# Conservation Through Aviculture

# Conservation Through Aviculture

ISBBC 2007

Proceedings of the IV International Symposium
on Breeding Birds in Captivity

Myles M. Lamont, Editor

Toronto, Ontario, Canada • September 12-16, 2007

ISBN 978-0-88839-731-7
Copyright © 2014 Myles M. Lamont

**Cataloguing data available from Library and Archives Canada**

All rights reserved. No part of this publication may be reproduced in any manner without permission of the editor. All images are © of the authors, reproduced with the kind permission of the authors and/or their representatives.

This publication is the result of a compilation of efforts by Hancock House Publishers, Disney's Animal Kingdom, The University of Missouri-Columbia and over 100 international authors.

Editor: Myles M. Lamont

Printed in the U.S.A- Lightning Source

*We acknowledge the financial support of the Government of Canada through the Book Publishing Industry Development Program (BPIDP) for our publishing activities.*

*Published simultaneously in Canada and the United States by*

**HANCOCK HOUSE PUBLISHERS LTD.**
19313 Zero Avenue, Surrey, B.C. Canada V3S 9R9
(604) 538-1114  Fax (604) 538-2262

**HANCOCK HOUSE PUBLISHERS**
1431 Harrison Avenue, Blaine, WA U.S.A 98230-5005
(604) 538-1114  Fax (604) 538-2262

**www.hancockhouse.com**

🦁 University of Missouri

*Sample citation:*

Bowden, CGR, C Böhm, MJR Jordan & KW Smith. 2014. Why is reintroduction of Northern Bald Ibis *Geronticus eremita* so complicated? An overview of recent progress and potential. In: Lamont MM, editor. Conservation Through Aviculture: ISBBC 2007. Proceedings of the IV International Symposium on Breeding Birds in Captivity; 2007 Sept 12–Sept 16; Toronto, Ontario, Canada. Surrey, BC, Canada: Hancock House Publishers; 2014. p. 38-46.

# Contents

FOREWORD .................................................................. 11

INTRODUCTION ............................................................ 12

ACKNOWLEDGEMENTS .................................................. 13

THE ISBBC AWARDS ...................................................... 478

## CONSERVATION BREEDING

ARTIFICIAL REARING OF THE ENDANGERED
TAKAHE IN NEW ZEALAND
*Martin Bell*.................................................................. 16

HAND-REARING THE NORTHLAND RACE OF
NORTH ISLAND BROWN KIWI AT AUCKLAND
ZOO, NEW ZEALAND
*Martin Bell*.................................................................. 27

WHY IS REINTRODUCTION OF NORTHERN
BALD IBIS SO COMPLICATED? AN OVERVIEW
OF RECENT PROGRESS AND POTENTIAL
*Christopher Bowden, Christiane Böhm, Mike Jordan
& Ken Smith*................................................................ 38

PROGRESS WITH A CONSERVATION
BREEDING PROGRAM TO SAVE THREE
CRITICALLY ENDANGERED GYPS VULTURES
FROM EXTINCTION
*Christopher Bowden, Vibhu Prakash, Nick Lindsay, Ram
Jakati, Richard Cuthbert, Asad Rahmani, Deborah Pain,
Rhys Green, Andrew Cunningham & Jemima Parry-Jones*....... 47

COLLABORATIVE EFFORT FOR THE LIGHT-FOOTED
CLAPPER RAIL PROPAGATION PROTOCOL
*Laurie Conrad, Charles Gailband, Judy St. Leger
& Richard Zembal* ........................................................ 55

CAPTIVE BREEDING, REARING, AND RELEASE
OF THE ATTWATER'S PRAIRIE CHICKENS AT
THE HOUSTON ZOO
*Mollie Coym*............................................................................................. 61

DEVELOPMENT OF CAPTIVE PROPAGATION
TECHNIQUES TO SUPPORT CONSERVATION OF
STELLER'S EIDERS IN ALASKA
*Tuula Hollmen, Heidi Cline & Nora Rojek*................................................. 69

THE PLACE OF CAPTIVE BREEDING IN THE
CONSERVATION STRATEGY OF THE ASIAN
HOUBARA BUSTARD
*Olivier Leon & Olivier Combreau*............................................................. 76

HAWAI`I ENDANGERED BIRD CONSERVATION
PROGRAM: A TOOL FOR THE RECOVERY OF
AN INSULAR AVIFAUNA
*Alan Lieberman*........................................................................................ 88

CAMPBELL ISLAND TEAL: THE SAVING OF A
SPECIES AND THE RESTORATION OF AN ISLAND
*Pete McClelland*........................................................................................ 93

THE VULTURE CAPTIVE BREEDING AND
RESTORATION PROJECT IN PAKISTAN
*Campbell Murn & Uzma Khan*.................................................................. 102

CAPTIVE BREEDING AND RELEASE OF RED
KITES TO HAMPSHIRE, ENGLAND
*Campbell Murn, Amy King, Samuel Hunt & Ashley Smith*......................... 108

CAPTIVE FIELD PROPAGATION AND EXPERIMENTAL
RELEASE OF EASTERN LOGGERHEAD SHRIKES
IN ONTARIO, CANADA
*Rina Nichols, Lance Woolaver, Elaine Williams,*
*Jessica Steiner & Ken Tuininga*................................................................. 116

SPECIES RECOVERY OF THE MAURITIUS FODY:
A MULTI-FACETED APPROACH TO SPECIES
CONSERVATION USING AVICULTURAL TECHNIQUES
AS KEY COMPONENTS OF RECOVERY
*Andrew Cristinacce, Richard Switzer, Amanda Ladkoo,*
*Lara Jordan, Andrew Owen, Markus Handschuh, Vanessa*
*Vencatasawmy, Frederique de Ravel Koenig, Carl Jones,*
*Diana Bell & Roger Wilkinson*................................................................... 127

THE CAPTIVE HUSBANDRY AND BREEDING OF THE
CRITICALLY ENDANGERED MONTSERRAT ORIOLE
*Gary Ward* .................................................. 142

BRITISH COLUMBIA WILDLIFE PARK CAPTIVE
BREEDING PROGRAM FOR BURROWING OWLS
*Paul Williams & Tyna McNair* .................................................. 158

FORTY YEARS OF BREEDING AMERICAN KESTRELS
IN CAPTIVITY: WHAT HAVE WE LEARNED?
*David Bird & Lina Bardo* .................................................. 169

# *AVICULTURE*

THE PAST, THE PRESENT AND THE FUTURE OF
NORTH ISLAND BROWN KIWI OUTSIDE OF NEW ZEALAND
*Kathy Brader* .................................................. 186

ESTABLISHING CAPTIVE POPULATIONS OF
ENDEMIC BIRD SPECIES FROM MADAGASCAR:
AN OVERVIEW OF THE EFFORTS OF THE VOGELPARK
WALSRODE FOUNDATION IN MADAGASCAR 1997–2007
*Simon Bruslund Jensen & Mario Perschke* .................................................. 191

THE BLACK COCKATOOS
*Neville Connors* .................................................. 199

THE HUSBANDRY AND CAPTIVE BREEDING
PROGRAM OF THE HORNED GUAN AT
AFRICAM SAFARI
*Juan Cornejo* .................................................. 209

PROPAGATION OF LIVE FOOD FOR BIRDS
AND OTHER VERTEBRATES
*Peter Karsten* .................................................. 219

PUBLIC AND PRIVATE SECTOR COLLABORATION:
AN OPPORTUNITY TO BUILD AND
PRESERVE AVIAN GENE POOLS
*Peter Karsten* .................................................. 227

FLAMINGOS IN CAPTIVITY:
THOUGHTS ON HOW AND WHY
*Catherine King* .................................................. 232

THE AVICULTURE OF HUMMINGBIRDS:
A HISTORICAL PERSPECTIVE
*Josef Lindholm, III* .................................................................... 252

THE STATE OF CAPTIVE WATERFOWL
*Michael Lubbock* ...................................................................... 311

EXPERIENCES WITH THE RESPLENDENT
QUETZAL IN NATURE AND CAPTIVITY:
A BRIEF HISTORICAL REVIEW
*Jesús Estudillo López* ................................................................. 324

INFECTIOUS AVIAN DNA TESTING:
SOME STATISTICAL ANALYSES
*Yuri Melekhovets, Tatiana Volossiouk*
*& Alexander Babakhanov* ............................................................. 332

IMPROVING THE KEEPING OF PARROTS BY
NEW CONCEPTS OF ENVIRONMENTAL
ENRICHMENT METHODS
*Rafael Zamora Padrón* ............................................................... 334

PREVENTION AND TREATMENT OF COMMON
MEDICAL CHALLENGES ASSOCIATED WITH
REARING AND BREEDING WATTLED CRANES
FOR A RECOVERY PROGRAM
*J.M. Pittman, M. Barrows & S.D. van der Spuy* ............................... 341

SOUTHERN CASSOWARY IN THE WILD
AND CAPTIVITY IN AUSTRALIA
*Liz Romer* ............................................................................... 353

FIRST CAPTIVE BREEDING OF THE
PHILIPPINE EAGLE-OWL
*Leo Jonathan, A. Suarez & Cristina Georgii* .................................... 363

FACING THE CHALLENGE OF SPIX'S MACAW
MANAGEMENT: OPTIMIZING THE HUSBANDRY,
VETERINARY CARE AND REPRODUCTIVE
MANAGEMENT OF SPIX'S MACAW AT THE
AL WABRA WILDLIFE PRESERVATION
*Ryan Watson, Richard Switzer, Simon Bruslund-Jensen*
*& Sven Hammer* ....................................................................... 370

# CONSERVATION & RE-INTRODUCTION

ROLE OF CAPTIVE BREEDING CENTRES IN
REDUCING IMPACTS OF BIRD TRADE IN
SUB-SAHARAN AFRICA
*Abdou Karekoona, Jason Beck & Raymond Katebaka* ........................... 392

CONSERVATION OF THE CRITICALLY ENDANGERED
NEGROS BLEEDING-HEART PIGEON ON THE ISLAND
OF NEGROS, PHILLIPINES
*Apolinario Cariño, Angelita Cadeliña, Rene Vendiola,
Jose Baldado, Charlie Fabre, Mercy Teves, Pavel Hospodarsky,
Emilia Lastica & Loujean Cerial* .......................................... 396

A COMPARISON OF BEHAVIOR AND
POST-RELEASE SURVIVAL OF PARENT-REARED
VERSUS HAND-REARED SAN CLEMENTE
LOGGERHEAD SHRIKES
*Susan Farabaugh, Ania Bukowinski, Susan Hammerly,
Christine Slocomb, Angela Sewell, Kathleen De Falco,
Lynne Neibaur & Jeremy Hodges* .......................................... 406

THE SCIENCE AND ART OF MANAGING
CAPTIVE BREEDING FOR RELEASE:
SAN CLEMENTE LOGGERHEAD SHRIKES
*Susan Hammerly, Susan Farabaugh, Tandora Grant,
Christine Slocomb, Angela Sewell, Kathleen De Falco,
Lynne Neibaur & Jeremy Hodges* .......................................... 407

POPULATION HISTORY AND MITOCHONDRIAL
GENE POLYMORPHISM IN BIRDS:
IMPLICATIONS FOR CONSERVATION
*Austin Hughes & Mary Ann Hughes* ........................................ 408

CONSERVATION EFFORTS TO RESTORE THE
CAPTIVE-BRED POPULATION OF HOUBARA
BUSTARD RELEASED IN THE WILD IN THE
KINGDOM OF SAUDI ARABIA
*M. Zafar-ul Islam, P. Mohammed Basheer, Moayyad Sher
Shah, Hajid al-Subai & Mohammad Shobrak* ................................. 417

THE ROLE OF SCIENCE IN AVIAN CONSERVATION:
EXAMPLES FROM PACIFIC ISLAND KINGFISHERS
*Dylan Kesler* ........................................................... 423

TRANSLOCATING SPECIES: THE VALUE OF
THE SITE VS. THE VALUE OF THE SPECIES
*Pete McClelland* ........................................................ 436

HOW DISEASE CAN AFFECT CAPTIVE BREEDING FOR CONSERVATION:
A NEW ZEALAND EXPERIENCE
*Kate McInnes, Richard Jakob-Hoff, Emily Sancha
& Jack Van Hal* .................................................... 441

POPULATION BIOLOGY: THE SCIENCE OF
POPULATION MANAGEMENT FOR CAPTIVITY,
REINTRODUCTION & CONSERVATION
*Colleen Lynch* ...................................................... 445

REARING TO RELEASE: MANAGING RISKS IN
THE REINTRODUCTION OF CAPTIVE-BRED BIRDS
*Philip Seddon & Yolanda van Heezik*................................. 454

CALIFORNIA CONDOR RECOVERY:
A WORK IN PROGRESS
*Michael Wallace*..................................................... 471

**ISBBC AWARD RECIPIENTS** ............................................ 478

**AUTHOR INDEX** ...................................................... 479

# Foreword

*Editors note: Written nearly 30 years ago, this foreword from the 1983 International Foundation for the Conservation of Birds Symposium in Los Angeles, still holds true today and describes with great accuracy the current state of conservation and aviculture, along with the goals and objectives of this symposium almost three decades later.*

Man has always destroyed plants and animals on which he must subsist; such is the law of Nature. Until the last century, however, the balance of life on earth had been fairly well preserved. It is true that vulnerable species, especially those restricted to islands' special habitats, had been exterminated during the 17th and 18th centuries; the Dodo and other birds in Mauritius and Reunion; the Great Auk; the Passenger Pigeon, the Labrador Duck in North America; the macaws of the Caribbean islands to mention only a few. But since, with the advent of the industrial age, the means of access to the most remote areas of the earth have changed the situation. Man is now well on the way to exterminating many forms of life, among which birds appear to be particularly threatened.

I became keenly interested in nature, particularly birds, at a very early age. In those days, there were still large numbers of them everywhere. But some abuses, causing excessive losses, were already worrying bird lovers. They all came from excessive shooting and trapping for commercial purposes, affecting mostly game birds, which were eaten on a large scale, and those whose beautiful feathers were used for decorating hats and other ornamental articles. Societies for bird protection had already been established in England, in Germany and in the United States, when, in 1910, we founded in France the French League for the Protection of Birds. I was its first secretary and later its president for 47 years.

But bird preservation was needed on a world level and the International Council for Bird Preservation (ICBP) was established in 1922, the founders, consisting of Gilbert Pearson, President of the Audubon Society, Lord Grey, Pieter van Tienhoven and myself. I was its president from 1938 until 1958, succeeded by Dillon Ripley. All these groups still are accomplishing a great deal, but much more is now needed to stop the appalling threats.

Human populations are not only increasing tremendously in numbers, but they also now penetrate even the remotest parts of the world to overexploit its resources at an alarming rate. Not only birds, conspicuous and attractive, are the victims to their greediness; they disappear as a consequence of the destruction of their necessary habitats. The indiscriminate use of pesticides has already caused appalling casualties, even at our very doors, as have also the various air and water pollutions. We certainly must try to stop the cause of such destruction and endeavor to establish natural parks and reserves where birds will be able to continue to live a normal existence. But due to physical, political and other adverse conditions, it is not always going to be possible to preserve and reestablish suitable sanctuaries.

If we really want to save many species of birds from extinction, we must also propagate them under control. Luckily, many of the most threatened species have proven to be easy to breed in captivity and to establish sufficient numbers in suitable accommodations where they can be protected, studied and reared under excellent conditions. Such popular groups as the ratites, cranes, waterfowl, game birds, pigeons, parrots, finches and other passerine birds are among them. Birds of prey are now successfully reared and released. Zoos and other public institutions, as well as private aviculturists, are working hard and successfully at it.

We all came here to learn about captive breeding and share our achievements, to benefit from our experiences and to work together to save the wonderful birds which still exist on earth so that future generations will still be able to enjoy them. It seems essential that sufficient stocks be established to insure their survival. It is my hope that from this international symposium will come a deeper commitment to conservation through captive breeding.

— Dr. Jean Delacour, 1983

# Introduction

As coordinators of the IV International Symposium on Breeding Birds in Captivity (ISBBC), we take great pleasure in welcoming all the speakers, delegates and guests to Toronto, Ontario, Canada.

Over the last 30 years there have only been three prior symposia to this one today. The first, held in Seattle in 1978, hosted by Jan Roger van Oosten, then director of Woodland Park Zoo. The following two were held in Los Angeles, California, the second in 1983 and the third in 1987, both sponsored by Gary Schulman. The 1987 symposium was held in commemoration of the death of Dr. Jean Delacour who passed away two years prior in November of 1985.

Twenty years since the gathering of the last symposium, we have again amassed a great wealth of knowledge, experience and talent in one room to discuss the issues and challenges facing avian conservation today. Much has changed over the last two decades in the progress of avicultural techniques, avian husbandry, medicine and nutrition, however we still face the same challenges today as we did twenty years ago in terms of a continued decline in both the *in-situ* and *ex-situ* populations of birds, perhaps at a heightened extent today than at any time previous.

It may seem somewhat disheartening to think that little progress has been made in bringing species extinction to a halt since the first symposium was held in 1978, however a great number of strides have been made with many species on a global scale as has been described in the following pages. New advances in technology have helped progress many aspects of avian medicine as well as refining processes such as cryopreservation, DNA profiling and phylogenetics. Great changes have also been made in captive husbandry instruments, particularly by incubator and brooder manufacturers as well as egg monitoring devices and remote viewing cameras, giving aviculturists the opportunity to gain insight into the processes of avian reproduction that wasn't possible even a decade ago.

Since the first symposium in 1978, a good number of captive breeding programs have been implemented on a global scale for a variety of species. Reintroduction biology is now being recognized by many conservation biologists as a key component of many conservation programs. Not long ago it was often regarded in a rather negative light due to the poor success rates and high expense that often accompanied early reintroduction programs. With the increased knowledge we have gained in avian biology in a variety of fashions along with advances in avicultural management techniques, captive breeding for conservation will likely continue to play a pivotal role in preventing extinctions for a great number of species.

Since the discussion of hosting a fourth symposium began in 2005, our aim for this event had always been to gather some of the worlds greatest and imaginative minds in the avicultural, zoological and conservation fields in order to once again bring this vast array of talents and experiences under one roof. It is our hope and our goal, that this event will provide a medium to exchange ideas, thoughts and provocative discussions in order to once again bring the avicultural field into the limelight and to exemplify the great work that has been, and is being, accomplished today. We hope that this symposium will provide that opportunity as it last did twenty years ago this year, and we share in the words of Jean Delacour from the 1983 symposium, "It is my hope that, from this international symposium will come a deeper commitment to conservation through captive breeding".

Myles M. Lamont &
David T. Longo
*September 12, 2007*

# Acknowledgements & Notes from the Editor

I must start this section off first with some apologies:

Firstly, to the authors of these proceedings who worked tirelessly and took valuable time away from their own individual projects and programs to partake in this event. Had it not been for your commitments to this symposium, these proceedings and to your programs, the avicultural community would be less for it.

Secondly, to the speakers, delegates and interested parties in these proceedings, I owe you my deepest and most sincerest of apologies for the delay in providing you with what you now hold in your hands. I could provide an entire chapter on the various hurdles and obstacles that have presented themselves over the last few years in the delay of this publication, many of which were beyond my own control, however none of those will make up for the wrong doing on my part for not having produced this publication much sooner. It of course was never my anticipation that the production of these proceedings would take so long, however none of that can be changed at this point, all that can be expressed is my utmost gratitude for your patience and support in this publication.

Thirdly, if I were to be completely honest with the reader, I would have to start by saying that this was a much greater undertaking, requiring much more effort and many more sleepless nights than I had ever envisioned. I can say with some certainty that had it not been for some youthful ignorance, my part in the coordination of this symposium would have been much smaller or completely non-existent. In hindsight, it was far too great a task for just two individuals to attempt to accomplish on their own, alas, it was, and I can only hope it was worth the effort.

As with the undertaking of anything of this nature, a great many people are to thank for their time, assistance and cooperation. I must first thank all the paper contributors for allowing us to print, what has been in many cases, a compilation of many years worth of efforts. Some of the chapters found within these proceedings hold information on captive avian reproduction that has never been published; my gratitude for allowing us to print this information through these proceedings cannot be overemphasized.

A few speakers were unable to provide manuscripts, either due to publication restrictions by employers or personal reasons, I thank you nonetheless for sharing your research with the delegates.

One of my biggest regrets in the delay of this publication is the loss of one our speakers and one of our ISBBC Award recipients: Jesús Estudillo López and Lynn Hall. Dr. Estudillo López was an avicultural great who needs no introduction. He lead the way in avicultural techniques for not only his work with cracids but also trogons, quetzals and many other species. It is my understanding that the ISBBC was the last conference he ever attended and I am greatly honoured he was able to attend and share his experiences with our delegates. Lynn Hall was nominated and awarded one of the Jean Delacour Avicultural Awards for his avicultural accomplishments, particularly with the Columbidae, along with his success in other avicultural pursuits. Both will be sadly missed and I have great regrets that neither were able to see these proceedings come to fruition.

I owe an immense amount of appreciation to two key individuals and institutions: Dr. Dylan Kesler of the University of Missouri-Columbia and Chelle Plasse of Disney Animal Kingdom. Both were instrumental in helping offset some of the financial burdens faced in this publication, I cannot thank you both enough. The late Neville Connors must also be thanked for his generous donation in making these proceedings possible.

I must also thank David Hancock of Hancock House Publishers for his continued encouragement, patience and advise on the development and printing of these proceedings; it is unlikely they would have been finalized without his efforts. I will also take the opportunity to thank him for his support on both a personal and professional level over the last ten years that we have been working together on a variety of projects and programs.

I would like to thank Josef Lindholm III for his advise, support and great literary contribution, not only in these proceedings, but of many over the years. I would also like to thank Chris Bowden, Phil Seddon, Peter Karsten and David Bird for their professional support with this publication. To Richard Switzer,

Gary Ward, Simon Bruslund-Jensen, Dick Schroeder, Scott Colomb, John Del Rio, Martin Vaillancourt and others too numerous to mention, my thanks for your kind words, advice and friendships over the duration of the production of these proceedings.

I of course cannot forget to thank my co-coordinator, David Longo, for his incredible patience and support over the last few years in awaiting for these proceedings to be finalized. Without his help and shared insanity, the symposium and these proceedings never would have been realized. Similarly, his mother, Clarice Longo was instrumental in helping organize finite details and providing support that had we not had, would have led to do the demise of the whole event.

Lastly, I need to thank you, as the reader, be it as an aviculturist, ornithologist, zoologist, veterinarian, conservationist or other avian enthusiast for taking an interest in the conservation of many of the worlds threatened birds. It is my hope that the culmination of the efforts and information enclosed in these pages may be used to not only educate, enlighten or clarify, but also perhaps to simply inspire the next generation of avian conservation practitioners. Lets us hope another twenty years won't pass before the next such gathering is attempted.

Myles M. Lamont

# Conservation Breeding

*Ex-situ* Propagation for
Conservation & Re-introduction

# ARTIFICIAL REARING OF THE ENDANGERED TAKAHE IN NEW ZEALAND *PORPHYRIO MANTELLI*

MARTIN BELL

Queenstown, New Zealand

## SUMMARY

The Takahe is a large flightless rail which inhabits alpine areas of the southern South Island. It is unique to New Zealand. Once considered to be extinct, it was rediscovered in Fiordland by ornithologist Dr. Geoffrey Orbell in 1948. At that time, the population was estimated to be about 200 pairs. By 1980, numbers had dropped to just 120 pairs. Sub-fossil remains suggest that Takahe once lived throughout the North and South Islands of New Zealand. The Takahe's recent decline has been attributed to competition for food with introduced Wapiti and Red deer and predation by the introduced stoat. Furthermore, their low reproductive rate has compounded the problem. Birds do not breed until they are three years old, and usually lay a single clutch of two eggs a season. If, as is often the case, the first clutch fails; it is rare that they attempt breeding a second time. Initially, conservation of Takahe focused on their management in the wild through stoat control, culling of deer and fertilizing tussock. Since 1958, 17 adults and six chicks have been taken to the National Wildlife Centre in the North Island for captive breeding but the results have been less than spectacular.

It wasn't until the 1980s that aviculture was seriously considered in the conservation of the Takahe. In January 1983, the first four Takahe eggs were flown from the Murchison Mountains in Fiordland to the nearby Te Anau Wildlife Park where they were hatched in incubators and hand-reared using puppets in a similar manner to the Californian Condor program. It was hoped that through the use of puppets resembling adult Takahe, and the use of taped calls, would significantly reduce potential mal-imprinting of the hand-reared birds. The aim was to rear birds suitable for release back into the wild. In addition, the removal of eggs from the wild early in the season would encourage birds to re-nest. This first (and somewhat crude) attempt was considered to be a success and resulted, in 1985, in the development of a specialised facility dedicated to hand-rearing Takahe for release into the

wild. Twenty-six years on, the Takahe hand-rearing program continues to contribute to the conservation of the species. Hand-reared Takahe can now be found breeding in Fiordland and on off-shore island sanctuaries. This paper looks in more detail at the history of hand-rearing Takahe, and the contribution that aviculture has played in the survival of this very endangered species.

## INTRODUCTION

My avicultural experience begins with the humble domestic chicken. As a boy, I loved animals but was limited to having just a small number of pets because we lived on a small suburban property. I had cats, mice, guinea pigs, and an assortment of bantams and chickens. My grandfather kept chickens for most of his adult life and my mother always said my interest in chickens must come from him. When I left school, my first job was at a commercial poultry hatchery where I worked for three years, learning a great deal about incubation. Little did I know that my experiences with chickens would one day have significance in the conservation of many species of New Zealand's endangered birds like the Takahe *Porphyrio mantelli*.

As I got older, I developed a keen interest in New Zealand wildlife. After accompanying my father to a wildlife trainee course in 1976, where he was teaching first aid, I thought I would like to become a wildlife officer. In 1980, I successfully applied for a traineeship as a wildlife officer specialising in aviculture at the National Wildlife Centre (locally known as Mt. Bruce) in the Wairarapa. It was here that my long-time involvement with the Takahe began. The rather dignified looking Takahe is a stocky bird with dark blue and green plumage and a vivid red beak. The impression of dignity is destroyed as soon as the Takahe breaks into its characteristic "lolloping" run, with stubby wings outspread and feathery white behind displayed for all to see. The National Wildlife Centre's promotional video describes New Zealand's most endangered rail

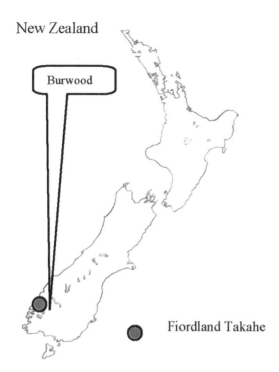

**Fig. 1.** *Location of Burwood.*

**Fig. 3.** *An adult Pukeko.*

as a 'Pukeko' *Porphyrio p. melanotus* with middle-aged spread. This comical and accurate analogy is closer to the truth than most people realise because the Takahe is actually a distant relative to the Pukeko.

The Takahe is a very vocal bird, almost continually making sounds; an impressive yodel, a soft 'thrum', a whine or delicately pitched hoot. The Takahe uses sound to advertise a territory; to maintain contact in the tall grasses; and to communicate with a chick in its egg. The Takahe is not a small bird and can weigh as much as 3.5kgs. Although fully formed, its wings are too small to haul the bird's considerable bulk of the ground. They are strictly used for balance and courtship displays. Like many other New Zealand birds, evolving in the absence of mammalian predators, the Takahe did not need to be able to fly. Like many other New Zealand birds, the introduction of those predators (especially the stoat) affected its population greatly.

During the summer months, Takahe feed primarily on alpine tussock, the Mid-ribbed Snow Tussock *Chionochloa pallens*, Curled Snow Tussock *Chionochloa crassiuscula* and the Broad-leaved Snow Tussock *Chionochloa flavescens*. They move below the snow line during winter to feed on *Hypolepis millefolium* fern rhizomes. They also feed on the less nutritious Red Tussock *Chionochloa rubra*. It takes considerable strength and dexterity to pull a tussock tiller (leaf) from its host. Takahe do it with ease, pulling the stalk out of the clump with their beaks and then holding it in the claws of one foot while nibbling the tender base.

I learnt how strong Takahe were during a routine examination of a bird at the National Wildlife Centre, when a bird grasped a piece of skin on the back of my hand in its beak and, rapidly twisted its head from side to side. The resulting sensation was not unlike accidentally pinching oneself with a pair of pliers. Needless to say I went to great lengths to ensure it didn't happen to me again.

In 1898, the last known living Takahe was reported to have been taken by a Mr. Ross' dog along the shore of the middle Fiord of lake Te Anau and for 50 years thereafter the Takahe was considered to be extinct. In 1948, it was rediscovered in the Murchison Mountains of Fiordland by Dr. Orbell. A survey carried out by the Wildlife Service (the Government Department responsible for wildlife management at that time) estimated the population to be around 400 birds. This number declined further in the years to follow. Today, the population is around 300 birds, half of which reside outside of Fiordland. Management of Takahe, since its rediscovery, has been multi-disciplined. Takahe have been; studied in the wild, their predators trapped, and their competitors (deer) shot.

**Fig. 2.** *A pair of adult Takahe.*

They have been studied in captivity, and hand-rearing has played an important role in boosting the size of the total population.

## 1981- Takahe Management Plan

The New Zealand Wildlife Service prepared a number of successive management plans for Takahe. In August 1981, the Wildlife Service produced a management plan which placed emphasis on the conservation of species in the wild; swapping fertile eggs from two-egg clutches with infertile eggs from other clutches so that each nest would contain at least one fertile egg. The plan also proposed to collect chicks that were at risk and to rear them in captivity using tapes and models of adult Takahe so that the birds' behavior would not be adversely modified by contact with humans. In September of 1981, Colin Roderik (a wildlife officer who had previously been based at the National Wildlife Centre and had considerable experience with Takahe) was asked to implement the captive management aspects of the plan. It was considered important to demonstrate that the procedure of rearing Takahe would work in order to justify the setting up of a more permanent facility.

## 1981 & 1982- A False Start

The captive rearing program began in the 1981/1982 breeding season. A total of five chicks of similar age, each from a two-chick brood were to be collected for the trial. Colin prepared specialised equipment; incubators, brooders and model 'surrogate' Takahe parents. By December 1st, 1981, all nests in the search areas had been located. A number of clutches were rearranged. Three infertile eggs were replaced and one fertile egg was taken in a portable incubator to the National Wildlife Centre where it was subsequently hatched and reared by a captive pair of Takahe. The planned collection of three chicks from the wild had to be cancelled because there was a risk that chicks might be chilled if handled in the wet and windy conditions. Further attempts to collect chicks was thwarted by a continuation of the bad weather and unavailability of helicopter transport. On the 17th of December 1981, five two-man parties were flown into the field to search those areas most likely to have two-chick broods. It was soon evident that the bad weather had been responsible for a number of chick fatalities and it was pointless to continue.

It was at this stage that one of the Wildlife Service Scientists, Roger Lavers, suggested that eggs rather than chicks be collected in the next season. This idea was supported by evidence that in tussock seeding years when food was plentiful, Takahe had been known to nest a second time. If eggs were removed early in the breeding season, there was a good chance they would re-nest and lay a second clutch.

It was therefore planned to re-survey the pairs in the "chick removal area" during January 1982, with a view to selecting eggs for transfer to the Te Anau Wildlife Park for incubation and hand-rearing. Measurements and weights of eggs were to be recorded and the eggs were to be candled to determine fertility and embryonic development. Because it was considered best to rear chicks of similar age in a small group so that they had social contact with each other, it was necessary to replace both fertile eggs from one clutch with one fertile egg from another two-egg clutch. A minimum of five eggs were to be collected in this way.

## 1982 & 1983 - The First Hand-Rearing of Takahe

Colin Roderick had assembled the necessary equipment which included; a large high sided plywood brooder with small feeding hatches, two force draft incubators, two heated pet blankets, two hand-made polystyrene brooding models (painted and feathered to look like Takahe), a similarly designed feeding head, tapes of Takahe calls and small speakers for the models. Colin, however, died unexpectedly in 1982 and the author was recruited to the project in December, 1982. [I was working as a wildlife avicultural trainee based at the National Wildlife Centre at the time and I travelled to Te Anau to undertake the work each summer for the next three breeding seasons; 82/83, 83/84, 84/85.] A small number of bantams were acquired as back-up to the incubators in case of power failure. (Almost every bantam in Te Anau was auditioned in

**Fig. 4.** *Trialing equipment with Pukeko chicks.*

**Fig. 5.** *First two Takahe eggs in incubator.*

order to find ten suited to long term sitting). A hen house was built and the four best sitters of the flock were moved into specially made nest boxes in the "Takahe rearing room" which was a converted garage. The bantams were let out of their boxes twice a day to feed, drink and stretch then they were returned to sit on their plastic eggs.

Because the equipment and techniques were untested I decided to trial it out on their nearest relative, the Pukeko. However, time did not permit me to rear the five Pukeko chicks beyond two weeks of age because of the arrival of the Takahe eggs. In January 1983, the Takahe eggs were collected. Equipped with pre-heated vacuum flasks for transporting eggs, two teams of two people were flown into the Murchison Mountains by helicopter, only a ten minute flight from Te Anau. [I was fortunate to be able to fly into Fiordland to help find and uplift the Takahe eggs. Often the helicopter could not land so we had to climb out of the machine as it hovered with one skid touching the ground. In these situations, entry and exit from the helicopter had to be quick and expertly timed, because the machine could be unexpectedly thrust into the ground at any time as a result of down draughts. The pilot, Bill Black was something of a legend in wildlife circles at the time.] On this trip, four eggs were collected and flown to Te Anau in separate pre-heated flasks. Uplifting of the eggs couldn't have been more expertly timed. Two of the eggs had started pipping. On arrival at the Te Anau Wildlife Park, the two hatching eggs were put in an incubator (Turn-X) equipped with small speakers in order to play Takahe parental calls. The incubator temperature was set at 37.5°C and a relative humidity of 60%.

The automatic turner was switched off because the eggs were soon to hatch. The other two eggs were placed under two broody bantams. During hatching, the eggs were watched closely. It was considered important to transfer the hatching eggs from the incubator to the model Takahe quickly, in case they mistakenly identified staff as their parents. A 24-hour vigil was maintained (with the help of lots of strong coffee). Within half-an-hour of the birds completely emerging from their shells, they were moved to the brooding model Takahe equipped with speaker and sitting on a warm heated pet blanket. Initially, a thermometer was used to check the temperature under the model didn't exceed 30°C. Although similar to Pukeko chicks; black, and downy, and perpetually noisy, Takahe chicks do not have the Pukeko's delicacy or its charming disproportionately large and rather whimsical feet; Takahe chicks are stocky and beaky.

**First Brooders**

The surrogate parent Takahe was a polystyrene shell covered in real Pukeko feathers, with a bright red beak. It sat on a vinyl electric blanket (marketed for cats and dogs) which provided the warmth for the chicks. The chicks could come and go from the model through archways cut in its body. Inside the model were "velcroed" removable strips of woollen sock material which hung down providing the birds with added warmth, comfort and security. Embedded under this lining was a small speaker (connected back to a tape deck outside the brooder), used to emit the different parental Takahe calls when required.

The brooder in which the model and chicks were kept was a very large ply-wood box (2.4m x 0.8m x 1.2m high) built on a stand. The brooder floor was lined with a removable artificial grass mat. The walls had strategically placed peep holes, and hand holes necessary for allowing the second artificial parent to be inserted at feed times. The feeding head was made of solid polystyrene, again covered in Pukeko feathers. A small speaker was held on the outside of this "solid" model and parental soliciting calls were played at feeding times. As anticipated, the Takahe chicks responded by running to the source of the calls. Their diet consisted of a variety of tussock tiller bases (just the very succulent end of 5mm), mixed with moistened 'Farex' (a commercial baby food) to a porridge like consistency easily fed using the model head. Other foods such as potato, clover, and carrots were added from time to time. Insects and commercial poultry mash were also fed to the chicks.

The chicks became mobile at the age of two days. They began to tug at objects. This included the plastic strands of the artificial grass matting. Their unrelenting interest prompted the introduction of handfuls of real grass and turf to try and divert their attention. The grass was eagerly accepted but was to cause unexpected problems (and subsequent fatalities), and it did not stop the birds from tugging at the artificial grass.

Even at this very young age, the chicks attempted to use their feet in the characteristically parrot-like fashion to pick up and hold their food when eating. To begin with, water was given to the chicks by dunking the model head and allowing the chicks to take the water droplets from it. This process was tedious, but after several days, the chicks began to take the water from the dish provided. Cleaning the artificial grass in the brooder was difficult because it was awkward to lift out. The most satisfactory way to clean it was with the vacuum cleaner. Simultaneously using the vacuum cleaner and playing the parent alarm call initially sent the chicks running for cover. Unfortunately, the chicks quickly got used to the vacuum cleaner and didn't bother to hide. Later this problem was solved by remodelling the brooder. The first three weeks of chick rearing was plain sailing, but then one chick stopped eating. This was most unusual because they normally feed continually. The following day the local vet diagnosed the bird as having a probable intestinal blockage. Faecal analysis confirmed that; a *Salmonella* related bacterium was present in the gut, so the bird was isolated from the others in a second stand-by brooder. It was treated with laxatives to try and clear the blockage and antibiotics to stem infection. The artificial grass was removed from the brooders because it was suspected to harbour stale food and bacteria, despite efforts that had been made to clean and sterilize it. There was also concern that the chick had perhaps ingested some of the fibres in the material and these might have caused the intestinal blockage. The chick subsequently died three days later.

## Modifications

After this early set-back, the brooders were modified. They were altered to make cleaning and servicing easier while ensuring that the chicks would not see their keepers. Each brooder was divided in half with a solid wall and fitted with a hinged door at one end. Each compartments front wall became a door providing better access for cleaning and they were each fitted with a one-way observation window (replacing the crude peep holes). Each side of the partition had an aluminium drawer which when fitted went halfway into the brooder compartment. Behind and raised slightly above each drawer was a wooden floor with a sunken removable water dish large enough for the chicks to drink from and bath in. The drawers were

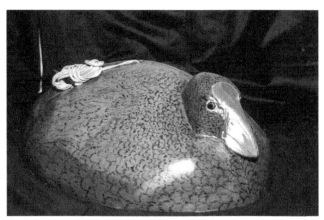

**Fig. 6.** *New fiberglass brooding model.*

filled with soil, leaf litter and clumps of grass that could be regularly changed. The birds were only kept in one side at a time. When their compartment required cleaning, the partition door was opened and they were called by their 'parent' to the clean compartment.

At four weeks of age the three remaining chicks were moved outside on fine days to a small secure pen. As each day passed, the chicks grew larger and stronger and new problems arose. The model parents became the targets of unrelenting attention. The adolescent birds began pulling feathers from them and pecking apart the polystyrene beaks when hungry (or bored). They also began to pull the speaker from the brooding model. The chicks became too big to get through the holes in their model parent's side and so rolled it over and simply climbed inside. It seemed that, even upside down, the model continued to represent warmth and security for the chicks but it was all too clear that both it and the feeding beak required redesigning.

Towards the end of March 1983, the three chicks fledged into dull versions of adult Takahe. These three birds (later christened 'Tussock', 'Snow', and 'Alpine') remained in captivity at the Te Anau wildlife Park and two of these (Alpine and Snow) went on to breed in November 1986, successfully hatching two chicks. 'Alpine' could still be seen at the Te Anau Wildlife Park in January 2007, at 24 years of age, almost twice the average life

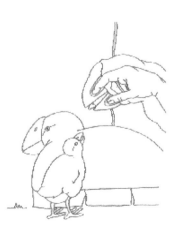

**Fig. 7.** *New brooder design.*

**Fig. 8.** *Wayne Hutchinson in the hide feeding chicks.*

span for wild Takahe.

## 1983/84 - Making Significant Improvements

In the following summer of 1983/84 the program continued and the author returned to Te Anau. Further improvements were made to the facilities and equipment at the Te Anau Wildlife Park. A steel kit-set garage was erected in a staff only area with sleeping facilities, brooders, incubators, and hot and cold running water. A large outdoor pen with predator proof solar powered fencing was erected and planted with tussocks to provide food and shelter. Changes were made to the brooders and new improved Takahe brooding models were made.

The new models were made entirely of hard wearing fibre glass and the speaker was glued firmly inside the head. There were no feather's, they were simply painted to resemble adult Takahe. They had a removable fringe of material hanging from their rims to the ground and a removable internal lining of soft cloth. They were hung over their nest areas so as the chicks grew the model would simply be raised to allow more head room. The feeding model was now a blue cloth hand puppet with two sturdy wooden mandibles. One went on the thumb and the other on the forefinger. The speaker was held in the palm of the hand, hidden inside the glove. The "tweezer-like" hand puppet could much more precisely pick up food. Both these new models worked well and are still used in unmodified form today. I later designed similar brooding models for the Black Stilt *Himantopus novaezealandiae* and New Zealand Fairy Tern *Sterna nerius davisae*. The Black Stilt model was more sophisticated having thermostatically controlled heating built in. At seven weeks of age, the six birds reared that season were moved outside with their surrogate model parent into small pens with a hide at one end from which they continued to be fed and weighed as before. Staff spent many hours crouched in the tiny hides feeding and observing the birds.

The small pens were positioned in the larger predator proof enclosure and the birds were eventually released into these when they had a full complement of feathers. A specially formulated, pellet food was prepared by Massey University in Palmerston North for Takahe as a supplement. This was provided to the birds in a hopper in the enclosure. It was always anticipated that the Takahe would require this supplementary pellet if released into areas outside Fiordland. The chicks were introduced to the pellets (in mini hoppers in the brooders) from the time they were small and they were readily accepted. On rare occasions when the birds caught sight of a human lurking about their enclosure, they would immediately hide, and unfamiliar noises or sudden movements elicited the same response. As they got close to full size, it became difficult to weigh them without a great deal of drama, so weighing was restricted to occasions when handling the birds was required for other reasons and ceased as each bird reached a kilogram in weight. At two months of age the surrogate parent was removed and they were no longer fed by models. It was gratifying to see that the birds were very wary of people as had been intended by the rearing method. Unfortunately, once they became accustomed to the presence of staff who came to fill their food hopper, they began to parade along the fence-line in greedy anticipation. To counteract this, they were fed at irregular times and small piles of food were put in random places around the enclosure. These six birds were sent to Maud Island in the Marlborough Sounds in April 1984. Initially, they were kept in small pens on the island to acclimatize them to their new home before being free to range over most of the island. Then in January 1987, the first Takahe chick was hatched on Maud Island (a second chick was hatched by another pair in February of the same year).

## 1984 & 1985

In November 1984, the author was permanently appointed to the Takahe program and began setting up a new captive rearing facility at Burwood Bush; although the existing facilities at the Te Anau Wildlife Park were to be used for the 1984/1985 season. The results were disappointing. A total of 11 eggs had been collected and were already being incubated when the author arrived to take over the work. Of these 11 eggs, only five hatched; and of these, four required interven-

**Fig. 9.** *Red Tussock reserve in Gorge Hill.*

tion to hatch. The other eggs died at varying stages of development. Suggested causes for the embryo deaths were varied but nothing was obvious at the time. On the 20th of December, two further eggs were incubated and subsequently hatched. One of these died two days after hatching as a result of localized peritonitis caused by an unabsorbed yolk sac. Early in December, one of the original five older chicks died as a result of an intestinal blockage. Just five birds were reared this season.

## 1985/86- Establishment of Takahe Rearing Facilities at 'Burwood Bush'

In 1985, more than 3,000 hectares of crown land administered by the Lands and Survey Department at Gorge Hill, 35 kilometers east of Te Anau was declared a Scientific Reserve. At the same time, it was made available to the Wildlife Service for Takahe management. The area consists of 1500ha of Red Tussock and 1570ha of remnant beech forest. The area became available following public submissions on a discussion paper released by the Department of Lands and Survey in Invercargill late in 1984. The area became known as 'Burwood Bush'.

There was an existing building built by Lands and Survey for their farmers on a ten hectare piece of land opposite the reserve. This area of land was also bequeathed to the Wildlife Service for the Takahe program. The brooder facilities were to be built on this site.

Takahe management staff set about establishing permanent rearing facilities which included; an incubator/brooder room, workshop, staff quarters, fuel storage shed, and the first ten hectare predator proof enclosure on the Red Tussock Reserve. Takahe would be hand-reared at the brooder facility then transferred to specially fenced enclosures on the Red Tussock Reserve. The majority of these birds would later be released into the wild in Fiordland but some would be kept as breeding stock on the reserve.

## Predator Proof Enclosure

The enclosure (known as Block 'A') was predominantly covered in red tussock but included a patch of beech forest. The specialised fence was constructed using round wooden posts (120cm above ground) and 10mm galvanized chicken netting. The netting went into the ground 10cm and then out 50cm at a right angle to prevent digging animals getting in. Outriggers were embedded in the top of each post at a slight angle; these supported three solar powered wires. Power output from the solar panel was sufficient to electrify the wires and so deter predators. Solar power was necessary because of the isolation of the area (approximately 5km from the nearest mains power supply). Another two electrified wires ran along the perimeter of the fence supported by outriggers just above the ground. On completion of the fence, the enclosure was extensively trapped using 450 'Gin' traps. Hedgehogs, possums, cats, ferrets and stoats were the main targets.

Trapping around the outside perimeter of the enclosure continues to this day. A second enclosure (Block 'B'), with a fence similar in structure to the first was completed in May 1986 and a third (Block 'C') was completed in February 1987.

## Incubator/Brooder Room

In October 1985, the program built a 1700 sq ft soundproof incubator/brooder room was sufficiently completed in time for the 1985/1986 breeding season. It was designed to rear up to 16 birds a year. The building consists of a temperature controlled incubator room with sleeping quarters, a separate toilet, storage room, kitchen, and

**Fig. 10.** *Predator proof fence design.*

**Fig. 11.** *Incubator/brooder building at Burwood.*

brooder room.

Four large wooden brooders were built against the outside walls, two on each side, each divided in two and equipped with one way glass. Glass fibre floors were fabricated; shaped as an artificial landscape and incorporating strategically placed drainage pipes. The inside walls were painted with marine paint, all to make cleaning quicker and easier. Lighting was controlled with dimmer switches for simulating dusk and dawn. On the outside walls of each brooder compartment was a door operated from outside the brooder which gave access to an outside pen packed with tussock. These pens are solid walled (Hardiflex) and have electric wires around the top to prevent predators getting in. Pellet hoppers, filled from the outside were installed in the outside walls.

### 1985 & 1986

Twelve chicks were reared from the 16 fertile eggs received that year. On 21 March 1986, the Burwood Bush facilities were officially opened by the Minister of Internal Affairs, Peter Tapsell; and special guest Dr. Orbell. The opening coincided with the release of the first six Burwood-reared Takahe, into the fenced area (Block A) of the Red Tussock Reserve. The birds were equipped with small transmitter back-packs so they could be monitored from a distance without being disturbed. The other six birds were released into the same area four days later. On release, none of the birds showed any signs of imprinting, all running and hiding in the tussock as soon as their boxes were opened. The birds were supplementary fed Takahe pellets which were provided in special hoppers.

### 1986 & 1987

The 1986/1987 season was disappointing; only three chicks were successfully reared from 17 eggs received. Two eggs were infertile and ten died during incubation. Unlike previous years, the eggs were received much earlier in incubation and the cause of death could be the result of one or a number of factors difficult to isolate.

Five chicks hatched but two suffered intestinal blockages. The first was given castor oil and required force feeding. Three days later its condition had worsened so, after consulting the local vet, surgery was performed to remove the blockage. Ether was used to anesthetize the chick for the hour-long operation. Although the operation was successful, the chick died several hours later as a result of its undernourished and stressed condition. The second chick developed the symptoms ten days later. It was initially given a feline laxative (Kat-A-Lax) which was unsuccessful so before the bird lost too much condition surgery was performed and a large fibrous blockage was removed from the intestines. The chick survived the operation but died two days later. The autopsy revealed there was another blockage in the intestine and gizzard which had gone unnoticed in surgery. The blockages were caused by long pieces of tussock and grass, not the artificial food provided.

### 1987 to 2007- Methods Refined

I left the program in November 1986 to take up a new position at the National Wildlife Centre and my involvement in the Takahe program gradually tapered off. Very little has been changed at Burwood Bush in the last 20 years. The original fibre glass models are still in use, as are the original brooders. The incubators which are now 'Curfew' replacing the old 'Turn-X'.

| Table 1 | |
|---|---|
| **NUMBERS OF TAKAHE AT THE END OF 2005/06** | |
| Murchisons Mountains core census area | 165 |
| Fiordland beyond core census area (Stuart Mountains) | 3 |
| Islands | 86 |
| Burwood Bush[1] | 34 |
| Display (Mt Bruce at Te Anau Wildlife Park) | 7 |
| **Total**[2] | **295** |

[1] *including six breeding pair*
[2] *not including birds under a year of age*

**1989- Fern Rhizomes**

The hand-reared Takahe could not be shown how to dig up the *Hypolepis* fern rhizome (which is an important food source for wild birds during the winter) using puppets. They would need to be able to do this if they were to be released into Fiordland. The first training to dig up and feed on *Hypolepis* began in winter 1989. Each juvenile group was enclosed within an area of *Hypolepis* for one to two weeks, until they discovered and began feeding on the rhizomes. This procedure was very successful, but their confinement to small pens was often stressful, especially during poor weather.

**1990 to 1993- Adult Birds as Tutors**

The following year, a group of juveniles was penned with an older, experienced bird, and rapidly learnt the feeding technique. Since 1993, all juvenile Takahe at Burwood have learnt how to feed on *Hypolepis* from either one experienced adult or a pair of Takahe and no longer require penning. The juveniles learn quickly to feed on *Hypolepis* by association, watching and occasionally being fed by the adults, soon after the first snowfall. Presently four adult pairs each care for up to six juveniles from the age of two to ten months. The extended family spends the winter together.

**1996- Foster Parents**

In 1996, it was decided to use real adult Takahe as partial foster parents to the puppet reared birds. From the time the chicks hatch until they are eight weeks of age they spend time rotating between puppet Takahe and real Takahe.

**Stage at Which Eggs are Collected**

Preference still goes to bringing eggs out late in incubation rather than early and the hatching results are better. Because so few eggs are available for artificial incubation each year, the program has not had the luxury of experimenting with incubation techniques to improve hatchability of eggs taken earlier in incubation.

**Intestinal Blockages**

One recurring problem was the loss of chicks as a result of intestinal blockages. The common factors in each case were the chicks were less than two weeks old and the blockages were always caused by long strands of tussock and grass. These long pieces were picked up when scattered as supplementary food. These young birds had not yet consistently learnt to hold the food in their feet and bite small off at a time as adults do. They just swallowed the long strands whole. The tussock tillers fed to the chicks in the artificial diet were always cut short as they would be if fed by the parents. Mills, Lavers and Lee (1984) stated in their paper "*The Takahe-a relic of the Pleistocene grassland avifauna in New Zealand*", New Zealand Journal of Ecology, that the gastric system of the Takahe is unspecialised (monogastric with a small caecum), which means that they cannot digest fibre. The problem was solved by not giving the younger chicks access to long tillers of tussock and grass.

Today Burwood Bush aims to hand-rear up to 17 birds per season. There are six breeding pairs held at Burwood Bush in the predator proof enclosures on the Red Tussock Reserve.

**Numbers of Takahe at the end of 2005/06 season:**

**Islands That Have Takahe:**

North Island
Tiritiri Matangi (Hauraki Gulf)
Kapiti (Wellington coast)
Mana (Wellington coast)

South Island
Maud (Marlborough Sounds)

Captive Display:
National Wildlife Centre (Mt Bruce)   3
Te Anau Wildlife Park                 4
*(including "Alpine")*

**A Successful Program?**

Between 1987 and 1992, 58 yearling Takahe were released into the Stuart Mountains of Fiordland adjacent to the Murchison Mountains. They were released in small groups of two to eight birds. Radio transmitters were used to monitor 24% of them.

Up to 1993, results showed that at least 13 of those (22%) survived their first winter in the wild, and two breeding pairs formed, one of these producing a yearling in two different breeding seasons. A wider survey in 1995 found only eight birds (14%) including one unbanded (probable offspring). To date further occasional sightings have been spread as widely as 27km from the release site, over an area of approximately 35,000 hectares. Because of the poor results, no further birds were released in the Stuart Mountains. Releases focused on the Murchison Mountains of Fiordland. Survival of captive-reared birds released into Murchison Mountains was more promising and looks similar to wild-reared birds. The Department of Conservation is currently in the process of comparing breeding output over their lifetime, which is obviously an important consideration to the net gain to the wild population. The total number of captive-reared birds released between 1982 and 2006 has been 256 and 26 of these have been put on islands (J. Maxwell pers. comm. 2007).

**The First Hand-Reared Takahe Alpine and Snow's Descendants:**

As at end of 2004/05 season this pairing had either 39 or 42 descendants (depending on who was the true parent of three of them in later generations). At that point in time the islands population numbered 83, so a significant contribution, although not all their descendants were still alive at that point.

## DISCUSSION

It has never been established whether Takahe behaviour would be adversely modified if they were hand-reared without the elaborate 'puppet show'. Anecdotal observations over the years suggest that at least some of the Takahe behaviour may be innate; such as their instinct to hide. However, there have been other observations that suggest Takahe learn some important behaviour from more experienced birds; such as feeding on *Hypolepis* ferns. In the wild, Takahe chicks have been observed being fed by their parents up to five months of age and remain around their parents for up to 18 months suggesting that parents play an important role in chick learning and survival.

"The majority of chicks reared at Burwood over the past 13 years have been hand-reared from hatching through to independence. Nineteen chicks (up to three weeks old) collected for hand-rearing from the wild and a few partially parent-reared chicks at Burwood have displayed noticeably different behaviour compared to chicks that have been entirely hand-reared. The partially parent-reared chicks appear to have a more relaxed temperament, are less aggressive or competitive towards their siblings and learn feeding techniques faster, for example, feeding directly from a dish rather than relying on being hand-fed. They also hold grass tillers with their feet, in the usual adult manner, about two weeks earlier than hand-fed chicks.

There also appear to be differences between partially parent-reared and totally hand-reared juveniles during the winter. The parent-reared chicks tend to feed on *Hypolepis* rhizomes earlier and appear to be in better physical condition after winter than their entirely hand-reared cohorts. Since 1996, the chick rearing procedure has attempted to include parent-rearing for a minimum of five days within the chicks' first three weeks of life. One chick at a time is transferred from the brooder to an adult pair and back to a brooder again. Advantages are birds are better adapted behaviorally, as well as a reduction in Takahe hand-rearing labour (Eason & Willans 2001).

Past experience from Mt. Bruce suggested that early mal-imprinting of Takahe chicks had a negative effect on reproductive behaviour (Reid 1977). A strongly human-imprinted male on Maud Island failed to associate with other Takahe for five years and has not shown any reproductive behaviour despite being paired in captivity for more than a decade. Two Takahe reared by Pukeko on Mana Island (Bunin & Jamieson 1996) failed to associate with other Takahe by age four and six years respectively. An attempted forced pairing of one bird to a parent-reared Takahe and its later transfer to Pukeko-free Maud Island at age five, has had no effect as yet.

The graph shows the total population has increased between the 1981/1982 (118) and 2005/2006 (295), an increase of 177 birds over 25 years. It clearly shows the most significant contribution to the population is the island inhabitants. The majority of these birds are hand-reared birds and their descendants. It is clear how important the hand-rearing program has been to

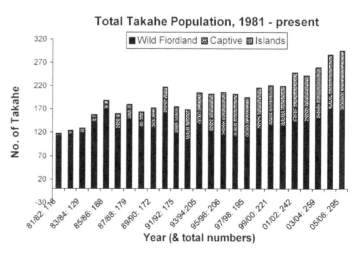

the overall population increase. The graph also shows that the wild Fiordland population has fluctuated over the 25 years but never exceeded 170 birds. There has been some research on nutritional recovery of tussock that has been grazed by deer. The results have shown that it takes many years for tussock to get back to normal nutritional levels. This would suggest that the wild population of Takahe in Fiordland may increase in the long term if deer were completely absent.

Recent research by Catherine Grueber (2007) for her MSc thesis (Zoology, University of Otago, supervisor Ian Jamieson) has highlighted the level of inbreeding occurring amongst the island Takahe population. Previous research by Ian Jamieson and Christine Ryan has showed that such inbreeding is undesirable. It has already started to have an effect on the breeding efficiency of some females. (When the inbreeding coefficient of females was compared with their breeding fitness the more inbred females showed reduced egg fertility, hatching success and fledging success compared with non inbred females (Jamieson & Ryan 2001; Jamieson et al 2003)). The islands currently available for Takahe are each nearing capacity. Jane Maxwell, Takahe ecologist, with the Department of Conservation states in her report of winter 2005; that Takahe management needs to focus on good quality production rather than high productivity. She states it will be more important to produce fewer birds with desirable genes than many birds which we may not wish to breed from in future. One thing is for certain, the hand-rearing of Takahe has been productive and is likely to continue for the foreseeable future.

## Acknowledgements

I'd like to thank my wife Paula Bell for editing this paper and Jane Maxwell (Takahe ecologist) and Linda Kilduff (Officer-in charge of 'Burwood Bush') for supplying me with up to date information on the program.

## References

BUNIN JS & IG Jamieson. 1996. Responses to a model predator of New Zealands endangered Takahe and its closest relative, the Pukeko. Conservation Biology 10(5): 1463-1466.

EASON D & M Willams. 2001. Captive rearing: a management tool for the recovery of the endangered Takahe. Pages 80-95 in The Takahe: Fifty Years of Conservation Management and Research, W.G. Lee & I.G. Jamieson (eds), University of Otago Press, Dunedin.

GRUEBER C. 2007. Unpublished M.Sci thesis (Zoology, University of Otago, supervisor Ian Jamieson).

JAMIESON IG & CJ Ryan. 2001. Closure of the debate over the merits of translocating Takahe to predator free islands. Pages 96-113 in The Takahe: Fifty Years of Conservation Management and Research, W.G. Lee & I.G. Jamieson (eds), University of Otago Press, Dunedin.

JAMIESON IG, MS Roy and M Lettink. 2003. Sex-specific consequences of recent inbreeding in an ancestrally inbred population of New Zealand Takahe. Conservation Biology 17(3): 1-10.

MAXWELL J. 2005. Objectives for Takahe Program to reduce the effects of inbreeding in the Takahe population outside of Fiordland. Unpublished Report for the Department of Conservation.

MILLS JA, RB Lavers and WG Lee. 1984. The Takahe- a relic of the Pleistocene grassland avifauna in New Zealand. New Zealand Journal of Ecology 7: 55-70.

REID BE. 1977. Takahe at Mt. Bruce. Wildlife-A Review. Wellington Dept. of Internal Affairs 8: 56-73.

# HAND-REARING THE NORTHLAND RACE OF NORTH ISLAND BROWN KIWI AT AUCKLAND ZOO, NEW ZEALAND *APTERYX MANTELLI*

MARTIN BELL

Queensland, New Zealand

## SUMMARY

The Northland race of the North Island Brown Kiwi has declined significantly in southern Northland in the last 30 years as a result of land clearance, and the effects of introduced mammalian predators, more specifically the stoat and cat. The Northland Brown Kiwi is now in danger of extinction. The first trial release of captive-reared North Island Brown Kiwi chicks by the Department of Conservation took place in July 1995. The birds were released onto predator-free islands and coped well with the transition from captivity to the wild. However, the challenge was to see if captive-reared birds could be successfully released into areas, on the mainland in the presence of predators. In December 1995, Auckland Zoo was approached by the Department of Conservation and asked to participate in 'Operation Nest Egg'. The zoo would incubate wild Kiwi eggs, hatch the chicks and rear them to an age considered suitable for release back into the wild. Fifteen eggs were collected between September 11th, 1996 and December 20th, 1996 from 12 different nests on private property, 30 minutes north of Whangarei. The eggs were transported back to Auckland Zoo in polystyrene 'coolers' packed with shredded paper. The egg ages were estimated to be between ten days and 52 days into incubation. Twelve birds were reared (this included two birds that were collected as very young chicks) and all 12 were released back into the wild. On May 20th, 1997, four of the birds were released into disused burrows in 'Hodges Bush' and the other two birds were released in nearby 'Poroa Bush'. Later in the same year, the remaining six birds were released into the same areas. This paper looks in detail at the contribution Auckland Zoo made to 'Operation Nest Egg' and the continuing conservation of the Northern race of North Island Brown Kiwi.

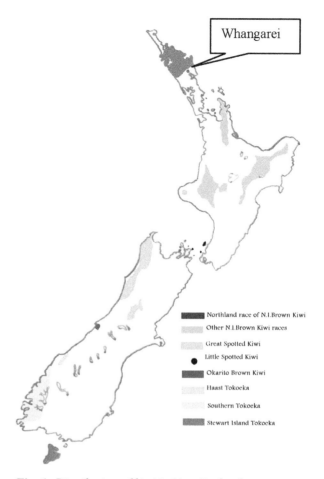

Fig. 1. *Distribution of kiwi in New Zealand*

## INTRODUCTION

The Kiwi is the smallest member of a group of flightless birds known as ratites which include the cassowary, ostrich, rhea, emu and New Zealand's extinct moa. Ratites are an ancient bird group having evolved at the time of the dinosaurs about 80 million years ago. Five species of Kiwi are now recognized: the North Island Brown Kiwi *Apteryx mantelli*, the Haast Tokoeka *Apteryx australis* 'Haast', Southern Tokoeka *Apteryx australis*, Little Spotted Kiwi *Apteryx owenii* and Great Spotted Kiwi *Apteryx haastii*.

The Okarito Brown Kiwi *Apteryx rowi*, until recently, was considered a sub-species of the North Island Brown Kiwi but is now considered a species unto itself. The Tokoeka is divided into three races; 'Haast Tokoeka', restricted to a small area of the West Coast of the South Island, the 'Southern Tokoeka' found in Fiordland and the 'Stewart Island Tokoeka', only on Stewart Island. The North Island Brown Kiwi has been further split into four distinct populations; the

Northland race, Coromandel (Coromandel Peninsula), Western (Taranaki, Wanganui, and Tongoriro) and Eastern (Hawkes Bay, Urewera and Bay of Plenty).

The kiwi has evolved in a land devoid of terrestrial mammalian species (except for three species of bat). It has taken on the role of a small mammal; it is a bird with no tail, it has vestigial wings, feathers more like hair than feathers, cat like whiskers, two functioning ovaries, marrow filled bones, nostrils at the tip of its long bill and a cooler mammalian blood temperature of 38°C. The kiwi has a highly developed sense of smell with the olfactory lobe (which controls the sense of smell) being larger and more developed than other birds. It is believed that the Kiwi became nocturnal to evade the large raptors that are now extinct. Invertebrates, their main food source, are more active at night. Male kiwi are smaller than females in all species. Perhaps the best known feature of the kiwi's physiology is the egg which is equivalent to six domestic hen's eggs and weighs up to one fifth of the female's body weight. North Island Brown Kiwi lay one or two eggs in a clutch. After laying the first egg, the female goes off to build up condition, sometimes laying a second egg in the same nest three weeks later. Although the kiwi has two functioning ovaries, production of the second egg is delayed because it would be physically impossible for her to carry both at the same time. In captivity, female kiwi have been observed standing in water just prior to egg laying (perhaps to relieve the pressure on their abdomens).

Once the female has laid the egg, she will stay with it for 24 to 48 hours, then leaves it to be incubated by the male. The male North Island Brown Kiwi incubates the eggs from the time the first egg is laid. The eggs vary in size from 300 to 500 grams. The only other species of kiwi where the male is the sole incubator is the Little Spotted. Incubation varies between the species but the North Island Brown Kiwi incuba-

**Fig. 2.** *The Northern Brown Kiwi.*

**Fig. 3.** *Day old Kiwi chick.*

tion ranges from 68 days (as witnessed by the author at Auckland Zoo), to 85 days. Breeding begins in winter (May/June) and extends to late summer (February). The male and female of the closely related Okarito Brown Kiwi share incubation with the male sitting on the egg during the day and the female at night.

Kiwi chicks hatch into fully developed miniatures of the adults and require limited parental care. They are intrinsically built for life on their own. Because of their long beaks and lack of an egg tooth, the chicks stab at the shell in one place and simultaneously kick with their powerful legs to shatter their shells. They hatch with large internal yolk sacs which they live off for the first few days of their lives. North Island Brown Kiwi chicks begin foraging between two and five days of age but return to be brooded by the adult males during the day. At three to five weeks of age they are fully independent. Because kiwi chicks are unattended for so much of the time, mortality from predation is high.

In the wild, kiwi live between ten and twenty years. In captivity, they have been known to live in excess of 40 years. All kiwi are monogamous and pair for life unless their mates die. Males are considered sexually mature at 18 months of age but females may take three to five years to reach sexual maturity. Kiwi are territorial with territories ranging from as little as 1.6 hectares to 100 hectares. Kiwi defend their territories fiercely, even to the death. In territorial disputes they loudly snap their bills, hiss and growl and lash out with their very powerful legs and claws. Many nesting and roosting burrows will be used in their territory. A bonded pair of Kiwi spend just some of their time sleeping in the same burrow. The female North Island Brown Kiwi has a loud hoarse 'crrruik' call while the male has a high pitched ascending whistle making them easy to distinguish without being seen. Kiwi feed at night, and typically have their bill to the ground sniffing for food. Their manner can be likened to a blind man tapping his cane in front of him. When the Kiwi senses a worm or other food in the ground, it will push its beak to the hilt. Valves in the nostrils prevent unwanted soil entering. Their main diet is earthworms of which there are 192 species in New Zealand. Kiwi are omnivorous, opportunistic and inquisitive feeders. They have been known to eat Koura (a native fresh water crayfish), frogs and fish. A common problem in captivity is Kiwi ingesting unusual objects such as nails, keys, wire and plastic sometimes resulting in illness or death. Kiwi drink water by lowering their bills horizontally and tipping back, not by using their bills like straws.

Although Kiwi evolved in the absence of mammalian predators; today they are suffering at the hands of introduced predators such as the stoat, ferret, wild pig, cat, and dog. The most significant of these is the Stoat *Mustela erminea* which takes the more vulnerable kiwi chicks. Despite its small size, the stoat is an effective predator and a prolific breeder. Dogs that are allowed to range free in kiwi areas have been responsible for randomly killing adult kiwi. The most notorious case recorded was a German Shepherd which killed 500 North Island Brown Kiwi in the Waitangi Forest in 1987. The owner of the dog was never located. Loss of habitat through land clearance is also affecting kiwi populations. Neville Peat (1999) states in his book that, "During the twentieth century, the total kiwi population has fallen by more than 90 percent - a staggering decline. The present rate of decline, in order of six percent a year, means that in the absence of effective conservation measures, the population will halve every decade. The key to the recovery of kiwi is the survival of the chicks and juveniles against the onslaught of stoats and other predators."

For most of the twentieth century, kiwi management was very much ad hoc because as a whole, kiwi were not considered to be at risk. Research findings by people like John McLennan started to raise alarm bells. Then, in 1991, the first Kiwi Recovery Plan was prepared by the Department of Conservation (DOC) by Butler and McLennan (1991); and since then has been implemented by D.O.C in conjunction with the Bank of New Zealand and the Forest and Bird Protection Society. The plan covered a five year period and was aimed at halting the decline of the kiwi. A revised plan was prepared and took effect on July 1996 then the latest plan covered 2006 to 2016.

One strategy in the plan was, 'Operation Nest Egg', developed to assist the wild populations of Kiwi. This began in 1995, and involved the removal of eggs from monitored nests, artificial incubation and hatching in captivity. This was to be done at approved institutions that had extensive captive management experience. The chicks were to be reared to a size at which they could defend themselves from stoats, (800g+) before being returned to where they came from. Removing eggs early in the season would mean there was a better chance that pairs would re-nest; potentially doubling chick output. The first trial release of captive reared North Island Brown Kiwi chicks took place in July 1995 by D.O.C staff member, Rogan Colbourne, onto offshore islands. These birds have coped well with the transition from captivity to the wild; however, the real test is to see whether captive reared birds can be successfully released into areas on the mainland in the presence of predators. A number of institutions around the country became involved in Operation Nest Egg, focusing on the populations in their area.

**Auckland Zoo Approached**

In late December 1995, Auckland Zoo was approached by D.O.C and asked if they would be prepared to participate in 'Operation Nest Egg'. This would involve incubating wild Northland Kiwi eggs, hatching the chicks and rearing them to an age considered suitable for release. The Northland race of the North Island Brown Kiwi has been in rapid decline, especially in southern Northland in the last 20 years (Robertson 2003) as a result of land clearance, and the effects of introduced mammalian predators. According to D.O.C; at the (then current) rate of decline of 5.8% per annum (McLennan et al 1996), the Northland Brown Kiwi would be extinct in ten years without intervention.

**First Eggs for Auckland Zoo**

**Fig. 4.** *Auckland Zoo bird breeding facility.*

Fifteen eggs were collected between September 11th and December 20th, 1996 from 12 different nests on the 'Hodges' property, 30 minutes north of Whangarei. The eggs were transported back to Auckland Zoo by vehicle. The eggs were transported in plastic bodied polystyrene 'coolers' packed with shredded paper. A warm (30°C) hot water bottle was placed at the bottom of the cooler. This was used to prevent egg chilling, rather than provide a constant accurate incubation temperature (kiwi are known to leave their developing eggs unattended for hours at a time). The eggs were in transit for up to four hours or more depending on the location of the nest. The egg ages were estimated to be between ten days and 52 days into incubation.

On arrival at Auckland Zoo, the eggs were taken to the 'Bird Breeding Facility' incubation room in a quarantined area of the zoo, where they would not be contaminated by other animals. White gumboots, coats, and Virkon solution footbaths were provided. Once unpacked, the eggs were checked for any damage and then cleaned. This was done by submerging them in a mixture of warm water (38-40°C) and 'Chickguard' solution (1% dilution rate) for no longer than five seconds (the water had to be warmer than the egg to prevent the solution entering the shell).

The eggs were individually weighed, measured and marked with graphite pencil for identification. Then each egg was candled by staff at the zoo to get a more accurate age of the embryo and its condition. Using a graphite pencil, the air cell was drawn on the egg along with the date. Every few days at candling, the advanced air cell would be marked on the egg. Before they were put in the incubator, they were placed on a flat surface to roll to their natural stop position, then a line was drawn down the centre on the top as a reference point for turning. Until 56 days into incubation, the eggs were turned four times a day, 90

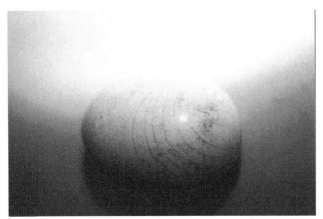

Fig. 5. *Kiwi egg with air cell marked in pencil.*

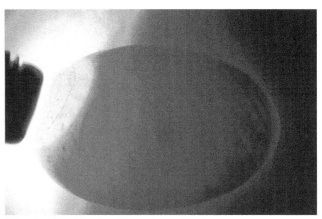

Fig. 6. *Candling kiwi egg (early DIS).*

degrees at a time. They were turned twice in one direction (180° total), then back so the line was at the top again. Then turning would go the other way. The line would never be more than 45° from the top position. Manual turning was done to avoid unnecessary jarring. The turning of Brown Kiwi eggs is a relatively new innovation recommended by Rogan Colbourne (DOC Kiwi Recovery Group staff) from his research on Brown Kiwi incubation.

During incubation, the eggs were weighed daily at 7:00am and removed from the incubator and placed on the bench for one hour to cool down. The cooling was done because there was a lot of anecdotal evidence that the North Island Brown Kiwi leave their eggs unattended for hours at a time. The weights and densities were recorded and 'density loss' over the previous 24hrs was calculated. Temperature and humidity were recorded up to four times each day. Eggs were candled periodically in order to make a visual check on embryonic status.

A Brinsea Hatchmaster 'A' incubator (with automatic turner switched off) was used to incubate the eggs until they were a minimum of 56 days into incubation, then the eggs were moved to a Brinsea Hatchmaker incubator for the remainder of the incubation. These were both still air incubators which better simulated the temperature gradient of natural incubation. Eggs of similar age were incubated together. Incubation was carried out in a temperature controlled room in the Bird Breeding Facility, maintained at 25°C. Humidity in the room was kept at a satisfactory level by a portable dehumidifier. Auckland could get very hot and humid during summer which would otherwise affect the normal operation of the incubators.

The target incubation temperature for the first 56 days was between 36°C and 36.5°C (actual range 34.7°C - 36.5°C). From 56 days to pipping, the target temperature was 35.5°C (actual range 34.5°C -

## Table 1
### INDIVIDUAL INCUBATION RESULTS

| ARKS # | Egg no. | Arrival date | No. days incubation ex-situ | Weight on arrival (g) | Length (mm) | Width (mm) | Incubator | Outcome (Hatch/Died in shell/Infertile |
|---|---|---|---|---|---|---|---|---|
| 960119 | 50a | 17/08/96 | 26 | 409.04 | 120.82 | 79.41 | B | H |
| 960136 | 50b | 17/08/96 | 37 | 465.79 | 130.20 | 80.45 | B | H |
| None | 13a | 17/08/96 | 26 | 420.75 | 119.53 | 80.33 | B | DIS |
| None | 58a | | | 396.02 | 119.70 | 78.05 | | INF |
| None | G/W | | | 401.50 | 122.60 | 75.70 | | INF |
| None | 58b | 17/08/96 | 29 | 431.85 | 126.56 | 78.74 | B | DIS |
| 960147 | 13b | 17/08/96 | 54 | 444.87 | 121.70 | 81.49 | B | H |
| 960138 | 88 | 27/09/96 | 0 | chick | | | | H |
| 960159 | X97/NYO | 16/10/96 | 0 | pipping | | | B | H |
| 960174 | Chick | 16/10/96 | 0 | chick | | | | H |
| 960171 | X11N/Yel | 16/10/96 | 18 | 369.54 | 123.58 | 76.46 | B | H |
| 960173 | 99Na/BW | 31/10/96 | 10 | 441.44 | 125.80 | 82.5 | B | H |
| 960172 | 99Nb/BW | 31/10/96 | 21 | 427.39 | 124.60 | 81.60 | B | H |
| 960192 | 98N/Red | 31/10/96 | 31 | 373.45 | 117.36 | 77.84 | B | H |
| 970004 | 26(2) | 06/12/96 | 34 | 461.49 | 124.93 | 82.28 | B | H |
| 970005 | 69(2) | 06/1296 | 39 | 408.47 | 126.87 | 78.25 | B | H* |
| 970013 | wild | 20/12/96 | 35 | 425.71 | 125.19 | 79.72 | B | H |

*This egg arrived with a hair line crack*

37.5°C). Humidity was maintained between 50% and 60%.

Egg #69(2), second egg of the two egg clutch, arrived at the zoo with a hair-line crack. This meant that weight loss was going to be significantly increased. To compensate for the increased weight loss the egg was incubated separately and the humidity in the incubator was increased. The monitoring records indicated that this was indeed working. Although there was some discussion on whether to seal the crack with 'nail varnish', this was not done.

In addition to the 15 eggs collected, two chicks were transferred to Auckland Zoo; one estimated to be four to five days of age, and one at less than one day of age. The one day old chick was hatching when uplifted and transported back to the vehicle inside D.O.C staff member Pat Miller's shirt next to his body for warmth because the weather was wet and cool and the chick had not yet dried out. The chick was then transported back to Auckland Zoo in a cardboard box on top of a hot water bottle covered with newspaper and cloth material. The older, more robust chick did not require the same treatment and was just transported in a cardboard pet-box.

### The Results

Of the 15 eggs, two were infertile. Two embryo's died in the egg before hatching. Autopsies carried out on the dead embryo's did not reveal any deformities and causes for their deaths could not be certain. However, the embryo in one egg (13a) was measured and it was clear that the embryo was less developed than expected for its age. Its estimated age on arrival could have been wrong. From the time of this eggs arrival at the

## Table 2
### INCUBATION SUMMARY

| | |
|---|---|
| Total number of eggs received (including 2 chicks) | 17 |
| Number hatched (including 2 chicks) | 13 |
| Number died in shell | 2 |
| Number infertile | 2 |

zoo, it was incubated as a 50 day embryo at the lower temperature. As a result, this may have contributed to the embryo's stunted growth and ultimate death. The remaining 11 fertile eggs hatched, including egg #69(2). However, the chick from egg #69(2) hatched with a small piece of external yolk sac. There were 13 chicks to be reared for release.

*Ex-situ* incubation ranged from 18 days to 54 days and these extremes hatched without any problems. Artificial incubation of freshly laid eggs through to a successful hatch has rarely been done. More research is required to do this regularly. As is usual with endangered species, the relatively few eggs available are too important to be experimented with. An overall hatch rate of 73.3% was achieved (of fertile eggs, 84.6%). Table 2, below summarizes the results.

## Chick Rearing

Three to four days elapsed between internal pipping and complete hatch. Once the chicks bills entered the air cell (determined by candling), they were moved to the Brinsea Hatchmaker incubator. The temperature was decreased to 35°C and the humidity increased to 80% or more. When the eggs were almost completely hatched, most were transferred to the 'Bell' zoo built wooden cabinet style incubator/ brooder (based on the successful design by Barry Rowe, formerly of Otorohanga Kiwi House). The few that weren't moved to this incubator were left in the Brinsea Hatchmaker; only because there was insufficient room. This homemade still air incubator had a pull-out basket drawer lined with flannelette material. It was ideal for checking the chicks without having to handle them. Here, they spent the next five to seven days while they lived on the contents of their retracted yolk sacs and gained strength. The temperature was reduced to 27°C by day two or three, and ambient humidity was maintained. The chicks were kept on cloth 'Nappies' in the incubator/brooder and these were changed daily to maintain good hygiene. Each chick's umbilicus was dabbed with iodine solution to clean and dry it, helping to prevent infection.

**Fig. 8.** *Inside kiwi brooder.*

The chick which hatched from egg #69(2) (ARKS #970005) was force fed sooner than normal because there was concern that it was small and light and wouldn't have sufficient yolk reserves. However, the bird was alert and on its feet. It was tube fed 'Ensure Plus', and 'Vytrate' primarily for hydration. It was also force fed small amounts of artificial food (see diet). By day 13, it became apparent that its abdomen was very hard and distended so the chick underwent emergency surgery (carried out by Veterinarian Kerry Rose) to remove its yolk sac. The chick survived the operation but died four hours later. In hind sight, the bird probably had sufficient yolk at hatching and the early feeding may have stopped the bird absorbing the yolk contents. The other chicks were moved to one of two large portable brooders between days five and seven, following an assessment of their strength and alertness. One to four chicks were kept together at any one time. The brooders were of chipboard construction, elevated from the ground on a detachable frame with castors to assist in moving. They were 1750mm long x 635mm wide x 420mm high. At one end of each brooder was a closed in heated compartment with a

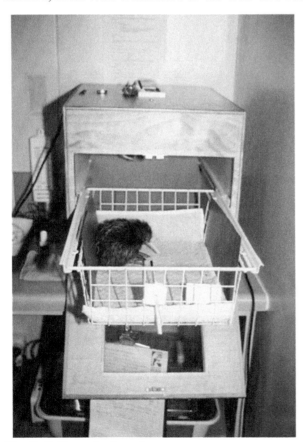

**Fig. 7.** *Bell wooden brooder.*

### Table 3
### THE AGE & WEIGHT OF CHICKS WHEN THEY BEGAN TO EAT

| ARKS # | Hatch Weight | First night of self feed | Last weight before feeding | Weight loss from hatch to self feeding (g) | % Weight loss |
|---|---|---|---|---|---|
| 960119 | 320.40 | 14 | 250.52 | 69.88 | 21.81% |
| 960136 | 359.82 | 9 | 306.25 | 53.57 | 14.88% |
| 960147 | 363.87 | 9 | 271.62 | 92.25 | 25.35% |
| 960138 | 277.36 | 12 | 220.56 | 56.80 | 20.47% |
| 960159 | 352.59 | 11 | 264.47 | 88.12 | 24.99% |
| 960174 | 283.75 (~4-5days old) | 14 | 252.08 | N/A | N/A |
| 960171 | 308.35 | 14 | 228.61 | 79.74 | 25.86% |
| 960173 | 366.06 | 38 | 195.50 | 170.56 | 46.59% |
| 960172 | 357.17 | 9 | 279.53 | 77.64 | 21.73% |
| 960192 | 290.07 | 14 | 224.92 | 65.15 | 22.46% |
| 970004 | 375.38 | 11 | 276.27 | 99.11 | 26.40% |
| 970013 | 335.24 | 13 | 240.45 | 94.79 | 28.27% |

small doorway for the birds to enter and exit through. The doorway was covered by loosely hanging strips of shade cloth to give the chicks a sense of security. Brooding heat was provided by way of a 'Elstein 100 watt infrared heat bulb (which does not give off light). The temperature was crudely regulated by a 'Honeywell' thermostat and was recorded up to four times a day. It ranged from 19.9°C to 30.4°C.

The entire floor area of the brooders was covered in peat moss to which live earth worms were added on an irregular basis. Some leaf litter was added to the heated compartment for nesting.

The peat moss was changed between each batch of chicks (approximately two to four weeks). Initially, the brooders were not covered; but when one chick jumped out on to the brooder room floor, wooden frames, covered in 10mm wire chicken mesh were built as a covers for them. Fortunately, the chick did not suffer any injuries. The brooders were located in the 'Kakapo' brooder room in the Bird Breeding Facility next to the incubator room. Although air-conditioning was available, it was not used so that the chicks would harden to ambient air temperatures. However, room temperatures were monitored and recorded.

As younger chicks came along, the older chicks were moved from the portable brooders to one of three larger brooders attached to the brooder room wall which provided direct access to an outside pen. These brooders (initially designed for Kakapo) are one divided into three by two removable solid partitions. Each compartment (830mm wide x 815mm deep x 980mm high) consists of a light controlled by a dimmer switch, a one-way glass window for viewing, a drawer base to hold the substrate and all surfaces are made of easy clean Formica. The design is based on that used at Burwood Bush for Takahe. Each compartment also has an access door to a fully enclosed outside enclosure ($4m^2$). By the time the chicks were moved into these brooders, they were no longer provided with heat but were given a simple cardboard box in which to sleep. The box was regularly replaced with a clean one.

The door to the outside enclosure was opened one or two days after they were moved to the brooder. With the exception of one chick (#970013), all others vacated the brooder on the first or second night and from then on chose to sleep in the artificial burrows provided outside. These burrows were made of wood and were often not to any specific design. However,

Fig. 9. *Kiwi chick examination.*

the ideal burrow had square sleeping quarters (29cm high and 24cm square with a tunnel 70cm long and 19cm square). Access for staff was from the back of the box or the top by way of a hinged door. The bottom of the sleeping quarters and the tunnel were open to the earth. For convenience, chicks were often kept in groups of four. Without exception, the chicks chose to sleep together in the same burrow. The chosen burrow did not change unless a different one was imposed on them by zoo staff who were instructed to swap dirty burrows with cleaned burrows as the need arose.

## Feeding

Between five and six days of age (the first day in the portable brooder), the chicks were force fed small quantities of the artificial food up to four times a day. This was so that they would acquire a taste for the artificial diet. The chick's weight was recorded before and after feeding. From this point, artificial food was available to them each night. To begin with, 50g was provided and this gradually increased to 200g by the time they were 300 days old. Once the chicks were consistently putting on weight and taking artificial food from the dish, force feeding discontinued. The chicks began feeding on their own between nine and 14 days of age (see Table 3). One stubborn chick didn't start eating on its own until it was 38 days old. The mean percentage of weight loss between hatch weight and last weight before feeding on their own is 24.3%.

Fig. 10. *Owners of 'Hodges Woods' holding kiwi to be released.*

## Diet

The artificial diet consisted of; lean beef heart, finely diced tofu, peas and corn kernels, diced banana, soaked sultanas, wheat germ, yeast flakes, specially formulated Kiwi Premix #178 and vegetable oil. The beef heart and tofu were cut into julienne strips of a size that would be manageable by the chicks (starting at 15mm long). All ingredients were mixed together. Earthworms (*Lumbricus* sp) were provided ad lib once they were consistently eating the artificial diet. The chicks would flick their food around when eating, and it was therefore important to remove any food from the soil. Food containers were placed on sand trays to

### Table 4
**BIRDS RELEASED ON THE MAY 20, 1997**

| ARKS # | Age at Release (days) | Weight (g) | Last Bill Measurement (mm) | Sex (by weight & measurements) | DOC band # |
|---|---|---|---|---|---|
| 960119 | 251 | 1.491 | 82.1 | female | R47389 |
| 960136 | 239 | 1.151 | 73.8 | male | R47388 |
| 960173 | 192 | 1.228 | 70.4 | male | R43466 |
| 960171 | 199 | 1.338 | 80.3 | female | R43465 |
| 960192 | 175 | 1.307 | 70.5 | male | R43464 |
| 960172 | 181 | 2.285 | 80.6 | female | R47390 |

help catch uneaten food.

Fresh drinking water was always available in dishes wide enough for kiwi to lower their bills in horizontally. They often put pieces of food in the water before eating it.

## Chick Rearing Outcome

Of the 13 chicks, 12 survived to be released. The only chick to die was #970005 at 13 days of age following surgery to remove the yolk sac. All remaining chicks were reared without major incident. Two chicks that hatched in the Brinsea Hatchmaker were noted to have bruised bills as a result of poking them into the mesh floor of the incubator. This was later avoided by placing paper towels on the mesh floor so the birds could not poke their bills through the mesh.

## First Release

One month prior to the birds release, they underwent examinations for parasites (particularly Coccidia) and disease. Coccidiosis had been detected in Operation Nest Egg birds. Blood samples were taken, along with cloacal and oral swabs. It was important these birds posed no threat to the wild population. Some birds had tapeworm (a commonly found parasite in wild kiwi) and they were treated and cleared of the parasite before release.

On May 18$^{th}$ 1997, leg transmitters were taped to five of the birds and the sixth bird received its transmitter on the morning of May 20$^{th}$ (in order for the procedure to be caught on film for the TV3 documentary 'Inside New Zealand'). The transmitters are expected to have a life span of 17 months. Weights and measurements taken of the birds suggested there were three males and three females. All birds were over one kilogram which meant they had a good chance of defending themselves against stoats.

On May 20$^{th}$ 1997, the six birds fitted with transmitters were released into the wild on the mainland. Four of the birds were released into vacated disused burrows in 'Hodges Bush' and the other two birds were released in 'Poroa Bush' nearby. Two birds were released into each burrow. These were pairings that were housed together during their pre-release isolation at Auckland Zoo. The remaining six birds underwent thorough examinations and treatments before they were released in the same areas later in 1997.

## Summary of Results of the 1996 & 1997 Season

Overall, 12 chicks were reared for release from 17 eggs/chicks taken from the wild, a success rate of 70.58% (if we exclude the two infertile eggs, the percentage increases to 80%).

Clearly, the objective to rear kiwi in captivity from eggs and chicks taken from the wild was very successful, especially considering that some of these eggs and chicks would have perished if left in the wild. There would be an advantage in being able to incubate freshly laid eggs which are vulnerable to misadventure in the wild. In 1976, at the Otorohanga Kiwi House, a dummy egg with a thermometer was placed under a brooding male North Island Brown Kiwi to determine a more accurate incubation temperature (Rowe 1978). It was found that the top of a naturally incubated kiwi egg reached a temperature of 37°C. This exercise was repeated in 1994 by Rogan Colbourne with similar results. It is still a rare event for eggs to hatch when artificially incubated from 'point of lay'. Although, Auckland Zoo has incubated and hatched two eggs that were collected freshly from a captive pair of Kiwi (Nelson, A. unpubl. data). Research undertaken by Bassett and Potter (1998), indicated that far too many kiwi embryos are dying under artificial incubation, and they suggested that this was a result of sub-optimal incubation regimes.

Up to 2006, Auckland Zoo has contributed 130 birds to the Northland population as a result of its involvement in Operation Nest Egg. Since 1998, Auckland Zoo has used Motuora Island (85ha) in the Hauraki Gulf as a safe creche (free of stoats). Releasing chicks back to the mainland once they are large enough to defend themselves against stoats. Initially, all young Kiwi were radio tagged before release on the island. After two of the first 15 released died because their transmitters became entangled, all chicks have been released with 'Trovan' identification implants. Once the chicks have reached sub-adult size they are released back on the mainland with transmitters. Auckland Zoo's participation in 'Operation Nest Egg' continues today.

## The Special Egg of 1998 & 1999 Season

In the 1998/1999 season, one egg that was brought to the zoo for hatching, when candled, revealed a well developed embryo (67 to 75 days into incubation) but the air cell was malpositioned at the side of the egg (instead of off-centre at the blunt end). There was no way of knowing what position the embryo was in by candling alone. If the embryo was malpositioned, human intervention would probably be required. Kiwi eggs visibly rock from side to side once the embryo has developed beyond three weeks (pers. obsv.), so it was soon established that this embryo was alive and well. After consulting with the zoo veterinarian, Richard Jakob-Hoff, it was decided to X-ray the egg. This showed that the chick was in the normal position. However to get a better idea of its age the vet suggested that the Auckland Hospital be approached and asked if they would do an Ultra Scan and/or an MRI scan of the egg. The hospital was intrigued enough to say they would do the scans. On September 11, 1998, eight days after its arrival at the zoo, the egg was taken to the Auckland Hospital where it first underwent the Ultra Scan. Unfortunately the density of the shell made the Ultra Scan ineffective. The following MRI clearly showed the chick in the egg. Despite its unusually positioned air cell, the chick was in the normal position and had an external yolk-sac which suggested it was not yet ready to hatch. There was concern that; when and if the chick tried hatching it might suffocate in the amniotic fluid or be overcome by carbon monoxide. The egg was taken back to the zoo and put back in the incubator. Any assistance given to the chick had to be timed perfectly to have a successful hatch. There was also the possibility the chick would hatch without assistance.

Thirteen days after its arrival, the chicks' activity had reduced to the point where intervention became necessary. It was X-rayed at the zoo to see where the bill was positioned, then the vet began an exploratory operation with the assistance of avicultural assistant, Kelly Cosgrave (this X-ray showed an increase in the chicks bone calcification, a sign it was close to hatching). Small pieces of shell and membrane were removed to expose the tip of the bill. It was then left for two more days to see if it would finish hatching on its own. However, on September 18th at 10:15pm a painstaking, two hour assistance hatch began which resulted in a healthy but weak kiwi chick. Iodine solution was swabbed on the umbilicus. This chick was successfully reared and went on to be released into the wild (Cosgrave 1999).

### Haast Tokoeka - worth a mention

The Haast Tokoeka is one of New Zealand's most endangered Kiwi. Two to three hundred are believed to exist. They are only found in a small area about 25 kilometers south of Haast on the west coast of the South Island. They are mainly found from the foot of the Haast range up to an altitude of 1700m above sea level. The main threats are the same as for other Kiwi.

In 2003/2004 the Kiwi Recovery Team decided to trial Operation Nest Egg with the Haast Tokoeka for the first time. The nearest captive institution with kiwi expertise was the kiwi and Birdlife Park in Queenstown, 200kms south. The author was enlisted to oversee the program, having had experience with the Northland Brown Kiwi. Facilities and procedures at the Kiwi & Birdlife Park were modelled on those used at Auckland Zoo. It was planned that D.O.C field staff would lift up to five viable eggs from the wild and deliver them by portable incubator to the Kiwi and Birdlife Park. The plan was to hatch the chicks and rear them for several weeks before taking them to Burwood Bush (Takahe rearing facility). A total of three eggs were transferred on August 28th, 2003 and November 4th, 2004 to the Kiwi and Birdlife Park. Two eggs were hatched (named 'Tititea' and 'Wilson'). They were both sent to Burwood Bush Takahe rearing facility once they were eating well on their own to continue growing to 1kg in weight, before being returned to Haast. Unfortunately, Wilson died at Burwood Bush having lost condition. There was concern that this bird could not find enough to eat. However, Tititea was successfully released back at Haast. The third egg died just prior to hatching. This egg had been artificially incubated from day one. A fourth egg which was pipping, was collected later in the season. This chick died following an assisted hatch. The egg had been abandoned and was very cold when uplifted. It was initially thought to be dead on arrival at the park. Chilling may have been the ultimate cause of its death.

Despite the small number reared, the first trial was declared a success so was continued the following

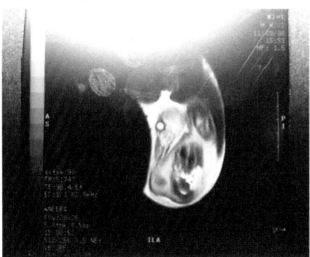

**Fig. 11.** *MRI scan of embryo.*

**Fig. 12.** *Tokoeka chick in brooder.*

year (2004/2005). A chick was taken to Queenstown where it was reared for two weeks. At 400g, it was released on Centre Island on lake Te Anau where it stayed until it reached 1250g. Then on April 5$^{th}$, 2005 it was released back at Haast. One egg was taken to Queenstown on October 28$^{th}$, 2004 and successfully hatched on November 20$^{th}$, 2004, however, this chick died several weeks later.

The following year, the Department of Conservation decided to continue the ONE program for the Haast Tokoeka at the Willowbank Wildlife Reserve in Christchurch and there it continues today.

## ACKNOWLEDGEMENTS

Acknowledgements go to; Dr. Richard Jakob-Hoff (Auckland Zoo), Chris Hibbard (Auckland Zoo), Kelly Cosgrave (Auckland Zoo), Sandra Rice (Auckland Zoo), Ann William's (Auckland Zoo), Kirsty Chalmers (Auckland Zoo), Chris Smutts-Kennedy (DOC Hamilton), Rogan Colbourne (Kiwi Recovery Group), Pat Miller (DOC Whangarei), Tracey Johnson (Captive Coordinator), Wayne Johnson TVNZ, and the 'Hodges', who have all contributed to the success of 'Operation Nest Egg' at Auckland Zoo. Bevan Cameron (Kiwi & Birdlife Park), Paul Wilson (Kiwi & Birdlife Park), Matt Wong (Kiwi & Birdlife Park), Kristy (Kiwi & Birdlife Park), Paul Vanklink (DOC Haast), Phil Tisch (DOC Haast).

## REFERENCES

BUTLER D & J McLennan. 1991: Kiwi recovery plan. Threatened Species Recovery Plan 2. Department of Conservation, Wellington.

BASSETT SM & MA Potter. 1998. Kiwi embryo mortality. Proceedings of the New Zealand Conservation Management Group Kiwi Workshop, 1998.

COLBOURNE R, S Bassett, T Billing, H McCormick, J McLennan, A Nelson and H Robertson. 2005. The development of Operation Nest Egg as a tool in the conservation management of kiwi. Science for Conservation 259, (24p).

COSGRAVE K. 1999. Kiwi- A hard egg to crack. ARAZPA/ASZK conference proceedings, Alice Springs, N.T, Australia.

JOHNSON T. 1996. Husbandry Manual for North Island Brown Kiwi.

PEAT N. 1999. Kiwi New Zealands Remarkable Bird. Random House, Auckland, New Zealand.

PEAT N. 2006. Kiwi the people's bird. Otago University Press, Dunedin, New Zealand.

ROBERTSON H. 2003. Kiwi Recovery Plan, 1996 - 2006. Threatened Species Recovery Plan 50.

ROWE B. 1978. Incubation temperatures of the North Island Brown Kiwi. Notornis 25(3): 213-217.

## AUTHOR BIOGRAPHY

Martin has been involved in the captive management of New Zealand wildlife since 1980. His interest in birds began with keeping chickens as a child and he worked at a commercial chicken hatchery for three years after leaving school. He later went on to work for the Wildlife Service, the Department of Conservation, Auckland Zoo and the Kiwi and Birdlife Park in Queenstown. He has specialised in the captive management of a variety of endangered species including; Takahe, Kiwi, Blue Duck, New Zealand Dotterel and the New Zealand Fairy Tern. He was responsible for pioneering specialist techniques and equipment for hand-rearing Takahe, the Black Stilt, and the Fairy Tern; specifically to prepare them for release into the wild. Martin's most recent contribution to conservation was the trial incubation and rearing of the very rare "Haast Tokoeka" Kiwi. Today he has come full circle; enjoying, with his family in Queenstown, the company of a small number of domestic chickens.

# WHY IS REINTRODUCTION OF NORTHERN BALD IBIS *GERONTICUS EREMITA* SO COMPLICATED? AN OVERVIEW OF RECENT PROGRESS AND POTENTIAL

CHRISTOPHER G.R. BOWDEN[1], CHRISTIANE BÖHM[2], MIKE J.R. JORDAN[3] & KEN W. SMITH[1]

[1] Royal Society for the Protection of Birds, Bedforshire, UK
[2] Alpenzoo Innsbruck-Tirol, Weiherburggasse 37a, A-6020 Innsbruck, Austria.
[3] North of England Zoological Society, Chester Zoo Chester, UK.

## SUMMARY

The Northern Bald Ibis is classed as Critically Endangered by IUCN, with just over 100 breeding pairs left in the wild, and most of the top conservation priorities relate to *in-situ* threats in Morocco and Syria. There is a significant captive population, presenting the theoretical potential for reintroduction to parts of the former range. Due to the social complexity of the species and a variety of other factors, early release attempts were unsuccessful, and the actual potential for reintroduction has proved to be very difficult. However, there has been very significant progress towards this over the past ten years. There is now a proven (if highly intensive) method developed for establishing a sedentary population, but most areas where the species occurred would have had migratory populations, and progress with developing a way to reestablish migratory behaviour is still more complex and is at a more experimental stage. Less intensive options are also being tested. There is a very diverse set of organisations and interests in the species, all of whom have relevant expertise right across the spectrum. These include *in-situ* conservationists, government bodies, NGOs, zoo and captive experts, as well as behavioral biologists. Focusing and coordinating efforts between the diverse players involved, and at the same time keeping the conservation priorities for the species firmly in mind has been the key objective of the International Advisory Group for Northern Bald Ibis (IAGNBI), an independent group that has been effectively pursuing these aims since 1999. IAGNBI produces newsletters, holds meetings, produces very detailed meeting reports and has contributed heavily to the recently produced Species Action Plan.

## INTRODUCTION

The Northern Bald Ibis *Geronticus eremita* (NBI) is classified as "Critically Endangered" because of its small range and population. It has undergone a long history of decline over at least four centuries, having been formerly distributed over much of north and northeast Africa and the Middle East. It even occurred throughout the European Alps where it disappeared over 400 years ago. The main population now occurs in Morocco and now numbers just over 100 breeding pairs. In addition, a small relict population of two pairs persists in Syria where it was rediscovered in 2002. These represent the remaining wild western and eastern populations that are regarded as genetically distinct. In Turkey, a small semi-wild population is maintained at Birecik, the site of a former wild breeding colony, but this has lost its migratory habit and requires active intervention each year to ensure the birds survive the winter. However, it represents an important genetic resource for a time when reintroduction methodology has been developed further. A further 1300 birds of western population origin are held in studbook-registered collections, mainly in European zoos.

Conservation efforts and Species Action Plan priorities (Jimenez-Armesto et al 2006) focus on first securing the remaining Moroccan wild birds, by gaining understanding of their ecology and needs and implementing a program of local support measures for these. This has been quite successful recently, with the population showing some recovery from a low of 59 pairs in 1997, but the 108 breeding pairs are still confined to essentially two sites, so they are potentially very vulnerable to any one-off disasters, which could dramatically change the situation. In Syria, the situation is still more fragile, but efforts have concentrated on raising the profile of the bird, ensuring it is fully protected and monitored at the breeding sites and investigating the migratory behaviour. With such a low population, the Syrian situation is critical. As far as we know these birds are the last in the Middle East to retain the migratory habit, and in 2006 it was confirmed for the first time, using satellite tagging, that the birds wintered in Ethiopia with brief stops in Saudi Arabia, Yemen and Eritrea (Peske 2007; Serra et al 2007; J. Lindsell pers. comm. 2007).

Northern Bald Ibis is a highly socially complex species, breeding colonially and usually feeding in groups. The remaining wild birds are consequently very concentrated in just three sites, making them even more susceptible to one-off incidents than the low population figures might suggest.

Concerted efforts by the zoo community since the 1960s have ensured the establishment of breeding populations in many zoos. All captive birds (with the exception of around 30 held in Turkish zoos) are of Moroccan origin. Breeding and keeping is no longer a problem, due to the very considerable progress that has been made with feeding, aviary design and other features, ensuring that captive birds can breed and survive successfully. Efforts to release birds back into the wild have so far been less successful, with a number of notable failures in the early days although a more considered and intensive approach over the last decade is showing that potential methods are likely to become available.

The following summary outlines the plight of the species over the past hundred years:

- 1900-1920 - 50 colonies in Morocco, four in Algeria, c.25 colonies in Syria & Turkey, + frequent records in Eritrea/Yemen.
- 1950-60 - Massive declines documented especially in Turkey (& Morocco)
- 1985 - Critical status highlighted (Collar & Stuart 1985 - African Red Data Book) as already down to only four to five colonies and declines continuing.
- 1989 - Extinction in Turkey and two other Moroccan sites left only two colonies.
- 1997 - Moroccan population low of 59 pairs after 1996 mortality (Bowden et al 2003; Touti et al 1999)
- 2002 - Two remnant pairs rediscovered in Syria (Serra et al 2003)
- Morocco population 1998 to 2007 currently undergoing small expansion to 108 pairs (El Bekkay & Oubrou 2007), but Eastern population in Syria remains at just two breeding pairs (Serra & Peske 2007).
- Turkish semi-wild population at Birecik currently increasing (18 pairs in 2007 pairs), but in need of appraising habitat viability in that area (Ozbagdatli 2007).

## Captive Breeding Success

Captive breeding was first attempted in the late 1950s and was initially problematic. Success was greatly improved particularly by changing food, but also with better husbandry and aviary design. There is currently a very substantial captive NBI population, which is managed through three different studbooks. These are North America- since 1990, currently 130 birds; Japan since 1992- 120 birds; and Europe since 1988- 900 birds (Böhm 2006). The European studbook is not only the one with most birds, but also holds the main body of founders. All captive birds within the studbook originate from Moroccan stock, comprising 150 birds that were originally imported from the wild. There are also a substantial number of unregistered birds held which bring the total number of birds in captivity closer to an estimated 2000.

## The Origin and Characteristics of the Registered EEP Population

The first reported imports of NBI from Morocco into Europe date back to 1949-1954 when birds were brought to Basel Zoo. A total of 71 imports are recorded into the NBI Endangered Species Program (NBI EEP), the majority of which had arrived before 1960 (42 birds), and the latest of which were in the 1970s (29 birds). Forty-nine of the imported birds survived at least two years and so this is an indication of the likely size of the founder population (ie: birds that are likely to have bred). Having said this, there are only six definite breeding records involving founder birds.

In 1934, several birds (of unknown origin) were brought to Berlin Zoo from Palestine and one more from Birecik, Turkey in 1973 to Basel Zoo. The exact fate of the former is unknown but they are unlikely to have bred. The bird from Birecik died within a few days. Therefore, no birds of the eastern population survived in the European zoos. Two thirds of the EEP population is younger than ten years, and the population is gradually increasing. Breeding success is actively managed and suppressed in most of those colonies in accordance with studbook recommendations (Böhm 2006).

## Potential Health Concerns of Captive NBI

Apart from susceptibility to avian tuberculosis, which has seriously reduced some captive populations, NBI have few special concerns. The one exception may be a distinctive "skin problem" which has appeared in several separate collections (Quevedo & von Houwald 2003). It appears as a chronic ulcerative dermatitis characterized by the loss of feathers, rawness, and ulceration. It is found on the back, neck, and underside of the wings. Although some birds died quickly, others lived for quite a long time with those areas of bare skin exposed. Some individuals have recovered and feathers regrown. Recent evidence now suggests this may be a result of zinc or other mineral deficiencies (M. Quevedo pers. comm). Further, there is an indication that stress (transfers, low ranking in the hierarchy, overcrowded aviaries, etc.) may trigger the problem, and it does highlight the need for caution and strict veterinary protocols when releasing birds to the wild.

A final particular trait to mention for NBI is their habit of swallowing any small objects, including nails and wire, that they find. This is a particular problem in captive conditions where very significant numbers have died because of a punctured digestive tract.

## THE HISTORY OF RELEASE METHODOLOGY WITH REFERENCE TO THE DEVELOPMENT OF A VIABLE REINTRODUCTION TECHNIQUE

### Early Reintroduction Trials in Israel and Elsewhere

Various preliminary release attempts were carried out which highlighted the fact that for NBI, release is far from being a straightforward issue. Four separate trials were carried out in Israel between 1983 and 1987. These ranged from the releasing of 34 birds, mainly adults while they were actively breeding, to releasing newly fledged juveniles, and then releasing 16 wing-clipped juveniles and finally a group of six hand-reared fledglings. Despite the very negative results from all of these trials, it was well documented by Mendelssohn (1994), which paved the way for further progress to tackle the difficulties, particularly the progress that was later made in Austria. The birds from all of the Israeli work became disorientated, often emaciated, and the social bonding was apparently very weak. The only signs of success came from one or two of the hand-reared birds in the last of these trials, which gave some important leads for later work. More details are given in Pegoraro (2003).

Other proposals were developed in Spain and Morocco in the early 1990s, but these were not implemented largely due to complications of funding and methodology. One in Italy, in 1991 and 1992 was originally planning to use a hand-rearing method, but due to the large number of people needed to do this, and limited resources available, it became a less intensive method, and in the end, the birds were apparently never released. There was also an early proposal to release birds within the Souss-Massa area, but this was not taken forwards on the grounds that methodology was not sufficiently developed at that time, and it would have been inappropriate to carry out a release so close to the only wild population.

The other ongoing experiment through the 1980s and 1990s was the soft-release of the semi-wild birds at Birecik. Each year the birds are recaptured into the aviaries in July and August, to prevent them disappearing over the winter period. But each year, a number have gone missing, sometimes groups disappearing together, such that a total of over 200 birds have disappeared over the 25 year period. Whilst many of these are simply local mortalities, it is clear that at least some have been effectively released. Unfortunately, monitoring has not been systematic for most of the period concerned, but it seems fairly clear that no birds have survived to return the following spring.

### Alpenzoo/Austrian Preliminary Work and Breakthrough Using Hand-rearing Techniques

This was the first use of intensive hand-rearing method, carried out at Alpenzoo in 1991. It led to strong social bonding within the group of nestlings, and illustrated that close social contact with a restricted number of human foster-parents could be effective. It attracted a lot of attention as hand-rearing was not formerly considered a suitable tool in this field due to other potential problems associated with imprinting on humans (Pegoraro 1996). Although the trial was not continued to the following year, it laid the foundations for development of this technique by the Konrad Lorenz Institute work.

### Behavioural Projects at the Konrad Lorenz Institute & Waldrapp Team

These experimental projects started in 1997 with the aim of establishing a local colony of semi-tame birds at the Konrad Lorenz Forschungsstelle in Grünau, Austria. These birds were to be used as a model for investigating the mechanisms of sociality in birds and later, to investigate techniques for establishing NBI colonies. Over the first four years of the project, 43 hand-reared birds were released and these birds bred themselves for the first time in 2002 and again in subsequent years (Kotrschal 2007). The project utilized the hand-rearing techniques that had previously proved successful and have established a sedentary, free-flying colony of NBI, albeit in habitat unsuitable for a sedentary colony and therefore intervention is required in the winter for the birds to survive.

This project has had considerable success and a high degree of scientific monitoring and knowledge sharing and has contributed very substantially to the current concepts of NBI reintroduction.

As a related project, a second group was established in 2002 at the nearby airstrip at Scharnstein with a view to investigating migration tradition and physiology utilizing NBI following ultralight aircraft. This project has achieved considerable success and in 2004 the first successful autumn migration occurred - the birds followed the ultralight aircraft from Austria to an over-wintering area in Italy (Fritz 2007). By the 2006 season, no birds had completed the return journey although several birds made long distance flights (up to 300km) north towards their original fledging site in Austria. This work is very much ongoing with further progress already being reported in 2007.

An important (but unfortunately very intensive) feature of the hand-rearing method used by both of these projects is that the hand-rearing is only carried out by a very small number of 'ibis-mothers' who talk to the birds and bond very much as individuals with the individual birds. This seems to be a feature needed to make that bond to the group so strong.

**Bechar el Kheir Project - Mezguitem, Morocco**

This project was initiated in 1997 and in 2000 construction of breeding aviaries and ancillary buildings commenced. The project was originally a GTZ project together with the Moroccan Government, but soon transferred to an independent conglomerate of zoos headed by Munich Zoo. The first NBI arrived at the station at the end of 2000, with the aim of releasing birds in 2003 and continuing for ten years (with the release of a total of 200 birds). By 2006, the project had still not established a productive captive colony that would facilitate release; the first successful year of breeding was 2006 and by the end of that year the aviaries housed 19 NBI (Fritz & Pfistermueller 2007).

The project has been controversial since its inception and attracted considerable criticism, mainly for its failure to disclose information and engage with IAGNBI or the IUCN, and the Moroccan Government also has growing concerns with regards to how appropriate the work is. Concerns have been raised over the apparent lack of health screening of birds and of the unsuitability of the site chosen for release. It seems unlikely that NBI could establish a sedentary colony at the chosen site and so would be forced into winter movements with the inherent risks of contacting the existing wild populations at Souss Massa.

An evaluation of this project is urgently outstanding and the likelihood of birds being released and the project's future role remaining in question.

**Spanish Release Programs (Proyecto Eremita)**

Reintroduction of NBI in Spain was first considered in 1991, when birds were established in Jerez Zoo with a view to release at Almeria. This project did not progress further for a variety of reasons, and it was not until 1999 that in a revised plan, 'Proyecto Eremita' was formulated. This project proceeded in 2002 as a joint venture of Jerez Zoo and the Andalusian Government.

Proyecto Eremita has been from its conception an experimental reintroduction designed to test techniques for establishing a primarily sedentary colony in suitable habitat in La Janda, Cadiz, SW Spain. The first release of 23 NBI occurred in 2004 and birds have been released each year since then (a total of 73 birds up until the end of 2006), with releases continuing in 2007. One quite separate technique was tested, rearing together with Cattle Egrets *Bubulcus ibis* (which was not successful) and other methods tried have focused on making the methodology less intensive and thereby less costly than the hand-rearing in Austria (Quevedo et al 2007). A distinctive feature of the hand-rearing technique is the use of uniform T-shirts and an ibis helmet worn by staff. By trying multiple variations within the group each year, it has not been easy to interpret results, and the dispersal and losses of most of the birds are a sign that there is clearly not a stable group as yet, apparently due to weaker bonding with the group or site. Therefore, success has been mixed and at times somewhat controversial with some birds known to have dispersed to Morocco.

Steps to prevent the heavy losses and dispersion of the birds are being taken by closing them in at the critical dispersal period. Nevertheless, the program has carried out extensive health screening and post-release monitoring and has engaged at every opportunity with IAGNBI and the IUCN SSC Reintroduction Specialist Group to discuss its plans. Some useful methodologies are been tested, with emphasis on reducing the costs involved. The latest step is attempting the integration of parent-reared and hand-reared birds at the time of release.

## Turkish 'Soft Releasing' and Syria Emergency Supplementation Plan

As mentioned above, birds had been 'escaping' in late summer (shortly before being enclosed for the winter) for years from the Birecik semi-wild colony in Turkey, although these have never been proved to return the following spring (Arihan 1998; Ozbagdatli 2007). Improved monitoring, husbandry and aviaries have lead to a sustained increase in the number of birds, particularly over the past four years, and plans are being developed to satellite tag selected birds to discover the fate and movements of birds allowed to remain free-flying.

In addition, the situation with the remaining Syrian wild population is so critical that despite the lack of a proven technique, an 'Emergency Plan' involving a small-scale supplementation with a Birecik-origin juvenile has been developed. This would only be put into action in response to a further decline of the remaining two pairs. The details of this proposed method are still being discussed by IAGNBI and the authorities involved - concerns relating to possible disruption to the remaining wild birds mean that only one or maximum two juvenile birds would be supplemented in a given year. There is a key need to test such a plan elsewhere before carrying it out in Syria, but there may not be time left to do this. Of course, there are additional practical and administrative hurdles involved in implementing such a plan, which are being tackled and preparatory work is already underway in case this last resort action is implemented in the near future.

## The Creation of the Advisory Group (IAGNBI), its role and functions

The International Advisory Group for Northern Bald Ibis was created in 1999 at a workshop in Agadir (Anon 1999) which brought together all of the main NBI players, and produced some guidelines on key actions needed for the Moroccan population, and criteria for any potential release trials. It has representation from a wide range of the players involved, (full designations and responsibilities given in Appendix I, together with terms of reference) and has subsequently held two major meetings (2003 and 2006) with well documented reports as outputs (Böhm et al eds. 2003; Böhm et al eds. 2007). It has, in addition to e-mail discussions within the group, it has produced three very detailed newsletters (Böhm ed. 2004; Bowden ed. 2001; Bowden ed. 2003) and contributed very substantially to the Species Action Plan, which was largely funded by AEWA (Jimenez-Armesto et al 2006). The group has been recognised by BirdLife and IUCN as the authority for the species, and AEWA has first approached IAGNBI for the Action Plan workshop and indeed to take responsibility for its implementation. Whether the latter function will be formalized is still under discussion, and not yet clear due to the diversity of the group, and potential problems associated with any fund-raising role it might develop.

## Key Priorities From SAP

The Species Action Plan resulting from a meeting held in Madrid in 2004, came up with the following priorities for the species (Table 1). It is notable that the highest priority actions for the SAP are almost exclusively *in-situ* issues for the remaining wild populations, ie: six of the seven critical priorities, with the seventh being the need for coordination of efforts through IAGNBI. Even the high priority actions are mostly concerned with safety precautions for the wild birds and further *in-situ* measures, with just the careful management of a captive population being the main *ex-situ* action. The development of release techniques for reintroduction comes among the medium importance measures. This reflects the lower urgency for this work, but of course, this whole area is one which can potentially be progressed more readily than some of the *in-situ* challenges, but it often comes down to the availability of potential resources, and appeal to potential funders. It is a very real challenge to find ways to channel more resources to addressing the higher priorities that are needed for the species, and one that IAGNBI is well placed to influence.

## Discussion

The level of interest in NBI is high and still growing for a variety of reasons, not just because of its critical conservation status and the increasing threats associated with these, despite the recent slight recovery in numbers of the main wild population in Morocco to just over 100 breeding pairs. There is also the historical interest, eg: having its own Egyptian hieroglyphic symbol, and the story of the former European Alps populations that disappeared 400 years ago (Pegoraro & Föger 1999). Then there is the accessibility of the species to the public in numerous zoo collections; the recent activity and progress with release and ultralight

## Table 1

**RANKED RESULTS OF THE INTERNATIONAL SPECIES ACTION PLAN FOR NORTHERN BALD IBIS**

**Critical Importance:**
1. Building on or near to NBI breeding and feeding sites restricted.
2. Hunting stopped
3. Agriculture and grazing regimes maintained/altered to provide suitable feeding areas.
4. Risk of intoxication reduced
5. Collection of firewood controlled to prevent destruction or degradation of NBI feeding areas.
6. Socio-economic factors driving land use changes investigated and addressed in partnership with local communities and stakeholders.
7. The conservation of the NBI through international coordination and cooperation promoted by the International Advisory Group for the Northern Bald Ibis (IAGNBI)

**High importance:**
1. Breeding success, inter- and intra-specific competition, and predation monitored at all exiting breeding colonies.
2. Provision of uncontaminated fresh water sources close to breeding sites maintained and improved
3. Risk of infection disease reduced
4. A comprehensive health screening conducted on all birds prior to reintroduction/release experiments.
5. A captive population maintained with health, inbreeding and age structure managed.
6. Habitat requirements, food availability and foraging ecology in the current range and release trial sites researched and compared.

**Medium importance:**
1. The level of genetic variation within the captive, semi- wild and wild populations assessed.
2. Techniques for the establishment of new colonies by reintroduction investigated.
3. Disturbance by military firing range reduced

**Low importance:**
1. The impact of the introduction of new birds to existing breeding colonies researched in captivity during the breeding season.
2. Discarded fishing line and other potentially dangerous debris to be collected and disposed of safely.
3. Risks reduced related to electric wires and collision
4. Reservoir construction affecting feeding and breeding sites controlled.

migration trials; and now the latest satellite tagging results for the Syrian birds, revealing the migration route for the eastern population. All of these elements have combined to create a lot of good will and potential for publicity and funding for the species, and presents a challenge that needs to be channeled carefully for the maximum benefit of the species.

The range of organisations, individuals and issues involved in conserving NBI is extraordinarily diverse, and the main line of communication for coordinating efforts has become through the IAGNBI, together with the recent production of an international SAP, which has further formalised the agreed priorities.

Although the development of release methodologies for eventual use in full-scale reintroduction work may not be the highest priority for the species, it is undoubtedly still very important, and an area that can be addressed by a far wider range of institutions than are directly involved in the *in-situ* measures. The social complexity of the species is a particular challenge, which makes the whole process less straightforward than for most birds. Progress with techniques has been very significant over the past five to ten years, with the major progress so far being development of a method (admittedly intensive) for establishing a sedentary population, as well as the encouraging signs for developing a method to establish a migratory one. The next steps will involve identifying more clearly where these methods would best be applied, and at the same time ensure that the wild populations receive the attention they need with growing development pressures threatening to undermine the recent positive trends. Issues such as: defining the former range; what constitutes an introduction versus a reintroduction; how close to any remaining wild populations a reintroduction should be carried out; and just how serious the risks presented by reintroduced birds are if they come into contact with wild birds; are all potentially hot topics that make decisions on where future reintroductions should be and observing IUCN reintroduction guidelines quite difficult. If the key players can continue to develop links, and the network can ensure that the remaining wild populations receive the necessary resources they need to persist, then this will be a huge contribution towards the ultimate conservation of the species, where IAGNBI can play a very important coordinating role. There will always be potential tensions that can be caused by greater funding potential for appealing release programs compared with the apparently more mundane and ongoing issues at sites where the wild population remain, but good communication and cooperation between programs can go a long way to preventing this from causing problems.

For the remaining wild NBI populations, it will be the engagement of local people and identifying ways in which they can gain some benefit and pride from the continuing presence of NBI, that will be the most important priority for the long-term future of the species. National NBI species action plans for Morocco, Syria and Turkey are all planned in the near future which will undoubtedly address these issues and the specific threats in each case as a top priority. Once a release methodology has been refined further, we can consider in more detail where it could most effectively be implemented.

*Note: This paper outlines the position in 2007 and subsequently, largely because of satellite tagging results from Turkey and Syria, there have been major increases in our understanding of the birds in the Middle East. The plan is now to attempt to supplement the Syrian birds after recent breeding failures of the two remaining wild pairs in 2008 and 2009. IAGNBI now has an active web site www.iagnbi.org where further updates can be obtained.*

## References

ANON 1999. International workshop on a strategy for the rehabilitation of the Northern Bald Ibis. Agadir, March 8-12 1999; Min. Chargé des Eaux et Forêts, Maroc; BirdLife International/Royal Society for the Protection of Birds, Deutsche Gesellschaft für Technische Zusammenarbeit (GTZ). Workshop report: RSPB, Sandy, UK. 47pp.

ARIHAN O. 1998. Recent information on the occurrence of the Northern Bald Ibis *Geronticus eremita* in Turkey. Turna 1:10-15.

BÖHM C. (ed). 1999. Northern Bald Ibis Geronticus eremita, 2nd EEP Studbook 1999: 52-64. Alpenzoo, Innsbruck-Tyrol.

BÖHM C. 2006. Northern Bald Ibis Geronticus eremita, 3rd EEP Studbook, Alpenzoo, Innsbruck-Tyrol pp 64.

BÖHM C. (ed). 2004. IAGNBI Newsletter No. 3 (electronic newsletter, pp 72).

BÖHM C, CGR Bowden & MJR Jordan (eds). 2003 Northern Bald Ibis conservation and reintroduction workshop. Proceedings of the International advisory Group for Northern Bald Ibis (IAGNBI) meeting Alpenzoo, Innsbruck-Tirol, July 2003. pp81 ISBN 1-901930-44-0.

BÖHM C, CGR Bowden, MJR Jordan & C King (eds). 2007. Northern Bald Ibis Conservation and Reintroduction workshop. Proceedings of 2nd Meeting of International Advisory Group for Northern Bald Ibis (IAGNBI), Vejer, Spain, September 2006. pp 124. ISBN 978-1-905601-00-4.

BOWDEN CGR (ed.). 2001. IAGNBI Newsletter No. 1 (electronic newsletter, pp 40).

BOWDEN CGR (ed.). 2003. IAGNBI Newsletter No. 2 (electronic newsletter, pp 49)

BOWDEN CGR, A Aghnaj, KW Smith and M Ribi. 2003. The status and recent breeding performance of the critically endangered Northern Bald Ibis *Geronticus eremita* population on the Atlantic coast of Morocco. Ibis 145: 419-431.

COLLAR NJ & SN Stuart. 1985. Threatened Birds of Africa and Related Islands: the ICBP/IUCN Red Data Book. Cambridge, UK: International Council for Bird Preservation and International Union for Conservation of Nature and Natural Resources.

EL BEKKAY M & W Oubrou. 2007. NBI conservation project in Souss-Massa region. In: Böhm, C, Bowden, CGR, Jordan, M. & King, C. (eds). (2007). Northern Bald Ibis Conservation and Reintroduction workshop. Proceedings of 2nd Meeting of International Advisory Group for Northern Bald Ibis (IAGNBI), Vejer, Spain. September 2006.

FRITZ J. 2007. The Scharnstein Waldrapp Ibis Migration-Project after four years: the birds leave the microlites behind. In: Böhm, C, Bowden, CGR, Jordan M & King C (eds). (2007). Northern Bald Ibis Conservation and Reintroduction workshop. Proceedings of 2nd Meeting of International Advisory Group for Northern Bald Ibis (IAGNBI), Vejer, Spain, September 2006.

FRITZ J & R Pfistermueller. 2007. Travel report November 2006 Station Bechar el Kheir Morocco. In: Böhm, C, Bowden CGR, Jordan M & King C (eds). (2007). Northern Bald Ibis Conservation and Reintroduction workshop. Proceedings of 2nd Meeting of International Advisory Group for Northern Bald Ibis (IAGNBI), Vejer, Spain. September 2006.

JIMENEZ-ARMESTO M, C Böhm & C Bowden (Compilers). 2006. International Single Species Action Plan for the Conservation of the Northern Bald Ibis *Geronticus eremita*. AEWA Technical Series No. 10. Bonn, Germany.

KOTRSCHAL K. 2007. Konrad Lorenz Forschungsstelle: NBI Project 1997-2006: an update. In: Böhm C, Bowden CGR, Jordan M & C King (eds.). (2007). Northern Bald Ibis Conservation and Reintroduction workshop. Proceedings of 2nd Meeting of International Advisory Group for Northern Bald Ibis (IAGNBI), Vejer, Spain. September 2006.

MENDELSSOHN H. 1994. Experimental release of Waldrapp Ibis *Geronticus eremita*: an unsuccessful trial. Int. Zoo Yearbook 33:79-85.

ÖZBAGDATLI N. 2007. NBI at the Birecik breeding centre. In: Böhm, C, Bowden, CGR, Jordan, M. & King, C. (eds.). (2007). Northern Bald Ibis Conservation and Reintroduction workshop. Proceedings of 2nd Meeting of International Advisory Group for Northern Bald Ibis (IAGNBI), Vejer, Spain. September 2006.

PEGORARO K. 1996. Der Waldrapp. Vom Ibis, den man fur einen Raben hielt. Wiesbaden.

PEGORARO K. 2003. Release trials of NBI: an overview. In: Böhm C, Bowden CGR & Jordan MJR (eds). 2003. Northern Bald Ibis conservation and reintroduction workshop. Proceedings of the International advisory Group for Northern Bald Ibis (IAGNBI) meeting Alpenzoo, Innsbruck-Tirol, July 2003.

PEGORARO K & M Föger. 1999. The Northern Bald Ibis *Geronticus eremita* in Europe: A Historical Review In: Northern Bald Ibis *Geronticus eremita*. 2nd EEP Studbook (Böhm C. ed.) Alpenzoo Innsbruck-Tirol:10-20.

PESKE L. 2007. Satellite tagging of 3 NBI in Syria in 2006. In: Böhm C, Bowden CGR, Jordan M & C. King (eds). 2007. Northern Bald Ibis Conservation and Reintroduction workshop. Proceedings of 2nd Meeting of International Advisory Group for Northern Bald Ibis (IAGNBI), Vejer, Spain. September 2006.

QUEVEDO M & von Houwald. 2003 Skin problems in Northern Bald Ibis. In: Böhm C, Bowden, CGR, & Jordan MJR (eds.) 2003. Northern Bald Ibis conservation and reintroduction workshop. Proceedings of the International advisory Group for Northern Bald Ibis (IAGNBI) meeting Alpenzoo, Innsbruck-Tirol, July 2003.

QUEVEDO M, JM Lopez Vazques & EA Prieto. 2007. Update of Proyecto eremita. In: Böhm, C, Bowden, CGR, Jordan M & King C (eds.). (2007). Northern Bald Ibis Conservation and Reintroduction workshop. Proceedings of 2nd Meeting of International Advisory Group for Northern Bald Ibis (IAGNBI), Vejer, Spain. September 2006.

SERRA G, M Abdallah, A Abdallah, G Al Qaim, T Fayed, A Assaed & D. Williamson. 2003. Discovery of a relict breeding colony of Northern Bald Ibis *Geronticus eremita* in Syria: still in time to save the eastern population? Oryx 38(1): 1-7.

SERRA G & L Peske. 2007. NBI conservation efforts in Syria 2002-2006: Results and lessons learnt In: Böhm, C, Bowden, CGR, Jordan M & King C (eds.). (2007). Northern Bald Ibis Conservation and Reintroduction workshop. Proceedings of 2nd Meeting of International Advisory Group for Northern Bald Ibis (IAGNBI), Vejer, Spain. September 2006.

SERRA G, L Peske & M Wondafresh. 2007. First survey of Eastern NBI wintering on the Ethiopian highlands: Field mission report 14 November - 1 December 2006. In: Böhm, C, Bowden, CGR, Jordan M & King C (eds.). (2007). Northern Bald Ibis Conservation and Reintroduction workshop. Proceedings of 2nd Meeting of International Advisory Group for Northern Bald Ibis (IAGNBI), Vejer, Spain. September 2006.

TOUTI J, Oumellouk F., Bowden CGR, Kirkwood JK, and KW Smith. 1999. Mortality incident in Northern Bald Ibis *Geronticus eremita* in Morocco in May 1996. Oryx 33: 160-167.

## AUTHOR BIOGRAPHY

Chris is now International Species Recovery Officer for RSPB. He has worked for RSPB for 18 years, much of that time on Northern Bald Ibis (involvement for 13 years, included living in Morocco for over five years) and he chairs the International Advisory Group for the species (IAGNBI). He carried out detailed breeding ecology research on Woodlark and Nightjar in UK in the 1980s, but since then has worked on international projects, including four years interlude for BirdLife International in Cameroon. Apart from the ibis work, he has been involved with other Critically Endangered species including Montserrat Oriole, Ibadan Malimbe and Jerdon's Courser. The past three years, Chris has headed RSPB's South Asian vulture program, mainly focusing on India.

## Appendix 1

**INTERNATIONAL ADVISORY GROUP FOR NORTHERN BALD IBIS (IAGNBI) MISSION STATEMENT, TERMS OF REFERENCE AND POSTS/COMPOSITION.**

IAGNBI was created on 12th March 1999 at the International workshop on a strategy for the rehabilitation of the Northern Bald Ibis held in Agadir, Morocco. The primary objectives of the committee were to ensure international co-ordination and co-operation on Northern Bald Ibis projects.

**Mission statement:**
"Promoting the conservation of the NBI through international co-ordination and co-operation"
Terms of Reference

- Focusing attention on the priority conservation problems
- Facilitating communication and cooperation between concerned groups
- Encouraging applied scientific research to close gaps of knowledge on NBI and updating what the most urgent are
- Acting as the implementing partner of the SAP for the NI for AEWA (subject to AEWA agreement)
- Produce release guidelines for the NI
- Review propositions for all NI release/re introduction projects/trials in relation to release guidelines produced for the species
- Support fund raising for the priority projects
- Produce regular newsletters
- Liaison with AEWA

**Committee composition:**

| | |
|---|---|
| Research Biology | - Chris Bowden (Chair) |
| | - Karin Pegoraro |
| | - Kurt Kotrschal |
| Captive Population | - Christiane Boehm (Secretary) |
| | - Cathy King |
| IUCN/Reintroduction | - Mike Jordan |
| Veterinary | - Miguel Quevedo |
| | - Andrew Cunningham |
| Moroccan Population | - Mohammed El Bekkay |
| | - Mohammed Ribi |
| Syrian Population | - Gianluca Serra |
| Turkish Population | - Taner Hatipoglu |
| | - Nurettin Ozbagdatli |
| Algeria | - Amina Fellous |
| Ethiopia | - to be appointed |

# PROGRESS WITH A CONSERVATION BREEDING PROGRAM TO SAVE THREE CRITICALLY ENDANGERED *GYPS* VULTURES FROM EXTINCTION

Christopher G.R. Bowden[1], Vibhu Prakash[2], Nick Lindsay[3], Ram M. Jakati[4], Richard J. Cuthbert[1], Asad Rahmani[2], Deborah J. Pain[1], Rhys E. Green[1], Andrew A. Cunningham[3] & Jemima Parry-Jones[5]

1 Royal Society for the Protection of Birds,
  The Lodge, Sandy, Beds, SG19 2DL, UK
2 Bombay Natural History Society, Hornbill House,
  Shaheed Bhagat Singh Road, Mumbai, 400023, India
3 Zoological Society of London, Regents Park,
  London, NW1 4RY, UK.
4 Haryana Forest Department, Van Bhawan, sector 6,
  Panchkula, 134109, Haryana, India
5 International Centre for Birds of Prey, Little Orchard
  Farm, Eardisland, Herefordshire HR6 9AS, UK

## Summary

Three South Asian *Gyps* vulture species are critically endangered with extinction, despite having been abundant in the 1990s. The main cause of the declines is now established, being ingestion of a veterinary painkiller, diclofenac, from cattle carcasses, which is toxic to the vultures. Two key actions have been identified by the range state governments and IUCN; namely the eradication of diclofenac from the environment and establishing a conservation breeding and release program. The Bombay Natural History Society has taken up the challenge, and with support from a great diversity of players is addressing both of these issues. The breeding program is making considerable progress, and has established two centres: in Haryana and West Bengal, with a third, smaller facility in Assam recently receiving its first birds. There are currently 165 birds in the program, roughly divided between the three species most affected. Eleven birds are also held in Punjab Province, Pakistan at a WWF facility there, and there are plans for a small facility in Nepal. The overall target for South Asia is for at least 75 breeding pairs of each of the three *Gyps* species, and we hope to reach this within the next two years. Then it will be to get the birds breeding and preparing for the pre-release phases once there is a diclofenac-free environment. Meanwhile, we have the challenge of coordinating, resourcing and running the centres, and getting the required expertise and experience plugged into the program. We are establishing governing bodies and a technical advisory committee to help with this and summarise here how these are constituted. The further challenge is to get full ownership by the local and regional governments, NGOs and other players so that the ultimate goals of eradicating diclofenac, and breeding and releasing the vultures are successful. It is encouraging that the Indian Central Zoo Authority is now planning to resource vulture breeding facilities in close proximity to four leading zoos in India, and it is hoped that these will bring the program closer to the required targets.

## Introduction

Since the early to mid 1990s, the dramatic declines of three *Gyps* vultures across South Asia have been among the most widespread and rapid of any bird species on record. Declines are documented at 97% across India between 1993 and 2003 (Green et al 2004), and although slightly later for Pakistan, they have declined even faster there (Gilbert et al 2006). Formerly abundant across the region, the declines over the past twelve years have meant that in all likelihood, less than 1% remain, and the remnant populations are in most cases still probably declining by over 30% per year (Green et al 2004). The species most affected are Oriental White-backed Vulture *Gyps bengalensis* (OWBV), Long-billed Vulture *Gyps indicus* (LBV) and Slender-billed Vulture *Gyps tenuirostris* (SBV), and all three were classified by IUCN as Critically Endangered in 2000 (IUCN 2007). These birds are keystone species, playing an important cleaning role, for cattle meat in particular because it is not generally consumed by people for cultural reasons. The vultures thus reduce health risks to people of having carcasses rotting in the wider environment. There is also good evidence that feral dogs have increased in parallel with the vulture declines, presumably due to greater food abundance, and this presents an increase risk to humans of rabies, which largely results from feral dog bites (Pain et al 2003).

It was not until surveys undertaken in 1991-1993 were repeated in 2000 that the extent and rapidity of the vulture declines were realized, and detailed investigations into the cause started immediately. By 2003, the link with veterinary diclofenac, a non-steroidal anti-inflammatory drug (NSAID) administered to cattle as a painkiller and for a variety of ailments, was established in OWBV in Pakistan (Oaks et al 2004). Shortly afterwards, research indicated that the same primary cause was affecting OWBV and LBV across India and Nepal (Shultz et al 2004) and a South Asian Species Recovery Plans was drawn up (ISARPW 2004), outlining the key actions necessary to prevent extinction. Further, an IUCN resolution was passed in Bangkok in November 2004, stating the need for a conservation breeding program. The major actions outlined were to remove diclofenac from the environment as quickly as possible and, recognising that removing diclofenac may not be feasible before most or all populations have gone extinct, to instigate a major conservation breeding program for all three species.

## Feasibility and Scale of the Planned Conservation Breeding Program

The minimum target was set for getting 75 breeding pairs of each of the three species into the breeding program, and holding these at least three separate locations as fast as possible was agreed (ASARPW 2004; Watson 2004). The figure was arrived at by considering a number of factors and making certain assumptions. With the likely mortality, productivity, group sizes it will make the release of 100 vultures within ten years (Watson 2004). The practicality of getting birds in future, bio-security factors, minimum numbers needed at each centre as well as the needs for adequate genetic founders are all serious concerns. Having well scattered locations to reduce stochastic risks such as an outbreak of disease is an important precaution. With all of these considerations and bearing in mind that some centres (for practical and political reasons) will not be for all three species, a target of six centres across South Asia is a realistic figure that we are working towards. The proposed time scale for the centres has been set at fifteen years, with the aspiration to have birds ready for release within eight to ten years.

The first challenge is that only one of these species (OWBV) has previously been bred in captivity, and then only once. There has been no great incentive to do so until now, and in fact neither Long-billed nor Slender-billed Vultures have previously even been held in registered collections. However, thirteen species of vulture have been successfully captive bred, and this includes seven species in the genus *Gyps*, and most notably Eurasian Griffon *Gyps fulvus* which has been successfully bred and indeed has been the subject of highly successful reintroduction programs across Europe (Sarrazin et al 1994; Terrasse et al 1994). There will undoubtedly be some learning involved in developing the best methods for breeding each species, and an ongoing need for input from raptor breeding specialists.

The plan for six centres (with 25 pairs of each of the three species), for practical purposes has become the hope for at least four centres in India, and one each in Pakistan and Nepal. Currently we are still well short of this.

## Development of the Breeding Program

The first breeding centre was established at Pinjore, in Haryana, India in late 2001 and inaugurated in February 2003. This was originally set up for diagnosing the cause(s) of decline, building capacity in vulture husbandry, and with the possibility of a conservation breeding centre in mind should this become necessary. By early 2004, 30 OWBV and LBV were housed there. This has now been upgraded and expanded to become the first conservation breeding centre, and a second centre has been established at Raj Bhat Khawa, Buxa Tiger Reserve in West Bengal, inaugurated in September 2006.

The centres have been financed largely by RSPB and Darwin Initiative (UK government funds), with BNHS the main implementing body, and having a signed MoU with the state government. The Zoological Society of London and National Birds of Prey Trust have been the other main contributors, both with expertise and additional funding which have allowed the program to progress. The respective state governments have formally allocated two hectares for a period of 15 years, and it is now a key element for future success of the program for this state and central government support to increase further.

The need to react quickly was recognised by the organisations involved, and largely with Darwin Initiative, RSPB and NBPT funds, which together with their personnel and the key involvement and expertise of BNHS, Haryana State Government, and ZSL, the program was expanded considerably. Expansion of the Pinjore centre was the first major step, followed by the initiation of the West Bengal centre. One further centre has been agreed by Assam Forest Department and initiated at Rani Forest, Assam by BNHS with RSPB and Oriental Bird Club support and funds. This is primarily planned for 25 pairs of Slender-billed Vultures.

Key steps were taken by the Indian Ministry of Environment and Forests, both in granting the necessary permissions for the program to take place, and in the development of an Indian Government Vulture Action Plan (MoEF 2006), which categorically endorses the need for a breeding program as well as legislating for the removal of diclofenac as a veterinary drug. The Central Zoo Authority has also taken a very positive step, and are currently in the process of issuing grants to up to four Indian zoos to construct further breeding facilities. These are currently at an advanced planning stage.

In Pakistan, WWF Pakistan has established a facility together with Punjab Wildlife and Parks Department of the Provincial Government, which is planned to hold ten pairs of Oriental White-backed Vultures. This was inaugurated in April 2007, and has technical input from the Hawk Conservancy Trust, as well as some funds from the environment agency of UAE.

There have also been plans to construct a similar scale centre to the Pakistan facility in the terai of Nepal, close to Chitwan National Park. This would also be for a smaller number of Oriental White-backed Vultures. Political unrest has been a major hindrance to progress on this, despite funds being available from RSPB and ZSL. The proposal between the Government Department of National Parks, the National Trust for Nature Conservation (formerly the King Mahendra Trust for Nature Conservation) and Bird Conservation Nepal (BirdLife Partner for Nepal) is still at a planning stage. Meanwhile, some important *in-situ* initiatives have been tested, to try to remove diclofenac through awareness-raising and exchange programs of diclofenac with the safe alternative drug meloxicam.

One feature of the program is the broad diversity of international players and organisations willing to help with the program. The program already spans a number of specialist fields, and as it moves from the diagnosis and construction phases through the capture and housing, the husbandry and veterinary elements come to the fore. We are also beginning to consider the pre-release and release phases. Each has its own sets of skills and experience that are needed, and we need to ensure that the BNHS and coworkers receive adequate and timely training for each. The ZSL has been able to provide much of the veterinary skills, and NBPT on the husbandry side. The pre-release phases have started to draw upon The Peregrine Fund's experience with California Condor *Gymnogyps californianus*, and attention will be given to ensuring that other relevant programs will be brought more into this program in future.

**Aviary Design, Management and Progress on the Ground**

Pinjore is the most advanced of the centres, and now consists of three large 'colony aviaries', a series of nursery and holding aviaries, as well as hospital aviaries. There is also a well-equipped laboratory on site, as well as an office where remote control CCTV monitoring is carried out for the main aviaries.

The current vulture stock (as of July 2007) for each centre is given in Table 1. Just over 50% of the birds currently at the centres were collected as adults. Many were picked up sick in neighboring states, and successfully recovered after a few days or weeks of good care. Others were injured by kite strings in Ahmedabad during an annual kite festival there (the glass-coated strings designed for kite-fighting are very dangerous to passing birds, often seriously cutting their wings). A few adults were trapped using traditional trapping methods, which are very time consuming, especially now that vulture densities are so low.

**Table 1**

**CURRENT VULTURE STOCK (AS OF JULY 2007) FOR EACH BREEDING CENTRE**

|  | Pinjore Haryana | Raj Bhat Khawa (Buxa) West Bengal | Rani Forest Assam | Changamanga Punjab Pakistan |
|---|---|---|---|---|
| OWBV | 43 | 20 | 2 | 11 |
| LBV | 55 | 17 | - | - |
| SBV | 14 | 12 | 2 | - |
| Total | 112 | 49 | 4 | 11 |
| Total Capacity Planned (adults) | 150 | 150 | 50 | 20 |

**Fig. 1.** *Colony aviary at Pinjore.*

The remaining birds were collected as nestlings.

States that have played a key role in pro-actively sending sick/injured or trapped vultures (apart from the states Haryana, West Bengal and Assam with centres) include Madhya Pradesh, Gujarat, Rajasthan, Uttar Pradesh, Punjab and Orissa. Some states have been more willing than others to quickly grant Schedule I permissions for the transfer or catching of birds between states. This process takes time and effort, and unfortunately has in many cases been too slow to allow birds to be collected. It relies on highly proactive involvement by state chief wildlife wardens and other officials from both the sending and receiving states concerned.

Haryana was selected for the first centre largely due to the extremely positive attitude of the forest department at the outset, and the consequent relative speed of getting formal permission to establish the centre. This together with the available Darwin funding and the commitment of BNHS, RSPB, ZSL and NBPT has given the program a good start. This positive factor of state government attitude is key for all three states selected for BNHS vulture conservation breeding centres, in addition to seeking a scatter of centres across the range-state regions for bio-security reasons.

One area where faster progress may be possible in future is by 'double clutching' (or 'egg-pulling'), i.e. removing and artificially incubating the first egg, allowing the birds to re-lay and hence double the number of potential chicks per pair (all three species lay only one egg per year otherwise). This option will be tested once regular breeding is established. Hand-rearing the chicks has many potential difficulties, which will need careful management to avoid problems of imprinting on humans. These issues are fast approaching, and whilst they can be carefully planned for, they cannot be tackled until breeding is underway.

### Food Supply - a major issue

The captive vultures need to be provided with a food supply guaranteed to be free from diclofenac. Just one contaminated meal could kill all of the vultures in a breeding aviary. This has so far been a hugely expensive issue to address. So far, goat meat (already the

**Fig. 3.** *Aviary from the outside.*

most costly meat source) has been used as goats are far less likely to be administered diclofenac, but more importantly, they can be kept for a minimum of ten days to ensure that any diclofenac passes through the system before slaughtering on demand. Experiments show that diclofenac has a short residence time in goat tissue, with residues undetectable after 26 hours. In cow tissue, diclofenac residues are still detectable after 72 hours, but are not detectable after 7 days (Taggart et al 2006). Slaughtering of cows or even buffalo is not culturally acceptable in Haryana, and other options such as rabbits and stall-fed goats are being investigated, to help reduce costs in future. Regional differences will determine what is cost-effective and acceptable. Currently, vulture food constitutes over

**Fig. 2.** *Vultures in colony aviary from inside.*

half of the centre's running costs and this will grow with increasing numbers of birds.

## Recent Progress and Plans for Breeding

Of the birds currently held, two pairs of OWBV built nests and laid eggs in the 2005/2006 season. Neither egg hatched. In the 2006/2007 season, four eggs were laid, and two hatched. One survived just six days; the other survived over three weeks. The progress of OWBV compared to LBVs and SBVs is probably explained by the higher proportion of adult OWBVs relative to the other two species, although here was some pairing up of LBVs and preliminary nest building by two pairs in 2006/2007 season. There are plans to keep more established pairs out of the colony aviaries and keep them in individual aviaries during the coming season to potentially improve their chances of breeding.

## Framework for Running the Centres in India

Each centre will have a governing body, which meets at least annually, and potentially more as needed. The composition for Haryana is as follows:

Secretary of Forests, Govt. of Haryana (chair)
Principal Chief Conservator of Forests, Haryana
Chief Wildlife Warden, Haryana
Deputy Secretary of Forests, Haryana
Director, Bombay Natural History Society (BNHS)
Principal Scientist & Vulture Breeding Program Director, BNHS
Vulture Program Manager, RSPB
International Zoo Programs Director, ZSL
National Birds of Prey Trust Director & raptor breeding expert.
Professor, Dept Pathology, Hissar Vet College, Haryana

This committee effectively engages all parties, and helps assign responsibility for key tasks. It also provides a formal reporting system between the parties. A similar body is planned for the other centres. A technical advisory committee has also been created, and the intention is to formalize it to include central government in the near future. This ensures that there is regular consultation and input from raptor breeding and husbandry experts, and an opportunity for the team to draw on wider expertise. This is part of the overall capacity-building process for the program staff. The composition of the committee is as follows:

Director General Wildlife, Government of India (from 2007)
Chief Wildlife Warden, Haryana, RM Jakati (chair)
Chief Wildlife Warden Assam
Chief Wildlife Warden West Bengal (represented by Mr Lepcha)
Andhra Pradesh Forest Dept
Maharashtra Forest Department (Representative)
Director, Central Zoo Authority, BR Sharma
Indian Veterinary Research Institute, Dr Swarup
Wildlife Institute of India, BC Choudhary
Director, BNHS
Principal Scientist & Vulture Breeding Program Director, BNHS
Vulture Program Manager, RSPB, Chris Bowden

International Zoo Programs Director, ZSL, Nick

**Table 2**

**UPDATED NUMBERS OF GYPS VULTURES HELD FOR CAPTIVE BREEDING IN CENTRES RUN BY BNHS IN INDIA & IN NEPAL AND PAKISTAN 2009**

|  | Haryana | Bengal | Assam | Nepal | Pakistan | Total |
|---|---|---|---|---|---|---|
| OWBV | 53 | 48 | 19 | 43 | 14 | 177 |
| LBV | 54 | 17 | 0 | - | - | 71 |
| SBV | 14 | 12 | 9 | - | - | 35 |
| Total | 121 | 77 | 28 | - | - | 284 |

Lindsay
National Birds of Prey Trust Director & raptor breeding expert, Jemima Parry-Jones, with support from
Curator of Birds, ZSL, John Ellis
Vulture Program Research Manager, RSPB, Richard Cuthbert
Condor Program Director, The Peregrine Fund
IUCN Asia reintroduction specialist group, Mike Jordan
Chief Vet, ZSL, Andrew Routh

These committees will meet shortly before the breeding season to review the previous years progress in time to implement necessary changes. The first meeting, held in December 2006, was of an interim committee, as unfortunately the Director General for Wildlife was not able to attend. It is hoped that the committee will be formalized at this years meeting, once the post holder has been appointed.

Finally, Pinjore was formally registered with the Central Zoo Authority as a breeding and care centre, which further consolidates the status of the centre and makes it eligible for funding support (e.g. for aviary extensions). The CZA will carry out annual inspections as with all zoo facilities, making recommendations as appropriate; we expect other centres to follow in this respect.

**Long Term Plans**

The longer-term objective is to have 75 pairs of each species by 2009, which is the year we first anticipate successful breeding. Assuming methodology for breeding these species develops as expected, productivity should allow us to have released 100 birds by 2019, and we hope some of these will be earlier. We will only start releasing birds once diclofenac has been shown to be out of the environment for a defined potential release area, and once we have at least 30 additional birds of a given species available. There will be a pre-release phase, when the groups of birds selected for release are kept in temporary aviaries close to the proposed release area for a minimum period of several months. We anticipate this phase will start between 2014 and 2016.

Indian zoos are now planning vulture breeding facilities on a somewhat smaller scale to Pinjore, each proposed facility being for either one or two species (OWBV or LBV). These could play a crucial role in achieving our targets, and initial signs are that there is a willingness to integrate these centres into the current BNHS program. Two of the four zoos have requested that BNHS manages the centres, which would be quite separate from the main zoos. The additional financial burden of staffing the centres is a major challenge, but we hope that this can be overcome. If these come to fruition as hoped, and indeed if BNHS can take on the staffing then this will greatly assist the standardization and coordination of the program in India as a whole.

## Discussion

The challenges presented by this program are unprecedented in several respects: three species are involved, the husbandry requirements for these species are largely unknown, time is short due to the decline rates in the wild, there are considerable costs associated with guaranteeing a safe food supply, along with the administrative challenges of obtaining catching permissions in time. In addition, it is crucial that diclofenac is removed from the environment so that the birds can eventually be safely released. The BNHS vulture advocacy program spearheaded by Dr. Nita Shah has made major progress on this issue, and the Indian government has banned the manufacture of diclofenac for veterinary purposes. This example has been followed by Nepal and Pakistan who have banned the manufacture and importation of veterinary diclofenac. The successful safety-testing of a suitable alternative painkiller and NSAID, called meloxicam (Swan et al 2006; Swarup et al 2007), was an essential component for getting the diclofenac manufacturing ban in place. However, import and sale are still legal in India, and preliminary monitoring information suggests that the prevalence of diclofenac in dead livestock has not yet declined. There is still a job to do in getting this switch across to the veterinary and farming communities. There is consequently a long way to go before the drug will be effectively removed from the vultures environment.

The magnitude of the declines, and the environmental health role played by the vultures have attracted the attention of a wide audience, and the problem is taken seriously by the conservation and many parts of the wider community. Coordinating efforts and resources in an effective way remains a challenge. Obtaining the full support of all stakeholders, but especially at local and regional community level, for the key measures required, will be a key factor that will ultimately determine the projects success.

*Note: The program had its first breeding successes in 2008 with two Oriental White-back Vultures fledged at Pinjore, and a further three in 2009. Also in 2009, two Slender-billed Vultures were successfully fledged, one at Pinjore and one at Rajabhat Khawa. There has also been some significant progress with getting more founder stock - see Table 2 with updated figures for each centre including the new one at Chitwan in Nepal which is run by the National Trust for Nature Conservation.*

## ACKNOWLEDGEMENTS

A number of people have played hugely important roles in the progress of the program in India so far, and in reacting quickly enough to make things happen. The Haryana state Government PCCF Mr. Srivastava, Mr. Ujjwal Bhatacharya, Mr. S.S. Bist, Dr. B.R. Sharma, Mr. Malakar, Mr. Lepcha, have been particularly key. BNHS staff who have played their part include Dr. Nita Shah, Devojit Das, Sachin Ranade, Jeherul Islam, Nikita Prakash and the other centre staff. From ZSL, Dr. Andrew Routh and John Ellis are very closely involved and providing essential support, and at RSPB, Dr. Mark Avery and Ian Barber are likewise. Hem Sagar Baral and staff at Bird Conservation Nepal have been very closely involved in Nepal, as has Siddartha Bajracharya and the National Trust for Nature Conservation. In Pakistan, Dr. Uzma Khan has been leading the initiative for WWF Pakistan, with strong support from Hawk Conservancy Trust and Dr. Campbell Murn in particular. The timely provision of funds, support and resources from the Darwin Initiative, RSPB, NBPT, ZSL and the state governments of Haryana, west Bengal and Assam as well as the Ministry of Environment and Forests have been absolutely essential for the progress to date in each case.

## REFERENCES

ASARPW. 2004. Report on the International South Asian Recovery Plan Workshop. Buceros 9: 1-48.

GILBERT M, RT Watson, MZ Virani, JL Oaks, S Ahmed, MJI Chaudhry, M Arshad, S. Mahmood, A. Ali and AA. Khan. 2006. Rapid population declines and mortality clusters in three Oriental white-backed vulture *Gyps bengalensis* colonies in Pakistan due to diclofenac poisoning. Oryx 40: 388-399.

GREEN RE, I Newton, S Shultz, AA Cunningham, M Gilbert, DJ Pain & V. Prakash. 2004. Diclofenac poisoning as a cause of population declines across the Indian subcontinent. J. Appl. Ecol. 41:793-800.

GREEN RE, MA Taggart, D Das, DJ Pain, C Sashikumar, AA Cunningham and R. Cuthbert. 2006. Collapse of Asian vulture populations: risk of mortality from residues of the veterinary drug diclofenac in carcasses of treated cattle. J. Appl. Ecol. 43:949-956

IUCN. 2007. The IUCN Red List of Threatened Species, http://www.iucn.org.

MoEF. 2006. Proceedings of the International Conference on Vulture Conservation. Ministry of Environment and Forests, Government of India, New Delhi. pp 44.

OAKS JL, M Gilbert, MZ Virani, RT Watson, CU Meteyer, BA Rideout, HL Shivaprasad, S Ahmed, MJI Chaudry, M Arshad, S Mahmood, A Ali & AA Khan. 2004. Diclofenac residues as the cause of population decline of vultures in Pakistan. Nature 427: 630-633.

PAIN DJ, AA Cunningham, PF Donald, JW Duckworth, DC Houston, T Katzner, J Parry-Jones, C Poole, V Prakash V, P Round and R Timmins. 2003. *Gyps* vulture declines in Asia; temporospatial trends, causes and impacts. Conserv. Biol. 17(3): 661-671.

PRAKASH V. 1999. Status of vultures in Keoladeo National Park, Bharatpur, Rajasthan with special reference to population crash in *Gyps* species. J. Bombay Nat. Hist. Soc. 96: 365-378.

PRAKASH V, DJ Pain, AA Cunningham, PF Donald, N Prakash, A Verma, R Gargi, S Sivakumar and AR Rahmani. 2003. Catastrophic collapse of Indian white-backed *Gyps bengalensis* and long-billed *Gyps indicus* vulture populations. Biol. Conserv. 109: 381-390.

PRAKASH V, DJ Pain, AA Cunningham, PF Donald, N Prakash, A Verma, R Gargi, S Sivakumar and AR Rahmani. 2003. Catastrophic collapse of Indian white-backed *Gyps bengalensis* and long-billed *Gyps indicus* vulture populations. Biological Conservation 109(3):381-390.

SARRAZIN F, C Bagnolini, JL Pinna, E Danchin and J Clobert. 1994. High survival of Griffon vultures *Gyps fulvus fulvus* in a reintroduced population. Auk 111: 853-862.

SHULTZ S, HS Baral, S Charman, AA Cunningham, D Das, GR Ghalsasi, MS Goudar, RE Green, A Jones, P Nighot, DJ Pain and V Prakash. 2004. Diclofenac poisoning is widespread in declining vulture populations across the Indian subcontinent. Proc. R. Soc. Lond. B. (Suppl.) Lond. B 271 (Suppl. 6): S458-S460.

SWAN G, V Naidoo, R Cuthbert, RE Green, DJ Pain, D Swarup, V Prakash, M Taggart, L Bekker, D Das, J Diekmann, M Diekmann, E Killian, A Meharg, RC Patra, M Saini and K Wolter. 2006. Removing the Threat of Diclofenac to Critically Endangered Asian Vultures. 2006. Public Library of Science Biology. March 2006 4(3): e66.

SWARUP D, RC Patra, V Prakash, R Cuthbert, D Das, P Avari, DJ Pain, RE Green, AK Sharma, M Saini, D Das and M Taggart. 2007. The safety of meloxicam to critically endangered *Gyps* vultures and other scavenging birds in India. Animal Conserv. 10: 192-198.

TAGGART MA, R Cuthbert, D Das, DJ Pain, RE Green, S Shultz, AA Cunningham and AA Meharg. 2007. Diclofenac disposition in Indian cow and goat with reference to *Gyps* vulture population declines. Environ. Pollut. 147(1): 60-5.

TERRASSE M, C Bagnolini, J Bonnet, JL Pinna and F Sarrazin. 1994. Reintroduction of the Griffon Vulture *Gyps fulvus* in the Massif Central, France. Pp. 479-491 in BU Meyburg & RD Chancellor, (eds). Raptor conservation today. Berlin: WWGBP/The Pica Press.

WATSON R T. 2004. Report of the Kathmandu Summit Meeting: "A new environmental threat posed by the drug diclofenac" Godavari Village Resort, Kathmandu, Nepal.

# COLLABORATIVE EFFORT FOR THE LIGHT-FOOTED CLAPPER RAIL PROPAGATION PROTOCOL

Laurie Conrad[1], Charles Gailband[2], Judy St. Leger[1] & Richard Zembal[3]

[1] SeaWorld California
500 Sea World Dr., San Diego, CA 92109
[2] Chula Vista Nature Center 1000 Gunpowder Point Dr.
Chula Vista, CA 91910
[3] Clapper Rail Study Team, 24821 Buckboard Lane,
Laguna Hills, California 92653

Development and degradation of coastal wetlands has severely impacted the population on the Light-footed clapper rail in southern California (Zembal et al 2007).

## SUMMARY

The Light-footed Clapper Rail *Rallus longirostris levipes* is one of the most endangered coastal birds in California. The population had declined to fewer than 600 individuals in the 1980s, primarily as a result of coastal wetlands destruction and degradation. The range of this subspecies is from Santa Barbara to Imperial Beach California in approximately 23 distinct locations. U.S. Fish & Wildlife Service (USFW), Chula Vista Nature Center (CVNC), Sea World California (SWC), San Diego Wild Animal Park (SDWAP) and independent biologists have partnered to develop a Captive Propagation Protocol. Studies suggest detrimental demographic and genetic effects, attributed to habitat fragmentation, may hamper recovery of the species. To combat potential genetic bottlenecking in wild rail populations, captive breeding efforts were initiated in 1998 as part of a Light-footed Clapper Rail Propagation Protocol (Protocol). The focus of the breeding activity is to increase genetic diversity in isolated and fragmented populations by propagating and releasing birds derived from a genetically desirable population. The program has successfully bred wild-captured rail pairs since 2001. Both parent and artificial incubation/hand-rearing techniques have been used to fledge a total of 154 offspring which were released into nine distinct habitats that have been identified as rail habitat with depressed populations. Telemetry was used to monitor post release movements of some rails in 2005 and 2006. The Light-footed Clapper Rail population in California has reached a record high for the third consecutive year. A total of 438 rail pairs were identified in 19 marshes, which is an increase from 218 pairs in 2001. This increase has been the result of management efforts for the rails and major habitat restoration.

## INTRODUCTION

This collaborative project was designed to reduce the further decline of the rail, and even restore the population through a managed captive breeding/release program and public education. The goal of the protocol was to facilitate the introduction of new rail bloodlines to small, isolated rail populations in California through the practice of egg trans-relocation. The program intended to swap eggs between wild rails in isolated areas with eggs produced from captive rails trapped at Newport Bay. Newport Bay has the largest and presumably most genetically desirable bloodline, which is why this subpopulation was chosen as a source of breeding founders (Nusser et al 1996).

Two unrelated juvenile rails were captured from Newport Bay, California in the winter of 1998. The birds were transferred and housed together on exhibit at CVNC, located within the Sweetwater Marsh National Wildlife Refuge. The exhibit is approximately 6m x 6m with an artificial slough designed to mimic the high and low tides of the natural marsh. The exhibit was planted with Cord Grass *Spartina foliosa*, Pickleweed *Salicornia* sp., and a variety of marsh plants native to the area. The birds were also provided with several artificial nest structures to ensure they had a few undisturbed areas. The pair spent two springs together without any nesting success. The female sat in the artificial nest structure the second breeding season but the pair showed no other signs of breeding.

Two more rails were captured from Newport Bay in the fall of 2000 and transferred to CVNC. The original rail pair was broken-up and repaired with the newly acquired birds. The two pairs were housed in separate enclosures that shared a common wall. Both pairs produced eggs the following spring. Light-

**Fig. 1.** *The Light-footed Clapper Rail ranges between Santa Barbara to Tijuana River Estuary, Imperia Beach in the United States. There are also several subpopulations in Baja California, Mexico. The largest population in Mexico is San Qunitin.*

**Fig. 3.** *Captive rails will sometimes double and triple clutch. To increase egg production and chick survivability early clutch eggs/chicks are artificially incubated/hand-reared. Humidaire incubators are used for artificial incubation. Parameters are 99.0°F (37.2°C) dry bulb, 82-84°F (27.8-28.9°C) wet bulb.*

footed Clapper Rails are territorial; they denote their range by their distinct clappering vocalization. The territorial maintenance behaviors, such as clappering, were mostly absent from the single pair housed in isolation from other rails in the first two breeding seasons. The absence of other rails probably contributed to the lack of breeding productivity. Although this theory has not been thoroughly tested the captive breeding institutions house at least two rail pairs in close proximity to each other. Translocation of eggs from captive rails to wild nests was attempted several times but was successfully accomplished only once. Identifying wild nests at similar embryonic stages as the captive produced eggs proved very time intensive and difficult. One egg swap was attempted in 2001 when eggs produced at CVNC were taken to a suitable nest a Mugu Lagoon. The egg swap was unsuccessful because the nest with wild eggs was destroyed by a mammal mere hours before the planned swap.

The failed egg swap at Mugu Lagoon combined with the desire to have the captive pair double clutch led to the attempt to artificially incubate and hand-rear Light-footed Clapper Rails. The eggs intended for the egg swap were transferred to SWC where they were incubated in a Humidaire incubator with parameters set at 99.0°F (37.2°C) dry bulb, 82-84°F (27.8-28.9°C) wet bulb. The eggs hatched without incident and the chicks were hand-reared with every effort to avoid having them imprint on their human caregivers. Puppets were used for feedings and coveralls were worn by keepers to hide their presence when weighing the birds, cleaning the brooders/enclosures, and feeding. A recording of ambient marsh sounds was played in the chick's enclosures to drown out the sounds of keeper activities. The techniques employed proved sufficient to avoid the risks of habituation experienced when hand-rearing other precocial avian species. However, the techniques were relaxed after the chicks proved overstressed by the routine and necessary handling for banding prior to transfer to the proving enclosure at CVNC. The new protocols allow for some visual exposure to people while avoiding having the chicks associate their keepers as a food source.

The program also allows for chicks to be parent-reared. Rail pairs at CVNC have produced between one and three clutches per season. Early season clutches have experienced a higher degree of mortality of chicks less then three weeks of age. We cannot be certain of the cause(s) for early season losses but assume cooler weather played a contributing factor. Partial, sometimes entire early season clutches, are pulled for artificial incubation/hand-rearing when necessary as indicated by parent behavior and environmental conditions. Regardless of the rearing method, hand-reared or parent-reared, all rail chicks are transferred to a

**Fig. 2.** *Rails nest in inertial areas in coastal wetlands and usually produce six to ten eggs per clutch.*

proving enclosure between 40 to 60 days of age. The proving enclosure is a large pen constructed in wetland habitat on Sweetwater Marsh. The enclosure is densely vegetated with native plants that the rails will rely on for hiding and foraging after they are released. The young rails spend 30 to 50 days in the proving enclosure. At this stage of their development they are fed more live foods and native species of coastal invertebrates, similar to those they will find in the wild. Fresh frozen fishes and commercial pellets are reduced and eliminated form their diets. The rails are closely observed and evaluated for desirable survival skills during their time in the final enclosures. The birds are not released until they demonstrate the ability to find and consume a variety of natural, live prey items. They also must demonstrate appropriate hiding behaviors in the presence of raptors and other predators. Prior to release all rails are fitted with a numbered aluminum band and a colored band. The numbered bands are provided by USFW. The colored bands are plastic spiral bands that are fitted and the have the external seem melted with a butane soldering iron. The color bands are coded to represent the year of banding.

Band sightings on rails are difficult and rare due to their elusive behavior. Band returns of deceased birds have been equally rare. Select captive-reared rails were fitted with radio transmitters in 2005 and again in 2006. In 2005, two attachment methods were used: harnesses and a tail mounts. Both attachment methods functioned well with acceptable transmitter to bird weight ratios. However, in 2006, a smaller transmitter model for tail attachment was used exclusively. Rails were released and tracked at three locations over the two years, Mugu Lagoon, Sweetwater Marsh and San Elijo Lagoon. Most of the radio tagged birds at Sweetwater and Mugu Lagoon lost their signals. Birds whose signals were lost could have moved out of range, had their transmitter fall off into the water, or been carried out of range by a predator. Conversely, rails with radio tags released at San Elijo were tracked and seen for the full lifespan of the radio tag batteries, approximately 42 days. Only one of the rails at San Elijo dropped the transmitter before the battery died. That signal was investigated in an attempt to find the rail. Neither the bird nor the transmitter could be found in the dense marsh vegetation. The use of radio telemetry to track the captive-reared rails after release has proven that the techniques used for the protocol produce birds that are capable of surviving in the wild. There has been no evidence or study to determine if these birds are breeding in the wild. The study also

**Fig. 4.** *Chicks take several hours to dry at which time they are transferred from the hatcher to a brooder box.*

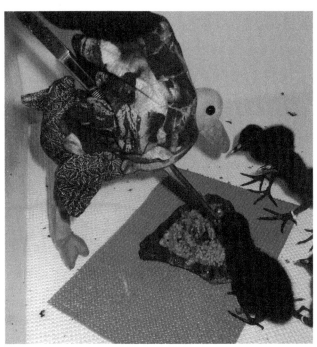

**Fig. 5.** *Hand-reared chicks are fed using a puppet while they are in the brooder box.*

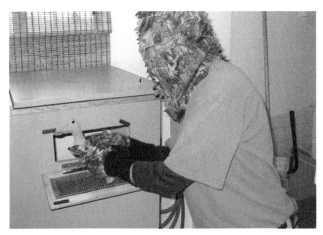

**Fig. 6.** *Chicks are housed in a brooder box for their first 7-10 days of life. Caregivers hide/disguise their presence while feeding chick in the brooder.*

**Fig. 7.** *At approximately 30 - 40 days of age all parent-reared and hand reared rails are housed in a conditioning/training enclosure for four to six weeks prior to release into the wild.*

**Fig. 8.** *All captive bred rails are banded with an aluminum numbered band and a plastic colored band to represent the year of release. The colored bands are plastic spiral bands that are fitted and then have the external seem melted with a butane soldering iron.*

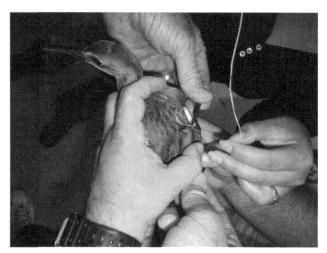

**Fig. 9.** *A radio transmitter weighing 6.7g being attached via a harness. The harness was held together by three cotton stitches that were intended to dissolve so the harness would detach from the bird after several weeks.*

suggests that captive-reared rails are not staying or surviving in some of the release areas. Some of these release sites, including Mugu and Sweetwater, have historically depressed rail populations both before and after release efforts. This might suggest a variety of other factors, including high densities of raptors/predators that might prevent some habitats from reaching a natural carrying capacity.

## DISCUSSION

The success of the breeding and education components of the program have done much to restore the numbers of Light-footed Clapper Rails in southern California while teaching people about the importance of coastal wetlands. Since 2001 more than 150 captive-bred rails have been released into nine distinct rail subpopulations. The wild California population of Light-footed clapper is at a historic high since annual surveys have been conducted since the 1980s. CVNC and SWC have developed programs and educational opportunities to increase public awareness of Light-footed Clapper Rail. CVNC exhibits the rail, allowing 40,000 to 60,000 guests a year to see the elusive, secretive bird up close. Exhibition has allowed students, bird keepers, and scientists including the USFWS to study the Clapper Rail. Details including reproductive behavior, molt patterns, and chick development have been studied and used to fill in many facets of unknown natural history. Sea World California has brought the story of the Light-footed Clapper Rail in to the homes of millions by featuring the rail and reintroduction efforts on their educational television series Shamu TV. Both SeaWorld and Bush Gardens have also facilitated having the rail featured on a segment of Jack Hanna's Animal Wise. Another method used to create public awareness is the involvement of local volunteers for Light-footed Clapper Rail studies and reintroduction efforts. Richard Zembal, program leader, has incorporated people who live near a wetland in a variety of activities, including: censuses, artificial nest construction and maintenance, rail releases, and even radio telemetry tracking. Integrating residents of nearby wetlands has resulted in increasing their sense of "ownership" of the resource and their willingness to be stewards and advocates for wetlands and the Light-footed Clapper Rail. The protocol has been refined each year. In 2007, the San Diego Wild Animal Park joined the program and produced eggs and offspring. Their highly significant contributions including housing additional founder pairs, chick production, and increasing the scope of the educational component are a great indicator of the potential for additional partners along the coast of Southern California.

The program has also provided expanded developmental opportunities for individuals working in a traditional zoological/avicultural setting. Aviculturists from SWC, CVNC, and SDWAP have been incorporated in a variety of field programs; including, nest searches, artificial nest construction, and call counts/censuses to support the program. Multiple benefactors including the SeaWorld Busch Gardens Conservation Fund, Port of San Diego and, allow for ongoing conservation of the Clapper Rail. More information and updates are available at www.clapperrail.com and www.swbgconservationfund.org.

## Conclusion

The program has shown that the captive propagation and release of Light-footed Clapper Rails is a viable means to increase the genetic diversity and boost the wild population. The techniques for breeding Light-footed Clapper Rails are now being used by San Diego Zoo's Wild Animal Park. Increased participation from other zoological institutions is being investigated for future program growth. However, a healthy Clapper Rail population is not solely dependant on the number of birds raised in captivity and returned to the wild; wetland conservation and restoration are vital to the future of this coastal bird. There are currently important environmental measures in place to protect and restore coastal wetlands. However, these restoration efforts take time and will not be possible without support from the general public. Continued conservation measures and education are critical to ensure a healthy future for Light-footed Clapper Rails.

## Acknowledgements

We thank Brian Collins of USFWS and Lyann Comrack of California Department of Fish and Game for their leadership in the recovery program. Also Jim Kelly, John Konecny, and Jim Robins for the many hours of searching for rail nests. Special acknowledgment goes to Susan Hoffman; the staff of the Chula Vista Nature Center, particularly Joyce Remp; the

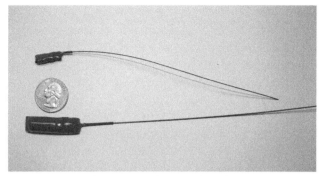

**Fig. 10.** *Several rails were fitted with radio transmitters in 2005 and 2006. In 2005, a larger transmitter model weighing 6.7g was used. This transmitter included a mortality indicator. In 2006, a smaller device without the mortality indicator was selected. The smaller unit weighed 1.1g.*

**Fig. 11.** *Radio-tagged rails were tracked with a handheld radio receiver made by Communications Specialists, Inc.*

**Fig. 12.** *Education about the Light-footed Clapper Rail and its association with coastal wetlands is an important component to securing this specie's future. SeaWorld California has set up the birds for breeding in their education facility to allow campers to be involved with the program.*

SeaWorld Avian Center team; the staff of the San Diego Wild Animal Park, particularly Michael Mace; and the Huntington Beach Wetlands Conservancy. This program is generously funded and supported by: the Department of the Navy under Cooperative Agreement N62473-06-LT-R0022, the City Of Chula Vista, SeaWorld California, the Unified Port of San Diego and SeaWorld Busch Gardens Conservation Fund.

## References

NUSSER, J.A., R.M. Gotto, D.B. Ledig, R.C. Fleischer and M.M. Miller. 1996. RAPD Analysis Reveals Low Genetic Variability in the Endangered Light-Footed Clapper Rail. Molecular Ecology 5(4): 463-472.

ZEMBAL, R., S. Hoffman and J. Konecny. 2007. Status and Distribution of the Light-footed Clapper Rail in California, 2007. San Diego: Clapper Rail Recovery Fund.

## Author Biography

As one of three assistant curators in the Bird Department at SeaWorld San Diego, Laurie Conrad oversees the day-to-day operations of the Avian Center. The Avian Center team members are responsible for the care and propagation of various bird species including toucans, flamingos, hornbills, parrots, waterfowl and Humboldt penguins. Conrad has been instrumental in getting SeaWorld into the endangered light-footed Clapper Rail recovery program in cooperation with U.S. Fish and Wildlife Service (USFWS) and also supervises the pelican rehabilitation and release program.

Conrad started her career at SeaWorld San Diego in 1989 as a bird keeper. From 1995 to 1996, she left SeaWorld and worked at the San Diego Zoo to continue learning about bird incubation and hand-rearing of many different bird species. She returned to SeaWorld in 1996 as a senior bird keeper and was promoted to supervisor in 1999, then assistant curator in 2004. In addition, she is a member of the American Zoological and Aquarium Association (AZA). She is the SeaWorld taxon advisory representative for Gruiformes and the North American studbook keeper for the Lesser Flamingo *Phoeniconaisis minor*.

Charles Gailband has worked with the Light-footed Clapper Rail at the Chula Vista Nature Center since 1990. He spent six years with the Aviculture Department at SeaWorld California, working with penguins, parrots, waterfowl, flamingos, and a variety of other birds. While at SeaWorld he was part of a team that developed techniques to hand-rear African Lesser and Caribbean Flamingos chicks. In 1999 Charles was hired by Chula Vista Nature Center to manage the bird collection and assist in developing techniques to breed and release the Light-footed Clapper Rail. Charles holds a BS in Liberal Studies with and emphasis in Science from National University.

# CAPTIVE BREEDING, REARING, AND RELEASE OF THE ATTWATER'S PRAIRIE CHICKENS AT THE HOUSTON ZOO

Mollie Coym

Houston Zoo
1513 N. MacGregor Way
Houston, TX 77030

## SUMMARY

The Attwater's Prairie Chicken is a species of grouse that once numbered at least a million along millions of acres of the coastal prairies of Texas and Louisiana. By the early 1900s, the Attwater's Prairie Chicken had vanished from Louisiana and their numbers in Texas were dwindling. In 1967, they were listed as endangered and in 1973 they were protected by the Endangered Species Act. Now these birds are down to just two small tracts of prairie land - The Attwater's Prairie Chicken National Wildlife Refuge (APCNWR) and The Nature Conservancy's Texas Prairie Preserve (TNC). The Houston Zoo works in cooperation with five other institutions as part of the Attwater's Prairie Chicken Recovery Team: Fossil Rim Wildlife Center, San Antonio Zoo, Sea World of San Antonio, Caldwell Zoo and Abilene Zoo. These facilities work in conjunction with the U.S. Fish and Wildlife Service and The Nature Conservancy in order to help the Attwater's Prairie Chicken population grow in the wild through captive breeding, rearing, and release into the wild.

## INTRODUCTION

### Natural History

Attwater's Prairie Chickens *Tympanuchus cupido attwateri* are members of the order Galliformes, family Phasianidae, subfamily Tetraoninae (grouse and relatives), and genus *Tympanuchus* (prairie chickens and sharp-tailed grouse). The Attwater's prairie chicken is considered to be one of three subspecies of the Greater Prairie Chicken which also includes the extinct Heath Hen *Tympanuchus cupido cupido* and the Greater Prairie Chicken *Tympanuchus cupido pinnatus* (BirdLife 2004). The Attwater's historic range includes millions of acres of the coastal prairies of southeast Texas and southwest Louisiana. As a result of habitat loss due to farming, industrialization and

pollution, they are currently restricted to two small prairie reserves, The Attwater's Prairie Chicken National Wildlife Refuge (APCNWR) and The Nature Conservancy's Texas Prairie Preserve (TNC) (Figure 1).

In the wild, adult Attwater's Prairie Chickens live approximately two to three years, in captivity they can live to about seven years old. The Attwater's mating ritual consists of courtship display "booming", that takes place in a lek, or booming ground. The males inflate the airsacs on their necks, extend their pinnae and tail feathers upward, and then drop their heads creating an "oo-la-woo" sound as they rapidly stomp their feet (Figure 2). Mating occurs between from February through mid-May. The hens make a depression in the ground filled with grass and feathers under the cover of grass clumps to create their nests. The eggs are a pale green color, the average clutch size is 12, and the incubation period is about 26 days long. The nests often fall victim to predators such as snakes, fire ants, and hawks;

**Fig. 1.** *US Fish and Wildlife Service census of the Attwater's Prairie Chicken in 2005. The range of the Attwater's Prairie Chicken has been reduced to two small refuges, the Attwater's Prairie Chicken National Wildlife Refuge in Eagle Lake, Texas and the Texas City Prairie Preserve in Texas City, Texas.*

heavy rains can also pose a threat to the success of a nest. The chicks hatch covered in bright yellow feathers with patches of brown and black. Chicks will stay with the hen for about six weeks. The chicks are fairly mobile and their diet consists mostly of insects early on; as they get older they begin to eat prairie grasses, seed, and plants as well (USFW 2004).

## Houston Zoo Breeding and Husbandry for Adult APC

A captive breeding program was developed for the Attwater's Prairie Chicken in 1993 in hopes to help boost the amount of prairie chickens in the wild. The Houston Zoo joined the effort in 1994. The zoo developed a holding and breeding area on grounds called "Boomtown" for the Attwater's Prairie Chickens, which served as their home until 2006. Due to further development of the zoo, it was determined that the prairie chickens needed a more natural, quieter, and more secluded home. Through a partnership with NASA's Johnson Space Center, the Houston Zoo was able to build 12-6m x 12m holding pens for breeding. During breeding season, the Attwater's are housed in the NASA pens, one pair per pen. Visual mesh barriers are placed along half of the length of the pens to prevent the males from becoming over stressed by the constant presence of another male. Large native clump grasses are maintained in each pen to provide cover and nesting areas for the hens. Additionally, unused Christmas trees are placed in each pen to make additional nest options available for the hens. They also have lean-tos and A-frames present for more durable cover.

The pairs are chosen by the Attwater's Prairie Chicken Recovery Team using genetic analysis. All of

**Fig. 2.** *Male Attwater's Prairie Chicken performing mating display, "booming," at the Houston Zoo Attwater's Prairie Chicken breeding and holding facility at NASA's Johnson Space Center.*

the parentage data is entered into a program called PM 2000 and the best matches are determined by the least amount of relation between birds. Birds are not paired together if their relation is more than half a percent for the sake of keeping the population as genetically diverse as possible. The birds are moved between the participating institutions each January. The Houston Zoo has participated in a diet study conducted by Mazuri Zoo Feeds to determine the optimum vitamin levels needed by the Attwater's Prairie Chickens. The Attwater's are all placed on a special, high Vitamin E test diet produced by Mazuri. It has been shown through diet studies over the years that higher Vitamin E levels lead to better egg production. In 2007, half of the breeding pairs were placed on 500 IU Vitamin E diet and the other half were placed on 1000 IU Vitamin E diet. Each pair had food measured out in the morning and the remaining food was measured in the afternoon to ensure that each pair had an equal amount of food available.

Fig. 3a

Fig. 3b

Fig. 3c

**Fig. 3.** *Nesting options of Houston Zoo APC Hens. Figure (3a) is an example of a clump grass nest, (3b) is an A-frame nest, and (3c) is a nest built underneath a Christmas tree.*

Fig. 4. *The brooder boxes are home to the Attwater's Prairie Chicken chicks until they have reached a weight of 25g.*

## Houston Zoo Egg Collection and Egg Data

When the males begin their booming displays, the keepers begin to search the pens thoroughly for eggs. This can be challenging; with all of the grass and Christmas trees, the hens have many options to choose from when making their nests. Some choose to build a traditional nest under the cover of clump grasses, similar to what would be seen in the wild. Some build their nests under A-frames. Others choose to build very secretive nests under Christmas trees (Figure 3).

Eggs are always handled wearing gloves; they are carefully removed from the nest and replaced with dummy eggs. When an egg is found it is given an egg log number and the information about which pen it was found in, where it was found in the pen, who the sire and dam are, and the condition of the egg are all recorded in the log. The egg is marked with its egg log number using a soft lead pencil. It is then placed in a cooler filled with egg crate foam for transport back to the Houston Zoo. Once back at the zoo, the egg width, length, and weight are all measured and recorded on an egg data sheet. This egg data sheet serves as a record of everything that happens to the egg. Once the measurements are taken the egg is placed into a storage cooler until it can be moved for incubation. All of the Attwater's Prairie Chicken eggs are carefully monitored for weight loss and any signs of distress throughout the incubation period. All eggs are candled and weighed twice a week to monitor development. Each egg is placed under a domestic chicken for a period of ten days.

The domestic chickens provide a more natural incubation than what our artificial incubators can provide. After this initial incubation, the eggs are placed in a Grumbach incubator, where the incubation staff carefully controls and monitors the humidity, temperature, and turning of the eggs. The egg is candled throughout the incubation process to ensure that the chick is making progress and not under and distress. When this chick is internally pipped, the egg is moved into the hatcher and is candled twice a day. Once, the external pip occurs, it takes approximately 24 hours before the chick is able to hatch. If any problems occur during the hatching process, the incubation staff

Fig. 5. *Attwater's Prairie Chicken stacker cage. The stacker cages house chicks with weights between 25g and 50g.*

## Table 1

**ATTWATER'S PRAIRIE CHICKEN FEEDING GUIDE**

| Week | Starter Pellets, (min offered) | | Salad, (max offered) | | MMW, (max offered) | | AFB Ratio | DMB Ratio |
|---|---|---|---|---|---|---|---|---|
| | Wgt, g | Tbls | Wgt, g | tsp (packed) | Wgt, g | No. | | |
| 1 | 4.00 | 0.5 | 1 | 0.25 | 0.75 | 15 | 70 : 17 : 13 | 90.5 : 2.5 : 7.0 |
| 2 | 8.59 | 0.75 | 2 | 0.5 | 0.45 | 9 | | |
| 3 | 15.88 | 1.5 | 4 | 1 | 1.00 | 18 | 78 : 18 : 4 | 95 : 2.5 : 2.5 |
| 4 | 25.06 | 2 | 5.7 | 1.25 | 1.05 | 21 | ↓ | ↓ |
| 5 | 28.27 | 2.5 | 6.03 | 1.5 | 1.20 | 24 | ↓ | ↓ |
| 6 | 37.15 | 3.0 | 8.09 | 2 | 1.5 | 30 | ↓ | ↓ |
| 7 | 47.51 | 4.0 | 8.09 | 2 | 1.5 | 30 | ↓ | ↓ |
| 8 | 55.0 | 4.5 | 8.09 | 2 | 1.5 | 30 | ↓ | ↓ |
| 9 | 77.28 | 6.0 | 8.09 | 2 | 1.5 | 30 | ↓ | ↓ |
| 10 | 84.49 | 7.0 | 8.09 | 2 | 1.5 | 30 | ↓ | ↓ |
| 11 | 102.26 | 8.0 | 8.09 | 2 | 1.5 | 30 | ↓ | ↓ |
| 12 | 104.26 | 8.5 | 8.09 | 2 | 1.5 | 30 | | |

*This guide determines how much food each chick receives. Mealworms and salad are maximum daily values, while starter values are minimums. Chicks have clean, accessible food at all times.*

assists the chick with hatching if necessary.

**Houston Zoo Husbandry for APC Chicks**

Once the chicks hatch, they are weighed, banded, and have their umbilicus swabbed. They remain in the hatcher until they are dry. Anything unusual about the chick is recorded on its data sheet. Chicks move from the hatcher to the APC Brooder Room into one of the brooder boxes (Figure 4). The brooder boxes are about 0.6m deep x 0.9m wide and about 0.3m tall wooden boxes designed to keep the chicks warm and easily observable during their first week or so of life. The brooder boxes are equipped with a florescent light, a ceramic heater, and a red heat bulb, each on independent switches so that the temperature can be adjusted to around 35°C on the bulb side and 37.2°C on the ceramic heater side. Thermometers are placed in the boxes for accurate temperature monitoring.

The brooder boxes also have removable panels on the top and sides of the box to allow for additional airflow for the chicks if needed. The bottom of each box is lined with shelf liner that is edged by folded pillowcases and mealworm bags on the outside. The chicks' food is placed on the pillowcases and mealworm bags which serve as a contrasting background for the food. Each box is additionally supplied with two water bottles containing a probiotic water solution and *Ligustrum* sp. branches for cover. The chicks remain in the brooder boxes until they weigh approximately 25g; once they have achieved that weight they are moved into stacker cages.

Attwater's chicks live in the stacker cages when they weigh between 25g and 50g (Figure 5). The stacker cages are about 0.6m deep by 0.9m wide and 0.6m tall wire mesh cages that allow the chicks to be outside and continuing to be in an easily observable and controlled environment. Each stacker cage is heated by two heat lamps, one containing a red heat bulb and the other containing a white heat bulb. The temperatures in the stacker cages are between 33.8°C and 36.6°C. Each cage is wrapped in plastic sheeting to keep the heat in and if temperatures are too hot, the lamps can be turned off. The stacker cages are set up in the same manner as the brooder boxes. The bottom is covered with shelf liner, folded pillowcase, and mealworm bags. There are thermometers for monitoring the temperature of the cage, *Ligustrum* sp. for cover, and two water bottles filled with tap water.

When the chicks reach approximately 50g, they are moved to holding pens (Figure 6). The bottom of each pen is covered in gravel and *Ligustrum* sp. or *Celtis* sp. branches and lean-tos for cover. Each pen also has a heat lamp and a thermometer to ensure that the chicks have temperatures ranging from 29.4°C to 32.2°C. Food and water are now provided on large, textured plastic trays.

The Attwater's Prairie Chicken chicks are placed

on a strict diet created by the Ft. Worth Zoo Nutrition Department that provides specifics about how much chicks eat of each starter, bugs, and salad mixture (Table 1). Chicks receive carefully measured amounts APC Starter Diet made by Mazuri, a salad mixture containing greens, apples, carrots, and peas, and either crickets or mealworms. This diet is measured out four times daily and the remaining food that is not eaten is measured the next morning. Strict records are kept about how much food each chick enclosure receives and have remaining each day. The hope is that the diet study will help us refine the diet given to the chicks to provide the most optimal nutrition.

Fig. 6. *Attwater's Prairie Chicken holding pen. Chicks are moved into the holding pens when they are over 50g.*

Each day the chicks are weighed and their behavior is observed to monitor their health and growth. The veterinary staff is notified of any chicks that have lost weight, have curly toes, rotating legs, or have been pecked and the chicks are treated accordingly. While the chicks are being weighed each morning, their enclosure is cleaned by another staff member. The remaining food is collected, weighed, and recorded; the water bottles and thermometers are removed and washed; the *Ligustrum* sp. is removed; the liners, pillowcases, and mealworm bags are removed and laundered; and the box is wiped clean and disinfected. The brooder box is then ready to be set up again. This process is then repeated for the stacker cages. Finally, the chicks in the pens are weighed. After they have been weighed, everything is removed from each pen for cleaning. While one keeper washes the food and water trays, another carefully rakes and sifts the gravel in each pen. Once the pen has been raked, the lean-tos are replaced. The *Ligustrum* sp. or *Celtis* sp. branches are either washed off and replaced in the pen or replaced with fresh branches. Clean food and water trays are then put in the newly cleaned pen.

**Feather Sexing**

Sex determination of the Attwater's Prairie Chicken is done before the chicks are released or transferred to another institution by examining the outer tail feathers. The outer tail feathers show different barring patterns between males and females. Males have been determined to have little to no barring across the outer few tail feathers. The females, however, have a great deal of barring on the outer tail feathers (Figure 7). Each bird's sex is determined in preparation for making breeding recommendations and determining whether the bird should be released or transferred to another facility in the Attwater's Prairie Chicken Recovery Team. Observing certain chick behaviors, such as booming and back stepping, also helps in determining the sex of the juvenile Attwater's Prairie Chickens.

**Release**

The goal of the Attwater's Prairie Chicken Recovery Team is to boost the wild populations. In order to achieve this goal, several Attwater's Prairie Chicken releases are held during the summer and fall. Some birds are released at the Attwater's Prairie Chicken National Wildlife Refuge in Eagle Lake, Texas, while others are transferred to the Nature Conservancy's Texas Prairie Preserve in Texas City, Texas. The strongest and healthiest birds are released in hopes that they will be able to survive the weather conditions and be able to hide themselves effectively from predators. All of the release birds have a radio collar placed around their neck for tracking purposes. The radio collars facilitate in monitoring the population and the survival of individual birds (TPWD 2005).

## Results

Results of egg production and survivability of the Attwater's Prairie Chickens bred in captivity have varied over the years due to many different variables; however the trends in recent years appear to be

Fig. 7. *Attwater's Prairie Chicken feather sexing. Figure 7a is an example of male tail feathers on a chick at 30 days old. Figure 7b is and example of female tail feathers on a chick at 30 days old.*

improving. As more facilities have become involved with the recovery of the Attwater's Prairie Chickens, overall the values of viable eggs, hatchability, and survivability to eight weeks have maintained or increased since the recovery project began in 1993 (Graphs 1, 2, & 3). In 2007, the twelve Attwater's Prairie Chicken hens at the Houston Zoo facility laid 166 eggs, of those eggs, 137 were viable, about 82.5 percent. From the 137 viable eggs, 107 eggs hatched, roughly 78 percent. At the time of this paper, 77.6 percent of chicks hatched had survived; these chicks ranged in age from 11 days to six and a half weeks, none of the chicks had yet reached the eight week mark.

The egg production of the Attwater's Prairie Chicken at the Houston Zoo was equal between the two diets in 2007. A total of 166 eggs were laid. Hens on the 500 IU vitamin E diet laid a total of 83 eggs, as did the hens on the 1000 IU diet, averaging 13.8 eggs per hen. However, only hens on the 1000 IU diet produced a second clutch of eggs. When the three low producing (two eggs or less) hens are not averaged into the data, the hens on the 1000 IU diet laid an average of 20.2 eggs, while the hens on the 500 IU diet laid an average of 16.2 eggs. Further diet studies will be conducted to conclusively determine how much vitamin E influences egg production (Graph 4). It is estimated that the recovery team will release approximately 40 Attwater's Prairie Chickens from the Houston Zoo into the wild this year. The recovery team's short term goal is to release at least 200 birds in a season with the combined efforts from all of the facilities participating in the recovery effort. In 2006, a total of 166 Attwater's Prairie Chickens were released into the wild.

## Discussion

The coastal prairie of Texas has diminished greatly in the last century consequently limiting the amount of land available to animals such as the Attwater's Prairie Chicken. Hunting has also contributed to the severe decline in their population. However, awareness and captive breeding may bring this bird back from the brink of extinction. Diet studies, genetic studies, and captive breeding efforts are being fine tuned in the effort to save the Attwater's Prairie Chicken from the fate of the extinct Heath Hen *Tympanuchus c. cupido*.

Since the Attwater's Prairie Chicken recovery project's conception in 1993, over 1000 birds have been released, maintaining the population. However,

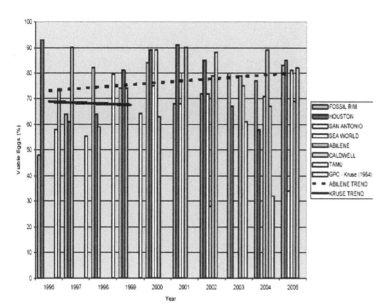

**Graph 1.** *Percentage of viable eggs produced at each facility from 1996 to 2005.*

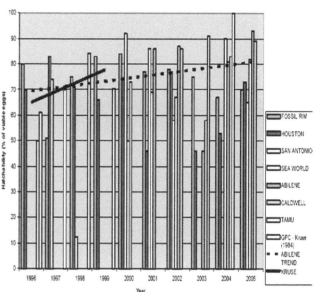

**Graph 2.** *Hatchability based on viable eggs produced at each facility from 1996 to 2005.*

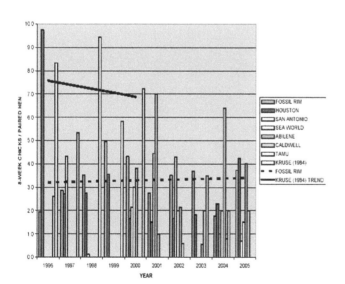

**Graph 3.** *Number of chicks to reach 8 weeks of age per paired hen at each facility from 1996 to 2005.*

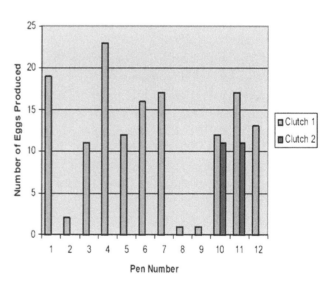

**Graph 4.** *Egg production per hen at the Houston Zoo in 2007. A total of 166 eggs were produced in 2007. Pens 1-6 were on the 500 IU vitamin E diet. Pens 7-12 were on the 1000 IU vitamin E diet. Egg production between the two diets was equal. However, when the three low production hens are removed, the hens on the 1000 IU vitamin E diet, on average, laid more eggs.*

the goal is to go beyond maintaining the population; the recovery team would like for the Attwater's Prairie Chicken to become a self sustaining population once again.

**Commercial Products Mentioned**

Corners Limited Custom Cages, 841 Gibson, Kalamazoo, MI 49001

Grumbach Incubators: distributed by Lyon Technologies, 1690 Brandywine Avenue, Chula Vista, CA 91911

Mazuri Zoo Feeds: manufactured by PMI Nutrition, PO Box 66812, St. Louis, MO 63113-6812
ProBios: manufactured by Bomac Vets Plus, Inc., 102 3rd Ave East, Knapp, WI 54749

Rubbermaid Shelf Liner: manufactured by Newell Rubbermaid, Inc., 29 East Stephenson Street, Freeport, IL 61032

## References

BIRDLIFE INTERNATIONAL. 2004. *Tympanuchus cupido*. In: IUCN 2006. 2006 IUCN Red List of Threatened Species. Downloaded on 08 May 2007. www.iucnredlist.org.
SARTORE J. 2002. A Chance to Survive. Lincoln, NE: Joel Sartore Photography.
TEXAS PARKS AND WILDLIFE DEPARTMENT. 2005. Adopt a Prairie Chicken Newsletter, Fall 2005. Austin, TX: Ed. Mark Klynn.
U.S. FISH AND WILDLIFE SERVICE. 2004. Attwater Prairie Chicken National Wildlife Refuge. Eagle Lake, TX. http://southwest.fws.gov.

## Author Biography

Mollie Coym is a zoo keeper who works full time with the Attwater's Prairie Chickens at the Houston Zoo. She graduated from the University of Houston with a Bachelor of Science Degree in Biology. She began her career with birds in 2003. She began volunteering for the Houston Zoo Bird Department in 2005. In 2006, Mollie began working there full time.

# DEVELOPMENT OF CAPTIVE PROPAGATION TECHNIQUES TO SUPPORT CONSERVATION OF STELLER'S EIDERS *POLYSTICTA STELLERI* IN ALASKA

Tuula Hollmen[1,2], Heidi Cline[1] & Nora Rojek[3]

1 Alaska SeaLife Center
2 University of Alaska Fairbanks, 301 Railway Avenue, Seward, AK 99664, USA
3 Fairbanks Fish and Wildlife Field Office, U.S. Fish and Wildlife Service, 101 12th Avenue, Room 110, Fairbanks, AK 99701, USA

## Summary

The Alaska-breeding population of Steller's Eiders *Polysticta stelleri* was classified as threatened under the U.S. Endangered Species Act in 1997, based on contraction in the species nesting range in Alaska. A Recovery Plan was published in 2002, providing guidance to recovery efforts for the species. Due to contracted range and limited nesting success of the Alaska-breeding Steller's Eiders, maintaining and enhancing breeding populations is considered vital for recovery efforts. To support recovery, current high priority recovery tasks include establishment of a captive flock at the Alaska SeaLife Center, development of artificial propagation techniques, and evaluation of field techniques to enhance egg survival using artificial incubation. Since 2003, several projects have been initiated to develop captive propagation techniques to support conservation efforts. The Alaska SeaLife Center established and maintains a captive flock of Steller's Eiders, and Steller's Eider eggs have been artificially incubated and reared in captivity. In 2005-2006, a pilot artificial incubation project was conducted in the field to reduce exposure of eggs to predation and enhance survival. Preliminary results show promise that artificial incubation techniques have potential to enhance productivity and support conservation of this threatened species.

## Introduction

The Alaska-breeding population of Steller's Eiders *Polysticta stelleri* was listed as threatened under the U.S. Endangered Species Act in 1997, based on contraction in the species nesting range in Alaska (U.S.F.W. 1997). A recovery team was formed and a recovery plan published in 2002, providing guidance to recovery efforts for the species (U.S.F.W. 2002).

The vicinity of Barrow on the Arctic Coastal Plain is the core northern breeding area of the threatened Alaska-breeding population of Steller's Eiders. A monitoring program was initiated in this area in 1991 (Rojek 2007). Since 1991, egg laying has been observed in only eight of 15 years and furthermore, Steller's Eiders were observed to experience poor success in breeding years. In years combined, only 15.5% of clutches monitored survived to hatching and only 17.6% of 17 monitored broods survived to fledging (Rojek 2007). Predation is likely the main cause of breeding failure in this population. Due to contracted range and limited nesting success of the Alaska-breeding Steller's Eiders, maintaining and enhancing breeding populations is considered vital for recovery efforts. In 2005, high priority recovery tasks were established to initiate development and refinement of captive breeding techniques to provide optional tools to support recovery efforts. Current high priority recovery tasks include establishment of a captive flock at the Alaska SeaLife Center (ASLC) in Seward, Alaska, development of artificial propagation techniques, and evaluation of field techniques to enhance egg survival using artificial incubation. Several collaborative projects have been initiated among the ASLC, the U.S. Fish and Wildlife Service (USFWS), and other partners, including the North Slope Borough of Alaska and Dry Creek Waterfowl in Port Angeles, Washington. In 2005-2006, the ASLC established a captive flock of Steller's Eiders of known geographic origin through opportunistic collection of eggs and rearing in captivity. Likewise in 2005-2006, a pilot project was conducted in the field, using artificial incubation of wild Steller's Eider eggs at field incubation facilities to reduce exposure of eggs to predation and thus, enhance survival. This pilot study was designed to evaluate feasibility of replacing Steller's Eider eggs with artificial eggs during early phases of incubation, incubating the Steller's Eiders eggs artificially, and returning the artificially incubated eggs back to nest just prior to hatching. This presentation is a preliminary summary and update of the artificial incubation and duckling rearing projects in the field and in captivity at the ASLC.

## Methods

**Field Methodologies: study area and nest surveys**

Steller's Eiders nest in Alaska near Barrow at the northwestern corner of the Alaska coastal plain (71°18'N, 156°40'W). The Barrow area is dominated by ice-wedge polygons, shallow oriented lakes, and drained lake basins, all underlain by continuous permafrost. Plant communities include upland meadow, wet meadow, marshes with emergent vegetation, and open water in large and small lakes and ponds (Bunnell et al 1975). Steller's Eiders nest intermittently at Barrow, with gaps of one to four years between nesting years (Quakenbush and Suydam 1999). Typically, onset of incubation (clutch completion) occurs during the second half of June to early July (Quakenbush et al 2004). In conjunction with the ongoing USFWS Steller's Eider project, a systematic ground-based breeding pair survey was performed in spring of 2005 and 2006 as soon as Steller's Eiders dispersed from large water bodies to breeding areas on the tundra. This survey was conducted within a 192km$^2$ study area that encompasses tundra within 4km of the Barrow road system and provided a census of Steller's Eider breeding pair distribution and abundance during the pre-nesting and early-nesting period. Following this survey, a nest search was conducted to estimate nesting density and distribution in the Barrow area. Nests were monitored and evaluated for inclusion in the egg substitution study and/or salvage and opportunistic collection for the captive breeding flock. Details on nest searching and monitoring methodology are described in annual field season reports (Rojek 2007).

**Field Methodologies: egg substitution pilot study and opportunistic collection**

After nest discovery, eggs were counted and candled to determine developmental stage. Powder-free latex or polyethylene gloves were worn during egg or hen handling. All materials contacting eggs or nests (e.g., artificial eggs, artificial incubator trays) were cleaned prior to contact and as necessary to maintain a clean environment using chlorhexidine solution. To minimize disturbance, nest visits were conducted either after hens accidentally flushed off nests or when hens were on incubation breaks.

Only nests with ≥5 eggs (indicating a complete clutch; (Quakenbush et al 2004) were considered for the egg substitution pilot project. Also, only every third active nest discovered was to be included, and a maximum number of nests included in a year would be three. Thus, with three active nests, one nest exchange would be conducted; with six active nests, two exchanges; and with ≥9 active nests, three exchanges. Priority was given to nests with eggs that were not developmentally advanced. Based on the number of active nests discovered in 2005 and 2006 that met the above conditions, eggs were exchanged in two nests in each of those years. All eggs in the clutch were removed and replaced with an equivalent number of artificial eggs that were the same size and color as Steller's Eider eggs. As a result of the experience from the first exchange, further egg exchanges only occurred while hens were on incubation breaks. After artificial eggs were placed in the nest, hens were monitored at every three days from a distance that would not cause flushing. Steller's Eider eggs were transported and incubated as described in the sections below. One to two days prior to external pipping, the Steller's Eider eggs were returned to the original nest and the artificial eggs were removed.

After pre-pipping eggs were returned, hens were monitored for acceptance of the eggs and to determine nest success. If a hen abandoned at any phase, the eggs were to be fostered into other nests with similar nest initiation dates (based on date of laying and/or egg development), including other replaced nests. If no such nests were available, clutches were to be transported to the ASLC for hatching and rearing. Hatching and dry brooding facilities (a heat lamp and a small brooder) were maintained in Barrow for backup purposes only. In 2005, abandoned eggs from other nests were salvaged, incubated as described above, and transported to the ASLC for hatching and captive rearing. In 2006, we planned for opportunistic egg collection to increase the size of the ASLC captive breeding flock. One or two eggs were collected from viable nests using the following criteria: one egg was collected from nests with three to five fertile eggs; two eggs were collected from nests with six fertile eggs. Fertility of eggs was determined by candling at time of collection. All field and incubation methods were the same as previously described for the artificial incubation pilot study.

**Egg Transport**

In 2005, eggs were hand carried from nest to a transport vehicle in soft, padded coolers with heat packs as heat source. Temperature was kept at 34.5°C - 37.5°C and vibration minimized during carriage. In 2006, compact electronic brooders (Dean's Animal Supply, St. Cloud, FL) were used instead, to facilitate more consistent temperature maintenance. During vehicle transport to field incubation facility, eggs were kept in portable electronic brooders. Electronic, battery operated portable brooders were used for airline transport from Barrow to Anchorage and further transport by car to the ASLC in Seward. Eggs were candled prior to departure and again upon arrival at ASLC.

**Artificial Incubation and Egg Monitoring**

An incubator with cool down timer (Grumbach Compact S84, Lyon Electronics, Chula Vista, CA) was maintained at the field facility and at the ASLC. The temperature was kept at 37.3°C and relative humidity at 60%. The incubator was programed to turn the eggs every two hours, and cool the eggs for 30min twice a day (at 7am and 5pm). Eggs were marked with nest and egg numbers and candled after temperature had stabilized in the incubator (in 2005) or at arrival (in 2006). Eggs were weighed and measured using a caliper at arrival. Embryonic development was monitored by candling and weighing eggs every three days. Average mass loss was calculated in 2005 for five eggs that hatched easily.

**Hatching and Rearing in Captivity**

After external pipping, a small piece of shell was removed from the edge of the hole to maintain an open airway during hatch process. The membranes of a hatching egg were kept moist by spraying with distilled water. The umbilicus of a hatchling was disinfected with diluted Betadine, and hatchlings were allowed to dry in the incubator for two to eight hours. Wet weight was recorded immediately after hatching, and dry weight, culmen length, and tarsus length was recorded when hatchling was moved from the incubator to the brooder.

In the brooder, supplementary heat was provided with infrared lamps and temperatures were initially maintained at 36°C at the warmest point of the brooder. Water was provided *ad libitum* and during the first one to two days, a pinch of food was sprinkled into the water each time the water was changed. Ducklings were fed with Mazuri (Purina Mills, St. Louis, MO) waterfowl starter food. Additionally, *Mysis, Daphnia*, and mosquito larvae were offered in water in small amounts. At two to three weeks of age, adult Mazuri sea duck formula was gradually mixed in with the starter food, and by approximately six weeks, ducklings were fed with adult formula. Ducklings were introduced to water after they started eating, and were housed in totes with constant access to water by approximately seven to ten days of age. Ducklings were introduced to salt water habitats at five to six weeks of age, by converting fresh water swimming pools into salt water (equivalent of local ocean salinity concentration) over a one week period.

**Health Monitoring and Disease Screening**

In 2006, screening for highly pathogenic avian influenza was conducted in the field prior to shipment of eggs by commercial airlines. Samples were collected from nests and eggs using standard virus swabbing techniques, and screened at the State of Alaska, Department of Environmental Conservation, Environmental Health Laboratory (Anchorage, Alaska). At the rearing facility at the ASLC, health monitoring involved daily monitoring of weight, growth, food consumption, and observations of behaviors and clinical signs of illness. Signs monitored included stationary or dropping weight, abnormal posture or locomotion, lethargy, soiled plumage, nasal or ocular discharge, runny droppings or stained vent area, sneezing or coughing, panting or heavy breathing, redness or swelling of umbilical area, and skin lesions or abnormalities.

Ducklings were weighed daily between 9:00 - 11:00am until two months of age, and then weekly until six months of age. To monitor structural growth, tarsus and bill length was measured every three days until two months of age, and then weekly until measurements did not change between two sessions.

Eggs and ducklings were maintained in an isolation facility during incubation and for a minimum of 30 days from hatch, and until results from disease screening and health evaluation were completed. After hatch, swabs for microbiological analyses were collected from egg membranes and ducklings, and screened for a suite of potential avian pathogens. Cell culture and egg inoculation methods were used for virus isolation attempts, and aerobic culture methods were used to screen for bacterial pathogens. At approximately one month of age, a blood sample was collected from each duckling, and evaluated for antibodies for a suite of potential avian pathogens and for routine hematology. If stationary or dropping weight or other signs of illness were observed during rearing and quarantine, a more thorough clinical health evaluation was performed and diagnostic testing conducted to evaluate the case and determine treatment.

## Results

### Egg Substitution Study

In 2005, the first egg substitution was conducted on July 3rd in conjunction with concurrent video monitoring project. After flushing the female from the nest, video equipment in proximity of the nest was readjusted, and seven fertile eggs were switched with seven artificial eggs. The nest was revisited on July 4th, and was found abandoned. The substitution protocol was revised so that further switches were to be conducted during incubation recess only. On July 6th, seven eggs from a second nest were substituted with seven artificial eggs. The nest was revisited on July 7th, and eggs were found warm although the female was off the nest. On July 8th, the female was confirmed back on the nest and incubating. On July 17th, the eggs were determined to be at 22 days of incubation, and were returned to the female during an incubation break. The female resumed incubation, and all eggs had hatched and the brood had departed from nest by July 24th. No further nests were available in 2005 to continue with the pilot substitution study. In 2006, the first nest with seven eggs was substituted with six artificial eggs on June 30th. The female was confirmed on the nest and incubating on July 1st, and five eider eggs were returned to the nest two days prior to hatch. All five hatched successfully. The second nest of eight eggs was substituted with six artificial eggs on July 2nd, and female was confirmed on the nest and incubating on July 3rd. Six eider eggs were returned to the nest two days prior to hatching. After approximately seven-and-a-half hours of monitoring, the female had not been observed to take incubation recesses, and was flushed to return the eggs. The female resumed incubation and all six eggs hatched successfully.

### Salvage and Opportunistic Collection of Eggs

In 2005, a total of 11 eggs were salvaged or opportunistically collected from four different nests. One of the eggs at 14 days of development had been abandoned for approximately ten hours prior to salvage, and produced a viable duckling. In 2006, a total of 19 eggs were opportunistically collected from 11 different nests. Four of these eggs were kept from two nests participating in the egg substitution study as described before. All opportunistic collections took place during incubation recesses. All of these females resumed incubation and 10 of 11 nests hatched successfully. One nest was predated at a later stage of incubation and did not hatch successfully.

### Egg Transport and Monitoring

Transport of eggs from Barrow to ASLC took approximately 12 hours. In 2006, one egg was slightly cloudy and abnormal in appearance although determined viable by candling prior to transport, but showed signs of embryonic death upon arrival. Dried fecal matter was observed on the shell of the egg at the time of collection. The cause of death was determined to be bacterial infection of embryonic membranes, preventing normal movement of the embryo and hatching process. Eggs lost an average of 11.6% of mass from start of incubation to internal pipping (n=5).

### Hatching and Rearing

Estimated artificial incubation range was 25-29 days in 2005 (n=9) and 24-28 days in 2006 (n=18). Hatching success was 82% in 2005 and 95% in 2006. Ducklings exhibited a tendency to filter feed their food during the first three weeks of life, and showed a preference to feed food items served on water. Frozen *Daphnia* and mosquito larvae served in the drinking water were readily consumed and encouraged filter feeding bouts during early age. Representative growth curves for mass and culmen/tarsal measurement up to 60 days of age are presented in Figure 1 and 2.

### Health Monitoring

No evidence of viruses, including highly pathogenic avian influenza, was found in the swabs or serum samples collected in the field or during isolation rearing at ASLC. Weight curves of one duckling in 2005 and five ducklings in 2006 were observed to deviate from expected growth pattern. Slower gain was typically first observed at approximately one week of age and, simultaneously, abnormal (watery or mucoid) droppings were observed. At this point, additional samples were collected for microbiological screening, and antimicrobial sensitivity testing. Commonly isolated organisms included *Escherichia coli, Acinetobacter sp, Enterococcus* sp. and *Staphylococcus* sp. Ducklings were successfully treated with five to seven days of oral trimethoprim-sulfa. One duckling did not show a favorable response to trimethoprim-sulfa treatment, and was treated with amoxicillin and supportive care of oral nutrition and multivitamin

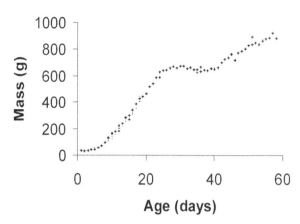

Fig. 1. *Daily mass of a representative Steller's Eider duckling from hatch to 60 days of age.*

supplements.

## DISCUSSION

The pilot study to use artificial incubation to enhance egg survival and productivity of Steller's Eiders showed promising results in the field. Using an approach to initially replace Steller's Eider eggs with artificial eggs during an incubation recess, we successfully conducted an egg substitution in three separate nests. The first attempt to substitute Steller's Eider eggs with artificial eggs was conducted in conjunction with other procedures near the nest after the female was flushed, possibly contributing to abandonment of the nest. When the egg switch was performed during an incubation recess, no abnormal responses to artificial eggs were observed. All of the three females participating in the pilot study appeared to readily accept the artificial eggs, maintained incubation routines, and later accepted their original eggs close to pipping and hatching. All eggs returned to their original nest hatched successfully under a female. Field incubation facilities were established and transport protocols developed and tested, both in the field and for a 12-hour transport from the field to the ASLC facility. One egg that was found inviable upon arrival at ASLC had likely died of bacterial contamination and infection of embryonic membranes. The slightly abnormal appearance observed prior to transport and visual fecal contamination on the shell suggested that the contamination had likely originated from earlier exposure in the nesting environment. Weight loss of eggs during incubation was similar to those reported earlier for other avian species, including sea ducks (Zicus et al 2004). Total incubation times were similar to those reported earlier for Steller's Eiders, although we estimated a four day range in each year in total incubation time. The range we found included slightly shorter and longer incubation periods than those reported earlier for free ranging Steller's Eiders (Solovieva 1997). Ducklings were reared using commercially available diet formulas enriched with invertebrate food items during early rearing. As shown in Figure 1, a plateau in weight gain was observed in many ducklings at approximately three to five weeks of age, but stationary or dropping weight at earlier age was observed in conjunction with clinical signs of gastrointestinal disease. Approximately 20% of ducklings reared during 2005-2006 showed weight loss and symptoms of gastrointestinal disease during first two weeks of life. The majority of these cases responded well to oral trimethoprim-sulfa treatment and resumed normal weight gain and growth. The onset of gastrointestinal symptoms during the first week of life coincided with resorption of yolk and may be related.

Many of the logistical challenges we encountered in the field related to the rare and listed status of the species, and the remoteness of the field location. The original pilot project was planned to occur in 2003, however, no nesting was observed during 2003-2004. In 2005, Steller's Eiders bred for the first time since 2000 in our study area near Barrow, and the pilot study was initiated after a three year waiting period.

Due to the rare and listed status, conservative approaches were taken when designing field protocols. To conserve wild populations, only a minority of wild nests would be participating in pilot protocols in any given year, resulting in relatively few opportunities to test protocols. Also, staff resource needs were high because significant staff time was allocated to the

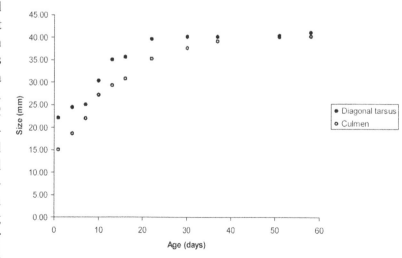

Fig. 2. *Daily tarsus and culmen length of a representative Steller's Eider duckling from hatch to 60 days of age.*

**Fig. 3.** *Wild Steller's Eider nest.*

**Fig. 4.** *Candling eggs in the field.*

**Fig. 5.** *Close up view of nest.*

monitoring of nests for incubation recesses. Additional cautionary approaches involved minimizing handling time of eggs for candling and limiting monitoring of success to document hatch result only, instead of further brood monitoring using radio telemetry. Remoteness of field sites provided some logistical challenges of transport from the nests to the field incubation facility and back, especially during adverse weather conditions. Due to cold temperatures, maintaining appropriate egg incubation temperatures was a challenge in the field, especially while waiting and monitoring the females to take an incubation recess when replacing artificial eggs with her originals. Also due to remoteness of the study area, transport time to the main facility was relatively long, even though regular commercial airline service provided a daily transport option. Additional logistical challenges included avoiding x-ray screening of viable embryos through airline security, and coordinating for rapid avian influenza screening prior to transport from the field. We conclude that our pilot efforts to enhance productivity of Steller's Eider using artificial incubation showed potential and promise to enhance productivity of Steller's Eiders, however, several logistical challenges would need to be addressed before a regular or larger scale program would be feasible. The project was supported by leverage from ongoing field projects to monitor nesting ecology of Steller's Eiders near Barrow. Without leverage and collaboration from the ongoing field project, the cost of the pilot studies would have been much greater. Furthermore, the long term monitoring project by USFWS provided baseline information to assist planning of this project. The technique we used was relatively labor intensive, and required significant field support for nest searching, monitoring of females, and operations at the field facility. Because the egg substitutions were conducted during an incubation recess, the field crew had to monitor nests to determine incubation recess schedules and time their nest visits accordingly. Sometimes, this resulted in hours of waiting time on the tundra. Whereas the protocol was successfully performed on three separate nests in our pilot study, a larger scale operation would likely require either a significant staffing resource or a modification of the protocol.

## Acknowledgements

We would like to acknowledge our partner and collaborator Arnold Schouten from Dry Creek Waterfowl, Port Angeles, Washington for his help and support throughout this project. Help from Debbie Schouten is also greatly appreciated. We would also like to recognize the contributions of our friend and colleague Paul Dye, who participated in the design and planning of this project but sadly, passed away before we were able to initiate the field project. The authors would like to acknowledge the Ukpeagvik Inupiat Corporation (UIC) for access to land and the Barrow Arctic Science Consortium and North Slope Borough Department of Wildlife Management for logistical support. Field assistance was provided by Rebecca Howard, Jessica Eden, Tasha DiMarzio, Amy Kilshaw, and the rest of the field crew. We also appreciate the generous support of Ted Swem of the U.S. Fish and Wildlife Service, Fairbanks Fish and Wildlife Field Office. At the Alaska SeaLife Center, we would like to acknowledge the help and support of avian husbandry staff (Tasha DiMarzio, Amy Kilshaw, and Mike Grue), eider research staff (Ann Riddle and David Safine), veterinary services staff (Dr. Pam Tuomi, Dr. Carrie Goertz, and Millie Gray), and other ASLC support. We would like to thank Dr. Bob Gerlach for AI testing and providing rapid diagnostics results to assist with field and transport logistics.

Funding for this project was provided by U.S. Fish and Wildlife Service, Alaska SeaLife Center, and the U.S. Air Force 611th Civil Engineer Squadron's Conservation Program.

## References

BUNNELL FL, SF Maclean and J Brown. 1975. Barrow, Alaska, USA. In: T Rosswall & OW Heal, (eds). Structure and Function of Tundra Ecosystems. Ecological Bulletin. (Stockholm) 20: 73-124.

QUAKENBUSH L & R Suydam. 1999. Periodic non-breeding of Steller's Eiders near Barrow, Alaska, with speculations on possible causes. Pages 34-40 in: RI Goudie, MR Petersen and GJ Robertson, (eds). Behaviour and ecology of sea ducks. Occasional Paper, No. 100, Canadian Wildlife Series, Ottawa, Ontario.

QUAKENBUSH L, R Suydam, T Obritschkewitsch and M Deering. 2004. Breeding biology of Steller's eiders *Polysticta stelleri* near Barrow, Alaska, 1991-99. Arctic 57(2): 166-182.

ROJEK NA. 2007. Breeding biology of Steller's eiders nesting near Barrow, Alaska, 2006. U.S. Fish and Wildlife Service, Fairbanks Fish and Wildlife Field Office, Fairbanks, Alaska. Technical Report. 53 pp.

SOLOVIEVA D. 1997. Timing, habitat use, and breeding biology of Steller's Eider in the Lena Delta, Russia. Pp. 35-39 in Proceedings from Steller's Eider workshop. Wetlands Int. Seaduck Specialist Group Bull. No. 7.

U.S. FISH AND WILDLIFE SERVICE. 1997. Endangered and threatened wildlife and plants; threatened status for the Alaska breeding population of the Steller's eider. Federal Register 62:31748-31757.

U.S. FISH AND WILDLIFE SERVICE. 2002. Steller's eider recovery plan. Fairbanks, Alaska.

ZICUS MC, DP Rave and MR Riggs. 2004. Factors influencing egg-mass loss for three species of waterfowl. Condor 106: 506-516.

# THE PLACE OF CAPTIVE BREEDING IN THE CONSERVATION STRATEGY OF THE ASIAN HOUBARA BUSTARD *CHLAMYDOTIS MACQUEENII*

OLIVIER LEON & OLIVIER COMBREAU

National Avian Research Centre, Po Box 45
553 Abu Dhabi, United Arab Emirates

## SUMMARY

The Asian Houbara Bustard *Chlamydotis macqueenii*, favorite quarry of Arabian falconry, is threatened by a menace of cultural origin. Hunted for decades on a sustainable scale, recent improvements of the living standards in the Arabian peninsula have changed the dimension on which the art of falconry is practiced. Concerned by the possible extinction of the species, dedicated organizations devoting their efforts to the conservation of the Houbara were recently created. In the UAE, the National Avian Research Centre has been studying the Asian species of Houbara for more than 15 years, with an ecology program monitoring wild populations and a captive breeding program aimed at restocking depleted populations. Houbara captive breeding is not simple. Contrary to other bustard species, Houbara does not breed well naturally in captive conditions, and is very sensitive to stress. Consequently, successful Houbara captive breeding not only has to use and adapt techniques developed in poultry aviculture, such as artificial insemination and artificial incubation, but also has to keep the bird tame from hatch to death, which makes it extremely demanding in terms of manpower. Being conservation oriented, the captive breeding of Houbara endeavors in maintaining the original genetic diversity of the founders. As a result, genetic diversity is handled by the management of each individual's mean kinship. Finally, when outside conditions are not suitable for triggering off and maintaining breeding, environmentally-controlled buildings are used. It poses the challenge of defining an appropriate light regime and temperature program for a flock which exhibits different responses for a given stimulus, because of an intrinsic population phenotypic polymorphism. Conservation will be achieved when sufficient numbers of genetically valuable individuals are produced. By adapting aviculture techniques to the particular case of a non-standardized flock, Houbara breeding centre will succeed and participate fully in the global conservation strategy of this species.

## INTRODUCTION

The captive breeding of Asian Houbara bustards *Chlamydotis macqueenii*, and hereafter Houbara, is a major tool to achieve the success of the global conservation strategy defined by NARC (Fox et al 2000; Combreau et al 2005). The bird benefits from a peculiar status in Arabian cultures where falconry is an art, and the Houbara the main subject of it. Until the middle of the 20th century, falconers used to catch falcons during their fall migration and use to train them quickly until they became ready to hunt the Houbara that came to winter in Arabia. Because of the scale on which this was done, and the relative protection of the breeding grounds, falconry was thought to be sustainable, and a balanced was achieved (Upton 1989).

The oil boom in the second half of the last century has had spectacular effects on Arabian human populations and the improvement of their standards of living. However, the consequent loss of habitat due to urbanization, the rising numbers of hunters who can afford their passion, the poaching of wild Houbara for falcon training purposes, and the ability to access more remote areas thanks to the progress in air and ground transportation have had severe consequences. Hunting traditionally took place on the wintering areas (Pakistan, Iran and Arabian Peninsula) but more and more Central Asian countries are attracting falconers to hunt. Central Asian states of the former Soviet Union are traditional breeding grounds of the migrant Asian Houbara and are essential stopovers for birds migrating further north and east to China and Mongolia. As these areas are used as breeding grounds by the

Houbara, the hunting takes place late in early autumn and targets breeders and juvenile. As a result most of the migrant Houbara populations, hunted once in these Central Asian states, are hunted again further south in Afghanistan, Iran or Pakistan. In the past, Houbara populations were safe once passed the frontiers of the former Soviet Union but currently there is no longer any available refuge (Combreau et al 2005). As a consequence, the pressure on Houbara populations has increased to an alarming extent.

Reports for the past 25 years have been suggesting a sharp decline in Asian Houbara populations (Collar 1980; Goriup 1997). More recent surveys based on satellite transmitted data all stress on the fact that extinction of the species could occur in a foreseeable future. A mortality assessment study showed that the mortality rate is more than 11 times higher in winter, when hunting occurs, than on the breeding grounds (Combreau et al 2001). A survey made over several years in the same regions showed striking decline in relative abundance from one year to another on the breeding grounds (Tourenq et al 2005). A recent review of the status of the Houbara conducted by Birdlife International for the International Union for the Conservation of Nature (IUCN) lead to an elevation of threat status from Low risk/near threatened to Vulnerable (Birdlife International 2004). The Houbara is also included in Appendix I of the Convention on International Trade in Endangered Species of Wild Fauna and Flora (CITES), in Appendix II of the Convention on Migratory Species (CMS) and in the Convention on Biological Diversity (CBD).

In its Global Strategy for the Conservation of the Falcons and Houbara, NARC has proposed the implementation of a series of conservation measures aimed at preserving the remaining wild Houbara populations in their ecological, migratory, physiological, and genetic diversity and integrity. The NARC Houbara strategy targets a substantial reduction in the hunting and poaching pressure on wild birds on a global scale through management of the breeding and hunting grounds and through management of the wild Houbara populations. To complement this, NARC proposes to produce Houbara in captivity for reintroduction or population reinforcement, put and take (release and hunting) and for falcon training purposes. To achieve its goals, NARC is articulated around three main programs: an Asian Houbara captive-breeding program, an Asian Houbara ecology program, and a bird rehabilitation program. NARC programs have a strong cohesion and form an inseparable entity. The purpose of this paper is to present the progress and challenges faced for breeding this species in captivity. It will be discussed of the role and place of a captive-breeding unit such as NARC's one within the existing Houbara conservation strategy.

**The Houbara Challenge**

Though tremendous progress has been made in the last ten years, producing Houbara in captivity remains a difficult and expensive activity. The NWRC in Saudi Arabia has pioneered Houbara captive-breeding (Saint-Jalme et al 1996). In Taif, captive-breeding was started with birds of the north-African and Asian species collected from the wild in Algeria and Pakistan respectively. Later, birds and techniques were exported to other centres including the NARC in UAE and the Emirates Centre for Wildlife Propagation (ECWP) in Morocco. Since NARC was set up in 1989, the production of Houbara Bustard has been low or even nil in 1997 and 1998. An analysis of the performance of captive breeding at NARC over years reveals that the low reproductive performance of Houbara at NARC can be explained mainly by two factors including the quality of the stock and the weather conditions prevailing in the UAE. The captive stock of Houbara was originally composed of birds donated to NARC from private collections. The origin of these birds was unclear, as most of them had been captured as adults in Pakistan, Iran or Oman. Whether the birds have had belonged to resident or migrant populations was unknown. Because of that heterogeneity and intolerance to human presence, the number of females which laid and the number of males which gave semen in the early days were extremely low. The sole solution offered to NARC for developing the captive breeding of Houbara was to obtain a homogenous pool of founders from known origin. For obvious practicality reasons, a new stock of birds was acquired from the National Wildlife Research Center of Taif (Saudi Arabia) in 1999 and 2001. Taif center's Houbara chicks originate from Balochistan and belong to the large south-central Asian resident Houbara population. It is from those chicks that the NARC captive-breeding has really started. In 2007, 780 birds were produced. However, the bird exhibits a number of characteristics which make it challenging to breed in captivity.

<u>The bird is cryptic and shy:</u> It is commonly acknowledged in any ecology survey on Houbara that the probability of detection is very low (Seddon et al 2001). In addition, it has a marked tendency to stress when held in captivity.

The bird also does not breed well under the harsh weather conditions prevailing in the UAE: Although some highly adapted populations of Houbara bustards may breed regularly in the Arabian Peninsula, the greater part of the Houbara population breed in much cooler conditions north of the 40th parallel. Consequently, attempts at breeding Asian Houbara in the UAE have always been hampered by high temperatures, this being complicated by the heterogeneity of the breeding pool. It seems also more and more obvious following NARC studies on the ecology of the wild populations that primary reproductive triggering factors in Houbara are population specific (NARC, unpublished data). Northern birds may respond well to an increasing photoperiod and low temperature whereas rainfall and subsequent atmospheric depression might be a sufficient cue to trigger breeding even under decreasing photoperiod in southern populations such as the Omani one. Therefore, it was decided as early as 1998 to put the birds inside climate-controlled buildings. This was an entirely new concept and a new challenge for the birds and bird care staff.

The bird does not breed well naturally in captivity: Until recently, breeding behavior of Houbara in the wild was hardly documented and sometimes contradictory (Lavee 1988; Ponomareva 1983). Recent studies show that, at least in eastern Kazakhstan, males are territorial and present a lekking behavior without the geographic distribution usually associated with leks (Riou et al in press), no protection offered by the male, no resource on the male site used by the female, and the male attracts the female by its displaying behavior only. However, the breeding behavior could be greatly depending on the habitat, and notably the available resources.

Natural breeding in captivity yielded very low results in terms of fertility (Saint Jalme et al 1996). The average fertility observed in pair bonded, small groups (two males/two females) and large groups (three to seven males/six to eight females) was never above 45%. Natural breeding data from NARC show a fertility of 18.2% in average on groups of around 30 birds. This is why all Houbara breeding centres are using artificial insemination (AI).

The bird has a low productivity: In the wild, a Houbara lays an average of three to four eggs per clutch (Combreau et al 2002). However this number might vary depending on the origin of the bird or the habitat condition found on the breeding ground. It is well described in domestic species that a brooding behaviour is negatively correlated to the number of eggs laid per female (Johnson 1986). On the other hand, removing the egg from the nest at the earliest after laying triggers off the female to lay again. Therefore in order to enhance productivity, and optimize the number of eggs an individual female can lay, Houbara captive breeding is based on egg pulling and artificial incubation (Saint Jalme et al 1996).

The bird is omnivorous: In the wild, Houbara have a very diverse feeding regimen (Collar 1996). They are mostly insectivorous, but depending on the habitat, they can feed on other arthropods, plants and small mammals as well. Hence, designing a suitable diet is a great challenge.

**The Asian Houbara Captive Breeding**

Houbara captive breeding units have implemented a series of techniques in order to handle the physiological, behavioural and ecological traits mentioned above.

**1. Adult care**

Besides feeding and monitoring for clinical signs and breeding behaviour, bird keepers devote their activities to maximize interactions with the birds. At NARC currently, one bird keeper handles around 80 birds daily. Because taming has to be continuously maintained and is always reversible, taming sessions are specifically scheduled in the daily routine, and starts from hatch until death. Adult taming is a great part of the success a Houbara captive breeding can achieve. Untamed birds are subject to trauma while trying to escape and flying in the cage, and are too intolerant to human presence to develop sexual activity. Crickets, mealworms and alfalfa *Medicago sativa* are the most commonly used taming devices.

**2. Artificial insemination**

Male collections and artificial insemination have been developed originally in NWRC in Taif, Saudi Arabia. In Missour, Morocco, the ECWP has enhanced those techniques. The semen is injected directly into the oviduct.

AI is a great tool for captive breeding management:

- It increases the average fertility. At NARC, the average fertility on inseminated eggs is significantly higher than that in natural breeding conditions (76% vs. 18.2% respectively (Leon et al

2006)).

- It allows the pairing of any female with any males with no logistic bottleneck, therefore allowing for proper genetic management.

- It allows for a better assessment of individual performances, by accessing to each bird's productivity.

To be carried out on a regular basis, several important factors have to be taken into account and dealt with efficiently: semen collection, semen storage, insemination planning and genetic management.

- Sperm has to be manually collected from males. Although well described in the poultry industry (Sauveur 1988), sperm collection is much less documented in non-domestic species. Contrary to chickens, sperm collection by abdominal massage has not been used successfully on Houbara so far. Instead, collections are made with a dummy on which the male copulates, every two days. The sperm is collected with a glass Petri dish, and transferred into straws or glass recipient, and sent to the lab for concentration assessment. Taming is obviously a key element for a male to be collectable.

- Suitable sperm storage technique has to be implemented. Sperm preservation is a key factor to success. Indeed, a full optimization of the sperm can be achieved, by allocating the sample to a genetically desirable female later. NARC is storing sperm by refrigeration or cryo-preservation, using the method described in Hartley et al (1999) and latter refined in the Emirates Centre for Wildlife Propagation (ECWP), Missour, Morocco (Challah, unpublished data). Much progress still needs to be made, as around half of the sperm do not survive to the cryo/thawing process.

- Females need to be planned for insemination. Houbara present a variable laying pattern that is mostly female dependent. This individual specificity makes it difficult for a regular insemination schedule to be implemented. A study showed that the best fertility results are obtained when females are inseminated twice with a minimum of 10 million active sperm three and six days before laying (Saint-Jalme et al 1994). To be workable, this method entails that the next laying can be inferred from previous laying history. Hence, each female is planned according to the pattern she had in previous and current seasons. Pattern is described as the average number of days between clutches and the average number of eggs per clutch. For first-time layers, the first clutch is left un-inseminated, the exception being if the female shows a sexual behaviour indicative of sexual maturity like a display. NARC aviculturists use a specifically developed computerized management system which shows the laying history of each female, can display the laying schedule of any female on demand, and records any relevant information concerning the insemination process for later analysis.

- Artificial insemination has to be carried out along with a proper genetic management, which is a pillar of the program's efficiency from a conservation point of view. NARC is using data from its own studbook to compute the genetic structure of the stock through Population Management 2000 software (Pollack et al 2002). The parentage of the birds is determined by microsatellite analysis, using 17 of the microsatellites initially developed for the North African subspecies and the great bustard *Otis tarda* (Chbel et al 2002; Lieckfeldt et al 2001). In the case of unknown parentage, a relatedness matrix between individuals is calculated using microsatellite data, and used as an input in Population management 2000. The mating is chosen based on the Mate Suitability Index (MSI) given by MateRx program (Ballou et al 1999). The MSI calculates a single value inferred from key genetic parameters of the pair which determine the genetic quality of the potential offspring.

### 3. Artificial incubation

Artificial incubation of Houbara eggs is a time consuming and technically difficult task. Progress has to be made in this field, as the difference between wild and captive hatchability is still significant: wild hatchability has been estimated at 84% (SD=23%), whereas 63% of incubated actually hatched at NARC in 2006 (Combreau et al 2002; Leon et al 2006). The incubation period is around 23 days. In the incubators, a common temperature and humidity regimen (37.5°C, 30% relative humidity) is applied for the first days after which a weight loss assessment is made. Like most incubation facilities, an average weight loss of

15% is targeted (Anderson 1979; The Game Conservancy's Advisory Service 1993). If the actual value differs from the target, the egg is moved to an incubator which is specifically prepared to hold eggs out of the normal range. The transfer to the hatchery occurs at day 20. Egg morphological characteristics seem to have an influence on hatchability. Notably, eggs laid by younger females present a significantly higher proportion of calcium aggregates, which is shown to impair hatchability (Leon et al 2006).

## 4. Chick rearing

Chick rearing is probably the most demanding activity in Houbara captive breeding. For 10 to 12 hours a day, one bird-keeper takes care of 20 to 25 chicks. Moreover, Houbara chicks are not autonomous before ten days of age minimum, before which they must fed by hand. They are kept by groups of five, handled and monitored for behavioral and health problem several times a day. Chick growth is monitored individually by regular weighing.

## 5. Feeding

For standardization and ease of use purposes, specific pellets have been developed and are given depending on the breeding status and/or age of the bird. Water is given ad libitum.
Five regimens are identified:

- First age, until 15-30 days.
- Grower, from 30 days to 8 months
- Maintenance, for adults in sexual rest.
- Breeding Female.
- Breeding Male.

(Table 1 shows the principle characteristics of Houbara feed.)

Complementary to pellets, a variable amount of live food is given to the birds for taming purpose: mealworms, crickets and alfalfa *Medicago sativa*. Around five to ten grams per bird per day of each item is offered per bird. Mice are used as medication device.

## 6. Environmental parameters

As mentioned earlier, reproduction in outside cages yielded poor results in the UAE because of harsh weather conditions, with temperatures as high as 35°C in March. To place the birds in a suitable environment, environmentally-controlled buildings have been developed.

First, greenhouse-like buildings with a pad-cooling system coupled to extractors were tried out. The birds were exposed to the natural photoperiod and the temperature could be cooled down in summer to 28-30°C when the outside temperature is around 45-48°C.

In a second time, closed buildings with air conditioning have been tried. This led to the issue of defining a suitable light regimen and temperature program able to trigger and maintain breeding, and synchronize the birds. Houbara has a seasonal reproduction (Collar 1996), and the first tries mimicked the conditions to

### Table 1
**MAIN CHARACTERISTICS OF HOUBARA FEEDING PELLETS ACCORDING TO BIRD'S AGE**

|  | Starter | Grower | Breeding Female | Breeding Male |
|---|---|---|---|---|
| Energy (kcal ME/kg) | 2700 | 2700 | 2400 | 2400 |
| Crude Protein | 26.2% | 23% | 23% | 22% |
| Crude Fat | 2.7% | 2.6% | 4% | 4% |
| Crude Fiber | 4.0% | 3.8% | 6.7% | 8.7% |
| Crude Ash | 7.1% | 6.6% | 12.2% | 7.4% |
| Calcium | 1.2% | 1.1% | 3% | 1% |

*Legend: kcal ME/kg: kilocalories of metabolisable energy per kilogram of pellet*

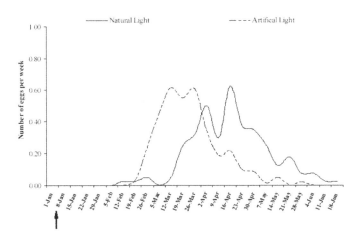

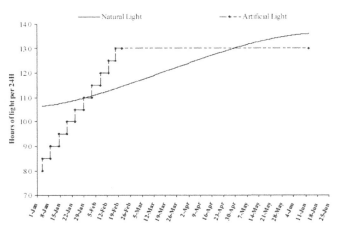

Fig. 1. *Differences in laying pattern for Houbara females exposed to artificial or natural light. Top: Laying curve. The arrow represents the start of light increase for the group exposed to artificial light. Bottom: Photoperiod pattern according to the type of light (natural vs. artificial).*

which birds were exposed in the wild, i.e. increase the photoperiod, increase temperature. Compared to birds housed in natural light buildings, it resulted in a much better synchronization of the laying and yielded better productivity per female. Figure 1 shows the difference in laying pattern between birds kept in natural light and artificial light conditions, and plots the photoperiod according to the type of light, natural vs. artificial. Table 2 shows the main environmental characteristics of the buildings depending on breeding conditions.

Light regimens are a key factor to breeding onset and duration (Johnson 1986). NARC has carried out a series of experiment in order to understand the influence of light and temperature on breeding onset in Asian Houbara. Notably, breeding can be triggered out of the normal breeding season, and the pace of light increase along with the apparent timing of temperature increase can influence the laying pattern (Leon et al 2007).

### Table 2

**ENVIRONMENTAL PARAMETER SETTINGS IN THE BREEDING BUILDINGS ACCORDING TO THE BREEDING STATUS**

| Breeding status | Light | Temperature |
|---|---|---|
| Sexual rest | 8 hours light/16 hours dark | 20°C |
| Transition to breeding conditions | 1) 8 hours light/16 hours dark | Decrease to 5°C |
|  | 2) Increase by 30 min-1h | Increase by 2-3°C |
| Breeding conditions | 13.5/10.5 light/dark | 15°C |
| Transition to sexual rest | No light | 28-30°C |

## 7. Cage design

A suitable cage has had to be designed. In breeding buildings, space has to be optimized to improve the cost effectiveness per m². At NARC, birds used to be housed in individual indoor cages of 16m² whereas surfaces of up to 25m² per individuals were used outdoor. After many refinements, trials and errors, Houbara are now housed in 4m² cages. Walls of the cage are made of soft net to limit traumatic injuries. Trials are being carried out in 1m² cages, and preliminary results are encouraging.

## 8. Biosecurity and preventative medicine

Strict hygiene controls are to be observed in all aspects of bird care and building operations. NARC has been built in a remote area to be purposely far from poultry farms. Each building has its dedicated laundry and shoes. Strict hands and feet washing are required. However, the objective is not to produce Specified-Pathogen-Free chicks, so there is no air pressure gradient and UV air disinfection. An environmental screening program is undertaken by the veterinary department on a regular basis to monitor the contamination status of the premises and equipment.

Captive birds are vaccinated against Newcastle disease and poxvirosis. A trial of different Newcastle vaccine types, strains and routes of administration showed that the inactivated/adjuvanted vaccine by intramuscular route gives the best results in terms of antibody persistence in adults and transferred immunity to the chick (Facon et al 2005).

To summarize, Houbara captive breeding is a difficult task which requires lots of manpower along with techniques found in poultry aviculture such as artificial insemination, semen assessment, artificial incubation, defining suitable lighting and temperature programs, and proper biosecurity measures. Houbara's sensitivity to stress requires a constant taming, which is why starting a captive breeding colony with adult birds from the wild is sure to fail. Constant adjustments and continuous research is needed in order to better understand the bird's reproductive physiology. In the UAE, because of the harsh weather conditions, the key achievement resulted in breeding birds in environmentally-controlled buildings. Not only does it open the door to a rationalized way of production, it also allows for proper research by standardizing the conditions for all birds. Future challenges remain in the optimization of building space, rationalizing the insemination process, improving the diet and switching to a group management insemination planning rather than individual one which is easier to handle on a large scale.

## NARC's Captive Breeding Results

Since 1998, NARC has made a number of progresses in the breeding techniques by importing what was successful in other centres and notably in ECWP, Morocco, and by adapting to the very peculiar context it has to evolve in: all-in system, and a restraint on the number of founders. NARC's yearly production since 1998 is shown in Figure 2. The increase in the number of chicks produced partly results from the improvement of females' productivity, as shown on Figure 3. There is a major effect of the age on the breeding performances. Younger females (aged one year old and less at the beginning of the season) are significantly less productive than older females as shown on Figure 4. The existence of a previous breeding experience has an effect on productivity. In Table 3, the productivity of two groups is compared: birds which had laid in previous seasons, and birds which never laid in previous seasons. To avoid any bias due to young age, birds aged a year or less have been excluded from the comparison. The effect is highly significant. This has been shown for outdoor bred birds (Van Heezik et al 2002), but is also described here for birds held in artificial conditions.

As far as the males are concerned, more than 95% of the males are displaying after two years of age. In 2006, 56% of the males have been collected at least once, and 75% of them gave at least one usable sample. The main bottleneck in Houbara male reproduction is the ability to collect usable sperm from a bird. This ability is mostly male-dependent: there is no collector effect as long as the collector is experienced (Leon et al 2006). However, a few males show a very clear preference for a particular collector.

Even if a male gives semen samples regularly usable, the consistency between collections for a given male can be an issue. Indeed, the variations in concen-

### Table 3

**EFFECT OF PREVIOUS BREEDING EXPERIENCES ON FEMALE PRODUCTIVITY**

| | Group 1 | Group 2 | Statistical Significance |
|---|---|---|---|
| Average number of eggs per laying female | 6.0 | 9.1 | p=0.001 |
| Average number of chicks per laying female | 3.6 | 5.8 | p=0.008 |
| Fertility/ Hatchability | 0.67/0.50 | 0.78/0.60 | NS, p=0.181/0.145 |

*Mann Whitney test has been used, statistical difference if p<0.05. Group 1: Females with no breeding history. Group 2: Previous layers*

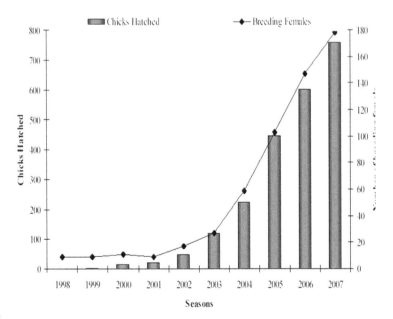

**Fig. 2.** *Evolution of NARC's yearly production 1998-2007.*

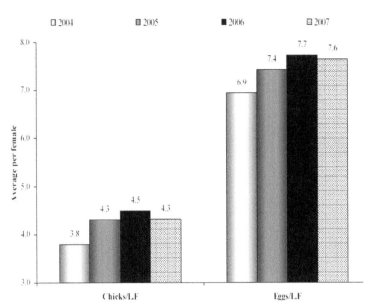

**Fig. 3.** *Evolution of females' productivity from 2004 to 2007 season.*
Legend: Chicks/L.F: Number of chicks per laying female.
Eggs/L.F: Number of eggs per laying female

trations from an ejaculate to another can be drastic. A trial has been performed to assess the effect of widening the gap between collections on males which semen concentrations decreased to the point of being unusable. The average number of sperm per ejaculate per male increased when the gap is widened. As shown in Figure 5, it is attributed to an increase in concentration, the volume being comparable whatever the gap.

Additionally, there is a well described age effect on performances. Young males (one-year old and less at the beginning of the season) are significantly performing worse than older males, as shown on Figure 5.

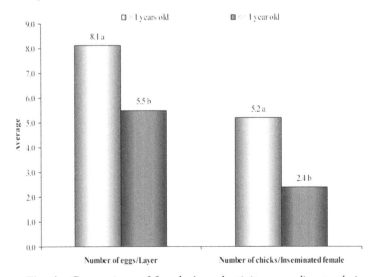

**Fig. 4.** *Comparison of females' productivity according to their age at the beginning of the season: effect on number of eggs laid and number of chicks produced (a not equal to b, p<0.05, Mann Whitney test).*

These encouraging results have been achieved after years of dedicated research. In order to cope with the numbers expected, Houbara breeding centres have adapted themselves by trying to apply poultry aviculture methods to the system. However, dealing with wild birds entails an absence of standardization of phenotype, unlike domestic poultry after years of genetic selection. This has a major impact for two main reasons. First, it implies that birds will not respond equally to a given stimulus. Although this variability can be handled when dealing with a small number of birds, it becomes very complicated with hundreds of them. Second, it impairs the interpretation of experimental results. In most models used to compute the effect of a particular variable, the associated $R^2$ (coefficient which expresses the percentage of the overall observed variation explained by the tested variable) are low, even if the effect is significant. The main conflict faced by large captive breeding structures is expressed here: the variability is the main goal to achieve from a conservation point of view, because genetic diversity is kept to its maximum, but it is counterproductive when trying to breed large numbers of birds under standardized conditions.

**Perspectives in Asian Houbara Bustard Conservation**

The future of Asian Houbara bustards lies on a ground of cultural and biological questions that need to be addressed. To efficiently deal with the challenge in all its diversity, a coordination of efforts undertaken by all dedicated centres is the key to success.

**1. Continue the development of a cost-effective mass production of captive bred Houbara.**

Houbara captive breeding has a major relevance in the conservation issue.

By developing subgroups by geographic origin, chicks of compatible genetic pattern can be produced and help restock declining wild populations. Genetic studies based on microsatellite data are currently being carried out to determine the genetic distances between individuals from different geographic origins and different migratory patterns (migratory vs. resident populations). Studies based on mitochondrial DNA have failed to show genetic divergence between populations (Pitra et al 2002; Pitra et al 2004). However, satellite tracking of Houbara has revealed that though birds are not faithful to their wintering ground, conversely all

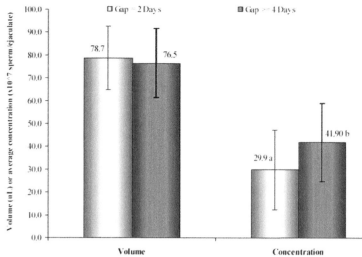

Fig. 5. *Comparison of males' results according to the gap between collections: effect on volume and concentration (average ± standard deviation). The difference in volume is not statistically different (p=0.611, Friedman test), but it is in concentration (a not equal to b, p=0.011, Friedman test).*

migrant Houbara followed so far by this technique have shown a strong fidelity to their breeding ground (NARC, unpublished data). This specificity indicates that the risk of local extinction on some of the breeding grounds is high in case of catastrophe (heavy poaching, hunting, lethal disease) whereas it is moderate in the wintering ground. To reduce the risk of local extinction that most long migrant Asiatic Houbara Bustard populations are currently subject to, captive breeding offers the possibility to keep and breed a small number of each important population of Houbara so as to maintain a live gene bank of the species. This can only be achieved with a proper genetic management based on the correct evaluation of the genetic value of each individual.

With a core of birds of mixed origin, a captive-bred alternative for falcon training purpose and for hunting can be developed, therefore protecting wild populations from massive hunting and poaching. It has been evaluated that 4,000-7,000 Houbara are illegally traded from Pakistan each year (Goriup 1997), but accurate information is missing. Ramadan-Jaradi (1989) estimated in the late 1980s that in the markets of the UAE alone, up to 2,000 wild-caught Houbara were being sold annually. Furthermore, large numbers of birds die during transport, as the birds are kept under unhygienic and crowded conditions (Bailey et al 1998). Conservationists are concerned that this live trade may have an impact on the status of the Houbara as high as that of falconry. This practise, generated by the high prices offered for live Houbara in the Arabian Peninsula, is extremely detrimental to the Houbara populations and no effort should be spared to fight it. The objective of UAE projects is to produce 10,000 Asian Houbara by 2015.

## 2. Continue the efforts in birds' rehabilitation.

Houbara illegally captured and smuggled in the middle-east countries are usually trapped on their wintering grounds, and may be kept with poultry under unhealthy conditions during several days or weeks before they are shipped to Arabian Gulf countries for falcon training purposes. Seized Houbara are generally in extremely poor condition, severely emaciated, suffering from traumatic injuries, carrying viral, bacterial and/or parasitic infections (Bailey et al 1998).

Two countries on the distribution range are involved with the rehabilitation of smuggled birds, Pakistan and the UAE. United Arab Emerates ratified the CITES convention in 1990 and has built up the legal framework to proceed with confiscation of Houbara and other birds illegally smuggled into the country. The hunting, capture, and transaction of Houbara in the UAE fall under a federal law that regulates the movement and possession of individuals of this species. NARC is the designated facility to receive Houbara confiscated by UAE cus.toms authorities, in compli-

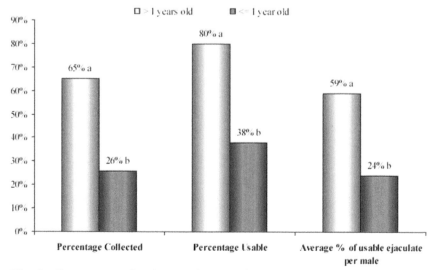

Fig. 6. *Comparison of males' results according to their age at the beginning of the season: effect on ability to be collected, ability to be used if collected, and average percentage of usable sample per male. A sample is usable when it contains a minimum of 15 million active sperm. (a not equal to b, p<0.05, Fisher Exact test).*

ance with internationally recognised IUCN Guidelines for the Placement of Confiscated Animals, which include the destroying of diseased and injured birds and release of rehabilitated birds back into the wild. To date, NARC has admitted about 1,100 confiscated Houbara at its quarantine station of which 50% survived and about 25% were released in UAE, Pakistan, and Iran. The remaining birds never recovered from injuries and diseases contracted during the smuggling process. In Pakistan, Houbara Foundation International (HFI) operates a rehabilitation center in Rahim Yar Khan (Punjab) for Houbara illegally trapped and confiscated. Birds admitted in this facility receive medical and nutritional attention before they are released back into suitable habitats. In the late 90s, Houbara were confiscated by the hundred and the Rahim Yar Khan rehabilitation centre was sometimes housing up to 1,500 Houbara. Again, the poor conditions of the birds lead to the death of at least 25% shortly after admission. Nowadays, the number of confiscated Houbara is smaller but still in the range of a few hundred every year.

## 3. Improve the knowledge on wild population ecology.

Tremendous work has been achieved in the ecology field of wild populations. It is unrealistic to undertake a captive breeding project on such a scale without having scientific facts on wild bird's behaviour, habitat selection, mating system, migration routes and population viability studies. At NARC, the ecology program aims to research and monitor wild Houbara throughout Asia and advise on conservation measures. Methods include satellite tracking of wild birds, regular surveys to monitor long-term trends in Houbara numbers and assessment of breeding success and mortality rates (Combreau et al 2001). Rehabilitated birds released in 2003 and equipped with satellite transmitters are still being followed today. Similarly, the majority of the captive bred chicks released in the wild have been followed by the same method which proved to be very accurate for assessing the survival rate and the overall success of the release program. Collaboration with local centres is also primordial. In Kazakhstan, in China and Pakistan, local institutes and foundations monitor twice a year several thousands of kilometers of transects to evaluate trends in Houbara populations.

## 4. Insist on public awareness

Numerous public awareness efforts have taken place in the middle-east in the last 15 years to explain the Houbara conservation issues to the public. Centres such as the NARC in Abu Dhabi, UAE, the NWRC in Taif, Saudi Arabia, and HFI in Lahore, Pakistan, have developed specific public awareness departments. Awareness strategies are adapted to local requirements and are targeted either to falconers or to the general public. Falconers, and especially wealthy falconers who are also decision makers in the countries involved must be partners of Houbara conservation strategies if any sort of success is to be expected. The Asian Houbara agreement under the Convention of Migratory Species specifically insisted on the need for the range countries to promote the creation of falconers clubs and association to coordinate activities and bring the Houbara conservation issue within the acknowledgement of all, and inviting falconers to be part of the conservationists. In the UAE, an Arabian falconer's club was created in 2001 to encourage the participation of falconers as an entity in Houbara conservation. The Arabian falconer's club aims at propagating a certain ethics of falconry by providing information and advice for its members on healthy principles of falconry, by organizing workshops and lectures on falconry-related matters, by publishing and distributing information on hunting legislations and guideline and by proposing substitute quarries for falcons training.

In Saudi Arabia, UAE and Pakistan, awareness is brought to the general public via specific school visits during which issues related to wildlife conservation are brought to the knowledge of students. In addition, NARC and NWRC are producing brochures, booklets, films, and keep public informed via regular press releases in local newspapers, and exhibitions related to hunting and falconry. In Pakistan, where the vast majority of people totally lack education on environmental issues, lectures and seminars in urban and rural areas as well as specific workshops are conducted by HFI.

## CONCLUSION

By establishing a thorough cooperation with all countries of the distribution range and facilitating exchanges between dedicated centres, by emphasizing research on ecology and captive breeding, the issue of Houbara conservation will be handled in an efficient and sustainable way, with significant positive effects in the long term. Houbara captive breeding, one of the different initiatives undertaken, has to deal with a large stock consisting of non standardized animals. The conflict between optimizing genetic diversity and defining environmental conditions suitable for all is the main great challenge each large captive breeding has to face. This is why a strict application of poultry techniques is meant to fail, because they handle a phenotype, not birds. Behind this fascinating challenge, the fate of the species is at stake.

The perceived decline of the Houbara Bustard in Asia has had positive consequences for conservation as a whole in the middle-east. Houbara decline and its potential disastrous consequences on Arabian falconry have prompted series of reactions such as the creation of specific Houbara centres in several middle-east countries. Besides addressing specifically Houbara and falcons issues, biologists in those centres soon pointed out to a vast range of environmental issues that had grown in parallel with the recent and massive development of those countries. The variety and the extent of environmental issues that necessitated specific undertakings prompted the establishment of the National Commission for Wildlife Conservation and Development (NCWCD) in Saudi Arabia in 1986 and that of the Environmental Research and Development Agency (ERWDA) in the UAE in 1996, now Environment Agency-Abu Dhabi. These agencies are currently involved in the promotion of environmental awareness of the local public, and participate to the education of the populations on key ecological issues, from waste recycling to water management and wildlife conservation.

## References

ANDERSON A. 1979. The incubation book. Spur Publications, Saiga Publishing Co. Ltd. Hindhead, Surrey, England.

BALLOU J, S Thompson, J Eearnhardt. 1999. MateRx. http://www.vortex9.org/materx.html.

BAILEY TA, JH Samour & TC Bailey. 1998. Hunted by Falcons, Protected by Falconry: Can the Houbara Bustard (*Chlamydotis undulata macpueenii*) Fly into the 21st Century? Journal of Avian Medicine and Surgery, 12 (3): 190-201.

BIRDLIFE INTERNATIONAL. 2004. *Chlamydotis undulata*. In: 2004 IUCN Red List of threatened species. IUCN 2004.

CHBEL F, D Broderick, Y Idaghdour, A Korrida and P McCormick. 2002. Molecular Ecology Notes. 2: 484-487.

COLLAR NJ. 1980. The world status of Houbara: a preliminary review. In: Proceedings of the Symposium on the Houbara bustard Chlamydotis undulata. Coles CL & Collar NJ Ed., Athens, Greece, 24 May 1979. Syndenhams Printers, Poole, 12pp.

COLLAR, NJ. 1996 Family Otididae (Bustards). Pp 240-2XX in: Handbook of the Birds of the World. Vol. 3. Hoatzin to Auks. Del hoyo J., Eliott A and Sargatal J. eds., Lynx edicions, Barcelona.

COMBREAU O, F Launay and M Lawrence. 2001. An assessment of annual mortality rates in adult-sized migrant Houbara bustards (*Chlamydotis [undulate] macqueenii*). Animal Conservation. 4:133-141.

COMBREAU O, J Qiao, M Lawrence, X Gao, J Yao, W Yang and F Launay. 2002. Breeding success in a Houbara Bustard *Chlamydotis [undulata] macqueenii* population on the eastern fringe of the Jungar Basin, People's Republic of China. Ibis. 144 (on-line): E45-E56.

COMBREAU O, F Launay and M Lawrence. 2005. Progress, challenges and perspectives in Houbara Bustard conservation in Asia. In: Ecology and conservation of steppe-land birds. Bota G., Morales M.B., Mañosa S., Camprodon J. eds. Lynx Edicions & Centre Tecnològic; Forestal de Catalunya, Barcelona.

FACON C, JL Guerin and F Lacroix. 2005. Assessment of Newcastle disease vaccination of Houbara bustard breeders. Journal of Wildlife Diseases. 41(4):768-774.

FOX N, M Al Bowardi, H MacDonald and F Launay. 2000. A global strategy for the conservation of falcons and Houbara. Environmental Research and Wildlife Development Agency, Abu Dhabi.

GORIUP PD. 1997. The world status of the Houbara bustard *Chlamydotis undulata*. Bird Conservation International. 7:775-779.

HARTLEY P, B Dawson, C Lindsay, P McCormick and G Wishart. 1999 Cryopreservation of Houbara semen: A pilot study. Zoo Biology. 18(2):147-152.

JOHNSON A. 1986. Reproduction in the Female. In: Avian Physiology. Pp 403-432. Springer-Verlag New York, Inc.

LAVEE D. 1988. Why is the Houbara *Chlamydotis undulata macqueenii* still an endangered species in Israel? Bustard Studies. 3:103-107.

LEON O, I Barcelo Llanes, A Brand, CJ McCarty, F Maozeka, J Howlett, A Toosy and O Combreau. 2006. Captive breeding season report. Environment Agency-Abu Dhabi.

LEON O & O Combreau. 2007. Modifying breeding season onset in Asian Houbara bustards (*Chlamydotis macqueenii*) by light and temperature program. Under Press. 2007 American Association of Zoo Veterinarians conference, Knoxville, USA.

LIECKFELDT D, A Schmidt & C Pitra. 2001. Isolation and characterization of microsatellite loci in the great bustard, *Otis tarda*. Molecular Ecology Notes. 1: 133-134.

PITRA C, D Lieckfeldt & S Frahnert & J Fickel. 2002. Phylogenetic relationships and ancestral areas of the bustards (Gruiformes:Otididae), inferred from mitochondrial DNA and nuclear intron sequences. Molecular Phylogenetics and Evolution, 23:63-74.

PITRA C, MA D'aloia, D Lieckfeldt & O Combreau. 2004. Genetic variation across the current range of the Asian Houbara bustard (*Chlamydotis undulata macqueenii*). Conservation Genetics 5(2):205-215.

POLLACK JP, R Lacy & JD Ballou. 2002. Population Management 2000. Chicago Zoological Society, Brookfield, IL.

PONMAREVA TS. 1983. Reproductive behavior and distribution of the Houbara bustard (*Chlamydotis macqueenii*) on the breeding grounds [Russian]. Zoologicheskii zhurnal 62:592-602

RAMADAN-JARADI G & MG Ramadan-Jaradi. 1989. Breeding the Houbara bustard at the Al Ain Zoo & Aquarium, Abu Dabi, UAE. Der Zoologische Garten 59:229-240.

RIOU S, O Combreau, M Lawrence & F Launay. 2007. Solitary display in the Asian Houbara bustard on the Kazakh steppe: resource-defence or self-advertisement? Ethology, in press.

SAINT-JALME M, P Gaucher & P Paillat. 1994. Artificial insemination in Houbara bustards (*Chlamydotis undulata*): influence of the number of spermatozoa and insemination frequency on fertility and ability to hatch. Journal of Reproduction and Fertility. 100: 93-103.

SAINT-JALME M & Y van Heezik. 1996. Propagation of the Houbara Bustard. Kegan Paul International ed. London, New York.

SAUVEUR B. 1988 Poultry reproduction and egg production [French]. INRA ed.

SEDDON P & Y van Heezik. 2001 Counting Houbara bustard in Northern Saudi Arabia: An assessment of methods. In: Counting Houbara Bustard. IUCN/SSC/Birdlife Working Group on the Houbara Bustard Ed.

THE GAME CONSERVANCY'S ADVISORY SERVICE. 1993. Egg production and incubation. The game conservancy publications, Fordingbridge, Hampshire, UK.

TOURENQ C, O Combreau O, M Lawrence, S Pole, A Spalton, G Xingji, M Al Baidani & F Launay. 2005. Alarming Houbara Bustard populations trends in Asia. Biological Conservation. 121:1-8.

VAN HEEZIK Y, M Saint-Jalme, S Hemon & P Seddon. 2002. Temperature and egg laying experience influence breeding performance of captive female Houbara bustards. Journal of Avian Biology 33(1):63-70.

UPTON J. 1989. The Houbara Bustard and the Arab falconer. Bustard Studies. 4: 174-176.

## Author Biography

Graduating from the national veterinary school of Toulouse, France, in 2002. Started as an assistant professor in the veterinary school of Toulouse, swine and poultry health, until September 2004. Responsible for teaching veterinary students about swine and poultry breeding techniques, diseases and pathology, and the general clinical approach in population medicine. Research activities were devoted to the descriptive epidemiology of *Chlamydophila psittacii* carriage in duck flocks in the southwest of France. Hired as a researcher in aviculture in the National Avian Research Centre in November 2004. Responsible for retrospective data analysis, protocol designs, current procedures improvements, as well as having hands on the captive breeding activities such as male collections, female inseminations, artificial incubation. Designed an enhanced genetic management for the optimization of the captive flock genetic diversity, and experimental protocols to explore Houbara reproductive physiology. I am now managing the captive breeding department, and responsible for organizing NARC's transition to large scale Houbara production.

# HAWAI`I ENDANGERED BIRD CONSERVATION PROGRAM: A TOOL FOR THE RECOVERY OF AN INSULAR AVIFAUNA

ALAN LIEBERMAN

Conservation Program Manager
Zoological Society of San Diego
15600 San Pasqual Valley Road
Escondido, CA 92027-7000

## SUMMARY

The Hawai`i Endangered Bird Conservation Program (HEBCP) is a unique conservation partnership, composed of the San Diego Zoo's CRES Department of Applied Conservation, government agencies (U.S. Fish and Wildlife Service and the State of Hawaii), and Hawai`i's private landowners, working together to recover 22 species of endangered Hawaiian forest birds. This program clearly demonstrates the significant role captive propagation can play in worldwide conservation efforts as well as providing a model for a collaborative, multidisciplinary approach to endangered species recovery.

The HEBCP is playing a pivotal role as an "endangered species bank account" for the reestablishment of endemic Hawaiian species of birds into managed habitat. Without this propagation effort, many of these endangered species might go extinct while their habitat is being identified, reserves are being designed, and management measures being implemented. The propagation effort, based on many years of technological advances in zoos and private aviculture, has saved at least one of these species from extinction and will continue to play a role in their recovery. Currently, the program manages nine Hawaiian species at the Maui and Keauhou Bird Conservation Centers, as well as three active release programs; the Puaiohi releases on Kauai, now completing its eighth year of releases (113 birds), the Palila (21 birds) in its third year on Mauna Kea Volcano and the Nene one three islands (346 birds). Successful hatchings have been accomplished with the `Alala, HI `Akepa, HI Creeper, Palila, Puaiohi, Nene, Maui Parrotbill, `Akohekohe, `Apapane, Common `Amakihi, `I`iwi, and `Oma`o.

## INTRODUCTION

Captive management forms one of the cornerstones of forest bird conservation in Hawai`i. To date, 14 Hawaiian taxa have been raised in captivity, exclusively in island facilities. The goal has been to learn how to meet species requirements in captivity and to maintain, raise, breed, and release native birds with the purpose of bolstering or re-introducing populations of endangered species. In rare instances, species have been brought into captivity while they were going extinct in the wild, e.g., `Alalâ and Po`o-uli. The path to saving forest birds by means of captive propagation has been made especially difficult because of the overall lack of comparable work on passerine birds elsewhere, necessitating the development of technology and expertise locally, and because of the great cost of building and staffing suitable aviary centers in the islands. Perhaps for these reasons, use of captive propagation has come late to Hawai`i and was not underway to save forest birds until the late 1980s, by which time many species most in need of such assistance had become extinct.

A variety of captive management options have been practiced in Hawai`i, and these serve as models for future work on endangered forest birds. Which species should be tackled first, and what strategy would work best, have been determined by a host of considerations including primarily the species' biology and the particulars of its conservation situation. In lieu of captive management, field support of dwindling wild populations and translocations of endangered birds to safer locations have been or could be preferable (e.g., Northwestern Hawaiian Island passerines). Captive management options successfully carried out in Hawai`i include: (1) rear-and-release of chicks from wild-collected eggs (suitable for a variety of species but generally not used as yet), (2) captive breeding for immediate release (e.g., Puaiohi and Palila), (3) captive-breeding for a self-sustaining captive population (e.g., Nênê and `Alalâ), (4) captive breeding for production (e.g., `Alalâ eventually), (5) emergency search-and-rescue (e.g., Po`o-uli), and (6) preemptive technique development for many species.

## The Program

The Hawaii Endangered Bird Conservation Program has hatched over 800 chicks of 14 taxa, and handled over 2,250 eggs of which over 1,050 have been viable (Table 1). Eight of these species are classified as Federally endangered, to include the `Alala *Corvus hawaiiensis*, Maui Parrotbill *Pseudonestor xanthophrys*, Hawai`i Creeper *Oreomystis mana*, Hawai`i `Akepa *Loxops c. coccineus*, `Akohekohe *Palmeria dolei*, Puaiohi *Myadestes palmeri*, Palila *Loxioides bailleui*, and Hawaiian Goose or Nene *Branta sandvicensis*. Five non-endangered native Hawaiian species; `Oma`o *Myadestes obscurus*, Hawai`i `Elepaio *Chasiempis sandwichensis*, `Apapane *Himatione sanguinea*, `I`iwi *Vestiaria coccinea*, and Common `Amakihi *Hemignathus virens virens* and *H. v. wilsoni*, were propagated to serve as surrogate models for the development of captive propagation and release technology.

The Program, run by the Zoological Society of San Diego, manages the two major avicultural centers in the State, one at Olinda, Maui Is., and the other at Keauhou, Hawai`i Is. Since 1993, the program has established the avicultural techniques to incubate, hatch, maintain, breed, rear, transport, acclimate, and release all taxa of endemic Hawaiian birds attempted to date. The Keauhou Bird Conservation Center on the Big Island of Hawai`i was inaugurated in 1996 and now includes an incubation and brooding building with laboratories, fledging aviaries, office space and neo-natal food preparation area. Additional structures include two forest bird buildings with 37 aviaries and bird kitchen, fifteen `Alala aviaries, a workshop and two caretakers' accommodations. Also in 1996, the program assumed the operations of the Maui Bird Conservation Center. This facility has areas for incubation and hand-rearing, 'Alala and Nene breeding complexes, and indoor-outdoor forest bird aviaries. The facility also serves as an incubation and neonatal area for the endangered Maui forest bird eggs that are brought from the field for captive management.

With the wild population of `Alala numbering less than ten individuals, the Hawai`i Endangered Bird Conservation Program joined with the `Alala Partnership (U.S. Fish and Wildlife Service, USGS-BRD and McCandless, Kealia and Kai Malino Ranches, Kamehameha Schools) in an intensive reintroduction program from 1993-1998. During the period, thirty-six `Alala were hatched and thirty-four survived to fledging. Twenty-seven `Alala were released into historical habitat in the South Kona District of Hawai`i. Twenty-five birds survived until independence (~120 days post-release). Although the long-term survivorship of the released `Alala was lower than first expected (21 of 27 did not survive to breeding age), through the release and monitoring program, biologists have been able to better identify the factors that limit the long-term `Alala survivorship in the native Hawaiian forests (predators, disease, etc.). In an effort to accelerate the recovery of the `Alala, the U.S. Fish and Wildlife Service is reviewing the options to establish an additional (alternative?) release sites with expanded and enhanced habitat restoration efforts to ensure the long-term survival of released 'Alala in the future. Although the wild population of `Alala is likely extinct, the captive population continues to grow and now numbers 52 birds, with the addition of three chicks reared in 2006. Once this flock reaches a level of genetic sustainability (approx. 75-80 birds), it is anticipated that releases will once again take place in secure and managed habitat.

Perhaps the most spectacular of Hawai`i's endemic avifauna is the sub-family of Honeycreepers (Drepanidinae). Reproductively isolated from the mainland populations and from each other on their respective Hawaiian islands, this group evolved into more than fifty unique species and subspecies. Many of these are now extinct, with the majority of the remaining taxa threatened with extinction. In order to test the effectiveness of captive-rearing and release strategies for this sub-family for future restoration efforts in Hawai'i, a pilot study was conducted with the Common `Amakihi in forests where introduced avian disease and mammalian predators were present. Methodology used resulted in the first successful artificial incubation, hatching and rearing of a Drepanidinae. Sixteen chicks were hatched (mean hatch weight = 1.4g) and reared. Two different release strategies were evaluated for small Honeycreepers; a ten to fourteen day acclimatization period in a hacking aviary ($4m^2$) in the native forest with subsequent food supplementation ("soft release") and a two day adjustment period in small field cages ($1m^2$) with food supplementation. Although all the birds survived the initial release and returned for food supplementation, twelve of the sixteen birds succumbed within thirty days to malaria infections, and four birds were not seen nor bodies recovered after fourteen days. This is a clear demonstration that irrespective of successful propagation techniques, recovery will not succeed unless mosquito-free, predator-controlled reintroduc-

## Table 1

**SUMMARY OF EGGS AND CHICKS HATCHED AND REARED IN CAPTIVITY: HAWAII ENDANGERED BIRD CONSERVATION PROGRAM, 1993-2006.**

| Species | Year | Total Eggs Collected/Laid | Eggs Viable at Collection | Chicks Hatched From Viable Eggs | Chicks Survived to Independence | % Hatch From Viable Eggs | % Survival Of Chicks |
|---|---|---|---|---|---|---|---|
| Hawai`i `Amakihi *Himatione v. virens* | 1994-1995 | 38 | 26 | 21 | 21 | 81 | 100 |
| Maui `Amakihi *H. v. wilsoni* | 1997-2000 | 11 | 1 | 1 | 1 | 100 | 100 |
| `I`iwi *Vestiaria coccinea* | 1995-2001 | 15 | 12 | 11 | 6 | 92 | 55 |
| `Ôma`o (*Myadestes obscurus*) | 1995-1996 | 36 | 29 | 27 | 25 | 94 | 93 |
| Hawai`i `Elepaio *Chasiempis sandwichensis* | 1995-2003 | 33 | 16 | 11 | 10 | 69 | 91 |
| Palila *Loxioides bailleui* | 1996-2006 | 125 | 91 | 71 | 51 | 76 | 74 |
| Puaiohi *Myadestes palmeri* | 1996-2006 | 885 | 271 | 227 | 176 | 86 | 77 |
| `Akohekohe *Palmeria dolei* | 1997 | 6 | 6 | 6 | 5 | 100 | 83 |
| Hawai`i Creeper *Oreomystis mana* | 1997-2006 | 36 | 19 | 17 | 14 | 89 | 82 |
| Maui Parrotbill *Pseudonestor xanthophrys* | 1997-2006 | 37 | 17 | 15 | 13 | 88 | 87 |
| `Apapane *Himatione sanguinea* | 1997 | 7 | 2 | 2 | 2 | 100 | 100 |
| Hawai`i `Akepa *Loxops c. coccineus* | 1998-2006 | 42 | 24 | 19 | 13 | 90 | 72 |
| `Akiapola`au *Hemignathus munroi* | 2001 | 1 | 0 | 0 | 0 | - | - |
| `Alala *Corvus hawaiiensis* | 1993-2006 | 390 | 162 | 84 | 71 | 55 | 85 |
| Nene *Branta sandvicensis* | 1998-2006 | 621 | 386 | 323 | 304* | 84 | 95 |
| Totals | | 2,283 | 1,062 | 835 | 712 | 79% | 85% |

\* Nênê goslings are precocial and unlike the passerine species in this table, are independent at hatch. For purposes of this metric, independence" for the Nênê is considered to be 30 days of age

tion sites are available or strategies are developed to decrease mortality in naïve Honeycreepers exposed to disease after release. However, the experience gained in the incubation and rearing of the Common `Amakihi has subsequently provided the technology to hatch and rear seven additional species of Honeycreepers, the smallest being the Hawai`i `Akepa with an adult weight of 9-11g and a hatch weight of 1.0g.

Very similar to the mainland Solitaires, five species of Hawaiian *Myadestes* Thrushes survived until very recently. However, it is now thought that only two of these species persist; the `Oma`o on the Big Island of Hawai`i and the Puaiohi on the island of Kauai. In 1995 and 1996, the first restoration attempt of a small Hawaiian passerine in disease-free, predator controlled habitat was made with the release of captive-reared `Oma`o, into the Pu`u Wa`awa`a Forest Reserve; habitat that has been without this species for nearly 100 years. In 1995, two birds were reintroduced as a preliminary test release and in 1996, twenty-three birds were released in cohorts numbering from two to seven birds. Of the twenty-five released birds, twenty-three are known to have survived 30 days (life of the transmitters). Follow-up surveys in 1997 and 1998 indicate that many released `Oma`o survived to sexual maturity and did successfully breed.

The Puaiohi is an endangered thrush, endemic to the island of Kauai and restricted to the Alaka`i Wilderness Area above elevations of 900m. Since 1995, this native Hawaiian Solitaire has been the focus of an aggressive recovery effort that has incorporated the funding, field efforts and the captive propagation and release expertise of several governmental and private agencies. In 1996 and 1997, fifteen Puaiohi eggs were collected from the wild, hatched and reared at the Keauhou Bird Conservation Center; becoming a captive breeding flock in 1998 and 1999. Since that time, over 250 Puaiohi have been hatched in captivity. Since 1999 (8 years; 1999-2006), 132 Puaiohi have been transported to Kauai, acclimatized for seven to fourteen days in hacking aviaries ($3m^2$), transmittered and soft-released from two release sites in the Alaka`i Wilderness Area. Supplemental food is offered at the hacking cages, but only a few of the birds return to feed. Although each of the birds carries a radio transmitter, it is extremely difficult terrain in which to track birds. However, many of the birds have been confirmed to survive into the breeding season with documentation of many successful breedings between captive and wild birds as well as captive-captive birds in the wild. For example, in the first year's release (1999) of fourteen birds, there were twenty-one nesting attempts from which seven chicks successfully fledged. This is the first release program for a passerine that has successfully incorporated all of the following techniques to include: collection of wild eggs, artificial incubation and hand-rearing, captive-breeding, release and subsequent breeding of the released birds in native habitat. This complete reintroduction scenario for the Puaiohi; from the wild to captivity and back to the wild, where breeding has been confirmed on several occasions, occurred over only three years time, a remarkably successful recovery action.

In 2003, the first release of captive bred Palila took place in Pu`u Mali on the north side of Mauna Kea Volcano on the island of Hawai`i. Currently the Palila survives in one population of the west side of Mauna Kea (Pu`u La`au). This release, coupled with a translocation effort by USGS-BRD in the same area of managed and recovering mamane forest will hopefully establish a resident, breeding population of Palila that is disjunct and independent of the main population on the other side of the mountain. Over the past three seasons (2003-2005), 21 Palila have been released from the captive flock. The majority of the birds are believed to survive in the managed area and the first successful breeding of a captive bird (with a wild translocated bird) in the wild was confirmed this past season (2006)... an exciting first for the recovery program.

The first 13 years of this program presents a more optimistic future for the beleaguered avifauna of the Hawaiian islands. As the captive flocks of the endangered species grow, and the techniques for rearing and release are refined, it is hoped that many of the endangered Hawaiian birds will benefit from restoration efforts. However, it must be emphasized that captive propagation and reintroduction is only one aspect of the ecosystem management tools required in Hawai`i to conserve and restore endangered native bird species.

## Author Biography

I have been in the zoo and conservation discipline for over thirty years; working for the San Diego Zoo, The Nature Conservancy, the Smithsonian Peace Corp Environmental Program and The Peregrine Fund. I have worked in the bird and reptile departments at the SDZ in Colombia working with caiman conservation and research in the Amazon rainforest; and for the past 13 years have been responsible for the design, development, construction and operation of the Hawai`i Endangered Bird Conservation Program. Most recently, in addition to managing the Hawai`i Program, I have assumed the supervision of CRES's San Clemente Island Shrike propagation effort as well as other translocation and propagation release projects managed by CRES.

# CAMPBELL ISLAND TEAL: THE SAVING OF A SPECIES AND THE RESTORATION OF AN ISLAND

PETE MCCLELLAND

Department of Conservation
P O Box 743, Invercargill, New Zealand

## SUMMARY

Seven-hundred kilometers south of mainland New Zealand lies 11,300ha Campbell Island, a windswept refuge for a diverse range of plants and animals including the small flightless Campbell Island Teal. Eliminated from the main island by introduced rats and cats, the teal survived for over 100 years on 26ha Dent Island, where a small population was rediscovered in 1975. Between 1984 and 1990, eleven birds were taken into captivity to establish a backup population in case of disaster on Dent. Breeding first occurred in late 1994 and, as the captive population grew, birds were released onto a holding island, where they quickly bred. The ultimate aim of the recovery program was to re-establish teal on Campbell Island, however, the presence of rats meant this was not possible until suitable eradication techniques were refined and tested on smaller islands. Trials on Campbell Island showed eradication was feasible, and in 2001 the largest rat eradication ever attempted was carried out. Logistically, the eradication was extremely challenging with four helicopters and 120 tonnes of bait. After two years, to enable any surviving rats time to increase to a detectable level, an intensive survey was undertaken. No sign of rats was found, and planning for reintroduction of teal commenced. Between 2004 & 2006 158 teal, both captive raised and from the holding island, were released. Monitoring in 2006 found that breeding had occurred and the population should rapidly recolonise given their dispersal behaviour. This paper describes two internationally significant conservation programs - the world's largest rat eradication, benefiting a wide range of species, and completion of the recovery plan for one of the world's rarest ducks. New Zealand's Department of Conservation is active in providing advice on island eradications around the world and continues to help refine the techniques for a wide range of situations.

## INTRODUCTION

The 11,300ha Campbell Island lies 700km south of the mainland of New Zealand (Figure 1). It contains a diverse range of flora and fauna, including six species of albatross, a range of burrowing seabirds, and three species of penguin, all of which breed on the island. It is also home to three endemic species or sub-species of land birds and a variety of plants, including a group of large leaved herbs loosely classed as 'megaherbs', which are endemic to the New Zealand Subantarctic. Campbell Island was discovered in 1810 by an adventurous sealing captain Frederick Hasselburgh in his vessel the Perseverance, after which the main harbour is named. The next 140 years saw the natural resources of the island and surrounding waters being exploited by sealers, whalers and farmers. This period also saw the introduction of a range of mammalian herbivores and predators which were to have major and long lasting impacts on the island.

The first species to arrive at the island was the Norway rat *Rattus norvegicus*, which is believed to have hitched a ride on a sailing ship soon after the islands discovery. This was followed by cats *Felis domesticus*, presumably to try and control the rats, as well as sheep and cattle for the farming attempts which lasted from 1895 to 1931 before they proved to be uneconomic. The island was declared a Flora and Fauna Reserve by the New Zealand Government in 1954, was then upgraded to Nature Reserve in 1978 and then in 1986 to National Nature Reserve, New Zealand's highest protected land status with entry by permit only. The sheep and cattle, along with the burning which took place to increase the area of grazing land, had a major impact on the island's vegetation, which is now recovering following the cessation of burning at the end of the farming era in 1931, and the removal of cattle in 1987, and sheep over three stages finishing in 1990 (Brown 2002).

Cats died out in the 1990s for unknown reasons, possibly related to the sheep eradication program, (e.g. changes in vegetation or loss of a winter food supply in the form of dead sheep). This left only the rats, which, in conjunction with the cats have had the greatest long term impact on the island by causing the local extinction, and in some cases total extinction, of a number of bird species. When the first scientific party visited the island in 1840 they found virtually no native land birds. Pipits *Anthus novaseelandiae*, a small ground dwelling passerine, occasionally made it to the main island from several of the smaller predator free islands and islets which fringe the coast of Campbell, but were quickly predated by the cats and rats. What other species had been present before the rats was largely a matter of speculation based on the species which had been found on the other subantarctic islands, especially on the Auckland's 250km to the North-East and Macquarie Island 700km to the South-West. These included snipe *Coenocorypha* sp., rail *Rallus* sp. and parakeets *Cyanoramphus* sp.

One species which provided much intrigue was a small flightless duck which had only been recorded twice, in 1886 and 1944, both times on the coast at North-West Bay, a large bay with a predominantly rocky coastline which opened to the North West. (Williams M. & Robertson C.J.R. 1997). Where these birds had come from was a mystery, although the limited evidence available pointed at 26ha Dent Island, 2km off the coast near the entrance to North West Bay (Figure 2). This was confirmed in 1975 when a small population, less than 100 and possibly less than 30 birds, was located on the island.

**Fig. 1.** *Map of New Zealand showing locations important to the Campbell Island teal recovery program.*

## Discussion

As the size of both Dent Island and the teal population meant it was in real danger of extinction from stochastic events such as landslips, disease or introduced predators, a decision was made to try and establish a backup population in captivity. This was done without a specific long term recovery objective other than to try an ensure the survival of the species, as no possible sites for future release had been identified and reintroduction to Campbell was unthinkable due to the presence of the rats. The confidence in being able to breed the birds in captivity was based on the knowledge which had been gained during the successful Brown Teal *Anas chlorotis* captive breeding program on the New Zealand mainland, which had produced over 2000 ducklings between 1964 and 2000 (Dumbell 2000).

In 1987, three male and one female Campbell Island Teal were caught and taken to the mainland. It had been hoped to catch up to five male and five female teal but the dense vegetation on Dent Island, along with an abundance of seabird burrows which the teal could easily escape down, meant that catching the birds was very difficult. Ten of the closely related, but more common, Auckland Island teal *Anas aucklandica* were also taken in to captivity at the same time to act as surrogates for the Campbell program. This included trialing aviary layouts and testing the fertility of the Campbell Island males (i.e. pairing Campbell Island males with receptive Auckland Island females to see if they produced fertile eggs, as well as confirming the suitability of such techniques as double clutching).

Due to the relatively low threat status of the Auckland Island species, it has self sustaining populations on at least four large islands, meant that there was no real benefit to the species itself in taking these birds into captivity, aside from confirming avicultural techniques that may be required in case of an unforeseen and unlikely crash in the population. Alternatively there was no real impact on the wild population from removing them and, by having a larger number of individuals to work with for pairing trials etc, they could provide information that was crucial to the conservation of the Campbell Island Teal in captivity.

Unfortunately, despite trialing different pairings of the four Campbell Island birds and different aviary lay outs, no breeding occurred within this small group. This was despite the Campbell Island males being willing and able to breed, as shown by fertile eggs being produced by male Campbell Island and female Auckland Island teal pairings. The lone Campbell Island female was not receptive possibly due to stress when flock mating was trialled with her and the three males. A further four male and three female Campbell Island teal were added to the captive population in 1990, but the next three years saw no change in the breeding status. In 1993, the Subantarctic Teal Recovery Plan (McClelland 1993), covering both Campbell Island and Auckland Island teal, was produced by the Department of Conservation. The plan set out as the long-term goal for the project the return of Campbell Island teal to Campbell Island, specifically through the removal of introduced predators from the island. At the time of the writing of the recovery plan it was recognised that the techniques needed to carry out the rat eradication, that was required before any reintroduction could be considered, were not yet well enough developed, and it could be a long time, if ever, before rat eradication on an island the size and with the logistical complexity of Campbell could be contemplated.

Taking this in to account the recovery plan gave the primary interim conservation objectives for Campbell Island Teal as:

a) *Maintain the wild population of Campbell Island teal on Dent Island*

Protecting the Dent Island population was carried out by, as much as possible, minimizing the risk from controllable human influences (e.g. disease, introduced predators, human induced habitat damage etc.), by restricting access for all but the highest priority projects - this effectively meant only one visit was permitted in the following 10 years and that was to complete objective b).

b) *Survey the islands around Campbell, for any further populations of Campbell Island teal*

As any additional populations could have affected not only the threat status of the species but also the genetic base, a survey of all the islands believed to be large enough to hold a self sustaining population of teal was carried out in 1997. Although no additional populations of teal were found, Campbell Island Snipe *Coenocorypha* undescribed sp. were discovered for the first time on Jacquemart Island during the course of this task. The resurvey of Dent found an apparent significant drop in the teal population on the island putting even more emphasis on the captive breeding program.

c) *Establish a captive breeding population of Campbell Island teal*

The process to establish a captive population was already under way, although at the time of the writing of the plan there had been no breeding.

d) *Establish an additional wild population of Campbell Island teal- in order to reduce the risk of extinction in the event of a decline in the Dent Island population*

Establishing a back up wild population involved selecting a suitable island which had suitable habitat; this was an issue since the only site we had to gauge it on directly was Dent, dense tussock and megaherbs, and there is nothing really like Dent anywhere else, or at least not outside the subantarctic. It was therefore compared to sites utilised by Auckland Island Teal on the islands they inhabit. The island had to be large enough to hold a self sustaining population of teal; be free of introduced predators; have no near relatives of the teal present, then or likely in the future- this eliminated many sites since the Brown Teal Recovery Group were also looking for suitable sites for their birds and they were the original teal present on the islands around the mainland; have acceptable biosecurity and security from human interference and be logistically feasible.

e) *Promote the removal of predators from Campbell Island.*

While at the time of the writing the plan it was generally regarded that an island the size of Campbell would never have rats removed, it was decided that it was important to have this as the long-term goal, as the previous ten years had already seen huge advances in rat eradications, and the teal was a prime candidate as a flagship species for the eradication both due to its charismatic nature and economic advantage as compared to an expensive captive breeding program.

The writing of the plan also lead to the establishment of the Subantarctic Teal Recovery Group. Recovery groups have a formally recognised role in the management of selected threatened species in New Zealand; they are set up to provide advice to the conservation managers and to try and ensure a coordinated management across the country for species which are present in more than one Conservancy (land management area). As Campbell Island Teal were only present in the wild in one conservancy (Southland) and the captive population at the National Wildlife Centre (Mt. Bruce) was in another, it meant that only these two areas, along with a science advisor, had to be represented in the group. The small number of people involved, along with the unified focus for the future, breeding sufficient birds to release in to the wild, made for a very straightforward decision making process which provided the advice based on the consensus of the group. Fortuitously, although largely coincidentally, not long after the establishment of the recovery group the first breeding in captivity occurred in the 1994/95 season. This is believed to have been due to a modified aviary lay out based on field research on the closely related Auckland Island teal (Williams 1995) and on observations from the previous years, that allowed flock mating without allowing any single male to dominate the breeding aviary.

The problem continued to be getting the wild caught female Campbell Island teal to breed, as nearly all the wild males bred successfully with captive Auckland Island Teal, and later with captive bred Campbell Island females when they became available. So while seven of the wild caught males are represented genetically in the captive population only one wild caught female ever bred. "Daisy" produced 24 ducklings during her 12 years in captivity, which in turn produced another 39 ducklings and so the captive population was established (Anon 2002). Beginning from a single female appears to have made little difference to the genetics of the captive population. Studies have shown that, presumably after going through bottlenecks when the original birds arrived at Campbell Island, and again when they established on Dent, the founder birds for the captive population were basically genetically identical anyway- i.e. microsatellite DNA fingerprinting showed that individuals were 82.3% to 95.5% genetically similar (Lambert et al 1997). Once the first pair bred successfully in captivity the captive population grew rapidly, both from additional breeding by the initial pair and from captive bred and wild caught males mating with captive bred females, which were far more receptive to the males than their wild caught counterparts. The primary aims of having the teal in captivity were first to establish a back up wild population and then, when possible, reintroduce the birds onto Campbell, with advocacy as a secondary, although important goal. As such it was decided that it was preferable to restrict the number of birds held on the mainland in order to manage both the genetics of the population and the disease risk for any future releases. As such a limited number of birds, those of least value genetically to the release program, were released to selected external institutions for advocacy purposes. Some of these institutions had assisted the program by holding surplus male Campbell Island Teal and or Auckland Island Teal and carrying out trial matings.

By the time the captive population was at the point where it could support releases to the wild it was recognised that the eradication of rats from Campbell was feasible, and it was a matter of when, rather than if, the eradication would take place. As such in addition to the previously listed criteria for selecting a release site, the site had to be suitable for recapturing of birds for later transfer to Campbell, and ideally as close as possible climatically and habitat wise to Campbell Island to allow the birds to adapt as much as possible to living in the wild before their eventual transfer to the harsh conditions they would find in the subantarctic. All the islands around New Zealand, and even some as far away as the Falklands were considered, but for a variety of reasons from politics to logistics to perceived habitat suitability and the presence of other teal species either then or in the future, a decision was made to use 1396ha Codfish Island/Whenua Hou as the holding site. The island holds a diverse range of habitats, and although in the past it had had Brown Teal present, they were deemed very unlikely to recolonise. It is relatively easy to access although isolated enough to provide a reasonable level of security. Biosecurity was already being managed as it is a primary site for the management of the critically endan-

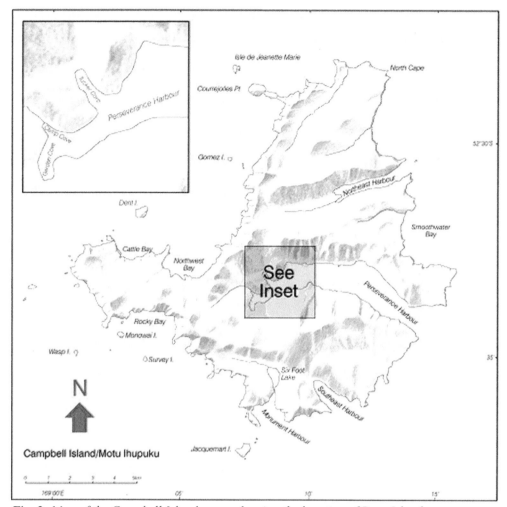

**Fig. 2.** *Map of the Campbell Island group showing the location of Dent Island.*

the eradication of rats from Campbell Island. Building on the information gained during an array of eradications on smaller islands, as well as some onsite trials to test bait types and baiting rates, a plan was developed to aerially spread 16mm (4g) cereal baits containing 20ppm of brodifacoum over the whole island. More than twice the size of the previous largest successful eradication and nearly five times the size of the largest aerial operation, Campbell provided a diverse range of problems that had to be worked through from non-target issues to helicopter storage. These were compounded by the isolation of the island and the fact that in order to minimise non-target issues, and maximise bait take by the rats the operation had to be done in the middle of the subantarctic winter. After setting up base in the disused meteorological station on Campbell, a team of 18, including four helicopter pilots, oversaw the unloading and storage of 120 tonnes of bait. In order to protect the bait from the weather and minimise handling a new system of transporting and storing the bait was developed, namely 1.2m$^2$ plywood "pods" which could be transported by forklift, crane or under a helicopter as the situation demanded. The pods were sealed so they could be left out in the open and flown to the required helicopter loading sites as required.

The island was divided into four blocks based on size and topography, with two loading sites being selected in each block to minimise ferrying time by the bait spreading helicopters, and the down time as they waited to be filled. One Squirrel helicopter ferried the pods of bait to the loading sites as it was required and three smaller Jet Ranger helicopters focussed on spreading the bait, with one machine primarily working on the cliffs, some of which rise to over 600m. Differential GPS allowed the pilots to fly the accurate flight path required to ensure that bait was laid in every

gered Kakapo *Strigops habroptilus* which provided the additional benefit of having a permanent base on the island for use during the releases, monitoring and future recapture programs. Having undergone intensive disease screening and a short period in individual holding pens on the island to regain condition lost during the transfer, 12 teal which had been bred at the National Wildlife Centre were released at two sites on Codfish Island/Whenua Hou in both 1999 and 2000. All birds survived the transfer and release and rapidly dispersed around the suitable habitat of their release sites.

Breeding occurred in the first summer, eight months after the release, and a monitoring program utilising radio transmitters to follow as many birds as possible allowed the first information on the bird's survival and productivity in the wild to be collected. (McClelland 2002). As more birds were bred on the island and the number of birds able to be relocated to have transmitters replaced, dropped, the monitoring was phased out and the birds were left to their own devices. At the same time planning was underway for

### Table 1

**WEIGHT AND WEIGHT CHANGES OF CAMPBELL ISLAND TEAL FROM PRE-TRANSFER TO RELEASE ONTO CAMPBELL ISLAND**

| | | NWC females (n=19) | | NWC males (n=14) | | CI (Codfish) females (n=7) | | CI males (n=15) | |
|---|---|---|---|---|---|---|---|---|---|
| | | mean | range | mean | range | mean | range | mean | range |
| NWC tx. attachment/ CI catch rate | g | 462 | 373 - 521 | 493 | 431 - 638 | 369 | 305 - 460 | 395 | 355 - 470 |
| Pre-transfer weight (31 Aug) | g | 462 | 429 - 513 | 481 | 423 - 551 | 342 | 325 - 365 | 424 | 390 - 460 |
| Campbell arrival weight (3 Sept) | g | 436 | 385 - 480 | 464 | 425 - 535 | 349 | 325 - 365 | 413 | 360 - 460 |
| % change over transfer | % | -6 | -10 - -6 | -4 | 0 - -3 | +2 | 0 - 0 | -3 | -8 - 0 |
| Campbell release weight | g | 405 | 360 - 450 | 441 | 400 - 495 | 369 | 305 - 460 | 395 | 350 - 470 |
| Release as % of pre-transfer rate | % | 88 | 84 - 88 | 92 | 95 - 90 | 108 | 94 - 126 | 93 | 90 - 102 |
| Release wt. as % of Codfish wt. | % | - | - | - | - | 100 | 100 - 100 | 100 | 99 - 100 |

possible rat territory- one missed rat could mean failure of the entire operation! Due to the normally bad weather at the island, where when the wind is not blowing it is often too cloudy to fly, three months had been allowed for the operation, but unusually good weather, for Campbell Island anyway, and a high level of planning to allow the best advantage to be taken of it, as well as extremely experienced and skilled pilots, meant the project was completed in less than one month.

The standard practise in New Zealand for confirming the success of a rat eradication is to wait two years to give any remaining rats the chance to breed up to detectable levels and then do a check to see if any rats are present. However the size of Campbell Island and the level of monitoring that would have to be done to be confident after two years meant that it was decided to wait five years before declaring success. However, in the meantime the captive teal population was growing and there was a strong push to release more birds as aviary space was limited.

This meant that in 2003 the first result monitoring for the eradication was carried out. When no sign of rats was found, giving a high level of confidence that the eradication had been successful, planning was started for the first teal release. A reintroduction plan was written (Seddon and Maloney 2003) with the aim of establishing at least two self-sustaining sub-populations of Campbell Island Teal on Campbell Island. The plan called for two releases of 60, then 40 birds over two years with intensive monitoring followed by a third release of a further 40 birds if required. The plan also set criteria for gauging the success of the reintroduction including survival rates and evidence of breeding. Unfortunately the resources required to follow the plan exactly were not available so the level of post release monitoring was significantly reduced to a week or less immediately post release and three to four weeks during the summers following the first two releases.

The size of Campbell Island, and the dispersal abilities of the teal meant that within logistical constraints the more birds which could be released the better. Therefore in order to provide the desired number of birds for the release a second institution - the Isaac Wildlife Trust's "Peacock Springs" which is a privately funded facility which is not open to the public and focuses on assisting the Department of Conservation with its captive breeding programs, was asked if it would participate in the program. Given the national priority of the species plus past experience with breeding Brown Teal, the Trust agreed and for the rest of the captive program they, and the National Wildlife Centre, worked closely to maximise production taking into account disease, genetics, etc. This involved moving birds between the two institutions as required. The first release on Campbell took place in September, 2004 with 50 birds, 28 captive bred and 22 wild birds from Codfish being released. This release was followed by 55 birds and then a further 54 birds in 2005 and 2006 respectively. These releases involved intensive disease screening and treatment for parasites leading up to the transfer. The deaths of three juvenile Kakapo to Erysipelas on Codfish Island/Whenua Hou led to the vaccination of all teal for that disease as a precautionary measure against low risk of transferring the disease to Campbell.

During the course of the program the opportunity was taken to carry out baseline disease screening on Campbell Island to confirm what diseases may have been present already. This involved sampling as many species as possible although the, most desirable species, as far as comparing the teal to, Mallard Ducks *Anas platyrynchus* were hard to catch. So only one individual could be sampled. Nothing unexpected turned up in those samples (Jacob-Hoff 2004; Potter 2005). All the Campbell Island Teal for release were bought together at the mainland departure point where they were crop fed and checked before being loaded into individual purpose built transfer crates. The crates were stacked into a ventilated room on the transport vessel for the 40 hour trip to the island. All the birds were crop fed approximately every 12 hours during the trip, as well as having ad lib access to food and water. This worked well with minimal loss of weight during the trip as shown in Table 1 (Gummer 2004).

Upon arrival at Campbell the birds were transferred in to small holding pens to give them a chance to clean and regain any weight lost during the trip - each bird had its own target weight - for wild caught birds this was the weight at time of capture, while for many of the captive raised birds it was an individual estimate of the desired weight as many birds were considered grossly over weight (25%+ over normal wild weight) which could cause problems with transmitter harness attachment! Under weight birds continued to be crop fed, although it was found that this was actually of minimal benefit so for the last release the birds were released as soon as possible after arrival - this did not appear to have any affect on survival. During the first two transfers all birds had backpack harness transmitters attached to allow survival and dispersal monitoring. This showed early survival, up to six months post release, to be basically equal for captive raised (71%) and wild-bred birds (71%) with no losses in the first two weeks but with a very wide dispersal - some birds moved over 5km. This dispersal and the rugged topography meant ongoing monitoring was difficult and while the confirmed survival rate for the first year was only 70%, no dead birds or dropped transmitters were found indicating a much higher survival rate.

Only one bird is known to have been lost following the transfer - a male who was predated by a Great Skua *Catharacta skua*. All the birds went through an initial naïve phase upon release where they would sit out in the open with little regard to potential predators, however after two days the innate survival instincts seemed to click in and it was rare to see a bird during daylight after that.

In 2006, young from at least four nesting attempts were seen during a monitoring trip, as well as one bird which was believed to have been bred the previous year. While this does not confirm that the teal have established a viable population, the initial signs are good and it is hoped that this will be confirmed in the southern summer of 2007/08. A population of Campbell Island Teal is still being held in captivity as a safeguard and for advocacy purposes with the management of the population to be done by the Conservation Management Group, an independent group of institutions and individuals involved in a wide range of captive breeding programs, under guidelines from DOC.

Teal were not the only species to benefit directly from the rat eradication, additional benefactors included two other endemic land birds, the Campbell Island Pipit *Anthus n. aucklandicus* and Campbell Island Snipe *Coenocorypha a. perseverance*. The latter was only discovered in 1997 and was restricted to a single 19ha island. Some people had advocated for the establishment of a captive snipe population with the aim of safeguarding the species and then releasing birds on to Campbell when it was predator free. However, this would potentially have put the species at risk, as the size of the wild population was unknown although given the area of habitat it had to be small - estimated at <50 (Barker et al 2005), and captive breeding techniques have yet to be developed for snipe so there was no surety that a captive population could be established. It was therefore decided by the Department of Conservation, to give the snipe a chance to re-establish themselves. An arbitrary and unofficial period of ten years was given for this to happen but the first breeding on the main island was recorded within five years of the rat eradication and it appears that the population is now on its way to being established.

Burrowing seabirds are also re-colonising the island with Grey-backed Storm Petrels *Oceanites nereis* and White-chinned Petrels *Procellaria aequinoctialis* both having been recorded breeding on the main island since the eradication. Other species including the Double-banded Plover *Charadrius bicinctus* and Bellbird *Anthornis melanura* have been blown to the island, presumably from the Auckland Islands but have yet to become established. A wide range of invertebrates as well as plants have also benefited with some inverts.

## Conclusion

The Campbell Island Teal project shows how captive breeding can not only be important in protecting a species in the short to medium term but also how, with good forward planning, it can play an integral role in re-establishing species back into the wild. This can be by providing birds either directly or via a secondary holding site. The most important thing for any reintroduction is the removal of the causal factors of the initial population loss. Often, as in the case of Campbell Island, it can sound theoretically easy, although logistically challenging (eg. removal of predators), while for many others it will not be so straightforward especially on most mainland sites. Precautions must be taken to minimise the risk of introducing disease to the new sites, which can, as in the case of Campbell Island Teal be done by managing the exposure of the birds to be released and screening of the birds prior to release, or by a higher level of intensive and often expensive screening.

Individual species such as Campbell Island Teal are often used as the flagship species for habitat restoration projects such as eradications. It is especially useful if they are a charismatic species, preferably large and cute with a high public profile. Managers can make good use of these species but should not forget that most habitat restoration projects, especially eradications, will benefit a wide range of species including invertebrates and plants, often with little or no additional input required. This has been the case on Campbell Island where there has been a dramatic recovery not only of the birds but also for the vegetation and invertebrates, with further recovery continuing over many years to come. The examples of Campbell Island Teal and Campbell Island Snipe show that there is no set cure for all species and whether they would benefit from a captive breeding program should be considered on a case by case basis, depending on their numbers, biology, source and future release locations and the risks that each action poses.

There is an inherent risk in both strategies; removing birds for a captive breeding population can put the source population at risk with no guarantee of establishing the captive population, and while not removing birds from a single vulnerable population also puts the species at risk, it is up to the managers to decide which approach to take based on the best available information at the time. Due to the low numbers of the subject species, the information is often based on research into closely related species. It is often a case of make haste slowly! As with all major projects, both the teal recovery and the Campbell eradication were true team efforts involving a large number of people over many years (more than 20 years in the case of the teal). It is through the dedication of teams such as this that conservation managers around the world will continue to restore habitats and save species.

## References

ADAMS L. 2005. Second transfer of Campbell Island teal to the subantarctic in September 2005. File NHS 03-12-06 Department of Conservation: Invercargill.

BARKER D, J Carroll, H Edmonds, J Fraser and C Miskelly. 2005. Discovery of a previously unknown *Coenocorypha* snipe in the Campbell group New Zealand subantarctic. Notornis vol. 52:143-145.

BROWN D. 2002. Island Pest Eradications New Zealand's subantarctic Islands and Codfish Island/Whenua Hou Nature Reserve 1984-1993. Unpublished report Department of Conservation: Invercargill.

DUMBELL G. 2000. Brown Teal Captive Management Plan - Department of Conservation Threatened Species Occasional Publication No 15. Department of Conservation.

GUMMER H. 2004. First transfer of Campbell Island teal to the Subantarctic in September 2004. File NHS 03-12-06 vol 1 Department of Conservation; Invercargill.

JAKOB-HOFF R. 2004. Report on Avian Disease Surveillance of Campbell Island, September 2004. File note NHS 03-12-06 Department of Conservation Invercargill.

LAMBERT DM, J Robins & A Harper. 1997. Parentage and genetic variation in Campbell Island teals Unpublished Report to Department of Conservation.

MCCLELLAND P. 1993. Subantarctic Teal Recovery Plan Department of Conservation Threatened species recovery plan No. 7 Department of Conservation.

MCCLELLAND P. 2002. An assessment of the success of a recently introduced population of Campbell Island Teal (*Anas nesiotis*) on Codfish Island (Whenua Hou Nature Reserve) and implications for returning teal to Campbell Island. MSc thesis; Lincoln University.

SEDDON P & R Maloney. 2003. Campbell Island Teal Reintroduction Plan DOC Science Internal Series 154 Department of Conservation; Wellington.

WILLIAMS M. 1995. Social structure dispersion and breeding of the Auckland Island Teal. Notornis 42: 219-262.

WILLIAMS M & CJR Robertson. 1997. The Campbell Island Teal *Anas aucklandica nesiotis*: history and revue; Wildfowl 47: 134:16.

## Author Biography

Pete has worked for the New Zealand Department of Conservation for over 20 years on a wide range of threatened species projects. His current position as Program Manager Outlying Islands in the Southern Islands Area sees him playing a key role in the management of New Zealand's five subantarctic island groups, including: protected species management, invasive species control, research permitting and visitor management. Over the last 10 years Pete has had an increased focus on rodent eradications, both within New Zealand and internationally, including managing the Campbell Island rat eradication, which at 11,300ha is the largest to date, a record that Pete hopes will soon be broken. His experience with rat eradications has lead to him giving advice in such places as the Galapagos, Aleutians and the Azores where he gets a great deal of personal satisfaction from seeing the techniques which have largely been developed by Department of Conservation staff being refined for each local situation. Pete's involvement with Campbell Island teal started in 1990 with the second trip to Dent Island to capture birds for a captive breeding program. This was followed by the writing of the recovery plan and leadership of the recovery group culminating in the recently completed reintroduction back to Campbell Island after the island was declared pest free. Pete considers it a privilege to not only have been involved in two internationally significant conservation projects, but to have the two projects so closely linked - the teal could not have been reintroduced unless the rats had been removed. Pete is a self confessed island nut but appreciates that working on islands has many advantages over the mainland where conservationists often have to keep battling away against the same issues just to hold their ground.

# THE VULTURE CAPTIVE BREEDING AND RESTORATION PROJECT IN PAKISTAN

CAMPBELL MURN[1] & UZMA KHAN[2]

[1] The Hawk Conservancy Trust, Andover, Hampshire, SP11 8DY, England
[2] WWF-Pakistan, Ferozepur Road, Lahore 54600, Pakistan

## SUMMARY

The rapid population declines of three *Gyps* vulture species in south Asia since the early 1990s has necessitated the development of captive management facilities and conservation breeding programs to protect core populations of the affected species. In 2004, WWF-Pakistan launched the Gyps Vulture Restoration Project, which will maintain a viable population of Oriental White-backed Vultures *Gyps bengalensis* in captivity and eventually produce vultures for release back to the wild. The captive breeding facility is located in Changa Manga forest, near Lahore, in the Punjab Province of Pakistan. Captive vultures have certain idiosyncrasies in their needs, and we discuss these in relation to their breeding and husbandry. In June 2007, the captive population at the facility comprised 11 birds (7.3.1). Additional vultures will join the project in 2007/2008, which has a target of up to 15 breeding pairs in addition to non-breeding juveniles and sub-adults. Herein, we describe the facility, husbandry regime, veterinary aspects and future plans for the project.

## INTRODUCTION

The unprecedented decline of *Gyps* vultures in south Asia since the 1990s has seen the introduction of a range of conservation initiatives. Significant among these has been the establishment of conservation breeding centres for the three species affected: Oriental White-backed Vulture *Gyps bengalensis*, Long-billed Vulture *G. indicus* and Slender-billed Vulture *G. tenuirostris*. The primary aim of these centres is to hold safely a core population of the species affected. Once the environment is safe for vultures, they will act as a source population for reintroductions or as supplementation to wild populations. However, removing from the environment the primary cause of the vulture declines, veterinary Diclofenac, is a significant task.

Despite the drug having been banned in three of the range countries (India, Pakistan and Nepal), it is unlikely to be removed quickly from the environment. Diclofenac has been used widely in Pakistan (Oakes et al 2004; Ahmed & Khan 2005) and India (Riseborough 2004), with recent unpublished reports confirming that it is still available - at least in remote areas where regulatory enforcement is low.

Even if Diclofenac is removed from circulation within the next five years, estimated mortality rates are currently up to 50% each year, suggesting that extinction, at least across most the range for these species, is considered likely (Green et al 2004; Shultz et al 2004). The task facing these conservation centres is therefore considerable. When compared to the size of the population declines, and the fact the declines are continuing, it appears immense. However, their importance cannot be underestimated. It is imperative that these centres manage their captive stock well, and have prolonged breeding success over many years, perhaps decades. Without such success, the prognosis for species survival is poor. The long term husbandry of the captive vultures, breeding success and the preparation of vultures for release requires involvement from a wide range of organisations and people. Their commitment for the duration and continued liaison with researchers working on wild populations is an essential part of the international conservation effort for south Asian *Gyps* vultures. This paper provides information on the captive management and breeding of *Gyps* vultures, particularly the Oriental White-backed Vulture *Gyps bengalensis*. We also describe Pakistan's first vulture breeding centre, recently established at Changa Manga in the Punjab Province of Pakistan. We outline the development and current status of the facility, and provide details of future plans for the *Gyps* vulture restoration project.

**Captive Management of Vultures**

Regardless of the species in question, there are many principles of captive management that apply. We do not intend to discuss these, but to highlight aspects of captive management relevant to gregarious vultures, particularly *Gyps* vultures.

Vultures are intelligent birds that learn quickly and dislike surprises, so routine and consistency are key elements of successful management. Owing to their exceptional eyesight, attention to detail and long memory, vultures quickly recognise individual keepers, regardless of what they are wearing. However, the use of a 'uniform' by keepers is important, but not from the perspective of trying to conceal the identity of an individual person, as this is misguided. A staff uniform can consist simply of a certain hat or colour shirt, with the key element being that the birds learn that when a keeper wearing the 'uniform' enters the enclosure, they do so only for feeding or maintenance work. Attempting any capturing, restraint or other activity that unsettles the vultures (such as veterinary procedures) whilst wearing the 'uniform' is an abuse of trust that can lead to vultures becoming permanently unsettled when anyone enters the enclosure. The opposite situation and what is desirable, is that the vultures will recognise keepers involved in routine tasks (such as feeding or pool cleaning) and be comfortable with them entering the enclosure at any time.

**Enclosure Design and Construction**

Compared to many captive avian species, vultures are large birds. They are generally long-lived and resilient. Despite the fact that the dimensions of enclosures is an area where opinion differs greatly, the size and weight of vultures dictates several recommended design parameters.

Height is important and should be a minimum of three meters. Overall dimensions for a pair of birds can be in the range of eight to ten meters long and six to eight meters wide, although larger enclosures are recommended for groups. Enclosures must have a mixture of covered and uncovered areas, as birds will often deliberately sit unprotected in the rain and will actively 'sunbathe' with wings outstretched. Enclosure materials should be robust with consideration given to potential for injury. Thin gauge wire is totally unacceptable. Welded mesh is a superior enclosure medium, and should be of sufficient gauge and mesh size (10gauge; 40-50mm) to avoid abrasion. Enclosure substrate can vary between sand, pea gravel, grass or a combination of these. Wood chips should be avoided, particularly in enclosed or confined areas due to risk of Aspergillosis. Vultures do not appear overly prone to geosedimentation problems when housed on sand and do not appear to deliberately eat stones.

Perch sizes should be variable. Although vultures can land and sit on surprisingly narrow perches, an enclosure should contain a number of perches 10-15cm in diameter sufficient for the number of birds accommodated. Care is required to ensure perch positions do not allow faeces to reach water baths or feeding areas. Large vultures kept on small, hard perches are more prone to bumblefoot (a chronic degenerative foot condition) - as are those where a significant part of the flooring is concrete. Astroturf™ or similar artificial material can aid weight dispersal on perches, although some vultures will destroy this material rapidly, particularly wild vultures. Rope wound around perches is also effective. Bathing, preening, drying and loafing around water are essential behaviours for vultures, and constitute a significant proportion of their daily activity. Interaction between individuals occurs at these times, and it can be a highly social activity. Fresh water is therefore essential and should be available in a bath at least one meter wide and 10-15cm deep. Vultures will often only bathe in or drink water that is changed virtually every day.

**Feeding**

The gregarious feeding behaviour of vultures must be taken into account for successful captive management. Interaction between birds during feeding is important for general social development, the establishment of a colony 'pecking order' and possibly pair formation. Food type and presentation is important, and ideally consists of large animal parts (e.g. jointed beef) or small to medium sized mammals (e.g. rat, rabbit). It is important that bone is given and that large bones are smashed to enable swallowing. An effective method of ensuring bone consumption is to sprinkle smashed bone pieces over meat. Nutritional osteodystrophy has been recorded in these birds when fed meat alone or when given bone in a form they cannot ingest. Calcium supplementation is also recommended for breeding/growing birds in the form of a calcium/vitamin D3 powder (eg. Nutrobal: Vetark, Winchester, UK). Feeding regimes should attempt to mimic wild conditions. At least one fasting day per week is recommended, and should follow a sizeable feed the previous day. It is not unusual for vultures to still exhibit a protruding crop more than 24 hours after a large feed. Within reason, carcass remains should be left in enclosures for the day. Scavenging and picking around carcass remains and the consumption of small food pieces around a feeding area is an important behavioural

### Table 1
**NORMAL BLOOD VALUES FROM CAPTIVE *GYPS BENGALENSIS***

| Parameter | Units | N | Mean | Range |
|---|---|---|---|---|
| Haemoglobin | g/dl | 7 | 16.21 | 14.6-18.5 |
| Packed Cell Volume | % | 7 | 49 | 40-55 |
| Red Blood Cells | $\times 10^{12}/l$ | 7 | 2.74 | 1.86-4.15 |
| Mean Cellular Volume | fl | 7 | 190 | 96-258 |
| Mean Cellular Haemoglobin Concentration | g/dl | 7 | 33.2 | 31.4-36.5 |
| Mean Cellular Haemoglobin | pg | 7 | 62.7 | 35.2-81.7 |
| White Blood Cells | $\times 10^9/l$ | 7 | 12.97 | 9.5-18.3 |
| Heterophils | % x $10^9/l$ | 77 | 7910.21 | 72-867.49-14.64 |
| Lymphocytes | % x $10^9/l$ | 77 | 141.89 | 4-190.38-3.33 |
| Monocytes | % x $10^9/l$ | 77 | 50.59 | 1-100.18-1.04 |
| Eosinophils | % x $10^9/l$ | 77 | 00 | 00 |
| Basophils | % x $10^9/l$ | 77 | 20.28 | 0-60-1.1 |
| Calcium; ionised | mmol/l | 3 | 1.11 | 1.00-1.18 |
| Calcium; total | mmol/l | 8 | 2.56 | 2.31-3.09 |
| Total Protein | g/l | 8 | 54.9 | 50.3-63.1 |
| Albumin | g/l | 8 | 17.0 | 15.9-17.9 |
| Globulin | g/l | 8 | 38.0 | 33.0-46.2 |
| Alpha Amylase | IU/l | 3 | 1342 | 1125-1535 |
| Aspartate Transaminase | IU/l | 8 | 842 | 319-2550 |
| Creatinine Kinase | IU/l | 8 | 320 | 165-659 |
| Sodium | mmol/l | 3 | 151 | 144-156 |
| Potassium | mmol/l | 3 | 5.4 | 4.9-6.2 |
| Bile Acids | umol/l | 3 | 51.79 | 40.3-74.2 |
| Uric Acid | umol/l | 8 | 997.9 | 181.06-1298.0 |

*Source: Chitty & Murn (2004)*

characteristic, and is an important activity for young or subordinate birds held within a colony. Care should be taken that birds do not become obese. Under usual conditions, if vultures do not feed almost immediately the food is presented, then overfeeding should be suspected and body condition monitored.

## Behavioural Characteristics and Breeding

Their gregarious nature means that, unless requiring veterinary attention, it is important to avoid keeping vultures in isolation. Such birds can become 'depressed' (lethargic, loss of appetite) and will fail to thrive. Despite the need for routine and stability, captive vultures are almost invariably interested in novel items. Given the opportunity, they will also invest significant amounts of time and energy in destructive behaviour. These tendencies are useful additions to management. Although feeding and bathing are important activities, the development of methods to keep captive vultures occupied requires additional time and thought. Enclosures should contain sufficient horizontal perches, in addition to post perches. Long horizontal perches are important for social behaviour and pair formation because they reduce conflict over single perch positions and allow birds to approach one another quietly by walking along the perch.

This is particularly the case for potential breeding pairs. In the wild, courtship behaviour often consists of close mutual soaring. In captivity, however, courtship can be subdued or not easily observed. Consideration of perch type and position, in addition to an experimental approach in this regard will yield results. Separate breeding enclosures are an essential part of management. Although wild Oriental White-backed Vultures breed in colonies, difficulties can arise when captive breeding pairs are in the same enclosure as non-breeding birds. To avoid these difficulties, an important technique is the availability of a large communal enclosure, which provides a venue for breeding pairs to form. Once established, breeding pairs require separate enclosures to avoid disturbance from unpaired birds during the breeding season. Females generally take main responsibility for nest building, although males will also bring material and add to the nest. Mating takes place either in the nest or on a preferred perch nearby. Copulation can be prolonged (compared with other avian species) and lasts between 10 and 30 seconds. The male will often bite the nape of the female's neck during copulation.

## Veterinary Aspects

For veterinary treatment purposes, the regular capture and restraint of a vulture should be a last resort. Due to their vomiting reaction, birds will lose condition quickly if captured regularly. During restraint, care must be taken not to throttle the bird. A vulture that is aware of the fact it is to be captured will usually vomit, and this reaction is likely to continue during restraint. Care must be taken that the bird does not aspirate regurgitated food resulting in choking or inhalation pneumonia. Treatment regimes should be as non-invasive as possible and the bird should be isolated for as short a time as possible. Vultures should be anaesthetised using isoflurane with induction by restraint and open mask. Breath holding is common making stabilisation difficult on occasions. They should therefore be intubated and placed on mechanical ventilation as soon as possible after induction. Trained veterinary personnel may take blood samples relatively easily by jugular or ulnar venipuncture. The following table outlines 'normal' blood values obtained from captive *Gyps bengalensis* at The Hawk Conservancy Trust, as part of a health screening program (Table 1). Adult birds (mixed males and females) were kept unfed for 12-24 hours before blood was taken. Heparinised samples were posted and analysed the next day. In general, drugs used in other raptors appear safe in vultures and they can be given by the same routes and methods. However, it should be noted that vultures are generally excellent at recognising drugs hidden in meat and rejecting them. It should also be noted that if on a course of oral drugs, birds should be handled and examined/treated before drugs are given in food. Otherwise regurgitation and loss of the drug may result.

## The *Gyps* Vulture Restoration Project in Pakistan

In 2004 WWF-Pakistan launched the *Gyps* Vulture Restoration Project in Pakistan. The immediate project objective is to conserve a viable population of *Gyps bengalensis* in a safe and secure environment. Once secured, the breeding potential of the captive population must be realised.

Additional project objectives include continued monitoring of wild populations, lobbying for the complete removal of Diclofenac from the environment and to build staff capacity for the eventual release of captive-bred vultures.

Plans for Pakistan's first conservation breeding facility for vultures began in 2005. Government approval, land allocation, facility design, fundraising and staff selection took place over the following 18 months. The project, run by WWF-Pakistan, is a partnership between WWF-Pakistan, the Punjab Wildlife and Parks Department, the Environment Agency, Abu Dhabi and the Hawk Conservancy Trust. WWF-Pakistan is the project manager and staff provider, whilst the Hawk Conservancy Trust has provided technical and training support and will contribute towards facility running costs into the future. The Environment Agency and WWF-US provided keystone funding for the facility construction.

## The Facility

Fig. 1. *Vulture holding enclosure.*

The vulture conservation centre is located in a secluded area of Changa Manga forest, which is approximately 80km southwest of Lahore. Government and local officials and project partners attended an official opening of the facility in April 2007. There is currently one large holding enclosure (Figure 1). It is 38m long, 6.5m high and increases in width from 14m to 27.5m. Construction materials consist of 150mm steel pole supports and welded steel frames on concrete bases. The walls and roof are chain link. The enclosure substrate is local soils and plants. Within the enclosure, perch types consist of live trees in addition to a number of artificial perches. One end of the enclosure contains a roosting/nesting ledge, which runs the width of the enclosure. This also provides shade and shelter for the birds. Additional shade cloth on the roof provides sun protection over a smaller area near the water pool. Heat stress in the vultures is evident on hot days. Head drooping is frequent at these times, and the vultures will spend increased amounts of time bathing and drying (Figure 2). Apart from environmental considerations, there are other challenges specific to the management of captive raptors and these necessitate staff training. A key component of the project is the development of staff skills through training and capacity building.

Within the enclosure, potential breeding birds have access to artificial nest sites (shallow woven baskets) and nests retrieved from the wild. These have proved popular already with the birds, and may provide an additional stimulus for courtship and/or breeding behaviour. Although a tree-nesting species in the wild, captive Oriental White-backed vultures have shown an occasional preference for nest ledges and artificial 'caves' (Chitty & Murn 2004). To accommodate this, the roosting/nesting ledge in the enclosure is fitted with partitions to provide separate areas for potential breeding pairs. There is an attached service building, storage facilities, and provision for office space (Figure 3). The facility currently has a capacity of approximately 30 vultures.

## Current Population and Future Plans

In June 2007, there were eleven vultures in the facility (seven males, 3 females, one unsexed). Five were collected as chicks from nests during the 2005/2006 breeding season, with one additional bird collected during the 2006/2007 season. The remaining five vultures are older, and are the remainder of the captive population used during Diclofenac toxicity testing work by the Peregrine Fund (Oakes et al. 2004). All birds have identification rings visible from a distance and microchip implants.

Future breeding potential with this small population is limited, as there are only three confirmed females in the group. The sex of the 2007 chick is currently unknown, but even if this bird is female, breeding potential is still low.

Clearly there is a need to increase the number of vul-

Fig. 2. *Vultures bathing during hot weather.*

**Fig. 3.** *Facility addition allowing for storage and office space.*

tures at the facility. To this end, trapping of wild vultures will take place in late 2007, following the monsoon season. Small populations and available food will undoubtedly make trapping attempts difficult; however the project aims to trap between 15 and 20 sub-adult birds. Second phase building will commence in the last quarter of 2007. The plans for phase two building include four breeding enclosures, plus additional infrastructure. This will include livestock paddocks, perimeter fencing, a facility for breeding rats for supplementary feeding and freezer rooms. Solar panels are being investigated for the electricity supply.

In the longer term, the construction of additional breeding enclosures is a primary goal. Only in this respect is the time scale favourable. It is likely to be many years before the environment is safe for the release of vultures back to the wild. However, unless the facility can reach a production capacity of at least 10-15 chicks per year, there will be limited potential for any release program in the future.

## CONCLUSION

The WWF *Gyps* vulture restoration project in Pakistan is part of a regional effort for the conservation of south Asian *Gyps* vultures. The project benefits from a dedicated staff team, solid governmental and partner support and a series of clear targets. Further development and the expansion of facilities will occur over future years, as there is an urgent need to increase the capacity of the facility.

There are no prospects for a rapid conclusion to the conservation of *Gyps* vultures in south Asia. Continuing the lines of communication and sharing information between Pakistan and India will be of benefit to all parties that are working together on this international conservation effort.

## REFERENCES

AHMED S & U Khan. 2005. Evaluation of the current status of Eurasian Griffon *Gyps fulvus* and Cinereous Vulture *Aegypius monachus* in central and northern Balochistan. WWF-Pakistan Technical report. 32pp.

CHITTY J & C Murn. 2004. Veterinary Aspects of Captive Old World Vultures. Proceedings of the 25th Annual Conference of the Association of Avian Veterinarians. New Orleans, Louisiana, USA. August 2004.

GREEN RE, I Newton, S Shultz, AA Cunningham, M Gilbert, DJ Pain and V Prakash. 2004. Diclofenac poisoning as a cause of vulture population declines across the Indian subcontinent. Journal of Applied Ecology 41: 793-800.

OAKS, JL, M Gilbert, MZ Virani, RT Watson, CU Meteyer, BA Rideout, HL Shivaprasad, S Ahmed, JL Chaudry, M Arshad, S Mahmood, A Ali and AA Khan. 2004. Diclofenac residues as the cause of population decline of vultures in Pakistan. Nature 427: 630-633.

RISEBROUGH RW. 2004. Fatal medicine for vultures. Nature 427: 596-597.

SHULTZ S, HS Baral, S Charman, AA Cunningham, D Das, GR Ghalsasi, MS Gouday, RE Green, A Jones, P Nighot, DJ Pain and V Prakash. 2004. Diclofenac poisoning is widespread in declining vulture populations across the Indian subcontinent. Proceedings of the Royal Society of London Biology: Biological Science (Suppl) 271: S458-S460. DOI 10.1098/rsbl.2004.0223.

# CAPTIVE BREEDING AND RELEASE OF RED KITES *MILVUS MILVUS* TO HAMPSHIRE, ENGLAND

CAMPBELL MURN, AMY KING, SAMUEL HUNT & ASHLEY SMITH

The Hawk Conservancy Trust, Andover, Hampshire, SP11 8DY, England

## SUMMARY

Since the early 1990s, populations of the Red Kite *Milvus milvus* have been re-established successfully in England and Scotland through translocation reintroduction programs. For other species, breed and release programs continue to be a successful method of reintroducing species, particularly raptors. Between 2003 and 2005, the Hawk Conservancy Trust released 12 kites to an area in northwest Hampshire, England. Four of these were captive bred, foster-reared and soft released in artificial nests. The remainder were mature ex-zoological birds and one wild-injured rehabilitated bird. Radio tracking revealed post release movements for captive bred kites of up to approximately 27km during the eight months following release. Interaction with powerlines killed two of the captive bred birds three weeks post-release and a third captive bred kite died from head injuries six months post release. The fourth captive bred kite ranged widely for two seasons, was observed in courtship flights with another kite near the release site in mid 2006 and remains local in 2007. Mature kites left the release area between two and 56 days, and travelled distances of between 0.4km and 12.2km during a continuous five month period of observation. Open areas were preferred, particularly large hedgerows surrounding arable or grassland fields. Independent foraging began nine days post release. The combined mortality rate for the larger group was 12.5% during the period of observation.

## INTRODUCTION

During the late 18th and 19th centuries the Red Kite *Milvus milvus*, once a numerous and widespread raptor in the United Kingdom, was exterminated in England and Scotland, primarily due to human persecution (Lovegrove 1990; Carter 2001). A relict population survived in Wales.

Following an improved public perception of the species and a major reduction in persecution, a Red Kite reintroduction program began in the United Kingdom in 1989. Kites translocated from Spain, Sweden and Germany were released to sites in northern Scotland and southern England (Carter et al 1999, Carter 2001). Using the same method, Red Kite populations were established successfully in three other locations in the United Kingdom in the period 1995-2000 (Wotton et al 2002). Two further release sites operated from 2003, bringing the total number of release sites to seven. Despite the success of the reintroduction program, Red Kites remained scarce in northwest Hampshire in southern England. Between 1995 and 2002 there were confirmed observations of only four kites in this vicinity. The Hawk Conservancy Trust (HCT) is a bird of prey conservation, research and rehabilitation centre based in Hampshire, England. Since inception in 1980, a long-term aim of the HCT has been to re-establish successfully a population of Red Kites in the local area. To achieve this aim, since July 2003, The Hawk Conservancy Trust has released 12 Red Kites into the wild. The initial release birds were captive bred. Subsequent releases consisted of captive stock birds and one rehabilitated wild bird. This report outlines the methods we used for breeding, rearing and releasing Red Kites to the wild. The subsequent release involving mature kites is described, in addition to post release observations, mortality factors and survival rates for all release birds.

## METHODS

### Housing, laying, hatching and rearing

Between 1996 and 2001, we obtained suitable stock to create four potential breeding pairs of Red Kites. Each pair was kept in a large semi-seclusion aviary with nest site choices of basket, platform and open box. The aviaries were 40' long, 20' high and 20' wide.

Previous experience with breeding kite species (Black Kite *Milvus migrans* and Brahminy Kite *Haliastur indus*) revealed the importance of large aviary size in a quiet environment. Captive male kites regularly perform courtship flights during the pre-breeding season, usually expressed as rapid and acrobatic flights, often in a circular pattern. The males also appear to demonstrate simulated foraging behaviour. Before egg laying, female kites will preferentially accept food from male kites, rather than collecting their own. The females clearly respond to this behav-

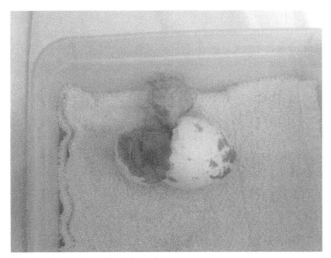

Fig. 1. *Red Kite chick hatching.*

iour, and appear to reject the advances of a male who has not exhibited the courtship and 'foraging' behaviours described. Large aviaries also provide space for separate roosts. We observed that male kites preferentially roost away from breeding females, who become intensely defensive of their selected nest site. Diet consisted of quail, rat, mice and chicken in rotation. Monitoring and regulating food intake is important as even mildly overweight birds fail to breed. We used occasional fasting days to increase interaction with food, with an aim of ensuring that females remained receptive to male overtures when providing food. In April 2003, a nine year old kite layed two fertile eggs in an open box nest site. This bird had produced eggs on three previous occasions, but each clutch had failed. Owing to this history, ten days after laying the eggs were transferred to a broody hen for incubation. At the end of the incubation period, which was 33 days, the pipping eggs were transferred to an incubator (Isolette C100 Infant Incubator) for hatching (Fig 1).

The newly hatched chicks (Fig 2 & 3) were hand-fed for four days on a diet of minced quail and chicken. The unknown rearing ability of the breeding pair dictated that surrogate parents would rear the chicks. We used a pair of European Buzzards *Buteo buteo* for this purpose. These surrogate parents had a long history of successful chick rearing, and proved to be exceptional parents to the chicks over 30 days until their removal prior to release. Surrogate rearing of kites by this species has been a successful method in the past (Carter 2001). Whilst the first clutch was incubating, the female kite returned to the nest and produced a second clutch after approximately three weeks. Incubation, hatching and rearing of this clutch followed the same methods as the first clutch prior to placement with the surrogate buzzards.

To test the rearing behaviour of the kites, the second clutch was replaced with a 12-day-old buzzard chick. Unfortunately, this proved unsuccessful, as the substitute chick disappeared after four days.

When the first pair of chicks was removed prior to release, the second pair of chicks replaced them in the foster aviary. To remove the risk of rejection by the foster parents and potential injury, the male buzzard was removed from the aviary as a precaution. Additionally, a mesh cage was constructed over the nest site to shield the chicks from any female aggression or rejection behaviour. The mesh gauge was of sufficient diameter to allow the passage of food to the chicks. Although the male buzzard remained in a separate enclosure, after two days the mesh was removed and the female buzzard reared the chicks successfully.

## Release

### Captive bred kites - 2003

The release of the captive bred kites was the culmination of many years of planning. Apart from the experiences of breeding birds for release and developing a suitable method for release, the HCT worked closely with local farmers and landowners, outlining the project, developing links and ensuring community involvement and support. In recognition of the local support for the project, the chicks were named after local farms surrounding the release site.

We used a soft release technique for all kites. Chicks were transferred to artificial nest sites ('hack' sites) in pairs when their flight feathers were half grown, at between 41 and 46 days of age (Fig 4). At this age the kites are fully feathered (usually achieved by +/- 30 days), but were unable to fly. Prior to release

Fig. 2. *Red Kite chick.*

**Fig. 3.** *Red Kite chick.*

all kites were fitted with a radio transmitter (Biotrack, Dorset, UK) attached using a backpack harness with perishable cotton stitches. Captive Black Kites used for flying demonstrations at the HCT enabled testing and modification of the harnesses before being fitted to the release kites. A rope and pulley system supplied food to the hack site remotely. Food consisted of local carrion types such as rabbit, rat and pheasant. Researchers, staff and volunteers monitored the kites during all daylight hours. Each pair of kites had a separate release site. Before release, all kites were health screened, treated for internal and external parasites, weighed and fitted with leg rings.

*Mature kites*

Following the reintroduction of captive bred kites to the district in 2003, the HCT decided to further supplement and reinforce the local population of kites in an attempt to maintain a stable population. Even as additional kites appeared in the district following the first reintroduction, the Trust concluded that further releases would sustain the small resident population, and introduce a source of genetic diversity. Eight adult birds (five females and three males) were released including one wild rehabilitated kite. The rehabilitated kite arrived at our bird of prey hospital with a blood disorder, and upon recovery exhibited leucism (partial loss of pigment in plumage and pigmented eyes). After an extended period of recovery from March to November 2005, this kite joined the release stock. Pre-release screening and identification was the same as for the captive bred kites. Radio transmitters were tail-mounted rather than attached via backpacks. The previous release highlighted that the battery life of the transmitters was much shorter than the longevity of the backpack harness. We expected therefore that the battery life of the transmitters would roughly correspond with the time at which the tail feathers would moult.

For three weeks before release, the kites lived in a large aviary, approximately 21m long, 8m wide and 9m high. This permitted the development of fitness in the birds before release. The aviary had a favourable aspect over the surrounding countryside and enabled the kites to assess their location and observe a nearby feeding station that would be their primary post-release support. The feeding station has been operating since the early 1990s. It supplements the diet of the local raptors, herons and corvids. At the beginning of December 2005, a section of the pre-release aviary was removed to allow the kites to leave. Over two days all the kites left the aviary, after which tracking commenced. The period of continuous tracking was five months, after which tracking efforts were opportunistic. We used an ICOM RC 310 receiver and a 3-element yagi to track the kites. For all kites, tracking efforts started each day at the release site unless the signal direction indicated that tracking should resume at the location of the previous day. Tracking usually continued until a bird was sighted. Precipitation and wind-speed reduced tracking efficiency, whilst damage to the antennas on the transmitters reduced the signal range.

The mature kites received supplementary feeding during their first few months in the wild. Their length of time in captivity suggested that it would be necessary to supplement their food during winter, whilst they developed foraging skills. Food was also available at the feeding station near the release site in the mornings and afternoons. The morning feed was gradually reduced and ceased completely six months

**Fig. 4.** *Sub-adult Red Kite.*

post release. For any kites that dispersed early, food was taken to their location; this was either quail or rabbit, simulating natural prey.

# RESULTS

**Post Release Observations**

*Captive bred kites*

Of the two captive bred pairs, the earliest departure from a hack site was by Kite 1 of Pair 1. This bird left the nest after only four days; however, the occurrence of a storm with strong winds on the night of the fourth day may have meant this was a premature departure. Early the next day it was found, grounded and wet, 20m from the base of the hack site tree. It was replaced in the hack site. Kite 2 from Pair 1 left the hack site after eight days, whilst the Kites 3 and 4 from Pair 2 made initial excursions from their hack site after seven days. In the days before leaving the hack sites, all kites spent much time exercising their wings and exploring the tree containing the hack site (Figure 5).

None of the kites moved very far (+/- 50m) from the hack site during their first movements. The hack site trees were both part of large hedgerows, which enabled the birds to make gliding flights along the hedgerow and land a short distance away. The kites would often return to the hack site by hopping from branch to branch along the hedgerow. Roosting positions were in either the hack site or less than 10m away. Food provision to the hack site gradually decreased to zero over a period of three weeks as successful foraging was observed. The kites had also begun to make use of the nearby feeding station.

During the late summer and autumn following release, the movements of Kite 2 revealed an interesting pattern. Weather conditions largely dictated movements. Specifically, a series of clear, warm days with thermal activity and winds from a particular direction saw the kite move away. When these conditions ceased (cooler temperatures, lower cloud and variable winds), the kite remained in one location for several days or weeks. A return to warmer conditions and an opposing wind direction would invariably see the return of the kite to the release site. This behaviour became predictable, to the point where a planned outing to resume tracking the kite would be postponed if the weather was warm and the wind direction was from the kite's position on the previous day. During these excursions, Kite 2 almost exclusively visited a specific area approximately 27km northeast of the release site. On occasions it would spend nearly two weeks in this location, making only small local movements, before a prevailing northeast wind and thermal conditions prompted a return to the release area. The kite clearly appeared to wait for specific weather conditions and wind directions to go to a specific location. In the spring of 2006, Kite 2 was performing courtship flights with one of the mature kites from the 2005 release program.

*Mature kites*

The movements of Kite 2 after fledging highlighted the importance of thermals and wind direction to kite dispersal behaviour. During the winter, these weather conditions are generally absent, compared with summer and autumn. Further, communal winter roosting and limited winter movement are recognised features of wild kite ecology (Carter, 2001). Thus, with an aim of holding the release kites near the supplementary feeding station for a longer period, a winter release was conducted.

Despite this, Mature Kite (MK) 6 and MK 8 remained within 0.9km of the release aviary and made use of the feeding station for only two days. The remaining kites made use of the feeding station, and all had dispersed from the release site by day 56. MK 4 was the only bird that dispersed from the release site and returned. This happened several times; the bird would return for up to a week before leaving again, generally staying within 12km of the release site. The remaining kites gradually increased the distances they moved away from the release site. Between 39 and 94 days post release, several kites increased the distances covered to between 6.7km and 12.2km from the release site. Observations of the mature kites feeding

**Fig. 5.** *Red Kite hack site.*

**Fig. 6.** *Post-release Red Kite chicks feeding alongside a Magpie.*

in the wild include MK 6 feeding on rabbit carcass nine days post release, and feeding alongside a Buzzard and a Magpie *Pica pica* at the same location (Fig 6). MK 2 preferred foraging across arable fields, and MK 7 fed from lamb remains for a number of days. MK 4, MK 6 and MK 7 interacted with wild kites on a number of occasions. These interactions included aggressive behaviour in addition to general associations. Within two weeks of release, all mature kites were accomplished fliers.

Preferred habitats for all kites were areas with large hedgerows surrounding arable or grassland fields. The kites generally avoided the centre of woodlands, although all mature kites frequently utilised woodland edges or large hedgerows (Fig 7). On one occasion were kites observed in a coniferous wood and a beech wood. The only habitat requirement often cited for Red Kites is a mixture of forest patches to breed and open areas to search for food (Seoane et al 2002; Hardey et al 2006). This corresponds with our observations that the mature kites preferred open habitats with large woodland hedges surrounding arable or grass fields.

## Mortality and survival

### Captive bred kites

After 18 and 17 days post-fledging Kite 1 and Kite 4 died from electrocution on a nearby powerline. Until this point, their foraging behaviour was increasing and the kites were beginning to roost in a small wood approximately 300m from the hack site.

Nearly five months (141 days) of surviving in the local countryside, Kite 3 was found near the release site. A veterinary examination revealed severe head trauma, possibly due to a collision or encounter with another raptor. The kite was euthanized after further examinations revealed widespread infection and abscess development in the thoracic cavity.

As noted previously, Kite 2 still frequents the release area (i.e. natal site) during 2007, and occasionally makes use of the feeding station, particularly during winter.

### Mature kites

Despite careful post release monitoring and supplementary feeding, the rehabilitated leucistic kite (MK 1) died ten days post release. There were no other confirmed fatalities during the period of post release monitoring, indicating a mortality rate during this period of 12.5%. Unfortunately, one of the radio tags was found inside the pre-release aviary two days after the release, thus making the kite untraceable with tracking equipment. This bird (MK 3) was occasionally identifiable using leg rings. At the end of January 2006, a radio tag was retrieved approximately 6.9km from the release aviary. There was no evidence to suggest the bird (MK 6) had died, and the most likely explanation is that the transmitter had detached from the tail feathers. Another radio tag was found at the end of June 2006 near the release aviary. The bird (MK 4) had lost several of its tail feathers and was thus recognisable at the feeding station. Results so far indicate a successful release, with no confirmed fatalities other than the leucistic kite. Sightings have been made of the birds feeding in the wild and significant behavioural observations include courtship behaviour of two pairs of kites within the release area.

**Fig. 7.** *Post-release Red Kite utilizing local hedgerows.*

## DISCUSSION

Mortality rates for raptors in their first year and up to breeding age are often based on ring returns and vary considerably. Broad mortality estimates across the family Accipitridae range between 60 and 90% for the period between fledging and adult plumage (Thiollay 1994). Radio tagging studies can be more revealing. Studies on large falcons using radio tagging indicate first year mortality rates of between 28% (McFadzen & Marzluff 1996) and 77% (Kenward et al 2007) for Prairie Falcons *Falco mexicanus* and Saker Falcons *Falco cherrug* respectively. Studies on Common Buzzards suggest relatively low first year mortality rates of between 25% and 40% (Kenward et al 2000).

It is thus clear that mortality rates in excess of 50%, whilst not certain, are not unusual for first year and/or pre-breeding age raptors. However compared with other accipitrid species in other locations, mortality rates for young Red Kites in Britain appear low and rarely exceed 50% (Wotton et al 2002). The 75% mortality of the released captive bred kites therefore appears high. However, the single electricity pylon that killed Kites 1 and 4 was inherently unsafe due to a small phase gap and earthed metal cross arm. As small consolation the pylon was eventually modified, but not before nearly a month had passed and Kite 4 had been killed (21 days after Kite 1). Although it is speculation, we believe that had these birds not been killed, their survival prospects were at least moderate. The early foraging and behaviour patterns exhibited by these kites before their deaths followed a similar pattern to Kite 2, which still survives four years post release. Clearly, if the pylon had been modified in less time the outcome for Kite 4 might have been positive.

The many observations of Kites 1 to 4 indicated that early in the fledging period these young birds were taking frequent short flights, presumably to exercise and improve their flying abilities. Their inexperience at this early stage, however, often forced them to land on almost any available structure; this behaviour is uncommon in adult birds, which possess the flight skills to perch on the branch of a tree when required. Compared to the captive bred kites the flight skills of the mature kites developed quickly after their release. They rarely made mistakes and possessed superior skills and coordination, despite having been in captivity for up to ten years. There was only one sighting of a mature kite using a pylon, and there were no fatalities resulting from the use of pylons by mature kites. The use of hacking to release young raptors into the wild is well-known to be successful. Indeed, the term has become widely used to denote any release scheme that employs release and supplementary feeding of juveniles. However, 'hacking' per se, refers to the release of pre-fledged juveniles in an artificial nest site that is closely monitored and fed. Historically, the aim of hacking was to re-capture the birds once they had learned to hunt effectively, and train them for use as falconry birds.

In any case, fledging and mortality rates of 'hacked' juvenile raptors are often comparable with wild juveniles. It is worth noting, however, that 'proper' hacking may not be the most suitable release method for captive bred specimens of species such as the Red Kite. There are two reasons for this. Firstly, as described, young kites make frequent short flights during their post fledging learning period. This often results in them making unexpected landings on potentially unsuitable structures (i.e. unsafe pylons). Holding young captive bred kites before release until they are fully flight fit may thus prove to be an effective release technique. Secondly, Red Kites are highly efficient scavengers that rely less on predation compared with other raptors (cf. Goshawk *Accipiter gentilis* and Peregrine *Falco peregrinus*). Other release programs for scavenging raptors, such as vultures, will always employ release aviaries that hold young birds until they are able to fly. It would appear that scavenging is an inherent behaviour in kites, which begins once a bird has mastered the necessary flight skills. These conclusions differ for Kite 3. Despite avoiding electrocution, Kite 3 died after 141 days, and field staff generally considered this kite to be the weakest of all the release birds. It is possible that Kite 3 was the victim of attack by buzzards, which regularly harassed and attacked the kites. The local corvid population often mobbed the young kites, and Kite 2 quickly became proficient at avoiding a mob of pursuing crows. Kite 3, however, was less successful and often had to land and defend itself from the nearest perch. It is therefore possible that Kite 3 may never have been an accomplished flier, even if it had been held in a pre-release aviary. In contrast to the captive bred kites, the lower mortality rate of the mature kite group is favourable. The only confirmed fatality was the leucistic kite, which although unfortunate, was not a surprise. Leucistic birds are rare, and may be at a disadvantage to conspecifics due to their conspicuous plumage and possible optic deficiencies. In light of our previous points regarding the development of flight skills in young kites and release strategies for captive bred kites, it is tempting to reflect upon the low post release mortality of the mature kites compared with the captive bred kites, and draw the conclusion that the release of captive bred nestlings via hacking is an inferior method. This project, however, provides insufficient grounds to support this conclusion, as it is unknown how the two electrocuted kites would have survived, but for the existence of an unsafe pylon near the release site. The post release behaviour of Kite 2 appears to conform with the recorded movements of young wild kites, whereby juvenile dispersal is followed by philopatric tendencies and a return to the natal locality (Newton et al 1994; Carter 2001). Similarly, post-fledging dependence on the hack site broadly corresponds with the behaviour shown by wild kites, which can be dependent upon their parents for three to four weeks (Carter 2001). However, the mature kites have not exhibited convincing philopatric behaviour. At the end of the observation period, only one kite (MK 4) remained in the area, and this bird was the only kite to leave the release site and return. This may be due to the existence of the feeding station, or possibly a type of release site fidelity. With the expiration of the radio tags, confirming the identity of every kite that appears at the feeding station is no longer possible. As a result, conclusions about the area of settlement for MK 4 remain speculative. It is possible that genetic composition affects the dispersal tendencies of kites, particularly birds from German populations that migrate to southern Europe (Carter 2001). As the release site was not the natal area of any of the mature kites, with four of the kites originating from German stock, the likelihood of them dispersing would appear high. Furthermore, the presence of an established local population tends to reduce dispersal distances (Carter 2001). This is another possible reason for the dispersal of the mature kites over a larger area. The Trust has no plans to continue reintroducing Red Kites, but with observations of courtship behaviour from the released kites, any subsequent breeding success and the presence of juveniles in the area may reduce the tendency of new birds to disperse. Those kites that dispersed early made use of the food taken to them, but never have any of the kites depended completely upon the supplementary food. Although having been captive birds all their lives, observations of feeding behaviour by the mature kites soon after their release demonstrates their innate foraging behaviour.

## CONCLUSION

The main objective of this release scheme was to re-establish a population of Red Kites into northwest Hampshire after an absence of over 150 years. In 2003 the Hawk Conservancy Trust successfully ran the first ever captive breed and release project for Red Kites in England. This program, in conjunction with subsequent releases in 2005, lead to the establishment of a small population of Red Kites in the target area. Additional birds have been attracted to the region, and Red Kites are seen virtually every day in 2007. There are no plans to continue the release program. With established kites in the area, and a species that is increasing nationally, the prospects for Red Kites in northwest Hampshire remain positive.

Education will play an important role in the future of Red Kites. The Hawk Conservancy Trust worked hard to ensure local support for the Red Kite reintroduction projects, and local landowners (and their children) occasionally participated in tracking the kites. Encouragingly, most landowners and farmers are now proud to have kites on their land, protecting them and monitoring their success. Increased public awareness through talks, general publicity and visitor information will continue to be a major activity of the Hawk Conservancy Trust.

The prospects for Red Kites in the UK appear extremely good, with increasing numbers at most of the release locations. They are still in a recovery phase however, and are 'Amber Listed' by the British Trust for Ornithology as a Species of Medium Conservation Concern.

## REFERENCES

CARTER I. 2001. The Red Kite. Arlequin Press. Chelmsford, England.

CARTER I, M McQuaid, N Snell and P Stevens. 1999. The Red Kite (*Milvus milvus*) reintroduction project: Modelling the impact of translocating kite young within England. Journal of Raptor Research 33: 251-254.

HARDEY J, H Crick, C Wernham, H Riley, B Etheridge and D Thompson. 2006. Raptors a field guide to survey and monitoring. Scottish Natural Heritage, Edinburgh.

KENWARD RE, SS Walls, KH Hodder, M Pahkala, SN Freeman and VR Simpson. 2000. The prevalence of non-breeders in raptor populations: evidence from rings, radio-tags and transect surveys. Oikos 91: 271-279.

KENWARD RE, T Katzner, M Wink, V Marcström, S Wall, M Karlbom, K Pfeffer, E Bragin, K Hodder and A Levin. 2007. Rapid sustainability modelling for raptors by radiotagging and DNA-fingerprinting. Journal of Wildlife Management 71: 238-245.

LOVEGROVE R. 1990. The Kite's Tale: The story of the Red Kite in Wales. RSPB, Sandy, Bedforshire.

MCFADZEN ME, JM Marzluff. 1996. Mortality of Prairie Falcons during the fledging-dependence period. The Condor 98: 791-800.

NEWTON I, PE Davis and D Moss. 1994. Philopatry and population growth of red kites, *Milvus milvus*, in Wales. Proceedings of the Royal Society of London, Series B 257: 317-323.

SEOANE J, J Vinuela, R y Diaz-Delgado and J Bustamante. 2003. The effects of land use and climate on red kite distribution in the Iberian peninsula. Biological Conservation 111: 401-414.

THIOLLAY J. 1994. Family Accipitridae (hawks and eagles). Pp. 52-105 in J. del Hoyo, A. Elliott, J. Sargatal, eds. Handbook of the Birds of the World. Vol. 2. New World Vultures to Guineafowl. Lynx Edicions, Barcelona.

WOTTON SR, I Carter, AV Cross, B Etheridge, N Snell, K Duffy, R Thorpe and RD Gregory. 2002. Breeding status of the Red Kite *Milvus milvus* in Britain in 2000. Bird Study 49: 278-286.

# CAPTIVE FIELD PROPAGATION AND EXPERIMENTAL RELEASE OF EASTERN LOGGERHEAD SHRIKES *LANIUS LUDOVICIANUS MIGRANS* IN ONTARIO, CANADA

Rina Nichols[1], Lance Woolaver[1], Elaine Williams[1], Jessica Steiner[1] & Ken Tuininga[2]

1 Wildlife Preservation Canada, 5420 Highway 6 North, Guelph, Ontario, N1E 3N7, Canada
2 Environment Canada, 4905 Dufferin Street, Toronto, Ontario M3H 5T4, Canada

## Summary

The eastern subspecies of the Loggerhead Shrike *Lanius ludovicianus migrans* is a raptor-like migratory passerine found in south central Canada and the eastern United States. Since the 1970s, the decline of the Eastern Loggerhead Shrike has been precipitous throughout most of its range with an estimated 30 breeding pairs remaining in Canada by 1995. A captive-breeding population was established in Ontario in 1997 due to concern for the decreasing wild population. In 2001, an experimental program was developed to trial field breeding enclosures and to develop release techniques in the event that a reintroduction was needed.

Currently, there are 24 field propagation and release enclosures at two field sites in southern Ontario. A total of 221 young shrikes have been released in Ontario since the program began in 2001. The program has had seven separate re-sightings of release birds including three birds that returned to the breeding grounds the following season to produce fledglings of their own in the wild. The development of large improved wintering and breeding enclosures and new management of pairings has greatly increased the number of fledglings produced annually, from 10 in 2001 to 129 in 2006. Productivity of field propagated pairs was similar to, and in some years greater than wild pairs, with double brooding much more likely to occur within the field enclosures. Field breeding and release enclosures allowed young shrikes to be raised in a natural setting by their own parents, where they learned critical survival skills including hunting of wild prey and predator avoidance. Due to the large size of the enclosures the young developed strong flight skills *in-situ*. Birds were "soft released" with provision of a recognizable food source while they learned to find wild food. The techniques developed were simple, flexible and cost-effective and produced strong young birds that had been raised by their own parents in their native habitat. These techniques have wide ranging applications for other reintroduction efforts and should, in many cases, be considered a viable alternative to *ex-situ* breeding, hand-rearing and/or translocations.

## Introduction

The eastern subspecies of the Loggerhead Shrike *Lanius ludovicianus migrans* is a raptor-like migratory passerine found in south central Canada and the eastern United States. Since the 1970s, the decline of the Eastern Loggerhead Shrike has been precipitous throughout most of its range with breeding populations in Canada and the northeastern States becoming nearly extirpated (Pruitt 2000). The reasons for the decline of shrikes in North America remain unclear, but suspected factors include habitat loss/fragmentation, pesticides, mortality along roads, climatic effects, and the threat of introduced diseases such as West Nile Virus. Very little is known about wintering areas, migratory routes or general wintering ecology of this subspecies, which is when most mortality is suspected to occur.

Between 1993 and 1995, the number of breeding pairs in Canada decreased by 70%, from 100 pairs to 30 pairs. There are currently an estimated 25 breeding pairs in Canada. The majority of known remaining wild pairs are found in the Carden and Napanee areas of southern Ontario (Figure 1), with only a few pairs in Manitoba (K. DeSmet pers. comm. 2007)

Due to concern over the rapid decline and range contraction of this subspecies, in 1997 when the wild population had decreased to only 18 known pairs, a captive-breeding population was established. A total of 43 nestlings were collected from wild nests in 1997-98 and taken into captivity as original founders. By 2001, the captive population had increased to approximately 100 birds. With breeding possible in captivity, and so few wild pairs remaining in Canada, an experimental field propagation and release program was initiated in 2001 to develop techniques in the event that a full-scale reintroduction was required. Propagation of captive birds in field enclosures constructed in shrike habitat was considered a viable means of increasing productivity while at the same time producing healthy fledglings, raised in as natural an environment as possible for release. At present, the program is releasing captive-raised young at two locations in southern Ontario. One is an area where shrikes have been recently extirpated, and the other is an area with a small existing breeding wild population. The project is therefore considered a Re-establishment and/or Supplementation program (IUCN 1998).

The field propagation and release program is part of a multidisciplinary effort toward shrike recovery. Complementary programs include wild population monitoring, habitat stewardship and restoration, public awareness, landowner involvement, and scientific research. This article focuses on the techniques that have been developed for the field propagation and release of Eastern Loggerhead Shrikes in Ontario, and summarizes the results of the program since its inception in 2001. Also discussed are important milestones, productivity comparison of field propagated pairs to wild pairs, and the potential of field propagation and release for other reintroduction programs.

**Fig. 1.** *Map of Ontario, Canada showing locations of the field propagation and release sites (Carden, Dyer's Bay, and Smiths Falls), the two wintering facilities (Toronto Zoo and Ingersoll) and the remaining wild populations (Napanee and Carden).*

## METHODS

### Field sites

Field propagation and release enclosures are currently situated in two areas of southern Ontario: the Carden Alvar (hereafter referred to as Carden) and Dyer's Bay (Figure 1). The habitat at both sites consists of cattle grazed fields separated by patches of mixed-wood forest and native short grassland, similar to the habitat being used by the remaining wild breeding shrikes in Ontario. Sites were selected within the historical range of the shrike in areas where breeding pairs either currently exist or had only recently been locally extirpated, in accordance with the IUCN Guidelines for Reintroductions (IUCN 1998). The program began in the Smiths Falls area, with releases occurring there from 2001-2004. In 2003, the program was expanded to include a second site at Dyer's Bay. Most recently, in 2005, the enclosures at Smiths Falls were moved to a new site in Carden. This move was made because Carden contains one of the last remaining wild breeding populations and has significant areas of short grassland that are protected from development. Currently, there are 24 field propagation and release enclosures at the field sites: 14 at Carden and 10 at Dyer's Bay.

### Enclosures

Field enclosure design has evolved over the years, based on the observations and recommendations of program staff. The first two years of the program, 2001-2002, concentrated mainly on the design and construction of field enclosures and development of field propagation and release techniques. During this time, manipulations of enclosure size, design, orientation and placement were carried out to determine ideal conditions for reducing stress to the birds during daily management and for maximizing productivity of breeding pairs.

All enclosures at present consist of either two or three units (Figure 2). Each basic unit was constructed of galvanized welded wire mesh (1.27 by 2.54cm), cedar, and marine plywood. Two general enclosure sizes and designs have been used: 1) The original enclosure design with units that measure 2.44m wide by 2.44m high by 3.66m long, joined by a flight corridor of welded wire mesh measuring 0.6 to 0.9m long and; 2) Second generation designs with units that measure 2.44m wide by 3.08m high by either 3.66m or 4.88m long. These enclosures were designed without a flight corridor as the avicultural staff found the corridors problematic, both during enclosure construction and during daily management of the birds (Nichols and Woolaver 2002). Instead of a tunnel, the second generation enclosures have a shared wall between units, with a window that can be opened to allow access between enclosures. All original enclosures have since been modified to eliminate the flight corridor and join units by a shared window. The most important change to the second generation enclosures has been the increase in height to 3.08m. Avicultural staff noted that pairs in these taller enclosures appeared less stressed during feeding times, and the extra height allowed the birds to raise two broods within the same enclosure at the same time (Nichols and Woolaver 2002).

Of the 24 field propagation and release enclosures currently in use, six are the original design without the flight corridors (one with a height of 3.08m) and the remaining 18 enclosures are variations of second generation design. Four of the most recently built second generation enclosures have larger release doors and shared wall doors (1.22m by 1.22m) compared to smaller doors (ranging from 0.30m to 0.61m) on the older enclosures. For more detail on other minor changes that have been made to the enclosure design see Temple 2003, Nichols et al 2004, Nichols and Steiner 2005, and Steiner 2006a. Enclosures were built as individual pre-constructed panel sections that were easily transported to the field and then bolted together at the site. Enclosures were placed at least 30m from the forest edge and provided clear views of open fields. Poultry wire mesh was anchored to the bottom and along the ground outside to discourage mammalian predators. Enclosures were built around a live Hawthorn bush *Crataegus* sp. to provide nesting cover and thorns for the shrikes to impale their prey, as they would in the wild. Each unit was also furnished with ample food, water, perching, impaling branches, nesting material, nest cups, and cover to facilitate bonding and breeding.

Wintering enclosures have also evolved over this time. From 1997-2002, the majority of birds were wintered at the Toronto Zoo in small (0.91m wide x 1.52m high x 1.22m long) indoor cages, with only 9 birds having access to large indoor/outdoor flights. Birds in the large indoor/outdoor flights tended to be less stressed, having space to escape to during daily husbandry routines. It was also noted that birds win-

**Fig. 2.** *Second generation field propagation and release enclosures with two (A) and three units (B)*

tered in the small indoor enclosures had decreased muscle tone and poorer body condition, which became very evident when they were transferred to the large breeding enclosures, with some birds being unable to fly the length or height of these larger spaces. In 2003, the addition of a wintering facility in Ingersoll allowed 30 birds to be housed in large indoor/outdoor flights, measuring 1.22m wide, with an indoor portion of 3.05m high x 3.05m long, and an outdoor section of 2.44m high x 3.05m long (6.10m length total). At the Toronto Zoo, 3-4 of the small indoor cages were joined together to create larger enclosures for indoor birds. By winter 2005, the number of large indoor/outdoor flights at the zoo was doubled to 18 and an additional 17 indoor/outdoor enclosures were built at Ingersoll, providing all but 5 birds with large enclosures and access to outside.

**Daily Husbandry**

Food items consisted of live invertebrates (crickets and mealworms) in large feeding corrals and dead vertebrates (thawed mice) in small clean dishes. Birds were provided with fresh food and water twice daily. Feeding routine and amounts varied depending on stage of breeding, age, and number of birds in the enclosure. For a more detailed account of husbandry techniques see Experimental Field Propagation Husbandry Protocols for Eastern Loggerhead Shrike (WPC 2007a). During all management stages, disturbance to the pairs and their offspring was kept to an absolute minimum. Staff were careful to be as quiet and inconspicuous as possible when near the enclosures.

**Management of Breeding Pairs**

Currently the Eastern Loggerhead Shrike captive population in Ontario consists of 77 adults; 35 females and 42 males. From mid September through late April, these birds were kept at captive facilities at the Toronto Zoo and at a shrike-specific over-wintering facility in Ingersoll, Ontario (Figure 1). From late April through mid September, selected pairs were relocated to the field propagation enclosures at the two Ontario field sites. Surplus birds remained at the over-wintering facilities. Selection of individual birds for the field propagation and release program was determined by an Eastern Loggerhead Shrike Captive Breeding Advisory Group. The studbook data was maintained by John Carnio (Canadian Association of Zoos and Aquariums) and analyzed using a genetic program (Single Population Analysis and Records Keeping System, SPARKS) provided by the International Species Information System (ISIS). This program developed a breeding matrix utilizing mean kinship coefficients and inbreeding coefficients as parameters to determine the best pairings. Previous breeding history and experience were also considered in pairing recommendations, with most pairings consisting of at least one proven breeder, and successful pairs being kept together as long as possible. One of the goals of the captive program is to maintain 90% of the genetic diversity of the original founders for 25 years. At present, the captive population is retaining 97.1% of the genetic diversity of the original wild founders (Carnio 2007). Offspring from pairs considered of highest genetic priority to the captive population were kept back, while young from lower priority pairs were considered candidates for release.

Selected pairs were transferred from their respective over-wintering facilities to the field propagation enclosures at the end of April to early May, closely approximating the timing of returning wild shrikes. Females and males were initially placed in different units of the same enclosure and separated by a door of wire mesh. Once a pair was observed mutual feeding (generally the male passing a food item to the female through the mesh) or exhibiting other signs of courtship, the doors were opened and both birds were allowed access to the full enclosure.

Detailed daily observations (minimum 20 minutes) were carried out on each pair, from their arrival at the field site through fledging of their young. Once a pair was observed nest building, the daily observation time was increased to one hour, to determine lay dates and the onset of incubation. Nests were inspected after a minimum of 3 days of incubation to determine clutch sizes. Females were never flushed from the nest and inspections made only after a female had left the nest on her own. Once hatched, nests were inspected after a minimum of 3 days to determine brood size. Daily observations were continued for the remainder of the breeding season, to ensure that adults were feeding their young.

In cases where the pair had successfully fledged one brood and the female was sitting on a second nest, the older fledglings were separated into the adjacent unit before hatch of the second nest, or earlier if they were interfering with the incubating female. This allowed the male to continue feeding the female on the nest uninterrupted, and to feed the older fledglings through the wire mesh of the shared wall, if they were not yet fully independent.

## Management of Fledglings and Release Birds

Once young fledged, they were observed daily to monitor their development and to ensure that they were being fed and taught to hunt on their own by their parents. Fledglings were separated from their parents when 37-49 days old, to approximate wild behaviour and to encourage the young birds to feed on their own. In some cases, younger fledglings or young not yet observed feeding on their own were separated into the unit next to their parents until they were observed independently feeding, before being moved to a release enclosure. Since parents could still feed the young through the wire mesh of the shared wall, this allowed for ongoing parental care whilst young were encouraged to hunt for themselves. In most cases, once the young had been observed feeding on their own, they were moved to a predetermined release enclosure where they were grouped with other release young of approximately the same age. A release enclosure was simply one of the field propagation enclosures that had become available, i.e. had housed an unsuccessful breeding pair that had been removed, or a successful pair where young had already been separated from their parents and their parents removed. Adult birds were taken back to the over-wintering facilities.

On the day of transfer to a release enclosure, fledglings were banded following the Eastern Loggerhead Shrike Banding Protocol (WPC 2007b). Morphometrics were taken and the young were banded with a silver ID band on one leg and a colour band on the other leg. Feathers were also collected for an ongoing genetics doctoral study (Chabot 2006). For birds that were considered of highest genetic priority to the captive population and therefore to be retained, a few drops of blood were collected on an avian DNA sexing card so that sex could be ascertained. These young were kept in a separate enclosure at the field site until they could be transferred to an over-wintering facility.

## Pre-release

After banding, fledglings from different broods were placed within different units of a release enclosure, so they could later be released as larger groups of 6-12 birds. Releasing young in larger groups of mixed broods approximated wild behaviour wherein young from different nests travel together post-fledging (C. Grooms pers. comm. 2007). Initially, broods were kept in separate units for 1-3 days before the shared wall was opened and all young allowed access to the entire enclosure. Young were observed closely, before and after opening of the shared walls, to ensure there were no signs of aggressive behaviour between birds.

Release birds were then kept within the release enclosures for 7-10 days. On each of these days, mealworms were placed on a shelf just inside the release doors at the intended time of the release. This encouraged shrikes to feed near the release doors and habituated the birds to staff approaching the enclosure at this time of day. Pre-release training of young also included the provision of live mice so the fledglings developed strong hunting skills and the ability to capture and impale live vertebrate prey. Young birds received a minimum of three weeks of live-mouse training before release. This included a minimum of two weeks training with their parents, followed by one week in the release enclosure. Fledgling groups received 1-2 live "hopper" mice per group per day during the training period, in addition to their regular insect and thawed mice diet. Daily observations ensured that all young were feeding on their own prior to release.

## Release

Releases of strong, healthy young were undertaken when the fledglings were 44-66 days of age. Release enclosures were selected for their availability and proximity of natural cover to the release door. Field propagation enclosures were transformed into release enclosures prior to receiving any young by placing shelves under the inside and outside of the release door, and by adding branches to the outside of the cage to provide perching sites near the release door. Feeding corrals were placed within each unit.

Releases occurred after 7-10 days of pre-release training within the release enclosure, once it was confirmed that all young were hunting on their own. On release day, birds were carefully coaxed into the release unit and all food removed from the enclosure in the late morning. In the afternoon, three feeding corrals were placed 5-15m outside the enclosure, in front of the release door and proximate to available perching sites. Releases began at 1600h. Bowls containing mealworms were placed on the outside shelf, food was placed in the outdoor corrals, and the release door was opened. Observers watched the release door from a hidden observation point at least 60m away, and recorded detailed behavioural notes during the release. Between 1900h and 1930h (approximately dusk), food on the shelf was moved inside, and the door was closed if any birds remained inside. Remaining birds were released as early as practical the following morning. If all birds had left the cage before dusk, the door was left open and food left on the shelf. All movements by program staff around the enclosures and surrounding habitat post-release were careful and deliberate, so as not to scare birds from the release site before they were ready to leave on their own.

Providing the newly released birds with supplemental food was an extremely important component of the release. The full diet was placed in the outside corrals for at least a week post-release, continuing with both morning and afternoon feeds. After a week post-release, the amount of supplemental food was modified according to how many shrikes were still in the area, and was gradually decreased over time. Staff and volunteers monitored the shrikes daily, until the birds had left the release site. Volunteers and staff continued to survey the surrounding area adjacent to the field sites until birds were no longer seen.

## RESULTS

### Observations

Daily observations were recorded using a behavioural ethogram (Chabot 2002a). Although this information has yet to be statistically analyzed, several noteworthy behavioural trends have been observed since 2001. Frequent daily activities involved natural shrike behaviour such as scanning fields for predators, alarm calling, perching and scanning, and hunting for live prey. The wire mesh of the enclosure allowed entry of wild prey and shrikes were frequently observed catching wild invertebrates in the air and on the ground. Shrikes were also able to catch wild vertebrate prey, including jumping mice (Zapus), snakes (Thamnophis and Diadophis), and frogs (Rana) that had entered the enclosures. Fledglings were observed exhibiting anti-predator behaviour immediately in response to parental alarm calls by flying under cover and remaining completely still. Young were also frequently observed intently watching and then mimicking parents as they hunted and impaled prey.

### Pairings and Productivity

Since 2001, the program has expanded with techniques and enclosure designs developed to maximize productivity of breeding pairs. The number of pairings (including re-matches) has increased substantially from three in 2001 to 33 in 2006 (Table 1). From 2001-2003 pairings were based solely on maximizing genetic diversity. In 2005 and 2006, breeding history and experience were also considered in pairing recommendations. Wintering enclosures have also been greatly improved over this time, with a shift towards housing the majority of birds in large flights with outdoor access rather than small indoor cages.

The number of fledglings produced each season has increased greatly over the years with the addition of larger wintering and breeding enclosures of improved design and with new management of pairings (Table 1).

The number of pairs double brooding has also increased significantly since 2001 (Table 1). Of the 18 pairs that have successfully fledged two broods since 2001, 16 were in the second generation enclosures (3.08m tall with a shared wall). Although it was initially suspected that 3-unit enclosures would be needed for double brooding to occur, pairs will also readily double brood in 2-unit enclosures, provided units were 3.08m tall.

### Productivity of Field Propagated Pairs Compared to Wild Pairs

Detailed monitoring of wild breeding pair productivity was also carried out each season. Data from the wild population in Carden and data from the field propagation program are presented below (Table 2). This comparison suggests that productivity of the field propagated pairs was similar to their wild counterparts, with more fledglings produced per nest by field propagated pairs in 2005 and 2006 than in the wild. Double brooding in particular, was much more likely to occur within the field enclosures.

### Releases

A total of 221 young shrikes have been released in Ontario since 2001. The number of young released each season has increased substantially from 10 in 2001 to 111 in 2006, with more fledglings released in Ontario in 2006 than all previous years combined (Table 1). All of the fledglings produced have been parent raised and released from the field sites. In 2004 and 2005, a few pairs remaining at the over-wintering facilities also bred, and their young were translocated to the field sites after fledging and released with similar aged young (see Table 1).

Releases occurred in the late afternoon, providing enough daylight for the shrikes to leave the enclosure and orient themselves before roosting nearby. The approaching dusk discouraged shrikes from taking long exploratory flights away from the release area.

### Table 1

**PRODUCTIVITY OF FIELD PROPAGATED PAIRS AND NUMBER OF FLEDGLINGS RELEASED TO THE WILD BETWEEN 2001 AND 2006.**

| Year | Number of original pairings / re-matches* | Number of pairs to fledge young | Number of pairs to double-brood | Number of fledglings produced | Number of fledglings released |
|---|---|---|---|---|---|
| 2001 | 3 | 3 | 0 | 10 | 10 |
| 2002 | 6 | 5 | 1 | 19 | 14 |
| 2003 | 20 / 7 | 3 | 0 | 4 | 0 |
| 2004 | 20 / 5 | 9 | 2 | 34 | 26  (6)** |
| 2005 | 20 / 13 | 9 | 5 | 53 | 49  (5)** |
| 2006 | 24 / 9 | 18 | 10 | 129 | 111 |

* if pairs were not exhibiting breeding signs or were unsuccessful breeders they were re-paired according to pairing recommendations
** additional young that were produced by pairs at the Ingersoll over-wintering facility and released at the field propagation and release sites.

All shrikes left the enclosure on their own and were never flushed out. Once released, the birds generally flew directly to nearby trees or the feeding corrals to feed. Some birds returned to the release doors to feed on mealworms, while a few were observed flying back into the release enclosure. A few birds were even seen "hawking" wild insects shortly after leaving the release cage. Shrikes were observed interacting and flying together within hours of the release and making short flights of less than 50m, before returning to the immediate area of the enclosure. Generally birds would roost just before dark within 20-50m of the enclosure.

For the majority of the releases, birds stayed around the release site area for several days post-release, but there was considerable variation with some birds gone by the next morning and some staying for several weeks. The shrikes initially relied heavily on the feeding corrals, with some birds feeding exclusively on the supplemental food for the first day or two. Typically the shrikes were observed hunting wild prey on their own within two days of being released. Newly released shrikes were often observed improving their flight and hunting skills by aggressively chasing any other nearby passerines and by catching wild insects on the wing. They also exhibited natural anti-predator behaviour by hiding under cover when predators such as hawks (*Accipiter* sp. and *Buteo* sp.) were in the area. In most cases, newly released birds stayed together as a group while perching in nearby trees, feeding from corrals, or chasing other passerines away from the release site.

### Sightings and Returns

The program has had seven separate sightings of release birds, including three (and potentially 4) birds that have returned to the breeding grounds the following season to produce young of their own in the wild. In 2004, a shrike released from Dyer's Bay was captured at Long Point Banding Observatory c.350kms south east of the release site, a month after it had been released. This bird was captured in a mist net and dropped a frog in the net when captured (J. McCracken pers. comm. 2006). In 2006, another young bird released in Dyer's Bay was observed hunting near Long Point at least 11 days after release (C. Wood pers. comm. 2006). Most recently, in March 2007, a 2006 release bird was spotted in West Union, Ohio, and remained there for almost a week (P. Whan pers. comm. 2007). This Ohio sighting was the first winter band recovery for this subspecies since banding began in 1999, with over 1000 shrikes banded to date.

Most encouraging for the future recovery of the

Eastern Loggerhead Shrike has been the return and integration of two female and one male release birds to the wild population in Carden. The first female had been released in Dyer's Bay in August 2004 and returned to the Carden area in late May of 2005. She paired with a wild male and successfully fledged five young in June, producing one of the highest wild broods of the season (Nichols & Steiner 2005). In 2006, a second female, released in 2005, was found tending a nest with a wild male in late May in Carden. She was later observed feeding a minimum of three fledglings (Steiner 2006a). In 2007, a male bird released in Carden in August 2006 was observed taking food to a brooding female in late May. Also in 2007, a 2006 release bird (sex undetermined) was observed around the field breeding enclosures at the Dyer's Bay site. At the time of writing, this bird has remained in the area for over two weeks but remains unpaired. Shrikes have not been confirmed in Grey/Bruce counties (which include Dyer's Bay) since 2002. The return of these birds to the breeding grounds translates into a 1.8% return rate, which is comparable to rates of natal philopatry seen in wild populations of migrant songbirds (J. McCracken pers. comm. 2007).

## DISCUSSION

A significant step in the development of the project has been the modification of field enclosure design from season to season. The addition of only 0.64m to the height and 1.22m to the length of each unit of the field propagation enclosures resulted in a substantial increase in the likelihood of double-brooding. The extra space and shared walls between units allowed the male to care for the first group of fledglings while the female produced a second brood. The flexibility of being able to modify enclosure design and management techniques as soon as challenges were encountered led to continuous improvement in methodology and a resulting increase in number of young produced each season.

Winter housing has also improved over the course of the program, with the majority of birds now being wintered in large flights with outdoor access. The extra space ensures that muscle tone and flight skills are maintained, while outdoor access may provide more natural synchronization to the seasons.

The number of fledglings produced has increased each year with the exception of the 2003 season, when productivity was extremely low. The reasons for this are not fully known, however 3 separate inoculations for West Nile Virus (WNV) were given to each bird very close to the onset of the breeding season. Since there were no other major changes to field propagation protocols in 2003, it is suspected that either the inoculation itself, and/or the stress of repeated catching-up and handling associated with the process, had a nega-

### Table 2
**PRODUCTIVITY OF WILD PAIRS IN THE CARDEN AREA AND FIELD PROPAGATED PAIRS FROM 2004 THROUGH 2006**

| Year | Mean clutch size (range) | | Mean brood size (range) | | Mean number of fledglings (range) | | Proportion of pairs to double brood | |
|---|---|---|---|---|---|---|---|---|
| | Wild | Captive | Wild | Captive | Wild | Captive | Wild | Captive |
| 2004 | 5.4 (3-7) n = 14 | 5.2 (4-7) n = 4 | 5.0 (3-6) n = 14 | 4.0 (3-5) n = 3 | 4.9 (3-6) n = 14 | 3.3 (1-5) n = 11 | 1/14 | 2/25 |
| 2005 | 5.2 (1-7) n = 14 | 5.3 (3-7) n = 12 | 4.9 (2-6) n = 14 | 4.3 (2-6) n = 13 | 3.0 (1-6) n = 14 | 4.2 (2-6) n = 13 | 1/14 | 5/32 |
| 2006 | N/A | 5.3 (2-7) n = 28 | N/A | 4.6 (2-7) n = 27 | 3.0 (3-4) n = 7 | 4.6 (2-7) n = 27 | 1/7 | 10/33 |

*Not all captive nests were approached to collect clutch and brood numbers in 2004. Wild nests were not approached to collect clutch or brood numbers in 2006.*

tive effect on breeding activity.

Release techniques were modified from successful reintroductions of Mauritius Pink Pigeons *Columba mayeri* and Echo Parakeets *Psittacula echo* (Jones et al 1992; Woolaver et al 2000). As a soft-release, it was critical that the release birds were provided with a recognizable food source while they learned to rely entirely on wild food. The young birds relied on the invertebrates in the corrals during the first week post-release until they adapted to hunting exclusively wild prey. Supplemental feeding has been a critical component of the San Clemente Loggerhead Shrike *L. l. mearnsi* reintroduction and is particularly important during the first week post-release (D. Brubaker pers. comm. 2007). It was also important that the enclosure remained a safe and positive location for the shrikes. Human activity around each enclosure was minimal and shrikes were never forced to leave the enclosure during the release but were allowed to explore on their own terms while learning to avoid predators, fly long distances and hunt outside of the enclosure. Pre-release training habituated the young to the approach of field staff to the release door in the early evening so that birds were not stressed at the time of the release.

There were many advantages to field propagation:

- Field enclosures were flexible and relatively easy to transport and set up in field conditions, allowing breeding to occur in the shrike's natural habitat.
- The mesh design of the enclosures allowed captive birds to hunt both invertebrate and vertebrate wild prey. This allowed development of natural behaviours and provided a nutritious, balanced and varied diet closely approximating that of wild shrikes.
- Due to the much higher likelihood of double brooding in field enclosures (30% of pairs in 2006), and the elimination of depredation risk, overall productivity of captive pairs was substantially higher than their wild counterparts.
- The adult shrikes raised their young to fledgling, eliminating any need for hand-rearing staff. This produced young birds for release with no possibility of imprinting or behavioural modifications, and also saved significant costs by reducing project salaries.
- Since young shrikes were raised in a natural setting by their own parents they learned critical survival skills such as predator avoidance and hunting. Due to the large size of the enclosures they developed strong flight skills *in-situ*. Release birds were extremely fit, and able to hunt for prey and avoid predators on their own upon release. Birds were exhibiting natural shrike behaviours, such as hunting and impaling wild prey, within the enclosures and immediately upon release.
- Young shrikes could be released as a group, which worked extremely well. The shrikes explored the release site, chased one another and other passerines that attempted to perch near the enclosure, and fed from the corrals and on wild prey together. This helped improve their flight skills and younger birds were able to learn skills from older ones.

**Challenges for a Migratory Passerine**

Carrying out releases for a migratory passerine poses the additional and significant challenge of extremely high mortality during migration, particularly during the first year. Band returns suggest that migratory passerine return rates are 1-10% for juveniles (J. McCracken pers. comm. 2007). Unlike successful releases of captive raised birds on islands, once release birds had left the release area we were unable to provide additional support in the form of supplemental food or predator control. To counter this high mortality, releases of any migratory passerine must be carried out in substantial numbers, just as wild productivity is high with wild migratory passerines, to ensure that some of the birds return the following spring.

As technology improves, shrikes fitted with radio-transmitters will allow monitoring of future releases to provide information on post-release survival, dispersion and migration of release birds after they have left the release site. A monitoring program post-release is a major recommendation of the IUCN's Reintroduction Guidelines (IUCN 1998) that is currently not being adequately satisfied by this release program. A radio-tracking study is planned for the 2007 season. Trials with dummy transmitters on captive shrikes in 2006/07 have had positive results, with no negative physical or behavioural effects on trial birds of various ages (Steiner 2006b).

The sightings and returns of release birds represent important milestones for the recovery of this subspecies in Ontario. The return rate for released shrikes was similar to that recorded in other small passerines (J. McCracken pers. comm. 2007), further evidence that the use of field propagation enclosures produced strong individual birds for release. The return and integration of field propagated shrikes to the wild breeding population is significant and establishes the use of field propagation and release enclosures as a viable, cost-effective component of a multi-disciplinary recovery effort for the Eastern Loggerhead Shrike.

**Potential for Other Re-introduction Projects**

Field propagation and release techniques similar to the ones currently being developed for the Eastern Loggerhead Shrike may be preferable to *ex-situ* captive breeding for many species, particularly those that are difficult or expensive to breed in large captive facilities. It may also be preferable for species that are poor candidates for translocation. Field propagation and release in this case was extremely straightforward and cost-effective, and produced skilled young birds that had been raised by their parents for the full breeding episode in their natural environment and were acclimatized to the release site. Parent-raising of young in field propagation enclosures *in-situ* and releasing the fledglings directly to the natal territory should be considered a viable alternative to *ex-situ* captive breeding, hand-rearing and/or translocations.

## Acknowledgements

We thank all the partners of the experimental field propagation and release effort, including the Avian Science and Conservation Centre at McGill University, Bird Studies Canada, Canadian Cattleman's Association, Canadian Wildlife Service, Canadian Association of Zoos and Aquariums, the Couchiching Conservancy, Bill Dobson and Linda Hynes, Bill and Barb McNair, the Ontario Ministry of Natural Resources, Ross Snider, Neil and Carolyn Turnbull, and the Toronto Zoo. Individuals that have provided significant contributions to the field propagation and release program include Robert Wenting, Amy Chabot, Tom Mason, Jon McCracken, Letitia McRitchie, Pete and Sue Read, Ian Ritchie, Murray Smith, and Merilee Temple. The field interns and volunteers, although too numerous to mention individually, have been the backbone of the project and their effort and enthusiasm deserve our gratitude.

## References

CARNIO J. 2007. Eastern Loggerhead Shrike Studbook Report 2006/2007. Unpublished Report to Eastern Loggerhead Shrike Recovery Team, Ken Tuininga (Canadian Wildlife Service - Ontario), Chair. 7 pp.

CHABOT A. 2002a. Fieldwork guidelines and protocols for the Eastern Loggerhead Shrike Recovery Team. Unpublished report to Eastern Loggerhead Shrike Recovery Team, Robert Wenting (Canadian Wildlife Service - ON Region), Chair. 45 pp.

CHABOT A. 2002b. Summary report on the experimental field propagation and release of Eastern loggerhead shrikes in 2001. Canadian Wildlife Service. Ottawa, Ontario. 15 pp.

CHABOT A. 2006. Demography, Migration and Population Genetics of the Loggerhead Shrike (*Lanius ludovicianus*) in eastern North America. PhD thesis, Queen's University, Kingston, Ontario. In progress.

IUCN. 1998. Guidelines for Reintroductions. Prepared by the IUCN/SSC Reintroduction Specialist group. IUCN, Switzerland and Cambridge, UK. 10 pp.

JONES CG, KJ Swinnerton, CJ Taylor & Y Mungroo. 1992. The release of captive-bred Pink pigeons *Columba mayeri* in native forest in Mauritius. A progress report July 1987-June 1992. Dodo 28: 92-125.

NICHOLS R, M Temple, & A Jones. 2005. Eastern Loggerhead Shrike in Ontario End of Season Report 2004: Field Monitoring and Experimental Captive Breeding and Release Programs. Unpublished report to Canadian Wildlife Service, Ontario Region. 78 pp.

NICHOLS R & J Steiner. 2006. Eastern Loggerhead Shrike in Ontario End of Season Report 2005: Field Monitoring and Experimental Captive Breeding and Release Programs. Unpublished report to Canadian Wildlife Service, Ontario Region. 162 pp.

PRUITT L. 2000. Loggerhead shrike status assessment. U.S. Fish and Wildlife Service. Bloomington, Indiana, U.S.A. 169 pp.

STEINER J. 2006a. Eastern Loggerhead Shrike in Ontario End of Season Report 2006: Field Monitoring and Experimental Captive Breeding and Release Programs. Unpublished report to Canadian Wildlife Service, Ontario Region. 179 pp.

STEINER J. 2006b. A trial of "dummy" radiotransmitters on juvenile captive-born eastern loggerhead shrike. Wildlife Preservation Canada, Guelph, Ontario, 13 pp.

TEMPLE M. 2003. Eastern Loggerhead Shrike in Ontario End of Season Report 2003: Field Monitoring and Experimental Captive Breeding and Release Programs. Unpublished report to Canadian Wildlife Service, Ontario Region. 90 pp.

WOOLAVER L & R Nichols. 2002. Captive field propagation and experimental release of Eastern loggerhead shrike in Ontario, 2002. Wildlife Preservation Trust Canada, Guelph, Ontario. 38 pp.

WOOLAVER L, CG Jones, K Swinnerton, K Murray, A Lalinde, D Birch, F de Ravel & E Ridgway. 2000. The release of captive bred Echo parakeets to the wild, Mauritius. Reintroduction News 19: 12-15.

WPC. 2007a. Experimental Field Propagation Husbandry Protocols for Eastern Loggerhead Shrike. Prepared for Ontario Eastern Loggerhead Shrike Recovery Program by Wildlife Preservation Canada. 45 pp.

WPC. 2007b. Eastern Loggerhead Shrike Banding Protocol. Prepared for Ontario Eastern Loggerhead Shrike Recovery Program by Wildlife Preservation Canada. 19 pp.

## Author Biography

Jessica Steiner holds a Bachelor of Science (Zoology) from the University of Guelph (Ontario, Canada) and a Masters of Applied Science in Wildlife Health and Population Management from the University of Sydney (Sydney, Australia). She has previously worked with various aspects of recovery programs for several endangered and threatened species, including the Puerto Rican Crested Toad, Black-Footed Ferret, and Eastern Missassauga Rattlesnake. In 2005, she was the recipient of Wildlife Preservation Canada's New Noah scholarship, which provides practical training and field experience in managing and conserving endangered species to young Canadian biologists. Through this opportunity, she received extensive training in endangered species management at the International Training Centre of the Durrell Wildlife Conservation Trust (Jersey, British Channel Islands), followed by intensive field work in Mauritius working with some of the world's most critically endangered passerines, the Mauritius Fody and Mauritius Olive White-Eye. Upon her return to Canada she quickly became heavily involved with the Eastern Loggerhead Shrike Recovery Program. In 2006, Jessica monitored the wild population of shrikes in the Carden Alvar and managed the captive-breeding and release program in that area. This year, Jessica co-ordinated recovery efforts for the eastern loggerhead shrike across Ontario, which include captive-breeding and release, wild population monitoring, habitat stewardship and restoration, public awareness and education, and landowner involvement.

# SPECIES RECOVERY OF THE MAURITIUS FODY *FOUDIA RUBRA*: A MULTI-FACETED APPROACH TO SPECIES CONSERVATION USING AVICULTURAL TECHNIQUES AS KEY COMPONENTS OF RECOVERY

Andrew Cristinacce[1,2], Richard Switzer[1,6§], Amanda Ladkoo[1], Lara Jordan[3], Andrew Owen[4], Markus Handschuh[4], Vanessa Vencatasawmy[5], Frederique de Ravel Koenig[1], Carl Jones[1,4], Diana Bell[2] & Roger Wilkinson[3]

§ Speaker
[1] Mauritian Wildlife Foundation, Grannum Road, Vacoas, Mauritius
[2] University of East Anglia, Norwich, NR4 7TJ, UK
[3] Chester Zoo / North of England Zoological Society, Caughall Rd, Upton, Chester, CH2 1LH, UK
[4] Durrell Wildlife Conservation Trust, Les Augrès Manor La Profonde Rue, Trinity, Jersey JE3 5BP, English Channel Islands, British Isles
[5] National Parks and Conservation Service, Mauritius
[6] Al Wabra Wildlife Preservation, P.O. Box 44069, Doha, Qatar

## Summary

The Mauritius Fody *Foudia rubra* is an endangered passerine, endemic to Mauritius. The recovery program for the Mauritius Fody has involved a multi-faceted approach to conservation, in which both wild management and avicultural techniques have been key components. An *in-situ* captive breeding program was initially established as an experimental strategy, involving the rescue and harvest of eggs and chicks from wild nests as founders. Subsequent nest harvests and rescues provided birds for release on a predator-free offshore island. The process of rescuing eggs and chicks, artificially incubating and raising the new recruits, and the subsequent release and monitoring of the birds on the offshore island was a collaborative effort involving Mauritian staff and aviculturists from British zoos. Fieldworkers also played key roles in the monitoring of nests in the source population and the establishment of the new population post-release. Since the species recovery effort began in earnest in 2002, the population of the Mauritius Fody has increased by an estimated 60%. The success of the program can be attributed to multiple management techniques, a flexible interface between the wild and captive populations and the involvement of aviculturists who brought valuable expertise for hands-on management of birds both in captivity and the wild.

## Introduction

The genus *Foudia* represents a group of sparrow-like birds that have radiated throughout the islands of the Indian Ocean. The Mauritius Fody is found in native and exotic forests, forest edges and heathland, where it feeds on both insects and nectar. The Mauritius Fody is particularly charismatic, bold and inquisitive; but it is not until the breeding season that the Fody lets its true colours show. At the start of the breeding season in September, the male Fody develops an outrageously scarlet head and neckline. Excited "plicks" and splutters accompany nest-building activities, resulting in a dome-shaped nest with entrance tunnel - similar to the nests of many estrildids, and slightly resembling the nests of their close relatives, the weavers of the African mainland.

Unfortunately, it is possible that the Fody's extravagant nesting behaviour contributes to its downfall. Like many inhabitants of small oceanic islands, the Fody displays great naivety towards introduced predators, in this case rats (particularly the arboreal *Rattus rattus* and the Crab-eating Macaque *Macaca fascicularis*). Jones (1996) states that at least 16 species of Mauritian native birds, 11 of which were endemic, have been extirpated from Mauritius since 1600. These extinctions are the result of widespread habitat degradation and the introduction of mammalian predators, and partly as a consequence of the high level of endemism that occurs on this oceanic island. Only nine species of native forest bird remain; several are threatened with extinction, some critically.

Nesting failure, due to predation from exotic mammals, is certainly the major causative factor in the Mauritius Fody's decline. Habitat destruction is a major factor in the decline of many Mauritian species, but ironically native habitat forest does not hold the key to the Fody's survival. In the face of exotic nest predation, a small patch of forestry plantation may be enabling the species to hang on a by a thread. Studies by Safford (1997b) showed that nesting success in the grove of Japanese Red Cedar *Cryptomeria japonica* is 46%. This contrasts starkly with nesting success in native trees, Pine, Eucalyptus and other exotic trees, which is only 4%. Recent data by Cristinacce et al. (in prep b.), shows an increased dependence upon Pine plantations for nest sites, although nest predation still appears to be high in these areas.

This phenomenon of exotic vegetation providing a refuge from exotic predators has been reported elsewhere. On the New Zealand mainland, the Mahoenui Giant Weta, an immense orthopteran (Deinacrida sp), has survived in the prickly exotic Gorse *Ulex europeaus*, which provides good cover and protection from potential predators (Sherley & Hayes 1993). In the early 1990s, the last nine wild Pink Pigeons *Columba mayeri* survived in a grove of exotic Cryptomeria, known as Pigeon Wood; it is exactly the same patch of Cryptomeria that offers the Fody its last stronghold. It is speculated that the lack of food items in the canopy of Cryptomeria, combined with the dense, prickly foliage, limit the optimistic foraging activity of potential nest predators.

Ongoing monitoring of the Fody population has shown that the population is stable (Cheke 1983; Safford 1997c & Nichols 2000), with the 2002-2003 census (Cristinacce et al in prep; Switzer et al 2003), estimating the population at 93-116 pairs. However, it is hypothesized that the 12-15 pairs living in the Cryptomeria grove acted as a source population, with youngsters dispersing to other areas which act as a sink population due to low recruitment. Crucially, the Fody's range was shrinking, and it was restricted to a small upland area of mostly exotic forest at the southern edge of the Black River Gorges National Park. Furthermore, the population is not significantly divided into sub-populations, and more than 90% of adult breeding birds occur in only one population unit.

Therefore, in 2002, the species was listed as critically endangered (50% chance of going extinct in 5 years). It triggered IUCN criteria B1a + bi, ii, iii, iv (but not v) and C2aii.

## The Decline of the Mauritius Fody - A Historical Perspective

Serious interest in the status and conservation of Mauritian passerines began with the Mascarene expedition of the British Ornithologist's Union in 1973. This expedition coincided with a forestry project between 1972-74, which destroyed a considerable proportion (28km$^2$) of the Fody's habitat. It has been estimated that this single forestry project at Les Mares was responsible for the loss of 200 pairs - over half of the pre-1972 population (Cheke 1983). As a follow-up to this initial study, Anthony Cheke investigated the ecology and status of the Mauritius Fody and other species in 1974-75. Census data from 1974-75 estimated the Mauritius Fody population as 247-260 pairs (Cheke 1983).

Roger Safford carried out a 4-year study between 1989-1983, and resulted in a PhD thesis (Safford 1994) and subsequent published papers (Safford 1997a, b, c, d & e). This study had a particular focus on the distribution, status and ecology and nesting success of the Fody. Safford mapped a total of 90 Mauritius Fody territories and the total population was extrapolated to 104-120 pairs (Safford 1994, 1997a).

Since then, the Mauritian Wildlife Foundation (MWF) had monitored the populations of the Fody, Olive White-eye *Zosterops chloronothos* and other passerines as a minor component of the Fauna Program, which has predominantly focused on intensive efforts to restore the Mauritius Kestrel *Falco punctatus*, Pink Pigeon and Echo Parakeet *Psittacula eques echo*, as well as the recovery of Rodrigues and Round Island fauna. More recently, studies had focused on the breeding ecology of native Mauritian passerines (Nichols 1999) and wild population monitoring of the Fody (Nichols et al 2000). The combination of all these studies had heightened the concern for the status of the Fody and other passerines, increasing their significance as priorities for species recovery programs. In 2001, the Mauritius Fody was identified as the first passerine warranting a consolidated species recovery effort.

## The Development of a Mauritius Fody Recovery Program

Of the three most threatened native Mauritian passerines, the Fody had been identified as the top priority for experimental management trials. Certainly, at the time more was known about the ecology, distribution and threats of the Fody than was known of the Olive

White-eye and Mascarene Paradise Flycatcher *Terpsiphone bourbonnensis desolata*. This certainly does not mean that the other two species were overlooked - indeed, their population was continually monitored, whilst scientific studies of ecology and distribution were undertaken.

Paradoxically, with a population between 200-250 birds, by Mauritian standards the Fody was still relatively abundant, and the population had by no means fallen to the low levels that the Kestrel, Pink Pigeon and Echo Parakeet had once sunk. This allowed MWF the opportunity to fine-tune management techniques, in the knowledge that the survival of the species did not hang on the breeding success of one or two key individuals. Additionally, although conservation management of the Fody was vitally important, it was anticipated that it would not necessitate the same intensity in the efforts to restore it.

## Goals of the Mauritius Fody Recovery Program

Simply, the greatest threat facing the Mauritius Fody is nest-predation by exotic mammals (e.g. *Rattus rattus* and *Macaca fascicularis*) (Safford 1997b). Overcoming this threat would involve either:

1. The total eradication of exotic predators over a large enough area to support a sustainable population of Fodies.

or

2. The provision of habitat to benefit the nesting success of the Fody in the face of exotic predators.

In order to target the ultimate threats to the Mauritius Fody, the most obvious strategy would have been to eliminate exotic predators from a sufficient area to support a sustainable population of Fodies. In reality this was an impossible task. Although representatives of New Zealand's Department of Conservation are becoming progressively more ambitious in the area of islands on which they attempt to eliminate rats and other predators, Mauritius has a dense population of 1.2 million, precluding the use of such methods.

Another possible tactic may have involved the establishment of a 'super-fence' to enclose a key patch of habitat, and prevent the passage of all mammalian predators. However, under the circumstances at the time, the development and management of such a 'super-fence' was simply unfeasible.

Unfortunately experimental measures to limit nest predation of wild nests in sink habitats, by a variety of strategies, had met with mixed results during the 2001-2002 season. The mixed results of these, combined with the high labour-intensity of the work, prompted us to focus on other options.

Without doubt, the top priority in the conservation of the Fody was the creation of habitat that would provide Fodies with high nesting success in the face of alien predators. Ironically, although the restoration of native forest is crucial to the recovery of many Mauritian species, it is not the solution to the Fody's problems. Cryptomeria has proven to benefit Fody nesting success, and a planting program for Cryptomeria (or a similar species such as Araucaria) was considered as a strategy to provide the necessary habitat. (Safford & Jones 1998; Carter & Bright 2002).

However, the deliberate planting of an exotic Cryptomeria plantation was controversial as a conservation measure. Subsequent data (Cristinacce et al. in prep b.) has shown that Cryptomeria has little benefit over Pine (35% compared to 34%), and Pine was already abundant in the surrounding area. Certainly, whatever strategy selected to improve the quality of habitat for Fody nesting success was likely to have a timeline of many decades.

Meanwhile it was essential to address proximate goals, so that the Fody population would be secure in the in the interim, albeit with management. Therefore The Mauritius Fody recovery program aimed to focus on increasing the population to a level which:

1. Is sustainable and genetically viable.
2. Would not require prolonged intensive management.

## Predator-free Islands

Some of the world's most successful bird recovery programs have focused on establishing populations on predator-free islands, including in New Zealand and Seychelles. Similarly, it was considered that the future survival of the Mauritius Fody might lie mid-term in the establishment of sub-populations on predator-free islands.

Île aux Aigrettes (IAA) is a small island (27 hectares) lying in Mahébourg Bay, approximately 0.8km off the south-eastern coast of Mauritius. Most importantly, IAA is an isolated (and therefore manageable) site, where alien mammalian predators have been eliminated and native lowland plants are once again regenerating. IAA has already proven a successful site for the establishment of a sub-population of the Pink Pigeon.

It was proposed to establish a sub-population of Fodies on IAA, as a major mid-term goal of MWF's recovery program for the species, beginning in 2003-2004. Unlike mainland Mauritius, IAA lacks mammalian predators of Fodies, thereby presenting the potential of a predator-free environment for the Fody to exist and reproduce. As a model for the restoration program, a closely-related species, the Seychelles Fody *Foudia sechellarum*, had been restored and/or translocated to predator-free islands, with great success. In particular, Cousin, an island of a similar size to IAA (27 hectares), supports very high numbers - approximately 1,000 (L. Wagner pers. comm.). Whilst it was acknowledged that there are significant differences between the Seychellois and Mauritian species, and it was uncertain how Mauritius Fodies would adapt to offshore island life, IAA surely possessed enormous potential.

Although ecological studies on IAA helped to predict how Mauritius Fodies might have adapted to the habitat on IAA, it would have been only possible to assess the suitability of IAA for Fodies by translocating an experimental population. Subsequent studies of Fody survivability, feeding and breeding ecology on Île aux Aigrettes in comparison to the uplands would help to assess how translocated Fodies would have fared.

During the 2003-2004 season, the release of a cohort of Fodies on IAA was intended on an experimental basis, to assess the survivability of the released Fodies and to attempt to develop translocation and release protocols. Initially a soft release was implemented, with subsequent post-release support and monitoring.

If IAA were to prove successful as a reintroduction site, the following major benefits would be achieved:

1. The population of Mauritius Fodies would increase.
2. The IAA population would be located in a predator-free refuge, when nest predation has been identified as the major limiting factor in the population's abundance.
3. Two separate sub-populations would reduce the impact of stochastic events or disease on the survival of the species.

Additionally, the release of birds on IAA presented many new and exciting major opportunities for study:

1. The release population would be ringed. This is the first time that an opportunity had arisen for a self-contained sub-population of Mauritius Fodies to be ringed in its entirety. This would provide details on life history (survival, longevity, age at maturity and first breeding), territoriality and mate loyalty. Subsequent generations could be ringed, providing the foundation of a detailed long-term study.
2. Research on breeding ecology and nesting success in predator-free conditions. For the first time in centuries, a sub-population of the Mauritius Fody would exist in the absence of exotic predators. All previous studies of breeding ecology and nesting success have been in the presence of exotic predators.
3. Research on the species in lowland coastal forest. Comparisons could be made with the population in the uplands.
4. Genetic studies of a newly created, ringed sub-population.

**Captive Management**

For threatened avifauna, techniques have included the harvesting of eggs and nestlings from the wild for rearing in captivity to increase the productivity of wild populations and provide individuals for reintroduction programs (Cade & Temple 1995; Jones et al 1995,; Jones 2004). Many of these programs have focused on birds of prey (Seddon et al 2005) and the *in-situ* rearing of endangered passerines seems to have been restricted to Hawaii (Kuehler et al 2000). Translocation of adult passerines onto islands has been successfully used in the Seychelles (Komdeur 1994, 1996) and New Zealand (Armstrong et al 2002; Dimond & Armstrong 2007; Armstrong & Ewen 2002).

With the goal of establishing a new population on IAA, captive-breeding was seen as the last resort. However, as with all "safety nets", it was prudent to ensure that this would be dependent upon fail-safe techniques (Snyder et al 1996). At this point, the Mauritius Fody had never reproduced successfully in captivity. Consequently, during the 2002-2003 season we focused on the experimental harvest of a number of chicks from wild nests in order to establish a captive population. This experience with the management of captive passerines would enable us to establish hand-rearing and captive husbandry protocols.

Initially the techniques attempted in the species recovery of the Fody were employed on an experimental basis. Prior to the start of the program, it was not possible to predict for certain which techniques would be most successful. The establishment of a captive population was to present opportunities for developing incubation, clutch- and brood-manipulation and veterinary techniques. Importantly, these techniques have great potential for management of the species in the wild - the successes of many bird conservation programs throughout the world have been dependent upon the application of such practices.

By maintaining a captive population in the Gerald Durrell Endemic Wildlife Sanctuary (GDEWS), it would also provide the first opportunity to study the impact of disease in the lowlands. In the Hawaiian Islands avian disease, particularly avian malaria, has decimated populations of native forest birds in the lowlands. This was a perfect opportunity to determine the effect of any transmissible pathogens present in local Mauritian avifauna that may have precluded the release of Mauritius Fodies in the lowlands.

Naturally, it was hoped that the captive population would reproduce successfully. Detailed observation of reproductive behaviors in a captive environment had the potential to improve our understanding of the reproductive effort of the species and the threats which impact it in the wild. Finally, it was hoped that the captive population, if successful, could even contribute offspring for release on IAA, minimizing the impact of removing new recruits from the wild mainland population. However, this was not the primary goal of the species recovery effort, and a full-scale captive breeding program was not the initial intention.

## METHODS

*(Methods adapted from Cristinacce et al. (in prep a))*

### Nest Surveys

Nests of Mauritius Fodies were located in upland southwest Mauritius by continuously observing adults carrying nesting material, in the act of nest building. Once located, nests were monitored at least once every two days by observing the nest from ground level. The behavior of the parents - such as frequency of visits to the nest, or silent, discrete behavior - was used to determine the stage of the nest. The age of the chicks was estimated from the time of the first observations of incubation at a nest (assuming an incubation period of 14 days) and the behavior of the parents.

### Table 1
**TIMELINE FOR MAURITIUS FODY RECOVERY**

| Fody breeding season | Targets |
|---|---|
| Sept 2001 - April 2003 | Experimental nest protection measures to ameliorate predation.<br>Population surveys. |
| Sept 2002 - April 2003 | Harvest of wild-hatched chicks for experimental hand-rearing trials.<br>Establishment of hand-reared birds in aviaries at GDEWS.<br>Population surveys. Nest monitoring studies in upland wild populations. |
| Sept 2003 - April 2004 | Harvest of wild-hatched chicks for experimental release program<br>Experimental release of Mauritius Fodies on Île aux Aigrettes<br>Post-release monitoring and population / ecological studies |
| Sept 2004 - April 2005 | Harvest of wild-hatched chicks for hand-rearing<br>Harvest of wild eggs for experimental incubation trials<br>Establishment of hand-reared birds on Île aux Aigrettes, with continued post-release support, monitoring and population / ecological studies<br>Captive-breeding trials at GDEWS |
| Sept 2005 - April 2006 | Harvest of wild eggs and chicks for hand-rearing<br>Further releases of wild- and captive-bred birds on Île aux Aigrettes<br>Opportunity for reproduction of Île aux Aigrettes sub-population<br>Studies of breeding ecology on Île aux Aigrettes |

## Nest Harvests

Due to the potential threat of nest predation, all nests were taken that could be harvested at the appropriate stage. During the 2005-06 season, nests were selected for harvest only if the pair had not contributed offspring to the gene pool previously. This strategy was intended to increase the genetic diversity of the released population on the offshore island. Pairs were assumed to remain in the same territory between seasons. During the earlier seasons of the program, nests were harvested when chicks were estimated to be around 5 days old, because at this age chicks had developed a strong begging response and were robust enough to cope with translocation; however their eyes were still unopened, and consequently they were considered to be more suitable to adapt to the captive environment and feeding process.

In later years (2004-06), it became possible to take eggs for artificial incubation and to rear chicks from hatching. Taking eggs instead of chicks has several advantages, because:

1. Wild nests were exposed to predation for a shorter duration.
2. Parents would be able to recycle and produce a replacement clutch earlier in the season.
3. Transferring eggs, rather than chicks, reduces the possibility of transferring disease from the wild to the captive breeding center (Ounsted 1991).

Harvests of chicks took place in the morning, allowing parents time to give chicks the first feeds of the day, but leaving enough time for chicks to adapt to the new hand-rearing regime before the end of the day. Before a harvest, the target nest was observed until a parent left the nest to ensure that the chicks had just been fed and typically they had been brooded by the female.

Only female Mauritius Fodies incubate and they leave the nest unattended for periods of around 20 minutes to feed (Safford, 1997c; pers obs.). As soon as the female left a nest, the tree was climbed and the whole nest was removed intact, placed in a tight plastic tub and lowered in a canvas bag to staff waiting on the ground. The chicks were quickly transported in their nest to a waiting jeep where they were driven to the rearing facility in Black River. The whole rescue process took approximately one hour and so it was not considered necessary to feed chicks or place them in a portable brooder for the translocation. The domed shape of the nest offered adequate thermal and physical insulation and leaving the chicks in the nest was considered less disruptive.

Harvests of eggs were undertaken at a mean clutch age of approximately 10 days of incubation. At this stage the vein growth of the eggs has reached 100% and the eggs are more resilient to change. (One nest was taken at earlier stage because it was considered to be under threat of immediate predation due to the presence of a group of Crab-eating Macaques around the nest tree). The eggs were then removed from the nest, candled and placed in a pre-heated flask. The flask contained a sponge base in which there were hollows for up to four eggs (the largest recorded clutch size). The eggs were then transported on foot from the forest to the nearest field station, where they were transferred to a portable incubator. The portable incubator was transferred to a truck at the nearest accessible location, without the temperature dropping below 30.0C, before being connected to the jeep battery as a power source.

## Artificial Incubation Parameters

Eggs were incubated at 37.2C and a mean relative humidity of 65%. Eggs were either turned 1800 by hand every two hours between 8.00 and 18.00, or automatically rolled for 15 minutes out of every 90 minutes. Once an egg internally pipped it was moved to a hatcher with a relative humidity of above 70%, and no turning mechanism.

## Hand-rearing & Weaning

Appendix Table A1 presents the hand-rearing diet regimen for Mauritius Fodies. The hand-rearing diet evolved throughout the program, as colonies of mice *Mus musculus* and cockroaches (sub-order: Blattodea) were progressively established at the breeding centre. Care was taken to ensure adequate hydration of the chicks - bee larvae is an excellent food item to assist in hydration; bee larvae were harvested from local hives, wild nests, or even nest cavities rustled from Echo Parakeets. Food items were also soaked in reconstituted lactated ringer's solution. Adequate sources of vitamins and minerals were also a consideration and supplements were added to the diet.

Chicks were fed every 60 minutes for the first four days, every 90 minutes between days 5 and 9 and every 120 minutes between days 10 and 18 (Appendix Table A1). The brooding temperature for chicks decreased accordingly with age and development. Chicks were weighed at the beginning of each day before the first feed, to monitor weight gain and development and to evaluate food mass targets.

Having fledged at approximately 18 days, fledglings were maintained in small fledging cages. Afterwards, the fledglings were maintained either in larger aviaries or were transferred to the pre-release aviaries at the release site on IAA. Birds destined for release were treated for parasites with an oral dose of Panacur (Intervet UK, Milton Keynes, UK). Fledglings were weaned either in outdoor aviaries at the captive breeding center, or in the pre-release aviaries.

## Husbandry for the Captive Breeding Population

Mauritius Fodies were kept in the outdoor aviaries at GDEWS in Black River. The aviaries measure approx 3m x 8m x 2.5m, with a small covered shelter at one end. The aviaries are vegetated with Bamboo spp, Hibiscus sp. shrubs, Pongamme *Milletia pinnata*, with Chinese Guava *Psidium cattleianum* branches for perching. During the breeding season Cryptomeria branches taken from the uplands were used for perching and to provide nesting sites. Each aviary contains a water basin for bathing and drinking, which was cleaned and refilled daily. Considerations for pest control included mousetraps baited with peanut butter and mixed seeds just outside aviaries, and ant moats around food dishes.

In 2003-04, each Fody pair was provided with one aviary. However, males often had to be separated from females due to over-vigorous chasing during courtship - a behavior frequently observed in the wild. Consequently, in following seasons, each pair was given access to three connecting open aviaries (approx 9m x 8m x 2.5m). Moss, guava rootlets, Isachne sp. grass, Pink Pigeon feathers (soaked in Virkon-S, then rinsed), rope fibers and coconut husks were placed in the aviaries for nesting material. Later it became clear that Fodies would use any fine grass species to build nests.

Food was prepared daily and consisted of a universal insectivorous mix, scrambled egg, diced fruit and vegetables and a sprinkle of a dietary supplement of either Nutrobal (Vetark Ltd, Winchester, UK), Avipro (Vetark Ltd) or Nekton Tonic I (Pforzheim, Germany) given on alternate days. Fruits and vegetables used were apples, grapes, bananas, mangoes, plums, pears, carrots, papaya and tomatoes. Additional fruit was spiked daily onto branches within the aviary. Each food bowl contained enough food to provide a constant supply for the captive Fodies. Avesnectar solution was provided daily in feeders attached to the aviary walls between 11.00 and 16.00h - for limited time to avoid fermentation. Live crickets or wax moth larvae were introduced at approximately 10.00 and 15.00h. Bottlebrush *Callistemon citrinus* flowers collected from the wild were placed in the aviaries for nectar feeding at irregular intervals. Feeding bowls were not placed in the aviary in which the pair had built their nest in order to minimize human disturbance.

Prior to the breeding season, red peppers, extra carrots and the supplement Nekton R-Beta were added to the food from June to provide extra red pigmentation for males molting into breeding plumage. This was discontinued as soon as the male's head and neck were fully red. Pairs of Fodies who were parent-rearing chicks, were given extra bee larvae and wax moth larvae every 2 hours from 06.00 until 16.00h; this was continued until the chicks had fledged and were feeding independently.

Early in the 2004-05 season, a brood of recently-hatched chicks died due to nest fly larvae infestation. Following this, each nest was treated with Carbaryl mixed with talcum powder at a concentration of 5%, after the first egg had been laid and again after the first egg had hatched. After incubation had started, the male was separated from the female to prevent disturbance. If a captive pair was successful at raising offspring, the chicks were transferred to the release aviaries on IAA, as soon as they had shown themselves to be independent.

## Estimating Survivability of Chicks

The Mayfield method (Mayfield 1961, 1975) was used to estimate the daily probability of survival of nests in captivity and in the wild with confidence intervals from Johnson (1979). The probability that chicks would have fledged from each nest in the wild was calculated using the formula for Mayfield's estimator of nesting success; $p = s^t$, where p= probability of a nest surviving to fledgling in the wild, s = the daily survival rate of nests in the upland population, t = estimated time (in days) until chicks would have fledged from the nest in the wild. The probability of the nest surviving to fledging in the wild was then used as an estimate of the probability of survival to fledging in the wild of each egg or chick in the harvested nest. The estimated probabilities of survival to fledging in the wild from all harvested eggs and chicks were summed to give an estimate of the number of chicks that would have been expected to fledge in the wild if no nests had been harvested.

## Releases of Mauritius Fodies on Île aux Aigrettes

In most cases, Fodies were transferred to Île aux Aigrettes just before or just after weaning. The young birds spent up to two weeks in pre-release cages, where they received a continuation of the typical captive diet. Within this period, the birds became accustomed to the vegetation, terrain and climate of IAA. Once their pre-release acclimation period was complete, the hatches of the pre-release aviaries were opened, to allow the Fodies access to the rest of the island. At this point the birds were free to come and go to and from the pre-release aviary.

The pre-release aviary also served the dual purpose of facilitating post-release monitoring and supplemental food provision. Although the birds were now essentially wild birds, every effort was made to ensure that post-release survival was maximized. Supplemental food was offered ad libitum. Additionally, it was possible to catch up a bird in the aviaries if it required hands-on management, such as veterinary care or ring alterations. If and when the Fodies returned to the aviaries, it was possible to monitor their survivorship. For those birds who failed to return to the aviaries for food, survivorship could only be assessed by searching the island. It quickly became apparent that it was necessary to cover the hatches with a 2-inch x 2-inch mesh grill which prevented the passage of larger birds such as Mynahs *Acridotheres tristis* and doves *Streptopelia* sp., but easily allowed the movement of Fodies.

## RESULTS

*(Results data and analysis adapted from Cristinacce et al. (in prep a))*

### Nest Harvests

In four breeding seasons of 2002-06, 29 Mauritius Fody nests were harvested from the mainland and four were rescued from Île aux Aigrettes. All of these were delivered safely to the hand-rearing facility except for one nest containing three eggs which was dropped during the harvest, smashing the eggs. These harvests provided a total of 25 viable eggs and 52 chicks (Table 2). Nests were harvested from 21 different pairs around Pigeon Wood and Les Mares Chasse. Fifteen pairs had one nest harvested, four pairs had two nests harvested and two pairs had three nests harvested.

### Artificial Incubation & Hand-rearing

Hatchability of the 25 eggs harvested or rescued was

**Table 2**

**SUMMARY OF SUCCESS OF NEST HARVESTS FROM THE WILD AND SUBSEQUENT ARTIFICIAL PROPAGATION, 2002-2006**

| Season | No. eggs successfully harvested | No. eggs hatched | % hatchability* | No. chicks harvested | No. chicks fledged | % survivability of chicks |
|---|---|---|---|---|---|---|
| 2002-03 | 0 | 0 | - | 14 | 14 | 100% |
| 2003-04 | 0 | 0 | - | 21 | 20 | 95% |
| 2004-05 | 8[1] | 8 | 100% | 15 | 23 | 100% |
| 2005-06 | 17[2] | 14 | 82% | 2 | 16 | 100% |
| Total | 25 | 22 | 88% | 52 | 73 | 99% |

[1] Includes one egg rescued from IAA
[2] Includes four eggs rescued from IAA
* Value for hatchability includes all eggs - fertile, infertile or unviable
(Cristinacce et al in prep a.)

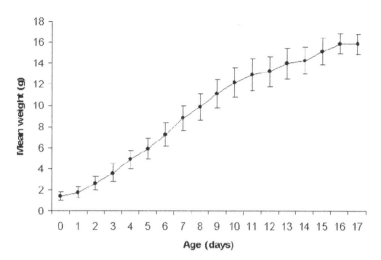

**Fig. 1.** *Mean daily weights of Mauritius Fody chicks reared from egg with standard error bars (n=22 days 0-12; n=21 day 13; n=13 day 14; n=8 day 15; n=5 day 16; n=3 day 17). (Cristinacce et al in prep a.)*

88% (n = 22). One egg died in late incubation and two died during hatching.

Survivability to fledging of the 74 chicks harvested or rescued, either initially as eggs or chicks, was 99% (n = 73). The one chick that died had a ruptured air sac, possibly caused during the harvesting process. Growth curves for the Fodies hand-reared from hatching are shown in Figure 1.

**Effectiveness of Harvesting & Rearing**

Using the Mayfield method, the mean probability that a harvested nest would have survived to fledging in the wild was 63% (Cristinacce et al in prep b). On the basis of this value, if the 75 viable eggs or chicks which were harvested from the mainland distribution had been left in the wild, the estimated number of fledglings that these nests would produce is 46.54. In comparison, the harvest and hand-rearing process successfully produced 69 fledglings. (A further 4 fledglings resulted from eggs salvaged from IAA.)

**Captive Breeding**

During the 2003-04 season, there was no successful breeding in the four pairs of one-year-old Fodies present in captivity. Some males had to be separated from females due to over-zealous courtship and aggressive behavior. However, one of the separated females built a nest and laid 3 infertile eggs. By the start of the 2004-05 season, there were only 3 established pairs surviving in captivity. One of these pairs was separated halfway through the season so the female could be paired with a different male. Fourteen breeding attempts produced 31 fledglings, but one died soon after fledging. Eight of these chicks fledged prematurely, possibly due to disturbance caused by ringing them at a late age (11 or 12 days old), so they had to be hand-reared until they could fly and feed themselves.

In the 2005-06 season there was just one pair of Fodies in captivity - the remainder of birds had been released on IAA. This last pair had one breeding attempt, resulting in three chicks.

Hatchability of the 58 eggs laid in captivity was 83% (n = 46). Of the 46 chicks that hatched, chick survivability to fledging was 74% (n = 34). (Table 3). Nesting success calculated using the Mayfield Method was 64%. Again, this is in comparison to a Mayfield value for nesting success in the wild as 63%.

## DISCUSSION

Certainly, the harvest of wild Fody nests in order to rescue or procure recruits for a new island population, or experimental captive breeding program, was a highly successful strategy. The production of 69 fledglings from the 20 eggs and 52 chicks, taken from the 29 nests harvested in upland areas, can be considered an excellent rearing result. Values of 88% for hatchability (including infertile eggs) and 99% for chick survival can be considered outstanding. This degree of success, especially working in a field environment, is a testimony to the aviculturists and field biologists involved in the process.

The Mayfield analysis supports the success of the harvest and artificial propagation effort. The successful fledging of 69 youngsters, in comparison to the 46.54 fledglings predicted if the nests had been left in their wild state, is emphatic. The rear and release strategy was selected over adult translocation for several reasons. Harvesting of nests has had little impact on the wild population - no decline in the upland population of Fodies throughout the timeline of the harvest and rear program has been reported. The effect of removing breeding adults from this remnant population was unknown, but was anticipated to have a much more significant impact on recruitment into the wild population. The great benefit of removing individuals as eggs or chicks was the parents' ability to recycle and produce subsequent successful clutches - Mauritius Fodies can begin nest building as soon as two days after a nest is predated (Safford 1997c). By taking nests at an early stage of incubation, the process is

### Table 3

**SUMMARY OF BREEDING ATTEMPTS OF MAURITIUS FODIES IN CAPTIVITY AT GDEWS (PARENTAL INCUBATION AND REARING)**

| Season | No. pairs | No. breeding attempts | No. eggs laid | No. eggs hatched | % hatchability* | No. chicks fledged | % survivability of chicks |
|---|---|---|---|---|---|---|---|
| 2003-04 | 4 | 1 | 3[1] | 0 | 0% | 0 | - |
| 2004-05 | 4[2] | 14 | 51 | 43 | 84% | 31[3] | 72% |
| 2005-06 | 1 | 1 | 4 | 3 | 75% | 3 | 100% |
| Total | - | 16 | 58 | 46 | 79% | 34 | 74% |

1 Infertile due to separation of male and female.
2 Includes three females. One pair was split up and the female was repaired to increase the genetic diversity of juveniles release d on IAA.
3 Eight chicks hand-reared after premature fledging.
* Value for hatchability includes all eggs - fertile and infertile
(Cristinacce et al. in prep a.)

increasingly more effective, by reducing the chances of egg predation in the wild and further increasing the parental capacity to recycle. (This approach must be counter-balanced with leaving eggs under parents for approximately the first quarter of the incubation period, to ensure that the clutch is complete and that the eggs receive an initial stimulus of natural incubation).

A second rationale for transferring captive-reared youngsters, rather than wild adults, reflects the significant differences between the coastal native forest habitat available on IAA, in comparison to the exotic tree plantation habitat of the remnant source population in upland Mauritius. Captive-reared juveniles were considered more likely to adapt to this new habitat and less likely to disperse from the island than adults with experience of upland habitats. Furthermore, it was anticipated that hand-reared youngsters would have lower levels of stress during the transfer and pre-release acclimation period, and survivability would benefit as a result. Sources of stress anticipated in the wild adults included the inevitable human disturbance and intra-specific aggression between territorial adults.

The experimental captive breeding program was also considered a success. Although first year mortality was high, the initial captive cohort of 14 birds was sufficient to establish a breeding population of 3 pairs. In the season at the end of their second year (2004-2005), these three breeding females together produced a total of 51 eggs and 31 chicks. High levels of fecundity were matched by 84% hatchability and 72% chick survivability - both very encouraging figures for parent-rearing attempts in pairs in their first year of breeding. The captive breeding trials showed that large numbers of individuals could be produced for a release program from a small number of pairs - the provision of appropriate, spacious housing was a key consideration.

This experimental program was also notable for the first successful breeding of the Mauritius Fody in captivity. (Cristinacce et al. 2006) Although the Rodrigues Fody *Foudia flavicans* has previously been propagated in captivity (Darby et al 1984), this is the first documented instance of a threatened Fody species successfully parent-rearing its own chicks to independence. However, the Madagascar Fody *Foudia madagascariensis* has regularly been bred in captivity (Brickell 2006). Furthermore, productivity and nesting success from captive pairs was high compared to wild populations both in upland Mauritius and on Île aux Aigrettes (Safford 1997b; Cristinacce unpublished data). The relative merits and problems associated with *in-situ* and *ex-situ* conservation management intervention for critically endangered taxa are well-documented (Snyder et al 1996; Jones 2004). Criticisms of *ex-situ* captive breeding programs include claims that it deflects attention and resources away from more urgent conservation problems such as habitat preservation and restoration (Rahbek 1993; Snyder et al 1996). However, captive breeding programs in the country of origin have been used to provide individuals for release (van Heezik et al 2005; Tweed et al 2003) and captive breeding has been used in one third of all programs that have recently saved a bird species from extinction (Butchart et al 2006). Furthermore, the release of large number of juveniles increases the chances of a successful release program (Griffith et al 1989; Wolf et al 1998). The experimental captive breeding program for the Mauritius Fody is a case study for an efficient and effective program to increase the overall population of a threatened species. Most importantly, the release program, using IAA as a predator-free refuge, has proven very successful. Between 2003-2006 the releases of captive-reared young Fodies have successfully established a new sub-population; the releases have been bolstered by successful breeding of the newly re-established population. Before the first releases in February 2003, IAA had no population of Mauritius Fodies. By the end of the breeding season in April 2007, the population on IAA was estimated at 147 birds (C. Jones pers. comm.). This represents a total population increase of approximately 60%. Both the population increase and the establishment of a second sub-population represent significant progress in securing the future of the species, and may eventually lead to a reclassification of IUCN/BirdLife status from Critical to Endangered.

One of the tremendous challenges that zoos have in the fulfilling their conservation mission is the hurdle of transferring captive-bred individuals to a strategically viable release program. The greatest obstacles are the logistical and bureaucratic processes, which make international transfer of animals for reintroduction programs, at appropriate ages, an almost impossible process. This is emphasized by a review of the some of the most notable avian conservation success stories, where reintroduced individuals have been sources from *in-situ*, rather than *ex-situ*, zoos or captive breeding centers. In North America, Peregrine Falcons *Falco peregrinus*, California Condors *Gymnogyps californius* and Whooping Cranes *Grus americana* and Nene *Branta sandvicensis* have originated almost entirely from facilities in the country of origin, even if not always technically "*in-situ*". In Mauritius, the great majority of released Mauritius Kestrels, Pink Pigeons and Echo Parakeets originated from GDEWS, and not from European zoos involved in the "safety net" captive breeding program.

In the cases of the Mauritian Kestrel, Pink Pigeon and Echo Parakeet, as well as the Mauritius Fody, a captive breeding program has only been one component of a multi-faceted approach. Management of eggs, chicks and nest sites in the wild, all employ avicultural techniques. There is tremendous benefit to a fluid interface between captivity and the wild, where field biologist have avicultural skills, aviculturists have field expertise and birds are considered as being within one entire management unit. The success of intensive efforts to conserve New Zealand's avifauna is also dependent upon a similar philosophy. As in the case of the Mauritius Fody, where eggs and chicks have been "rescued" if nests are considered to be at risk from nest predation, wild Echo Parakeet chicks are monitored frequently and rescued if chicks are considered to be at risk from disease, neglect or a competition for a scarcity of food. Although the techniques involved are different, the principle is the same - there is no distinction between wild and captive birds and biologists, and aviculturists have a vital role throughout. (Disease transmission is one consideration which is a notable exception to this generalization.) Furthermore, zoos can play a pivotal role in *in-situ* conservation programs through the provision of such avicultural expertise and the training of local staff - aviculturists from Chester Zoo and the Durrell Wildlife Conservation Trusts have played valuable roles in the hands-on management of the Fody. In doing so, the zoos are building the capacity of their own staff, building the capacity of the *in-situ* conservation program and ultimately benefiting species recovery.

**The Future**

Following the dramatic success of the establishment of the Mauritius Fody on Île aux Aigrettes recovery program, the focus of the Passerine Program for the Mauritian Wildlife Foundation and its collaborators is aimed at two new major goals:

1. The reintroduction of the Mauritius Fody on other offshore islands, including on the 250-hectare Round Island, where a major island restoration program is also being undertaken.
2. Utilizing and developing the techniques and strategies employed in the recovery program for the Fody, effort will now focus on a species recovery program for the Mauritius Olive White-eye, which is also identified as being in need of immediate conservation action.

## Acknowledgements

We are grateful to everyone who contributed to the success of this project: Financial support from: Ruth Smart Foundation, Chester Zoo / North of England Zoological Society, Hong Kong Bank (Mauritius), Thrigby Hall Wildlife Gardens. The Mauritian National Parks and Conservation Service, particularly Yousoof Mungroo. Valuable assistance in incubation, hand-rearing and captive husbandry was provided by Tracé Williams, Elise Kovac, Anne Williams, Marie-Michelle Hippolyte, Nadine Lemarque, Stephanie Lolith and Clare Jones. Special thanks to all the staff of the Mauritian Wildlife Foundation; in particular Ruth Cole, Franziska Hillig, Andrew Bowkett, Anna Reveleaux, Jason Malham, Paul Freeman, Tom Bodey and Shivananden Sawmy for their assistance in nest finding and harvesting.

## References

ARMSTRONG DP, RS Davidson, WJ Dimond, JK Perrott, I Castro, JG Ewen, R Griffiths and J Taylor. 2002. Population dynamics of reintroduced forest birds on New Zealand islands. Journal of Biogeography 29:609-621.

ARMSTRONG DP & JG Ewen. 2002. Dynamics and viability of a New Zealand robin population reintroduced to regenerating fragmented habitat. Conservation Biology 16:1074-1085.

BRICKELL N. 2006. Collated data on the Madagascar fody *Foudia madagascariensis*. Avi Mag 112:169-173.

BUTCHART SHM, AJ Stattersfield, NJ Collar. 2006. How many bird extinctions have we prevented? Oryx 40:266-278.

CADE TJ & SA Temple. 1995. Management of threatened bird species: evaluation of the hands-on approach. Ibis 137: 161-172.

CARTER SP & PW Bright. 2002. Habitat refuges as alternatives to predator control for the conservation of endangered Mauritian birds. In Veitch C.R. and Clout M.N. (eds) Turning the tide: the eradication of invasive species: 71-78. IUCN SSC Invasive Species Specialist Group, IUCN Gland and Cambridge.

CHEKE AS. 1983. Status and ecology of the Mauritius Fody *Foudia rubra*, an endangered species. National Geographic Society Research Reports 15: 43-56.

CHEKE AS. 1987. An ecological history of the Mascarene Islands with particular reference to extinctions and introductions of land vertebrates. In Diamond, AW (ed.) Studies of Mascarene Island birds, pp. 5-89. Cambridge University Press, Cambridge, UK.

COLE RE, CG Jones and V Tatayah. 2007. Work Proposal to Chester Zoo 2007-2008 Season. Unpublished report by Mauritian Wildlife Foundation.

CRISTINACCE A, A Ladkoo, E Kovac, L Jordan, A Morris, T Williams, F de Ravel Koenig, V Tatayah and CG Jones. 2006. First hand-rearing of Mauritian white-eyes. Avi Mag 112:150-160

CRISTINACCE A, A Ladkoo, RA Switzer, L Jordan, A Owen, V Vencatasawmy, F de Ravel Koenig, CG Jones and DJ Bell; in prep a. Captive breeding and rearing of critically endangered Mauritius Fodies *Foudia rubra* for reintroduction.

CRISTINACCE A, RA Switzer, RE Cole, CG Jones and DJ Bell.; in prep b. Exotic tree species as refuges for endemic island bird species: The case of the Mauritius Fody *Foudia rubra*.

DARBY PWH, DF Jeggo, and ME Redshaw. 1984. Breeding and management of the Rodrigues Fody *Foudia flavicans*. Dodo 21:109-126

DIMOND WJ & DP Armstrong. 2007. Adaptive harvesting of source populations for translocations: a case study with New Zealand robins. Conservation Biology 21:114-124.

GRIFFITH B, JM Scott, JW Carpenter, C Reed. 1989. Translocation as a species conservation tool. Science 245:477-480.

IUCN. 2006. IUCN Red List of Threatened Species. http://www.iucnredlist.org. Accessed on 10th of May 2007

JOHNSON DH. 1979. Estimating nest success: the Mayfield method and an alternative. Auk 96:651-661.

JONES CG. 2004. Conservation management of endangered birds. In Sutherland, WJ, Newton, I & Green, R (eds). Bird ecology and conservation. Oxford University Press, Oxford, UK.

JONES CG, W Heck, RE Lewis, Y Mungroo, G Slade, T Cade. 1995. The restoration of the Mauritius Kestrel *Falco punctatus* population. Ibis 137: 173-180.

KAPLAN EL & P Meier. 1958. Nonparametric estimation from incomplete observations. Journal of American Statistical Association 53: 457-481.

KOMDEUR J. 1994. Conserving the Seychelles Warbler *Acrocephalus sechellensis* by translocation from Cousin Island to the islands of Aride and Cousine. Biol Cons 67:143-152.

KOMDEUR J. 1996. Breeding of the Seychelles Magpie Robin *Copsychus sechellarum* and implications for its conservation. Ibis 138: 485-498.

KUEHLER C, P Harrity, A Lieberman and M Kuhn. 1995. Reintroduction of hand-reared Alala *Corvus hawaiiensis* in Hawaii. Oryx 29: 261-266.

KUEHLER C, A Lieberman, P Oesterle, T Powers, M Kuhn, J Kuhn, J Nelson, T Snetsinger, C Herrmann, P Harrity, E Tweed, S Fancy, B Woodworth, T Telfer. 2000. Development of restoration techniques for Hawaiian thrushes: Collection of wild eggs, artificial incubation, hand-rearing, captive-breeding, and reintroduction to the wild. Zoo Biology 19: 263-277.

MAYFIELD HF. 1961. Nesting success calculated from exposure. Wilson Bulletin 73: 255-261.

MAYFIELD HF. 1975. Suggestions for calculating nest success. Wilson Bull 87:456-466.

NICHOLS RK, P Philips, CG Jones and LG Woolaver. 2002. Status of the critically endangered Mauritius Fody *Foudia rubra* in 2001. Bulletin of the African Bird Club 9: 95-100

NICHOLS RK. 1999. Endangered passerines on Mauritius: management report 1998/1999. Unpublished report by the Mauritian Wildlife Foundation.

NICHOLS RK, P Phillips and F de R Koenig. 2000. Endangered passerines on Mauritius: management report 1999/2000. Unpublished report by the Mauritian Wildlife Foundation.

OUNSTED ML. 1991. Re-introducing birds: lessons to be learned for mammals. In Gipps, JHW (ed). Beyond captive breeding: Re-introducing endangered mammals to the wild. Symposia of the Zoological Society of London 62: 75-85.

RAHBEK C. 1993. Captive breeding- a useful tool in the preservation of biodiversity? Biodiversity and Conservation 2: 426-437.

SAFFORD RJ. 1994. Conservation of the forest-living native birds of Mauritius. PhD thesis, University of Kent, Canterbury, UK.

SAFFORD RJ. 1997a. A survey of the occurrence of native vegetation remnants on Mauritius in 1993. Biol Cons 80:181-188.

SAFFORD RJ. 1997b. Nesting success of the Mauritius Fody *Foudia rubra* in relation to its use of exotic trees as nest sites. Ibis 139: 555-559.

SAFFORD RJ. 1997c. The annual cycle and breeding behaviour of the Mauritius Fody *Foudia rubra*. Ostrich 68: 58-67.

SAFFORD RJ. 1997d. The destruction of source and sink habitats in the decline of the Mauritius fody, *Foudia rubra*, an endemic island bird. Biodiversity and Conservation 6: 513-527.

SAFFORD RJ. 1997e. Distribution studies on the forest-living native passerines of Mauritius. Biological Conservation 80: 189-198.

SAFFORD RJ & CG Jones. 1998. Strategies of landbird conservation on Mauritius. Conservation Biology 12(1): 169-176.

SEDDON PJ, PS Soorae and F Launay. 2005. Taxonomic bias in reintroduction projects. Animal Conservation 8: 51-58.

SHERLEY GH & LM Hayes. 1993. The conservation of a giant weta at Mahoenui, King Country: habitat use, and other aspects of its ecology. New Zealand Entemomlogist 16: 55-67.

SNYDER NFR, SR Derrickson, SR Beissinger, JW Wiley, TB Smith, WD Toone and B Miller. 1996. Limitations of captive breeding in endangered species recovery. Conservation Biology 10: 338-348.

TWEED EJ, JT Foster, BL Woodworth, P Oesterle, C Kuehler, A Lieberman, AT Powers, K Whitaker, WB Monahan, J Kellerman and T Telfer. 2003. Survival, dispersal, and home-range establishment of reintroduced captive-bred Puaiohi *Myadestes palmeri*. Biological Conservation 111:1-9.

VAN HEEZIK Y, P Lei, R Maloney and E Sancha. 2005. Captive breeding for reintroduction: influence of management practices and biological factors on survival of captive Kaki (black stilt). Zoo Bioliogy 24:459-474.

WHITMORE KD & JM Marzluff. 1998. Hand-rearing corvids for reintroduction: importance of feeding regime, nestling growth and dominance. Journal of Wildlife Management 62: 1460-1479.

WOLF CM, T Jr. Garland & B. Griffith. 1998. Predictors of avian and mammalian translocation success: reanalysis with phylogenetically independent contrasts. Biological Conservation 86: 243-255.

## Author Biography

Richard Switzer's special interest is the interface between captive breeding and the wild, using both avicultural and field techniques as strategies in avian species recovery. Since graduating in 2001 with an MSc in Wildlife Management and Conservation, he has been Passerine Program Coordinator for the Mauritian Wildlife Foundation and Facility Manager of the Maui Bird Conservation Center for the Zoological Society of San Diego. Early in 2007 he joined the ZSSD team in the collaborative project to capture and translocate a cohort of Rimatara/Kuhl's Lorikeet from Rimatara to Atiu. Since May 2007 he has been Bird Curator at the Al Wabra Wildlife Preservation in Qatar.

### Appendix Table A1
**HAND-REARING DIETS, FEEDING TIMES & BROODER TEMPERATURE OF MAURITIUS FODIES IN 2005-06**

| Day | 1 | 2 | 3 | 4 | 5 | 6 | 7 | 8 |
|---|---|---|---|---|---|---|---|---|
| Temp (C) | 37-36 | 36-35 | 35-34 | 34 | 33 | 32 | 31 | 30 |
| Intervals between feeds (mins) | 60 | 60 | 60 | 60 | 90 | 90 | 90 | 90 |
| Times | 5.00-21.00 | 5.00-21.00 | 5.30-21.00 | 5.30-21.00 | 5.30 + 6.30-20.00 | 5.30 + 6.30-20.00 | 5.30 + 6.30-20.00 | 5.30 + 6.30-20.00 |
| **Feed** | | | | | | | | |
| 1 | B | B | B (vb) | B (vb) | B (vb) | B (vb) | B (vb) | B (vb) |
| 2 | B (p) | B (n) | B (n) | B (n) | B (n) | B (n) | B (n) | B (n) |
| 3 | Bcg | BEP (nk) | BEP (nk) | BEP (nk) | BEP (nk) | BEP (nk) | BEP (nk) | BEP (nk) |
| 4 | B | B | B | B | M* | M* | M | M |
| 5 | B | Bcg | Bcg | Bcg | Bcg | Bcg | Bcg | Bcg |
| 6 | Bcg | BEP | BEP | BEP | BEP | BEP | BEP | BEP |
| 7 | B | B | B | B | M* | M* | M | M |
| 8 | B | Bcg | Bcg | Bcg | Bcg | Bcg | Bcg | Bcg |
| 9 | Bcg | BEP | BEP | BEP | BEP | BEP | BEP | BEP |
| 10 | B | B | B | B | M* | M* | M | M |
| 11 | B | Bcg | Bcg | Bcg | B | B | B | B |
| 12 | Bcg | BEP | BEP | BEP | | | | |
| 13 | B | B | B | B | | | | |
| 14 | B | Bcg | Bcg | Bcg | | | | |
| 15 | Bcg | BEP | BEP | BEP | | | | |
| 16 | B | B | B | B | | | | |
| 17 | B | B | B | B | | | | |
| 18 | B | B | | | | | | |

***Key***

*B= Bee; Cg:=Cricket gut only; vb=Vitamin B; BEP= Egg and Papaya; M*= Mouse internal organs; M=Mouse; n=Nutrobal (Vetark Ltd.); Cab=Cricket abdomens; WW= Wax moth larvae; p=Avipro (Vetark Ltd.); nk:=Nekton® I tonic*

| 9 | 10 | 11 | 12 | 13 | 14 | 15 | 16 | 17 | 18 |
|---|---|---|---|---|---|---|---|---|---|
| 29 | 28 | 27 | Not regulated | | | | | | |
| 90 | 120 | 120 | 120 | 120 | 120 | 120 | 120 | 120 | 120 |
| 5.30 + 6.30-20.00 | 5.30 + 6.30-20.00 | 5.30 + 6.30-20.00 | 5.30 + 6.30-20.00 | 5.30 + 6.30-20.00 | 5.30 + 6.30-20.00 | 5.30 + 6.30-20.00 | 5.30 + 6.30-20.00 | 5.30 + 6.30-20.00 | 5.30 + 6.30-20.00 |

| 9 | 10 | 11 | 12 | 13 | 14 | 15 | 16 | 17 | 18 |
|---|---|---|---|---|---|---|---|---|---|
| B (vb) | B (vb) | B (vb) | B (vb) | B (vb) | B (vb) | B (vb) | B | B | B |
| B (n) | B (n) | B (n) | B (n) | B (n) | B (n) | B (n) | B (n) | B (n) | B (n) |
| BEP (nk) | BEP (nk) | BEP (nk) | BEP (nk) | BEP (nk) | EP (nk) | EP (nk) | EP (nk) | EP (nk) | EP (nk) |
| M | M | M | M | M | MWW | MWW | MWW | MWW | MWW |
| Bcg | Bcg | Bcg | Bcg | Bcg | Bcab | Bcab | Bcab | Bcab | Bcab |
| BEP | BEP | BEP | BEP | BEP | EP | EP | EP | EP | EP |
| M | M | M | M | M | MWW | MWW | MWW | MWW | MWW |
| Bcg | Bcg | Bcg | Bcg | Bcg | Bcg | Bcab | Bcab | Bcab | Bcab |
| BEP | B | B | B | B | B | B | B | B | B |
| M | | | | | | | | | |
| B | | | | | | | | | |

# THE CAPTIVE HUSBANDRY AND BREEDING OF THE CRITICALLY ENDANGERED MONTSERRAT ORIOLE *ICTERUS OBERI*

Gary Ward

Durrell Wildlife Conservation Trust
Les Augrès Manor, La Profonde Rue, Trinity
Jersey, Channel Islands JE3 5BP

## Summary

Eight Montserrat Orioles were brought into captivity in 1998 with a view to establishing an understanding of the husbandry requirements, so that if a captive breeding initiative was needed to help save the species in Montserrat the necessary information would be available. Montserrat Orioles had never been maintained in captivity before this time and much information on the captive husbandry of this species has now been gathered. Since being in captivity the Montserrat Oriole has proved to be relatively easy to breed, and seems to be well suited to life in captivity as long as certain requirements are met. This paper will discuss the captive requirements for this species including some of the innovative solutions that have been made to solve the challenges involved with breeding Montserrat Orioles in captivity. Many of these captive management techniques are applicable to the husbandry of other passerine species.

## Introduction

As its name suggests, the Montserrat Oriole *Icterus oberi* is a resident endemic of Montserrat, a 39km² volcanic island of the Lesser Antilles in the eastern Caribbean. There has been substantial deforestation for agriculture across the island, which led to much of the remaining forest habitat being concentrated in the central highlands. In 1995 the Soufriere Hills volcano in the south of the island began to erupt, resulting in heavy ash falls and pyroclastic flows which have caused severe damage to much of the remaining forest. Several major eruptions have occurred since 1995 and the volcano remains very active. Prior to these eruptions the mountain forests in the south of the island represented 60% of the forest habitat occupied by the Montserrat Oriole. Following the destruction of much of the Soufriere Hills forest, the main oriole popula-

tion has been restricted to the Centre Hills. Subsequently the Montserrat Oriole population in the Centre Hill has suffered a further decline of an estimated 50% between the years of 1997 and 2002. The current population estimate is 200 to 800 birds with a declining trend. Owing to the rapid population decline and small range of the species it is now classified as Critically Endangered by the IUCN/BirdLife International (Hilton et al 2005).

The Montserrat Oriole is the only endemic bird species on Montserrat and is the national bird. The oriole is a popular icon on the island and has an economic value to the growing nature tourism industry. The Montserrat National Trust uses the Montserrat Oriole as their logo (Hilton et al 2005).

Extensive research has been carried out by the Royal Society for the Protection of Birds (RSPB) and the Montserrat Forestry Department into the reasons for the more recent population decline of the Montserrat Oriole in the Centre Hills. This has highlighted several limiting factors affecting the species besides the ongoing volcanic activity. Like many islands, Montserrat has been subject to the introduction of invasive species which have had a negative effect on the endemic fauna and flora. Rats *Ratus* spp. can be included in the suite of invasive species which are affecting the natural ecology of the island and are found in high densities in the Centre Hills forest. These are major predators of Montserrat Oriole eggs and chicks. The presence of other introduced species such as feral pigs *Sus scrofa*, feral cats *Felis catus* and goats *Capra hircus* as well as invasive plants raises concerns for the longer-term health of the forest. Interestingly the native Pearly-eyed Thrasher *Margarops fuscatus* has also been implemented in the decline of the oriole population due to its habit of predating the nests of other bird species. Drought can also affect the breeding success of the oriole and this problem has worsened with recent climate changes and the diversion of spring water to supply the human population. Ash falls following major eruptions can have a negative effect on the arthropod food supply of the orioles and evidence suggests that during breeding seasons when recent ash falls have occurred the orioles are unable to find enough insects to rear their chicks (Hilton et al 2005).

Because of the unpredictable nature of the volcano and other threats to the species, a consortium of conservation bodies known as the Montserrat Alliance (see Appendix 1) made a decision that Durrell Wildlife Conservation Trust would collect a small number of orioles and bring them to its headquarters in Jersey, U.K, to conduct husbandry and captive breeding trials. In 1998 an expedition was launched which saw eight birds captured on Montserrat and brought to Jersey: five males and three females (Owen 2000). This species was brought into captivity with a view to establishing an understanding of the husbandry requirements, so that if a captive breeding initiative was needed to help save the species in Montserrat that information would be available. Montserrat Orioles had never been maintained in captivity before this time and much information on husbandry and captive breeding of this species has now been gathered (Ward 2006).

## Taxonomy & Natural History

The Montserrat Oriole is a member of the family Icteridae, also known as New World Blackbirds. The Icteridae family includes some 103 species ranging throughout the Americas, from Alaska to Cape Horn, with a large component found in the Caribbean (Jaramillo & Burke 1999).

Within the family Icteridae is the genus *Icterus*, New World orioles, with 27 species including the Montserrat Oriole. This species is closely related to other single island endemics found in the Caribbean: the Martinique Oriole *I. bonana*, St. Lucia Oriole *I. laudabilis* and the Puerto Rican Black-cowled Oriole *I. dominicensis*. It is thought likely that the Montserrat Oriole has been isolated for over two million years (Hilton et al 2005).

## Distribution

The Montserrat Oriole is endemic to Montserrat, in the Lesser Antilles in the Caribbean. The main population is now restricted to the Centre Hills mountain range with a small remnant population found in a 1-2 km² forest patch in the south of the Soufriere Hills just 1km from the summit of the volcano (BirdLife 2007).

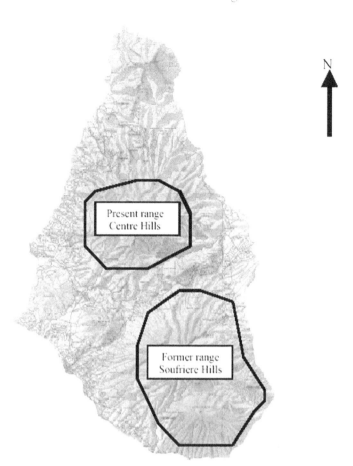

**Fig. 1.** *Map of Montserrat showing past and present range of the Montserrat Oriole* (Owen 2001).

## Habitat

Montserrat Orioles can be found in most forest types occurring between 150-900m. The species is found in higher densities in the wetter tropical mountain forest between 200-500m and avoids lowland dry forests and high altitude elfin forest. In the lower, drier altitudes they tend to inhabit gullies, known locally as Ghauts, where the forest habitat is damper. Most Montserrat

**Fig. 2.** *Montserrat Oriole habitat in the Centre Hills area. Photo by Will Masefield*

**Fig. 3.** *Adult male Montserrat oriole.*

Oriole territories are associated with stands of *Heliconia caribaea*, a plant that the birds use to build their nest in. The oriole is often found at the edges of forest clearings where small agricultural plots, such as banana plantations, occur. Data collected prior to the volcanic eruptions of the mid-90s suggests the oriole's strong-hold were forests of the bamboo *Bambusa vulgaris* found in the east of the Soufriere Hills (Hilton et al 2005).

## Description

Adult Male: The head, mantle, breast and wings are black. The black breast ends abruptly with the lower breast and belly a bright golden yellow. These golden yellow feathers are slightly tinged with tawny brown. The lower back and rump are lemon yellow and the tail is black. The legs and feet are bluish-grey and the bill is black with the base of the lower mandible pale grey. The eyes are dark reddish-brown.

Adult Female: Primarily olive-yellow over most of

**Fig. 4.** *Adult female Montserrat oriole.*

**Fig. 5.** *Two year old male Montserrat Oriole showing olive dappling on the nape.*

the body with a yellowish face and black lores. The crown is yellow-olive becoming olive on the nape and mantle. The wings are olive-brown and the tail is olive-yellow with a yellow wash. The bill, legs and feet are the same as the male (Owen 2001).

Juveniles: Similar to adult female but lacking the black lores. The juveniles appear more drab in appearance to the adult females. Immature males may exhibit a small number of black feathers on the throat. During the second moult at just over a year old the males acquire a mottled appearance as the black feathers on the head, nape and back grow through. Two year old males show olive dappling on the nape.

Length: 20-22cm.
Adult weight: 30-40g

## Housing Requirements

To date Montserrat Orioles at Jersey Zoo have been

**Fig. 6.** *The same male Montserrat Oriole as in Fig 5 during the second moult at just over one year old.*

kept behind two different sizes of square weld mesh wire netting: 25mm and 12mm. Both sizes of weld mesh exhibit advantages and disadvantages. The 25mm mesh has the main advantage of allowing less restricted visibility for the observer than that of the 12mm mesh. However, both sizes can be painted with matt, dark-coloured paint (black/green) which helps to reduce reflection and improve visibility into the aviary. 25mm mesh has the disadvantage of allowing easy access to vermin such as rodents and small wild birds. Montserrat Orioles also have the interesting behaviour of pushing their heads through the 25mm mesh in order to gain an unobstructed view of the aviary surroundings. There is a small risk of the birds coming to harm while doing this and could be potentially exposing themselves to predation from opportunistic predators. 12mm mesh is more suited to eliminating unwanted entry from vermin and only very small mice are capable of accessing the aviary through this size mesh.

## Shelter

The provision of an adequate indoor area for Montserrat Orioles is considered extremely critical for the welfare of these birds, due to the natural range of this species being confined to the tropical Caribbean island of Montserrat where the lowest yearly temperatures reach 24-25°C. In cold climates, the shelters should be provided with heating during the winter months - this can be provided with the use of heat lamps and/or bar heaters. If heat lamps are not used then the ambient temperature should be maintained above 15°C throughout the colder months of the year. The birds can be allowed access to the outside flight year-round though access pop holes with sliding doors. The pop holes incorporated into the aviary designs at Durrell Wildlife measure 30cm x 30cm and are fitted with sliding doors which can be adjusted to keep the gap as minimal as possible to reduce heat loss. The indoor areas provided are either block work or timber construction. The larger indoor areas incorporated into aviaries housing orioles at Durrell Wildlife (3m x 3.5m x 3m high) have been divided in two with 12mm weld mesh. Sliding pop holes and a keeper door are incorporated in the divides for management purposes. This feature is useful when introducing birds or when they need to be separated due to aggression. These sheds are fitted with florescent tube lights controlled by a time switch set to come on at 0700 hours and go off at 1900 hours. In winter the extra lighting within the indoor areas after dark encourages the birds to enter them and then roost indoors during these colder months.

Corrotherm sheeting is also used to cover parts of the roof over the outside flight, which provides shelter from heavy rain and is particularly beneficial if the birds choose to build their nests under it.

## Spatial Requirements

The smallest enclosures used for Montserrat Orioles are indoor floor cages measuring 2m x 1.2m x 2.45m high. It is recommended that only one bird be housed in this size cage and only for a short period of time. Orioles are prone to obesity and a large planted aviary where the birds can exercise by flying around and foraging is far more appropriate. Small outdoor holding aviaries are used to house individual orioles for longer time periods. These measure 2.5m x 2.3m x 2m high and incorporate a small indoor area measuring 2.5m x 1.2m x 2m high. The smallest size aviary used for housing breeding pairs is 4.7m x 2.4m x 2m high. Due to the low levels of aggression that can occur between the pair at any time of the year any aviary smaller than

### Table 2
**AVIARY SIZE IN RELATION TO HOUSING PAIRS/INDIVIDUALS**

| Aviary size | Recommended capacity |
| --- | --- |
| 2m x 1.2m x 1.45m high (indoor flight) | Should only be used for individual birds and for short periods of time |
| 2.5m x 2.3m x 2m high (outdoor flight) | Holding aviaries only used for individual adult birds or up to 3 juveniles |
| 4.7m x 2.4m x 2m high (outdoor flight) | Smallest size aviary suitable for housing a breeding pair (birds may need to be separated for short periods when aggression occurs) |
| 5m x 3.5m x 3.5m high (outdoor planted flight) | Ideal size aviary for housing a breeding pair compatibly throughout the year |

this for housing pairs of birds would be unsatisfactory. Even in this size aviary pairs often need to be separated for few days or weeks when aggression becomes too severe. Ideally a pair of Montserrat Orioles should be housed in a large planted aviary with plenty of space and cover for the birds to seek shelter from one another and with opportunities for exercise and foraging. The largest aviary size used for Montserrat Orioles at Durrell Wildlife is 9.5m x 3.8m x 3.6m high.

**Substrate/Enclosure Furnishings**

The substrate used in the indoor holding cages is either bare tiles or with a layer of wood shavings. Both substrates have proven satisfactory; however shavings are now preferred as this has eliminated the need for the daily cleaning associated with the use of tiles. The cages can now be spot-cleaned and fresh shavings added when necessary. The wood shavings also provide enrichment, as the birds will pick through the shavings searching for insects scatted on the floor by the keeper. In outside flights both sand and bark chip are used as substrates. However the sand is only used in off-display holding aviaries and the bark chip in the large on-display breeding aviaries. The bark chips provide an authentic looking forest floor simulation and will break down over time adding nutrients to the soil which is beneficial to plant growth. Fresh bark chip is added to the aviaries when needed (once or twice a year) as it decomposes. Sand is used as a substrate in the indoor areas of all aviaries. For perching natural branches are recommended for Montserrat Orioles as these can provide a variety of textures and diameters important for maintaining condition of the birds' feet. Dense twiggy branches are generally selected for orioles, incorporating a variety of horizontal and vertical angles, which the birds will readily clamber amongst. Cable ties can be used to attach the natural branches to the aviary walls and this provides a quick and easy way of securing the perching. Water features such as small waterfalls and creeks can also be incorporated into Montserrat Oriole exhibits. These add interest to the landscaping of the aviary and enrichment for the birds by providing them with a variety of bathing opportunities.

**Planting**

Montserrat Orioles benefit greatly from the inclusion of plants in their enclosure by providing environmental enrichment, nesting sites and by attracting insects. Plants obviously help to add authenticity to exhibits in recreating the natural forest habitat of the species. A huge variety of plant species have been used in the oriole aviaries with some been more useful to the birds than others. As a general rule any type of plant can be utilized by the orioles, at the very least for foraging for invertebrates. Montserrat Orioles are adapted to forging amongst dense vegetation and heavily planting their aviaries will allow the birds to make maximum use of their enclosure. In the wild the favoured nesting site for Montserrat Orioles is under the large leaves of Heliconia plants *Heliconia caribaea* (Owen 2000). At Jersey Zoo banana plants *Musa* sp. have been planted for the orioles as a suitable substitute for *Heliconia* and the birds have built nests in these on several occasions (D. Jeggo pers. comm. 2006). Bamboo also provides excellent nesting sites for orioles as a tight clump of bamboo has the ideal structure for the birds to attach their intricately weaved cup nest to. The orioles build these nests out of fine plant fibres, and plants that produce these fibres such as the Chusan Palm *Trachycarpus fortunei* are also useful. In holding aviaries the orioles are given potted plants of various kinds, within which insects can be thrown providing foraging enrichment for the birds. These potted plants can then be easily moved to a different aviary if necessary (See Appendix 2 for a table of plant species used in Montserrat Oriole aviaries).

**Hygiene**

Indoor holding cages with wood shavings for substrate are spot-cleaned when necessary (every 1-2 weeks) and fresh wood shavings are added. The indoor areas of outdoor aviaries can become rather soiled in a relatively short time especially during the winter months when the birds are using the inside areas more. Sand is used as a substrate in the indoor areas at Durrell Wildlife and when the build up of faeces and dropped food becomes too great (about every two weeks) the sand is sieved. This removes all the large particles such as dried faeces leaving clean sand that can then be spread back over the floor. Over time this sand will become too dusty and will need to be replaced. Outside flights require the minimum of cleaning other than spot-cleaning under roosting sites or raking over the bark chip or sand substrate. Faeces can be washed off plant leaves, logs and rockwork with the use of a garden hose and any ponds or creeks should be scrubbed and rinsed out on a weekly basis.

Orioles will clean their bills by wiping them on the perching and this can lead to a build up of a layer of

**Fig. 7.** *A typical inside area of an aviary showing the dense twiggy perching ideal for Montserrat Orioles.*

**Fig. 8.** *Heavily planted aviary housing a breeding pair of Montserrat Orioles.*

**Fig. 9.** *Holding aviary with potted plants.*

fruit and other uneaten food items on the perching, particularly near food dishes or where fruit is spiked. For this reason fresh clean perching should be provided when necessary.

After a certain length of time and always before birds are moved into a new aviary the walls and other surfaces should be scrubbed with a disinfectant.

## Captive Behaviour & Social Management

Montserrat Orioles are very inquisitive and agile birds and seem well suited to life in captivity. The birds will forage frequently around the aviary and amongst the vegetation in the search for insects or nesting material. They will use their bills to pry beneath loose bark or leaf litter in a similar fashion to birds of the family Sturnidae. Orioles show amazing abilities when clambering through the vegetation and are able to climb up and down vertical stems such as that of bamboo. They can also hang upside down under branching or the netting on the roof of the aviary. The orioles also have the ability to catch insects thrown to them on the wing. Montserrat Orioles will always hold insects in their feet while eating them. The birds will collect an insect, fly to a level perch and pin the insect to the perch with one foot while they proceed to devour it. This behaviour allows the orioles to tackle large insects such as adult locusts *Locusta migratoria*, which most other passerines of the same size would find difficult.

Preening is performed in typical passerine fashion and the birds will bring their feet over the top of a lowered wing to preen the back of the head. Bathing occurs at least once a day, usually just after fresh clean water has been provided. Sunbathing will also occur in patches of strong sunlight and the birds will lower themselves onto their front and spread out their wing and tail feathers whilst doing this.

## Social Behaviour & Management

Montserrat Orioles are an extremely territorial species and can show persistent aggression towards con-specifics, which can result in fatalities if the individuals are not separated. Both adult males and females should not be housed together with the same sex as severe injury or fatalities will result. Adult birds should only be kept in compatible pairs or separately and birds kept in pairs should be watched closely at all times for increased aggression. The male is usually the aggressor although in some instances the female can become dominant. Aggression between the pair usually only consists of minor squabbles over insects and scattering the insects in different areas of the aviary can easily mitigate this. A male will also show aggression towards the female during incubation when she leaves the nest for food. As long as the pair is housed in a large spacious planted aviary with plenty of areas to seek shelter they should be able to be kept safely together throughout the year.

Adult pairs or singletons should not be kept in aviaries immediately adjacent to those housing other orioles of the same sex. Males particularly will fight through the netting resulting in injuries to the head or feet. Even when visual screening has been put up between the flights the birds will inevitably find a small gap where they can still see each other and continue to fight. On one occasion two adult females were placed individually in adjacent floor cages divided with 25mm weld mesh. The birds had to be separated almost immediately due to fighting with one bird sustaining severe head injuries. When introducing a pair it is recommended to have the female established in the aviary first. The male can then be confined to the shed with visual contact before introduction. Montserrat Orioles don't seem to be too fussy when it comes to mates and there has so far not been a case where a pair has proven to be incompatible.

Juvenile birds should be removed from the breeding aviary as soon as they have become independent (approx. 6 weeks old) as the adult male will become aggressive towards them, especially if the pair has started to breed again. One individual male at Durrell Wildlife was notorious for killing his offspring and would even start to chase them before they were feeding for themselves. In this case the male was confined to one half of the shed to allow the female to continue to care for the fledglings.

Juvenile birds can be grouped together in spacious aviaries with individuals from several clutches from different pairs. The largest group of juveniles at Durrell Wildlife consisted of 5 birds but it is felt the groups could be larger if the aviary is of a good size and well planted. Juveniles can be kept together with little risk of aggression towards each other until the second moult, which occurs at approximately 1 year of age, at which time the birds will start to get their sub-adult plumage and become territorial.

## Enrichment

As already mentioned the best enrichment for Montserrat Orioles is providing them with a large heavily planted aviary. Insects can be scattered amongst the vegetation encouraging the birds to forage. Rotten logs that have been left in aviaries containing orioles have been found pulled apart, presumably by the birds looking for insects, although this has not been witnessed. The large flower stems of New Zealand flax can also be given to Montserrat Orioles and, even though these birds are not thought to eat nectar in the wild, they will readily feed on the nectar in the flax flowers. New Zealand flax has evolved to be pollinated by birds and the orioles' foreheads become covered in pollen while they are feeding on the nectar, giving the birds the appearance of having a golden crown. The orioles have also been seen feeding on other flowering species such as honeysuckle, bromeliads and canna lilies and these could also be added for enrichment purposes.

## Mixed Species Compatibility

To date Montserrat Orioles have only been tried in the same aviary with a few other bird species, and mixed species compatibility is an area that needs further investigation. One pair at Durrell Wildlife have bred successfully when housed with a single male Edward's Pheasant *Lophura edwardsi*, where both species showed indifference to one another. Also, in preparation for a planned Caribbean exhibit, a compatibility trial has been carried out by mixing a single male Montserrat Oriole with a single male St. Lucia Parrot *Amazona versicolor*. This has proven successful in a non-breeding situation and further trials with breeding pairs would be useful. Mixing orioles with other passerine species has not been fully investigated and the only case at Durrell Wildlife was when a pair of Pekin Robins *Leiothrix lutea* were introduced into a large planted aviary containing a single female Montserrat Oriole. The birds seemed compatible initially until the oriole started nest building, at which time she become aggressive towards the Pekin Robins and they had to be removed.

## Captive Diets

The following is a diet sheet for Montserrat Orioles kept at Durrell Wildlife Conservation Trust.

Main daily feed given between 10.30 and 11.00am.
*Insectivorous mixture*

- 2 parts Orlux Uni Patee (Universal mix)
- 2 parts Orlux Tropical Patee (Low iron mix)
- 2 parts finely grated hard-boiled egg
- 2 parts finely grated carrot

- 2 parts Wholemeal breed crumbs
- 1 part soaked NutriBird T16 original (Mynah pellets)

All ingredients are thoroughly mixed together

*Fruit mixture*

All fruit is chopped into approximately 5mm square cubes and placed into a plastic container with holes drilled into the bottom to allow the juices to drain out.

The fruit mix contains equal parts of the following.

- Papaya
- Mango
- Melon
- Plum
- Black grapes
- Kiwi fruit
- Tomato
- Pear
- Banana
- Green grapes

Other fruits such as pomegranates, passion fruit, peaches and nectarines are also added when available.

Both the insectivorous mixture and fruit mixture is offered in a 5cm diameter stainless steel dish (one dish per pair of birds). Two Wax Moth larvae *Galleria mellonella* per bird is also added to the food dish before being given out. This dish is attached in an elevated position within the aviary. In addition a small piece of fruit is spiked onto the perching when the food bowls are distributed. This spiked fruit is given in a rotational basis from the following selection: papaya, mango, melon, pear, passion fruit and kiwi fruit.

*Insect feeds given at 08.00 and 16.00 (winter) or 17.00 (summer)*

Either of the following species of insects are offered during these feeds:

- Small handful of brown crickets *Gryllus assimilis*
- Small handful of black crickets *G. bimaculatus*
- Locusts *Locusta migratoria*, one per bird (both extra large and fully winged adults)

The insects are scattered amongst the vegetation to encourage foraging and to mitigate aggression between the pair.

N.B. Montserrat Orioles are not given mealworms *Tenebrio molitor* due to their tendency to become

**Fig. 10.** *Freshly prepared oriole dishes ready for distribution to the aviaries.*

obese (mealworms have a high fat content)

## Supplements

The insectivorous mixture is liberally dusted with Avimix multivitamin supplement daily before the mixture is mixed together. All the insects given are gut-filled with a vitamin- and mineral-enriched insect food, as well as vegetables and greens. All insects that are given when parents are rearing chicks are also heavily dusted with Avimix.

## Drinking Water

Drinking water is supplied in ceramic bowls, which are scrubbed out and replenished with fresh water during the morning rounds each day. In aviaries containing water features such as small creeks or waterfalls, ceramic water bowls are not necessary as long as the water features are topped up with fresh water on a daily basis.

## Reproduction

Montserrat Orioles attain adult plumage at 2 years of age. However only one female to date has successfully produced offspring in her 2nd year (when paired to an adult wild-caught male). Young pairs seem to need at least one breeding season to perfect their nest building skills. At this age, rather than building the neat cup-shaped nest usually constructed by this species, the birds will practice their weaving skills by building platform-like structures. It is generally not until the 3rd or 4th year that the cup-shaped nests are perfected and egg laying will occur.

The breeding season for Montserrat Orioles at Durrell Wildlife is between May and September.

**Nest Building Requirements**

In the wild Montserrat Orioles weave their intricate cup-shaped nests out of fine plant fibres and only the female builds the nest. These nests are commonly found under the large leaves of *Heliconia*. In the past it was thought important to provide the orioles with a suitable substitute for *Heliconia* and banana was chosen and subsequently planted in the breeding aviaries. Nests have been constructed under the large leaves of the banana, however many more are positioned in other plants such as bamboo and more commonly attached to the netting on the roof of the aviary. The orioles will often build the nests in quite exposed places and regularly build in the shed of the aviary, attaching the nest to a wire netting division or light guard. The birds seem to prefer to build their nests as high as possible in the aviary with just a small gap between the nest and the aviary ceiling for the female to squeeze through. In some cases the females choose a nest site that lacks suitable structure for attaching the nest, such as an exposed wire mesh division within the shed. Once weaving has been started in these areas improvements to the site can be made by attaching a small twig to the netting, thus providing more structure for supporting the nest.

Unravelled Hessian rope fibres are given to the orioles as a substitute for the fine strands of plant fibres used to build nests in the wild. This has proven to be ideal nesting material and large amounts can be tied to the perching in the aviary for the female to collect. Lengths ranging from 10 to 30cm have proven to be most useful.

Dried cordyline leaves stripped into long fine strands will also be used by the orioles as nesting material (G. Harrison pers. comm. 2006).

**Clutch Size**

Normally 3 eggs are laid; however clutches of only one or two eggs have been produced. To date only one clutch has occurred when the parents were able to parent-rear all three chicks to independence.

**Incubation & Fledging Period**

Incubation is carried out solely by the female and commences once the last egg in the clutch has been laid. Incubation lasts for 13-14 days.

Montserrat Orioles fledge at about 14-16 days old

Fig. 11. *Montserrat Oriole nest in the wild suspended under a Heliconia caribea leaf.*
*Photo by Gerardo Garcia*

Fig. 12. *Captive Montserrat Oriole chicks in the nest.*

Fig. 13. *Montserrat Oriole nest inside a shed with a small twig attached for extra support.*

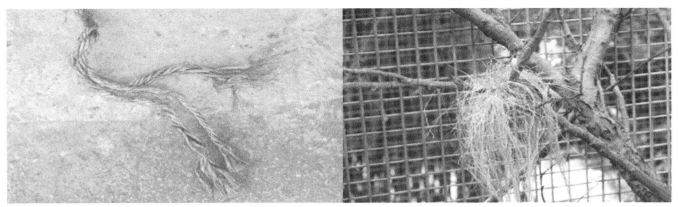

**Fig. 14.** *Ravelled and unravelled Hessian rope fibres material suspended in the aviary for the birds to collect.*

at which time they are fully feathered with dull grey-green feathers. The fledglings will continue to be fed by the parents until approximately 6 weeks old (Owen 2001).

**Chick Rearing Diet & Presentation**

During chick rearing the parents are given insects on an ad-lib basis. Soft bodied insects are offered while the chicks are still in the nest and these include the following:

- Brown crickets *Gryllus assimilis*
- Wax Moth larvae *Galleria mellonella*
- White skinned mealworms *Tenebrio molitor* (recently shed their skin)

When supplying insects ad-lib there is always going to be some wastage as the insects can move around (crickets in particular will tend to scurry into crevices such as those around doorframes, out of the reach of the birds). To mitigate this, the insects are offered in large ceramic bowls. This prevents the insects from crawling out due to the smooth sides of the bowl (Ward 2006). The bowls are placed on an elevated platform in a dry area of the aviary. All the insects given to birds rearing young are heavily dusted with Avimix multivitamin supplement.

**Age of Removal From Parents**

Adult male Montserrat Orioles can be come very aggressive towards their offspring once they have become independent. Thus it is important to remove the fledglings once they have been observed feeding for themselves. This should occur at approximately 6 weeks of age; however the chicks will continue to solicit food from the parents for several weeks after this (Owen 2001). Close observations should be kept on the adults during this time to ensure that the birds are not becoming aggressive and that no fatalities occur, especially if the parents have started another nesting attempt.

Ceramic bowl used for feeding insects to parents with chicks in the nest

**Hand-rearing**

Montserrat Orioles have proven to be very reliable parents and to date it has not been necessary to rescue chicks due to poor parenting on the part of the adults. However hand-rearing remained a very important aspect of the husbandry and captive breeding of this species that needed to be investigated. For this reason a hand-rearing trial was conducted in the breeding season of 2004. A clutch of 3 chicks was selected for the trial, of which two were removed for hand-rearing at 5 days old. The smallest and the largest chicks were removed leaving the middle-sized chick in the nest to be raised by the parents and to provide a comparison

**Fig. 15.** *Ceramic bowl used for feeding insects to parents with chicks in the nest.*

in the development between the hand reared and parent-reared birds. The chicks were kept in a small dish lined with crunched up tissue paper where they fitted snugly together which helped to prevent their legs from becoming splayed. The temperature was initially kept at 32°C, which quickly became too hot for the birds and they were soon being maintained at room temperature.

The hand-rearing diet consists of two parts finely chopped pinkie mice and one part finely chopped papaya, supplemented with Avimix multivitamin powder. This is a very simple diet and is quick and easy to prepare. The mixture was kept refrigerated between feeds and removed from the fridge 10 minutes before each feeding to allow it to warm up a little. Tweezers were used to feed the chicks suitable size pieces of this diet that they could readily swallow. Whistling, and tapping to create vibrations which simulate adult birds landing on the nest, was used to encourage the chicks to beg (Ward 2006). The chicks were started on two-hour feedings with the consistency of the pinkie/papaya mix becoming coarser as the chicks grew. Once the chicks stated to jump out of the nest bowl at day 14 they were moved to a small cage where they could be continued to be hand fed easily. Daily weights were taken of the chicks until 27 days old and morphometrics including the culmen, tarsometatarsus, gape width and beck height and width was taken daily until day 15. Weights of the chick remaining in the nest were taken on days 6, 8, 11, 13 and at 47 days old when this bird was removed from the parents and housed with the 2 hand reared birds. The weights show that the hand reared birds grew much faster than the parent-reared bird but at 47 days old the weights of all 3 chicks were in the same order as when the hand-

Fig. 17. *Montserrat oriole chicks being hand fed.*

rearing trial began (for complete hand-rearing daily record sheets see Appendix 3).

**Capture & Handling**

Montserrat Orioles should be captured with the use of the lightweight net that is deep enough to be twisted around once the bird has been caught to prevent it escaping. Mesh nets should not be used as the birds will easily get their heads, wings and legs stuck in the mesh and can be difficult to remove. Foam rubber should be fitted around the rim of the net to reduce the risk of injury if the bird flies into it. Once the oriole has been caught in the net the handler can then reach in and take hold of the bird by placing the hand over the back and the index and middle finger on each side of the neck. This will prevent the bird from flapping its wings and controls the head so the bird cannot attempt to bite. The hold should be firm but never tight and the bird must not have its breathing restricted in any way. Once the bird has been pulled from the net this hold will allow for easy examination of the bird's pectoral muscles to assess body condition. This hold is also useful for banding and checking the identification. The legs can be held with the other hand to hold them out of the way if the bird requires intra-muscular injections (Fig 20).

A second hold that can be used for orioles to allow for examination of the bird's back and wings is by holding the legs above the hocks with the fingers while securing the tarsus with the thumb (Fig 21). This hold is often used by bird ringers and is useful for taking a photograph without the body of the bird been obscured by the hands. There

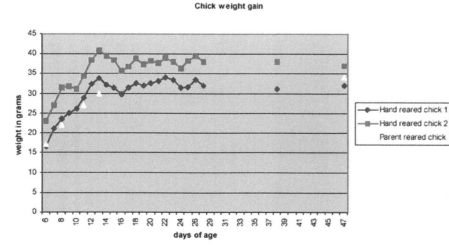

Fig. 16. *Graph of the weight gains of two hand-reared chicks compared to one parent-reared chick.*

Fig. 18. *Handling technique used for checking rings and intra-muscular injections.*

is a risk that the bird will escape if the hold on the legs is not secure as it will still be able to flap its wings.

## Health

The Montserrat Orioles at Durrell Wildlife undergo a bi-annual faecal screening examination for parasitology. Montserrat Orioles can be anaesthetised with isoflorine but should be considered as high anaesthetic risk.

*Pseudo-tuberculosis*

Several cases of Pseudo-tuberculosis have occurred in captive juvenile Montserrat Orioles. Infected birds show a lack of appetite and fluffed-up feathers. Small white lesions can be visible on the liver which can be

Fig. 19. *Handling technique often used for examining the back and wings.*

seen through the skin of the belly when the bird is in the hand. Antibiotic treatment should be given and if the symptoms have been identified early the prognosis is good. Any other orioles housed with the infected bird should also be treated with antibiotics to prevent further infections. Treatment can be given with either of the following two antibiotics.

Enrofloxacin (Baytril) 10% solution given in the drinking water. 100-200mg per litre (1-2ml per litre) of drinking water for 5 to 10 days

Ciprofloxacin given on drinking water.
100mg per litre (2ml per litre) of drinking water for 5 days

*Atoxoplasmosis and Coccidia*

The parasitic infection Atoxoplasmosis has occurred in both adult and fledgling orioles. Adult birds can be treated in the drinking water with the following:

Toltrazuril (Baycox)
25mg per litre of drinking water for 2 days with treatment repeated in 5 days time.

In recent years a significant percentage of fledglings have succumbed to this disease. Chicks can be treated pre-emptively whilst still in the nest with the first treatment at approximately 6 days old. The following dose rate applies:

Toltrazuril (Baycox) 12.5 mg per ml solution.
.01 ml per 10g body weight given orally for 2 days with treatment repeated in 5 days time.

Coccidia can be treated with Toltrazuril at the same dose rate as that given for Atoxoplasmosis.

## Routine Pre/Post Export Isolation

All orioles are quarantined before entering or leaving the collection and all birds require 3 clear faecal samples taken at 1 week intervals before leaving isolation (J. Lopez pers. comm. 2006).

## CONCLUSION

Montserrat Orioles have adapted well to captivity and will breed readily if the right captive conditions are provided. The information that has been gathered on the captive husbandry of Montserrat Orioles since the species was brought into captivity in 1998 would be more than sufficient to aid in the establishment of a captive breeding program should it become crucial to the survival of the species. However there is still much to be learnt about the captive management of Montserrat Orioles with areas such as artificial incubation, reasons for obesity and mixed species compatibility that still need further investigation.

## ACKNOWLEDGEMENTS

I would like to thank David Jeggo, Bird Department Head, Durrell Wildlife Conservation Trust and chairmen of the EAZA Passerine TAG, for his encouragement and advice in completing this paper.

I would wish to thank Gavin Harrison, Bird keeper at Edinburgh Zoo, for supplying me with information on the captive husbandry of Montserrat Orioles at Edinburgh Zoo. Javier Lopez, Head Veterinarian, Durrell Wildlife Conservation Trust, for information on the health requirements of Montserrat Orioles.

**Products Mentioned in Text**

Orlux Tropical Patee and Uni Pattee. nv ORLUX sa, Verbindingsstraat 20, B-8710 Wielsbeke-ooigem, Belgium. www.orlux.be

NutriBird T16 original. Versele-Larga NV, Kapellestraat 70, B-9800 Deinze, Belgium.

Avimix multivitamin powder. Vetark Animal Health, PO Box 60, Winchester, Hampshire SO23 9XN, UK

## REFERENCES

BIRDLIFE INTERNATIONAL. 2007. Species factsheet: Icterus oberi. Downloaded from http://www.birdlife.org on 21/5/2007

HILTON GM, GAL Gray, E Fergus, SM Sanders, DW Gibbons, Q Bloxam, C Clubbe and M Ivie (eds). 2005. Species Action Plan for the Montserrat Oriole *Icterus oberi*. Department of Agriculture, Montserrat.

JARAMILLO B & P Burke. 1999. New World Blackbirds, The Icterids. Christopher Helm Ltd, A and C Black, London, U.K.

OWEN A. 2000. The collection of eight Montserrat Orioles *Icterus oberi* and their establishment at Jersey Zoo. Dodo 36. Durrell Wildlife Conservation Trust. Jersey Is, U.K. Pp 51-61.

OWEN A. 2001. Species Management Guidelines - Montserrat Oriole, *Icterus oberi*. Durrell Wildlife Conservation Trust. Jersey Is, U.K.

WARD G. 2006. Passerine Husbandry and Conservation at Durrell Wildlife Conservation Trust. 2nd International Congress on Zookeeping, Congress Proceedings (CD rom).

## AUTHOR BIOGRAPHY

I was born and raised in New Zealand and have been keeping birds, and other animals, from an early age. I always knew that I wanted to work professionally with birds and achieved this when I started working at a small local zoo in my home town of New Plymouth. From there I moved to Auckland and spent time working with penguins at Kelly Tarlton's Underwater World and Antarctic Encounter. In 2000, I travelled to Cape Town, South Africa to volunteer in the rehabilitation of 20,000 oiled African penguins after the catastrophic "Treasure Spill". From here I was offered an internship with the International Bird Rescue and Research Centre in California, USA. On completion of this internship I became a member of the International Fund for Animal Welfare's oiled wildlife response team and have since responded to several major oil spills around the world. In 2002, I volunteered for six months with the Mauritius Wildlife Foundation, working on the Pink Pigeon recovery program. After this I joined the Durrell Wildlife Conservation Trust as a bird keeper and have since become senior keeper in charge of the softbill and pigeon section. My duties at Durrell include managing a large collection of softbills with a particular emphasis on passerines. I am also involved in Durrell's passerine conservation program have carried out fieldwork in the Seychelles and Mauritius.

## Appendix 1

### THE MONTSERRAT ALLIANCE

The Montserrat Alliance is a working group of conservation bodies including members from the following organisations:

- Montserrat Ministry of Agriculture, Trade and the Environment
- The Royal Society for the Protection of Birds
- Fauna and Flora International
- Durrell Wildlife Conservation Trust
- The World Wide Fund for Nature
- The Royal Botanical Gardens, Kew

## Appendix 2

### PLANT SPECIES USED IN MONTSERRAT ORIOLES AVIARIES AND THEIR USES

| Species | Plant Characteristics | Uses |
|---|---|---|
| Hardy Bananas *Musa basjoo* *Ensete ventricosum* | Probably the best species to use due to its hardiness. Important not to let this species dry out in summer and is prone to frost damage | Provided as a substitute for *Heliconia caribea* favoured by orioles to construct their nests in the wild, where they suspend them under the large leaves from the mid-rib |
| Palms Canary Island Date Palm *Phoenix canariensis* Chusan Palm *Trachycarpus fortunei* Mediterranean Fan Palm *Chamaerops humilis* Cabbage Palm *Cordyline australis* | (Technically not a palm) | Provide alternative perching and forest effect. The fine fibres produced at the stem of this palm can be used as nesting material by the birds |
| Tree ferns *Dicksonia antartica* | Suitable for outside aviaries as one of most hardy species in temperate regions | Excellent for the birds to forage in and adds a tropical forest effect |
| Evergreens Bay *Laurus nobilis* Elaegnus *Elaegnus x ebbingei* | | Both have been used to provide year round cover and seclusion |
| Shrubs Elder *Sambucus nigra* | | Generally a good structural plant providing good perching, seclusion and nest sites |
| Bamboos *Phyllostachys aurea* *P. bambusoides* *P. nigra* | 'Boryana' are all good species although most tall species are probably suitable | Provide perching and nest building opportunities |
| Climbers *Clamatis armandii* *Clematis montana* | All good climbers particularly *C. armandii* as it is evergreen | Ideal for the birds to forage in and provide suitable structures for nest building |
| Epiphytes Bromeliads *Billbergia nutans* and *B.x windii* | Both relatively hardy and survive the winter outdoors | Provide excellent enrichment as the orioles forage amongst them |
| New Zealand Flax *Phormium tenax* | This large clumping plant with rigid upright leaves | Produces 4m high flower stems providing nectar and pollen feeding enrichment for the orioles |

## Appendix 3

### MONTSERRAT ORIOLE 2004 HAND-REARING RECORDS AT DWCT

| ID No. | Date | Weight | Brooder Temp (°C) | Location | Comments |
|---|---|---|---|---|---|
| B5470 B5469 | 1/7 | 23g 16.57g | 32 | Nursery | Both chicks removed from the nest and placed in brooder in nursery for hand rearing. 6 days old. Third chick B5468 (weight= 17g) left in nest to be raised by the parents. Chicks fed on two parts finely chopped pinkies and one part finely chopped papaya every two hours between 06.00 and 22.00hrs |
| B5470 B5469 | 2/7 | 26.95g 21.2g | 27-29 | Nursery | Brooder turned off and maintained at room temperature with room heaters on. Day 7. |
| B5470 B5469 | 3/7 | 31.4g 23.6g | 25 | Nursery | Chick in nest being raised by parents B5468 weight= 22g |
| B5470 B5469 | 4/7 | 31.8g 24.97g | 25 | Nursery | Introduced thawed out frozen legless crickets to hand rearing diet. Day 9. |
| B5470 B5469 | 5/7 | 31.1g 26.0g | 24.5 | Nursery | Introduced thawed out frozen mealworms and wax moth larvae to hand rearing diet. Day 10. |
| B5470 B5469 | 6/7 | 34.3g 28.8g | 24 | Nursery | Chick in nest being raised by parents B5468 weight= 27g |
| B5470 B5469 | 7/7 | 38.4g 32.3g | 21 | Nursery | |
| B5470 B5469 | 8/7 | 40.7g 33.8g | 21 | Nursery | Chick in nest being raised by parents B5468 weight= 30g. B5470 observed making warning call. |
| B5470 B5469 | 9/7 | 39.5g 32.2g | 20 | Nursery | Both chicks fledged today and were moved to fledging cage in nursery. Day 14 |
| B5470 B5469 | 10/7 | 38.4g 31.5g | 20 | Fledging Cage | B5469 found to have lost nail, possibly caught in wire of fledgling cage. Fledglings fed five times during the day with first feed at 07.15hrs and last feed at 1900hrs. Diet consists of thawed out frozen crickets, live white mealworms, and live wax moth larvae with heads removed, chopped pinkie mice and chopped papaya. Fledglings also offered insectivorous mix and finely chopped fruit in a dish and a piece of spiked papaya. Water and some live insects also provided. Fledglings have not been observed feeding themselves at this stage. Both birds rung with plastic split rings |
| B5470 B5469 | 11/7 | 35.7g 29.8g | 20 | Fledging Cage | Fledglings being fed six times a day with first feed at 0800hrs and last feed at 2030hrs |
| B5470 B5469 | 12/7 | 36.65g 31.35g | 20 | Fledging Cage | Day 17 |
| B5470 B5469 | 13/7 | 38.72g 32.56g | 20 | Fledging Cage | Fledglings being fed seven times a day with first feed at 0820hrs and last feed at 2100hrs. B5470 seen drinking from water dish. |
| B5470 B5469 | 14/7 | 37.35g 32.05g | 20 | Fledging Cage | Day 19 |
| B5470 B5469 | 15/7 | 38.21g 32.57g | 20 | Fledging Cage | |
| B5470 B5469 | 16/7 | 37.74g 33.11g | 20 | Fledging Cage | Both fledglings seen feeding from spiked papaya. Day 21 |

| ID No. | Date | Weight | Brooder Temp (°C) | Location | Comments |
|---|---|---|---|---|---|
| B5470<br>B5469 | 17/7 | 38.9g<br>33.9g | 20 | Fledging Cage | |
| B5470<br>B5469 | 18/7 | 38.0g<br>33.5g | 20 | Fledging Cage | Fledglings moved to a raised cage in indoor bird room. Fledglings being fed five times a day with first feed at 0820hrs and last feed at 1800hrs. B5470 seen eating chopped piece of fruit |
| B5470<br>B5469 | 19/7 | 36.3g<br>31.5g | 25 | Bird Room (raised cage) | Fledglings fed eight times today with first feed at 0800hrs and last feed at 2030hrs. Started taking food from tweezers and swallowing voluntarily. Day 24. |
| B5470<br>B5469 | 20/7 | 38.2g<br>31.6g | | Bird Room (raised cage) | Fledglings fed seven times during the day with first feed at 0800hrs and last feed at 1845hrs |
| B5470<br>B5469 | 21/7 | 39.4g<br>33.5g | | Bird Room (raised cage) | Fledglings fed six times during the day with first feed at 0800hrs and last feed at 1815hrs. Both birds feeding from spiked papaya |
| B5470<br>B5469 | 22/7 | 38.1g<br>32.0g | | Bird Room (raised cage) | Fledglings fed five times a day with three hours between feeds. First feed at 0800hrs and last feed at 1845hrs. Stopped taking daily weights. Day 27 |
| B5470<br>B5469 | 24/7 | | | Bird Room (raised cage) | Fledglings fed four times today with first feed at 1030hrs and last feed at 1900hrs. Day 29 |
| B5470<br>B5469 | 25/7 | | | Bird Room (raised cage) | Fledglings fed four times today with first feed at 0830hrs and last feed at 1800hrs. B5469 seen using feet to hold crickets |
| B5470<br>B5469 | 26/7 | | | Bird Room (raised cage) | Fledglings fed five times today with first feed at 0900hrs and last feed at 1730hrs. Have eaten most of the crickets left in food dish overnight |
| B5470<br>B5469 | 27/7 | | | Bird Room (raised cage) | Fledglings fed three times today with feeds at 0830, 1300 and 1700hrs. Only taking very small amounts at each feed. Seen eating wax moth larvae. Day 32 |
| B5470<br>B5469 | 28/7 | | | Bird Room (raised cage) | Fledglings fed three times today at 0900, 1530 and 1844hrs |
| B5470<br>B5469 | 29/7 | | | Bird Room (raised cage) | Fledglings no longer being hand fed and are now feeding for themselves. Moved to large floor cage in bird room |
| B5470<br>B5469 | 2/8 | 38g<br>31g | | Bird Room (floor cage) | |

# BRITISH COLUMBIA WILDLIFE PARK CAPTIVE BREEDING PROGRAM FOR BURROWING OWLS *ATHENE CUNICULARIA*

Paul Williams & Tyna McNair

British Columbia Wildlife Park
9077 Dallas Drive, Kamloops, BC,
Canada V2C 6V1

## Summary

The Burrowing Owl *Athene cunicularia* a small ground dwelling owl, became virtually extirpated as a breeding species from British Columbia prior to the 1980s. In 1978, the Committee on the Status of Endangered Wildlife in Canada, listed the owl as a threatened species. In efforts to maintain biological diversity in the province, reintroduction through transplanting colonies of imported Burrowing Owls was initiated in 1983. The British Columbia Ministry of Environment recognized a need for a breeding facility in order to sustain a viable population within the BC region. It was thought that if successful, the breeding program could also potentially help populate other regions in Canada and the United States to down-list the Burrowing Owl status of endangered. In 1989, a Burrowing Owl breeding facility at the British Columbia Wildlife Park was initiated. Also at that time the province of BC developed a formal Burrowing Owl Recovery Team. The members of this team are: MOE, BCWP, Burrowing Owl Conservation Society of BC (Port Kells breeding facility), Stanley Park Conservation Society. To date BCWP has produced approximately 600 owls for release. Together with the owls from Port Kells, the program has released over 800 owls to the Kamloops Nicola valley region. There are over 700 artificial burrows that have been GPS mapped at the various release sites. This program has been operating for 15 years and will be reassessed in another 5 years for its feasibility for reintroduction and establishment of a self sustaining population of Burrowing Owls in the BC interior grasslands. This is a huge program that could not be undertaken without the support, partnership and cooperation of BC and Canadian government agencies, NGO's, corporate sponsors, landowners and all stack holders, volunteers, researchers and US government agencies.

## Introduction

The Burrowing Owl *Athene cunicularia*, a small ground dwelling owl, became virtually extirpated as a breeding species from British Columbia prior to the 1980s (Dunbar 1983; Wedgewood 1978). In 1978, the Committee on the Status of Endangered Wildlife in Canada (COSEWIC), listed the owl as a threatened species (Low 1995). In 1995, the Burrowing Owl was downgraded to endangered. The species was also designated endangered provincially under the BC Wildlife Act.

Although Burrowing Owls were never found in great abundance in the BC region, their range extended from the Cariboo-Chilcolton grasslands, down through Kamloops and the Okanagan area and various locations in the United States. In efforts to maintain biological diversity in the province, reintroduction through transplanting colonies of imported Burrowing Owls was initiated in 1983 (Dunbar 1983). The British Columbia Ministry of Environment (MOE), recognized a need for a breeding facility in order to sustain a viable population within the BC region. It was thought that if successful, the breeding program could also potentially help populate other regions in Canada and the United States to down-list the Burrowing Owl status of endangered. In 1989, a Burrowing Owl breeding facility at the British Columbia Wildlife Park (BCWP), was initiated.

### Natural History & Distribution

Burrowing Owls, also called gopher owl, rattlesnake owl and howdy owl, are distributed across southern Canada and the prairie provinces and throughout the western United States down to the Mexican border (Strode 1997). Isolated populations have also been located in suitable habitat near Florida, Cuba, Hispaniola and various other regions in Central and South America (Strode 1997). During the winter seasons the species migrate south during the September-October months and return to northern locations in

March-May. Based on banding recoveries, it is theorized that the Canadian owls migrate far south in a "leap-frog" migration (Strode 1997). The exact location of British Columbia's migrating Burrowing Owls is presently unknown (Leupin et al 1994).

## Habitat

Burrowing Owls occupy a unique habitat range unlike that of a typical owl. In contrast to a covered forest habitat, the Burrowing Owl prefers xeric to subxeric valleys which consist of low herbaceous vegetation (under 10cm) (Leupin & Low 1995). Low vegetation is a necessity for movement, hiding from predators, foraging and hunting efficiency (Johnsgard 1988). This dry open terrain also consists of existing burrows, natural perches and abundant food resources such as insects and rodents. This habitat is typical in the southern interior regions in BC and the prairie provinces. The distribution and abundance of Burrowing Owls is directly related to the availability of suitable habitat. The owls dwell in abandoned burrows of various ground dwelling mammals such as the Yellow-bellied Marmot *Marmota flaviventris*, American Badger *Taxidea taxus* and the Columbian Ground Squirrel *Spermophilus* spp. (Dunbar 1993; Leupin & Low 1995). The abundance of these burrowing mammals effects the species ecology by providing nesting chambers, sites for food storage, escape from enemies and shelter from temperature extremes (Johnsgard 1988).

## Characteristics & Appearance

The Burrowing Owl is unique in that it has a small rounded body with long, sparsely feathered legs. It is highly maneuverable with powerful talons for gripping. Adults, both male and female, are a rich sandy-brown color, thickly spotted with whites and buffs on the upper parts, while the underparts are whitish barred with brown (Blood 1993). The coloring of the owls allows for the greatest camouflage in short grasslands. The female owls tend to have slightly more barring and are darker than the male. This trait is thought to be an evolutionary adaptation since the female spends most of her time inside the burrow (Low 1995).

The average length of a Burrowing Owl is 24cm, with slight sex size dimorphism. On average, the males have longer tails, weigh 8g more and have 3-4 mm greater wing span than the female (Strode 1997). The average weight of a male is 158.6g, while the female averages 150.6 grams. The wing length for males averages 172.3mm, with the female at 170.3mm (Johnsgard 1988). From data collected At BCWP weights of adults can vary from as low as 150g at time of chicks fledging to as high as 210g in the fall. The differences are attributed to the owls foraging more in late fall to maintain a consistent body weight for winter.

The average life span of the Burrowing Owl is 5-8 years in wild, while captive bred owls with abundance of shelter and food, may live up to 10-1 2 years (D. Low pers. comm. 2007). The oldest recorded owl at BCWP was a male that live to 14 ½ years. The mortality of the owl is highly variable between specific populations and years with various factors contributing to success or mortality (D. Low pers. comm. 2007). Factors may include climate, abundance of prey and predation. Predators of the Burrowing Owl include raptors such as Red-tailed Hawks *Buteo jamaicensis*, Swainson's Hawk *Buteo swainsoni*, Great Horned Owl *Bubo virginianus*, Northern Harrier *Circus cyaneus*, Short-eared Owl *Asio flammeus* among various others. Other ground dwelling predators include Coyote *Canus lupus*, America Badgers, weasels *Mustela* sp., domestic dogs, foxes and snakes.

## Vocalization

The Burrowing Owl has a repertoire of a least 17 vocalization patterns with at least three by young owls (Johnsgard 1988). Adults produce a primary call, five calls associated with copulation and seven associated with nest defense and/or food begging (Strode 1997). The males primary song used in pair formation, pre-copulatory behavior and territorial defense consists of a double noted "coo-coo". The female calls associated with copulation are the "smack" and "eep". The male also utilizes the "tweeter" and "copulation song", while "warble" is utilized by both sexes. When threatened at the nest, adults will "chatter", "scream" and "chuck", while bobbing up and down, remaining erect (Strode 1997). The females' rasp tends to stimulate the male to begin foraging. The juveniles have three vocalizations along with bill snapping. Hunger or alarm will bring a low "eep" from chicks, and like the female will "rasp" to initiate foraging behaviors of adult male. When severely stressed the juveniles (and adult females), will produce a "rattlesnake rasp" which deters predators from entering the burrow. Along with the female making a contented "cluck-cluck" when the male is depositing food, most vocalization patterns are common for the owls at BCWP.

## Activity Patterns & Food Collection

The activity patterns of the Burrowing Owl vary during the time of year with the owl being mainly nocturnal during winter months and diurnal in summer (Strode 1997). In spring the owls are most active dusk until dawn, with reproductive activities occurring in late evening. Foraging and hunting occurs mainly in late afternoon until dusk; however the owl is highly opportunistic and may forage at any time if prey is available. Prey capture is done in various ways; ground foraging (insects), perch hunting, hovering and flycatching (Strode 1997). Prey is caught in talons and transferred to beak.

Burrowing Owls can usually be found living in loose colonies in the wild. Such groupings may be a response to local abundance of burrows and food, or an adaptation for mutual defense (Low 1995). Owls will aggressively defend the area around the nest burrow from intruders. This behavior has also been observed at BCWP where some of the male breeding pair owls would dive at the keepers head when they approached the nest chamber burrow opening.

Pellets collected in field studies by biologist Ernest Leupin (BCCF) in the Kamloops region, showed a diversity in prey dependent on the areas in which the pellets were collected. Burrowing Owl food caches include vertebrates such as amphibians, rodents, and birds. The prey base includes arthropods such as Coleopterans (carrion, ground and click beetles) and Orthopterans (crickets and grasshoppers), vertebrate prey including Masked Shrew *Sorex cinereus*, Deer Mice *Peromyscus maniculatus* and Meadow Vole *Microtus pennsylvanicus* (Leupin & Low 1995). Small birds and amphibians have also been collected from pellets, but are not often found. Burrowing Owls will consume 15% of their total body weight per day (Bryant, 1990). Undigested food (hair, bones, etc.) are regurgitated in form of 2-3 pellets (Johnsgard 1988). These pellets are of cylindrical shape, 3-4cm long and 1.5cm thick (Blood 1993). The daily food intake of owls at BCWP will be discussed with maintenance.

## Limiting Factors

The abundance of the Burrowing Owl is limited by the availability of suitable habitat and food resources (Dickinson et al 1994). Pest management of ground dwelling mammals and rodents has been found to be a primary concern of habitat decline. Further habitat loss is caused by agriculture, urban and industrial development, such as filling in of burrows for grazing livestock and plowing of native grasslands and farmlands (Bryant 1990). Other factors responsible for owl declines are uses of chemical pesticides and rodenticides, which is known to directly cause food loss or indirectly contaminate food resources. This contamination causes adult increased mortality and low reproductive success (Low 1995). Human caused mortality by shooting of owls or accidents are also a concern (Blood 1993). These negative factors along with natural predation and environmental hazards all attribute to the Burrowing Owl population decline.

## British Columbia Wildlife Park

### History

One of the first transplanting efforts of Burrowing Owls in British Columbia was made in 1980. The British Columbia Ministry of Environment (MOE) released an entire brood of Ontario Burrowing Owls in the Kamloops region. However, with the declining population of Burrowing Owls across Canada, MOE realized that in order to reestablish the owls as a breeding species, a facility was needed to ensure survival of individual owls along with increased numbers of Burrowing Owls for future release. Doug Jury (MOE) and Scott Mann (DMV), proposed the BCWP, as a suitable facility for captive breeding. The first owls arrived at BCWP in 1989 from Washington State and the Calgary Zoo. The first breeding occurred in 1991.

BCWP (established in 1966 and at that time called the Kamloops Wildlife Park or KWP) is a non-profit society with the motto and mandate of "conservation through education." BCWP is a CAZA (Canadian Association of Zoos and Aquariums) accredited zoo that mainly features native BC wildlife. In 2005 the Park entered into a formal Memorandum of Understanding (MOU) with the Ministry of Environment of BC. This has cemented a formal partnership with MOE in conservation and wildlife education matters concerning the province. Along with being part of the BC Burrowing Owl Recovery Team, the park is also directly involved in the public education of two other BC endangered species, the American Badger and the Woodland Caribou. BCWP is also a licensed wildlife rehabilitation centre and the only one in the Kamloops area. The park receives orphaned or injured wildlife from as far north as Prince George and as far south as Penticton.

## Recovery & Management

The Wildlife Park and the Kamloops region in general, was chosen as an ideal location for the breeding facility for many causal factors. The owls were originally more abundant in the Okanagan area, but due to increased development and orchard expansion, the Burrowing Owl population could not realistically be supported there. Kamloops provides suitable and abundant habitat for raising and releasing the owls, as well as the support system of BCWP and many local organizations and individuals such as landowners and ranchers. All of these factors allow the Burrowing Owls to achieve the greatest chance for survival. In 1988, by approval of MOE and cooperation with the University of British Columbia (UBC), BCWP was chosen as the official location for a breeding facility for Burrowing Owls. Initial funding was provided by MOE and the Provincial Conservation Assistance Fund (PCAF) which helped build the breeding facility in 1989. BCWP, with the support of MOE and numerous volunteers and donations, were dedicated to the success of the new facility.

The Burrowing Owl Recovery Team (BORT), was established in 1990 to plan, implement and supervise the recovery efforts for Burrowing Owls. The recovery team is governed by the BC Ministry of Environment. All of the Burrowing Owls in this program are owned by the province of BC. The other members of the team are as follows: The British Columbia Wildlife Park, The Burrowing Owl Conservation Society of BC (operating the Port Kells Breeding Facility) and the Stanley Park Ecological Society. A provincial recovery team and national recovery team, including Alberta, Saskatchewan and Manitoba, both aid in the recovery team. The goal of BORT is to achieve population and habitat objectives which allow for removal of the Burrowing Owl from the endangered species list. Objectives of the team include: increasing the number of suitable sites for release, reintroducing the owl into these areas, providing prime nesting habitat to increase breeding activity, providing protection for owls and public education, as well as supporting and maintaining programs to monitor and enhance the survival of the Burrowing Owl. In order to obtain these objectives and reestablish self-sustaining wild population in the Kamloops area, certain criteria must be met:

a. Aid with release efforts through the construction and maintenance of artificial burrows.
b. Monitor the behavior of released owls and to record patterns of burrow occupation.
c. Collect cast pellets from burrow entrances and use them to analyze food habits and seasonal dietary shifts.
d. Analyze macro and micro-habitats and determine burrow preference.

Burrowing Owl Mandate of British Columbia Wildlife Park with the Ministry of Environment

1. To secure a supply of releasable owls for MOE.
2. To expand the program by releasing increased number of Burrowing Owls into region.
3. To increase public awareness and support of wildlife enhancement projects, specifically those involving endangered species
4. In co-operation with other facilities within BC, to supply other regions with owls for release or breeding purposes.

## Breeding Facility Design

The construction of the original Burrowing Owl exhibit and breeding facility was completed in March 1990 with the help of KWP, MOE, BCCF and numerous volunteers and donators. The original pen design was drafted by Audrey, McKinnon & Partners with input from Scott Mann, Bob McGinn, Doug Jury and Steve Quiney. The facility design consists of a central solid wood constructed building 17 feet wide by 32 feet long with a concrete foundation and floor. This building is supplied with electric power for lights and for heat. Attached to the building and surrounding it are eight breeding pens, 12 feet wide by 20 feet long with an 8 foot ceiling height. Shade cloth was installed on any common wall separating breeding pens with the thought that this would aid in the relief of any territorial stress. The nest chambers for the breeding pens are located inside the main building where access is through an artificial tunnel to the outside breeding pen. Each breeding pen has an additional earth mound with a tunnel system and 3 perching areas. All of the artificial tunnels and burrows are made from different lengths of 6 inch big "O" weeping tile plastic tubing. They are perforated through out to aid in drainage during spring thaw and for heavy rains. This whole structure is surrounded by a continuous fly way that is 10 feet wide with an eight foot ceiling height. This is for flight exercise and is large enough for territorial development. The total length of the circular flyway is 368 feet and has numerous mounds with tunnel systems through out. Tunnel lengths are from 3 to 5 feet in length. All of the outside pen construction is with 4x4 post and 2x4 frame construction covered with 16 and 18 gauge 1 inch galvanized weld wire. There is a digging barrier around the whole structure as well as between each breeding pen. A guard fence was built 12 feet from the pen with one side completely blocked off from public view in order to allow for the birds to remain wild for release. An over head irrigation system was put in place to encourage proper growth of native grass and habitat. Educational signage and pictures are included for visitors which explain the owl program and reflect the work of BCWP, BORT, organizations, businesses and numerous volunteers. A boardwalk encloses the front and sides of the adult pen allowing for public viewing, yet allows the owls to remain private.

A new juvenile pen was constructed in 1996 against the backside flyway of the original pen. This new area was constructed for the juveniles to be overwintered in where they are separated into a male and female area. This will eliminate any juvenile breeding prior to the spring release and any aggressive competition from the adults. This separate juvenile pen was needed because of the increased production during the years prior to 1996. The juvenile pen measures 24 feet wide and 90 feet long and is divided down the middle for the male and female sides. Both male and female sides have 3 earth mounds with 3 separate nest chambers per mound. There are multiple perches through out. This pen is not accessible to the public for viewing.

In 2001 and 2002, there was a drastic decline in the production of the owls. The reasons for this were that park development was now too close to the breeding facility and the breeders were aging. A major park development was placed within 40 feet of the facility which resulted in extensive heavy equipment usage during the breeding season. This combined with the aging breeders (at least half were between 9 and 11 years of age) caused the normal production levels to be cut in half. Fertility rates and nesting success rates declined sharply. Two things needed to be done. Build a new larger facility in an isolated area of the park with no public viewing, and acquire enough new young breeders to replace the aging non productive ones. The park did not have enough non related owls in residence to increase production significantly.

In 2002, the park acquired a grant from Human Resources and Service Development Canada (HRSD) to start the development of a new breeding facility. The remainder of the cost for the facility was supplied by the park. This new facility is roughly twice the size (13,417 sq. ft.) of the original facility and is built on the same basic design as the original. There are 13 breeding pens and two separate juvenile pens. The male juvenile pen is attached to the south end of the facility and the female juvenile pen is attached to the east side of the facility. It was built to house 80 to 100 juveniles. This facility was completed in the fall of 2002 and was used for the 2003 breeding season. This facility design has been duplicated at the Port Kells breeding facility.

In both 2003 and 2005 the MOE of BC acquired an import permit for up to 7 wild caught juvenile Burrowing Owls from the Washington tri-cities area for each of those years. With the help of local Washington Fish and Wildlife officials and biologists, the BC recovery team was able to bring back six owls in 2003 and in 2005. This influx of unrelated young breeders has increased the production at the park dramatically as well as allowed an increased level of breeder owl exchanges between the park and the Port Kells breeding facility.

**Maintenance, Diet & Feeding**

The breeding facility ensures that there is an abundance of food for the young owls to be physically fit for release as well as to maintain the permanent breeders in optimal health. Generally there is over feeding to ensure there is limited competition and aggression. The captive owls are fed on a weekly schedule. All full grown owls are fed the equivalent of two food items per day. On Mondays, Wednesdays and Fridays the each owl is fed one mouse and one day-old chick. On Tuesdays and Thursdays each owl is fed two day-old chicks and approximately a dozen thawed crickets with a supplemental mineral powder. On Saturdays and Sundays each owl is fed two day-old chicks with supplemental mineral powder. The amount of mice is increased prior to egg laying of the paired breeders and in the first 10 days of the chicks hatching. Mice are the preferred food item of owls at BCWP. High costs of rearing and purchasing of mice limits the diet of the owls, but BCWP does recommend that mice and crickets be available more often if funding is prevalent. On average, an adult captive bird will consume 1-2 food items daily. When nesting, the male owl collects the food items and deposits them at the burrow entrance for the female to feed her young. The young will also consume 1-2 chicks daily, but are fed smaller amounts approximately every hour by the female. During the first five days after hatching, owl chicks will only consume one food item daily.

Maintenance of the Burrowing Owl exhibit is

continuous. The burrows and nesting boxes are spot cleaned three times per week and completely cleaned and lined with a new substrate three times a year. This is done after fledging in mid summer, prior to winter set up at the end of October and at breeder pairing time at the end of February. The substrate consists of 3-4 inches of top soil and peat moss. Bison dung is also place in burrow entrance for owls to line nest chambers. This is their preferred nesting material. Perches are cleaned on a regular basis. Regular maintenance prevents parasitic infestation and disease from bacteria build up. Periodically the juveniles have had problems with flea infestation, but this situation was overcome by removing the substrate in burrow more frequently (3 times a year) and using a flea spray 3 times per year. Although owls in general require very little water, the defrosted feed contains less moisture so the owls are supplied with fresh water for each separate enclosure. Debris and old feed is collected daily while new food is being deposited. Other maintenance such as raking enclosures and grass cutting is done as needed. Maintenance of pens is done quickly and efficiently to ensure the amount of human contact is limited, thus allowing the birds to remain as wild as possible. At BCWP only one person enters the pens to do the daily feeding and cleaning, while large maintenance may require two or more people.

Winter maintenance requires that the owls are well protected from the harsh climates. Feeding is increased to twice a day when the temperature is consistently below -10 degrees Celsius. The main reason for the increased fed is that the day-old chicks freeze in winter and become inedible. Many owls in an attempt to defrost the prey item, will crouch over the frozen chick and their chest becomes iced-up. This causes a decrease in the birds body temperature and can prove fatal if not noticed. Any birds showing an iced-chest are brought inside to the park hospital and kept there until healthy enough to return. Snow is removed at all burrow entrances as it falls and straw is placed there to keep them off the frozen ground as they enter and leave burrows. Outside nest chambers are lined with straw for added insulation as well.

Twice a year each individual bird is given a physical examination and weighed.

**Captive Breeding**

Mating in captivity occurs much like that of mating in wild. Burrowing Owls return from migration in early spring and begin the search for a mate. Pair bonding at BCWP usually begins in February. The males occupy a nesting site, usually one of the breeding pens, in which they will stand near the burrow entrance and utilize their primary song to attract females. In captive breeding the owls which have not already paired will be paired by staff in an appropriate pen in order to initiate the breeding process. Once enticed, courtship antics involving various postures and displays are made by both sexes. While singing, the male will bend forward, almost parallel to the ground with his primaries and secondaries held together over his back and white patches of the throat and brow fully displayed (Strode 1997). Other courtship behaviors include circular flight, billing and preening of head and face (Strode, 1988). Most often seasonal pairing is monogamous between the two owls. It is highly unlikely that wild owls will breed together for more than one season, but in captivity continuous pairing is common. Quite often at BCWP a previous years' pair will even use the same pen as past years for their territory. Both sexes are quite capable of breeding at one year of age, but the inexperience of copulatory behavior in a young male and the physical intrusion of an older male, lessen the chance of success. Forced pairing is used, where the breeding pairs are decided by park staff and placed in separate breeding pens at the end of February. As mentioned above, sometimes the owls will pair themselves and as long as they are not related and have had good success in the past, they are allowed to do this. The pair remains together in the breeding pen approximately until the end of June when the offspring fledge.

Copulation is done at the burrow, most often at dusk. The female will move towards or away from the singing male at the start of copulatory behavior. The male will then stop singing, stand and look down at the female with white patches exposed and feathers raised. The female will stand erect with exposed white patches. The male will then fly to the female and mount her, giving the primary call with or without the male warble and may terminate with a tweeter call. The male flaps his wings while mounted, probably for balance. He may scratch the female's head and both may bill nip. Copulation lasts 4-6 seconds and is done 1-3 times per evening.

**Owlets**

The clutch size for captive Burrowing Owls is the same for captive birds as wild birds, 6-12 eggs per clutch. The owls have the largest normal brood size of any North American owl (Weidensaul 1989). Egg laying at BCWP begins in mid March to early April. The eggs are white and weigh approximately 10.5g. The average length of an egg is 32.2 mm (n=21), average width is 25.7mm (n=21). Incubation takes approximately 24 days and is done solely by the female. Eggs are laid at a rate of 1.5 per day and are hatched in the order they were laid. At BCWP most females do not begin incubation until they have laid 5 to 8 eggs. Double clutches have been observed at BCWP on occasion and at one release location. If a female looses eggs or young early in spring, she will double clutch usually within 12-14 days. From analysis of breeding data at the park from 1991 to 2003, it was found that female reproductive success drops sharply after the 9th year of age. Clutch sizes decreased slightly while fertility rates dropped dramatically. Also nesting success rates decreased significantly due to abandoned nests or smashed eggs.

During incubation time, the female is highly secretive and rarely leaves the burrow. In the wild the male feeds the female and remains within 250m of the burrow (Johnsgard 1988) but is not permitted inside the burrow. If an intruder attacks, the male will sound alarm calls and swoop down to deter invaders. Each day when staff feed the owls, alarm calls can be heard by the males, as well as "rattlesnake rasps" from inside burrows from females and chicks. Although the chance of abandonment can occur in captive as well as wild situations, the male usually remains in the direct area of the burrow and continually collects food for the female and chicks.

Burrowing Owl chicks weigh approximately 8.9 grams upon hatching and are partially covered with a whitish down. After 24 hours, all chicks should have full stomachs and at 8 to 10 days their eyes fully open. It is at this stage that the chicks show evasive behavior and produce "eeps" for food and "rattlesnake rasps" if disturbed, although some chicks have been observed "rasping" before the eyes are open. Nest observations at BCWP are made every second day with no apparent stress to the female, who quickly runs up the burrow when the nest chamber is opened. Egg production, egg weights and dimensions, behavior, owlets weights and banding of broods are all recorded during observations. Because Burrowing Owls exhibit asynchronous hatching, if there is a large clutch size, the last chick to hatch usually is in danger of being trampled by its much larger siblings. If the chick is able to eat and grow through the 8-10 days, it will be of equal size to its siblings by 14 days. If owlets are in danger of not surviving, they are pulled by staff from the nest and hand-raised until the 4 week stage when it can eat whole food items on its' own, then fostered back to the same brood. BCWP has observed no problems with these fostered chicks integrating back into the brood and remaining wild. Hand-raising requires a special diet for the owlets.

Day 1: owlet is not fed for 24 hours.
Day 2-10: owlet fed deboned mouse with no stomach or intestines but with internal organs and some skin and water.
Day 10-14: owlet fed whole mouse cut up with bones and water.
Day 14- 21: owlet fed finely chopped mixture of chicks and mice in water.
Day 21- 28: owlet fed cut up mixture of chicks and mice in dish, no longer needs to be hand fed.
Day 28+: owlet feeding on whole food items.

At 14 days the chicks acquire feathers, but still lack the full feathers and barring of adults. As the chicks become less dependent, the female will leave the nest and begin to forage for food. When the nest becomes crowded, the chicks will venture to the entrance of the burrow. At 10-21 days the chicks will emerge from the burrow, but remain cautiously close to the entrance. Being unable to fly, the chicks can be seen stretching and flapping their wings as if preparing themselves for flight.

By 5 weeks the chicks will begin to fly outside the nest in the breeding pen and independently collect food. By 6-8 weeks owls have fledged, developing full feathers and are able to independently fly and forage. It is at this stage that the owls are processed: banded, weighed, blood samples taken for DNA sexing and the breeding pens are opened for access to the flyway. Clutch survival in the wild is food based with the highest mortality being juveniles. In captivity, the owls are given sufficient supply of food for survival and thus juveniles have high fledgling success at the park. Once the owls have surpassed the critical wintering stage, the owls are ready for release as yearlings.

**Banding & DNA Testing**

Chicks that have fledged are banded with an "in house" plastic spiral pigeon bandette. These bands are color coded and number sequentially. The color and leg correspond to the female parent of each brood. There are 12 possible color and leg combinations with the type of bands the park uses. The banding system aids greatly in identifying broods when all the owls have access to the common flyway as well as simplifying record keeping.

During the processing stage of fledglings, blood samples are taken by way of the toe clip method. Two samples of blood from each owl are placed on a piece of blotter paper and labeled to identify the owl. To insure there is no contamination of samples, all staff wear latex gloves while processing and toe clip equipment is cleaned with alcohol between each use. The samples are left to dry for a full half hour covered, before being placed in individual envelopes. These samples are sent to Thompson Rivers University (TRU) in Kamloops for testing.

TRU professor, Mairi Mackay, has designed a sex determination test adapted from "Sex identification in birds using two CHD genes", described by Richard Griffiths in the Proceedings of the Royal Society of London, 1996. The DNA is isolated from the blood. Using primers described in Griffiths' paper, the polymerase chain reaction (PCR) is used to amplify a 110 base pair fragment of DNA from the genes. Male birds carry two autosomal copies of this gene, while female birds carry a third copy on the female-specific W chromosome. The fragment derived from the autosomal genes can be cut with an enzyme (HaeIII) into two smaller pieces, while the fragments derived from the W chromosome remains uncut. The different sized fragments are separated by electrophoresis in a polyacrylamide gel and visualized by staining. DNA from female birds can be distinguished from male birds by the appearance of a 110 base pair fragment Sex determination is a necessity for separating the sexes to prevent breeding before release and for successful bonding for released owls.

**Pre-Release Training**

Pre-release training with live prey is done with the juveniles twice a year. It is done once in the fall when the juveniles are still with the adults prior to there separation at the end of October and once again in the individual juvenile pens for the month of April prior to the wild release. Shallow childrens wading pools are placed in the flyway in the fall and some of the regular food items are placed in them. This allows the owls to get used to the pools as a feeding area. As soon as they are eating out of the pools live mice are placed in them. This process allows the juveniles to see the adults hunt from these pools and will more quickly engage in this activity. This process is repeated again in April when the juveniles are separated from the adults and by sex. A successful capture consists of an owl flying down to capture a prey in talons, flying to a perch and biting of the prey's head. Most of the owls have no difficulties catching live prey and pre-release training is accomplished relatively easily.

**Release of Burrowing Owls**

**Criteria**

Sex determination of the owls is an extremely important factor before releasing the owls. In October, the juvenile sexes are separated to prevent pair bonding and egg laying at the facility. Until 1996, Mairi MacKay was still perfecting the techniques of the DNA sex determination test, therefore separating of the sexes was done by observation and physical characteristics. All owls are examined prior to release to assess general condition and body weight. If body weight is lower than 150g the owl is held back.

The month in which the juvenile owls are released is dependent on the weather patterns throughout the winter. If a particularly hard winter was prevalent, the owls may be released as late as May. Generally the owls can't be released before the end of April, even if it had been a mild winter because of the annual hawk migration through most of the release areas. Another factor which determines time of release is the abundance of prey and the predominance of predators (Leupin & Low 1995). Owls are released in single paired clusters in an attempt to increase breeding success. Genetic viability is also increased by ensuring that the owls released are not siblings. Release pairs are held in crates together over night before the release to help establish the pair bond. All these factors will limit or increase the success of a release and possibilities of breeding success in the wild.

Each owl is banded with a standard metal US Fish and Wildlife band and a metal bi-colored and numbered anodized band. The green over black coloring of these bands is specific to BC Burrowing Owls. Intensive monitoring after release is necessary in order to collect data on breeding and migration. Radio collaring of owls has been sporadic in the past due to financial constraints for the recovery team. For the last two years Aimee Mitchel, a masters student from UBC, has been radio collaring juvenile released owls as well as their offspring to track movements and dispersal. The radio collars are placed on the owls 24 hours before the release and can last for three months. Aimee is also studying how effective soft releases are in regards to early survival rates. This is done by placing wood or pvc tubing framed cages over the release burrows of selected pairs and keeping them there until egg laying is initiated. Supplemental feeding is a must in these cases. Early data suggests that the soft release cages significantly increases the chances of survival as well as nesting success.

**Release Locations**

Release sites are predetermined by members of BORT, after the site has proven to meet the criteria of BORT. They are all located in the Kamloops and Nicola valley area. The land must be abundant in prey and suitable habitat for the owls. It is critical that the land chosen is not considered for any further development (E. Leupin pers. comm. 2007). Many local landowners and ranchers are dedicated to helping monitor and protect the area, thus increasing the habitat for the owls. Each year existing burrows are prepared and new burrows constructed in release locations. To date over 700 artificial burrows have been installed at the various release sites, and most if not all have GPS coordinates. Annual releases occurred at most chosen locations, but if a location from a previous year proves to have negative factors such as high predation or consistently low prey abundance, the location will be abandoned.

Each location is chosen for its vegetation composition, structure, topographical features (slope, orientation) and drainage (Leupin & Low 1995). Suitable release sites include moderately grazed grasslands. Aspects vary from moderate through gentle rolling slopes with suitable native grasses and moderate food supply. Native grasses may include Bluebunch Wheatgrass, Crested Wheatgrass, Ponderosa Pine, Junegrass, Rough Fescue, Needle and Thread and Rabbit Brush (Leupin et al 1994; Leupin & Low 1996). Vegetation must be low or will inhibit the owls' ability to hunt. Virtually all sites have only artificial burrows.

Artificial burrows are a necessity due to habitat loss. The burrows that are constructed at release sites differ greatly from the facility. While the same plastic piping is used, the nesting chambers are made from three 4 gallon pails, which allow for easy access to monitor the owls. The bottom pail is the nest chamber while the top pail acts as the receptacle for the third insert pail. The third pail is filled with dirt for insulation and covers the access hole from the top pail into the nest chamber pail. The burrows sites are chosen to allow for development, drainage, prey availability and low predation (Dickinson et al 1994). Satellite burrows are also created near the artificial burrow. These burrows are created much like the artificial burrows, but are used as secondary burrows. Burrowing Owls use the satellite burrows to distract predators or intruders away from the main burrow site by entering the "false" burrow. Burrow occupancy is determined by pellets, scats and feathers. Owls will not be released into a burrow known to be inhabited by another Burrowing Owl in the area.

After migration the owls will normally return to the same site as previous year, but rarely the same burrow, so release sites and burrows are carefully monitored and kept available for returning owls.

**Education at BCWP**

Over the years the role of zoos has changed from a place to visit to a place to learn. Zoos are now directly responsible for helping educate the public and care for endangered species. As stated before, the BCWP motto is "conservation through education" and education is a key component of the success of the BC captive breeding and release program for Burrowing Owls. Areas that need to be discussed and understood by the public include grasslands conservation and the habitat needs of the species that live there. This will not only help the Burrowing Owl but will help other rare and endangered species such as the Badger and Sharp tailed grouse. The park has had two education Burrowing Owls. The first owl was called "Scout" and was known by name to thousands of students and visitors of the park from the Kamloops area. Scout was born in 1992 and had to be pulled from the nest to be hand raised as he was the last to hatch from his clutch and too small to fend for himself. Scout lived for 14 ½ years and was actively used in education programs for 12 of those years. He was also actively reproducing with his mate "Sage" until he was 12. The second owl is called "Quitemie" and was born in 2003. She too had to be pulled from her nest and hand raised as the runt of the brood. Quitemie has taken over where Scout left off. The use of a live animal that people can view close up, touch and form an emotional link to, is invaluable when trying to deliver critical information such as grasslands conservation. The understanding of this concept is central to the sustained use of our grasslands resources by all stake holders.

## DISCUSSION

During the seventeen years BCWP has been involved in this program many things have been learned, improved and fine tuned. All have helped to improve the success of the recovery program.

In the original breeding facility the tunnel systems were made of wood which only lasted a few years before they decayed and collapsed. This resulted in the use of the plastic perforated tubing at the facility as well as at the release sites. Wooden nest chambers were used at the release sites as well and also decayed quickly. The plastic bucket system now used has been improved upon each year as to bucket size and material type as well as connecting systems. Wooden nest chambers are still used at the breeding facility as they are easy to replace in a captive situation and provide more insulation in the winter. The use of shade cloth to add as a sight barrier to adjacent pens reduced stress levels in birds and increased nesting success. Record keeping systems have been developed as well as maintenance schedules. Hand-rearing of chicks that needed to be pulled from nests, and the best fostering time has been determined. Replacement times for aging breeders have been determined from breeding record analysis. Forced pairing was incorporated in to the breeding protocol in 1995 due to low production rates in 1994 and has increased rates. Both breeding facilities are large enough and have enough founding birds now to meet the BORT goal of 100 owls released each year. We have met that goal in 2006 and 2007. The banding systems used has been vastly improved upon. All owls must have the metal US fish and Wildlife band. Originally the owls were also banded with a plastic colored and number band like the "in house" bands. These bands fade and become brittle with age and made identifying birds in the field difficult. We now use the metal bi-colored bands with large numbers for easy identification. The use of soft release techniques has improved the survivability and nesting success of release birds but is costly and very labor intensive. Continued radio collaring of adults and juveniles will assist in determining habitat range and use.

## Conclusion

To date BCWP and the other members of the recovery team have fulfilled the following short-term goals:
- Initiating a captive breeding program
- Establishing two breeding facilities
- Conservation, assessment and protection of grasslands habitat
- Artificial burrow installation
- Yearly release and monitoring of juvenile owls from facilities
- Public education and publicity for the Burrowing Owl

To date BCWP has produced almost 600 owls for the release program. Over 800 have been released by both facilities combined. This effort takes a tremendous amount of financial and physical support. Funding for the program has always been a problem and will continue to be the major stumbling block. The continued support of both breeding facilities, public and corporate sponsors, ranchers and landowners, field crews as well as a large volunteer base is critical to the success of this program. Monitoring of the owls when they migrate south is something that must be added to the program to determine their real survivability. We know that 2 owls had migrated as far as California in the 15 years that the program has been working. Nothing is known about the other owls other than the return birds each spring. We must continue to partner with all stake holders of the local grasslands to maintain its' sustainability and work with other recovery teams to pool resources.

The BC Burrowing Owl Recovery Program will reassess in another 5 years to determine if it is feasible to establish a viable self sustaining Burrowing Owl population to the interior BC grasslands.

If it is determined that is not, at least we have done the ground work for other areas in North America that may have to resort to this extreme type of program when their endangered populations reach critical levels.

## References

BLANK RI. 1993. Reintroduction of Burrowing Owls into British Columbia. Unpublished Report. University of BC.

BLOOD DA. 1993. Wildlife at Risk - Burrowing Owl. Published Report. BC Ministry of Environment, Lands and Parks.

BRYANT AA. 1990. A Recovery Plan for the Burrowing Owls in British Columbia. Unpublished Report. BC Ministry of Environment, Lands and Parks.

DICKINSON T, E Leupin & M Murphy. 1994. An Ecological Assessment of Sites Used for Reintroduction of Burrowing Owls near Kamloops in 1993. Unpublished Report. BC Ministry of Environment, Land and Parks.

DUNBAR David L. 1983. Preliminary Recovery Plan for Burrowing Owls in British Columbia. Unpublished Report. BC Wildlife Branch. Victoria, BC.

JOHNSGARD PA. 1988. North American Owls; Biology and Natural History. Published by Smithsonian Institute. Washington, DC.

LEUPIN E, D Low, M Murphy. 1994. Burrowing Owl Releases in the Thompson-Nicola Region: Habitat assessment and management recommendations-1994. Unpublished Report. Ministry of Environment, Lands and Parks.

LEUPIN E, D Low D. 1995. Burrowing Owl Releases in the Thompson-Nicola Region: Habitat assessment and management recommendations-Unpublished Report. Ministry of Environment, Lands and Parks.

LEUPIN E, D Low. 1996. Burrowing Owl Releases in the Thompson-Nicola Region: Habitat assessment and management recommendations-1996. Unpublished Report. Ministry of Environment, Lands and Parks.

LOW D. 1995. Burrowing Owl. Unpublished Report. Royal British Columbia Museum with Ministry of Environment, Lands and Parks. https://rbcm1.gov.bc.ca/End_Species/es_franc/species/burowl.htm#risk.

STRODE Y. 1997. North American Regional Studbook for the Burrowing Owl. First Edition Racine Zoological Gardens. Racine, WI.

WEDGEWOOD JA. 1978. The Status of the Burrowing Owl in Canada. Published Report Committee on the Status of Endangered Species in Canada. Toronto, Ca.

WEIDENSAUl S. 1989. North American Birds of Prey. Quintet Publishing. New York, N.Y

## Author Biography

Graduating with a Bachelor of Science from the University of Manitoba in Winnipeg in 1978, Paul then worked at Assiniboine Park Zoo as a Zookeeper both seasonally and full time from 1974 to 1985. He began working for the Fish and Wildlife department in northern Alberta as a Habitat Technician in 1986. From 1990-1994 he worked at Calgary as a zookeeper and then moved to British Columbia Wildlife Park from 1994 to 2001 as a keeper. In 2001, he was promoted to animal care supervisor and has been running the Burrowing Owl breeding program since that time as well as being a director on the Burrowing Owl Conservation Society of B.C.

# FORTY YEARS OF BREEDING AMERICAN KESTRELS *FALCO SPARVERIUS* IN CAPTIVITY: WHAT HAVE WE LEARNED?

David M. Bird & Lina Bardo

Avian Science and Conservation Centre,
McGill University, 21,111 Lakeshore Road,
Ste. Anne de Bellevue, Quebec H9X 3V9 Canada

## Summary

The American Kestrel *Falco sparverius* is the smallest and most numerous North American falcon, and the only kestrel species to occur in the western hemisphere. This cavity-nesting, open-habitat-dwelling raptor is sexually dichromatic and can be aged at 12 days, ranges in size from 80 to 165g with females generally 10 to 30% larger than males, and is sexually mature at one year of age. As a result of the ease of breeding and housing American Kestrels in a captive setting, these falcons have been used for laboratory research over the past four decades, particularly in the field of toxicology but also for studies on reproduction, physiology, morphology, behavior, veterinary medicine, nutrition, and parasitology. In this paper, we describe the characteristics that make the American Kestrel such a useful and indeed invaluable animal for captive research. It also contains a compilation, though by no means exhaustive, of studies that have been conducted on captive kestrels with emphasis placed on research conducted after the original review evaluating the kestrel as a laboratory research animal by the senior author in 1985. With many generations of American Kestrels having been bred at the Avian Science and Conservation Centre of McGill University for 35 years and a local nest-box program existing for at least a dozen years, a very exciting and perhaps unique opportunity has been provided to investigate the impacts, if any, of long-term captive breeding on various aspects of the biology of a wild, semi-altricial avian species.

## Introduction

The American Kestrel *Falco sparverius* is the smallest and most numerous North American falcon, and the only kestrel species to occur in the western hemisphere (Willoughby & Cade 1964; Smallwood & Bird 2002). It can be found from Alaska to South America, and in the northern portions of its range the species is known to be migratory, though with considerable variation between individual populations (Smallwood & Bird 2002). This cavity-nesting, open-habitat-dwelling raptor is sexually dichromatic and can be aged at 12 days, ranges in size from 80 to 165 g with (females generally 10 to 30% larger than males, and is sexually mature at one year of age (Bird 1982; Smallwood & Bird 2002). As a result of the ease of breeding and housing American Kestrels in a captive setting, these falcons have been used for laboratory research over the past four decades, particularly in the field of toxicology but also for studies on reproduction, physiology, morphology, behavior, veterinary medicine, nutrition, and parasitology. The value of kestrels in laboratory research became especially known after various studies conducted on captive kestrels confirmed that DDE, a metabolite of DDT, could cause thinning in eggshells and thus, reproductive failure in birds (e.g. Wiemeyer & Porter 1970; Bird et al 1983; Blus et al 1997). The species subsequently served as a model for management practices for endangered Peregrine Falcons *F. peregrinus* when DDT caused their populations to decline drastically (Bird 1985). American Kestrels were used to study sperm production in falcons and to develop artificial insemination techniques in raptors, and were the first species to have been bred successfully from cryopreserved sperm (Bird et al 1976; Bird & Laguë 1977; Bird 1985; Brock & Bird 1991; Gee et al 1993). Reproduction, fertility, clutch size, lay dates, egg characteristics, effects of renesting and success of artificial incubation and hand-rearing have also been studied in the American Kestrel (e.g. Bird and Buckland

1976; Bird & Laguë 1982a, b, c). The American Kestrel was among the first avian species, and certainly the first raptor, studied to demonstrate the presence of sperm storage glands in the oviduct (Bakst & Bird 1987).

In this paper, we describe the characteristics that make the American Kestrel such a useful and indeed invaluable animal for captive research. It also contains a compilation, though by no means exhaustive, of studies that have been conducted on captive kestrels with emphasis placed on research conducted after the original review evaluating the kestrel as a laboratory research animal by Bird in 1985. Since that time, intensive research has been conducted, and is still ongoing, on such major environmental contaminants as polychlorinated biphenyls (PCBs) and polybrominated diphenyl ethers (PBDEs), as well as on heavy metals, pesticides and herbicides, electromagnetic fields, hormones and growth, morphology, behavior and diseases.

The vast majority of both past and ongoing research on the American Kestrel is aimed at extrapolating the findings to this species in the wild. Naturally, it has always been assumed in various published findings that captive-bred kestrels are little or no different from their free-ranging cousins. Thus, we deem it important to investigate whether long-term captive breeding has altered the make-up of the American Kestrel in some way(s), e.g. physiology, morphology, behaviour, etc. We therefore conclude this paper with a description of a newly initiated doctoral study to determine the effects of long-term captive breeding on this species and a discussion on how any changes might impact upon studies published in the past and future.

**Valuable Attributes of the American Kestrel**

The American Kestrel is considered a useful model for wildlife research in captivity for a variety of reasons. The species' wide geographic range and habitat variability, as well as its ability to live in close proximity to human populations, make it a relevant species to use as a model in various toxicological and ecological research throughout much of the continent of North America (Smallwood & Bird 2002; Smits et al 2002). The kestrel is also a well-studied bird, providing a large amount of available literature for consultation during research (Wiemeyer & Lincer 1987).

The American Kestrel is the only raptor species to date that can be produced in sufficient numbers in captivity and with a fairly rapid turnover for toxicological studies (Lowe & Stendell 1991). Similarly, large captive populations allow one to acquire more meaningful statistical results for studies in a variety of research fields (e.g. Bird & Rehder 1981). Kestrels breed readily in captivity if a pair is provided with a suitable nest box (Willoughby & Cade 1964). Up to 90% of captive kestrels paired in any given year have been shown to produce eggs, and a second clutch can readily be produced if the first clutch is removed within a week of laying the final egg (Bird 1982). As many as 26 eggs can be laid upon removal of each fresh egg laid (Bird & Buckland 1976). Females can also be inseminated artificially (Bird et al 1976). Egg fertility ranges between 65 and 80%, and hatchability of eggs ranges between 60 and 70% for natural or artificial incubation. Hatchlings can be reared by parents or hand-reared in groups without resulting abnormal behavior and they can be sexed by plumage color by 12 days of age (Bird 1982).

To further add value to the American Kestrel as a research animal, at least two booklets on aging nestling American Kestrels have been published (Griggs & Steenhof 1993; Klucsarits & Rusbuldt 2007), as well as two studies on aging developing embryos (Bird et al 1984; Pisenti et al 2001).

Cannibalism whereby a parent consumes its own nestlings or a nestling eats its own siblings, respectively, most commonly at an early age, has been observed in both captive and wild kestrels and does not appear to be a function of a captive environment (Bortolotti et al 1991). A study on wild kestrel pairs by Bortolotti et al (1991) did not find a strong correlation between human disturbance and cannibalism events, though food abundance was negatively correlated with cannibalism rates.

Due to its small size and relatively non-aggressive temperament, kestrels can be housed in large same-sex flight pens (6.7 x 6.7 x 2.7m; L x W x H) of up to 30 birds with minimal aggression between birds during non-breeding seasons, and winter losses are usually less than 10% (Bird 1982; Smallwood & Bird 2002). This allows for the maintenance of high numbers of study birds at one facility at any one time (Bird & Rehder 1981; Wiemeyer & Lincer 1987). Interestingly, a similar attempt to keep European kestrels *F. tinnunculus* met with failure because of severe physical aggression among the individuals (D. Bird, unpubl. obs.).

Captive populations have been successfully sustained on a daily diet of frozen-thawed day old cockerels (Bird 1982) with no noticeable nutritional impacts as yet, as well as on a commercially available carnivore diet such Bird-of-Prey Diet (Quinn et al 2005).

Though kestrels are capable of inflicting bites and scratches, they can be captured and handled easily by trained personnel.

Toxicological research can be conducted on captive birds without concern over results being masked by other unknown environmental contaminants, as can be the case when studying wild birds (Lowe & Stendell 1991; Fernie et al 2001a). Confounding effects of variable environmental conditions such as food availability can be controlled in a captive setting (Bortolotti et al 2003), and behavior is easier to observe in a captive environment, especially for a far-ranging species such as a raptor (Fisher et al 2001). Blood samples of up to 10% total blood volume can be taken from individual, healthy birds on a weekly basis for an extended period for analyses without any negative side-effects to the birds or changes to their packed cell volume (Rehder et al 1982, Rehder & Bird 1983). Captive birds have known pedigrees and documented clinical histories that can be consulted for reproduction or for any anomalies in research results. Most important, using captive bred birds for research reduces the need to remove birds from the wild for research (Bird & Rehder 1981).

Since the American Kestrel is deemed to be a relatively common species, this facilitates comparison between wild and captive birds and permits the exchange of eggs or young to enhance the genetics of the captive colony and prevent inbreeding. Nest boxes can be built and established in favorable kestrel habitat to permit safe and easy access to wild kestrel pairs, which readily use them (Smallwood & Bird 2002). Disturbance of nest boxes for monitoring and measurements of adults and nestlings appears to have little effect on the success of the nesting birds (Wiemeyer & Lincer 1987; Smallwood & Bird 2002).

Although a captive colony had been set up for behavioral studies at Cornell University in the early sixties (Willoughby & Cade 1964), two well-known captive colonies of American Kestrels in North America have risen to prominence over the last four decades. The Avian Science and Conservation Centre (ASCC), formerly the Macdonald Raptor Research Centre, located on the Macdonald campus of McGill University in Ste. Anne de Bellevue, Quebec has maintained a kestrel colony since 1972 (Bird 1982), and the Patuxent Wildlife Research Centre run by the U.S. Fish and Wildlife Service in Laurel, Maryland has housed kestrels mainly for toxicological studies since around 1968 (Wiemeyer & Lincer 1987). Access to colony birds at the ASCC has and continues to be extended to scientists at other research institutions. At least two other modest colonies established in the U.S. and composed of kestrels purchased from the ASCC were initiated for studies on impacts of selenium (e.g. Santolo et al 1999), embryological research (Pisenti et al 2001), and hormonal research (e.g. Sockman et al 2001, 2004). In some cases, kestrels have been taken from the wild at egg, nestling and/or adult stages to be used in laboratory research. Even wild kestrels deemed unreleasable have been borrowed from rehabilitation centers and clinics for research (e.g. Akaki & Duke 1999), though their genealogical and clinical histories and their ages are not as readily available as they are for captive-raised birds (Bird & Rehder 1981).

**Major Environmental Contaminants**

The American Kestrel continues to play a prominent role in toxicological research, especially in products involving organochlorines. As mentioned earlier, research on this species was critical to making the case against the use of DDT because of its secondary poisoning hazards (e.g. eggshell thinning in some species of birds leading to serious reproductive failure [Wiemeyer & Porter 1970]). While DDT and other organochlorine-based pesticides were banned from use in the early to mid-seventies, their persistence in the environment allows them to continue to pollute the food chain all over the world. Moreover, other classes of pesticides have taken their place. Finally, not all organochlorines have been employed as pesticides (see below).

**Organochlorines - PCBs**

The American Kestrel has been used fairly recently (e.g. Fernie et al 2000a) as a model to study the effects of polychlorinated biphenyls (PCBs), another major environmental contaminant of concern, on birds. PCBs are organochlorines once used in a variety of products ranging from electrical insulation to plasticizers before they were banned in the U.S. in 1978 (Lowe & Stendell 1991). PCBs are lipophilic, degrade slowly, and tend to bioaccumulate within the food chain, which makes them of particular concern to top predators such as raptors (Fernie et al 2000a). Ingestion of environmentally significant concentrations of PCBs has been shown to affect physiology, immunomodulation, behavior and reproduction of kestrels (Fernie et al 2003a).

The effects of PCBs on liver and kidney function and sperm quantity and quality have been studied in the past (Lincer & Peakall 1970; Bird et al 1983;

Elliott et al 1991, 1997), but more recently, PCBs have also been implicated in the coloration of birds. Coloration is often used for sexual signaling in various animal species, including birds (Bortolotti et al 2003a). Normally, male birds with higher carotenoid concentrations have brighter coloration, and are favored by females as better mates. Carotenoids affect immune response and other biological systems in birds, thus coloration becomes a signal of a bird's health status. The correlation between carotenoids in the blood and coloration on fleshy portions of the body did not hold true for PCB-dosed male kestrels during the breeding season (Bortolotti et al 2003a). A loss of correlation between coloration and carotenoid concentration could conceivably result in false assessments of potential mates by females and could lead to poor nesting success. PCBs also appear to affect the normal development of red pigmentation in the irides of kestrels as they age by suppressing it, though the consequences of this side effect have yet to be determined (Bortolotti et al 2003b).

Immunotoxicology, the study of immune function in relation to environmental contaminants, is becoming more common as an indicator of the effects of contaminant levels on wildlife (Smits et al 2002). The immune response of kestrels was affected by PCB exposure, both immediately after dosing and over the long term (Smits et al 2002). Effects were sex-linked, with males having depressed corticosterone levels and decreased antibody production, and both sexes having a reduction in thyroid hormone, even with in ovo exposure (Quinn et al 2002; Smits et al 2002; Love et al 2003a). The depression of corticosterone levels could have serious implications to a bird's response to stressors in the environment, such as predators.

Since contaminants often affect bird behavior, behavioral observations of birds undergoing toxicological studies have been encouraged in addition to the use of physiological examinations (Fisher et al 2006a). During direct dietary exposure to PCBs, aggressive courtship displays between kestrel pairs were noted, as well as clutch abandonment and cracking of eggs by parents (Fernie et al 2003a). Male kestrels exposed to PCBs displayed more sexual behaviors, though this did not contribute to better nesting success (Fisher et al 2001). PCB-exposed parents appeared to be less coordinated in terms of sharing incubation, which could result in poor hatching success (Fisher et al 2006a).

Effects of PCBs on the reproductive abilities of adult kestrels and on their offspring have been quite variable and not always easily explained. Kestrels dosed with environmentally relevant levels of PCBs through their regular diet exhibited delays in clutch initiation, smaller clutches, higher levels of infertility, reduced hatching success, greater hatchling mortality and reduced fledgling success (Fernie et al 2001b). Exposure to PCBs also resulted in longer incubation times, though this might have been the result of delayed clutch initiation (Fisher et al 2006a). Altered size of brood patches have been detected in captive kestrels exposed to PCBs, which indicates a possible impact on the endocrine system (Fisher et al 2006a, b). The endocrine system is thought to control incubation behavior and brood patch development in birds. Since organochlorines are fat-oluble, they are expected to concentrate in lipid-rich egg yolk (Fernie et al 2000a). Eggs from treated birds contained environmentally significant concentrations of PCBs (Fernie et al 2001b). PCBs caused greater variability in egg size within individual clutches, as well as heavier yolks and reduced albumen concentrations, resulting in more lipids and less protein available to developing embryos, which could have an effect on later survival of the nestlings (Fernie et al 2000a). Whether PCBs cause eggshell thinning and a reduction in shell thickness index remains controversial (see Lowe & Stendell 1991; Peakall & Lincer 1996).

Deformities followed by death were seen more often in nestlings of dosed kestrels than in those of control pairs (Fernie et al 2003a). Nestlings that died soon after hatching were abnormally small (Fernie et al 2003b), and hatchlings in general were more likely to die when hatched from PCB-exposed parents than from control pairs. The growth of nestlings from PCB-dosed pairs was affected both during the year of dosing and in the following reproductive season when the adults were no longer exposed to the chemicals. Changes in nestlings were sex-specific: females grew to heavier weights with longer bones, while males were lighter weight with shorter bones than control nestlings. Males were younger at fledging and had longer primaries, while female nestlings took longer to achieve their maximum size. Changes to growth rate and bone size could have implications for survival after fledging, and may indicate that PCBs alter the hypothalamic and pituitary regions of the brain that affect growth (Fernie et al 2003b).

Though many of these effects were not seen over the long term (i.e. in the reproductive season after dosing had ceased), the rate of multiple deformities per nest increased dramatically the year after exposure to PCBs had been terminated (Fernie et al 2003a). PCBs contain both non-persistent and persistent congeners. Non-persistent congeners are usually present in an organism immediately after dosing but are cleared from biological systems soon after exposure stops, while persistent congeners will be present in the system long after exposure has ceased (Fernie et al 2003a). Non-persistent PCB congeners have been shown to have varied and immediate effects on avian reproduction in terms of adult behavior and developmental abnormalities that are not seen again after exposure ceases, but persistent congeners have also been shown to affect laying patterns, development and deformity rates in kestrels (Fernie et al 2001a, 2003a). Deformities were even seen in offspring from males dosed in previous years bred with control females (Fernie et al 2003a). This suggests that PCBs result in changes at the genetic level since contaminants cannot be transmitted directly to nestlings from their sires (i.e. nutrients are only provided to embryos by females through the eggs) (Fernie et al 2003a).

PCBs have been shown to have an effect on kestrels exposed to the chemicals in ovo through their parents (Fernie et al 2001a). In ovo-exposed males had poorer hatching and fledgling success during reproduction, while females showed delayed clutch initiation, greater failure to lay eggs, and smaller clutches (Fernie et al 2001a). It is possible that in ovo exposure is more detrimental to birds than dietary exposure, since egg laying was not suppressed in females fed PCBs directly (Fernie et al 2001a). The female nestlings produced by in ovo-exposed kestrels had larger overall size but slower growth and fledging rates, while males were also heavier with shorter bones and grew and fledged more quickly than control nestlings (Fernie et al 2003c). Hatchlings from in ovo males had delayed growth. This was found to be partially due to male parental care (in ovo-exposed males ignored nestling food-begging calls). Thyroid hormones of the nestlings were also affected (Fernie et al 2003c). Eggs of in ovo females had a hundred-fold decrease in PCB concentrations compared to those of females exposed to dietary PCBs, but the concentrations were still sufficient to affect nestlings (Fernie et al 2003c).

Overall, the effects of PCBs on captive kestrels support the Great Lakes embryo mortality, edema, deformities syndrome (GLEMEDS) hypothesis, which suggests that growth and survival of embryos and nestlings is affected by PCBs and other contaminants in the Great Lakes region (Fernie et al 2003a). Studies done by direct exposure of embryos in eggs to PCBs also produced poorer hatching success, lower hatching weights and bone lengths, and deformities in bill structure, as well as edema and external yolk sacs (Hoffman et al 1998).

**Polybrominated Diphenyl Ethers**

Polybrominated diphenyl ethers (PBDEs) form part of the class of brominated flame retardants (Fernie et al 2006). Large quantities of PBDEs, over half of which are produced in North America, are used worldwide in products ranging from polyurethane foam to textiles. PBDEs are lipophilic, bioaccumulate in the food chain and have been found in animals around the world, with some of the highest concentrations found in peregrine falcons (Fernie et al 2006). Within the next decade, concentrations of PBDEs will be similar to those of PCBs in the Great Lakes region (Fernie et al 2006). PBDEs are suspected to alter neurobehavioral development and thyroid function, though their effects have yet to be properly studied in wildlife.

Research into the effects of PBDEs on kestrels is currently ongoing at both the ASCC and the Patuxent Wildlife Research Centre. Kestrel eggs dosed with environmentally relevant levels of PBDEs produced nestlings that ate more, gained weight more quickly and grew to larger sizes in weight and bone length than control nestlings (Fernie et al 2006). Increased size could put more stress on bone structure and increase energetic costs of living, which could be detrimental to their survival (Fernie et al 2006). The nestlings also had a reduced immune response and structural changes in the spleen, bursa and thymus, which could reduce the birds' abilities to fight off infection (Fernie et al 2005a). Thyroid hormone and retinol concentrations were also decreased (Fernie et al 2005b). Effects to the thyroid system can alter such biological functions as growth and metabolism, and reduced retinol levels can delay growth, cause deformities during development, and result in oxidative stress (Fernie et al 2005b).

**Metals**

*Lead*

Concern was raised over the risk of lead poisoning in raptors by direct exposure to lead through ingestion of lead shot from game species and from indirect exposure to lead biologically incorporated into prey animals (Hoffman et al 1985a,b). Raptors near urban areas, such as kestrels, are also at risk of lead poisoning because their prey species are often contaminated. At the time these studies were conducted, endangered peregrine falcons were being released into urban environments as part of a population recovery program, thus concern for their survival was expressed (Hoffman et al 1985a,b). Kestrel nestlings were used as a model to test the effects of lead poisoning on altricial (or semi-altricial) bird species such as raptors. High concentrations of dietary lead increased nestling mortality, and growth was impaired in surviving nestlings. Hematological effects were also noted, as were changes to brain, liver and kidney functions. Altricial nestlings were judged to be more sensitive to lead poisoning than nestlings of precocial species or adult birds exposed to lead, either from direct poisoning with lead powder or from biologically incorporated lead in prey items (Franson et al 1983; Custer et al 1984; Hoffman et al 1985a,b). Lead does bioaccumulate in adult kestrels though, and lead shot has been shown to cause clinical poisoning (Custer et al 1984).

*Selenium*

Selenium is found naturally in the environment, but elevated levels of the element can be found in areas of agricultural drainage and have been associated with mortality, wasting and poor reproduction in aquatic bird species (Santolo et al 1999; Yamamoto & Santolo 2000). Studies on selenomethionine, a highly toxic selenium compound, focused primarily on waterfowl and domestic species before kestrels were exposed to the compound to determine its effects on a wild terrestrial bird species such as a raptor. Impacts upon the health of dosed kestrels were not apparent in initial studies, though the project helped demonstrate that blood-sampling was an effective, non-lethal technique of measuring selenium concentrations in birds (Yamamoto et al 1998). Long-term studies indicated that kestrels fed environmentally relevant levels of selenium failed to increase body fat before winter months, which could affect survival during migration of wild birds and spring reproductive efforts (Yamamoto & Santolo 2000). Results suggested that elevated selenium levels due to agricultural drainage reduce fertility and possibly egg size in kestrels, though the impact upon kestrels appears to be less severe than to waterfowl (Santolo et al 1999).

**Herbicides, Pesticides, Avicides & Secondary Poisoning**

Kestrels are likely to come into contact with various herbicides, pesticides and avicides in the wild given that their preferred hunting grounds include agricultural and urban habitats (Smallwood & Bird 2002). Their nestlings would also be at risk of exposure, both from food offered to them by their parents and from transfer of bioaccumulated chemicals in females through egg yolk. Captive kestrels have been used in studies of such herbicides as paraquat (Hoffman et al 1985c) and those belonging to the diphenyl ether class of chemicals (Hoffman et al 1991). They have also been used to demonstrate the danger that organophosphorus (e.g. fenthion) and carbamate (e.g. carbofuran) insecticides represent to bird species, particularly to raptors (Vyas et al 1998). Fenthion is also used as an avicide against pest bird species, including species that form components of raptor diets, such as the European Starling *Sturnus vulgaris* (Hunt et al 1991). This raises the potential risk of secondary poisoning to raptors feeding on poisoned prey. Captive studies using American Kestrels and House Sparrows *Passer domesticus* demonstrated that by killing and consuming fenthion-treated prey, kestrels were affected not only by secondary poisoning, but by direct contact with the avicide still present on the plumage of the prey item (Hunt et al 1991).

The pesticide, Kelthane, also received a great deal of attention from researchers due to its organochlorine components (MacLellan et al 1997; Wiemeyer et al 2001). Dicofol (1,1-bis (P-chlorophenyl)-2,2,2-trichloroethanol), a component of the organochlorine pesticide, Kelthane, is used most frequently to control cotton and citrus mites, and is similar in structure to DDT (MacLellan et al 1997; Wiemeyer et al 2001). It contained significant concentrations of DDE and DDT-related compounds until the U.S. Environmental Protection Agency changed its regulations in the 1980s, a decade after most other organochlorines were banned or became strictly regulated (MacLellan et al 1996). Effects of dicofol on kestrels included eggshell thinning at high concentrations (Schwarzbach et al 1991; Wiemeyer et al 2001), reduced reproductive success (MacLellan et al 1996), feminization of male offspring (MacLellan et al 1996) and reduced competitive and reproductive skills in males (MacLellan et al 1997).

**Other Toxicological Studies**

Captive kestrels were used to determine if white phosphorus, used by the military as an obscurant and for range-finding artillery, could result in secondary poisoning of raptors at artillery practice sites (Sparling & Federoff 1997). White phosphorus, which is lipophilic and can accumulate in waterfowl habitat, kills or incapacitates large numbers of waterfowl every year. Concern was raised that the birds suffering effects of the toxicant would be attractive to raptors. The kestrel study suggested that both direct and secondary poisoning with white phosphorus can kill or severely compromise raptors (Sparling & Federoff 1997).

Other toxicological studies conducted on American Kestrels include those on the effects of sodium fluoride (Bird et al 1992) and of organohalogens such as halogenated dimethyl bipyrroles (HDBPs) (Tittlemier et al 2003). Kestrels were also used as a representative study species in a study to determine whether the Ixtoc I oil well blowout in Mexico in 1979 presented a risk to migrating Peregrine Falcons (Pattee et al 1982). Kestrels were employed to help determine a non-lethal means of reducing the impact of raptors on such economic species as Red Grouse *Lagopus lagopus*, since raptors can be considered pests to certain species of economic interest to humans (Nicholls et al 2000). Specifically, it was demonstrated that kestrels showed greater avoidance to novel coloring on food than to chemical food deterrents, which may have future applications to lessen raptor predation on certain economically valued species such as domestic pigeons *Columba livia* (Nicholls et al 2000).

**Models and Bioindicators**

Studies on captive birds with controlled chemical dosing over known time periods can be used to model the effects of environmental contaminants in the wild (e.g. Fernie et al 2003a,b,c). For example, data collected on captive adult and juvenile kestrels exposed to PCBs have been used to create bioaccumulation models of the chemical in different life stages of birds. This allows biologists to extrapolate exposure levels to populations of birds in the wild (e.g. Drouillard et al 2001; Drouillard et al 2007). Similar research was conducted on kestrels using dietary methylmercury (Nacci et al 2005). Captive birds were fed various doses of mercury to help create a model of the effects of mercury on birds. The goal of this research is to create a model that can be used to monitor Common Loons *Gavia immer* as a bioindicator species in the wild (Nacci et al 2005).

Using American Kestrels and laboratory mice as a model, the impact of rodenticides, in particular chlorophacinone (Radvanyi et al 1988), has also been revealed to be a serious secondary poisoning hazard to raptorial birds. Similarly, American Kestrels and house sparrows have been used as a model to show that raptors actively select fenthion-poisoned prey, at least in a laboratory testing (Hunt et al 1992).

**Electromagnetic Fields**

Many birds, particularly raptors, use power lines for perching, hunting and nesting (Fernie et al 2000a). Nest platforms and nest boxes have even been placed on transmission towers and on hydroelectric poles for the benefit of raptor species, including the American Kestrel (Fernie & Bird 1999). These birds are exposed to electromagnetic fields (EMFs) produced by power lines, but little was known about the potential effect of EMFs on birds. However, EMFs have been shown to not only affect embryo development, but also impact upon their reproductive success (Fernie et al 2000b). Exposed breeding pairs had higher levels of fertility but lower hatching success than pairs not exposed to environmentally significant levels of EMFs. Eggs produced by EMF-exposed pairs were larger, had greater volume, thinner shells, and contained larger, longer embryos, which may indicate impacts upon thyroid and growth hormones (Fernie et al 2000b). Fledgling success was also higher in EMF nestlings and was proposed to be the result of heavier, larger nestlings, though growth was delayed in males (Fernie and Bird 2000).

EMF-exposed breeding adults were more active during the courtship period, and males were more alert during the incubation period, though this did not seem to affect hatching or fledgling success directly since feeding rates of young were not significantly different from control pairs (Fernie et al 2000b). Short-term exposure to EMFs has also been shown to decrease plasma proteins, hematocrits and carotenoids in males, which suggests a change in the birds' immune response (Fernie & Bird 2001). Males also suffered from higher levels of oxidative stress, which is known to result in cancer formation, occurrence of neurodegenerative diseases, and aging in mammals (Fernie & Bird 2001). Further effects on breeding males included increased body mass without increased food intake and the earlier onset of molt (Fernie & Bird 1999). The earlier onset of molt might indicate that birds interpret EMFs as light, thus exposure to EMFs potentially increased the length of their photoperiod and advanced their molt period (Fernie et al 1999). Melatonin levels were also shifted so that levels in EMF birds early in the breeding season were similar to those of control males later in the breeding season (Fernie et al 1999). Exposed adults also produced fledglings with reduced melatonin levels. Melatonin acts on various physiological and behavioral functions in birds, including temperature regulation, metabolism, plumage color and migration. Results also suggested that males are more sensitive to environmental changes such as EMFs than females (Fernie et al 1999).

**Hormone Studies**

Before studying hormones in kestrels, blood hormone concentrations for birds were modeled primarily from domestic bird species (Rehder et al 1986). Kestrels have been used to study the effects of renesting, molt and weight on the concentrations of corticosterone, estrone, estradiol-17ß and progesterone in females during the reproductive and winter seasons (Rehder et al 1984, 1986). These were the first known studies of the above hormones in a raptor species (Rehder et al. 1984). Seasonal androgen levels in breeding and non-breeding male kestrels were similarly studied (Rehder et al 1988). Growth hormone affects the overall growth and metabolism of growing birds (Lacombe et al 1993). Its concentrations are usually highest just after hatching and during the peak growth period of nestlings. Though it has been well studied in domestic birds, little research has been done on wild bird species, and in particular on the effects of food restriction on growth in wild nestlings. Kestrels were used in growth hormone studies as a representative altricial bird species. Food restriction was found to increase the concentration of the hormone in nestling kestrels during the peak growth period and to delay growth (Lacombe et al 1993). It was postulated that the level of growth hormone increased in food-restricted birds to ensure that sufficient energy was supplied to vital organs.

The adrenal cortex produces corticosterone, a steroid hormone that is released in response to acute stressors and also to seasonal events such as migration or reproduction (Love et al 2003a). This hormone can alter a bird's behavior in response to a stressor, allowing it to adapt to and survive the event. Studies have been conducted on the effects of age, asynchrony of hatch, hatching order and growth on the corticosterone levels of kestrel nestlings to determine if differences in hormone levels affected competition between siblings (Love et al 2003a,b). The effects of repeated handling and restraint were studied in adult and nestling kestrels in an attempt to determine if adrenocortical studies upon wild and laboratory birds exposed to repeated handling provided a valid assessment of a bird's stress state or if handling interfered with results (Love et al 2003c).

Manipulations of photoperiod and food availability were performed to determine how such stressors affected plasma-prolactin levels in laying female kestrels, since hormone levels in females can influence hormone levels in yolk of eggs and thus, reproductive success (Sockman et al 2001). Using the same kestrels, Sockman et al (2004), determined the concentrations of prolactin and testosterone in male kestrels in relation to time during the breeding season and to transition from sexual and courtship behaviors to parental behaviors. More recently, the effect of the thyroid hormone, thyroxine, on the molt and plumage color of feathers has also been studied in captive kestrels (Quinn et al 2005).

**Disease and Parasites**

Kestrels have been used in research to determine causative agents of disease and as potential models for disease spread in wild birds. For example, kestrels were one of the species used to determine the source of avian vacuolar myelinopathy (AVM), a neurologic disease recently found in wild birds in the U.S. (Rocke et al 2005). Kestrels also have been used for research on West Nile virus (WNV), a fairly new but now widespread disease in North America (Nemeth et al 2006; D. Bird, unpubl. data). The transmission routes and effects of the virus on raptors were studied, in addition to determining potential risk to raptor handlers from their birds. Discoveries such as the fact that raptors can be infected with WNV via oral route, and that many raptors do not develop clinical signs of infection but instead develop lesions that might not be attributed to the virus upon inspection, can be useful for management of raptor species (Nemeth et al 2006).

To determine the effects of parasite infection on reproductive success in raptors, captive kestrels were infected with *Trichinella pseudospiralis*, a parasite that occurs in birds and is transmitted through the ingestion of infected meat, potentially a primary mode of infection in raptors (Saumier et al 1986). Infected males were found to be less competitive than uninfected males for mates and nest boxes (Henderson et al 1995). The reproductive success of infected birds was also affected; egg-laying was delayed, infected females were less likely to lay replacement eggs if their initial clutch was lost, and egg breakage and reduced nestling survival were related to reduced parental care (Saumier et al 1986). Non-breeding birds infected with the parasite showed reduced mobility as larvae migrated into the musculature of the birds (Saumier et al 1988). Infected birds walked instead of flew, reduced their exercise and preening, and were prone to remaining on the ground. Reduced ability to fly could make kestrels more susceptible to predators and less able to hunt, while reduced preening might reduce feather quality, water-proofing, and ability to thermoregulate (Saumier et al 1988).

**Effect of Diet on Growth**

Growth of nestlings is affected by such factors as diet quality and intake rates (Negro et al 1994). Captive kestrels have been used to study these aspects of growth in a controlled laboratory setting. The growth rate of nestlings subjected to short-term food deprivation was studied to determine what mechanisms nestlings possess to cope with natural food fluctuations experienced in the wild (Negro et al 1994). Continuous food restrictions were evaluated to determine their effect on the growth and development of nestlings (Lacombe et al 1994). Effects of food limitation were also investigated using brood size manipulation to determine if feeding abilities of parents limited the size of brood they were able to rear (Gard & Bird 1992).

Effects of different diet types on growth of American Kestrel nestlings have been studied, which has implications for captive-rearing programs. For example, Lavigne et al (1994) found that captive nestlings raised on a diet of day-old cockerels exhibited growth patterns better than those reared on a laboratory mouse diet and similar to those of wild nestlings.

Taylor et al (1991) also used American Kestrels and laboratory mice bred for their high fat content to demonstrate nutritional and energetic implications for wild raptors consuming starving prey.

**Morphology, Function & Behavior**

Captive kestrels have been used in studies of bird morphology, including research on musculature (Meyers 1992a,b) hind limb morphology (Ward et al 2002), and structure of the gastrointestinal tract (Reynhout et al 1997). Renal function under the effects of different diet regimes was studied in kestrels to determine the influence of high-protein diets on kidney function (Lyons & Goldstein 2002). Embryonic development was investigated using captive kestrel eggs to obtain a detailed description of the normal development of raptor embryos (Bird et al 1984; Pisenti et al 2001). Such information can be used in future research to identify abnormalities in embryos as caused by pollutants, diseases, and other environmental perturbations. Integument color and carotenoid levels in captive kestrels were examined for their variation throughout the seasons and to determine if color could be used for sexual signaling between birds during the breeding season (Negro et al 1998). Captive kestrels have been used in visual development studies, as well as in a series of eye structure, visual acuity, sensitivity and search image tests (Mueller 1971; Frost et al 1990; Andison et al 1992; Gaffney & Hodos 2003; Ghim & Hodos 2006). Chitin digestibility and its use as a potential source of carbohydrates for raptors was tested under laboratory conditions (Akaki & Duke 1998, 1999), as was the preference for certain parts of prey and the timing, rate and number of regurgitated pellets produced by kestrels on different diets (Balgooyen 1971; Duke et al 1996). Fault-bar formation as a result of stressors in growing kestrels has also been studied (Negro et al 1994).

Numerous behavioral studies have been done on captive kestrels. Some examples include mate choice studies (Duncan & Bird 1989; Henderson et al 1995), social behavior in fledgling kestrels (Agostini et al 1996), and fledgling play behavior (Negro et al 1996). Captive kestrels have even been used to look at competitive dominance of larger female nestlings and its effect on male nestling survival (Anderson et al 1993, 1997).

**Effects of Long-term Captive Breeding**

As indicated by the aforementioned brief review, the American Kestrel was first bred in captivity for research in the early sixties, but consistently in two major colonies for almost forty years. This raises the question though, as to whether the results of studies on these captive birds, especially those bred over many generations, can be legitimately extrapolated to their wild ancestors. In other words, are we dealing with a fully domesticated species that bears less resemblance to its wild counterpart? Is it really valid to employ wild birds raised and kept in captivity as "white mice" for free-ranging members of the same species?

Domestication, a term often used to describe intentional selection for certain traits such as those exhibited by farm, laboratory and companion animals, can also be defined as a process by which an animal adapts to an environment created by humans over time (Price 1999). Though the goal of a captive-breeding program for wild animals is to maintain as much as possible the animal's original wild state, captive environments differ from wild ones, and the potential for variability within a captive population may be reduced due to a small founding population (Wright et al 2006; Zeder et al 2006). In captive environments diet, mate selection, daily schedules, group size and composition, environmental characteristics and even the geographic location of the animal population are all controlled by humans (Price 1999; Zeder et al 2006). Adaptations to such conditions can occur over two time periods: short-term and long-term, and can result in changes to an animal's behavior, morphology and physiology (Kunzl & Sachser 1999; Price 1999). Short-term changes occur within an animal's lifetime and are a phenotypic response to the animal's immediate environment (Price 1999). Long-term changes occur over several generations and are the result of genetic changes (Price 1999). The rate of these changes can be affected by three factors: artificial selection by humans, whether intentional or not, natural selection in a captive environment, and the relaxation of natural selection to such things as predation and food stresses (Price 1999; Schutz et al 2001; Hakansson & Jensen 2005).

The alteration of behavioral traits is thought to be one of the first signs of domestication (Fleming & Einum 1997). Domestication or long-term captivity may not result in a loss or addition of behavioral patterns, but they can potentially change the frequency with which those behaviors are exhibited (Price 1999; Hakansson & Jensen 2005). Behavioral changes can include altered aggression levels and reduced activity levels (Kunzl & Sachser 1999; Wright et al 2006). Alteration in stress thresholds and increased rates of courtship displays and non-aggressive social interactions are also common changes to captive animal behaviors. Housing in same-sex or same-age groups or crowded or more isolated conditions than usually found in natural situations may deny young animals access to learning experiences essential for proper development (Price 1999; Hakansson & Jensen 2005). This may also lead to the formation of hierarchies in species where such social structures do not normally exist (Price 1999).

Captivity may deny animals key experiences during development that can result in a reduced ability to cope with predation, hunting and food searching (Price 1999). The costs of exhibiting anti-predator behaviors in a predator-free environment may no longer be outweighed by the benefits and the trait can become neutral instead of selected for (Wright et al 2006). Access to plentiful resources reduces or negates an animal's need to expend energy searching for such resources. Animals that normally exhibit contrafreeloading, expending energy to discover novel food sources when less energy demanding food sources are readily available, which is an adaptation to natural environments where food sources are not predictable, may lose this trait in captivity where diets are plentiful and often uniform (Price 1999; Schutz et al 2001).

Morphological and physiological traits commonly exhibited by animals undergoing domestication include faster growth, smaller brain size, changes in the sizes of internal organs, reduced activity level, prolonged retention of juvenile behaviors, earlier sexual maturity, multiple reproductive cycles per year and changes in clutch or litter size (Belyaev et al 1984; Jackson & Diamond 1996; Kunzl & Sachser 1999; Price 1999; Jensen 2006). Animals may also attain larger or smaller sizes, have different fat and muscle ratios and different colors or feather or fur textures (Jensen 2006). Changes can occur over few generations; the adult mass of Japanese Quail *Coturnix coturnix japonica* was doubled in 40 generations of selective breeding for commercial production (Jackson & Diamond 1996). New generations of animals brought into captivity may also have higher stress responses (increased endocrine, epinephrine and adrenal responses) to a captive environment and to handling than animals bred in captivity for long periods (Kunzl & Sachser 1999). These traits reoccur in many domesticated species, suggesting they are generalized adaptations to long-term captivity, and they can occur over as little as 10 to 20 generations (Jensen 2006). Some of these traits can be confounded with the effects of a captive environment; a more consistent and nutritious diet can also affect growth and reproduction, as can high stress levels in animals newly introduced to captive conditions (Hakansson & Jensen 2005). Such changes in captive populations of kestrels might make them unsuitable models for further wildlife research since they would no longer be considered a representation of a wild species. Studies are currently underway at the Avian Science and Conservation Centre to determine if any marked differences exist between wild kestrels and kestrels that have been bred in captivity for over 30 generations, and if so, what steps might be taken to reverse or reduce these effects so that captive colonies can still contribute to essential scientific wildlife research.

## CONCLUSION

Since the early sixties, the American Kestrel, in both captive and wild settings, has provided scientists with many and varied opportunities to study the biology and behavior of raptors and birds in general. They have been invaluable in toxicological research, providing solid, irrefutable data on secondary poisoning hazards that have led government agencies to place bans on certain chemical pesticides and industrial by-products. Thus, the important role played by this little falcon could be having a significant impact on the lives of not only free-ranging raptors, but also on other forms of wildlife, and ultimately on humans. With many generations of American Kestrels having been bred at the Avian Science and Conservation Centre for 35 years and a local nest-box program existing for at least a dozen years, a very exciting and perhaps unique opportunity has been provided to investigate the impacts, if any, of long-term captive breeding on various aspects of the biology of a wild, semi-altricial avian species. If there is a significant impact, it may, in fact, diminish the value of the American Kestrel as a useful model for laboratory research aimed at extrapolating the findings to kestrels in the wild.

## REFERENCES

AGOSTINI N, DM Bird and JJ Negro. 1996. Social behavior of captive fledgling American Kestrels (*Falco sparverius*). Journal of Raptor Research 30: 240-241.

AKAKI C and GE Duke. 1998. Egestion of chitin in pellets of American Kestrels and eastern screech owls. Journal of Raptor Research 32: 286-289.

AKAKI C and GE Duke. 1999. Apparent chitin digestibilities in the eastern screech owl (*Otus asio*) and the American Kestrel (*Falco sparverius*). Journal of Experimental Zoology 283: 387-393.

ANDERSON DJ, C Budde, V Apanius, JE Martinez Gomez, DM Bird and WW Weathers. 1993. Prey size influences female competitive dominance in nestling American Kestrels (*Falco sparverius*). Ecology 74: 367-376.

ANDERSON DJ, J Reeve and DM Bird. 1997. Sexually dimorphic eggs, nestling growth and sibling competition in American Kestrels *Falco sparverius*. Functional Ecology 11: 331-335.

ANDISON ME, JG Sivak and DM Bird. 1992. The refractive development of the eye of the American Kestrel (*Falco sparverius*): a new avian model. Journal of Comparative Physiology A 170: 565-574.

BALGOOYEN TG. 1971. Pellet regurgitation by captive sparrow hawks (*Falco sparverius*). Condor 73: 382-385.

BAKST, MR and DM Bird. 1987. Localization of oviductal sperm-storage tubules in the American Kestrel (*Falco sparverius*). Auk 104:321-324.

BELYAEV DK, IZ Plyusnina and LN Trut. 1985. Domestication in the silver fox (*Vulpes fulvus* DESM): Changes in physiological boundaries of the sensitive period of primary socialization. Applied Animal Behaviour Science 13: 359-370.

BIRD DM. 1982. The American Kestrel as a laboratory research animal. Nature 299: 300-301.

BIRD DM. 1985. Evaluation of the American Kestrel (*Falco sparverius*) as a laboratory research animal. In Archibald J, J Ditchfield and HC Roswell, ed. 1985. The Contribution of Laboratory Animal Science to the Welfare of Man and Animals. 8th ICLAS/CALAS Symposium Vancouver 1983, New York. pp 3-9.

BIRD DM and RB Buckland. 1976. The onset and duration of fertility in the American Kestrel. Canadian Journal of Zoology 54: 1595-1597.

BIRD DM, D Carrière and D Lacombe. 1992. The effect of dietary sodium fluoride on internal organs, breast muscle, and bones in captive American Kestrels (*Falco sparverius*). Archives of Environmental Contamination and Toxicology 22: 242-246.

BIRD DM and PC Laguë. 1977. Semen production of the American Kestrel. Canadian Journal of Zoology 55: 1351-1358.

BIRD DM and PC Laguë. 1982a. Influence of forced renesting, seasonal date of laying and female characteristics on clutch size and egg traits in captive American Kestrels. Canadian Journal of Zoology 60: 71-79.

BIRD DM and PC Laguë. 1982b. Fertility, egg weight loss, hatchability, and fledgling success in replacement clutches of captive kestrels. Canadian Journal of Zoology 60: 80-88.

BIRD DM and PC Laguë. 1982c. Influence of forced renesting and hand-rearing on growth of young captive American Kestrels. Canadian Journal of Zoology 60: 89-96.

BIRD DM, PC Laguë and RB Buckland. 1976. Artificial insemination vs. natural mating in captive American Kestrels. Canadian Journal of Zoology 54: 1183-1191.

BIRD DM, DB Peakall and DS Miller. 1983. Enzymatic changes in the oviduct associated with DDE-induced eggshell thinning in the kestrel, *Falco sparverius*. Bulletin of Environmental Contamination and Toxicology 31: 22-24.

BIRD DM and NB Rehder. 1981. The science of captive breeding of falcons. Avicultural Magazine 87: 208-212.

BIRD DM, P Tucker, GA Fox and PC Laguë. 1983. Synergistic effects of Aroclor 1254 and Mirex on the semen characteristics of American Kestrels. Archives of Environmental Contamination and Toxicology 12: 633-639.

BIRD DM, J Gauthier, and V Montpetit. 1984. Embryonic growth of American Kestrels. Auk 101:392-396.

BLUS LJ, SN Wiemeyer and CM Bunck. 1997. Clarification of effects of DDE on shell thickness, size, mass, and shape of avian eggs. Environmental Pollution 95: 67-74.

BORTOLOTTI GR, KJ Fernie and JE Smits. 2003. Carotenoid concentration of American Kestrels (*Falco sparverius*) disrupted by experimental exposure to PCBs. Functional Ecology 17: 651-657.

BORTOLOTTI GR, JE Smits and DM Bird. 2003. Iris color of American Kestrels varies with age, sex, and exposure to PCBs. Physiological and Biochemical Zoology 76: 99-104.

BORTOLOTTI GR, KL Wiebe and WM Iko. 1991. Cannibalism of nestling American Kestrels by their parents and siblings. Canadian Journal of Zoology 69: 1447-1453.

BROCK MK and DM Bird. 1991. Prefreeze and postthaw effects of glycerol and dimethylacetamide on motility and fertilizing ability of American Kestrel (*Falco sparverius*) spermatozoa. Journal of Zoo and Wildlife Medicine 22: 453-459.

CUSTER TW, JC Franson and OH Pattee. 1984. Tissue lead distribution and hematologic effects in American Kestrels (*Falco sparverius*) fed biologically incorporated lead. Journal of Wildlife Diseases 20: 39-43.

DROUILLARD KG, KJ Fernie, RJ Letcher, LJ Shutt, M Whitehead, W Gebink and DM Bird. 2007. Bioaccumulation and biotransformation of 61 polychlorinated biphenyl and four polybrominated diphenyl congeners in juvenile American Kestrels (Falco sparverius). Environmental Toxicology and Chemistry 26: 313-324.

DROUILLARD KG, KJ Fernie, JE Smits, GR Bortolotti, DM Bird and RJ Norstrom. 2001. Bioaccumulation and toxicokinetics of 42 polychlorinated biphenyl congeners in American Kestrels (*Falco sparverius*). Environmental Toxicology and Chemistry 20: 2514-2522.

DUKE GE, AL TerEick, JK Reynhout, DM Bird and AE Place. 1996. Variability among individual American Kestrels (*Falco sparverius*) in parts of day-old chicks eaten, pellet size, and pellet egestion frequency. Journal of Raptor Research 30: 213-218.

DUKE GE, J Reynhout, AL Tereick, AE Place and DM Bird. 1997. Gastrointestinal morphology and motility in American Kestrels receiving high or low fat diets. Condor 99: 123-131.

DUNCAN, J and DM Bird. 1989. The influence of relatedness and display effort on the mate choice of captive female kestrels. Animal Behaviour 37:112-117.

ELLIOTT JE, SW Kennedy, D Jeffrey and L Shutt. 1991. Polychlorinated biphenyl (PCB) effects on hepatic mixed function oxidases and porphyria in birds - II. American Kestrel. Comparative Biochemistry and Physiology C 99: 141-145.

ELLIOTT JE, SW Kennedy and A Lorenzen. 1997. Comparative toxicity of polychlorinated biphenyls to Japanese quail (*Coturnix c. japonica*) and American Kestrels (*Falco sparverius*). Journal of Toxicology and Environmental Health 51: 57-75.

FERNIE KJ and DM Bird. 1999. Effects of electromagnetic fields on body mass and food-intake of American Kestrels. Condor 101: 616-621.

FERNIE KJ and DM Bird. 2000. Effects of electromagnetic fields on the growth of nestling American Kestrels. Condor 102: 461-465.

FERNIE KJ and DM Bird. 2001. Evidence of oxidative stress in American Kestrels exposed to electromagnetic fields. Environmental Research Section A 86: 198-207.

FERNIE KJ, NJ Leonard and DM Bird. 2000a. Behavior of free-ranging and captive American Kestrels under electromagnetic fields. Journal of Toxicology and Environmental Health Part A 59: 597-603.

FERNIE K, G Bortolotti and J Smits. 2003a. Reproductive abnormalities, teratogenicity, and developmental problems in American Kestrels (*Falco sparverius*) exposed to polychlorinated biphenyls. Journal of Toxicology and Environmental Health Part A 66: 2089-2103.

FERNIE K, J Smits and G Bortolotti. 2003b. Developmental toxicity of in ovo exposure to polychlorinated biphenyls: I. Immediate and subsequent effects on first-generation nestling American Kestrels (Falco sparverius). Environmental Toxicology and Chemistry 22: 554-560.

FERNIE KJ, DM Bird and D Petitclerc. 1999. Effects of electromagnetic fields on photophasic circulating melatonin levels in American Kestrels. Environmental Health Perspectives 107: 901-904.

FERNIE KJ, JE Smits, GR Bortolotti and DM Bird. 2001a. In ovo exposure to polychlorinated biphenyls: Reproductive effects on second-generation American Kestrels. Archives of Environmental Contamination and Toxicology 40: 544-550.

FERNIE KJ, JE Smits, GR Bortolotti and DM Bird. 2001b. Reproduction success of American Kestrels exposed to dietary polychlorinated biphenyls. Environmental Toxicology and Chemistry 20: 776-781.

FERNIE KJ, DM Bird, RD Dawson and PC Laguë. 2000b. Effects of electromagnetic fields on the reproductive success of American Kestrels. Physiological and Biochemical Zoology 73: 60-65.

FERNIE KJ, GR Bortolotti, K Drouillard, J Smits and T Marchant. 2003c. Developmental toxicity of in ovo exposure to polychlorinated biphenyls: II. Effects of maternal or paternal exposure on second-generation nestling American Kestrels. Environmental Toxicology and Chemistry 22: 2688-2694.

FERNIE KJ, GR Bortolotti, JE Smits, J Wilson, KG Drouillard and DM Bird. 2000c. Changes in egg composition of American Kestrels exposed to dietary polychlorinated biphenyls. Journal of Toxicology and Environmental Health Part A 60: 291-303.

FERNIE KJ, G Mayne, LJ Shutt, C Pekarik, KA Grasman, RJ Letcher and K Drouillard. 2005a. Evidence of immunomodulation in nestling American Kestrels (*Falco sparverius*) exposed to environmentally relevant PBDEs. Environmental Pollution 138: 485-493.

FERNIE KJ, LJ Shutt, G Mayne, D Hoffman, RJ Letcher, KG Drouillard and IJ Ritchie. 2005b. Exposure to polybrominated diphenyl ethers (PBDEs): changes in thyroid, vitamin A, glutathione homeostasis, and oxidative stress in American Kestrels (*Falco sparverius*). Toxicological Sciences 88: 375-383.

FERNIE KJ, LJ Shutt, IJ Ritchie, RL Letcher, K Drouillard and DM Bird. 2006. Changes in the growth, but not the survival, of American Kestrels (*Falco sparverius*) exposed to environmentally relevant polybrominated diphenyl ethers. Journal of Toxicology and Environmental Health Part A 69: 1541-1554.

FISHER SA, GR Bortolotti, KJ Fernie, DM Bird and JE Smits. 2006a. Behavioral variation and its consequences during incubation for American Kestrels exposed to polychlorinated biphenyls. Ecotoxicology and Environmental Safety 63: 226-235.

FISHER SA, GR Bortolotti, KJ Fernie, DM Bird and JE Smits. 2006b. Brood patches of American Kestrels altered by experimental exposure to PCBs. Journal of Toxicology and Environmental Health Part A 69: 1603-1612.

FISHER SA, GR Bortolotti, KJ Fernie, JE Smits, TA Marchant, KG Drouillard and DM Bird. 2001. Courtship behavior of captive American Kestrels (*Falco sparverius*) exposed to polychlorinated biphenyls. Archives of Environmental Contamination and Toxicology 41: 215-220.

FLEMING IA and S Einum. 1997. Experimental tests of genetic divergence of farmed from wild Atlantic salmon due to domestication. ICES Journal of Marine Science 54: 1051-1063.

FRANSON JC, L Sileo, OH Pattee and JF Moore. 1983. Effects of chronic dietary lead in American Kestrels *(Falco sparverius)*. Journal of Wildlife Diseases 19: 110-113.

FROST BJ, LZ Wise, B Morgan and D Bird. 1990. Retinotopic representation of the bifoviate eye of the kestrel (*Falco sparverius*) on the optic tectum. Visual Neuroscience 5: 231-239.

GAFFNEY MF and W Hodos. 2003. The visual acuity and refractive state of the American Kestrel (*Falco sparverius*). Vision Research 43: 2053-2059.

GARD NW and DM Bird. 1992. Nestling growth and fledgling success in manipulated American Kestrel broods. Canadian Journal of Zoology 70: 2421-2425.

GEE GF, CA Morrell, JC Franson and OH Pattee 1993. Cryopreservation of American Kestrel semen with dimethylsulfoxide. Journal of Raptor Research 27: 21-25.

GHIM MM and W Hodos. 2006. Spatial contrast sensitivity of birds. Journal of Comparative Physiology A 192: 523-534.

GRIGGS GR and K Steenhof. 1993. Photographic guide for aging nestling American Kestrels. U.S. Department of the Interior, Bureau of Land Management, Boise, Idaho.

HAKANSSON J and P Jensen. 2005. Behavioural and morphological variation between captive populations of red junglefowl (*Gallus gallus*) - possible implications for conservation. Biological Conservation 122: 431-439.

HENDERSON D, DM Bird, ME Rau and JJ Negro. 1995. Mate choice in captive American Kestrels, *Falco sparverius*, parasitized by a nematode, *Trichinella pseudospiralis*. Ethology 101: 112-120.

HOFFMAN DJ, JC Franson, OH Pattee and CM Bunck. 1985a. Survival, growth and histopathological effects of paraquat ingestion in nestling American Kestrels (*Falco sparverius*). Archives of Environmental Contamination and Toxicology 14: 495-500.

HOFFMAN DJ, JC Franson, OH Pattee, CM Bunck and A Anderson. 1985b. Survival, growth, and accumulation of ingested lead in nestling American Kestrels (*Falco sparverius*). Archives of Environmental Contamination and Toxicology 14: 89-94.

HOFFMAN DJ, JC Franson, OH Pattee, CM Bunck and HC Murray. 1985c. Biochemical and hematological effects of lead ingestion in nestling American Kestrels (*Falco sparverius*). Comparative Biochemistry and Physiology C 80: 431-439.

HOFFMAN DJ, MJ Melancon, PN Klein, JD Eisemann and JW Spann. 1998. Comparative developmental toxicity of planar polychlorinated biphenyl congeners in chickens, American Kestrels and common terns. Environmental Toxicology and Chemistry 17: 747-757.

HOFFMAN DJ, JW Spann, LJ LeCaptain, CM Bunck and BA Rattner. 1991. Developmental toxicity of diphenyl ether herbicides in nestling American Kestrels. Journal of Toxicology and Environmental Health 34: 323-336.

HUNT KA, DM Bird, P Mineau and L Shutt. 1991. Secondary poisoning hazard of fenthion to American Kestrels. Archives of Environmental Contamination and Toxicology 21: 84-90.

HUNT K.A., D.M. Bird, P. Mineau, and L. Shutt. 1992. Selective predation of fenthion-exposed prey by American Kestrels. Animal Behaviour 43:971-976.

JACKSON S and J Diamond. 1996. Metabolic and digestive responses to artificial selection in chickens. Evolution 50: 1638-1650.

JENSEN P. 2006. Domestication - from behaviour to genes and back again. Applied Animal Behaviour Science 97: 3-15.

KLUCSARITS, JR and JJ Rusbuldt. 2007. A photographic timeline of Hawk Mountain Sanctuary's American Kestrel nestlings. Privately Published by the Authors. Lulu.com.

KUNZL C and N Sachser. 1999. The behavioral endocrinology of domestication: A comparison between the domestic guinea pig (*Cavia aperea f. porcellus*) and its wild ancestor, the cavy (*Cavia aperea*). Hormones and Behavior 35: 28-37.

LACOMBE D and DM Bird. 1994. Influence of reduced food availability on growth of captive American Kestrels. Canadian Journal of Zoology 72: 2084-2089.

LACOMBE D, DM Bird, CG Scanes and KA Hibbard. 1993. The effect of restricted feeding on plasma growth hormone (GH) concentrations in growing American Kestrels. Condor 95: 559-567.

LAVIGNE AJ, DM Bird, JJ Negro and D Lacombe. 1994. Growth of hand-reared American Kestrels I. the effect of two different diets. Growth, Development and Aging

58: 191-201.

LINCER JL and DB Peakall. 1970. Metabolic effects of polychlorinated biphenyls in the American Kestrel. Nature 228: 783-784.

LOVE OP, DM Bird and LJ Shutt. 2003a. Corticosterone levels during post-natal development in captive American Kestrels (*Falco sparverius*). General and Comparative Endocrinology 130: 135-141.

LOVE OP, DM Bird and LJ Shutt. 2003b. Plasma corticosterone in American Kestrel siblings: effects of age, hatching order, and hatching asynchrony. Hormones and Behavior 43: 480-488.

LOVE OP, LJ Shutt, JS Silfies and DM Bird. 2003c. Repeated restraint and sampling results in reduced corticosterone levels in developing and adult captive American Kestrels (*Falco sparverius*). Physiological and Biochemical Zoology 76: 753-761.

LOVE OP, LJ Shutt, JS Silfies, GR Bortolotti, JEG Smits and DM Bird. 2003d. Effects of dietary PCB exposure on adrenocortical function in captive American Kestrels (*Falco sparverius*). Ecotoxicology 12: 199-208.

LOWE TP and RC Stendell. 1991. Eggshell modifications in captive American Kestrels resulting from Aroclor® 1248 in the diet. Archives of Environmental Contamination and Toxicology 20: 519-522.

LYONS ME and DL Goldstein. 2002. Osmoregulation by nestling and adult American Kestrels (*Falco sparverius*). Auk 119: 426-436.

MACLELLAN KNM, DM Bird, DM Fry and JL Cowles. 1996. Reproductive and morphological effects of o,p'-dicofol on two generations of captive American Kestrels. Archives of Environmental Contamination and Toxicology 30: 364-372.

MACLELLAN KNM, DM Bird, LJ Shutt and DM Fry. 1997. Behavior of captive American Kestrels hatched from o,p'dicofol exposed females. Archives of Environmental Contamination and Toxicology 32: 411-415.

MEYERS RA. 1992. Morphology of the shoulder musculature of the American Kestrels, *Falco sparverius* (Aves), with implication for gliding flight. Zoomorphology 112: 91-103.

MEYERS RA. 1992. The morphological basis of folded-wing posture in the American Kestrel, *Falco sparverius*. Anatomical Record 232: 493-498.

MUELLER HC. 1971. Oddity and specific searching image more important than conspicuousness in prey selection. Nature 233: 345-346.

NACCI D, M Pelletier, J Lake, R Bennett, J Nichols, R Haebler, J Grear, A Kuhn, J Copeland, M Nicholson, S Walters and WR Munns Jr. 2005. An approach to predict risks to wildlife populations from mercury and other stressors. Ecotoxicology 14: 283-293.

NEGRO JJ, KL Bildstein and DM Bird. 1994. Effects of food deprivation and handling stress on fault-bar formation in nestling American Kestrels. Ardea 82: 263-267.

NEGRO JJ, GR Bortolotti, JL Tella, KJ Fernie and DM Bird. 1998. Regulation of integumentary colour and plasma carotenoids in American Kestrels consistent with sexual selection theory. Functional Ecology 12: 307-312.

NEGRO JJ, J Bustamante, J Milward and DM Bird. 1996. Captive fledgling American Kestrels prefer to play with objects resembling natural prey. Animal Behaviour 52: 707-714.

NEGRO JJ, A Chastin and DM Bird. 1994. Effects of short-term food deprivation on growth of hand-reared American Kestrels. Condor 96: 749-760.

NEMETH N, D Gould, R Bowen and N Komar. 2006. Natural and experimental West Nile virus infection in five raptor species. Journal of Wildlife Diseases 42: 1-13.

NICHOLLS MK, OP Love and DM Bird. 2000. An evaluation of methyl anthranilate, aminoacetophenone, and unfamiliar coloration as feeding repellents to American Kestrels. Journal of Raptor Research 34: 311-318.

PATTEE OH and JC Franson. 1982. Short-term effects of oil ingestion on American Kestrels (*Falco sparverius*). Journal of Wildlife Diseases 18: 235-241.

PISENTI JM, GM Santolo, JT Yamamoto and AA Morzenti. 2001. Embryonic development of the American Kestrel (*Falco sparverius*): external criteria for staging. Journal of Raptor Research 35: 194-206.

PRICE EO. 1999. Behavioral development in animals undergoing domestication. Applied Animal Behaviour Science 65: 245-271.

QUINN MJ Jr, JB French Jr, FMA McNabb and MA Ottinger. 2002. The effects of polychlorinated biphenyls (Aroclor 1242) on thyroxine, estradiol, molt, and plumage characteristics in the American Kestrel (*Falco sparverius*). Environmental Toxicology and Chemistry 21: 1417-1422.

QUINN MJ Jr, JB French Jr, FMA McNabb and MA Ottinger. 2005. The role of thyroxine on the production of plumage in the American Kestrel (*Falco sparverius*). Journal of Raptor Research 39: 84-88.

RADVANYI, A., P. Weaver, C. Massari, D. Bird, and E. Broughton. 1988. Effects of chlorophacinone on captive kestrels. Bulletin of Environmental Contamination and Toxicology. 41:441-448.

REHDER NB and DM Bird. 1983. Annual profiles of blood packed cell volumes of captive American Kestrels. Canadian Journal of Zoology 61: 2550-2555.

REHDER NB, DM Bird and PC Laguë. 1982. Variations in blood packed cell volume of captive American Kestrels. Comparative Biochemistry and Physiology A 72: 105-109.

REHDER NB, DM Bird and PC Laguë. 1984. Simultaneous quantification of progesterone, estrone, estradiol-17ß and corticosterone in female American Kestrel plasma. Steroids 43: 371-383.

REHDER NB, DM Bird and PC Laguë. 1986. Variations in plasma corticosterone, estrone, estradiol-17ß, and progesterone concentrations with forced renesting, molt, and body weight of captive American Kestrels. General and Comparative Endocrinology 62: 386-393.

REHDER NB, DM Bird and LM Sanford. 1988. Plasma androgen levels and body weights for breeding and non-breeding male American Kestrels. Condor 90: 555-560.

ROCKE TE, NJ Thomas, CU Meteyer, CF Quist, JR Fischer, T Augspurger and SE Ward. 2005. Attempts to identify the source of avian vacuolar myelinopathy for waterbirds. Journal of Wildlife Diseases 41: 163-170.

SANTOLO GM, JT Yamamoto, JM Pisenti and BW Wilson. 1999. Selenium accumulation and effects on reproduction in captive American Kestrels fed selenomethionine. Journal of Wildlife Management 63: 502-511.

SAUMIER MD, ME Rau and DM Bird. 1986. The effects of Trichinella pseudospiralis infection on the reproductive success of captive American Kestrels (*Falco sparverius*). Canadian Journal of Zoology 64: 2123-2125.

SAUMIER MD, ME Rau and DM Bird. 1988. The influence of Trichinella pseudospiralis infection on the behaviour of captive, nonbreeding American Kestrels (*Falco sparverius*). Canadian Journal of Zoology 66: 1685-1692.

SCHUTZ KE, B Forkman and P Jensen. 2001. Domestication effects on foraging strategy, social behaviour and different fear responses: A comparison between the red junglefowl (Gallus gallus) and a modern layer strain. Applied Animal Behaviour Science 74: 1-14.

SCHWARZBACH SE, DM Fry, BE Rosson and DM Bird. 1991. Metabolism and storage of p,p'dicofol in American Kestrels (*Falco sparverius*) with comparisons to ring neck doves (Streptopelia risoria). Archives of Environmental Contamination and Toxicology 20: 206-210.

SHUTT, LJ and DM Bird. 1985. Influence of nestling experience on nest-type selection in captive kestrels. Animal Behaviour 33:1028-1031.

SILEO L, JC Franson, DL Graham, CH Domermuth, BA Rattner and OH Pattee. 1983. Hemorrhagic enteritis in captive American Kestrels (*Falco sparverius*). Journal of Wildlife Diseases 19: 244-247.

SMALLWOOD JA and DM Bird. 2002. American Kestrel (*Falco sparverius*). In Poole A and F Gill, eds. The Birds of North America, No. 602. The Birds of North America, Inc., Philadelphia, PA.

SMITS JE, KJ Fernie, GR Bortolotti and TA Marchant. 2002. Thyroid hormone suppression and cell-mediated immunomodulation in American Kestrels (*Falco sparverius*) exposed to PCBs. Archives of Environmental Contamination and Toxicology 43: 338-344.

SOCKMAN KW, H Schwabl and PJ Sharp. 2001. Regulation of yolk-androgen concentrations by plasma prolactin in the American Kestrel. Hormones and Behavior 40: 462-471.

SOCKMAN KW, H Schwabl and PJ Sharp. 2004. Removing the confound of time in investigating the regulation of serial behaviours: testosterone, prolactin and the transition from sexual to parental activity in male American Kestrels. Animal Behaviour 67: 1151-1161.

SPARLING DW and NE Federoff. 1997. Secondary poisoning of kestrels by white phosphorus. Ecotoxicology 6: 239-247.

TAYLOR R, S Temple, and DM Bird. 1991. Nutritional and energetic implications for raptors consuming starving prey. Auk 108:716-718.

TITTLEMIER SA, JA Duffe, AD Dallaire, DM Bird and RJ Norstrom. 2003. Reproductive and morphological effects of halogenated dimethyl bipyrroles on captive American Kestrels (*Falco sparverius*). Environmental Toxicology and Chemistry 22: 1497-1506.

VYAS NB, LA Thiele and SC Garland. 1998. Possible mechanisms for sensitivity to organophosphorus and carbamate insecticides in eastern screech-owls and American Kestrels. Comparative Biochemistry and Physiology Part C 120: 151-157.

WARD AB, PD Weigl and RM Conroy. 2002. Functional morphology of raptor hindlimbs: implications for resource partitioning. Auk 119: 1052-1063.

WIEMEYER SN, DR Clark Jr, JW Spann, AA Belisle and CM Bunck. 2001. Dicofol residues in eggs and carcasses of captive American Kestrels. Environmental Toxicology and Chemistry 20: 2848-2851.

WIEMEYER SN and JL Lincer. 1987. The use of kestrels in toxicology. Pages 165-178 In Bird DM. and R Bowman, eds. 1987. The Ancestral Kestrel. Raptor Research Foundation. Raptor Research Reports Number 6.

WIEMEYER SN and RD Porter. 1970. DDE thins eggshells of captive American Kestrels. Nature 227: 737-738.

WILLOUGHBY EJ and TJ Cade. 1964. Breeding behavior of the American Kestrel (sparrow hawk). Living Bird 3:75-96.

WRIGHT D, R Nakamichi, J Krause and RK Butlin. 2006. QTL analysis of behavioral and morphological differentiation between wild and laboratory zebrafish (Danio rerio). Behavior Genetics 36: 271-284.

YAMAMOTO JT and GM Santolo. 2000. Body condition effects in American Kestrels fed selenomethionine. Journal of Wildlife Diseases 36: 646-652.

YAMAMOTO JT, GM Santolo and BW Wilson. 1998. Selenium accumulation in captive American Kestrels (*Falco sparverius*) fed selenomethionine and naturally incorporated selenium. Environmental Toxicology and Chemistry 17: 2494-2497.

ZEDER MA, E Emshwiller, BD Smith and DG Bradley. 2006. Documenting domestication: The intersection of genetics and archaeology. Trends in Genetics 22: 139-155.

## Author Biography

David is regarded as one of the world's leading experts on birds of prey and he is often consulted by governments, universities, funding bodies, corporations, and the general public for his expertise. Appointed as the curator of the Macdonald Raptor Research Centre, David completed his M.Sc. in 1976. After obtaining his Ph.D. in 1978, he became Director of what is now called the Avian Science and Conservation Centre. As a Full Professor of Wildlife Biology, he teaches several courses in ornithology, fish and wildlife management, scientific communication, and wildlife conservation. He has co-authored over 150 scientific papers on birds of prey, supervised 37 graduate students to completion, and is currently supervising nine. Much of his research has focused on the American Kestrel. He has maintained a colony of this species for research species at McGill University for 35 years, keeping as many as 500 at a time. David has served as President (and Vice-President twice) of the Raptor Research Foundation Inc. (RRF). He also was elected Vice-President of the Society of Canadian Ornithologists twice and is currently the President-Elect. He is an elected Fellow of the American Ornithologists' Union and an elected member representing Canada on the prestigious International Ornithological Committee.

Over the last 30 years, David has given countless talks all over North America and made innumerable radio and television appearances both in Montreal and across Canada. He has written and co-edited seven books, including City Critters: How to Live with Urban Wildlife, Bird's Eye-View: A Practical Compendium for Bird-Lovers, and The Bird Almanac: The Ultimate Guide to Facts and Figures on the World's Birds. He is also a regular columnist on birds for The Gazette of Montreal and Bird Watcher's Digest magazine.

# Aviculture

*Ex-situ* Management, Theory & History

# THE PAST, THE PRESENT AND FUTURE OF NORTH ISLAND BROWN KIWI OUTSIDE OF NEW ZEALAND *APTERYX MANTELLI*

Kathy Brader

Smithsonian National Zoological Park
3001 Connecticut Ave. NW
Washington, DC 20008

## INTRODUCTION

*The Past*

The first documented Kiwi sent outside of New Zealand was a North Island Brown Kiwi, *Apteryx mantelli*, a female sent to the Zoological Society of London in 1851 where she laid several eggs and lived for several years. Before 1872, more birds were sent over to London including more N.I. Brown's, Southern Tokoeka *Apteryx australis*, Little Spotted *A. owenii* and Great Spotted Kiwi *A. haastii*. North Island Brown Kiwi first came to the United States in 1906 to the Smithsonian National Zoological Park.

Since then only North Island Brown Kiwi have been kept outside of New Zealand and only sporadically, mostly in Europe and the United States, Jurong Bird Park and Japan have also kept kiwi. There have been 18 different institutions outside of New Zealand that have kept kiwi; the first hatching of kiwi outside of New Zealand happened in 1975 at the Smithsonian National Zoo. This bird, a male is still on display at the zoo. The next hatching that occurred was at the San Diego Zoological Park in 1983, a female. San Diego went on to hatch an impressive number of birds, twenty-two in all, unfortunately, only five still survive. The last hatchings at San Diego were in 1999, none of which survived. Frankfurt Zoo in Germany started keeping kiwi in 1978 and their first hatching in 1986, which the chick did not survive. The first surviving kiwi was a male hatched in 1987. Since then Frankfurt has hatched out 32 kiwi, out of which 16 still survive and Berlin has now added to the population.

*The Present*

Currently there are five zoos in North American and two in Germany that currently house kiwi (Smithsonian National Zoo, Conservation and Research Center, Front Royal, VA., Columbus Zoo, San Diego Zoo and the San Diego Wild Animal Park, Frankfurt Zoo and the Berlin Zoo). The population now stands at 39 birds, which are broken down to 21 males, 15 females and 3 unknowns (this does not include the two recently hatched birds at Frankfurt this year).

Fortunately, due to intense conservation efforts on the part of New Zealand to save these birds in the wild, protocols for incubation, rearing and the keeping of kiwi has improved our understanding of these birds. We are now better equipped to improve our breeding of kiwi kept outside of New Zealand. If we want to keep kiwi in our collections, we have to work together; with this in mind I traveled to New Zealand in 2005 to work and study with New Zealand's Department of Conservation and various New Zealand zoos.

To this end, I contacted the International Studbook Keepers in New Zealand with the hope of generating an interest in helping out the overseas zoos with getting birds in better positions for breeding. The end result was the that the International Studbook has decided to drop the overseas birds from their studbook and become a Regional Studbook for New Zealand birds only. This opened up the opportunity for the overseas zoos to manage our birds separately. This will actually work well for the remaining zoos to develop our own management plans. In the fall of 2006, I applied for and received the Studbook for the North America and Europe North Island Brown Kiwi, the new studbook will be published soon.

The genetic data was analyzed using the data that was current as of December 2006 (since then two more chicks have hatched at Frankfurt Zoo, and two at the Berlin Zoo). I was able to track down some information from New Zealand on birds that came into the overseas zoos, but some records were vague. Therefore, we had to make assumptions for these birds. I was able to determine that the parentage for these questionable birds had parents that were wild caught. We decided to count these birds as founders. The known percentage known before assumptions was 45.8% and afterwards was 100%.

Genetic data exports were based on the living populations on AZA only populations and AZA and Europe birds together. Given estimated population parameters and the population size suggested by the TAG space survey, projections indicate that genetic goals (90% GD or < 10% loss of GD over 100 years) will not be met. Improving population parameters (increasing target size and effective population size, and breeding existing potential founders) does not result in projections meeting these genetic goals

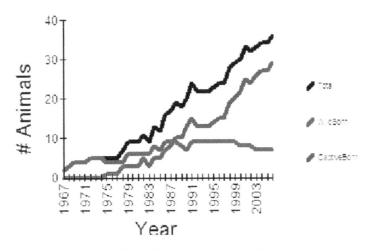

Fig. 1. *Census of AZA + European kiwi population.*

according to PM2000 projections (Strategy D). The addition of the European population does not improve the population genetically at first but does provide additional birds and additional spaces that can help the population grow and maintain gene diversity better in the long term. Including the European population, in addition to improving the growth rate, would allow the bi-regional population to maintain genetic diversity within 10% of its current level for 100 years (Strategy H). Additional founders may be available and would benefit the population. Researching and completing the pedigrees of unknown parentage and excluded birds may also improve the genetic outlook of this population. This looks great on paper but being of more practical use, is to look at our birds as individuals and their individual histories to know if they can be used as breeders. Some of the potential (including some wild caught) males are birds in their 30s; some have had the opportunity to breed with several different females over the years with no fertility. I seriously doubt that this will change and though kiwis have a projected long life span (possibly 60 years) and we can give the birds several seasons with a female, the odds are not good for obtaining offspring. This makes it all the clearer on how carefully we must look at our future pairings to ensure our potential genetic diversity and successful breeding.

Presently the last few years have been rewarding in the European zoo community and at the Smithsonian National Zoo where we have hatched our second chick in the zoo's history; we also currently have a fertile egg in the incubator currently (as of May 2007). We follow the protocol for egg incubation and hand-rearing that is recommended by Operation Nest Egg. This information has been invaluable to our current success. Kiwi incubation is an art all to itself, unlike any other bird species.

## Table 1
**DEMOGRAPHY & GENETICS OF BROWN KIWI IN AMERICAN AND EUROPEAN ZOOS**

| | Number of holding institutions | N | Estimated future holding capacity | T | l | GD (%) | $N_e/N$ | % known before assumptions | % known after assumptions |
|---|---|---|---|---|---|---|---|---|---|
| AZA | 4 | 17 (12.5.0) | 33 | 13.2 | 1.015 | 90.94 | 0.2824 | 45.8% | 100% |
| NA & Eur | 9 | 36 (21.13.2) | 67 | 13.2 | 1.015 | 87.01 | 0.3535 | 45.8% | 100% |

*N - Current population size*
*Estimated 5 year holding capacity was obtained from the Ratite TAG's space survey*
*T - Generation time (years)*
*l - Potential population growth rate based on historic data for this species (l = 1.0, 0% growth)*
*GD - Estimated current gene diversity of captive population*
*Ne/N - Ratio of effective population size to actual population size.*
*% Known - proportion of descendant population with known pedigree.*

## Table 2

**FUTURE PROJECTION FOR THE NORTH AMERICAN KIWI POPULATION**

|    | Projection strategy for AZA | % GD at 100 years | Years to 90% GD | Years to 10% loss GD | Tested target population size |
|----|-----------------------------|-------------------|-----------------|----------------------|-------------------------------|
| A. | Baseline | 54.6 | 0 | 16 | 33 |
| B. | Recruit 3 existing potential founders | 57.33 | 1 | 21 | 33 |
| C. | Recruit 3 existing potential founders; Increase growth rate to 1.05 | 59.11 | 2 | 28 | 33 |
| D. | Recruit 3 existing potential founders; Increase growth rate to 1.05; increase Target size to 75 | 72.04 | 2 | 42 | 75 |

### Incubation protocol

The incubation protocol is split between four age-related stages. Following is a breakdown of the parameters followed throughout each stage. Turning is done in four 45° stages spaced evenly throughout the day.

<u>Day one:</u> the egg is turned 90° clockwise (to the right) in 2 stages, and 90° anti-clockwise in two stages. (This will bring the top centre line of the egg back to 12 o'clock on the final turn).

<u>Day two:</u> the egg is turned 90° anti-clockwise (to the left) in two stages, and 90° clockwise in two stages. This 2-day cycle is repeated as necessary.

*Stage 1: day 0-9*

| | |
|---|---|
| Incubator | Still air Brinsea Hatchmaster |
| Temp | 36.0 - 36.5°C. |
| Relative Humidity | 60 - 65% |
| Turning | None |
| Cooling | 1 hour per day, 8am-9am. |

*Stage 2: day 10-55*

| | |
|---|---|
| Incubator | Still air Brinsea Hatchmaster |
| Temp | 36.0 - 36.5°C. |
| Relative Humidity | 60 - 65% |
| Turning | 4 x 45° daily |
| Cooling | 1 hour per day, 8am-9am. |

*Stage 3: day 56 to internal pip (around day 75)*

| | |
|---|---|
| Incubator | Still air Brinsea Hatchmaster |
| Temp | 35.5 - 36.0°C. |
| Relative Humidity | 55 - 60% |
| Turning | none |
| Cooling | none |

*Stage 4: internal pip to hatch (around days 75-78)*

| | |
|---|---|
| Incubator | Still air Brinsea Polyhatch |
| Temp | 35.5°C. |
| Relative Humidity | >60% |
| Turning | none |
| Cooling | none |

*(Prepared by Deidre Vercoe, updated by Andrew Nelson, Native Fauna Section, Auckland Zoological Park, March 2003)*

### Discussion

*The Future*

This is where we are currently and where we would like to go, the question is how do we improve our chances of obtaining our goals? With our improved understanding of the husbandry issues, we can improve our immediate goal of our future chick's survival and the breeding of our birds. However, we will need to bring birds in from New Zealand in the future to

### Table 3
**FUTURE PROJECTION FOR THE NORTH AMERICAN & EUROPEAN KIWI POPULATION**

|   | Projection strategy for AZA and Europe | % GD at 100 years | Years to 90% GD | Years to 10% loss GD | Tested target population size |
|---|---|---|---|---|---|
| A. | Baseline | 71.78 | 0 | 34 | 67 |
| B. | Recruit 5 existing potential founders | 74.86 | 0 | 82 | 67 |
| C. | Recruit 5 existing potential founders; Increase growth rate to 1.05 | 75.7 | 0 | 89 | 67 |
| D. | Recruit 5 existing potential founders; Increase growth rate to 1.05; increase target size to 100 | 80.75 | 0 | > 100 | 100 |

obtain our population target goals. The best way to go about this is to work together to show our commitment to kiwi conservation to the decision makers in New Zealand that we are serious about keeping a healthy population within the zoo communities. What would be the benefit to New Zealand in this? The answers I believe are two fold: firstly, is that if there were ever a devastating loss (such as disease outbreak) of the North Island Brown Kiwi in New Zealand there would be a healthy reserve of kiwi that could be used as a back up. This strategy became important in the US with the near disaster of the Whooping Crane loss. Secondly, and most important to me personally, is the opportunity of allowing people from all over the world to be able to actually see kiwi and come to appreciate what a unique species the kiwi is. Most people will never travel to New Zealand and see kiwi, this is a way for the world to learn what is so special about this bird and expound the great conservation work being done there to save them. At SNZP, we have been doing an education program called Meet A Kiwi for the last 17 years with an adult male and more recently our newest member, a chick that hatched here last February called Manaia. The response to this program from the public is astounding. We have people that come from all over the world that come for a chance to meet a kiwi up close. On the zoo's website, we have a section on the kiwi and the chick (including a webcam on the chick) and it comes in second most visited site behind the Panda pages.

### CONCLUSION

In conclusion, the status of kiwi outside of New Zealnad is not in badf shape and there are core groups of people that are determined to improve our care and expand our populations. By having kiwi in our bird collections, allows us to exhibit and educate our visitors on one of natures most unique treasures. The publicity generated by hatching a kiwi chick in our collections will get more visitors and press coverage to our institutions and further educate the public on this unique species. I believe that we should continue to have kiwi in our collections. The benefit is not just our visitor's pleasure but hopefully to assist the work being done in New Zealand to help ensure that kiwi remain on this planet for future generations.

### ACKNOWLEDGEMENTS

I would like to add a personal thank you to the staff at SNZP for their support of the kiwi project and the staff at the other kiwi zoos for their amazing work and the cooperation to the betterment of our kiwi.

## References

BILLING T. 1998. Kiwi Workshop Proceedings. Auckland Zoo, January 1998. New Zealand.

BRADER, K. December 2005. Animal Keepers Forum. The Journal of the American Association of Zoo Keepers, Inc. Vol. 32(12):578-584.

DEPARTMENT OF CONSERVATION. 1996-2006. Kiwi Recovery Plan. Wellington, New Zealand.

DEPARTMENT OF CONSERVATION. 2004. Captive Management plan for Kiwi. Wellington, New Zealand.

GREZELEWSKI D. 2000. Kiwi-icon in trouble. New Zealand Geographic. Number 45:66-97. Australian Consolidated Press, Auckland, New Zealand.

HOLMES T, L Hankey and S Oliver. 1999. Captive Management Plan for North Island Brown Kiwi. New Zealand Conservation Management Group. New Zealand.

JACOBS W. 2002. The New Zealand Experience. New Holland Publishers. Auckland, New Zealand.

JOHNSON T. 1996. Husbandry Manual for North Island Brown Kiwi.

KANZE E. 1992. Notes from New Zealand. Henry Holt and Company. New York.

ORBELL M. 2003. Birds of Aotearoa. Reed Publishing Ltd. Auckland, New Zealand.

PEAT N. 1990. The Incredible Kiwi. A Wild South Book. Auckland New Zealand.

PEAT N. 1999. Kiwi New Zealands Remarkable Bird. Random House. New Zealand.

PEAT N. 2000. Stewart Island. University of Otago Press. Dunedin. New Zealand.

PEAT N. 2006. Kiwi The People's Bird. Otago University Press. Dunedin. New Zealand.

WICKER R. 2002. Keeping the Kiwi and raising Kiwi in the Zoological Garden at the Frankfurt on the Main. The Zoological Garden Urban & Fisher Publishing House.

WILSON KJ. 2004. Flight of the Huia. Canterbury University Press. Christchurch, New Zealand.

WOLFE R. 2000. The People's Bird. New Zealand Geographic. Number 45:12-20. Australian Consolidated Press, Auckland, New Zealand

# ESTABLISHING CAPTIVE POPULATIONS OF ENDEMIC BIRD SPECIES FROM MADAGASCAR – AN OVERVIEW OF THE EFFORTS OF THE VOGELPARK WALSRODE FOUNDATION IN MADAGASCAR 1997 - 2007

Simon Bruslund Jensen & Mario Perschke

Vogelpark Walsrode Fonds e.V.
Am Rieselbach, D-29664 Walsrode, Germany

## Summary

The Vogelpark Walsrode Foundation was created to coordinate the bird conservation efforts of the Walsrode Birdpark. Apart form Madagascar several other projects are supported by the foundation. Project Tsimbasasa aims to develop the National Zoological Garden in Madagascar (PBZT). Enabling the staff and facilities to display and breed endemic avifauna and create the first breeding programs for several species. From 1997 - 2006 facilities of the park was renovated by combining Malagasy building techniques with modern enclosure design. Staff received training in avicultural management both in Madagascar and at the Walsrode Birdpark.

Annual expeditions to Madagascar served opportunities to follow-up on the progress and to perform field collecting trips. Strategies were developed with the Government of Madagascar and PBZT to minimize the impact on wild bird populations. In effect only young birds were collected and then hand-reared in the field with a minimum of loses these birds were subsequently established at the PBZT or Walsrode Birdpark in Germany. Vohidrazana Rainforest Corridor in the lowlands of Eastern Madagascar is still patched with intact rainforest and these forests contain some of the highest rates of endemism. This project concerns the unprotected area between two national parks. It is hoped to develop an awareness program for local communities with the aim to ensure that the national parks have continuous forest cover between them.

Tsarasaotra Heronry is one of the last known mixed heron colonies that are known to contain Madagascar Pond-herons and is located on a private property. The owner has agreed too support conservation initiaves and efforts are ongoing to renovate the nesting island which is in serious danger of being destroyed through erosion. In addition, WWF and RAMSAR are now showing interest in the area and an awareness campaign is planned in the heavy populated areas surrounding the park.

An association for the protection of water birds in the Bombetoka Bay was founded by an initiative of the Vogelpark Walsrode Foundation. The goal is the establishment of a conservation area and creating awareness. Local fishermen have agreed to patrol and protect the area. The first signs of increased tourism to the area have confirmed the benefits for the people in the region. Twenty-two young Blue-eyed Ibises were confiscated at the local meat marked, two are kept at PBZT and used for educational purposes and the remaining 20 birds were exported to Walsrode Birdpark were they will found a conservation breeding program.

## Introduction

The year 2007 has been declared Madagascar campaign by the European Association of Zoos and Aquaria's (EAZA) in recognition of the vast amount of environmental and social problems and at the same time the extraordinarily high degree of unique and endemic fauna and flora existing on this huge island making it one of Earths absolute conservation hot spots.

Walsrode Birdpark has been working in Madagascar for the past decade and has experienced the problems existing there first hand. In efforts to create awareness, develop *in-situ* conservation projects and create *ex-situ* safety populations we have been working with bird species that are little known to science and often has never been managed in aviculture before.

The efforts in Madagascar are a story of hard work, personal dedication and sacrifice. This presentation is a summary of the efforts, successes and failures performed in the past but also a look at what is still needed to be done in the future. All mixed with stories about the personal experiences we have made in Madagascar.

Walsrode Birdpark is a zoo that specializes in birds and extraordinary flowering gardens. Located in Northern Germany the park was founded in 1962 and today maintains one of the largest and certainly most diverse bird collections in the world. At Walsrode birds and their environment is being taken very seriously and the park has been involved in bird conservation since the very beginning in various ways including reintroductions, found-raising and conservation breeding.

The Vogelpark Walsrode Foundation (VWF) was created and registered as a charity in order to manage and organize the various conservation efforts of the birdpark and to ensure that donations would find their way directly into conservation work. Administrative, personal and office facilities for the foundation work are provided by the birdpark leaving the foundation with an absolute minimum of administrative expenses. In turn the work of the foundation is presented in the park with the dual benefit of creating more awareness about conservation and environmental issues as well as generating further funding. Individual donations and the spare change collected in the park is the primary income of the foundation however other foundations also provide funding for various projects through the VWF. The purpose of the foundation is to support bird conservation worldwide, an emphasis is made to projects that require hands on management and projects that benefit not only from financial support but also from the accumulated experience of the birdpark. It is a very proactive organization with a strong initiative that often supports important projects with very short notice and with the capacity to react on small scale avian environmental emergencies worldwide.

The organization functions in the form of an exclusive association for major sponsors and scientific advisors decision making is done by a small elected board of directors and the charity status is proofed by the Federal Government of Germany. Some of the long-term commitments and current projects apart from Madagascar include the Cloud Ambassadors program for Horned Guans in Central America, the Kagu conservation breeding center in New Caledonia, Golden Plover reintroductions in Germany and the Grenada Dove initiative in Grenada. Although often taking initiative the foundation rarely works entirely alone and often aims to involve as many same-minded organizations as possible in order to achieve a common goal. The Madagascar efforts has been supported by numerous individuals and organizations over the years including the San Diego Zoological Society, Wuppertal Zoological Gardens, Zurich Zoological Gardens, Landau Zoological Gardens, Berlin Zoological Gardens, Cologne Zoological Gardens, Heidelberg Zoological Gardens, Tokyo-Ueno Zoological Gardens, The Zoological Society for Conservation of Species and Populations, The IBIS-Ring, Mr. Luuc Van Havare and Mr. Serve Lemmens and many more.

**Project Tsimbasasa**

Our activities in Madagascar started with the urge for help from the Parc Botanique et Zoologique de Tsimbazaza (PBZT) in the capital of Madagascar, Antananarivo that had continuously failed to establish captive populations of endemic avifauna as had been requested by the ministry of higher education that was running the facility.

Funding and advice was initially provided but it was soon clear that help needed to be given on site in Madagascar in order to be effective; initial agreements also opened possibilities to develop parallel captive populations both in Madagascar and abroad. The VWF took on the task to develop the bird management at the PBZT in conjunction with the their own staff. Training was provided for biologists and keepers at Walsrode Birdpark in Germany. Theory and practice on population management as well as practical avicultural procedures were discussed and later implemented together in Madagascar. Every year from 1997-2006 another section of the bird facilities of the park were renovated by combining Malagasy building techniques with modern enclosure design, at times with a rather unconventional but practical outcome. The redesigned facilities provided the basis for several breeding successes by solving simple problems such as predator access. The annual expeditions to Madagascar served two main purposes: One was the planning and supervision of the construction efforts and follow up on the training programs and providing consulting service. The second was the field trips for surveying and collecting wild birds for the program. Strategies were developed with the Government of Madagascar and field biologists of the PBZT to minimize the impact on wild bird populations.

Pre-surveys, suitable locations and population sizes were evaluated prior to anything else. For most species, only nestlings were collected meaning that first task in the actual collecting trips was to locate sufficient amounts of active nests- this was on many occasions a very daunting task. Standing in a glowing hot spiny forest of the south or in dense wet rainforests of eastern Madagascar, the prospect of finding only a single nest often seemed an insurmountable task. Only with the massive efforts of skilled field staff and local guides made the location of nests possible.

The second task then was to survey the nests and to remove young at a stage where it would be possible to hand-rear them, leaving the parents to produce a second clutch. The only problem was that the collecting camps typically were located several days travel on foot from the nearest city, leaving the only option of building up a hand-rearing station in the field with enough gear to rear the young birds on site, including generators to run heating lamps and tents to protect the birds from predators such as snakes, Fossa's or dogs as well as enough equipment to provide all team members with food and hygienic material for up to 40 days in the field, all of this naturally within the frames of a tight budget. The work with hand-rearing the young birds in the field was performed by experienced staff from the Walsrode Birdpark and methods were developed over the years. Only a certain amount of the rearing food could be brought into the field and a major component therefore consisted of insects that were collected purposefully in the field. In some instances local children provided help but in most cases the majority of time in the field was spent collecting insects. The third task consisted of transporting the birds back to PBZT to the point where enough young birds had been collected to establish a founding population. This often proved rather difficult with the delicate and valuable cargo, in particular as we were usually several days away with young birds that had to be fed every few hours.

Once safely back at PBZT the young were stabilized and reared until independence. At this point the export to Europe was prepared for some and others were established at the PBZT. Only a certain quota of birds were allowed to be collected each year and government officials followed the field work and subsequent acclimatization at PBZT in order to verify this. All birds that were allowed to go abroad were initially quarantined at Walsrode Birdpark and from there then distributed to other participating parks or set up for breeding at Walsrode. A large off exhibit breeding centre has been dedicated to the birds from Madagascar and some species have in the meantime bred to several generations. The aim is to establish self-sustaining captive populations that functions as safety net in addition to generating awareness and funding for conservation work in Madagascar.

The species that where collected was carefully selected based on the assumption that it would be possible to establish captive populations of that particular species and on the basis that they were endemic to Madagascar and to some degree endangered through the loss of suitable habitat. In total, 12 different species have been collected, many more also fulfil the selection criteria but due to resource restrictions it is just not possible to establish populations of all.

One of the first species to be collected was the Madagascar Crested Ibis *Lophotibis cristata* this ibis is a solitary forest dweller and pairs make their nests high up in trees. Collecting the young certainly proved a challenge to the climbing abilities of the participants of the field expeditions. In 1998, four birds were collected and in year 2000 another five birds. A tenth bird was imported in 2001 and meanwhile all ten birds have reproduced. Several young have been placed at other zoos including in the US and Japan.

The Crested Coua *Coua cristata* was the first representative of the Silky-Cuckoo's that we worked with. This entire genus with ten living species is endemic to Madagascar, with three additional species having already gone extinct since the arrival of man to the island. This group of cuckoos are not parasitic but instead build their own nests of one or two eggs. The chick develops quickly and the species seems to adapt well to captivity, however it also seems to have a very short natural lifespan. Four birds were imported in 1998 and another four in 2000. Meanwhile 65 offspring have been produced by the Walsrode Birdpark alone, currently in the third generation. Birds have also been made available for several other institutions that financially support the project and some have even been exported back to Madagascar in order to be paired with unrelated birds at BPZT.

Sicklebilled-Vanga's *Falculea palliate* certainly proved to be one of the more challenging species. This group of relatives to the shrikes are also endemic to Madagascar and several of the species have developed specialized bills. They live in family bands that furiously defend their nest that is placed high in a thorny tree. The young are fairly easy to rear but to get the birds to breed in captivity proved much more difficult than expected. Two were imported in 1998, six in 2000 and another six in 2002. Only thee birds have been bred in Walsrode since, leaving fairly little hope that this species is likely to be established in captivity.

The import of Long-tailed Ground-roller *Uratelornis chimaera* was pursued by misfortune, although seven healthy birds that had been hand-reared in the field were imported in 2002 they all caught an infection within six months and only a single bird survived. The species have bred successfully in the aviaries at PBZT proving that there are good chances to establish this species. They have a distribution that is limited to a small stretch of dry forest in the south-western part of the island. This unique forest is known as the Spiny Forest and virtually all plant species in the region has thorns.

Equally misfortunate was the Subdesert Mesite *Monias benschi* that also only occurs in the spiny forest of the south-west. Very little is known about this group and the taxonomically placement of the three known species is still uncertain. They seem to live in very complex social groups where females are dominant. Groups of more than 20 birds rear only one or two young at the time and defend large territories. Only two nests could be detected within a very large range. The young proved very care-intensive to rear, and due to the low number of young birds collected it was decided to capture a number of adults for the PBZT. In the aviaries in PBZT the females proved to be very dominant and the males were systematically killed. Three hand-reared birds were imported to Walsrode a male and two females but unfortunately all died from the same infection that affected the Long-tailed Ground-rollers.

The Sakalava Weaver-bird *Ploceus sakalava* is an endemic species from the southern and western parts of Madagascar. Unlike the Fodies, they breed in colonies. Five young were hand-reared in Madagascar and 20 adult birds were purchased at a local market in Antananarivo; these birds were also imported in 2002 and are currently breeding in the aviaries of the Walsrode Birdpark.

The first Blue Coua's *Coua caerula* were imported in 2003 when four young birds arrived to Walsrode, another eight arrived in 2004 they were all collected as fledglings. This species is strongly tied to the dense rainforests in the east and have a nervous disposition. They resemble Turaco's quite remarkably with very similar movements and appearance. Initially this species proved extremely difficult to rear in captivity; of several eggs that hatched in 2005 and 2006 only a single young was reared successfully. In 2007, the breakthrough came and eight young were reared to independence. Temperatures and changes in the diet composition seem to have made the difference. So far six founder birds have been represented and careful efforts are now made to ensure that all founder birds will be represented equally well in the future populations. Currently this species is still maintained only at the Vogelpark Walsrode but with the recent breeding success and development of protocols the path is open for distributing this some-what delicate species to other experienced facilities as well.

The Pitta-like Ground-roller *Atelornis pittoides* has also proven difficult to breed. Nine sub-adult birds arrived in 2003 and another six in 2004. They quickly proved unsuitable for close quarters and individual birds are very aggressive towards each other. In the wild, this bird lives solitarily and pairs only meet during the breeding season and even then it's rare to see more than one bird at the nest site at the same time. They have bred in the huge Masoala Hall of the Zurich Zoo as well as Walsrode; they only seem thrive in very large exhibits.

In total, 12 Red-fronted Coua's *Coua reynaudii* were imported in 2003 and 2004, all were passed on to other facilities after ending the quarantine, namely Zurich and Wuppertal Zoos, in order to leave more space for other species at the breeding-centre in Walsrode. Until today none have been bred as far as we know.

The Madagascar Cuckoo-roller *Leptosomus discolour* is a highly unusual bird their loud calls are actually frequently heard in forested areas, where the birds occupy the treetops and hunt for insects and reptiles above the canopy. Six birds were imported in 2004 and another two in 2005 they were all hand-reared young and took two years to develop fully adult plumage. Eggs were laid in 2006 and 2007 but still no young were reared. Walsrode Birdpark remains the only facility in the world where this species maintained successfully over time and we are confident that the birds will proceed to breed as well. It has been determined that pairs do cause each other significant stress with their loud calls.

The Giant Coua *Coua gigas* is the largest of the surviving Silky-cuckoo's they are ground-dwellers similar to the well known American Road-runner. They live in relatively dry forests of the south and west. Five juvenile birds were imported in January 2006 and they quickly proved to be potentially very aggressive towards each other. A few eggs were laid already that same year but as the female were not compatible with any of the males the eggs remained infertile. In 2007, a number of efforts were made to bring a pair together for mating and it was some what surprising to find that one pair suddenly accepted each others company; at the time of writing a fertile egg is in the incubator. We realize that we still have a lot to learn about the social structures of the Giant Coua.

Finally, the Madagascar Blue Pigeon *Alectroenas madagascariensis* was imported in 2004 and 2006. Some of these fruit-eating pigeons were hand-reared from the egg in Madagascar prior to the transport to Walsrode. Seven pairs have been established out of which four have bred until now. We are very optimistic that this attractive species has a good chance to be established in captivity and the first offspring pairs have just recently been sent to other facilities.

## Discussion

From 1998 to 2006, birds were collected together with the PBZT under the Tsimbasasa Project from all the major endangered forest-habitats of Madagascar, the tall Monsoon-forests of the west, the dry spiny forests of the south and the wet tropical rainforests of the east. The island of Madagascar was once completely covered with forest, the initial human settlements started colonizing and deforesting the high plateau in the centre of Madagascar after reaching the island almost simultaneously from Africa in the west and Asia in the east about 2000 years ago. Today the highland forests are completely lost and remaining forests in the remoter regions are disappearing at an alarming rate. Human population growth in Madagascar is among the highest in the world leaving only little space undisturbed.

Initial field expeditions in 1998 and 2000 were supervised by the biologist Bernd Marcordes, in 2002 Simon Bruslund-Jensen headed the expedition in the south and in the following years the work in the field were performed by staff from PBZT with support from Mario Perschke and Bernd Marcordes from the Vogelpark Walsrode Foundation.

Mario Perschke's interest in Madagascar started as a primate keeper at the East Berlin Wild Animal Park and he started travelling regularly to Madagascar in 1995. Here he supported the development of the lemur section at PBZT as a volunteer and fundraiser until he moved permanently to Madagascar in 1999. His inspiring dedication for all Madagascar's wildlife was the reason that he also started supporting the efforts of Walsrode Birdpark. From 2002, he was employed by the VWF and received training in hand-rearing birds in order to continue the Tsimbasasa Project, however, he was also employed to investigate and perform other conservation work in Madagascar.

### Vohidrazana Rainforest Corridor

The lowlands of Eastern Madagascar are still patched with intact rainforest and these forests contain some of the highest known numbers of different endemic species per square kilometre any where in the world as well as some of the highest diversity figures known. A few national parks attract large numbers of tourists and still remain a safe heaven for many species. Among others the Indri *Indri indri*, the largest of the lemurs, roam these forests in monogamous pairs. This species is so adapted to a life in the trees that they never come down to the ground voluntarily. It is heart breaking to see how these large primates get caught in small island-like patches of forest that they are unable to leave, some starve as their forest-islands become too small to support them.

The two main National Reserves are Périnet and Mantady, they have a few kilometres between them and that area is called Vohidrazana. We visited this area for the first time in 2001 and determined that although satellite images leave the impression of a relative undisturbed forest much of the area has already been destroyed with slash and burn farming. After the cleared areas are abandoned by the farmer the areas are completely taken over by an, as ornamental garden plant, introduced thorny bush from Tanzania *Lantana camara*, leaving the forests no chance to regenerate. We also investigated the demographics of local settlements discovering that a very large percentage of the population was younger than 20 years of age, leaving high likelyhood for exponential population growth. The forest of Vohidrazana is currently being intensively used and no efforts are currently being made to change this; the VWF is looking to find partners and funding to start an awareness campaign in the area, educating people in sustainable use of the forest and the possibilities of eco-tourism. Long-term, we hope that larger areas could receive protected or buffer-zone status.

The Madagascar Pond-heron *Ardeola idae* have long been considered to be a relatively common bird, found even in the centre of Antananarivo. In recent years massive reductions in overall numbers and breeding colonies have been detected. They breed in large multi-species heronries and are one of the more shy species. During non-breeding they loose their brilliant snow-white plumage and attain a buff and brown plumage with white wings. In winter the birds curiously enough migrate to East Africa. The closely related African Pond-heron *Ardeola ralloides* look very similar in their non-breeding plumage but the breeding plumage is different with a buff coat. Numbers of this latter species, that is distributed throughout Africa, have been observed to be growing steadily in most of Madagascar.

Major concerns arose in 2001, when mixed pairs of the two species where noticed for the first time. One hypothesis is that the deforested landscapes of Madagascar encourages the expansion of the African

Pond-heron, native to open habitats in the African savannah rather than the Madagascar Pond-heron that has likely evolved in a forested habitat. The landscape changes of Madagascar causes the two closely related species to meet and compete for space.

After identifying the first mixed pairs, sightings of potentially intermediate birds increased excessively. As it was not fully confirmed that the two species actually hybridised with each other, determining this was considered a priority. Twenty young where collected from different nests that had been previously carefully observed and were known to be attended by Madagascar Pond-herons. These young birds were exported to Germany and based on their initial juvenile plumage no massive deviations could be detected. Blood was drawn for DNA examinations and laboratory work is being performed by Dr. Per Ericson at the Swedish Museum of Natural History. Initial results are very confusing, indicating a high degree of hybridizing within the population. Also the appearance of the young birds as they attained breeding plumage in their second year showed clearly that the young from even otherwise pure-looking parents showed variation and several characteristics shared with the African Pond-heron. These results indicate that the conservation status of the Madagascar Pond-heron must be re-evaluated considering this new threat. Meanwhile African Pond-herons have colonized the entire island and it may be questionable if any genetically pure Madagascar Pond-herons still exist.

**Tsarasaotra Heronry Renovation**

Currently Madagascar Pond-herons are only confirmed to breed at very few sites in the country. One is at the Tsarasaotra Park, on private property about four kilometers to the North-East of the capital Antananarivo. This is likely to be the only colony of this species in the highlands and possibly one of the last in Madagascar. The area is partly developed with housing, but also has a large area with trees and two fairly large lakes. The larger lake Alarobia has a small island which is the epicenter of the heronry. In addition to the two species of pond-herons, another six species of herons and egrets breed at this site in large mixed colonies, including Night-heron *Nycticorax nycticorax*, Green-backed Heron *Butorides striatus*, Cattle Egret *Bubulcus ibis*, Great White Egret *Casmerodius albus*, Black Egret *Egretta ardesiaca* and Dimorphic Egret *Egretta dimorpha*. The Madagascar Pond-heron is the most poorly represented species with only about 20 pairs. Due to the pressure from the high number of birds the island vegetation have been suffering for years, most of the trees and large portions of ground cover has been destroyed by the birds activities and as a result the island has started to erode.

Since 2001, the island and heronry have been observed by a group of concerned people, initiated by the VWF, that include staff from PBZT and members of the family that own the property. Also WWF and RAMSAR are now showing interest in the area and an awareness campaign is planned in the heavy populated areas surrounding the park. The most urgent matters are a restoration to ensure that the island itself will continue to exist. In August to September 2006, Mario Perschke and the VWF set up a team of seven labours and started the renovation work. Anchored poles and sandbags ensured the outline of the island and in total 2.8 tonnes of soil were moved to the island by hand and in small boats. Due to continued breeding of some species the work had to be performed gently with a minimum amount of disturbance. Planting ground cover plants initially proved unsuccessful as very large groups of wintering waterfowl especially whistling-ducks caused considerable damage to the new plants. Placement of artificial nesting platforms proved very successful and all 130 baskets were occupied with herons during the following breeding season, although no Madagascar Pond-herons were observed to use artificial nesting baskets, we still hope that this will help to increase the carrying capacity of the heronry and thereby support the Madagascar Pond-herons as well.

**Project Voronosy and the Blue-eyed Ibis *Threskiornis bernieri***

The 148,200ha large Bombetoka Bay is located at the mouth of the Betsiboka River northwestern Madagascar, some 20km south of the provincial capital of Mahajanga. About twenty small mangrove islets are located within the bay of Bombetoka, an area rich in various water birds. The conservation status of the endemic Blue-eyed Ibis, which were recently split away from the Sacred Ibis complex and given species status based on DNA studies, is becoming increasingly critical. The species distribution is restricted to a few remaining intact mangrove areas along Madagascar's west coast. The declining population is estimated to compromise no more than a 1,000 pairs.

Unless effective conservation measures are taken soon, this species may certainly face extinction. In addition to habitat destruction due to felling mangrove-trees, the collection of chicks for consumption has been identified as the main threat. The meat of the young birds is highly valued by local communities and is one of the few protein alternatives to fish that is available in the region. The Blue-eyed Ibis is a good flagship species candidate for this important area as the efforts to protect it from extinction will include protection of its mangrove habitat which is also home to many other species. The islands in Bombetoka Bay harbour the single largest existing population of Blue-eyed Ibis and were therefore chosen for this program.

In March 2005, an association for the protection of water birds in the Bombetoka Bay called Voronosy, was founded by a small group of scientists and keepers of the PBZT, Mario Perschke of the VWF was elected president. The goal of Voronosy is the establishment of a conservation area for the Blue-eyed Ibis and other endangered water birds in the Bombetoka Bay. An official collaboration contract was signed between all involved parties, including Malagasy government officials, in September 2005, giving the area a preliminary protected status.

The ongoing and planned actions for the conservation project for Bombetoka Bay consist of four different modules:

1) Education and development of awareness in the local communities.
2) Humanitarian aid that supports local communities in efforts to protect their own natural resources.
3) Definition and supervision of the protected areas.
4) The establishment of ecotourism at the site.

The inhabitants of the five villages which are located in close proximity, pose a considerable threat to the birds in the area through the collecting and consumption of chicks, and are directly involved in the conservation project. These five villages will be the main targets for awareness activities and humanitarian aid projects. It is planned to distribute and put up posters in the villages situated in the project area, informing the locals about the conservation area's location, the threats to the Blue-eyed Ibis and the benefits for the local people of conserving the mangrove forests. The schools receive special attention and all local teachers have become members of the Voronosy association. Posters with nature and species conservation topics will be designed and displayed, especially concerning the mangrove habitat and its inhabitants.

Regional associations of fishermen have been formed in each of the five villages. These associations have assumed responsibility for guarding the mangrove islands that are under protection. In turn they will receive practical help to increase their catch, a boat engine is provided to each of these associations, their members can use this to reach the fishing grounds, provided that it is also used for regular patrols of the protected areas. The members of the respective local association have dedicated themselves to practise these control cruises alternately and to submit detailed reports. To ensure professional continuance, a permanent guarding-group of eight wardens will be established. For each village one person will be appointed to organise, document and control the conservation actions.

The erection of a large publicity board at the Route Nationale 4 is planned. For passing tourists this board is to call attention to the sightseeing potential of Bombetoka Bay's water bird populations. The establishment of a tourist office is planned and one room to accommodate visitors to the site has already been constructed and has been accommodating visitors since March 2006. An entrance fee will be requested from every visitor, the money will be transferred to the fund of the respective local associations. Boat cruises for tourists can be offered alternately by the members of the local associations who have to transfer a defined percentage of their income from these trips to the project. These and other revenues will be managed jointly by the association and the community. The financial profits will be beneficial to the community projects and to the purchase of fuel for the supervision of the conservation area.

The local communities and the endangered bird fauna of Bombetoka Bay will equally benefit from the Voronosy conservation project. Twenty-two young Blue-eyed Ibises were confiscated at the local meat market in the early stages of the project. Two of the birds are kept at PBZT and are used for educational purposes in a newly built aviary that convey the story about Voronosy to the visitors of the zoo. The remaining 20 birds were exported to Walsrode Birdpark where they will form the basis of a breeding program. At Walsrode the birds have adapted well to captivity and as they reach sexual maturity next year, the expectations are high that they will produce plenty of young.

## CONCLUSION

Recent discoveries of the once thought extinct species, such as the four Madagascar Pochard that were recently rediscovered, gives a little shimmer of hope for Madagascar's fauna. Nevertheless the situation for many species and habitats are more than desperate. The ecological consequences of the habitat and species loss that goes on in many parts of the world cannot be foreseen. It is important that we all give our best attempts to counteract this and also pass the word to others that we have a world full of creatures worth saving. Aviculture is in many ways an appreciation of nature, although not being conservation in itself, many conservation techniques can be used directly in conservation work or to generate the interest and thereby also the needed funding.

On June 24th 2007, Mario Perschke, employee of the Vogelpark Walsrode Foundation and coordinator of the Madagascar projects was found dead in the Mahajanga National Park after having been missing for several weeks. His death leaves us with a huge gap in the organizations efforts and abilities to continue the sometimes difficult work in Madagascar. His life and dedication inspires us to continue at all costs. Future efforts in Madagascar will include a continuation of the Tsimbasasa Project in particular, to maintain the ongoing development of PBZT and to supplement those species that have already been collected in the past but where the genetic diversity is too low to support the captive populations.

## Author Biography

Simon Bruslund-Jensen has been working in the avicultural field since he was a teenager. He has worked at several facilities in his native Denmark before he went abroad in 1999 and worked with the Walsrode Birdpark in Germany with the responsibility for special breeding programs including Madagascar and Birds-of-Paradise. In 2002, he transferred to Al Wabra Wildlife Preservation in Qatar and worked as curator of birds, with responsibility for all bird related programs including breeding programs for Lear's Macaw and Spix's Macaws as well as Birds-of-Paradise. Since March 2007, he is back in Germany and acting as General Curator and Head of Scientific and Veterinary Management resumed responsibility for the collections at Walsrode Birdpark.

# THE AUSTRALIAN BLACK COCKATOOS: A REVIEW

NEVILLE CONNORS

Casuarina Parrot Gardens
Grafton, NSW, 2460, Australia

## Summary

The parrots of the world are a varied and dynamic lot and the cockatoo group are instantly recognisable. Just about everyone on the planet knows of them however a sector of this group are a little less known at least up to more recent times. The "Black Cockatoos" are one of the latter groups to be understood and consolidated in aviculture. Thirty years ago and more they were predominately kept in zoos and animal parks, there were a few dedicated and successful Black Cockatoo breeders in Australia but they were simply not a popular group of birds. This was mostly because of a belief perpetuated by a small group within the zoo system and repeated by the odd incompetent author that they were unsuitable to be kept in captivity. Quite the opposite is the case as they are delightful avicultural subjects, they are easily kept in good condition and have a willingness to reproduce. So much has been learnt over the past few decades that the myth and mystery has mostly been dispelled.

## Introduction

The four *Calyptorhynchus* species and the Palm Cockatoo make up the Black Cockatoos. In Australia, unfortunately the Palm is only seen in a couple of zoos and they are unlikely to reproduce. Fortunately they are popular in many countries and are being reproduced at an ever increasing and impressive rate. The remaining four species are well established in Australian aviaries and some are breeding well in several European collections.

### Threats to Survival of Black Cockatoos in their Native Habitat

### Fires

Australia is a large, hot and at times a very dry continent. Each summer some parts of this country will suffer the ravages of bush fires. These prove costly to property, the rural sector with farm crops, grazing paddocks and livestock being destroyed. Dwellings, farm buildings and human life as well may, and do, succumb to this devastation. The area that receives the least acknowledgement is the loss of native animals and the all important bushlands that they require for their survival.

To dismiss bush fires as a naturally occurring phenomenon is incorrect. It is a fact that fires can commence via activity from electrical storms and have done since time began. However, in our modern era many more factors contribute to the starting of such fires. Deliberate lighting is common, and obscure occurrences such as sparks from power lines and hot exhaust systems of motor engines driving through grass paddocks and of course the discarding of cigarettes all have played major roles in recent times. Aside from the direct loss and injury to the animals, including of course baby parrots still in their nests, there is insurmountable damage inflicted upon trees that provide food and hollows that are suitable for nesting cockatoos among many others. These hollows take decades to form and sadly hollow trees burn extremely well. In very recent times a bush fire of gigantic proportions went through the main breeding areas of the Gang Gang Cockatoo at a time of the year that would have trapped a great percentage of their young in the nest. *Calyptorhynchus* Cockatoos fortunately mostly breed throughout the cooler parts of the year thus not affected as badly directly but very much so by the loss of the food and nesting trees.

### Climatic Change

Perhaps the scientific community are not in total agreement regarding global warming but it is hard to believe that something is not happening. To me is seems logical that with the number of people that live on our planet and the amount of energy they use for heating, lighting and transport alone must surely alter what has been the norm. Unprecedented heat waves, droughts, cold periods are commonly reported

I feel compelled to report on something that we experienced literally in our own back yard. In something close to 150 years of record keeping of temperatures in the area where I reside, the hottest has been 44.6C. Just a few summers ago we experienced a day that touched 48C despite having sprinklers and misters operating we still lost about a half dozen birds. As the area where we have aviaries is set amidst gardens, shrubs and trees, and with the sprinklers being on, quite a few wild native birds apparently decided to sit out the heat wave in the aviary areas that is about five acres. I was shocked to find a number of these wild birds also died. It is my first experience of being confronted with something so seriously threatening. What is the cut off point for survival of different species? Would another degree or two of temperature mean the demise of a species within the affected area?

Such heat waves could take out a single, or several species in a single day, a terrifying thought! I don't believe I am an alarmist by nature but to pick up wild dead King Parrots, Rainbow Lorikeets, doves and even a magpie really shook me. (Now we are rebuilding aviaries with insulted panels to hopefully avoid such an occurrence)

## Land Clearance

Australia, as is the rest of the world is developing. Development equals clearing land for farming or for housing, roads etc. This of course thins the number of nesting trees, food sources and privacy for secretive breeding species. It also creates pockets and corridors that unnaturally squeeze the various populations together meaning those who do not compete so well, simply are lost. This will continue and is another example of how important it is for aviculture to secure many of these affected species.

## Nest Predation

Unknown but significant numbers are lost to predation. Again it is inaccurate to say this is normal. New introduced predators such as foxes, rats and feral cats take a tremendous toll on native wildlife including all of the cockatoo species by cats. Other creatures also take both eggs and nestlings. Monitors, (Goannas) pythons and opossums perform this task also other birds such as wood ducks have been known to "flatten" nestlings and take the log for their own nest. Galahs and Little Corellas also are guilty of nest robbing! As Black Cockatoos leave their young throughout the day at a very early stage any invasion of their nest log usually is successful. Additionally, illegal trapping or nest robbing is close to non-existent today, in part due to the successful propagation of these species in captivity providing a more valuable and adaptable pet or aviary bird.

## Characteristics

### Taxonomy

The classification of all black cockatoo species follows a hierarchical division where all parrots are divided into two families. Cacatuidae for cockatoo species and Psittacidae for other parrot species. Black cockatoos are further divided into subfamilies: Calyptorhynchinae (Glossy, Red-tailed, White-tailed and Yellow-tailed Black Cockatoos), Microglossinae (Palm Cockatoo) and Cacatuidae (Gang Gang Cockatoo). Subsequent divisions then classify the subfamilies into genera followed by species and in some instances, subspecies if warranted.

### Subspecies Classification

A subspecies is the division of a species into groups of related but significantly different members of a species complex that may have become initially separated from other groups of that species by a genetic or physical barrier. Evolutionary changes may occur, as well as changes in physical or behavioral appearance among other characteristics. Often this will lead to behavioural differences and if barriers disappear these birds may find interbreeding difficult. The identification of subspecies is often a cause for debate. Some species have undergone extensive and comprehensive research, sometimes coupled with genetic clarification, in this case separate status appears to be justifiably warranted.

### Species Classification

There are four species of Calyptorhynchus (Black) Cockatoos, currently there are twelve subspecies (including the four nominate):

Yellow-tailed Black Cockatoo

*Calyptorhynchus funerus funereus*
*Calyptorhynchus f. xanthanotus*

White-tailed Black Cockatoo

*Calyptorhynchus baudinii*
*Calyptorhynchus latirostris*

Glossy Black Cockatoo

*Calyptorhynchus lathami lathami*
*Calyptorhynchus l. halmaturinus*
*Calyptorhynchus l. erebus*

Red-tailed Black Cockatoo

*Calyptorhynchus banksii banksii*
*Calyptorhynchus b. samuli*
*Calyptorhynchus b. macrorhynchus*
*Calyptorhynchus b. naso*
*Calyptorhynchus b. graptogyne*

The Glossy Black *C. l halmaturinus* from Kangaroo Island South Australia is said to be a little larger than the nominate bird, therefore granted a subspecies title. In my opinion this is a very doubtful claim as Glossy's anywhere differ substantially in size. Again, in my opinion, the so called Glossy *C. l. erebus* is nothing more than a larger version of the nominate race.

Lastly, the Red-tailed *C. b. graptogyne,* was once more, in my opinion, a title granted for a purpose rather then a warranted one. The risk of the Red-tailed loosing protected status in much of Australia due to their abundance was very real and wildlife law enforcement officers literally called for a rare and endangered category for the Red-tailed. The obvious was confronting them when up popped *C. b. graptogyne.* Convenience or coincidence? This named group that reside in North West Victoria are today seemingly isolated and are not abundant in this part of the land however any species is usually quite rare at the extremity of their range.

The Palm Cockatoo that is also found in a small area of Cape York is currently divided into the nominate and four sub species: *Probosciger aterrimus* (nominate) and *P. a. goliath, P .a. stenolophus, P .a. intermedius* and *P. a. macgillivrayi.*

## Distribution

Collectively the four *Calyptorhynchus* species cover pretty much all of the Australian continent that is capable of sustaining birdlife.

## RED-TAILED BLACK COCKATOO

Are an extremely abundant species and have the widest distribution of any of the black cockatoos. Found across the top or northern part of Australia from the Pacific ocean front to the shores of the Indian Ocean. They are also found in large areas across the mid and southern parts of Australia, from western NSW to the south-western corner of western Australia. They reside in vast areas with the exclusion of the island state of Tasmania. The separation of some of these groups has led to the classification of a number of subspecies, although there is "overlapping" of these types in a few areas.

### Breeding

This cockatoo is a great bird for the new aviculturist as they are hardy, easy to care for, and keen to go to nest. Generally it is accepted that birds commence breeding from their fourth year, however like many species of psitticines, the more domesticated they become, the younger they seem to be capable of breeding, and there are some instances of three year old's successfully reproducing. The breeding period is quite variable as this species is found almost all over mainland Australia, hence with so many weather variables an aviculturist keeping this species in the north of Australia may experience differing breeding times to a similar situation at the other end of the continent, or to breeders from the northern hemisphere. However whatever the time, when the cock bird begins his "mating call" and starts parading up and down the perch with a series of hops, lowering his head and fanning his tail, the expectation of an egg is nigh.

A single egg only is laid and if removed for incubation the hen may re-lay. This can be repeated several times for the season. A number of instances have been recorded where this has reached into the double digits, in such cases it is paramount to supply the hen with calcium, as this species lays a large egg and calcium depletion is inevitable. If the egg remains with the hen she alone incubates for a period of approximately 30 days. The hatched chick is covered with yellow down and has a light pink/bone coloured beak. The young should climb to the top of the nest at 11 to 13 weeks of age. Generally if the young haven't climbed out by about 90 days then a problem, usually a calcium/dietary related, one exists. The juvenile usually sits at the top of the log (another good reason to have an opened top nest) for a day or so before taking its maiden flight. The period for weaning is around two months so if parent-reared they should remain in the aviary for at least that long. Young will often beg for a feed for a longer period even though they are eating themselves.

## WHITE-TAILED BLACK COCKATOO

Confined to the south-western corner of Australia where two types are found. One *C. baudini* inhabits the richer forested very corner of the land and has a longer mandible suitable for the extraction of many of the forest seeds from their region. Further north the shorter billed White-tailed *C. latirostris* inhabits more open land and its dietary requirements differ enough apparently explaining the difference between the two.

These are the genuinely most threatened of all the Black Cockatoos with a number of critical factors applying pressure on the populations.

### Breeding

Of the four *Calyptorhynchus* species the White-tailed has to date proven the most difficult to reproduce in aviaries. It must be accepted, aviculturists still have to gain more knowledge on this bird. It is felt that the trigger lays within the diet rather than any other factor.

This species is more willing to attempt to breed if brought into "condition". A diet feed all year round without too much alteration may see your pair of birds coast from one season to the next without an attempt to reproduce or perhaps some displaying may take place but no true results. Natural nesting takes place in late winter/early spring or after the intense heat of mid-summer.

For aviculturists in Australia's east the majority of breeding appears to be at the end of the summer around March. However, we have had eggs in November and December on occasion, invariably after a supply of banksias cones. Again the mere sight of the cones has the pairs excited. On the eastern side of Australia, prior to Christmas, the coastal banksias ripen and are devoured by the endemic Yellow-tailed *C. f. funereus*. We collect and feed out to both species and since they are all within hearing range of each other, it doesn't take long to hear both species displaying to their mate. Certainly the supply of pinecones will have a similar effect.

The male fans his tail and struts up and down the perch bowing with his crest raised delivering a variety of chatter like shrill squeals. Copulation, as with all cockatoos sees the male mount the female with both feet secured to her back. If aware of being observed the pairs will often break off their mating, unlike many of the *Cacatua* species that seemingly enjoy prolonging this process; White-tailed copulation is a relatively short process.

Two eggs form a normal clutch generally laid some four to seven days apart. The first egg is invariably larger than the second egg and if only one egg is laid then it is a larger egg of a similar size to that of the first. Usually only one chick will survive in both the wild and in the aviary particularly if there is some considerable gap between the laying and corresponding hatching of the eggs.

The rearing of two has been recorded in both captivity and in the wild; success must be more achievable if the eggs are laid close together. In aviculture the accepted practice is to remove one for hand-rearing.

## YELLOW-TAILED BLACK COCKATOO

Found along the eastern side of Australia from the Pacific coastline to a couple of hundred kilometres inland. They are found north of the tropic of Capricorn and south to Tasmania. The western extremity of their range is the south eastern corner of the state of South Australia. The southern subspecies *C. xanthanotus* is smaller with come colour differences and vocal calls to that of the larger northern nominate bird *C. f. funereus*. An intermediate sized bird is found around the Melbourne area of Southern Victoria where the two overlap. Although the size is intermediate the birds in colouration are that of *C. f. funereus* with the exception of the yellow patches on the tail feathers.

### Breeding

Pairs will commence reproducing from their fourth year of life, and some males certainly display to their mates earlier than this. Undoubtedly, they would be capable of breeding or certainly fertilizing a receptive hen. The nominate *C f. funereus* may commence breeding as early as the Christmas period through to July and occasionally even after that. The *C. f. xanthanotus* is willing to commence breeding during Australia's spring period around September to December or else will begin closer to the Easter period during Australia's autumn.

The hens will usually lay two eggs and in our experience and from information gleaned from other breeders there has been a break of four to nine days between them. The hen incubates the eggs commencing with the laying of the first egg for a duration of some 28 to 31 days and the size of the two eggs vary and therefore the weights of the chicks at hatching also vary. In almost every instance the second chick will perish if left for parent-rearing, although in both aviculture and with observations of wild Yellow-tailed occasionally the two may successfully fledge. It is however recommended to take at least one in either the egg stage or as a chick for hand-rearing. Whilst sitting, the hen as a rule comes off the nest three times a day, early in the morning again at around noon and lastly just as the light is fading at the days end. The babies have long and a rather deep yellow coloured down at hatching and it will remain in the nest for some eighty to ninety days or even longer. The young bird will generally sit on top of its nest for a couple of days prior to launching into its first flight. They will sometimes back down the nest at the approach of people until they are apparently secure with their surroundings.

## GLOSSY-BLACK COCKATOO

A similar distribution to that of the Yellow-tailed, the Glossy-black is also found from the shores of the eastern coast land to a few hundred kilometres inland that encompasses dramatically different terrains throughout their range. Their northern populations also reside north of the tropic of Capricorn and south into the state of Victoria (they are not found on the Island state of Tasmania). There is however, an isolated population on Kangaroo Island in South Australia. Two dubious subspecies titles currently exist for the Glossy-black Cockatoo. Personally I do not believe they differ from what is referred to as the nominate bird.

### Breeding

Breeding commences as early as February so the aviary should be totally furnished at least by the commencement of the calendar year. Copulation usually takes place in the morning although we have witnessed the occasional afternoon mating. The hen positions herself with her tail raised and her wings dropped. When the females come into breeding condition they undergo a real personality change, their usually soft and confiding nature becoming quite intolerant, sullen and distant.

The cock treads with both feet on the hens back with his wings fully opened, and he will then typically dismount and fly to the other end of the aviary then back again to repeat the mating. This performance may be repeated three or four times. All this is preceded by a lengthy display from the cock bird including a lot of vocalization, head bopping, jumps along the perch and tail fanning. Unlike the accepted approximate four years for the other *Calyptorhynchus* species, the Glossy's may reproduce as early as two years of age. It is not unusual for hens to lay at two years and we know of a couple of instances where a two-year-old male has fertilized an egg. Although a clutch is just a single egg, quite often the hen will lay and not brood, and some period later she will re-lay and sit. This is probably why there is some confusion in various literatures as to how many eggs are laid to form a clutch. If the egg is removed for incubation the pair will more than likely attempt to try again. The following is an account from one pair during the 1999 season: The first egg hatched on the 18th of April and the baby was removed at two weeks of age to be hand reared. They laid again and the egg was taken for incubation successfully hatching on the 30th June, they laid yet again and that egg was also removed to be hand reared hatching on the 14th of August. They in the meantime went down once more laying on the 10th of August, this was left for parent-rearing however it died hatching in September. Despite a rather late start to the season and leaving the first egg for the full incubation plus a couple of weeks we still finished up with three young healthy Glossy's for the year from one pair and with a little more fortune it could have been four.

Whilst incubating the hen sits tightly and usually only vacates the nest twice daily, once in the morning and again prior to sunset. The cock calls to his mate and she usually replies a couple of times from within the nest and then scrambles out to be fed by him and to drink. Prevailing weather seems to dictate how long the hen remains away from the nest. When entering the nest no matter what the entrance is like the Glossy-black hen will turn around and climb down to her egg traveling backwards, it is quite exceptional for the male to enter the nest.

The incubation time is usually 28 to 30 days however reports outside that time frame are no doubt accurate, further study will be of interest in this area taking into account the time of year and weather conditions. Some observations that we have made have been valuable when calculating brooding procedures for artificial rearing. After hatch, one hen we recorded leaving the nest only briefly on the first day at approximately

4pm and on the second she left the nest from 3:30pm until 5:15pm. This absence would be a concern for a single chick of another species, however the baby Glossy is covered with a thick, long and fluffy yellow down keeping it well insulated. Young remain in the nest for up to 100 days so if they have not vacated by around that period of time then there may have, or there has been some development problem. If the parent birds are hand reared then they are tolerant of nest inspection.

## Avicultural Requirements

### Nest Logs or Boxes

In Australia, most keepers of these cockatoos favour the usage of natural hard wood hollows. So long as the trees are already downed for some reason then that is fine however to fell a tree just for the express purpose of obtaining a nest log for an aviary must be frowned upon. The word hard is emphasized for a reason. Eucalyptus trees are a very hard solid timber and when hollowed out by termites and wood grubs are of course many decades old, so the timber is well seasoned. Black Cockatoos, especially the Yellow and White-tailed's can chew like no other bird on the planet. Any accessible wood will become "chips" in a heart-beat.

Therefore, nest boxes need to be flashed with bent sheet metal so as to deny the cockatoos any edges that will commence their "chomping". To deny them anything to chew within that nest is not the correct direction either as they actually deposit their eggs on substrate that they have manufactured.

These chips are dropped into the nest base and are of uniform size. By fixing slabs of Oregon pine near or around the top and entrance will satisfy this all important requirement. By leaving the top off the box or log and fixing the sacrificial slabs to that entrance area will be sufficient. Only the hen incubates the egg and she will back down into the nest to the point where the egg will be. In the wild, nest depths vary considerably however the domestically bred birds are more than happy with a nest of approximately three feet or a little more with an internal width of 35cm or just a little greater. If the nest is hung in a vertical position or close to that angle then it should be readily accepted. As they select nest sites in the open areas that are exposed to the elements in an aviary they enjoy it in an exposed area receiving the morning sun. If hung just under the roofing cover and down from the ceiling (roof) at a distance of around a meter to the top of the nest rim. An inspection door at a level just above the egg is helpful.

### The Aviary

Keeping in mind that this group of birds are chewers and are destructive if given the opportunity to be so, then the construction needs to be of steel and mesh of a quality and gauge that will repel the inevitable attacks upon it- two-and-a-half centimeter square and a gauge of 2.5 to 3mm will suffice. In Australia we have an ongoing problem with heavy metal poisoning. The manufactures have varied the amounts of lead and zinc over the decades but the problem hasn't gone away and indeed as a lot of our mesh comes from China now and the problem is certainly existent.

The obvious solution is unfortunately an expensive one, that is to use stainless steel mesh if you share the same problem as we do in Australia. Size is a matter of choice, finance and/or commitment. Although it is not up to any authors to try and dictate how large an aviary should be, they are a bird that seemingly needs to fly about quite a bit. It is beautiful to watch them flying around a good sized aviary and they willingly reproduce in such enclosures. So as a rule of thumb just a bit larger then you can afford should be suitable. We have housed and bred them in aviaries as short as only about 3.5 x 1.5m wide by a little over 2m high, in the case of the smaller Glossy-black, however the larger Blacks should have something of greater proportions. Like all parrots they are much more comfortable if their nest log and perches are above head height of their keeper or observers. Again perches need to be continually replaced no matter what type of branch is used as this frightens the cockatoos to a certain extent. Perch longevity can be extended by a couple of methods. If vertical branches of a substantial girth are dug into the ground, and the top of such being a little lower then the front and back perches so as not to obstruct their flight between the two, will see the birds spend countless hours on the vertical stump doing their wood chip thing. The second is to supply leafy branches of a safe plant/tree for them to rip to pieces. This can be done in whatever manner is easier for the keeper.

Depending on the climatic conditions of where one is situated, this can play a huge part in the design of an aviary, aside from space, suitable sunlight and shade must be mandatory and if in the colder parts of the world then the indoor, outdoor scenario that is afforded to other robust psittacines would need to be applied to the Black Cockatoo group as well.

Although we have a lot of suspended aviaries we do not house the blacks in such enclosures. Some breeders do and so long as they are large enough, then I have no doubt that they would work okay. When housing Black Cockatoos in banks of aviaries, thought should be given to other species that are to be housed in the same complex. One has to be confident that neighboring pairs will not inflict injury on each other or suffer stress and that a new fledgling may safely hang onto a side or dividing wall without fear of harassment. Separating the flights with double mesh will eliminate the problem. However, selection of neighbors is still important, as intimidation by aggressive species or individuals is a stress factor that the birds can do without.

**Food and Water Bowls**

Heavy ceramic bowls of generous proportions should be used in preference to any other material. It keeps the water cooler than metal containers and providing it is heavy enough the cockatoos will not turn it over.

Stainless steel bowls can be used so long as they are anchored firmly by one form or another. If the bowls are not secured, the birds bathing and splashing will empty the majority of their water out of the bowl, and once it is light enough most cockatoos will have little trouble tipping the remainder out, therefore denying themselves water until next checked by their keeper. It is beneficial to be able to access the bowls by not having to physically enter the actual aviary, if an escape flight or walk way is used then the feeding station may be serviced from there, also some similar adaptation for outside individual aviaries. This of course expedites the servicing procedure and eliminates stressing and distracting the pairs that will occur to varying degrees if the feeder is required to enter the aviary. Also if the pair are a little too imprinted to humans, then this servicing from outside helps reduce any contact time which may aid the "re-wilding" of the over-tame cockatoo.

**Aggression**

Like all cockatoos. the Blacks can also become seriously hostile towards one another. This usually occurs whilst a seemingly frustrated male wishes to commence breeding. Although the hen appears to go along with the idea the male still may launch into a vicious attack that may cause injury or death. Separating the aggressor will deny the pair the opportunity to reproduce and although that is a better alternative to that of loosing one, there are a couple of measures that can be performed.

If a wing of the offending male is clipped just enough to impair his flight to a degree that he cannot catch the hen, then breeding should still take place. Another method is to use electrical tape, wound around about 5/6 flight feathers that are bunched together. This tape may be removed without damage to the flight feathers once the violent moods have abated or disappeared. Obviously it is unnatural for an animal to kill the mate it is wishing to reproduce with and it is often just a few days where this situation prevails, however to observe and not take any measures may be a regrettable in-action. These birds, even established breeding pairs, should not be placed into one container for transport. A small cage will most likely have the same affect, increasing the chance of a serious fight between the pair or birds of the same sex. A separate container is a must.

**Handling**

Possibly only three words are required Handle With Care! That says it all really! If one is obtaining a pair of cockatoos there are some necessary items that you will require from time to time. A hand net of appropriate size and strength should be on hand prior to obtaining the birds. Once the bird is caught in the net it is easier and safer to bring it to the ground or floor, where the bird will mostly remain in an upright position allowing a hold to be taken on the bird by securing the back of his head through the net, while the other hand should encircle the wings securing the feet at the same time. With the head and wings held firm, the bird cannot bite the handler nor injure its wings, however if the feet are not restrained then a claw can inflict a certain amount of pain. The use of a towel can help if one isn't too confident with this exercise.

It is advisable to practice "a hold" on a smaller species. Learn by watching someone who is competent, and by taking the time to do this it will not be too difficult to get the right idea. Bird shop proprietors are often expert at this art as they perform this procedure many times on a daily basis.

**Food**

Sunflower seed is the staple diet for Black Cockatoos in aviaries. The majority of these cockatoos do not suffer from obesity or fatty tumors that would be inappropriate for say *Cacatua* cockatoos. Nuts such as almonds, pecan, brazil, etc are relished as are a few fresh peanuts, greens like spinach, silver-beet, green peas, etc are also enjoyed so long as the birds were educated as young ones to eat such food types. Sprouted seeds are relished as is beetroot with only the odd individual eating pieces of fruit such as apple or oranges. Most juveniles will give almost anything a go and this is the most important stage to set the pattern for their future. On a daily basis we give a piece of wet wholemeal bread with vitamins and calcium if it is required sprinkled on it. Invariably it is this food that is selected first by almost every pair and always if they have young in the nest, it is however only a small portion! It still has to be said that breeding results or at least to induce pairs to breed are improved if appropriate native food types are supplied. Once an egg is laid then it doesn't matter much. Yellow and White-tailed's are passionate about banksia cones in particular and the natural food source for Glossy Blacks, the Casuarina seeds, works for some Glossy's but doesn't seem to be as obvious as is the case with the banksias and the two Light-tailed types.

Red-tailed's seem willing to breed on just a normal diet and must be considered easy birds to cater for in aviaries. We have not used pellets however if it was to replace just the seed content of the diet then why not? A lot of work has been done with commercial foods and this will be an area that Blacks will most likely benefit from. In Australia, they have been bred in such numbers that the market place for them is at the least complete and we simply do not wish to breed them in great numbers unless of course export of such species is allowed.

**Hand-reared Versus Parent-reared Birds**

Many opinions have been expressed regarding which of the two are best to be retained for breeding purposes. Initially when the original black cockatoo stock was derived from the wild, breeding results were by and large few and far between. There were successes but production was slow.

This has again been demonstrated in more recent times with the harvesting of Red-tailed Blacks in the Northern Territory. The captured birds have been reproducing but at quite a comparatively slower rate than pairs that are several generations removed from the wild. With the availability of modern hand-rearing formulas healthy good-sized birds are being produced and often several from the one pair each season as opposed to just a single chick if left for parent-rearing. It is at the time of weaning that a deliberate and definite decision should be made as to the future of the cockatoo youngster whether it is to become a breeder or a pet.

If the desire is to develop the bird for breeding then only sufficient attention should be given to feeding, to complete the weaning and then move the bird into an aviary, preferably in the company of like birds and also preferably away from continuous human contact. It is rewarding to have the young birds quiet enough to take favorite morsels from their owners but it is intelligent to spend just a minimal amount of time in doing so and to remember they are birds and not feathered pets, so resisting mushy conversation may pay dividends for the future. This particularly applies to male birds, as very quiet hens are often okay for breeding.

Hand-reared but not silly tame birds make for very steady breeders that are less likely to suffer stress from such actions as nest inspection or replacing perches etc. Parent-reared birds may be steady particularly if bred from quiet parents however often they may be surprisingly flighty and slow to adapt to changes such as altering their aviary or even just some of the aviary furniture. They are more affected by outside stresses such as strange dogs and children. If illness or injury requires treatment then this type of bird is more stressed by the required handling than a hand-reared one might be.

**Reverting Tame Birds for Breeding**

This is a question that comes up quite frequently as people have a bird that thinks its a human but they wish to breed it. Unfortunately, as the damage was done in the birds informative weeks it is a difficult task if at all possible. Over tame male Black Cockatoos live just to see and connect with their owners or sometimes just any person, even when placed in an aviary with a group of non-tame like birds, they often totally ignore their peers, calling and parading for the attention of a person. If placed in a situation where the bird is again in attendance with like birds but totally out of view even when being fed and without any vocal contact from his keepers, this behavior may slowly be altered and even reversed. However, as the bird itself is totally happy about believing it is human, or a companion of a human then if breeding is desired it may be better having the tame bird as a pet and obtaining a new bird for breeding.

**Black Cockatoos as Pets**

This is not much of an option for people outside of Australia however in years to come it will be. By and large the Blacks make endearing and charming pets. Their ability to mimic however is almost non existent and as mentioned males of Red-tailed and Glossy are generally obsessive and demanding. The White and Yellow-taileds are extremely interesting and lots of fun with personalities that will charm most, however they are extremely destructive and would require intense supervision if kept in the household.

## Conclusion

The ISBBC symposium theme of "Conservation Through Aviculture" must make us all ponder upon this title and the future of the species we possess or that is under our care. With the world's human population increasing at an uncontrollable rate and climate changes becoming a reality, the future for many living animals and plants will surely be altered, severely threatened and worse, within most of our lifetimes.

Conservationists have been active for many years with countless projects undertaken throughout our planet. Many have been admirable and successful, many have been flaky, mercenary and abject failures, but everyone at least among the educated sectors of mankind are aware of the importance of conserving all that is! I believe, in aviculture and particularly within private aviculture, we have taken a most important and gigantic leap forward from unfortunately, the futility of trying to conserve absolutely everything as it is or more to the point as it was. We have become the world leaders in preservation. It is simply too risky for many species to try to leave them in ever reducing pockets of various habitats as conservation havens, which sadly are under an ever increasing amount of pressure. Unfortunately, many instances exist where that luxury has simply by-passed us and the last stand, that of preserving species, is of paramount importance.

Aviculturists are already actively doing this with numerous, but far from all species of birds. Perhaps quite unwittingly people are, or have learnt enough that under controlled conditions these birds can survive into the future. We, as aviculturists, can be and are, criticized for many aspects of our interest, hobby, and profession, however, we are still reproducing many of the world's gems without hand outs from the tax payer's coffers but via our own extraordinary efforts.

Instead of suffering opposition from government bodies, animal activists, misinformed wildlife agencies, misguided conservationists, and even worse various groups that have developed out of our own throng, aviculture will need to showcase just what we have and are offering to the very existence of many of our planets species. These groups through ignorance often view keepers and breeders of various animals in low esteem, we need to turn this train of thought into a pathway of logical progression, where not only can the mankind of the world today, but of tomorrows as well, mingle with surviving species.

## Author Biography

My journey into aviculture began in my early twenties. As a child I kept finches, budgerigars, king quail and a couple of rosellas. I do not consider that part of my life as really partaking in genuine aviculture. So from the first pair of birds (actually two cock superb parrots) to the present day's substantial collection has been a fantastic, wonderful and interesting journey. With my wife Enid or "Noddy" we have blended the raising of a family and the breeding of parrots in particular, into a busy but rewarding lifestyle.

In 1980, we purchased some bush land and relocated ourselves and aviaries along with about twenty pairs of parrots. We named our piece of paradise "Casuarina" and immediately set about expanding our collection. In 1990, we took the plunge to becoming full time bird breeders and added "parrot gardens" to this name. Although perhaps known mostly from experiences with the black cockatoos, we maintain in excess of 70 species of Psittacines and one large planted aviary that holds softbills, finches, doves, etc. Our aviary complexes are forever evolving with a number of different designs that suit the array of species we house. Set in the north of NSW, heat presents more problems than does the cold. The planning of the various structures have been a blend of factors, weather protection, privacy and user friendly for the good of both birds and their keepers. It is particularly difficult to say what has been the most rewarding achievement within the world of aviculture, however successfully having bred all of the Australian cockatoos and three exotics (Moluccan, Umbrella and Citron crests) is perhaps it.

Just breeding anything that we previously hadn't is satisfaction within itself. The other area of satisfaction is having been a founder and continuous organizer of the highly successful AVES bi-annual International Parrot conventions since 1993. To bring parrot breeders from around the world to Australia for the knowledge, fellowship and enjoyment is extremely rewarding. Speaking at various venues has been another area of involvement and I consider myself fortunate to have spoken at numerous avicultural club meetings and conventions around Australia along with presentations in NZ, USA, UK, Canada and Tenerife. Although I do not find writing a labour of love, my wife and I have contributed articles to a number of avicultural magazines in a variety of countries. We wrote the "Guide to Black Cockatoo" book that was released in 2005. Hopefully I can manage to continue this lifestyle until the final fall from the perch.

# THE HUSBANDRY AND CAPTIVE BREEDING PROGRAM OF THE HORNED GUAN *OREOPHASIS DERBIANUS* AT AFRICAM SAFARI

JUAN CORNEJO

Bird Departement- Africam Safari
11 Oriente 2407, CP 72007
Puebla, PUE, México

## SUMMARY

The Horned Guan is endemic to Chiapas, in the South West of Mexico and Guatemala. It's considered globally endangered by BirdLife International and of Immediate Conservation Priority by the IUCN Cracid Specialist Group. Its wild population has been estimated at 1000 individuals. The earliest records of Horned Guans in captivity come from the 1970s, and it was probably not until 1994 that the first successful captive reproduction took place. The last edition of the international studbook, current to December 31, 2006, keeps record of a historic population of 99 individuals, 61 of which are still alive and distributed between 11 different institutions in Germany, Guatemala, Mexico, UAE and Portugal. Five of these institutions have been successful in producing offspring, with a total of 46 birds bred, 82% in the last three years. Only 47% of the potential founders have reproduced in captivity and 27% of captive born birds belongs to the same pair. It has not yet been bred from the second generation (F2).

Africam Safari, located in Puebla (Mexico), initiated a breeding program for the Horned Guan five years ago; until the end of 2006 a total of 17 individuals have been bred from three wild caught pairs, and holds a total population of 23 individuals. Laying has occurred between December and May, with a peak in March. Double and even triple cluching has been achieved, and all incubation has been done by foster species and artificially. Detailed information about the incubation, rearing parameters, chick development, sexing techniques, housing, diet, and breeding behaviour are covered henceforth.

## INTRODUCTION

The Horned Guan *Oreophasis derbianus* is an endemic species from Chiapas, in the Southeast of Mexico

and southwestern Guatemala. It inhabits the cloud forest of the Sierra Madre de Chiapas, at elevations between 1200-2500m.

It is considered globally endangered by BirdLife International (2000), and of Immediate Conservation Priority by the IUCN Cracid Specialist Group (Brooks and Strahl 2000). Habitat alteration, hunting and illegal trade have been generally identified as the most important threats. Its biology has been well documented in the last decade, and summarized in González-García et al (2001), and in Delacour & Amadon (2004).

The first report of Horned Guans in captivity comes from a letter to Delacour in 1975 (Haynes 1975), were a private collection in Jalisco, Mexico where it states that they house three wild born birds, two obtained as eggs and another as a chick. In 1976, two chicks where captured and brought to the ZooMAT in Chiapas, Mexico (Alvarez del Toro 1976). Between 1982 and 1983 a total of four eggs were collected and hatched in captivity under domestic turkeys in a private collection in Mexico (Estudillo-Lopez 1986). In the 1990s different zoos and private collections acquired Horned Guans and the population in captivity has been slowly growing since then. It is still considered a very rare species in aviculture. The first successful reproduction in captivity probably did not occur until 1994 in the private collection "La Siberia", near Mexico City (F. Acevedo, pers. comm.).

Africam Safari is a private zoological institution founded in 1972, and located in Puebla, Mexico, at an altitude of 2200m. It received a group of wild caught (3.4) and captive born (1.5) Horned Guans from Fundacion Ara in 2000, before this institution became defunct. Since then, it has committed to the conservation of the aforementioned species, by exchanging and obtaining additional animals, investing in infrastructure, training, research, etc. A successful breeding program was initiated that leads the *ex-situ* conservation efforts.

Fig. 1. *Adult Horned Guan.*

## Captive Population Status

The last edition of the Horned Guan International Studbook, current to December 31, 2006 (Cornejo 2007), records a historic population of 90 individuals. Of these 61, 30 males, 19 females and 12 of unknown sex, are still alive and distributed between ten different institutions. There are nine birds in Europe, two in the USA, three in Guatemala, and 47 in Mexico (Table 1). In addition to these, there are around 50 more individuals in captivity, at non-participating institutions, both in Mexico and Guatemala. Breeding has been achieved in six of the registered institutions, with a total of 49 birds hatched since 1990, 77% of them in the last three years. But only three institutions have been consistent in breeding for more than one year (Africam Safari, Leon Zoo and M. Leal private collection).

In spite of the breeding success over the last few years, the total population size has not grown considerably because of the high mortality (Table 2 and Fig. 2). None of the captive bred birds have produced offspring to date, therefore the population is free of inbreeding. Only 42% of the potential founders have reproduced in captivity, and the founders' representation is highly biased towards one, which has produced 32% of the captive born birds (Fig. 3).

A genetic study of the relatedness of the founders in the studbook is underway at the Instituto de Biologia of the National University of Mexico. The results of

### Table 1

**THE TOTAL NUMBER OF HORNED GUANS HELD AT EACH OF THE INSTITUTIONS REGISTERED IN THE INTERNATIONAL STUDBOOK INCLUDING BIRDS HATCHED FROM 1990 TO DECEMBER 31, 2006.**

| Country | Institution | Hatches | Individuals |
|---|---|---|---|
| Germany | Walsrode | 0 | 1.1.0 (2) |
| Guatemala | La Aurora Zoo | 0 | 2.1.0 (3) |
| Mexico | Africam Safari | 22 | 13.8.2 (23) |
| Mexico | Fauna de Mexico | 0 | 1.2.0 (3) |
| Mexico | Fundacion Ara* | 6 | 0.0.0 (0) |
| Mexico | Guadalajara Zoo | 0 | 1.1.3 (5) |
| Mexico | Leon Zoo | 7 | 1.1.3 (5) |
| Mexico | Reavyfeex | 2 | 4.2.0 (6) |
| Mexico | ZooMAT | 1 | 4.1.0 (5) |
| Portugal | M. Leal | 11 | 2.1.4 (7) |
| USA | Saint Louis Zoo | 0 | 1.1.0 (2) |
| | TOTAL | 49 | 30.19.12 (61) |

\* no longer in operation

this research will be extremely valuable in order to minimize the genetic diversity loss by pairing the wild caught birds in accordance to their mean kinship values.

**Husbandry at Africam Safari**

*Diet*

The Horned Guan is a specialized herbivorous bird that feeds mainly on fruits and green leaves. It has been reported that it consumes more than 50 species of plants (González-García 2005; Montes 2005). There is no consensus on what the optimal diet for the Horned Guan in captivity is, as it is clearly demonstrated by the large differences, both quantitative and qualitative, in the diets presented in the different institutions that hold the species. At Africam Safari, Horned Guans are offered a morning dish of fruit and vegetables, and chicken pellet are available *ad libitum*. The diet is composed of nine different vegetables and commercial chicken pellets [(Tovar et al 2006); see Table 3]. A total of 324.3g of food per kg of live bird weight is offered, and an average of 38.2% of this is ingested. Avocado is consumed to the greatest extent (>96% of that offered), followed by grapes (85.9%) and bananas (77.7%).

The consumption of tomatoes, cantaloupe, and cabbage are minimal. Since there is no previous information about the nutritional requirements of the Horned Guan, and considering that the animals in the study were in good health and have bred, we can assume that most of the nutritional values of the consumed diet may meet the as-yet-undefined nutritional requirements of the species [(Tovar et al 2006); see Table 4]. Like other fruit and leaf-eaters (Pryor 2003), the Horned Guan, might have lower nutritional requirements for protein as compared to domestic poultry. Additionally, the fat content of the diet might be critically factor as a primary energy source (Witmer and Van Soest 1998).

*Housing*

Africam Safari's main breeding aviaries measure 6 x 7m and 3m high, with the 2m of the back sheltered. They are well planted, to provide visual barriers and adequate perches. The substrate is a mixture of volcanic gravel and leaf-litter. They enjoy sun as well as sand bathing, so available space and resources should be provided for these purposes. Different pairs are able to hear each others calls but they are not within visual contact. They can be mixed in their enclosure with smaller birds such as conures and quails. Outside the breeding season they tolerate the presence of other individuals and can be held in mixed groups if enough space is available. During the breeding season males can be aggressive towards females, and will fight with other males. Caution should always be taken when introducing new ani-

### Table 2

**POPULATION TRENDS OF THE HORNED GUAN INTERNATIONAL STUDBOOK POPULATION FROM 2004- 2006**

|  | 2004 | 2005 | 2006 |
|---|---|---|---|
| Initial population size | 19.19.0 (38) | 27.28.3 (58) | 30.27.6 (63) |
| Hatches | 8.7.3 (18) | 0.0.11 (11) | 0.0.9 (9) |
| Deaths | 1.2.0 (3) | 0.3.8 (11) | 0.8.3 (11) |
| Final population size | 26.24.3 (53) | 28.26.6 (60) | 30.19.12 (61) |

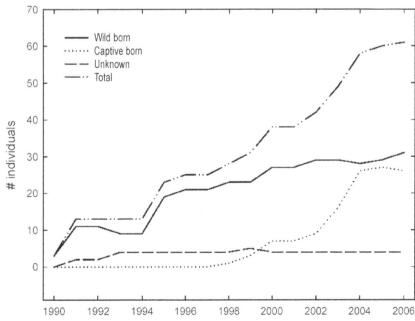

**Fig. 2.** *Historic census of the international studbook population of the Horned Guan.*

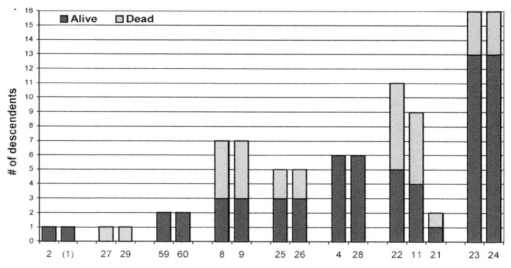

Fig. 3. Founders from the international studbook population of Horned Guans and the number of their descendants.

mals, and a previous period of visual contact is highly recommended. Although the Horned Guan is reported to be polygynous in the wild (González-García 1995), the best breeding results in captivity have been obtained with pairs. In the cases where a male has been put together with two females, one of the females is usually ignored and in some cases even attacked.

### Breeding at Africam Safari

From 2002 to 2007, Africam Safari's breeding program has produced a total of 29 chicks from 62 eggs and five different pairs.

*Sexing*

Horned Guans are not sexually dimorphic. Sexing can be done by surgical methods, genetic analysis, identi-

### Table 3

**AMOUNTS AND CONSUMPTION OF DIETARY INGREDIENTS FED TO HORNED GUANS AT AFRICAM SAFARI, PUEBLA, MEXICO**

| Africam Safari | Diet offered | Diet Consumed | | | |
|---|---|---|---|---|---|
| | g/kg LV | g/kg LV | s.d. | % Offered | % Consumed |
| Avocado | 21.7 | 20.9 | 2.4 | 96.4 | 16.9 |
| Avocado skin | 5.5 | 1.7 | 1.0 | 30.9 | 1.4 |
| Lettuce | 19.2 | 12.3 | 4.0 | 64.0 | 10.0 |
| Alfalfa, fresh | 9.6 | 5.2 | 3.1 | 53.6 | 4.2 |
| Tomato | 19.2 | 0.7 | 0.9 | 3.6 | 0.6 |
| Cantaloupe | 31.3 | 4.0 | 4.7 | 12.7 | 3.2 |
| Papaya | 32.5 | 11.5 | 11.2 | 35.4 | 9.3 |
| Banana | 36.1 | 28.0 | 9.5 | 77.7 | 22.6 |
| Grape | 16.9 | 14.5 | 4.2 | 85.9 | 11.7 |
| Cabbage | 98.6 | 14.8 | 22.9 | 15.0 | 12.0 |
| Chicken Pellets | 33.7 | 10.3 | 6.1 | 30.6 | 8.3 |
| Total | 324.3 | 123.9 | 84.8 | 38.2 | 100.0 |

n=5 pair

## Table 4

### DIETARY NUTRIENTS OFFERED AND CONSUMED BY HORNED GUANS IN AFRICAM SAFARI

|  | Offered | Consumed |
|---|---|---|
| Water % | 78.0 | 74.0 |
| Crude protein % | 10.6* | 8.5* |
| Crude fat % | 10.3 | 17.2 |
| Ash % | 7.9 | 7.6 |
| Crude fiber % | 8.2 | 9.4 |
| NFE % | 62.9 | 57.1 |
| K % | 0.56 | 0.53 |
| Na % | 0.21 | 0.14 |
| Ca % | 0.65 | 0.46* |
| P % | 0.33 | 0.23* |
| Ca:P | 2.0:1 | 2.0:1 |
| Mg % | 0.19 | 0.15 |
| Fe mg/kg | 47.51* | 327.92 |
| Mn mg/kg | 68.09 | 47.01* |
| Cu mkg/kg | 3.8* | 2.7* |
| Zn mg/k | 40.57 | 27.48 |

*Below the nutritional requirements of poultry (NRC 2004). All concentrations (except water) reported on a dry matter basis. N=5 pair.*

**Fig. 4.** *Part of Africam Safari's Horned Guan breeding complex.*

fying vocalizations (Gonzalez-García 1995), and by averting the intromittent organ of the males (Fig. 5). This technique has been proven successful at Africam Safari with birds as young as one month old, but it is not conclusive when trying to determine the sex of females.

### Breeding age

The earliest breeding age is four years for both sexes. It is plausible that it could be earlier, as during laparoscopies we have observed that males and females have active gonads at two years of age. The average reproductive age for the males and females in the studbook is nine years, and the maximum reproductive age is 19 for males and 14 for females.

### Nesting

The nest of the Horned Guans in the wild is a shallow depression on the epiphytic vegetation that grows on tree branches (González-García 1995). In Africam Safari they have used both, wood boxes (60 x 60 x 35cm deep) and raffia baskets (80cm in diameter), filled with moss, and with packed roots of aquatic

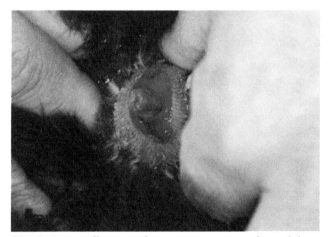

**Fig. 5.** *A partially averted intromittent organ of an adult male Horned Guan.*

**Fig. 6.** *Horned Guan utilizing a nesting basket.*

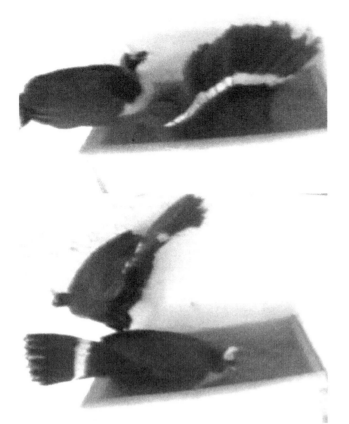

**Fig. 7.** *Photo capture from a CCTV recording of a breeding pair of Horned Guans in an elevated nest platform. In the first image, the female is on the left perching on the edge with the male digging in the substrate and fanning his tail. The second image is of the female inspecting the inside of the nest with the male perched on the platform rim.*

grasses. This latter material has proved to be a good nesting substrate, with a consistency more similar to that found in the wild. The best results have been obtained when the nest is placed high in the enclosure, in a corner under a shelter, but nests on branches have also been accepted by the birds. A pair at Africam Safari has been laying for three consecutive years on the floor of the enclosure, where they have been building a precarious nest between some vegetation. Also, there have been some instances of females laying from the perch, mainly in young birds.

Before laying, the male frequently visits the nest to prepare it by digging in the substrate. The female also participates in this task but with less intensity. In the presence of the female, the male will work on the nest in an inclined position: laying on his chest, with the neck and head resting on the substrate, and the tail fanned in an almost vertical position while facing her. If she decides to enter the nest, he will perch on the side of the nest with his back towards her and the tail fanned, sheltering her, and stays in this position for the duration of her work in the nest (Fig. 7). While these events take place, the male usually picks up small pieces of nest material with his beak and offers them to the female, who usually accepts them.

*Eggs*

According to Gómez de Silva et al (1999), the breeding season in the wild begins at the end of October and continues until May. At Africam Safari, the earliest eggs are laid in December, the last in July, with a main peak in March (Fig. 8). Horned Guans clutches consist of two eggs, laid with one day in between. The average measurements of 42 eggs are: 5.9 ± 0.2cm wide (max. 6.6cm, min. 5.3cm) and a length of 8.7 ± 0.2cm (max. 9.3cm, min. 8.3cm). For 55 eggs the average fresh weight was 165.7 ± 9.9g (max. 189.8g, min. 139.3g). Fresh weight can be calculated as 0.552 x length x width$^2$. If the first clutch is removed, a replacement clutch may be laid in an average of 34.2 ± 8.8 days later (max. 48 days, min. 16 days; n=13). With the goal of increasing productivity, pairs at Africam Safari are encouraged to double clutch, therefore, all eggs are removed soon after being laid. The maximum number of clutches per female in a single year obtained at Africam Safari has been four. Excluding the first year, the average fertility has been 74% (Fig. 11). At the beginning of the breeding program, artificial incubation was not successful. Although for last three years, and through a combination of artificial incubators (Grumbach and Humidaire) and domestic turkeys, 90% of viable eggs were successfully hatched (Fig. 10). Eggs were artificially incubated at 37.2°C. The average incubation time is 33.7 ± 1.0 days (max. 35, min. 32, n=25), with an average weight loss of 17.1 ±

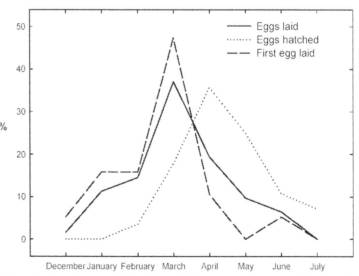

**Fig. 8.** *Distribution of the Horned Guans eggs laid (n=62), and chicks hatched (n=29) at Africam Safari from 2002 to 2007.*

2.6% (max. 23.2, min. 10.3, n=25). Internal pip occurs 3.3 ± 0.5 days before hatching (max. 4, min. 3, n=7).

## Chick Rearing & Growth

At Leon Zoo (Mexico) parent-rearing has been successful. They're experience (A. Ordaz pers. comm.) indicates that the chicks stay at the nest only for a few hours after hatching, and after that they jump to the floor where the female will look after and protect them by covering them under her wings and her chest feathers. The male does not participate in the rearing. The chicks are able to feed by themselves as early as one week old, but the female regurgitates for them until they are five months old.

No attempt at parent-rearing has been done so far at Africam Safari, so the chick data presented here comes from hand-reared chicks. Following the protocols for other cracids, Horned Guans do not pose any special difficulties to rear artificially. A diet of fresh alfalfa, chicken pellets, and soft fruits is offered for the first weeks, and is progressively changed to the adult diet. For the first few days, it is important to encourage them to eat and drink by offering food and water with the tip of ones fingers. Invertebrates such as mealworms will be accepted by the chicks. At hatching, chicks weigh an average 107.4 ± 7.8g (max. 120g, min. 87.9, n=26), and after three days of loosing an average of 4% each day, they start increasing their weight at a rate between 2-6% every day for the first two months (Fig 12 A & B). Adult weight is not achieved until approximately a year and a half after hatching (males 2212 ± 98g, max. 2324g, min. 2140g, n=3 wild born; females 1940 ± 138g, max. 2075g, min 1749g, n=4 wild born) (Fig. 13). Both culmen and tarsus adult size is achieved after the six months of age (Fig 13). The horn, which can measure more than 4.5cm in adults (measured from the front of the base to the tip), starts growing at the age of three months and takes more than three years to obtain its final size (Fig. 14). It starts as two seperate structures that later merge into a single horn (Fig. 16). The slight difference in adults' weights, and the preliminary data in the horns' measurements, suggest a dimorphism difference in body size as stated by Vaurier 1968, but more data is needed to run statistical analysis.

**Fig. 9.** *Eggs of Horned Guan at external piping.*

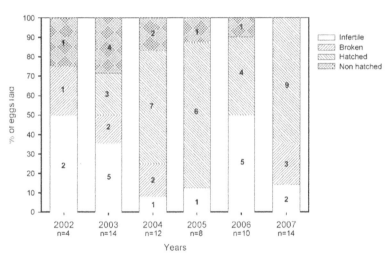

**Fig. 10.** *Results of the Horned Guan eggs laid at Africam Safari, by year.*

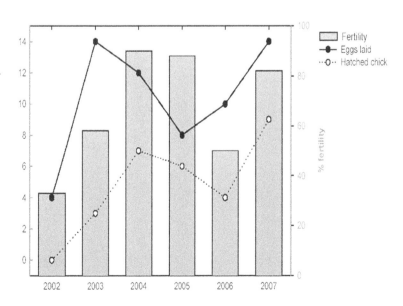

**Fig. 11.** *Number of Horned Guan eggs laid, chicks hatched, and % of eggs fertile, by year at Africam Safari.*

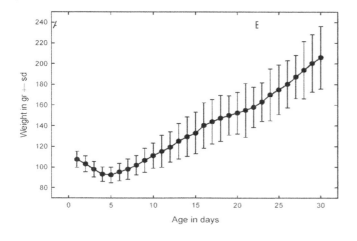

**Fig. 12a.** *Horned Guan growth curve for the first 30 days of life (n= 26).*

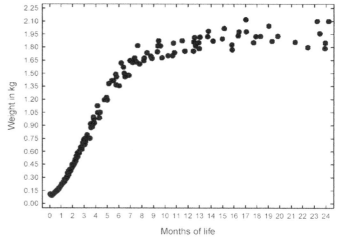

**Fig. 12b.** *Weight curve for the first two years of life (n=17).*

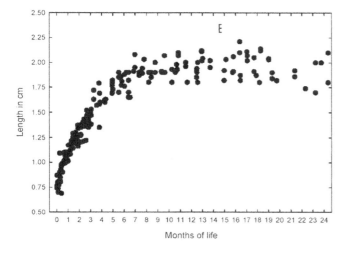

**Fig. 13a.** *Horned Guan culmen growth curve for the first two years of life (n=17).*

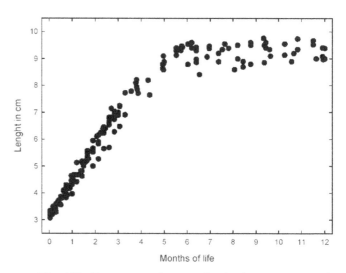

**Fig. 13b.** *Tarsus growth curve for the first two years of life (n=17).*

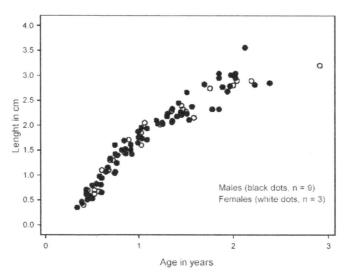

**Fig. 14.** *Horned Guan horn length curve for the first three years of life.*

**Fig. 15.** *Four day old Horned Guan chick.*

**Fig. 16.** *Horned Guan 14 weeks old young showing the beginning of the horn development.*

**Fig. 17.** *Comparison of the horn aspect of a one year old (left) and two year old (right) Horned Guan.*

**Fig. 18.** *Four year old Horned Guan with fully grown horn.*

## CONCLUSION

Our knowledge of the aviculture of the Horned Guan has increased dramatically over the last few years, nevertheless, current breeding achievements are not enough to maintain a viable captive population in the long term. To achieve this we need to increase the effective size of the population, and equalize the representation of the wild caught individuals. It is imperative to establish an active international captive breeding network that cooperates by sharing information and exchanging animals. We also need to establish husbandry guidelines in order to improve husbandry management, and train staff at different institutions, to insure consistency and improve the chances of successful captive breeding programs. The Cloud Ambassadors Program started in 2006 as a collaborative breeding program, supports field projects for the conservation of the Horned Guan and its habitat. Captive-bred birds are sent on loan to institutions willing to participate in the captive breeding efforts and to contribute with funds. As of June 2007, a total of two pairs have been sent from Mexico to zoological institutions in UAE and Europe.

The International Committee for the Conservation of the Horned Guan and its habitat initiative started after the Horned Guan PHVA in 2002, has already organized the I, II and III International Symposium for the Conservation of the Horned Guan, and will meet again in April 2008 at ZooMAT Chiapas, Mexico.

## Acknowledgements

The success in the Horned Guan breeding program at Africam Safari would not have been possible, nor the information presented here, without the dedication and persistence of Africam Safari's bird department staff, special thaks to Miguel de la Cruz, as well as the support of Frank C. Camacho, General Curator of Africam Safari.

## References

ALVAREZ DEL TORO M. 1976. Datos biológicos del pavón Oreophasis derbianus G.R. Gray. Univ. Aut. de Chiapas 1.

BIRDLIFE INTERNATIONAL. 2000. Threatened birds of the World. Barcelona and Cambridge, UK: Lynx Edicions and Birdlife International.

BROOKS DM & SD Strahl. (Compilers). 2000. Curassows, Guans and Chachalacas. Status Survey and Conservation Action Plan for Cracids 2000-2004. IUCN/SSC Cracid Specialist Group. IUCN, Gland, Switzerland and Cambridg, UK. Viii + 182 pp.

CORNEJO J. 2007. Studbook Internacional de *Oreophasis derbianus*. CRIO-Africam Safari. Puebla, Mexico.

DELACOUR J & D Amadon. 2004. Curassows and Related Birds. Second Edition. Lynx Edictions and The National Museum of Natural History, Barcelona and New York.

ESTUDILLO-LOPEZ J. 1986. Notes on rare cracids in the wild and in captivity. Journal of the World Pheasant Association 11: 53-66.

GÓMEZ DE SILVA GH, F González-García and MP Casillas-Trejo. 1999. Birds of the Upper Cloud Forest of El Triunfo, Chiapas, Mexico. Ornithologia Neotropical. 10(1): 1-26.

GONZALEZ-GARCÍA F. 1995. Reproductive Biology and Vocalizations of the Horned Guan *Oreophasis derbianus* in Mexico. Condor 97: 415-426.

GONZALEZ-GARCÍA F. 2005. Distribución, Densidad y Estado Poblacional del Pavón (*Oreophasis derbianus*) en México. En: Rivas et al. (eds.) Memorias del II Simposio Internacional para la Conservación de Oreophasis derbianus. Suchitepéquez, Guatemala.

GONZALEZ-GARCÍA F, DM Brooks & S Strahl. 2001. Historia natural y estado de conservación de los crácidos en México y Centroamérica. Pp. 1-50. In: Brooks, D.M. & Gonzalez-Garcia, F. (Eds.) Cracid Ecology and Conservation in the New Millenium. Miscellaneous Publications 2. Houston Museum of Natural Science, Houston, Texas.

HAYNES MH. 1975. News and Views. Avicultural Magazine 81(4).

MONTES L. 2005. Vegetación asociada a las aves (Pavo de Cacho, *Oreophasis derbianus*) en la reserva Los Tarrales, Patulul, Suchitepequez, Guatemala. Instituto Técnico de Capacitación y Productividad. INTECAP, Guatemala.

NRC. 1994. Nutrient Requirements of Poultry: Ninth Revised Edition. National Research Council, National Academy Press, Washington, D.C.

PRYOR GS. 2003. Protein requirements of three species of parrots with distinct dietary specializations. Zoo Biology 22(2): 163-177.

TOVAR G, J Cornejo and E Dierenfeld. 2006. Dieta y nutrición del Pavón (*Oreophasis derbianus*) en tres zoológicos mexicanos. Pp. 43-53. In: Memorias del III Simposium Internacional sobre Oreophasis derbianus. Cornejo, J. y Secaira, E. (Eds.). Comité Internacional para la Conservación de *Oreophasis derbianus* y su Hábitat. Veracruz, México.

VAURIER C. 1968. Taxonomy of the Cracidae (Aves). Bulletin of the American Museum of Natural History 138: 133-259.

WITMER MC & PJ Van Soest. 1998. Contrasting digestive strategies of fruit-eating birds. Functional Ecology 12(5): 728-741.

# PROPAGATION OF LIVE FOOD FOR BIRDS AND OTHER VERTEBRATES

Peter Karsten

Silver Pine Aviaries
5741 Stanehill Place
Denman Island, BC, Canada
V0R 1T0

## SUMMARY

Of the earth's 10,000 species of birds, more than 1,200 are considered to be in danger of extinction. Multi-generational breeding in *ex-situ* to sustain a diversity of avian species in aviculture for the future must offset the "consumptive" acquisition of birds from the wild. It is paramount that we hone our skills in the technology of captive breeding. The consistent breeding of insectivorous birds is dependent on an adequate supply of good quality, live insects. The consumption of live insects is greatest during the last days before fledging. A pair of Pekin Robins with three chicks may require up to 400 insects per day or more, depending on the size of the food insects. In most cases, there is a need to have insects at various stages of their development to meet the needs of the breeding colony. In addition, the food insects should be gut-loaded prior to feeding to correct nutrient imbalances. These needs are best met by breeding live food on location. Unfortunately, many aviculturists shy away from softbill breeding or simply maintain softbills without prospects of reproduction. The perception that it is difficult and messy to propagate insects and the cost of purchasing commercial live food has been a deterrent to pursue softbill breeding.

## INTRODUCTION

I have bred softbills for the last eight years on a Gulf Island off the west coast of Canada, in a region with limited availability and a high cost of insects for companion animal food. Crickets and waxworms are sold between eight to ten cents each and mealworms for one to two cents apiece. Bulk orders of mealworms are more affordable at $30 to $40 for 5000 (0.6 to 0.8 cent each). Ten pairs of softbills with their progeny, at various stages, consume approximately 600 waxworms, 600 crickets, 1,000 mealworms and 2,000 buf-

falo worms for a total of 4,200 insects per day. In my case, the cost of the aggregate in the retail market would be >$100 per day, while the average cost to produce the same amount of food insects is approximately $5.00 per day, not reflecting my labour of approximately one hour per day and the capital investment. Besides the problems with availability and quality in my location, the purchase of insects would be unaffordable. Operating efficiency and simplification of insect breeding regimes have been explored to reduce cost and time commitments, and so has the presentation of the live food to substantially reduce the need for frequent feeding of chick-rearing birds.

## To Breed or Not to Breed- That is the Question!

The adequate live food supply is often the limiting factor in successfully breeding softbills. In many cases, the on-site propagation of insects is more economical and reliable for bird breeders with large and diverse collections. Larvae and nymphs are required of different size and stage of development to accommodate young birds from hatching to weaning age. Nutritional quality is paramount for the proper development and maintenance of the birds. This can be a problem with insects obtained from commercial sources, which may have been stored at low temperature, to stall metamorphosis, and given inadequate or no food. "Empty" insects are just that empty of good nutrition.

On-site insect breeding demands attention to detail and a time commitment. In my estimation, the time allotment to care for live food colonies vs. a collection of 15-20 breeding pairs of softbills is approximately 1: 2, which translates into 60 vs. 120 minutes per day. It is important to devise low maintenance methods and simple food formulas to keep the work load down. An aviculturist must assess the pros and cons of breeding food insects in-house. If good quality insects can be purchased with assurance and at reasonable cost, then the day-to-day maintenance of stored live insects becomes the focus, which still requires good knowledge of the technology.

**What Live Food to Breed?**

The selection of food animals to breed depends on the animal inventory. African Rose Beetle, House Cricket, Lesser Mealworm Beetle, Darkling Beetle, *Zophoba*, Locust, Stick Insect, Greater Waxmoth and Whiteworms are discussed, however some were only tested for efficiency and ease of operation and not chosen as the core insect species for propagation. There are other food insects such as various species of flies, cockroaches and mammals, which I do not bred for the softbills of the families: Sylviidae (babblers), Paridae (tits) and Musicapidae (thrush relatives) in my collection. Fruit flies were bred for *Zosterops*, while they raised young and Honeycreepers for their maintenance diet. The relative workload and ease of propagation is reviewed in some detail below.

**Location**

Food insects love warm homes and some are highly invasive, hence propagating insects in "out-buildings", which are detached from the main residence, is highly recommended. The birdhouse kitchen, workshop or a garage are good locations. The ideal solution is a dedicated "bug room" with temperature controls to allow keeping cultures on shelves. Such room should be well insulated to heat it with an alternate emergency heat source (catalytic tent heater) in case of prolonged power outages, to ensure survival of the colonies. An insulated and centralized bug room saves energy costs and operational steps.

**Mealworms & Relatives**

Superworm *Zophoba morio*; beetle: to 25-28mm, larva: to 55 mm.
Mealworm *Trenebrio molitor*; beetle: 15-18mm, larva: to 28mm.
Buffaloworm (Lesser mealworm) *Amphitobius diaperinus*; beetle: to 6.5mm, larva: to 13mm.

*SUPERWORM (Zophoba morio)*

The larvae are called "Superworms" in the pet trade. The imago is a tropical South American beetle. Larvae feed on plant matter in the leaf litter on the forest floor. The culture medium is made of peat moss, sawdust, leaf and bark mulch (3:3:1). It is placed in plastic containers to 12cm depth and should be kept slightly moist, but loose to allow for air circulation. Target temperature is 27-29°C and RH 60-90%. Females lay eggs into narrow cracks of roots and bark to protect them from other imagoes, which may consume the eggs. Generation time is about four months at 27°C and 12 months at 18°C. Larvae feed primarily on plant and animal matter, such as lettuce, carrots and fruit. Additional food is provided for the beetles in the form of moistened cat and dog chow. Suitable culture containers are covered, screened, 25cm high plastic storage boxes. Wood is not suitable as the larvae will eat there way through it. Cultures are prone to mite infestation, which is hard to mitigate since the medium and food supply needs to be kept moist. I found mite invasions difficult to control. Mites covered the entire body of the larvae and are difficult to clean off.

The medium must be regularly sifted to remove droppings and to collect eggs and hatched larvae. I consider the culture rather labor-intense and since adult larvae are more suited for softbill species larger than Pekin Robins, I opted out of breeding Superworms.

*MEALWORMS (Trenebrio molitor)*

Mealworms are a widely used live food. They are very suitable for the aviculturist, provided the larvae are raised and stored on proper diets. Good availability and relatively inexpensive to purchase makes them a convenient food species. Purchased mealworms must be "gut-loaded" for several days to improve the food value.

They are easy to care for, but relatively slow to propagate. Generation time is about three months at 27°C and five months at 20°C. It is important to realize that mealworms have a highly inverted Ca:P ratio (1:16 vs. desirable 2:1), that their chitin of the exoskeleton negatively affects vitamin B complex absorption and that wheat bran contains phytic acid, which binds to calcium and diminishes calcium availability. Over supply in diet is real concern. Excessive protein levels lead to gout in adults and rapid growth in juveniles causing rickets and nerve damage due to vitamin B deficiencies.

Vitamin/mineral supplements are required to correct calcium imbalance, minerals and vitamins must be added. Some birds "bang" insects against the perch to kill them, causing supplements to fall off. A trace of margarine can be added to a portion of mealworms for better adhesion of the powder. Different products have different adhesion properties, however I found fine powder with high sugar content that sticks well, suffocates mealworms. I have liquefied additives to be injected into food animals, including poultry layer pellets in the medium and adding supplements to honey recipes improves the calcium content of food insects.

## Culture

I use a medium of wheat bran, poultry layer pellets, carrots and/or lettuce in approximate volume parts 8:1.8:0.2. I keep the medium dry enough to avoid clumping, mold and flour mite proliferation. If mites occur, ensure that the moist food is totally consumed before new vegetable matter is added. Containers (plastic storage boxes) must be covered if they contain beetles, which can fly short distances. The ideal temperature range is 26-28°C to keep generation time short (see Table 1). Larvae can be fed at all stages as required. Pupae are at times neglected by adults, which are feeding young. To encourage fully-grown larvae to pupate, I place folded, brown paper on top of the medium when the larvae start wandering around on the surface. The folded paper aids pupation and hatching of beetles. The culture should not be disturbed until the beetle have merged to be set up for breeding, at which time the beetles are transferred to their own container with 5-6cm of medium. I place folder paper on the top where a carrot is placed for moisture. This reduces mold and mite problems. I harvest eggs by sifting the beetle containers every seven to ten days.

I noted that eggs hatch better at high humidity and found placing the culture in the freezer chest (see cricket breeding) to be beneficial. The environment does not sustain mites in my experience and has a RH of about 70%.

## Equipment

Plastic storage boxes, 28x40cm and 15-23cm high are well suited to hold larvae and beetle colonies. A set of sieves are needed and can quickly be made out of flowerpots by gluing in hardware cloth of different size openings (2-5mm) onto the cut-away bottom. A discarded refrigerator is an ideal insect brooder. The insulation makes it economical to operate and it is easy to keep clean. It should have a twin (back up if one burns out) bulb light fixture and a thermostat to keep a selected temperature range. In a "bug room", which keeps a high ambient temperature, mealworm and or other insect boxes can be placed on shelves.

## Harvesting & Food Presentation

Maturing larvae will come to the surface to pupate and gather cleanly between folds of paper, which can then be picked up to pour the larvae cleanly into a container. In this way only mature larvae are removed to optimize yield, without disturbing the medium. Smaller larvae must be sifted from the medium by sieves of appropriate size. Mealworms are easily contained and can be presented in small glazed dishes or plastic storage boxes, used as "self-feeders", from which they cannot escape.

### *LESSER MEALWORM (Amphitobius diaperinus)*

The medium is the same as for mealworms, but regular additions of halved apples instead of carrots and presoaked poultry pellets, which are placed on top of the medium, are necessary to keep the medium slightly moist. Suitable additional foods are presoaked trout chow and dog/cat chow, since the larvae are more "carnivorous" than mealworms. Mite infestations do not occur in active culture. Beetles fly well and larvae run and hide quickly. These beetles are prone to infest other insect cultures, particularly cricket cultures. Propagation in a separate building is highly recommended. Beetles do not need to be sifted out and transferred to a separate breeding container. The medium is periodically sifted to remove feces.

I use 23cm high plastic storage boxes with screened covers to manage this very active insect. Temperature is best close to 27°C. If above 28°C, the insects appear to become stressed and super active to generate even more heat. Generation time is much shorter than that of mealworms, six weeks at 27°C.

## Harvesting & Presentation

The larvae are easily harvested once they mature and migrate to folded paper, which is placed on the medium for that purpose. The paper folds, containing larvae, pupae and beetles, are emptied into a plastic dish with 2.6mm holes in the bottom. The light sensitive larvae will immediately crawl through the holes and fall into a second container placed below. The beetles and pupae are neatly kept back and returned to the culture. Most birds do not consume beetles. This method harvests mature larvae and leaves most of the growing, small-sized larvae in the medium.

They are best presented for feeding in low, inwardly curved glass jars (small shoe polish or caviar type jars).

## STICK INSECTS

They are best bred in plastic aquariums. The Indian Stick Insect *Carausius morosus* is the most commonly used species. Up to 80mm long and 5mm diameter. Blackberry leaves and some other plants can be used as food. The 3mm long eggs are dropped to the ground. At 22-24°C, eggs hatch in three to four months. A female lays for six to twelve months, one to two eggs per night. Adult size is reached in four to five months. The insect is nocturnal and not detected by many birds. Stick insects are slow to produce and labor intense. Constant supply of green blackberry leaves is required.

## LOCUSTS

I bred the Egyptian Locust *Locusta migratoria* and the Desert Locust *Schistocerca gregaria*. The imago is 40-80mm long, depending on species. Eggs hatch in 12 to 16 days at 30-35°C. Nymphs grow to adult size in four weeks and shed their exoskeleton several times in the process. They are reproductive at four weeks of age and have a eight to nine weeks generation time. Grasses and other vegetables are fed. Locusts are diurnal. They need high temperatures, are labour intense, hard to handle and not eaten by all (small) softbills. Crickets are a much better choice.

## CRICKETS

House Crickets *Acheta domestica* up to 21mm long. Mediterranean Cricket *Gryllus bimaculatus*, up to 30mm long. Crickets are valuable live food with better nutrient balance than mealworms or waxworms (see Table 1). They are easy to propagate all year round in large numbers and relatively inexpensive to raise. I offer crickets in >25cm high "self-feeders" to provide live crickets for self-service over several hours. This method is ideal for having live food available for periods of time, when their parents feed chicks and the keeper can't be present. Crickets are tolerant to temperature changes, have a relative short generation time and are relished by a wide range of bird species (tits are an exception).

### Breeding Equipment

Plastic storage boxes with a 25 to 40 watt light bulb as heat source. Temperature is gauged by selecting the right size light bulb or by setting the ambient temperature in the bug room. Smooth walled, vented boxes 60 x 80 x 40cm are adequate to raise 500 crickets a week with the use of small back-up hatching containers. Hatching containers are small plastic boxes (aquariums) to hold two to four floral foam blocks. Egg cartons offer good living space to spread out the insects, which need good airflow. Discarded deep freezer chests are most efficient in terms of operating ease and energy conservation. (T. Pagel pers. comm. 2001)

A one cubic meter unit can produce >1000 crickets per week and can be considered a "Cricket Factory". To set it up, a pair of heat lamps, which are regulated by a thermostat, are mounted under the lid. The elevated platform of the compressor compartment serves a high shelve to keep hatching containers out of reach of adult crickets. The lid is kept slightly ajar for ventilation. The smooth walls are very easy to keep clean, which is important to prevent crickets from crawling up on the walls and to escapes.

Females have 10-16mm long ovipositor. Crickets will lay eggs into moist soil/ peat moss/sand mixture and in other moist substrates with air circulation. Floral foam, sold for wet flower arrangement, is ideal. (7.5 x 7.5 x 10cm foam blocks). It has no organic substances to attract mold, absorbs plenty of water and is porous to allow air exchanges. A floral foam block can hold >1000 eggs. Eggs are injected by the female into the moist floral foam. One female can lay 200 to 300 eggs over a period of 10 to 12 weeks.

To prepare the floral foam, a depression is cut into the top to control water flow when water is added. The egg-laden foam must be removed as soon as the crickets begin to dig out the eggs, within 48 hours, if well stocked, or they will dismantle the foam. Depending on humidity in the environment, the foam is watered daily or every second day. The foam must stay slightly moist at all times. Check (weight) saturation by lifting the foam. A tray with a layer of paper towel keeps the food dry in the culture in case the foam is over watered. Do not create freestanding water, since small crickets drown in it in large numbers.

Eggs hatch in 10 to 14 days at 26-30°C. The warmer it is, the faster the incubation and growth. The hatching nymphs are about 2mm long and white until their exoskeleton hardens and turns brown. The nymphs are fed with bran and goldfish flakes. Water is provided via a slice of an orange or lettuce. At 25°C they mature in six weeks; at 30-32°C they mature in four weeks. Temperatures >32°C are not beneficial. The foam is left in the rearing container as a watering device (in addition to the fruit). Crickets need good air circulation. Do not overcrowd. Move >8mm crickets to a more spacious containers or split up cultures.

Crickets eat a wide variety of food items. Basic food mixture: poultry layer pellets, wheat bran, ground puppy/kitten chow (>20% protein). Ratio: 2:8:2. Coarse components are ground up (i.e. coffee grinder). Water is offered through a constant supply of fresh fruit and vegetables, oranges work best. Crickets are harvested/sorted by shaking the egg cartons over a smooth-walled, >25cm deep container with a screen bottom. They can be selected by size with various sized sieves. Live feeding to birds is done in >23cm deep plastic boxes. It can have a landing perch to entice the birds to enter the box.

## GREATER WAXMOTH (Galleria mellonella)

Moths are 10-14mm, larvae up to 36mm long and pupae 8-10mm long. Cocoons 14-16mm long. Generation time is six to eight weeks depending on temperature. The ideal culture temperature range is 28-30°C. Eggs hatch in six to nine days. The opaque egg mass becomes transparent when hatching. The larvae grow to full size in 18 to 24 days. They will migrate to the lid and other surfaces to spin their cocoon. Some will stay in the medium and create pupating sites at its margins. The cocoon is spun in two to three days and the pupa rests for about 12 days to hatch into moths. Female lays up to 800 eggs in her lifetime. Moth lives about three weeks.

The Lesser Waxmoth *Achroea grisella* is not as productive, it produces fewer eggs and is smaller in body size.

## Equipment

An insulated insect brooder cabinet should have twin heat lamps, regulated by a thermostat. A recycled refrigerator works best to breed waxworms. Escaping larvae do not damage the walls, it is easily cleaned, well-insulated and serves other cultures as space allows, (i.e. storage for floral foam with cricket eggs in the door). Two 5-6cm vent holes are cut into the door at the top and bottom.

It is helpful to place a water dish in the fridge to improve humidity, which prevents the honey/cereal medium from drying out. The cultures are set up in two to six liter plastic storage boxes, which must have screened lids, with a screened area of no less than 50%. Metal fly-screen is glued into the lid for larger larvae and fuel filter screens for larvae hatching boxes.

## Food Preparation

Waxmoths were formerly raised on discarded honeycombs. Later, complicated formulas were developed as substitutions for honeycombs. Many recipes include expensive glycerin, beeswax, bee pollen etc, which do not seem to be critical additions.

Waxworms can be propagated on a simple formula of chick-starter, wheat bran, wheat germ and liquid honey (2:3:1:1 in volume parts). Small amounts of whey and brewer's yeast are good additions. I have also added ground up trout chow to offer more protein. The honey may be increased and chick-starter decreased if the mix tends to go dry (in dry environments). The liquid honey is kept fluid for easy mixing by keeping it in the warm brooder environment. The mix is worked into a moist, crumbly medium with a sturdy wooden spatula. If the medium dries up and or the honey crystallizes, the larvae will not survive or hatch. Very slight misting with water will reconstitute moisture levels, but mold must be prevented. A vitamin/mineral concentrate can be mixed into the honey prior to mixing the honey into the formula to "gut-load" the larvae.

## Culture

Eggs are collected from the rim of the container, where the moths deposit egg-strips containing hundreds of eggs. Eggs are deposited in narrow cracks between the lid and rim. The egg-strips are lifted off with a knife and placed on top of the food mix in a hatching container. I add a piece of tissue paper to place the eggs on. It allows for better monitoring of the hatching process. Hatching larvae are practically invisible to the naked eye, but their webbings soon become evident. Eggs can be collected every three to four days until moths die off.

## Larvae

Waxworms are hairless caterpillars, not worms or maggots, and must have relatively dry environments, good air circulation and slightly moist food. Lids must fit tightly, since the larvae will escape to find hiding places to pupate or in search of food if the medium dried out or is consumed. Good food supply and darkness keeps the larvae in place. Waxworms are simply harvested by pushing them off the underside of the lid with a paintbrush; it leaves the cocoons in place. The same container produces larvae of mature size over several days. One must leave enough larvae to pupate for the next generation of moths. It is important to divide overcrowded containers or they will build up excessive heat. Re-feed if necessary by adding new medium on top.

**Pupae**

As mentioned, waxworms will form cocoons under the lid and or in the medium unless they escape. Escaped larvae become a nuisance, as they chew a shallow depression at the pupation site into the substrate. Keep them away from your books! The cocoons have a slot at one end to help the moth emerge. It can be opened only at this end to remove the pupa, by grasping two of the three flaps and pulling them down to free the pupa. Waxmoth pupae (live) are eagerly fed by softbills to their chicks, while dead and pre-frozen insects are usually rejected. Pupae can be used to inject medication and supplements.

**Moths**

Moths are nocturnal, inconspicuous and not detected by most birds unless they move. They mate within hours after hatching, lay eggs for up to three weeks and then die off. To prevent moths from flying or scurrying away during handling in the containers, place them in a cool place (refrigerator) which will causes them become docile. Softbills relish moths.

*AFRICAN ROSE BEETLE (Pachnoda butana)*

The tropical African Flower or Rose Beetle is 22-25mm long and 13-15mm wide. The larvae are up to 50mm long and up to 10mm in diameter. Generation time at 28-30°C is only 10 to 11 weeks. Female can produce 70 eggs in clutches of 8 to 12. The round, white eggs are 1.5-1.8mm in diameter. Males live for two months, females up to five months. Mature larvae build solid cocoons out of saliva, soil and plant matter.

**Culture**

They are bred in >25 cm high, plastic containers, with screened lids. Beetles can fly with sudden starts. A converted fridge is ideal for a breeding unit. Medium consists of decayed wood, woodland humus, peat moss, fallen leaves and sand to equal parts. It should be about 15cm deep in the container. Beetles feed on flowers, fruits and leaves. Bananas are relished. An occasional addition of a hard-boiled egg stimulates reproduction and growth.

The larvae feed on decayed wood (alder, oak, beech, willow, etc.) but also fruit, especially bananas.

### Table 1
**FOOD ANALYSIS IN % OF INSECTS AS FED**

| Constituents | Adults Cricket | Nymphs | Mealworms | Waxworm | Buffalo* |
|---|---|---|---|---|---|
| Moisture | 69.2 | 77.1 | 61.9 | 58.5 | 63.0 |
| Ash | 1.2 | 1.1 | 0.9 | 1.2 | 1.4 |
| Protein | 20.5 | 15.4 | 18.7 | 14.1 | **21.6** |
| Fat | 6.8 | 3.3 | 13.4 | **24.9** | 10.6 |
| Ca | 0.041 | 0.027 | **0.017** | 0.024 | 0.020 |
| P | 0.295 | 0.252 | 0.285 | 0.195 | 0.230 |
| Ca:P | 1:7.2 | 1:9.3 | **1:16.8** | 1:8.1 | 1:11.5 |
| Fibre | 3.2 | 2.2 | 2.5 | 3.4 | 5.0 |

\* *Data except for buffalo courtesy of Mark D. Finke Ph.D., Scottsdale, Arizona, USA*

\* *Bold highlighting poor nutritional values*

A box 28 x 40 x 25cm can house up to 50 beetles. It can produce 40 to 50 larvae per week. The medium is periodically partly renewed. Females lay the eggs close to the bottom so these strata should not be disturbed.

## WHITEWORMS (Enchytraeus albidus)

Whiteworms are true segmented worms (phylum Annelida). They do not undergo a metamorphosis like insects. Whiteworms or Enchytraeids can be found in many compost heaps from spring to fall. The worms grow to about 40 mm in length with a 0.7mm diameter and can be propagated all year around indoors. Not all softbills feed on them, but it is excellent food for amphibian and fish.

## Culture

The medium must be porous, moist and high in humus content. Old, weathered horse dung is an ideal medium to grow white worms. It can be used in plastic pails with lids and a screen to keep other insects out. Mold and mites interfere with cultures; hence the soil should not get wet and compacted. Keep the culture in a shaded cool place <20°C. Good culture temperatures lies between 15-20°C.

## Food & Harvesting

For food, white bread, dipped in water before placing it on top of the soil works well, since it does not become moldy as quickly as other moist foods (Bread contains mold inhibitor). Good foods that attract large numbers to move to the surface and expedite growth are pieces of low-fat cheeses, hard-boiled eggs and small amounts of liver sausage in the peel, but avoid grease built-up.

The worms dislike light and will feed best under a slate/tile or a clump of sphagnum moss. It is best to feed worms with a clump of soil to keep them fixed in place and prevent them from drying out too quickly in the food dish.

### Table 2
**INSECT CULTURE PARAMETERS**

| Culture | Generation Time | Temp. | Work Load | Space Req. | Yield | Usage |
|---|---|---|---|---|---|---|
| Mealworms | 12 weeks | 27 C | low | medium | high | universal |
| Zophobas | 16 weeks | 27 C | high | medium | low | limited* |
| Buffalo | 6 weeks | 27 C | low | low | very high | universal |
| Stick insect | 32 weeks | 22 C | high | low | low | limited |
| Locust | 9 weeks | 32 C | very high | high | medium | limited |
| House cricket | 6 weeks | 28 C | medium | medium | very high | universal** |
| Waxmoth | 6 weeks | 28 C | medium | low | very high | universal |
| Afr. Rose beetle | 11 weeks | 28 C | very low | low | low | universal |

\* Too large for small ssp. of softbills if fully grown.
\*\* Tits of the genus Parus do not eat crickets (author's observation)

## References

FINKE MD. 2002. Complete Nutrient Composition of Commercially Raised Invertebrates Used for Insectivores. Zoo Biology 22: 269-285.

FRIEDRICH U & W und Volland. 1990. Futtertierzucht: Lebendfutter für Vivarientiere. Verlag Eugen Ulmer, Stuttgart.

KARSTEN P. 2002 Pekin Robins: Information on Care and Breeding. The AFA Watchbird 29(2): 56-68.

KARSTEN P. 2007. Pekin Robins and Small Softbills: Management and Breeding. Hancock House Publishers Ltd., Surrey, B.C. Canada, ISBN 0-88839-606.

## Author Biography

Peter Karsten, born November 1937 in Germany, immigrated to Canada in 1962, studied agriculture and business management in Germany, Sweden and Canada. He held the position of Director of the Calgary Zoo and Botanical Garden and Executive Director of the Calgary Zoological Society from 1974 to 1994. He worked in the field of agriculture, horticulture and zoological gardens in Germany, Sweden and Canada since 1955. His special interest has been zoo design, animal husbandry and species conservation. He was the first elected President of the Canadian Association of Zoos and Aquariums (1975/76), President of the American Zoo and Aquarium Association (1983/84) and President of the World Association of Zoos and Aquariums (1992/93). He published numerous technical papers including an internationally applied zookeeper training program. Peter Karsten provided on site consultant services to nearly 20 zoological gardens and wildlife parks in Canada, the USA, Mexico, Chile, Japan and Korea with emphasis on developing long range plans and facility design plans. He enjoys painting wildlife and teaching watercolour techniques. He now breeds rare species of birds in his own aviaries and published a book on bird breeding in 2007: Pekin robins and small softbills - management and breeding.

# PUBLIC AND PRIVATE SECTOR COLLABORATION: AN OPPORTUNITY TO BUILD AND PRESERVE AVIAN GENE POOLS

PETER KARSTEN

Silver Pine Aviaries
5741 Stanehill Place
Denman Island, BC, Camada
V0R 1T0

## SUMMARY

*Only human passion can save the diversity of species in aviculture for the future.*

More than 10% of the world's bird species are endangered today. Some species have been secured through *ex-situ* breeding programs and many more must be given that attention. In 1993, ISIS listed 2040 species of birds held by 425 instructions or 4.8 species on average per institution. In 2007, close to 650 institutions held 2257 species and per institution. This indicates a decline in avian biodiversity per zoological institution. The concerted effort by a single breeder produced greater number of Pekin Robins over a number of years than the 50 institutions, which were holding this species in their collection and submitted inventory data to ISIS. The resources of a private breeder could be combined with the resources of a zoological garden(s) to develop long-term conservation strategies to establish self-sustaining gene pools of birds in aviculture. Potentially, offspring could be used for reintroduction *in-situ*. This initiative requires agreements and protocols to be worked out between the private and public sector to overcome constraints. The collaboration will undoubtedly lead to a better utilization and preservation of bird species. Dialogue to on this topic is urgent and necessary to take advantage of this promising opportunity in conservation.

## INTRODUCTION

It is inconceivable that we will enjoy the wide variety of birds in *ex-situ*, which we have today, unless we invest in innovation, resources and longterm commitment to breed them for multi-generations. Passion and altruistic efforts will be fundamental to achieve this.

Monetary returns on investments will be sporadic at best. The most successful species conservation initiatives yield some economic rewards to help sustain conservation action. This is not always possible and other solutions must be found. It is my believe that some species, which have an uncertain future in aviculture, can be preserved through collaborative breeding programs between the stake holders of the public and private sector.

I worked 30 years within the zoological community and many years connected to it beyond that, to know the limitation of a public institution to pursue intensive *ex-situ* breeding of passerine bird species. Space and operating budget constraints set a high threshold for the breeding of many bird species, especially highly territorial species with specialized food requirements. In most cases the cost of producing progeny exceeded the acquisition cost of wild-caught specimens. This is changing dramatically. The recent avian influenza outbreaks closed the international transport of live birds between many countries, species protection legislation becoming more stringent and is advancing world wide, changing societal values challenge the keeping of wild animals in human care and many wild populations of birds are literally losing ground due to various pressures. William Conway of the Wildlife Conservation Society, stated two decades ago that it was wildlife spaces that had become endangered, not simply species. In the context of species conservation this encompasses both *in-situ* and *ex-situ* spaces.

The notion to find more wildlife spaces leads us to the private sector. Many species of hookbills and hardbills are secure now and can be preserved within the companion animal population. Potential wildlife spaces reach into the millions; and what is significant, they are generally not sustained on a cost benefit basis, but on passion for the animal itself. There is, of course, discussion about recovering costs and perhaps even "making money", because that is our human vision and measure of success, at least outwardly, to justify the keeping of wild animals at a cost. Net profit is not a prerequisite for conservation breeding and therein lies the opportunity. There ought to be more dialogue to compare notes between the private and the professional animal keepers to collaborate and forge long range conservation plans. Zoological institutions are driven by compassion as well and supported by altruistic contributions from the community at large. Capital investments are only marginally earned by a zoo through visitor revenue and almost all operating budgets receive tax base support. Wildlife conservation work is generally not secured by profitability from within.

**The Economics of Breeding Birds**

The short answer is: there is none. Species which require a distinct breeding territory with specific habitats demand more space and labour than species, which can be bred in colonies or a community aviary.

It is difficult to breed softbills below the market price. Example: I have been breeding Pekin Robins *Leiothrix lutea* since 1999. In 2005/2006, I maintained eight pairs of Pekin Robins, one pair of Silver-eared Mesia *Leiothrix argentauris* and a pair each of European Robins *Erithacius rubecula* and Blue Tits *Parus caeruleus*. I bred some 30 offspring at an average cost of $133 per bird, labour not included and conversely at $633 each with labour included. The current market price in Canada for these species is about $200 to $250 each.

**Operating Costs**

The capital investment for 20 aviary units is $27,000 in material cost; with a total of 2,185 square feet. ($12.36/sq ft) and 19,000 cubic board feet ($1.42/cbft). Including labour cost would raise these figures at least five-fold. Considering a life expectancy of 15 years for the structures, the yearly rate of amortization is $1800.

Annual operating cost:
| | |
|---|---|
| Capital investment/cost | $1,800 |
| Utilities/food, etc. $6/day | $2,190 |
| Labour @ $15/hr for 1000 hrs/yr | $15,000 |
| Total per year | *$18,990* |

| | |
|---|---|
| $18,990 / 30 birds bred | $633/bird |
| $3,990 (excluding labour) | $133/bird |

I can afford to ignore the labour component and replace it with passion or cost of recreation. I started breeding Pekin Robins in 1999 with one pair, added nine more founders and bred over 85 offspring to independence (2006), some of which are now third generation birds. Annual reproduction success (number of adults vs. weaned offspring) averages 92.8% between 1999 and 2006. The assumption can be made that the breeding program is self-sustaining and able to support satellite programs. Some of the birds are still reproducing at over ten years of age. Juveniles can breed within six months. Longevity of Pekin Robins may reach 20 years. Approximately ten Pekin Robins are relocated to other bird keepers each year.

While the above program has good results and presents a valuable conservation resource, it is unsupported by its revenue. The loss/profit relationship would be less favourable in breeding Pekin Robins in a public zoo. The breeding of softbills finds its justification in the context of a modern zoo's mandate to minimize remov-

**Table 1**

**PEKIN ROBIN *Leiothrix lutea* BREEDING ACTIVITY IN ZOOLOGICAL INSTITUTIONS**

| Year | 1993 | 1998 | 2003 | 2007 |
|---|---|---|---|---|
| Total population | 223 | 345 | 401 | 144 |
| Increase (decrease) vs. previous year | base | 35 % | 14% | (64%) |
| Males | 57 | 75 | 86 | 57 |
| Females | 52 | 59 | 75 | 35 |
| Sex unknown | 114 | 211 | 240 | 52 |
| Progeny last 12 months | 11 | 5 | 5 | 16 |
| % of total population. | 2.24 | 1.44 | 1.24 | 11 |
| Number of institutions | 48 | 46 | 55 | 59 |

*(data extracted from ISIS 2007, www.isis.org)*

## Table 2
**SILVER-EARED MESIA *Leiothrix argentauris* BREEDING ACTIVITY IN ZOOLOGICAL INSTITUTIONS**

| Year | 1993 | 1998 | 2007 |
|---|---|---|---|
| Total population | 127 | 186 | 160 |
| Increase (decrease) vs. previous year. | base | 32% | (14%) |
| Males | 55 | 50 | 28 |
| Females | 46 | 46 | 18 |
| Sex unknown | 26 | 100 | 114 |
| Progeny last 12 months | 24 | 5 | 8 |
| % of total population | 18.9 | 2.68 | 5 |
| Number of institution | 36 | 40 | 21 |

*Note the high "sex unknown" figure, despite clear dimorphism of the species. This species is protected by C.I.T.E.S as well and the specimens were clearly not acquired for breeding purposes. Unless the pattern changes, this species will likely decline faster than the Pekin Robin in zoological institutions.*

## Table 3
**WHITE-RUMPED SHAMA *Copsychus malabaricus* BREEDING ACTIVITY IN ZOOLOGICAL INSTITUTIONS**

| Year | 1993 | 1998 | 2007 |
|---|---|---|---|
| Total population | 71 | 104 | 144 |
| Increase (decrease) vs. previous year | base | 32% | 28 % |
| Males | 35 | 66 | 57 |
| Females | 26 | 19 | 35 |
| Sex unknown | 10 | 19 | 52 |
| Progeny last 12 months | 17 | 21 | 16 |
| % of total population | 24 | 20 | 11.1 |
| Number of institutions | 28 | 33 | 50 |

## Table 4
**PARAMETER COMPARISONS BETWEEN PUBLIC AND PRIVATE FACILITIES**

| Parameter | Public institution | Private breeder |
|---|---|---|
| Capital cost | high | low with own labour |
| Operating cost - labour | high and inflexible | low and flexible |
| Facility maintenance | high standard and high cost | low (no public services) |
| Technical expertise | high | variable and limited |
| Veterinary/ research support | high | low to none |
| Availability of space | low | high for selected spp. |
| Disturbance/business activity | high | low |
| Access to founders | high | low to nil, depending on sp. |
| Perpetuity of facility operation | high | low to none (situational) |
| Emergency keeper back up | good | problematic |
| Keeper hours per day on site | limited | unlimited |

al of animals from the wild and its role to serve research, education and recreation. Certain zoo functions generate visitor revenue to off set conservation work. The difficulty to commit to breeding softbills is evident in a relative low reproductive rate in zoo inventory data submitted to ISIS (International Species Information System, Eagan, USA).

## Breeding of Softbills in Zoological Institutions

### Pekin Robin

The influx of specimens shortly after 1998 may relate to confiscated wild-caught birds, which were transferred to non-profit institutions after the Pekin Robin became listed on C.I.T.E.S, Appendix II as of September, 1997. The high number of "sex unknown" birds support that notion. With an average longevity of less than ten years the population would naturally decline by 2007. Breeding success is notably rising, however still insufficient to secure the *ex-situ* population in the long term. Despite very marginal dimorphism this species has fewer "sexed unknown" specimen's than Silver-eared Mesia and White-rumped Shama *Copsychus malabaricus*.

The White-rumped Shama enjoyed better attention to *ex-situ* breeding than the genus *Leiothrix* (Table 3). Despite dimorphism, the number of "sex unknown" birds is rising, reflecting low attempts are made in breeding. Zoological gardens often receive unsolicited birds from private owners or, as noted above, confiscated birds from government agencies (customs authorities). These specimens represent an invaluable resource for *ex-situ* breeding programs beyond their home institutions.

## Collaboration Offers Opportunities

A cursory review of advantages and disadvantages between private breeders and non-profit institutions (Table 4) reveals a promising scenario for collaboration to strengthen the breeding results in aviculture. By pooling the resources between the two sectors, a number of the disadvantages and constraints can be mitigated to have better opportunities to build and preserve avian gene pools in *ex-situ*.

## DISCUSSION

### Potential Positive Scenario

A breeding consortium can be established between zoological gardens and private breeders. The private breeders would propagate relevant species of birds on his/her premises to provide specialized breeding environments, alleviate high production cost and focus on the specific technique of breeding to obtain high and consistent reproductive success. Genetically over-represented stock and non-breeding specimens could be transferred to public zoological gardens for education and public display in attractive community exhibits etc, to free up breeding spaces. The zoo would serve as a gene bank in case specimens must be re-entered into the active breeding population. The zoo, as a non-profit public education/research facility can import species covered by C.I.T.E.S. to inject new founders into the gene pool, while the private individual can only do this with restrictions and great difficulty. The imported specimens would be loaned to the breeder under contract and remain to be the property of the zoo.

Benefits between the collaborators include:

- Optimal breeding success of genetically enhanced and managed populations over the long term under SSP/EEP type guidelines.
- High rewards to the breeder and institution through conservation achievements.
- Assured long term availability of relevant species for public education, study and recreation.
- Assistance by the zoo to relocate a private inventory of birds in case the breeder can no longer operate the program.
- Collaborative research/study opportunity in an "off-exhibit" environment.
- Sharing of experience and knowledge.

Constraints and deterrents:

- Policies by zoos and zoo associations confine acquisition and sales, loans and trading of animals to accredited organisations and those maintaining institutional membership in an association to ensure animal standards are met.
- Private breeder may feel a loss of independence.
- No guarantees that the private breeder will or can participate over the long term.
- Establishment of trust relationships take time.

Steps to overcome constraints:

- All animal movements between parties must be documented and, where applicable, done with signed agreements (breeding loan, exhibit loan etc.).
- All specimens are permanently marked (leg bands) and listed in a studbook or similar register.
- The breeder's facility undergoes a form of accreditation.
- Provision of mutual site visit privileges and exchange status of reports.

**Common Breeding Loan Arrangements**

The breeding loans are usually set up to pair up single specimens or to introduce new bloodlines. In other cases, birds may be loaned out to give the recipient an opportunity to gain experience or to provide safe off-site storage. Breeding loan agreements are often made for multiple years.

Resulting offspring are normally shared by giving the aviary owner/operator, who is keeping and breeding the birds, half of the successfully raised, weaned offspring. The other half is divided between the owner of the male and female. If the pair has different owners, they each receive one quarter. If the breeder owns one bird he/she gains 3/4 of the offspring. The number offspring is not always divisible by four, in which case the distribution lines up as follows: first the breeder, then the owner of the hen, followed by the owner of the male. The formula is carried over to the following years to even out the distribution. If the pair is breeding successfully in one season it will likely breed again in the next. It is in the best interest of all parties to keep a pair intact for as long as it is productive and there is a need for the genetic combination of the progeny. It is understood that the breeder is making his/her best effort to maintain and breed the birds and possible losses are accepted as part of the course. Shipping costs are generally born by the recipient of the birds.

The breeder identifies the offspring with closed leg bands and keeps accurate records of breeding success and bloodlines. The birds are registered in a studbook or in-house registrar.

## References

FLESNESS N. 2007. International Species Information System (ISIS), Eagan, USA.

KARSTEN P. 2007. Pekin Robins and small softbills: management and breeding. Hancock House Publishers Ltd., Surrey, B.C. Canada, ISBN 0-88839-606-6.

# FLAMINGOS IN CAPTIVITY: THOUGHTS ON HOW AND WHY

Catherine King

Fuengirola Zoo, Avenida Camilo José Cela 6
29640 Fuengirola, Spain

## Summary

BirdLife International has categorized the conservation status of one of the six flamingo species as vulnerable and three as near-threatened. Flamingos live in highly specialized habitats, some of which are likely to be severely impacted in the near future with changing land use and weather patterns. Because flamingos are often difficult to approach closely in the wild, captive flamingos can greatly aid in understanding aspects of their natural history. The EAZA Ciconiiformes and Phoenicopteriformes TAG undertook two surveys to assess the status of flamingos in European zoos and then to explore how willing zoos are to change the management of their colonies. Flamingos are present in approximately 70% of European zoos. With more than 6,700 flamingos in European zoos, 92% of which are in the genus *Phoenicopterus*. Chilean Flamingos have the highest chick mortality and Caribbean Flamingos the highest non-chick mortality, followed by the Lesser Flamingo. The only two flamingos with more births than deaths are the Chilean and Greater Flamingos. A point system is described here to give some idea of the importance of different factors, both collectively and individually, in breeding flamingos. Size of colony is particularly important, and many colonies in Europe are too small to likely breed. When zoos were asked if they would consider making changes in their management, many were willing to increase colony and/or enclosure size, but not to reduce them if this was an option suggested. However, a number of zoos holding multiple species colonies indicated they were willing to reduce the number of species to one.

Social behaviours and partner choice appear to be different in wild and captive birds. This may well be a consequence of numbers, rather than captivity. It would be useful to study these behaviours in small, sedentary colonies of wild flamingos to confirm whether this is true, as understanding the dynamics of small colonies could become useful for conservation of flamingos in the future when and if breeding populations decrease in size and become more fragmented.

## Introduction

### Why are Flamingos in Captivity?

People love flamingos. The flamingo enclosure in a zoo is often strategically placed at the front of a zoo, near the entrance, because we know that flamingos please people. Approximately 70% of EAZA zoos have flamingos. There were at least 6,701 flamingos in 157 zoos in Europe on January 1, 2005, the date of the last EAZA census (King & van Weeren 2005). As I will frequently be referring to this census, I should now explain that while the International Species Information System (ISIS) is believed to fairly accurately reflect numbers of animals in the North American zoo region, it is much less reliable for Europe, because many European zoos are just beginning to use it. As North American and Europe are the only regions for which I have good zoo data, these are the only ones I will refer to.

Flamingos are not only popular in zoos, many private breeders of waterfowl and Ciconiiformes in Europe tend to have flamingos as well. Why are people so attracted to flamingos? Probably primarily because they are very pretty and easy to see, and usually are held in groups, so there are many to look at. We found at Rotterdam Zoo, where I worked as biologist for 16 years, that sharing the drama of what actually happens in a colony with the visitors can personify the birds, and arouse even more interest in them. Our volunteer observers referred to the Greater Flamingo colony as "their own Peyton Place", and for good reason. The flamingos squabble, break up with old partners to find new ones, they may leave new mates for previous ones, they may copulate with another flamingo on the sly, or form trios or homosexual relationships. Flamingos that are unlucky in love may harass others that are happily paired, sometimes even kidnapping an egg or offspring. Anything can happen in a flamingo colony, and while flamingos look alike, and luckily are enough alike that we can formulate general management guidelines, they are also very individual in their responses to situations, and show great behavioural plasticity in some areas.

**Fascinating Flamingo Facts**

Flamingos are among the most ancient groups of birds. The "modern" flamingo is 30 million years old, and older forms date back at least 50 million years. Flamingos caught the attention of humans early on: cave paintings of flamingos in Spain date back to around 5000 BC. The first scientific description of a flamingo was made in 414 BC by Aristophanes. Ancient Egyptians used a silhouette of the flamingo to symbolize the color red, and the flamingo represented the reincarnation of the sun god Ra. The flamingo even had a place in early Christianity, as many Christians thought that the flamingo was the legendary Phoenix, the immortal red bird reborn from the embers of its own funeral pyre, symbolically related to resurrection of Christ (del Hoyo et al 1992).

It is ironic that flamingos are associated with tranquil, tropical paradises when some flamingos live in the most hostile environments "known to bird". Despite their fragile appearance they can prosper in wetlands that no other vertebrate, including fish, can tolerate. Flamingos can withstand high levels of chlorides, sodium carbonate, sulphates and fluoride in water that may reach 68°C with a pH of 10.5 and a salt content double that of seawater. Leslie Brown, the famous Kenyan ornithologist, ended up in the hospital having to have skin grafts after his first attempt to walk through Lake Natron with inadequate foot and leg protection to more closely observe the Lesser Flamingo colony in the middle of that lake (Brown 1959). This tolerance for incredibly harsh conditions allows flamingos to exploit their food resources (mainly plankton, blue-green algae, diatoms and aquatic invertebrates) with little fear of predation or competition. Nonetheless, obtaining enough food, especially if there is a chick to be fed, can occupy much of a flamingo's day and also night (Rendon-Martos et al 2000).

Flamingos achieve their bright coloration though absorption of caretenoid pigments in their diet. Flamingos are famous for being the only birds to feed with their head upside down, filtering small particles through baleen whale-like lamellae using a piston-like movement of the tongue. It was recently discovered that flamingos have blood-filled sinuses on either side and behind the tongue that probably expand to aid the flamingos in their filter-feeding. Such erectile tissue has not been found in birds before (Beckman 2006). Physical adaptations to feeding in water of different depths include reduced feathering on the tibia, webbing between the front three toes, and a very long neck and legs. In fact, flamingos have the longest legs and neck relative to body size of any group of birds. The length of the neck is achieved through elongation of the cervical vertebrae rather than increased number: a swan has 25 cervical vertebrae while a flamingo only has 17 (del Hoyo et al 1998).

Flamingos are not truly migratory, and although there are some sedentary populations most flamingos are considered nomadic, moving around to find suitable conditions. Flamingos in temperate regions tend to have a true nesting season, while those in the tropics and subtropics are more opportunistic. This is probably why tropical flamingos form huge displaying groups throughout the year, to be in a state of readiness should a breeding opportunity present itself. Flamingo colony size can vary from just a few tens, such as recorded for Caribbean Flamingos in the Galapagos (Sprunt 1975) to a million, as was observed for Lesser Flamingos in the Etosha Pan (Berry 1972). Flamingos mainly nest on islands surrounded by water, and competition for a site to build a nest can be intense. As many as five nests per $m^2$ have been counted in Lesser Flamingo colonies (de Hoyo et al 1998).

Flamingos, along with pigeons, are the only birds known to feed their young "crop milk". In flamingos, this secretion is derived from modified salivary glands (Lang 1963). Unlike pigeons, flamingos are able to produce crop milk long before and after the chick is expected to hatch. Foster flamingo parents were able to start feeding an adopted chick after just 18 days of incubation and also after more than 90 consecutive days of incubation (King 2000). Once the young leave the nest they join groups of chicks chaperoned by a few adults. Parents are able to recognize their young in crèches of thousands of chicks.

A question that is frequently asked is: Why do flamingos stand on one leg? I do not think that question has been definitively answered, but it may be for stability (N. Jarrett pers. comm. 2006), as another aspect of the harshness of flamingo environments is the gale-like winds flamingos often encounter.

**The Flamingo Species**

The systematic relationship of flamingos with other birds remains murky and I will not even begin on this topic. To add to the confusion there is also much shuffling of the six flamingo taxa among genera within the family Phoenicopteridae. I will use the same taxonomy that BirdLife International currently does (BirdLife International 2007).

The Greater Flamingo *Phoenicopterus roseus* has the widest distribution of any flamingo. It ranges north to France, south to the tip of Africa, and east to Kazakhstan, India and Sri Lanka. It is also clearly the best studied flamingo. A ringing program using individually identifiable colored rings has been ongoing in the Camargue (France) since 1977 (20,947 flamingos ringed), coupled with intensive observation during the breeding season for many of those years (Johnson 2000). There have been other ringing programs carried out in Mediterranean countries, including Spain (since 1986, 16,335 flamingos ringed), Italy (since 1994, 1,847 flamingos ringed), Turkey (since 2003, 717 flamingos ringed), Sardinia (since 1999, 2,717 flamingos ringed) and Algeria (only in 2006, 208 flamingos ringed). A total of 41,974 flamingos have been ringed in the last 40 years (1977 through 2006) (Béchet 2006). These ringing programs have revealed much about the biology and movements of the flamingos in these locations. The conservation status of the Greater Flamingo is considered to be of Least Concern by BirdlLife International (2007). While not a very common bird in North American zoos (446 in 12 zoos listed in ISIS, 2007), the Greater Flamingo is much more frequently found in European zoos, with 2376 in 82 zoos in 2005 (King & van Weeren 2005).

The Lesser Flamingo *Phoeniconaias minor* occurs regularly in 29 countries from West Africa, across sub-Saharan Africa and along the southwest Asian coast to South Asia, and occurs as a vagrant in 23 additional countries. While this sounds like a wide distribution, closer scrutiny has revealed that confirmed regular breeding is confined to just five sites in four countries: Makgadikgadi Pans in Botswana, Etosha Pan in Namibia, Lake Natron in Tanzania and Zinzuwadia and Purabcheria salt pans in India. Of these five sites, only three, Etosha Pan and the two sites in India, are officially protected. More than 95% of the non-breeding population is concentrated on just 22 sites within 10 core countries (Childress et al 2007). Ironically, even though the Lesser Flamingo is the most numerous flamingo (population estimated at 2,220,000-3,240,000) this concentration of large numbers in just a few places leaves it vulnerable to local problems, and it is classified as Near Threatened by BirdLife International (2007). It is also included in the Convention on International Trade in Endangered Species of Wild Fauna and Flora (CITES) Appendix II, and Convention on Migratory Species (CMS) Appendix II.

The major threats to the survival of the Lesser Flamingo are loss and/or degradation of its specialized habitat because of changes in water quality and hydrology (Childress et al 2007). For example, there are plans to log the area around Lake Natron and to build a hydropower plant. Both of these activities could greatly change the lake's salinity levels, which could impact breeding. Extraction of salt and soda ash, and the disruption of Lesser Flamingos few breeding colonies by nearby human activities are also serious threats. Other threats include disruption of nesting colonies by predators, poisoning, disease, harvesting of eggs and live birds, human disturbance at non-breeding sites, and competition for food and breeding sites. An action plan is being developed for the Lesser Flamingo by the Flamingo Specialist Group in cooperation with CMS and Agreement on the Conservation of African-Eurasian Migratory Waterbirds (AEWA; Childress et al 2007).

Lesser Flamingos are not very common in zoos, with only 416 in 17 North American zoos reported in ISIS (2007) and 428 found in 23 European zoos in 2005 (King & van Weeren 2005). Only three European zoos have successfully bred Lesser Flamingos in the last ten years, each one once, but there are a couple of private breeders that have had more consistent success. The first North American captive breeding occurred in 1989 at Sea World San Diego, and Fort Worth Zoo is the only North American zoo now consistently breeding this species, even hatching F2 chicks in 2007.

The Caribbean Flamingo *Phoenicopterus ruber* most closely approaches the idyllic, brightly colored flamingo associated with warm weather and palm trees. True to its common name, it is found throughout the Caribbean region. Its global population is believed to number 850,000 to 880,000 individuals, and it is considered of Least Concern by BirdLife International (2007). There were 1336 Caribbean Flamingos in 56 European zoos in 2005 (King & van Weeren 2005), and ISIS (2007) indicates that 1676 Caribbean flamingos are held in 56 North American zoos.

The Andean Flamingo *Phoenicoparrus andinus* mainly occurs on the high Andean plateaus of Peru, Chile, Bolivia and Argentina. A global population of only 34,000 was estimated in 1997, a sharp decrease from the 1980s. Therefore, BirdLife International (2007) has categorized this species as vulnerable. It is also listed on CITES Appendix II and CMS Appendix I and II. Thousands of eggs have been collected annually for human consumption until recently, when nesting sites became better protected. Mining activities, unfavourable water-levels (owing to weather and manipulation), erosion of nest-sites and human disturbance and hunting may also affect productivity. Conservation programs are now in place in all four range countries. No zoos in North America and only five in Europe, hold this species, and the total European population is around 40 birds. The Andean Flamingo has only bred in captivity a few times, the last breeding occurred at Wildfowl and Wetlands Trust in Slimbridge several years ago.

The James, or Puna Flamingo, *Phoenicoparrus jamesi* also mainly occurs on the high Andean plateaus of Peru, Chile, Bolivia and Argentina. Egg-collecting and hunting were intensive during the 20th century, and the population probably declined rapidly then. It has started to increase, presumably owing to the success of international and national conservation programs organized in all four countries. A coordinated census in 2005 estimated the population to be 100,000 birds. BirdLife International has categorized this species as Near Threatened and it is on CITES Appendix II and CMS Appendix I and II. There are no James Flamingos in North America, and only two zoos hold them in Europe. The first hatching in captivity occurred in Berlin Zoo in 1989, and a second chick hatched in 2001. The small colony of seven birds kept with six Andeans continued to produce eggs but a third chick did not hatch until 3 July, 2006. The chick was euthanized due to problem with legs in November, 2006 (B. Blaszkiewitz pers. comm. 2006).

The Chilean Flamingo *Phoenicopterus chilensis* breeds in Peru, Bolivia, Argentina, Chile and perhaps erratically in Paraguay (at least one breeding record), with a few wintering in Uruguay and south-east Brazil. The global population is estimated to be no more than 200,000, and is undoubtedly declining. Egg-collectors have been responsible for the partial or complete failure of some colonies in recent years. Extraction of water for irrigation projects threatens the most important breeding site in Mar Chiquita in Argentina. Extensive habitat alteration as a consequence of mining, hunting and tourism-related disturbance is also a threat. The Chilean Flamingo is considered Near Threatened by BirdLife International, and is on CITES Appendix II and CMS Appendix II. Chilean flamingos are relatively well represented in North American zoos (1621 individuals in 62 zoos; ISIS 2007) and European zoos (2467 in 84 zoos) (King & van Weeren 2005).

## How Can Captive Flamingos Contribute to *In-situ* Conservation?

It is unlikely that the captive population will be important as a genetic reserve, a role that many captive animal populations are claimed to fill. However, because flamingos in the wild are most often difficult to approach and observe closely, captive populations can be used to gain information needed to understand and manage wild populations. Captive flamingos can serve as flagship species for their ecosystems, and their presence in zoos can link zoo and field professionals.

## Life History Statistics

To date the oldest living wild flamingos known were a male Greater Flamingo that was 40 years and 23 days of age (Johnson 1998) and a Lesser Flamingo that was 40 years and approximately nine months of age (Childress 2004). We know from flamingos in captivity that the potential is even longer. A male Greater Flamingo at Basle Zoo bred successfully at minimally 57 years of age (Studer-Thiersch 1998) and a female now at least 70 years of age is still alive. This female laid an egg when she was at least 60 years old (A. Studer-Thiersch pers. comm. 2007). A Greater Flamingo at Brookfield Zoo lived 60 years and 10 months (Shannon 1996).

## Communication & Reproductive Research

Analysis of ring information on Greater Flamingos nesting in the Camargue indicates that these birds, which nest in the thousands, do not usually repair with the partner from the previous year, or even from the same year if a bird makes two breeding attempts in the same season (Cezilly & Johnson 1995). Contrarily, flamingos in captivity tend to maintain partnerships for several years. The mean number of partners that 26 Greater Flamingos had over a ten year period was 3.1 (S.D. 1.4, range 1-7) at Rotterdam Zoo (King 2006). It is probable that, similar to captivity (e.g. Pickering 1992) as colony size increases, rate of partner change increases, and that small colonies in the wild are more monogamous than larger colonies. This could simply be because the flamingos that were paired the year before have a more similar physiological cycle (Studer-Thiersch 2000a). It has also been observed that flamingos that arrive together at a zoo tend to pair with other birds from that same shipment (King 2000), which could be for the same reason. Understanding the dynamics of mate choice and the effect of size of colonies could be important in maintaining sound genetic bases in small wild flamingo populations in the future.

Studies indicate that individual aggression and dominance interactions could have a negative impact on feeding time, and thereby fitness, of free-ranging flamingos (Bildstein et al 1991; Schmitz & Baldasarre 1992). Studies on dominance and agonistic behaviours in captivity can help elucidate relative importance of factors that can influence the outcome of agonistic encounters, e.g. size, age, reproductive status, sex, familiarity with the surroundings (King 2000).

Voice recognition is extremely important in flamingos, as parents must be able to find each other and to find their young in the crèche. Studies involving transfer of eggs and chicks at various stages of hatching to unfamiliar nests in captivity could help to discern when recognition starts. An investigation of vocalization development of parent and foster-reared chicks could provide insight into genetic and behavioural components of vocalisation acquisition. Non-vocal cues that pairs as well as parents and young use to find each other might be better studied in captivity (King 2000).

Cézilly et al (1997) concluded that Greater Flamingos nesting in the Camargue have an age-assortive mating system, with directional preference towards older and more experienced birds. They suggested that differences in performance of displays between individuals could provide proximate cues for assessing age. Testing this hypothesis by measuring qualitative and quantitative variation in display performance could be more easily achieved in captive populations.

## Morphology, Physiology, Endocrinology & Feeding Studies

Any study that involves measuring a physical or behavioural value can be more easily undertaken with captive flamingos. Investigations on sexual and taxonomic differences in several morphological parameters have been conducted with captive flamingos (Richter & Bourne 1990; Richter et al 1991; Studer-Thiersch 1986). Captive flamingos have been used as a material for studies of filter-feeding structures and mechanisms (Beckman 2006; Jenkins 1957; Zweers et al 1995), metabolism of carotenoids (Fox 1975), molt (Shannon 2000), vocalizations (Boylan 2000), vision (Martin et al 2005), crop milk (Lang 1963; Studer-Thiersch 1966) and normal blood chemistry values and values for comparative taxonomic cytochemistry (Hagey et al 1990; Péindo et al 1992). It has been proposed that flamingos can manipulate the gender of their young, with males hatching earlier in the season and females later, but finding a morphological indicator of age of juveniles that is independent of sex would be helpful in studying this topic (Bertault et al 2000), a goal that could be more easily accomplished using captive flamingos. Bildstein (1990) cites flamingos as an example in a paper he wrote advocating the use of zoo collections in studies of feeding ecology and conservation biology in wading birds.

## Development of Management Techniques

Management techniques for free-ranging flamingo populations can often be tested and adapted using captive populations. Marking methods, and transmitters for radio or satellite tracking can be easily tested. Use of artificial nests to encourage free-ranging flamingos to colonize a particular breeding area was adopted from zoos, and has been successful (Rendon & Johnson 1996). Studies related to disease such as avian influenza and response to medications and vaccinations can be studied much more easily in captive birds.

## Captive Flamingos as Flagships for Their Ecosystems

Flamingos can be very useful tools for illustrating problems in their ecosystem and potential solutions. For example, the area around Lake Nakuru was deforested, primarily for conversion to crops. This caused erosion, pollution, and, coupled with global warming, a great decrease in rainfall. As the lake dried up so did the tourist business, because there were no longer many flamingos for the tourists to see. Tourism was very important to the local economy, and now the Nakuru community is planting sapling trees, in the hopes that reforestation will bring back the flamingos and their major source of income (Planet Ark 2007).

Zoos working with flamingos are better placed to give financial and logistic support to field work. For example Dallas Zoo, represented by Chris Brown, the AZA Ciconiiformes TAG Chair, has been working since 1999 on a project in the Yucatan, Mexico to better understand the natural history of Caribbean Flamingos. Hundreds of young flamingos are banded, sexed and crop samples taken each year to gather data needed to create sound action plans. The Dallas Zoo's role is primarily supportive; channelling money, providing the leg bands, equipment (scales, band applicators), veterinary services, staff expertise with flamingo husbandry, means for DNA sexing and nutritional analysis of the crop samples. The zoo also provides education support and materials to the NGO, Ninos Y Crias, for this organization to educate children of the region about the flamingos and wetlands habitat preservation. Ninos Y Crias and the Ria Lagartos Reserve staff coordinate the banding day preparation (i.e. they apply to the Mexican government for the permit to band the flamingos, gather local people, build the corral, and initiate the herding of the chicks into the corral). During the remainder of the year they protect the flamingos on both coasts, make observations and notations of banded birds, and have just completed an artificial nest area similar to the one built in the Camargue in France. The birds are currently using the site (C. Brown pers. comm. 2007).

## Management of Flamingos in EAZA

### The Fabulous Flamingo Questionnaire

Because flamingos are some of the most commonly held birds in zoos, the EAZA Ciconiiformes and Phoenicopteriformes TAG considers optimizing their management as a high priority. Therefore, in 2005 we undertook a project to thoroughly assess the situation in Europe. We sent the "Fabulous Flamingo Questionnaire" to zoos asking questions about current colony size as of January 1, 2005 and plans for over five years ending in January 1, 2010, colony composition (mixed or single flamingo species in flock), other species in enclosure, number of enclosures and future enclosure plans, enclosure capacity, breeding results, mortality (first month and older), predation, age of colony, record keeping and ringing, past and future plans for acquisition, potential for supplying flamingos to other zoos in the next five years, and whether zoos are concerned about inbreeding in the colonies.

We approached 282 zoos, of which 230 (82%) responded. In total 157, (68%) of the responding zoos held flamingos, but four of these zoos' responses were received too late to be included in further analyses. The questionnaire covered all species, as well as hybrids since a few zoos are collecting the hybrids from other zoos to hold them together, out of harms way. Some of the analyses were only done on the three *Phoenicopterus* species, as these were the only ones in the EAZA collection plan at that time. The collection plan is intended to give guidance to EAZA zoos in selection of species for institutional collection plans. The summary of the survey results are presented here.

a. Most zoos (80%) holding flamingos had one flamingo enclosure (Table 1). Many (59) of the 153 zoos had a mixed (multiple flamingo) species flock. Of these, 27 were planning to separate the species before 2010. Most respondents reported that they held flamingos in mixed species flocks to encourage breeding and/or because there was no other enclosure. Some commented that it was not a problem, as different species do not interbreed. I feel compelled to interject here that this is a misconception. All the species can, and may, interbreed: even Lesser Flamingos and *Phoenicoparrus* flamingos have been known to hybridize with *Phoenicopterus* flamingos. Taking the approach that pulling eggs from hybrid pairs will solve this problem is also not valid. Flamingos that appear to be totally monogamous may still perform extra-pair copulations, as we know from cases where the partner could not possibly fertilize an egg, but nonetheless the egg is fertile.

b. Proportionally, the Caribbean Flamingo was the species most frequently held in mixed species flocks, and the Greater Flamingo the least frequently (Figure 1).

c. Chilean and Greater Flamingos are by far the most common flamingos in EAZA zoos. Together the three *Phoenicopterus* flamingos, *P. ruber*, *P. roseus* and *P. chilensis* constitute 92% of all flamingos in EAZA zoos. Zoos generally want to hold more flamingos in 2010 then they have now (Table 2).

d. The number of zoos that had any breeding success at all (at least one chick hatched) in the five years between 2000 and 2004 combined was similar for the three *Phoenicopterus* species, and was less than 45 % for all species.

- Andean Flamingo: no zoos (0%) with breeding results (4 zoos in total)
- James Flamingo: no zoos (0%) with breeding results (1 zoo in total)
- Lesser Flamingo: 1 (4%) zoo with breeding results (23 zoos in total)
- Chilean Flamingo: 37 (44%) zoos with breeding results (84 zoos in total)
- Greater Flamingo: 34 (43%) zoos with breeding results (80 zoos in total)

- Caribbean Flamingo: 23 (40%) zoos with breeding results (58 zoos in total)

e. The size of the colony affected reproduction in three ways:

1) The larger the colony, the greater the chance for any reproduction at all during the five year period 2000-2004. A regression analysis was done to look at the relationship between group size and incidence of any breeding (at least one chick hatched between 2000-2004). The R square was 0.696, meaning that 69.6% of the variance between group size and breeding results is explained by a linear connection between these two (Figure 2).

2) The larger the colony the more consistent the breeding was over years within the five year period in the colonies that bred. It can be seen in Table 3, that there was very little breeding in colonies of 20 individuals or less, even though many flamingo colonies (52% for the Caribbean Flamingo and 75% for the Lesser Flamingo) numbered less than 20. Breeding became more consistent, (i.e. the average colony of that size that bred, bred more of the five years, as colony size increased). It is interesting to note that five of the six categories in which the average colony bred more than four of the five years were single species situations. Even though there were more Caribbean Flamingos in mixed flock situations, the breeding was better in single species flocks.

3) The larger the colony, the greater the average number of young produced per individual per year, in colonies that bred at least once in the five year period (Figure 3). All three *Phoenicopterus* species showed a generally increasing trend here (the larger the flock the greater the number of young), although in some categories the paucity of colonies probably skewed results. For example, there were only four breeding colonies of Chilean Flamingos in the 41-50 size category (which had less young per year than the two adjacent categories), while there were 16 colonies in the 51-100 size category.

f. Most zoos that hatched chicks reared chicks, but even in successful zoos the mean number of chicks hatched and reared was not that high between 2000-2004 (Table 4). With the exception of the Chilean Flamingo, chick mortality is usually relatively low (Figure 4). It may be that Chilean Flamingo chick mortality is higher because this species typically nests very late in the year, sometimes as late as October. While it and the other South American species are quite cold-hardy, the damp cold and low amount of sunshine throughout much of Europe in the autumn could be factors in chick mortality. Leg problems are

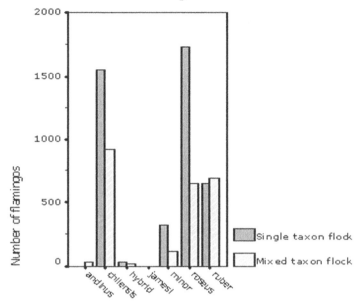

**Fig. 1.** *The total numbers of the different flamingos in mixed and single species flocks on January 1, 2005 (King & van Weeren 2005).*

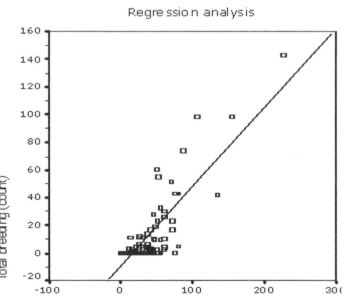

**Fig. 2.** *Relation between group size and breeding results of Phoenicoptrus flamingos in European zoos between 2000-2004. R square = 0.696 (at least one chick hatched between 2000 and 2004, n=133 single species flocks).*
*Sig. = 0.0000, the regression comparison is with 99.9% reliability significant for the whole population (King & van Weeren 2005).*

## Table 1

**AN OVERVIEW OF THE ENCLOSURES (N=153 ZOOS) IN EUROPEAN ZOOS AS OF JANUARY 1, 2005.**

| Number of Enclosures | Number of Zoos | Single Species Flocks | Mixed Species Flocks |
|---|---|---|---|
| 1 | 122 | 82 | 40 |
| 2 | 29 | 40 | 18 |
| 3 | 1 | 3 | 0 |
| 4 | 0 | 0 | 0 |
| 5 | 1 | 4 | 1 |

*(King & van Weeren 2005)*

often seen (N. Jarrett pers. comm. 2007) and may be due to inadequate Vitamin D3 synthesis (J. Nijboer pers. comm.).

g. Caribbean Flamingos > 30 days of age have the highest mortality rate of any species of flamingo in European zoos, fairly closely followed by Lesser Flamingos (Table 5). While flamingos in general are touted as being very cold-hardy, these two species are the most tropical of the flamingos, and it may be that they are less well adapted to the northern European climate then has been assumed.

h. Predation is not a great source of mortality in most years (Figure 5), which was surprising to me as my impression was that many zoos have fox problems. However some zoos mentioned that they bring the flamingos in at night because of predation threats.

i. Table 6 shows the percentages of single species colonies of different size categories of the species in the regional collection plan. It can be seen that there were many single species colonies of less than 21 flamingos in 2005, and only 13%-33% above 40. The expected colony sizes would be larger in 2010, although expected colony size of Lesser Flamingos remains particularly problematic.

j. Table 7 shows the number of flamingos that were produced in the last five years, the number that the zoos said they thought they could provide to other zoos in the next five years and the number that would be needed to meet the expected number in 2010. The number of flamingos needed to meet 2010 goals far out number what will be available from other zoos for all species, based on both what the zoos estimated and results from the past five years. Only the Greater and Chilean Flamingos have an increasing population based on birth and death rates.

## Table 2

**THE TOTAL NUMBER OF FLAMINGOS ON JANUARY 1, 2005 AND THE EXPECTED NUMBER ON JANUARY 1, 2010 IN EUROPEAN ZOOS**

|  | Andean | James | Chilean | Caribbean | Greater | Lesser |
|---|---|---|---|---|---|---|
| 01-01-2005 | 34 | 3 | 2467 | 1336 | 2376 | 428 |
| 01-01-2010 | 29 | 3 | 3414 | 2107 | 3505 | 520 |

*(King & van Weeren 2005)*

k. When asked how zoos planned to acquire the flamingos needed to meet 2010 goals most said that they planned to breed the flamingos themselves or acquire the from other zoos (Figure 6). Given the statistics shown here, this is unrealistic unless breeding of all species drastically improves and mortality of the tropical species and young Chilean Flamingos declines.

The Fabulous Flamingo Follow Up Questionnaire was designed to individually ask zoos to adhere to management recommendations made by the TAG, i.e:

1) zoos that only hold flamingos for exhibition should hold at least 20 individuals for well-being reasons;
2) zoos should hold only single species flocks;
3) zoos that want to breed flamingos should have a flock of more than 40 individuals to improve/stimulate breeding result.

Choices were given, e.g. "Your zoo reported having 30 Greater Flamingos. Would you consider reducing the number to 20-25 (minimum recommended for well being reasons) and giving the others to a zoo that wants to breed them? Or would you consider increasing the number to more than 40 to increase your own chance of breeding?

All 157 zoos that had responded to the Fabulous Flamingo questionnaire were approached, of which 129 (82%) responded to the follow up questionnaire. Only the four most commonly held flamingo species, the Greater Flamingo, the Chilean Flamingo, the Caribbean Flamingo and the Lesser Flamingo were included in the follow-up survey.

This was the first time that an attempt had been made to consider flamingo management on a European-wide level. We had no idea how the zoos would react, but the comments and response rate indicate that the zoos were very receptive to this approach. It can be concluded from the results that many zoos are willing to cooperate and to consider changing their current situation if necessary to improve flamingo management. Only 27% of the zoos that were asked to consider a making a major change in flamingo management (enlarge enclosure to hold more animals, stop holding mixed-species flocks, or hold less flamingos where appropriate) would not consider changing their management.

Zoos were generally reluctant to hold less flamingos or stop holding them entirely in single-species situations, as only six (15%) of 39 zoos chose this option. Zoos were more willing to stop holding one or more species in mixed-species situations in which they would still be working with one species, and usually trying to increase numbers of this species, as 11 (37%) of 30 zoos chose this possibility. Zoos making flamingos available to other zoos generally preferred loans or trades as possible transaction types, which would ease the financial restraints on receiving zoos.

Zoos clearly prefer to increase numbers of flamingos to achieve the TAG population goals rather than to reduce numbers. Half of the 34 zoos with insufficient space to hold a breeding group of more than 40 birds would like to enlarge enclosures to hold more individuals, and all of the six zoos that already have adequate facilities but smaller groups than recommended did want to increase group size when asked in this context. Forty percent of the zoos with mixed-species enclosures would prefer to build a new enclosure to accommodate two single species flocks in total, as opposed to giving up a species or not changing the situation. More than half of the zoos that would consider enlarging an existing enclosure or adding another enclosure anticipated being able to make these changes within five years.

Zoos were especially reluctant to stop working with Lesser Flamingos to enable these flamingos to be housed in larger groups. Prior to this survey Lesser Flamingos were not included in the European regional collection plan because the EAZA TAG did not want to encourage importation of Lesser Flamingos into Europe given the mortality and breeding problems with this species. However, we reversed this decision in 2006 following this survey, as it is clear that zoos wish to continue working with this species, but there is no hope that the population will become sustainable without much more cooperation among institutions and solving management problems. Therefore, the EAZA TAG is setting up a working group to achieve these goals. Similarly, the AZA counterpart TAG recently held a workshop on Lesser Flamingo husbandry (C. Brown pers. comm.), and we will work together to tackle this issue.

The Fabulous Flamingo Questionnaire already indicated a large deficit in flamingos needed to achieve collective institutional goals, with an estimated deficit of 2,401 flamingos based on what zoos thought they could supply and would need between 2005-2010, and an estimated deficit of 2,716 based on population growth between 2000-2004 for the four species discussed here. While it is wonderful that so many zoos reported in the follow up survey that they are prepared

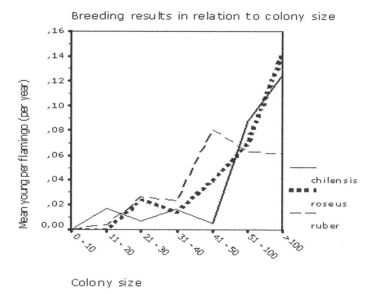

Fig. 3. *Mean number of young per year relative to colony size category for zoos that hatched 1 or more young between 2000-2004 (King & van Weeren 2005).*

to change their institutional goals to meet the TAG's three basic management targets, the deficit then becomes much larger still.

Half of the respondents from the 16 zoos with breeding colonies that had stated in the Fabulous Flamingo Questionnaire they were unable supply any flamingos to other zoos were willing to consider supplying flamingos in this follow up survey. Although most (83.3%) of the zoos already willing to supply flamingos thought they could increase the number, breeding of flamingos in European zoos can in no way match the need for flamingos. Less than 10% of the zoos wanting flamingos are willing to take them at still human-dependant stages, which may give zoos that are willing to do so a distinct advantage in acquiring flamingos.

The reduction of species by zoos now holding multiple species will ease the deficit some, but not much as in many cases the zoos need to trade for individuals to increase the size of the flock of the remaining species. Many zoos indicated that they were planning on finding flamingos by watching the EAZA surplus and wanted list. As few flamingos are offered on this list, a more active approach is necessary. Together with the report from this survey (King et al 2006) a database was made with a list of which flamingos individual zoos wanted to acquire or distribute. This was sent to all zoos participating in this project, and a number of recipients reported that they had used the list to initiate exchanges.

Almost all zoos that plan to increase flamingo group sizes through reproduction at their own zoos do not breed flamingos now, or breed only a few (56 of 59), but they are confident that breeding will improve in the next years. While we hope they are correct in their prognosis, it is clear that few of these zoos will actually be able to increase numbers quickly at this point, and we know that generally the larger the colony numbers the better the chance of breeding. However numbers are not the only factor, and even zoos that are unsuccessful in acquiring new flamingos at this point should still do their best to optimize other factors to improve the likelihood of flamingo breeding.

### **Flamingo Husbandry**

Joint AZA and EAZA Husbandry Guidelines have been produced for flamingos (Brown and King 2005), but I want to emphasize some aspects of husbandry here.

### Individual Identification & Sexing of Flamingos

Any zoo or private person that seriously wants to work with flamingos should be able to identify individuals at a distance, as this essential to understanding what is happening within the colony and being able to troubleshoot or otherwise alter the situation to improve success. While some zoos are reluctant to invest the time needed to observe a breeding colony, 15 to 30 minutes of daily observation by a practiced observer is all that is needed to monitor most colonies.

The EAZA executive office offers a service to EAZA zoos, in which the zoos can order coded rings in a choice of five colors, rather than looking for a

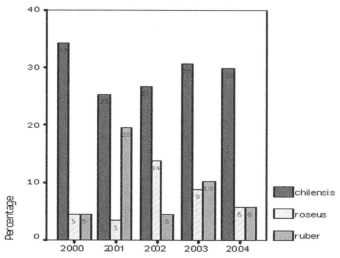

Fig. 4. *Percentage of young flamingos that died in the first month in European zoos between 2000-2004 (King & van Weeren 2005).*

### Table 3

**MEAN NUMBER OF YEARS THE COLONY HATCHED AT LEAST ONE CHICK BETWEEN 2000 AND 2004 (FOR ZOOS THAT HATCHED 1 OR MORE YOUNG) PER COLONY SIZE CATEGORY**

|  |  | 0 - 10 | 11 - 20 | 21 - 30 | 31 - 40 | 41 - 50 | 51 - 100 | > 100 |
|---|---|---|---|---|---|---|---|---|
| Andean | s | 0 | 0 | 0 | 0 | 0 | 0 | 0 |
|  | m | 0 | 0 | 0 | 0 | 0 | 0 | 0 |
| Chilean | s | 0 | 2 | **3** | *5* | 1 | **3.6** | *5* |
|  | m | 0 | 0 | 1.7 | 1.8 | **2.7** | **2.2** | **2.2** |
| Hybrid | s | 0 | 0 | 0 | 0 | 0 | 0 | 0 |
|  | m | 0 | 0 | 0 | 0 | 0 | 1 | 1 |
| James | s | 0 | 0 | 0 | 0 | 0 | 0 | 0 |
|  | m | 0 | 0 | 0 | 0 | 0 | 0 | 0 |
| Lesser | s | 0 | 0 | 0 | 0 | 0 | 0 | 0 |
|  | m | 0 | 0 | 0 | 0 | 0 | 0 | 1 |
| Greater | s | 0 | 0 | **2.7** | 1.8 | 1.5 | **3.1** | *4.5* |
|  | m | 0 | 0 | 2 | 1.7 | **3** | **3** | *4.5* |
| Caribbean | s | 0 | 2 | 1.3 | 2 | **3.7** | *5* | *5* |
|  | m | 0 | 0 | 0 | 2 | **2.5** | **3.5** | **3.5** |

*Bold letters: >2 years to 4 years, bold, italicized letters: > 4 years to 5 years.*
*(King & van Weeren 2005)*

distributor themselves. There are also instructions on the EAZA website for applying the rings. The service has been extremely successful, with almost 21,000 rings sold since the service began in 2000 (M. Los pers. comm.). Of these rings, 7,051 were of the size used for the *Phoenicopterus* flamingos (19mm inner diameter and 37mm height) and 1,148 were the size for Lesser Flamingos (17mm inner diameter and 33mm height). As there are fewer than 500 Lesser Flamingos in European zoos, the rings available for them are either also being used for similarly sized species or holders really want to keep a large number of reserve rings.

Many managers assume that a group of flamingos will have a sex ratio of roughly 1:1. This is often not the case, particularly regarding Lesser Flamingos (ISIS 2007), where the European population is 135.52.91 and the North American population is 247.132.68. It is essential that zoos that want to breed flamingos know the sex ratio of their colonies, and actively strive for a 1:1 sex ratio to maximize results.

### A Point System for Breeding Flamingos

There are a number of factors that are important for breeding flamingos, but it seems that if a certain number of criteria (not necessarily always the same ones) are met, the flamingos will breed. My theory is that this could be seen as a point system. For example, it may be necessary to have 27 of 39 points to breed flamingos, and points accumulated above the minimum can only increase rate of breeding success. Factors could be assigned a point value indicating their importance (Table 8). I am certain that others might argue with the point values I have given to factors, and they may be right. This is not meant to be a real formula, but only a way to present this idea, and to give zoos something concrete to work towards.

Colony size might have a value of seven points, as I think it is indisputably the most important factor for optimizing breeding. If it is impossible to gather together a large, single species group of flamingos, there are some tactics to try to achieve the same effect. Holding the birds closely together can help, although a compromise needs to be made between reducing area to increase density, and having sufficient area to display. Furthermore, it is important that flamingos have ample room to move around as reduced movement means reduced blood circulation, which means increased chance of foot lesions. Some zoos have reported success using mirrors to increase perceived colony size. However, as zoos tend to make multiple changes at the same it is difficult to establish whether it is the mirrors, or any single factor, that made a dif-

### Table 4

**NUMBER OF ZOOS HATCHING AND REARING FLAMINGOS AND NUMBER OF YOUNG REARED TO OLDER THAN ONE MONTH BETWEEN 2000-2004**

|  | N zoos reared/bred | Total number of young in five years | Mean number of young per year | Mean number of young reared per zoo that reared young per year | |
|---|---|---|---|---|---|
|  |  |  |  | Mean | SD |
| Andean | 0 | 0 | 0 | 0 | 0 |
| Chilean | 36/37 | 409 | 81.8 | 2.3 | 3.3 |
| James | 0 | 0 | 0 | 0 | 0 |
| Lesser | 1 | 1 | 0.2 | 0.2 | 0 |
| Greater | 31/34 | 602 | 120.4 | 3.9 | 6.2 |

*(King & van Weeren 2005)*

ference. Again it may be the summation of all the factors that made the difference.

Flamingos need to feel secure to breed. The most secure place may be within a couple of meters from people as long as the flamingos know that the people cannot reach them. More serious threats are larger animals held in their enclosure and predators, most commonly foxes in Europe. Bringing flamingos indoors every night to protect them from predation can negatively influence breeding. Flamingo enclosures situated where there is potential predation should be covered or hot-wired. Flamingos that are protected from predation but still can see potential predators may have an unsafe feeling. Surrounding the enclosure with vegetation may help, but best is to provide islands in water that the flamingos perceive as safe. Flamingos that do not feel safe tend to bunch together rather than spreading out. Safety deserves a point value of at least five.

Colony sex ratio does not seem important in whether a colony breeds or not, but it can affect breeding success. More atypical partnerships (not male-female pairs) occur with uneven sex ratios (King 2006), which may result in reduced fertility and/or more unrest in the colony, leading to lower reproductive success. Sex ratio may be assigned a value of two points.

Like sex ratio, the wing condition of the males (full-flighted, pinioned, wing-clipped, etc.) probably does not have much affect on whether a breeding attempt occurs or not, although there are not enough full-flighted groups to analyze this. Wing-condition clearly influences copulation ability of males. The response to flight restraint by males and the effect of flight restraint on males is individual. Many traditionally-pinioned males have difficulty springing up on the female. Some males respond by not even try to spring up on the female, while others continue to try even when never successful. Some males seem to be Olympic jumpers and are not much affected. The size of the male relative to the female can be an important factor in success. If a male is much taller than the female he may not have to jump at all to come into the correct position for cloacal contact. Fertility is not a 100% reliable measure of effect, as eggs can be fertilized by extra-pair copulations. Based on observations in several zoos, three-fourths to one-half of tradition-

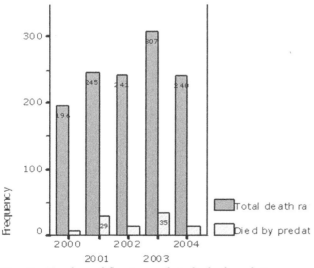

Fig. 5. *Number of flamingos that died of predation compared with total mortality in European zoos between 2000-2004 (King & van Weeren 2005).*

ally pinioned males in a colony are not able to copulate (King 2000; unpublished data). The effect of the different forms of flight restraint vary, with long-pinioning (taking only a few feathers off) being the least problematic and traditional pinioning the most. Full-flighted males have a value of 3, long-pinioned males have a value of 2.5, wing-clipped or tenectomized males have a value of 2 and traditionally-pinioned males have a value of 1.

Easily accessible water areas with gradual, smooth banks and primarily shallow bottoms (e.g. < 30cm deep) are features in flamingo exhibits that are lacking surprisingly often. This is very disheartening as flamingos are water birds, and deserve to be housed in enclosures in which they can be in their element. Foraging behaviour is essential to flamingos, and if there is no water flamingos will continue to make trampling movements with the feet (movements that function to stir up food in water) and bill foraging movements on dry land. Flamingos usually roost in water, as it presumably confers a feeling of safety. Much displaying is also performed in water. Water, especially if the basin bottom is smooth or covered with a layer of organic material, is good for the soles of the feet, as flamingos often have foot lesion problems in captivity. Appropriate water would certainly be worth 4 points. Ideally, appropriate water areas should constitute at least half of the enclosure (Brown & King 2005).

A study of four flamingo enclosures in western Europe (Greene 2005) showed that the flamingos in

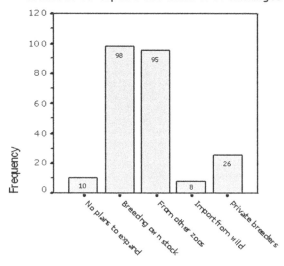

Fig. 6. *Plans to expand the number of flamingos in European zoos (King & van Weeren 2005).*

the one with much sun and easily accessible water areas had the best breeding, used the largest proportion of their enclosure (including spending half the time in the water) and were more active than the flamingos in the other enclosures. Another group that did not have access to much water but did have good access to mud in which flamingos also forage, was the second most active group. Preference for sun (versus shade) clearly affects use of enclosures in northern Europe (Greene 2005; Studer-Thiersch 2000a) and at least in that part of the world might have a value of four points. Studies

### Table 5

CHICK AND NON-CHICK MORTALITY BY NUMBER AND PERCENTAGE OF JANUARY 1, 2005 POPULATION

|  | Total number of flamingos | Total number of deaths 2000-2004 | | % deaths of the total population on 1.1.2005 | |
| --- | --- | --- | --- | --- | --- |
|  | January 1 2005 | = 30 days | >30 days | = 30 days | > 30 days |
| Andean | 39 | 0 | 5 | 0 | 12.8 |
| Chilean | 2774 | 172 | 307 | 6.2 | 11.1 |
| James | 3 | 0 | 0 | 0 | 0 |
| Lesser | 507 | 0 | 79 | 0 | 16.1 |
| Greater | 2700 | 47 | 324 | 1.7 | 12.0 |
| Caribbean | 1597 | 25 | 261 | 1.6 | 16.3 |

*(King & van Weeren 2005)*

### Table 6

**THE NUMBER OF COLONIES OF DIFFERENT SIZE GROUPS FOR SINGLE SPECIES FLOCKS IN 2005 AND THE EXPECTED NUMBER IN 2010**

|  | 1-20 2005 | 1-20 2010 | 21-40 2005 | 21-40 2010 | 41+ 2005 | 41+ 2010 |
|---|---|---|---|---|---|---|
| Chilean | 15 (33%) | 6 (11%) | 15 (33%) | 29 (49%) | 15 (33%) | 24 (39%) |
| Caribbean | 11 (52%) | 7 (18%) | 5 (24%) | 16 (42%) | 5 (24%) | 15 (27%) |
| Greater | 18 (35%) | 11 (17%) | 21 (40%) | 24 (38%) | 13 (25%) | 28 (44%) |
| Lesser | 12 (75%) | 5 (31%) | 2 (13%) | 9 (56%) | 2 (13%) | 2 (13%) |

*(King & van Weeren 2005)*

of colonies in warm climates should indicate if this is an important exhibit feature universally.

Other aspects of weather are also important. It has been frequently noticed that a few days of rain, cold and clouds early in the breeding cycle can shut reproductive activities off, at least temporarily, and sometimes for the season. Warmth may be particularly important for Lesser Flamingos. The influence of weather on continuation of reproductive activities becomes less important once nesting has advanced to the incubation stage (pers. obs.), although bad weather can influence success. Weather might have a point value of three. While photo period is not usually seen as an issue in breeding flamingos it can be a factor. Prolonged photo period (e.g. 24 hours a day) has stimulated reproduction in Caribbean and Greater Flamingos housed indoors (J. Lammers & B. Hiddinga pers. comm.). Photo period may also be particularly important in breeding of Lesser Flamingos in captivity (Harteman 2006). Photoperiod may deserve a point value of two.

Flamingos prefer long stretches of mostly unobstructed area for performing the "marching" display, this can be land, water or both. Too many obstructions can cause them to break off the displays prematurely, which can lead to reduced stimulation and synchronization, thereby reducing breeding success. Too hilly a terrain also impedes displays. Proper terrain is awarded a point value of two. On the other hand presence of an occasional soft visual barrier (e.g. a shrub, or large clump of grass or a stand of bamboo) and spreading of important resources (i.e. bathing area, feeding area, nesting area, and loafing area) requires that flamingos occasionally need to seek each other out, which can increase synchrony (Studer-Thiersch 2000a). Spreading of resources and strategically placed barriers might also count as two points.

A soft substrate enhances condition of the feet and should always be available. As mentioned above, water is recommended as a substrate. Short vegetation is also good, and matting or another type of flooring that has resilience should be used indoors. Substrate might count as 1 point for breeding, but from a well-being standpoint it would count as more.

Having a good nesting area (damp, pliable) available at the right time (when the birds are increasing display rates), can definitely be stimulating. In some cases presence of artificial nest mounds or even placing an egg on one or more artificial or flamingo built nests can even be more stimulating. A suitable nesting area may deserve a value of four points.

### Egg & Chick Manipulation

Manipulation of eggs is sometimes undertaken to increase number of (fertile) eggs produced, to safeguard eggs during incubation, or to give flamingos that do not have fertile eggs an opportunity to foster rear a chick. Some flamingo managers are adamantly against entering the colony during the nesting period to manipulate eggs or chicks, and feel that such disturbance will have a negative effect on reproduction. Contrarily, other zoo managers enter the colony and pull or switch eggs without any problems. "Colony intervention" was initiated at Rotterdam Zoo in 1993. The keepers were at first a bit worried about the effects, but an analysis of egg breakage in that first year showed that there was more egg breakage in the five days proceeding entering the colony than after. The keepers now also enter the colony regularly to microchip and pinion young two to five days of age, and the flamingos rarely bother to stand up when the keepers put their hands in the nest.

There are some things that should be kept in mind if egg manipulation is undertaken:

Keepers should attempt to always enter the colony from the same place, moving in a slow, non-threatening way to minimize disturbance. Generally flamingos are not terribly selective about the eggs that they incubate, wooden eggs and goose eggs have worked well in some zoos. Synchrony of the flock is important; therefore it is better to pull a number of eggs at the same time if the intention is to encourage the birds to recycle.

In some cases where only a few pairs produce fertile eggs, their first (and perhaps more) clutches are fostered to other flamingos that lay infertile eggs so that the pairs producing fertile eggs will recycle. Flamingos in captivity breed with similarly aged cohorts (same or one year apart) if there is a choice, at least in the first pairing (King 2006). Therefore, it should be kept in mind that soon the colony will have a number of similarly aged siblings that are quite likely to breed with each other unless zoos trade young. Trading young all of one sex (e.g. only males) is the most effective way to ensure that the flamingos breed with members of another colony. Trading of flamingos (or just sending them to another institution) in cases where inbreeding is likely to occur should be undertaken in the first two years, as some captive flamingos may breed at two years of age.

How many times a flamingo is encouraged to recycle varies with the philosophy of the manager. Even flamingos that lose eggs by natural events (e.g. colony fighting) may lay as many as six eggs a year (A. Studer-Thiersch pers. comm.). Theoretically, if the diet is good, the nutritional burden of laying so many eggs should not be that great on the female. However, I personally cannot believe that it is good for a female stress-wise, and I would not feel comfortable with encouraging a female to lay more than three eggs a year.

Flamingos are generally good parents, and parenting skills are fairly hard-wired in them, unlike storks, which usually require experience to learn how to rear young. However, end of the season chicks are often at a disadvantage over earlier hatched chicks. The parents may not be properly motivated to care for the chick because the rest of the colony is doing something else. This may be a problem if only one or two chicks hatch earlier in the year as well. It has also been observed that end of the season chicks have more leg problems, as discussed above. The survivorship of end of the season chicks, especially Chilean Flamingos that breed late in the year, can be improved by hand-rearing in some cases. Some zoos also opt to hand-rear if there have been chick predation problems in the past. There is no indication that hand-rearing negatively influences flamingo social behaviour or breeding. Of course it is easier on the chicks if more than

### Table 7
**GENERAL OVERVIEW OF POTENTIAL SUPPLY AND DEMAND FOR FLAMINGOS BETWEEN 2005-2010**

|  | Supply that zoos think they can deliver (total over five years) | Real surplus (birth - death) based on 2000-2004 collectively | Real surplus as % of the 1.1.2005 population | Flamingos needed (2010 - 2005) |
|---|---|---|---|---|
| Andean | 5 | -5 | -12.8 | -5 |
| Chilean | 220 | 102 | + 4.31 | 925 |
| James | 0 | -1 | - 2.5 | 0 |
| Lesser | 5 | -67 | - 13.5 | 192 |
| Greater | 245 | 259 | + 12.2 | 1070 |
| Caribbean | 101 | -38 | - 2.8 | 785 |

*(King & van Weeren 2005)*

### Table 8

**POTENTIAL POINT VALUES INDICATING RELATIVE IMPORTANCE OF FEATURES IN REPRODUCTION OF FLAMINGOS**

| Feature | Points |
|---|---|
| Colony size | 7 |
| Security | 5 |
| Appropriate water areas | 4 |
| Sun | 4 |
| Nest area | 4 |
| Weather | 3 |
| Wing condition males | 3 |
| Spread resources, soft visual barriers | 2 |
| Large unobstructed, level areas | 2 |
| Colony composition | 2 |
| Photo period | 2 |
| Substrate | 1 |
|  | 39 |

one young is hand-reared at a time, and introducing the young to the colony as quickly as possible can ease the transition. A single hand-reared chick was placed with the colony at Rotterdam Zoo during the day when only one month of age, and later he went on to be a reliable breeder. If it is necessary to remove a chick, or one of its parents, from the colony for health problems temporarily, it is quite likely that the parent-young relationship will not be disturbed, particularly if the chick is a couple of weeks old or older. Parents readily recognize their young and resume normal care and feeding, even after separations lasting several weeks.

## Colony & Individual Behaviour

After many discussions with other observers of captive flamingo colonies I can only conclude that I will never see it all, and flamingos will continue to surprise me. I think that behaviour of any given colony is influenced by the history of the colony (how long it has been together as a whole group, how many of the flamingos were acquired together, etc.), sex ratio and individual characteristics. A number of behavioural topics were studied with the colony at Rotterdam Zoo over the last 15 years (1992 through 2006). This colony initially consisted of both greater and Caribbean Flamingos and numbered 37 birds in 1992. The maximum number was reached in 2004 with 133 flamingos and dropped to 93 (47.46) when the 40 Caribbean flamingos left the colony in 2005. I certainly do not have the time to expound on the results of all these studies here, even though I would like to, so I will just discuss some findings on atypical relationships and flamingo fights and dominance. It would be very interesting to compare the results at Rotterdam Zoo to studies at other zoos to establish whether the observations can be generalized.

Atypical relationships, e.g. trios, quartets, same-sex pairs are quite common in captive flamingo colonies (King 2006). It is also clear that the public is fascinated with this aspect of flamingo behaviour, as illustrated at Rotterdam Zoo. Two males in the colony formed a pair bond in 1992 and disrupted nesting flamingos until they stole an egg from another pair and sat peaceably on it. The next year they were given an egg right away, so that they would not disturb others. Five fertile Chilean Flamingo eggs were received that year from Amsterdam Zoo to give to foster parents, as otherwise the chicks would have to be hand-reared. One of these eggs was given to the two males, and they did not miss a beat, rearing the chick with no problem at all. This occurred during what is called "the cucumber season" in Holland and when a press release about this pair appeared it was a hit, making two pages in France's most popular newspaper and the world news on the radio. We did interviews with journalists from several different countries and of course- gay newspapers. Colleagues were jealous of all the attention Rotterdam Zoo received, as after all they had gay flamingos too!

Eventually because of all the publicity I ended up writing a chapter for a book on homosexuality in flamingos from an evolutionary perspective, and only then really began thinking about why atypical relationships occur so frequently in captive flamingos. It may be that because flamingos pair with the bird they are most synchronized with, in a captive colony that just might be a bird of the same sex, because there are other factors that influence synchrony and there are not so many choices. Being paired, even to a flamingo of the same sex, would confer better access to resources. Another advantage would be that it keeps the flamingos in the game during the breeding season (King 2006). Breeding success of wild flamingos increases with age, at least to a certain point, because experienced flamingos are better at getting nest sites (Cézilly et al 1997). A pair can be defined as flamingos that sleep side by side, simultaneously carry out activities, call together and form coalitions during aggressive encounters (Studer-Thiersch 1975). While this is a helpful definition, flamingos seem to have a continuum of relationship possibilities that depend on the individual and the situation. Certainly trios and quartets may also fit the definition of a pair. There are different types of trios, as pointed out by Shannon (2000), and trios can sometimes be hard to separate from male-female pairs or same-sex pairs. Flamingo relationships are fluid, they can move from one type of atypical relationship to another, and back to a male-female pair. Flamingos maintain relationships with birds other than the current partner(s), even over years; they may switch among partners over the years, possibly within a small group. Although the data still need to be thoroughly analyzed, it appears that some flamingos at Rotterdam Zoo will not recycle in a given year unless some of their nesting buddies are also breeding then. There are also "coalition partners", flamingos that do not interact with each other much except to support each other during a fight, and "copulation partners", flamingos that copulate together but do not interact much otherwise, but may be previous or a future breeding partner.

To test the theory that having a partner helps win fights and provides better access to resources even during the non-breeding season, we studied fighting behaviour at Rotterdam Zoo. Fights were defined as any aggressive encounter, ranging from one bird displacing each other to physical fighting. We found that flamingos greatly increase the chance of winning a fight when they fight together with another flamingo- this may be their own partner, but it may also be a "coalition" partner. Male flamingos fight more than female flamingos (probably not a surprise) and younger flamingos (less than ten years of age) fight more than older flamingos during the non-breeding season. During the breeding season the breeding birds fight the most, regardless of age (de Ridder 2005; Stolze & King 2006). Interestingly, while females do not fight as much as males, we found in a long term study of dominance of individuals (based on how many fights each flamingo engages in relative to the number of fights they win), more females were dominant than males, and often a dominant female was paired to a subordinate male. Dominance rank was not entirely stable, although the lowest ranking bird in the colony remained the lowest ranking (Stolze et al 2006).

## Discussion

There are enough individuals of the Greater, Lesser, Chilean and Caribbean flamingos in captivity in Europe to maintain long-term genetically viable populations. However, intense cooperation between zoos to optimize colony size and composition is needed to maximize breeding potential. European zoos have largely indicated that they are ready and willing to make some changes in this area. However, even if EAZA zoos do greatly increase cooperation and improve their management, they are a long way from collectively achieving their goals, as there simply are not that many flamingos.

It is fair to wonder whether so many zoos really need to hold so many flamingos; possibly less zoos could hold more per zoo. The reality is that few zoos will agree to stop holding flamingos completely. At this point the TAG has several strategies to try to maximize the potential.

1) Convince as many zoos as possible to hold only one species and to trade with other zoos to increase numbers of the remaining species held to >40 if breeding is a goal. (Almost all zoos holding flamingos said they wanted to breed them);
2) Continue trying to persuade zoos with few flamingos (especially less than ten) to send these to a zoo better placed to hold them;
3) Encourage zoos that are willing to distribute some flamingos to give these to zoos that are closer to achieving the recommended breeding group number (> 40) and have enclosures suitable for flamingos;
4) Encourage zoos that do breed flamingos to make more available to other zoos;
5) Make it as easy as possible for zoos to arrange transfers by maintaining a current database on available and wanted flamingos, showing all the flamingos that these zoos have, so that choices can be better made.

Unfortunately, promoting these five strategies to the degree necessary to accomplish them is a time consuming, labor intensive task. We are still exploring means to achieve this. Many zoos have long viewed flamingos as highly desirable decoration, but have not adequately addressed their enclosure needs. Zoos need to evaluate their flamingo enclosures and consider which changes can be made to improve the flamingos' environment. The point system presented in this paper was proposed in order to help zoos see which factors they do and do not have control over, and to help define priorities.

Several zoo managers have recently requested information on holding flamingos in covered enclosures, especially walk through aviaries. Covered enclosures have distinct advantages; the flamingos can be held full-winged, which is bound to improve fertility and full-winged birds are less likely to fall when they trip or stumble because they have better balance. As trauma is the greatest death cause, holding flamingos full-winged could improve their life expectancy. Because the flamingos are able to move with more balance, their feeling of security may improve. Covering the enclosure, if done properly, should also prevent predation on the flamingos.

The drawbacks to a covered enclosure are costs and visibility of the structure (only from the outside if it is a walkthrough aviary). Of course if predation is a real risk, the enclosure would pay for itself quickly in potential lives saved. There are several possibilities for designing flamingo enclosures that reduce its drawback of visibility and/or costs. Several zoos in Japan cover their flamingo enclosures with microfilament that is barely noticeable, but keeps the flamingos in and herons out. It can also be easily hotwired. San Antonio Zoo has an open viewing area on one side that is closed at night to prevent escapes and predation. Rheine Zoo in Germany just constructed an enclosure with high enough barriers (10m) that it is believed the flamingos cannot escape. The possibilities to hold flamingos full-flighted are there, and I firmly believe that where there is a will there is a way.

Because there are so many flamingos in captivity and so much interest in them, we really do have the data to guide captive flamingo management. The real question is, how does what we know about captive flamingos apply to their wild counterparts? If flamingos normally nest in such huge numbers and move around so much (apparently independently of other individuals) why are they capable of such long term, identifiable relationships? It may be that flamingos did not always form such huge groups (even now there are some small, sedentary populations) and/or individual relationships may be equally advantageous in larger groups where breeding is opportunistic, as it could allow the birds to begin breeding more quickly. There is some evidence that some degree of more personal relationships do develop within large breeding groups of wild flamingos; the flamingos that a pair nests with is not random. Studer-Thiersch (2000b), observed that wild flamingos, like their captive counterparts, form small displaying groups of 15-20 individuals within larger populations. While it has not been possible yet to prove that these birds go on to nest together, there are small groups of flamingos that are synchronized in time of egg laying within a large breeding colony that probably do consist of these smaller displaying groups.

## CONCLUSION

While none of the six flamingo species are now considered threatened, the very specific environments that flamingos dwell in, put flamingos in a precarious situation in our rapidly changing world. Flamingos are present in captivity in impressive numbers, and captive populations can serve their wild counterparts through their availability for research to better understand flamingo biology and ecology, and to test techniques for field studies. Flamingos are very attractive to people, and thus can serve as flagship species for their environment. We know much about how flamingos should be managed in captivity. We also know that managing them well will require intensive cooperation, some sacrifices and some investments. But because of the extreme popularity of flamingos some zoos now seem willing to take the necessary measures, and hopefully more will follow. Obviously there is still much to be explored concerning flamingo behaviour. Understanding sociality among captive birds can only further illustrate the behavioural constraints and plasticity possible within the behavioural repertoire of wild flamingos. An ability to predict how flamingos might behave if reduced to small populations can only be advantageous in their long-term conservation.

## REFERENCES

BECHET A. 2006. Summary of greater flamingo banding in the Mediterranean region from 1977 to 2006. La Sambuc: Station Biologique de la Tour du Valat.

BECKMAN M. 2006. Science Now daily news: 31 October-2006.http://sciencenow.sciencemag.org/cgi/content/full/2006/1031/2?etoc

BERRY HH. 1972. Flamingo breeding on the Etosha Pan, southwest Africa, during 1971. Madoqua (Series 1), 5, 5-31.

BERTAULT G, M Raymond, F Cézilly & AR Johnson. 2000. Evidence of seasonal sex ratio manipulation in the greater flamingo. Waterbirds 2, (Special Publication 1), 20-25.

BILDSTEIN KL. 1990. The use of zoo collections in studies of feeding ecology and conservation biology of wading birds (Aves: Ciconiiformes). AAZPA 1990 Regional Proceedings, 353-360.

BILDSTEIN KL, PC Frederick & MG Spalding. 1991. Feeding patterns and aggressive behaviour in juvenile and adult American flamingos. Condor 93: 916-925.

BIRDLIFE INTERNATIONAL. 2007. Species factsheet: *Phoenicoparrus andinus*. Downloaded from http://www.birdlife.org on 27/5/2007.

BROWN LH. 1959. The mystery of the flamingos. London: Country Life.

BROWN C & C King. 2005. Flamingo husbandry guidelines. A Joint Effort of the AZA and EAZA in Cooperation With WWT. Dallas: Dallas Zoo.

CÉZILLY F & AR Johnson. 1995. Re-mating between and within breeding seasons in the greater flamingo. Phoenicopterus ruber roseus. Ibis 137: 543-546.

CÉZILLY F, V Boy, CJ Tourenq & AR Johnson. 1997. Age-assortive pairing in the Greater Flamingo. *Phoenicopterus ruber roseus*. Ibis 139: 331-336.

CHILDRESS B. 2004. Remarkable Lesser Flamingo recovery. Lanioturdus 37: 3-4.

CHILDRESS B, S Nagy, & B Hughes. 2007. International Single Species Action Plan for the Conservation of the Lesser Flamingo *Phoenicopterus minor*. First draft 31 January 2007.

DEL HOYO J, A Elliot & J Sargatal (Eds.). 1992. Handbook of the birds of the world. Vol. 1. Barcelona: Lynx Edicions.

DE RIDDER A. 2005. Flamingo Boulevard. Rotterdam: Rotterdam Zoo.

FOX DL. 1975. Caretenoids in pigmentation. In J. Kear & N. Duplaix-Hall (Eds). Flamingos (pp.162-182). Berkhamsted: Poyser.

GREENE T. 2005. Flamingo's in de ruimte. Onderzoek naar het verblijfgebruik van flamingo's in vier Nederlandse dierentuinen. Rotterdam: Rotterdam Zoo (In Dutch).

HAGEY LR, HT Schteingart, HT Ton-nu, SS Rossi, D Odell & AF Hofman. 1990. Beta-phocalacholic cid in the bile: biochemical evidence that the flamingo is related to an ancient goose. Condor 92: 593-597.

INTERNATIONAL Species Information System (ISIS). 2007. http://www.isis.org/CmsHome/content/MyHome (Accessed May 15 2007).

JENKIN PM. 1957. The filter-feeding and food of flamingos (Phoenicopteri). Philosophical Transactions of the Royal Society of London, Series B 240: 401-493.

JOHNSON AR. 1998. A noteworthy recovery. Flamingo Specialist Newsletter 9: 12.

JOHNSON A. 2000. An overview of the greater flamingo ringing program in the Camargue (southern France) and some aspects of the species' breeding biology using marked individuals. Waterbirds 23, (Special Publication 1), 2-8.

HARTEMAN J. 2006. Adviesen voor successvolle kweek van Phoeniconias minor. Dronten: Stoas Hogeschool. (In Dutch).

KING CE. 2000. Captive flamingo populations and opportunities for research in zoos. Waterbirds 23 (Special Publication 1): 142-149.

KING CE. 2006. Pink flamingos: atypical partnerships and sexual activity in colonial breeding birds. In V. Sommer & P.L. Vasey (Eds.). Homosexual behaviour in animals: an evolutionary perspective (pp. 77-106). Cambridge: Cambridge University Press.

KING CE & L van Weeren. 2005. Captive flamingo management on a European level. Rotterdam: Rotterdam Zoo.

KING CE, C Breedijk, W de Bruin, P Kok, S Sanders & A Tuit. 2006. Managing flamingos collectively and individually in Europe. EAZA Flamingo Regional-Institutional Collection Plan. Rotterdam: Rotterdam Zoo.

LANG EM. 1963. Flamingos raise their young on a liquid containing blood. Experentia, 15: 532-533.

MARTIN GR, N Jarrett, P Tovey & CR White. 2005. Field vision in flamingos: chick-feeding versus filter-feeding. Naturwissenschaften, 92: 351-354.

PÉINDO VI, FJ Polo, G Viscor & J Palomeque. 1992. Hematology and blood chemistry values for several flamingo species. Avian Pathology 21: 55-64.

PICKERING SPC. 1992. The comparative breeding biology of flamingos Phoenicopteridae at the Wildlife and Wetlands Trust Centre, Slimbridge. International Zoo Yearbook 31: 139-146.

PLANET ARK. 2007. http://www.planetark.com/avantgo/dailynewsstory.cfm?newsid=41524, 30-4-2007.

RENDON MM & AR Johnson. 1996. Management of nesting sites for Greater Flamingos. Colonial Waterbirds 19 (Special publication 1): 167-183.

RENDON-MARTÓS M, JM Vargas, A Garrido & JM Ramírez. 2000. Nocturnal movements of breeding greater flamingos in Spain. Waterbirds 23 (Special Publication 1): 9-19.

RICHTER NA & GR Bourne. 1990. Sexing Greater Flamingos by weight and linear measurements. Zoo Biology 9: 317-323.

RICHTER NA, GR Bourne & EN Diebold. 1991. Gender determination by body weight and linear measurements in American and Chilean flamingos, previously surgically sexed: within-sex comparison to greater flamingo measurements. Zoo Biology 10: 425-431.

SCHMITZ RA & GA Baldassarre. 1992. Contest asymmetry and multiple bird conflicts during foraging among nonbreeding American flamingos in Yucatan, Mexico. Condor 94: 254-259.

SHANNON PW. 1996. North American Regional Studbook for the Greater Flamingos (*Phoenicopterus ruber roseus*). Published under the auspices of the American Zoo and Aquarium Association.

SHANNON PW. 2000. Social and reproductive relationships of captive Caribbean Flamingos. Waterbirds 23 (Special Publication 1): 173-178.

SPRUNT A. 1975. The Caribbean. In J. Kear & N. Duplaix-Hall (Eds.) Flamingos (pp. 65-74). Berhkhamsted: Poyser.

STOLZE M & C King. 2006. Fighting flamingos: a study on fighting and aggression in flamingos at Rotterdam Zoo. Rotterdam: Rotterdam Zoo.

STOLZE M, S Saus & C King. 2006. Dominance in flamingos, a study on the social hierarchy in greater and Caribbean flamingos at Rotterdam Zoo. Rotterdam: Rotterdam Zoo.

STUDER-THIERSCH A. 1966. Altes en neues über das Futterungssekret der Flamingos *Phoenicopterus ruber*. Ornithologische Beobachter 63: 85-89 (in German).

STUDER-THIERSCH A. 1975. Basle Zoo. In J. Kear & N. Duplaix-Hall (Eds). Flamingos (pp. 121-130). Berhkhamsted: Poyser.

STUDER-THIERSCH A. 1986. Tarsus length as an indicator of sex in the flamingo genus Phoenicopterus. International Zoo Yearbook 24/25: 240-243.

STUDER-THIERSCH A. 1998. Basle Zoo's veteran flamingos. Flamingo Specialist Newsletter 8: 33.

STUDER-THIERSCH A. 2000a. Behavioural demand on a new exhibit for Greater Flamingos a the Basle Zoo, Switzerland. Waterbirds 23 (Special Publication 1): 185-192.

STUDER-THIERSCH A. 2000b. What 19 years of observation on captive greater flamingos suggests about adaptations to breeding under irregular conditions. Waterbirds 23 (Special Publication 1): 150-159.

ZWEERS G, F de Jong, H Berkhoudt & JC Van Den Berge. 1995. Filter feeding in flamingos. Condor 97: 297-324.

## Author Biography

Following graduation with a B.S. in Wildlife Biology at Kansas State University in 1980, I worked as a zookeeper in the bird department at San Antonio Zoo. The collection of long-legged wading birds there then was very impressive, and I was hooked. After gaining a few years of practical experience I went on to get a master's degree in Zoology in 1988 at Oklahoma State University. My master's thesis was an ethological comparison of three species of Ciconia, the White Stork *Ciconia ciconia*, the Oriental White Stork *Ciconia boyciana* and the Maguari Stork *Ciconia maguari* in captivity. Subsequent to my move to the Netherlands I began working as the bird biologist at Rotterdam Zoo in 1990, and remained there until moving to Spain in 2006. I began chairing the European Association of Zoos and Aquaria (EAZA) Ciconiiformes and Phoenicopteriformes Taxon Advisory Group (TAG) when the first EAZA TAGs were initiated in 1992, and continue to fulfil this function. Additionally, I am the studbook keeper for the Marabou Stork *Leptoptilos crumeniferus* ESB, and for the Von der Decken's Hornbills *Tockus d. deckeni* ESB, as well as coordinator for the Oriental White Stork European Endangered Species Program (EEP).

*A Dedication*

*My interest in flamingos dates back almost 27 years, when I became a zoo keeper at the San Antonio Zoo. San Antonio Zoo was one of the first zoos to breed Caribbean flamingos (in 1957). Breeding of this species was going strong when I worked there - and it still is. The flamingos have the good luck that Jeff Perry, a long-time bird department employee, not only resembles a flamingo physically because he is so tall and slender, he can also think like a flamingo. Jeff was one of the first zoo people to use coded color rings for individual identification of flamingos, and he helped pioneer many flamingo management techniques. I took what I learned from Jeff about flamingos with me when I went to Europe. Through my role as the European Association of Zoos and Aquaria (EAZA) Ciconiiformes and Phoenicopteriformes Taxon Advisory Group (TAG) Chair, I have been able to spread his expertise further, and to build on it. I dedicate this paper to Jeff Perry, and am forever grateful to him for showing me how fascinating trying to understand what is happening in a flamingo colony can be.*

# THE AVICULTURE OF HUMMINGBIRDS: A HISTORICAL PERSPECTIVE

JOSEF LINDHOLM, III

The Dallas World Aquarium
1801 N Griffin St, Dallas, TX 75202

## SUMMARY

An historical exploration of the aviculture of hummingbirds reveals that a fairly small and closely connected circle of individuals, institutions and businesses exerted a major influence on its development. and progress. Drawing extensively from primary sources, an examination of the evolution of husbandry, exhibition, and propagation of hummingbirds is presented, as well as a discussion of the rise and decline of commerce in them in various countries. Persons whose achievements are presented include John Gould, Major Albert Pam, Alfred Ezra, the Marquis de Segur, Jean Delacour, Cecil Webb, Lee Crandall, David Seth-Smith, Charles Cordier, August Fockelmann, Axel Reventlow, Anthony Chaplin, Captain Richard de Quincey, Eric Kinsey, the Marquis Yamashina, Percy Hastings, Raymond Sawyer, Oliver Griswold, Ettiene Beraut, Augusto Ruschi, K.C. Lint, Alex Isenberg, Lena Scamell, Walter Scheithauer, Tony Mobbs, Rod Elgar, Karl Schuchmann, and Jack Roovers.

Among institutions discussed are the London Zoo, the Bronx Zoo, Zoologischer Garten Berlin, Brookfield Zoo, Copenhagen Zoo, Cleveland Zoo, the San Diego Zoo, Wuppertal Zoo, Heidelberg Zoo, the Arizona Sonora Desert Museum, Butterfly World, and The Dallas World Aquarium.

## INTRODUCTION

*To be fully comprehended, the history of birds in captivity must encompass people and their relationships as much as it does birds.*

### The Brothers Pam

Jean Delacour (1966a) published his autobiography, *The Living Air*, when he was 76. Looking back over a lifetime of friendships, he wrote: "Major Albert Pam  was my friend since the beginning of the First World War and throughout the years we took great pleasure in each other's company. I spent many happy days at his fine Adam house, Wormleybury, in Hertfordshire, and he was very often with me at Cleres, not to speak of the almost forgotten days at Villers. Later on, until he died in 1955, he came to see me in Los Angeles every other year and we enjoyed Californian gardens together. He was primarily a connoisseur of plants and, particularly, an authority on amaryllids, and his gardens were full of rarities. He also kept wallabies and water-deer, cranes, flamingos, and lots of wild geese and ducks on his large lake, but difficulties gradually obliged him to reduce their numbers, and finally to dispose of them; many were used to restock Cleres." One achievement of Major Pam, which Captain Delacour did not mention was, perhaps, one the Major preferred to forget.

David Seth-Smith was, among his other responsibilities, to serve as the Zoological Society of London's Curator of Birds from 1909 to 1939. In 1906, however, he was not yet formally associated with the London Zoo but was instead acting in his capacity as Honorary Editor of the *Avicultural Magazine* when he wrote the following notes: "The arrival of a living Humming Bird at the Zoological Gardens is an event of great avicultural interest and importance, though it is somewhat ancient history now, and most of our members have read all about it in the daily and weekly papers. This bird, a specimen of the Bolivian Violet-eared Humming Bird *Petasophora iolata* [Sparkling Violet-ear *Colibri coruscans*], arrived at the Gardens on November 25th last [1905]. During the train journey from Southampton every effort was made to keep the bird warm, by placing foot-warmers, etc. round the cage, but in spite of all care, on arrival it was in a state of collapse from the cold: but it immediately recovered when placed in a warm place by the hot-water pipes in the Insect House. At a meeting of the Zoological Society on November 28th, Captain Pam gave an interesting account of his attempt to import living Humming Birds. These are caught by the

Natives with limed twigs, and many attempts have been made to keep them in captivity in Venezuela, but with very limited success. Capt. Pam fed his Humming Birds on sugar and water to which was added a certain amount of extract of meat as a substitute for insects: and this treatment succeeded well. He started from Venezuela with some half-dozen of these birds, all of which had lived with him for some time before starting. On arrival at Southampton one only survived. This bird lived for a fortnight in the Insect House, but, in spite of every attention, succumbed during the recent damp foggy weather. Captain Pam deserves the hearty congratulations of aviculturists for his success, though this was far less than he deserved to achieve. Success in importing these delicate denizens of the tropics is only to be attained as the result of infinite pains and unremitting attention to their every need during the voyage home" (Seth-Smith 1906a).

A few months later, David Seth-Smith had more to write (1906b), "During the past month [June 1906] two fine collections of South American birds have reached the Zoological Gardens, one brought home from Venezuela by Captain Pam, including four Humming Birds, all of which died within a day or two of their arrival, a Sun-bittern, Tanagers, Hangnest, Sugar-birds and others..."(Seth-Smith, 1906b).

Finally, by now speaking from experience, Seth-Smith wrote this report "The Zoological Society is indebted to Captain Albert Pam for many very valuable and rare South American birds, but the most wonderful collection of all arrived at the Gardens on May 27th [1907] and consisted of no less than twenty living Humming Birds, which were at once placed in two special cases in the Insect House. These comprised no less than four species, all from Venezuela; the Emerald Humming-Bird *Eucephala coerulea*, the Venezuelan Amazilia Humming-Bird *Amazilia feliciae*, the Oenone *Chyrsuronia oenone* and Prevost's Humming-Bird *Tampornis* [sic] *prevosti* [The Green-breasted Mango *Anthracothorax prevostii*]. Our members should not lose any time before inspecting this extraordinary collection, for these curious, and in many ways very insect-like creature, are short-lived however carefully they are tended..."(Seth-Smith, 1907a).

David Seth-Smith's pessimism was all too warranted, as witnessed by Albert Pam himself (then 32 years old), and his brother Hugo, who assisted him on the Venezuelan expeditions: "Last year the Editor wrote some remarks on the subject of the few Humming-Birds which were brought back from South America for the Zoological Society. Since that first trial another small consignment was brought over the in the month of June of last year, and last May [1907] we again succeeded in landing twenty Humming-birds, which were duly deposited in the Society's Gardens on the 27th of that month. The death of the last survivor of this consignment has caused us to review the facts which these three trials have brought out, and also the experience gained by us in the feeding and care of these small birds, which has been largely increased since the first article appeared in this Magazine on the subject. The last consignment of twenty birds was put on board the mail steamer in La Guayra in good condition, and the last bird to survive died after having been in captivity in London between four and five weeks. Inasmuch as this bird was caught about a fortnight before it left Venezuela, and the voyage took sixteen days, we can roughly state that it was in captivity for about nine weeks. It has now therefore been clearly proved that Humming-Birds, under favourable conditions, can be kept in captivity for a short time even in this country; but we have regretfully come to the conclusion that, owing to the climatic conditions which obtain in England, and the uncertainty of the weather, it is not to be expected that these little birds can be kept in a cage for any considerable period in England. ...The great difficulty of keeping these birds in small cages is this, in a house such as the Insect house, in which they were placed in the Zoological Gardens, they must be protected against the draughts caused by the doors being continually opened and shut, and the temperatures being kept fairly high caused the air to be too 'stuffy" and the ventilation deficient.

It will therefore be seen that in is practically impossible in this country to obtain the ideal conditions which would permit of Humming-Birds being seen in captivity in all their splendour and colouration, and we have therefore decided that unless some new scheme can be devised with would give these birds more natural conditions in which to live, we must give up our efforts to introduce them into this country." (Pam & Pam 1907).

The Brothers Pam had certainly not plunged recklessly into these experiments: "We had carefully discussed the question of the food with Dr. Chalmers Mitchell [Secretary of the Zoological Society], and had come to the conclusion that the solution of molasses, sugar, with the vegetable extract Marmite added did not contain a sufficient quantity of albuminoids to sustain life for any length of time. We therefore decided on the following plan. Every other day we added to a solution of the molasses sugar a quantity of well beat up white of egg and we found that the birds readily took this food and did very well on it. When the birds arrived in the Zoological Gardens we also gave them a small quantity of sweetened milk which they took very greedily indeed, by were are unable to say where this food was really beneficial or otherwise. In the former consignments we had experienced considerable difficulty owing to the fact that the birds, in taking the thick solution of molasses sugar, very often dipped their feather into the feeding glass, ad owing to the sticky nature of the solution were thus often prevented from flying properly.... In order to eliminate this danger we arranged for the glasses containing the solution to be fixed outside the cages with just sufficient opening for the birds to put their beaks through, and thus drink without any possibility of making their feathers sticky. This answered very well indeed, after birds had got used to finding their food, but caused a considerable mortality before they had learnt how to get at it. (Pam & Pam 1907).

They concluded the primary obstacle to keeping hummingbirds in England was lack of sunlight and fresh air: "In conclusion, we may repeat that the Humming-Bird is, above all, a creature of sunshine and liberty: in its habits it resembles a tropical butterfly much more than a bird, and the ideal conditions which would be necessary to keep the former in captivity are also required in the case of Humming-Birds.. For those who have seen these beautiful birds free in their native state, it can but be a sad sight to note their ruffled plumage and dejected appearance in captivity and it has made us very miserable to watch them slowly dying for the want of that sunshine which is so essential to their life. All lovers of animals will therefore understand our reasons when we say that we shall not try again to import Hummingbirds into England" (Pam & Pam 1907).

To this lugubrious account Editor Seth Smith (1907b) appended: "We feel sure that aviculturists in general will be very grateful to Messrs. A. and H. Pam for the amount of trouble they have taken to introduce living Humming-Birds, and to give us in this country an opportunity of studying these wonderful birds. The last consignment which arrived unfortunately came in for one of the most sunless Junes on record. Had they been favoured with the amount of sunshine that one expects during an English summer, and perhaps if they had had a somewhat less stuffy cage, they might have lived longer. At any rate we do not think the experiment can be regarded as quite a failure, and, although we quite understand Messrs. Pam's reason for not wishing to repeat the experiment, still we believe that under more favourable conditions Humming-Birds could be kept in this country for some time."

It is interesting to note David Seth-Smith and Albert Pam went on to maintain a professional relationship. For at least part of Seth-Smith's tenure as Curator of Birds and Mammals, Major Pam was Treasurer of the Zoological Society of London (Crandall 1939). His friend Delacour was eventually able to persuade him to construct an aviary for small birds at Wormlebury in 1922 (Pam 1923). He limited its inhabitants to finches.

*Attempts*

The Pam Brother's 1905 Violet-ear is generally considered the first hummingbird displayed in a public zoo. It was only while this paper was in preparation, that I became aware of evidence to the contrary. Marvin Jones (1928-2006) Registrar Emeritus to the San Diego Zoological Society accumulated, over most of his life, an unparalleled collection of data pertaining to animals in captivity. His archives are now in my care. Among his notes, I found a list of hummingbirds kept at one time or another by the Philadelphia Zoo. The date of first accession for Anna's Hummingbird *Calypte anna* is given, without comment, as 1874, the year the zoo opened. This is not entirely impossible. Anna's Hummingbird is the common hummingbird of the San Francisco Bay Area, and the Transcontinental Railroad was completed in 1869. The zoo opened with great fanfare, fifteen years after the incorporation of the Philadelphia Zoological Society, and it is not hard to imagine someone attempting a grand gesture.

I contacted Beth Bahner, Philadelphia's Animal Collections Manager and Registrar, who spent a considerable time searching through reports and records from the 1870s. She found no mention of Anna's Hummingbirds, but found something else quite fascinating instead: "I did find a Ruby-throated Hummingbird acquired 4 June, 1875, from PUBLIC, Kirkwood, NJ. We also have an old ledger page (in the same writing that dates back to opening) that suggests we had 194 Ruby-throated Hummingbirds in the collection between 1873-1900... At least I found something of interest". Of course *Archilochus colubris* is native to the Eastern Seaboard. How, and if, any of these were exhibited remains a mystery.

There has been some misunderstanding about London Zoo's pioneering exhibit. As will be seen, the 1905 Violet-ear was not the "first live hummingbird to arrive in Europe" (Scheithauer 1967). Another misconception was perpetuated in a 1970s traveling display on the history of zoos, emanating from the

Smithsonian Institution and shown at various zoos around the country. There was included a picture of Dickensian men and women peering at small glass cases full of hummingbirds on ornate stands in a lofty room. The same picture is featured in an essay entitled "Aviaries Past and Present" in a generally excellent book on zoos, captioned: "A striking innovation in the 1850s was the London Zoo's first hummingbird house. The birds were displayed in glass cases." (Anon. 1975). It is true that this building stood in the London Zoo in 1851 (attracting "over eighty thousand visitors" (Tree 1991). However, the hummingbirds were mounted specimens, part of the collection that John Gould was accumulating from skin dealers and his own commissioned collectors as he prepared his monumental Monograph of the Trochilidae, whose 25 parts, totaling 360 plates, were published from 1849 through 1861 (Tree 1991).

As it happened, John Gould did import a live hummingbird to England. While in Washington DC in 1857, he experimented with a Ruby-throated Hummingbird: "A *Trochilus colubris* captured for me by some friends in Washington immediately afterwards partook of some saccharine food that was presented to it, and in two hours it pumped the fluid of a little bottle whenever I offered it: and in this way it lived with me a constant companion for several days, traveling in a little, thin gauzy bag distended by a slender piece of whale bone and suspended to the button of my coat. It was only necessary for me to take the little bottle from my pocket to induce it to thrust its spiny bill through the gauze, protrude its lengthened tongue down the neck of the bottle, and pump up the fluid until it was satiated; it would then retire to the bottom of its little house, preen its wings and tail feathers, and seem quite content." It is apparently certain this bird-in-a-bag accompanied Gould on his visits to the eminent Smithsonian scientists Joseph Henry, and Spencer Fullerton Baird, and probable that it went on Gold's visit to the White House, to see President Buchanan (Tree 1991).

It is not recorded what became of this bird, but after journeying all the way to Montreal, Gould decided to take two more Ruby-throats with him when he departed New York: "The specimens... were as docile and fearless as a great moth or any other insect would be under similar treatment. The little cage in which they lived was 12 inches long by 7 inches wide and 8 inches high. In this was placed a diminutive twig of a tree, and, suspended to the side, a glass vial which I daily supplied with saccharine matter in the form of sugar of honey and water, with the addition of the yolk of an unboiled egg. Upon this food they appeared to thrive and be happy during the voyage along the seaboard of America and across the Atlantic, until they arrived within the influence of the climate of Europe. Off the western part of Ireland symptoms of drooping, unmistakably exhibited themselves; but, although they never fully rallied, I... succeeded in bringing one of them alive to London, where it died on the second day after its arrival at my house." (Tree 1991).

Regarding one possible previous arrival of living Hummingbirds in England, David Seth-Smith (1906a) commented: "There is an earlier record given by Latham, that all practical aviculturists will accept with reserve". Dr. John Latham (1740-1837) published his General History of Birds in ten installments, from 1821 to 1824 (Hopkinson 1940a). Concerning the Jamaican Mango, his account, which Seth-Smith met with skepticism, runs as follows: "... the circumstance of keeping Humming-birds alive in their own climate by means of sugar and water is well authenticated, and the following fact is also well attested. A young gentlemen, a few days before he sailed from Jamaica for England, met with a female Humming-bird sitting on the nest and eggs, and cutting off the twig, be brought all together on board. The bird became sufficiently tame, so as to suffer herself to be fed with honey and water during the passage, and hatched two young ones. The mother, however, did not long survive, but the young were brought to England and continued for some time in the possession of Lady Hammond... These little creatures readily took honey from the lips of Lady H. with their bills; one did not live long. But the other survived at least two months from the time of their survival" (Hopkinson 1940b).

David Seth-Smith (1906a) was aware of two other Nineteenth Century importations of hummingbirds to Britain. The eminent aviculturist Hubert Astley informed him that the "late Mr. Cholmondeley of Condover in Shropshire had a number of Humming Birds alive many years ago". An inquiry from Seth-Smith to "Miss Mary Cholmondeley, the well-known authoress" resulted in the following correspondence: 'My only regret is that I can give you but little information respecting the Humming-Birds which my uncle, the late Mr. Cholmondely of Condover imported in large numbers. My father, the Rev. R.H. Cholmondeley, brother to Mr. Cholmondeley of Condover, remembers the importation. We believe them to have been genuine Humming-Birds- on this point we have never had any doubt. They mostly either died on the journey or soon after arrival. One of my sisters remembers seeing some of them alive at

Condover, but only for a short time... The great difficulty was the food. I have heard my uncle speak of the difficulty, but not of how he solved it. But as, in spite of repeated trials the birds always died I fear he did not solve it" (Seth-Smith 1906a). The Reverand Astley (1915) added that these birds were kept around 1878, that "a lot of humming birds" were "kept in a conservatory, but as their food was not then understood, they died after a while; being given insects, the supply of which could not be sufficiently maintained."

The other report received by Seth-Smith (1906a) came from none other than Reginald Innes Pocock (1863-1947) who for years simultaneously held the positions of Superintendent and Curator of Mammals at the London Zoo, and was regarded as an authority on spiders, millipedes, and the taxonomy of mammals. He was also (with Seth Smith) honorary editor of the Avicultural Magazine from 1920-1923. A printing error in Seth-Smith's (1906a) article obscures what year in the 1890s Pocock's note in a December 9th Field was published, but it appears as follows; "Since the question of the importation of Humming-Birds is under discussion it seems a pity not to record the fact that in July, 1894, eleven specimens of *Trochilus cornutus*, which I presume to be the *Heliactin cornuta* of Sclater and Salvin's list, were brought over on the ss. Nile and bought by Mr. Arthur W. Arrowsmith, who sold them to the late Mr. Erskine Allon, of Gray's Inn." Nothing further is recorded regarding this reported importation of Horned Sun Gems, and it would be highly suspect were it not for Mr. Pocock's impeccable credentials.

There exist accounts of Nineteenth Century imports of hummingbirds to the European Continent. Hubert Astley (1915) states that "in 1878 there were numbers exhibited at the Paris Exhibition and offered for sale in small cages" and "some twenty humming birds were kept by a woman in Paris, perhaps some of those that were in the 1878 exhibition; and six years later 17 out of the 20 were still alive." No details are offered. The Reverend Astley goes on to mention that "as far back as June, 1876, [Alphonse] Milne Edwards [French mammalogist and ornithologist, who among other things, described the Giant Panda and the Pere David's Deer] quoted the case of Monsieur Vallet who kept about 50 humming birds, including five to six species, flying in a large aviary, but he refused to divulge the secret by which the birds were fed." Astley (1915) also mentions "one of the members of the family of Rothschild in Paris also received some, but they lived a fortnight: that was again in 1878". Finally, he states: "In 1885, a French lady received two from Vera Cruz, one of which lived for a year, and the other a little over that period", and concludes: "It seems a pity that when Mr. Pam imported some a few year ago, it was apparently not known at the London Zoological Gardens how these lovely birds had been successfully kept in captivity in Paris many years before." Jean Delacour (1926a) mentions: "... in his excellent book, Aviculture-Passeres, the Marquis de Brisay entertains us with various attempts, some of which appear to have been very successful." I have been unable to find anything more about this book. The Merquess appears to have been active in the last decades of the Nineteenth Century. It is also recorded that two arrived in Hamburg in 1876 (Schuchmann 1999).

While no details exist as to how the above-mentioned French hummingbirds might have been kept alive for year, there is one detailed account of hummingbirds kept in Italy for two months that seems tantalizingly plausible. Hubert Astley (1906) published this brief note in the Avicultural Magazine: "When paying a visit in March to Professor Giacinto Martorelli, the head of the Natural History Museum at Milan..., he told me that some few years ago several Humming Birds were kept in Milan in captivity for several months, and that they used to hover fearlessly round one's face and feed from the hand. Professor Martorelli himself saw them, and described it as the most lovely sight imaginable, adding that the beautiful little creatures remained in perfect health for some considerable time."

The Reverend Astley shortly received a reply from a Florentine aviculturist G.T. Baldelli (1906): "... In the last issue of the Avicultural Magazine, Mr. Astley mentions some Humming Birds that lived some time in Milan. I am able to add a few particulars that, I am sure, will interest your readers. On or two months ago a Genoese bird-dealer wrote to ask whether I would allow a Milanese lady to see my collection of birds and especially my Humming Birds. I wrote at once that I was perfectly willing to show all the birds I had, but that I had no Humming Birds, for they could not live in captivity. Shortly afterwards the lady called on me, and her first exclamation was 'How can you say that Humming Birds cannot live in captivity? I have kept them for some time.' Naturally I plied her with questions and she told me that when she lived in the Argentine Republic she found a nest of unfledged Humming Birds and placed it in a trap-cage. The mother fluttered round it in despair till she found the entrance to it, when she soon settled on the nest and was captured with her young. They were fed on sugar melted in water, but the lady soon found out they needed insects, and hit upon an ingenious device for cultivating small midges for them. She kept a barrel of fruit in the room, and as the fruit fermented with the heat of the room it gave birth to numberless midges, which the birds devoured. The young ones flourished and grew till they could not be distinguished from the parent birds, and lived happily for several months. She succeeded in bring them over to Milan, where they lived for two months more and were quite tame. She showed me a series of photographs of the lovely tiny creatures perched on her finger or hovering over a flower, and told me how she had nursed them and how fearless and tame they were. They could not stand the motion of a carriage, and she had to walk from the ship to the Hotel, holding them in the hollow of her hand… I have read in an old book that Budgerigars were very delicate birds, that it was almost impossible to bring them to Europe alive, as they could neither stand the journey nor the climate. Nowadays Budgerigars are among the hardiest of birds. Who knows but that in a few years time Humming Birds may be not only brought over, but bred and reared in Europe?"

*Mellin's Food*

While Signor Baldelli's optimism was rather premature, fully-documented sustained aviculture of hummingbirds was to be achieved less than a decade later.

In December, 1913, Hubert Astley recorded: "Herr August Fockelmann received in the beginning of November three Humming Birds, which only lived, a few days, and he sent me the bodies to inspect…" Both species were Cuban. Two were Cuban Emeralds *Chlorostilbon ricordii*. The third was a specimen of the world's smallest bird, the Bee Hummingbird *Mellisuga helenae*: "…One could hardly believe it to be a bird, and how this minute creature lasted out through severe ocean storms… to Germany, puzzles one. Herr Fockelmann still has hopes, when storms do not rage and when the sun shines more warmly, of successfully importing some of these wonderful little birds" (Astley 1913). August Fockelmann, based out of Hamburg, was a leading importer of cagebirds, his career ending in the Second World War.

In July, 1914, Fockelmann made another attempt at importing Cuban Hummingbirds. The Reverend Astley (1914), allowing himself a couple of Clergyman's jokes, describes this as follows: "Not long ago Herr August Fockelmann imported in July, six or seven of the beautiful little Emerald Green Humming Bird *Sporadinis ricordi*, which Mr. A. Ezra took charge of,

as they were unfit to continue the voyage from London to Hamburg, having been improperly fed so that all were almost in a dying condition, their feathers sticky (not blest!) with milk and honey. After constant nursing, not only by day but also by night, Mr. Ezra successfully brought one of the lovely little things back to health and strength, it having been a living skeleton when he received it. I saw it last month, figuratively falling on my knees to it. When the cage door is opened, this little "Hummer" darts out looking like an emerald dragon fly, whirring and hovering close to one with complete fearlessness, and the next moment away across the room to sip honey from flowers in a vase… Absolutely a gem, and causing the 10th Commandment to be broken to smithereens! Well! If a tiny thing like that can be brought back from the jaws of death, where many another species looked upon as far more robust would have succumbed, it gives one just cause to look forward to Humming Birds being kept quite easily, so long as they are imported properly and can be kept in a temperature not lower than 65 degrees."

Alfred Ezra (1944b) told this story himself thirty years later: "A dealer in Germany telegraphed to me to say that he had sent me five Humming Birds, and asked me to go and meet the small cargo boat in the Thames. Needless to say, I was thrilled at the idea of owning a live Humming Bird and rushed off to the boat at once. I had great difficulty in boarding it, but did so eventually with the help of a rope ladder. I was taken into the cabin, where I saw a small cage with five almost dead Humming Birds, not even strong enough to perch, but huddled up together on the bottom of the cage. I told the man in charge that I did not want birds which were in such a hopeless condition, but as he had taken so much trouble I gave him five pounds and took one home. By a good fire in the bird room, I warmed four very soft handkerchiefs, and then proceeded to wash the bird in warm water into which I had put a few drops of brandy. When wet, the bird was no bigger than a bumblebee. After getting all the sticky stuff off, I dried him in the warm handkerchiefs, and put him into a large cage. He at once started humming and fed on the wing. The wretched condition he was in was due to his having been fed with a twig on sugar and water, which ran all over his feathers and made them quite sticky and hard when dry. The other four birds were acquired by some other people, but only lived for a couple of days. I had no more trouble with this bird, and he moulted out perfectly twice, and was shown in wonderful condition at the International Cage Bird Show in [November] 1914… A very good colored drawing of this lovely bird, by Roland Green, appeared in our Magazine in 1915, on page five."

There would be no more German ships with Cuban hummingbirds in the Thames for a long time. Days before Alfred Ezra rescued this specimen, Arch-Duke Ferdinand of Austria was assassinated in Sarajevo on 28 June, 1914, and on 4 August, England entered the First World War by declaring war against Germany (The day after Germany declared war on France).

This did not prevent Alfred Ezra from acquiring another hummingbird in short order: "My second Humming Bird was the Garnet-throated, *Eulampis jugularis* [Purple-throated Carib], which was sent to me from France in charge of a friend. I met my friend in Victoria Station, and rushed home with the precious bird. When I unpacked the small cage in which it traveled the bird appeared absolutely lifeless, and, as far as I could make out, quite dead. I was standing by a fire and admiring his lovely colours, when I suddenly felt his heart beating. Taking him quite close to the fire, the bird, who was only torpid from the cold, soon recovered, and after a sip or two of his food, into which I had put a couple of drops of Brandy, was as active as ever, and never gave me any more trouble. This bird was also shown at the International Cage Bird Show in 1914 in perfect condition. I had him for two years, when I stupidly let him out of my window and I never saw him again" (Ezra 1944b).

This Carib had been collected by a person who remained rather mysterious to members of the Avicultural Society, styling himself as "A French Member of the Society" (1915a & b) in a two-part article for the Avicultural Magazine, an extremely detailed account of his success in bring hummingbirds to Paris, and maintaining them there. This master aviculturist eventually proved to be the Marqis de Segur (Delacour 1926 & 1955). His decision to launch a collecting expedition in February, 1914, had been prompted by his success in maintaining a collection of African sunbirds, using a formula originally devised by Alfred Ezra, who had experimented with sunbirds from India (Astley 1915; Delacour 1955; Finn 1912). Scion of a family that for centuries had served and advised the Caliphs in Baghdad, Alfred Ezra was born in India, and his brother, Sir David, also an aviculturist, was several times Sheriff [Mayor] of Calcutta. Ezra's pioneering formula consisted of honey, condensed milk, and Mellin's Food.

I had presumed Mellin's Food to be some sort of vitamin supplement, but was unaware of its exact purpose until I came into possession of the 23rd edition of the Illustrated Official Guide to the London Zoological Gardens in Regent's Park, published in 1925. The end papers and covers are full of advertisements, and the entire back cover, decorated with pictures of toy animals entering an ark, and a beaming infant, is devoted to Mellin's Food - "The Food that Feeds". Under the title "The Food for Your Baby", the reader is informed that "Cow' Milk alone is insufficient for baby. By the addition of Mellin's Food it is rendered more nutritious, easier to digest, and fit for the most delicate babe from birth. "Mixed in the correct proportions for infants of different ages, Mellin's Food provides a natural and perfectly balanced diet. Begin with Mellin's to-day and your baby will grow into a sturdy, merry, and healthy child.".

In March, 1914, the Marquis obtained two collections of hummingbirds on the island of Guadeloupe, representing the three resident species ("A French Member of the Society" 1915a). His specimens were primarily Purple-throated and Green-throated Caribs *Eulampis holosericeus*, but, after a ten day crossing, he also landed a single Antillean Crested Hummingbird *Orthorhynchus cristatus*. Of the birds that reached France, this was the first to die, during a difficult moult, after seven months in Paris ("A French Member of the Society", 1915b). His other hummingbirds went through their moults without issue, only to become extremely aggressive. Up to that point he had maintained several together, but after losses, was constrained to keep each in a separate cage. At this point, late in 1914, he presented Alfred Ezra with the aforementioned Purple-throated Carib. He also gave hummingbirds to a then 24 year old Jean Delacour, as well as initiating correspondence between Delacour and Ezra, resulting in a life-long friendship (Delacour 1955). The Marquis initially fed his hummingbirds the same way he fed his sunbirds, with metal containers with perforated lids, but by November of 1914 had switched to open glass containers.

It is of course all too well known that Jean Delacour's aviaries at Villers-Bretonneux were totally destroyed in 1918, when the Germans finally broke through the Western Front. In the meantime, however, Paris was far from the battle zone, and the Marquis continued his Hummingbird aviculture. In 1915 he made another expedition, this time to Venezuela, from whence he returned with five species of Hummingbirds, three of the genus Amazillia, as well as Black-throated Mangos *Anthracothorax nigricollis* and Ruby-Topaz *Chrysolampis mosquitus*. A Glittering-throated Emerald *Amazillia fimbriata* from this shipment lived "three years in France" (Delacour 1926a), which would constitute the captive longevity for more than a decade. The Marquis died in 1922 (Delacour 1922a), but that same year, Jacques Berlioz, Curator of Birds at the National Museum of Natural History, Paris found "a small collection of living Humming Birds in Count de Segur's aviary. Each bird was kept alone in a special cage, and several times every day was given a few minutes of freedom in a very large room" (Berlioz 1957). Jean Delacour (1926a), without mentioning a date or source, credits the Marquis with possessing the Cuban Bee Hummingbird, the first actual success following Fockelmann's above-noted 1913 attempt.

*Devil's Island*

Having established his new collection at Cleres, which he purchased in 1919, Jean Delacour embarked on his first collecting trip, a voyage to the Caribbean and South America which extended from October, 1921 to April, 1922 (Delacour 1966a).

Although he visited Guadeloupe and twice made two week visits to Martinque where he collected a number of living birds, he declined to acquire any hummingbirds there "as my late friend the Marquis de Segur had brought the three species to France" (Delacour 1922a), choosing instead to save his efforts for French Guiana.

His adventures in French Guiana are vividly and amusingly documented in two accounts (Delacour 1922b,c; 1966a), both very well worth reading. St. Laurent (The administrative center of "Devil's Island") became his base of operations, where Frank Fooks, his long-time aviculturist, cared for the growing collection. Delacour employed "a charming ex-crook who had become a fairly good collector of birds and butterflies. Commercial catching of the large blue Morpho, so numerous in French Guiana, had become one of the main occupations available to liberated convicts, and it had been built up into quite a business. We told our man that we mainly wanted live birds, particularly hummingbirds. As the use of firearms was not permitted, he had become a great expert with the blowpipe. He ingeniously replaced his normal hard clay pellets with soft ones, and he stunned the birds and brought them to us to be revived" (Delacour 1966a).

"In those days the method of caring for captive hummingbirds was still primitive, but we did well with them. Our man brought us a good many, including a few large Topazas, perhaps the most beautiful of all with their coat of glittering scarlet and amber and their long, X-shaped tail. They usually came to us unconscious. We took them in our hands and tried to make them feed. The whole point was to induce them to swallow a little of a rather thin mixture of honey, Mellin's Food and condensed milk diluted in hot water, the feed that Alfred Ezra had used so successfully for his sunbirds in India and in England... We had to plunge its bill into the mixture up to the nostrils, but after the first taste, if there still was even a spark of life in the bird, it would start eating with avidity. Until they realized that the bottle with a small, red-coloured spot was the counterpart of the blooms in which they usually fed, we had to catch the birds and make them drink every ten minutes, but most of the hummers soon found out how to feed. We caught a number of interesting birds in our conventional traps, but we could not really compete with our crook for results" (Delacour 1966a).

It is interesting to note a somewhat more vigorous account of this acclimatization which Delacour (1926a) had written forty years before the preceding report: "My birds were all captured by means of a pea-shooter loaded with a little ball of soft earth; this hit the Humming Birds, which fell to the ground stunned; they were at once picked up and brought into camp within two hours. They generally appeared to be lifeless. Every Hummingbird had then to be held in the hand and its beak plunged into a tepid mixture of Mellin's Food, sugar, honey, and milk (fresh or condensed), it usually began to drink at once; but some were stubborn; they were conquered by putting their beaks into my mouth and breathing gently on them. Should this not succeed there was another infallible plan: to plunge the whole beak and nostrils into the liquid; the bird begins to suffocate, puts out its tongue, and having tasted the mixture willy-nilly begins to drink greedily. Humming Birds are so gluttonous that even when newly caught and held in the hand they will begin to drink the delicious mixture as soon as they have tasted it." In the absence of Mellin's food, "phosphatine" was substituted (Delacour 1922c). Delacour (1926) considered "the best feeding vessel... [to be] a covered one, with a hole in the lid; the birds quickly discover and accustom themselves to use it."

"Thanks to constant attention we did not loose more than ten percent of our Humming-birds, and on leaving St. Laurent we took with us thirty birds..." (Delacour 1922c). Twenty-one reached France alive "in perfect condition" (Delacour 1926a). Eighteen were Common Wood Nymphs *Thalurania furcata*. There was a pair each of Crimson-Topaz *Topaza pella* and White-necked Jacobin *Florisuga mellivora* and a single Gray-breasted Sabre-wing *Campylopterus largipennis*. To Delacour's (1926a) knowledge, this brought the number of species certainly imported to

England or France to seventeen.

At the time of their importation, Delacour considered it essential that each specimen be caged separately: "... It is impossible to keep two in the same cage; they must be kept singly from the moment they are caught. But it is not necessary to give them a large cage, as their way of flying enables them to take exercise while, so to speak, remaining in one place.... Humming Bird cages should be furnished with only one, or two perches, so that they can fly without fraying their wings; they should be very slight. The floor should be covered with blotting paper. They should be gently sprayed with water every day, or given a wetted plant on which they can bathe. It is of the first importance that their plumage should not get soiled..." (Delacour 1926). In addition to Ezra's formula, to which he suggested adding "a little powdered meat or 'marmite'", Delacour (1926) recommended Drosophila, "easily bred in a vessel containing potato and red wine".

*Pernambuco*

In 1928, Delacour converted a pre-existing 40 foot long greenhouse at Cleres into an aviary with a netted twelve foot long viewing area, described in detail in the Avicultural Magazine (Delacour 1928). A variety of tanagers, honeycreepers, and sunbirds, with a Hooded Pitta (from Alfred Ezra) and a European Bluethroat were shortly established there, and several months later, pairs of Amythest Starlings and Little King Birds of Paradise were added (Delacour, 1929). However, there were no hummingbirds, as none were available. Delacour (1928) could only imagine they would "very likely do exceedingly well".

Likewise, hummingbirds were no longer included in Alfred Ezra's magnificent collection by 1925, when Jean Delacour (1926) described it in great detail. (Some time in the decade preceding his death in 1925, Hubert Astley included hummingbirds of unrecorded species in his aviary at Brinsop Court (Delacour 1925a & b) ). When the London Zoo opened its new bird house to much fanfare (Prestwich 1928) there were no hummingbirds among the fantastic array of soft-bills, listed in their entirety by Arthur Prestwich (1928).

There had been hummingbirds at London the year before, one of the many donations of John Spedan Lewis, the department store magnate (Seth-Smith, 1927). They had been collected in Costa Rica by Cecil Webb. Webb had begun his career as an importer of living birds in South Africa in 1919 (Webb 1954), and quickly established himself as a field collector of difficult African softbills, especially sunbirds, adapting Ezra's Mellin's Food formula, with the addition of fish liver oil (Webb 1954 p. 19). In 1927, the same year he made an expedition to French Indo-China on behalf of Jean Delacour, he made his first expedition to the New World, returning from Costa Rica on 21 August. For some reason, he did not see fit to include this trip in his otherwise richly documented autobiography (Webb 1954), making only bare mention in the foreward. Seventeen hummingbirds of five species, all believed new to aviculture (Finn 1928) were delivered to the London Zoo, where they were again exhibited in the Insect House. All were dead before April 1928, when a detailed article by the prolific author Frank Finn (1928) appeared in the Avicultural Magazine, describing some of these birds. [Finn (1912) also published observations on the Pam brother's 1908 shipment].

Frank Finn's (1928) observations prompted a note from J.B. Housden (1928), a founding member of the Avicultural Society, on a recent visit to New York, where he had seen a pair of Cuban Emeralds, resident at the Bronx Zoo for eleven months. In October, 1931, a letter from Lee F. Crandall, Curator of Birds at the Bronx Zoo, was published in the Avicultural Magazine: "... We have just lost our Ricord's Emerald Hummingbird, *Ricordia r. ricordii*, which lived here for 3 years, 4 months and 5 days. This certainly is the best American record... I am sorry to trouble you in such a trivial mateer, but wonder if you know off-hand what the best European records are..." (Crandall 1931a). To which David Seth-Smith (1931b) replied "This is certainly a splendid achievement and a record". The other specimen had lived for two years (Crandall 1939).

In response to a request from Mr. Seth-Smith for further information, Crandall (1931b) replied: "We change the food for our Humming Birds early in the morning, sometimes at noon in very hot weather, and again just before five o'clock. This last feeding is left in the cage overnight, and even in hot weather is still sufficiently fresh to be harmless to the birds before it is renewed, early in the morning. We use the usual mixture of condensed milk, Mellin's Food and honey, diluted in water. We try to keep the temperature of our building somewhere around 70 degrees, but as, of course, there is some fluctuation, we have introduced a small resistance unit into the Humming Bird's cage to make sure that there is sufficient warmth. I really attribute the loss of the specimen about which I wrote you to the failure of this resistance unit during a sudden chilly night. The bird was comatose the following morning, and continued efforts failed to revive him.

When cut flowers are available, we keep a few in the cage at all times, and during the winter months, replace them with a flowering begonia or something of the sort. When the weather is warm enough, we place the cage out of doors in the direct sunshine for an hour or two, and in the winter have made some use of an ultraviolet lamp. I am under the impression that the great amount of sunshine with which we are blessed is a most important factor in keeping Humming Birds alive. We now have a very nice specimen of Gould's Violet-eared Humming Bird, which we have had about six weeks. It is doing very nicely at present, but not having had previous experience with the species, know nothing of its potential longevity."

The New York Record would not stand for long. In 1932, Zoologischer Garten Berlin commenced keeping hummingbirds, and in 1936, three had been there "more than three and a quarter years" (G. Steinbacher 1936). These birds were a Swallow-tailed Humminbird *Eupetomena macroura*, a Black Jacobin *Melanotrochilus fuscus*, and a Plain-bellied Emerald *Amazillia leucogaster*. "Quite a large number" of other birds at Berlin in 1936 had been kept for more than a year and a half, a total of seventeen species, named by Assistant to the Director Georg Steinbacher (1936), having been exhibited there from 1932. As will be discussed, these were almost all from Brazil, but the Cuban Bee Hummingbird (on Stenbacher's list as "Trochilus helenae") was included as well. Hummingbirds were displayed in a large indoor aviary and "in three large cages, the fronts of which consist of glass panes." These were maintained at a constant temperature of 60-70F, and lit by "lamps mounted at each side in order to show off the varying colours of the birds". In winter the cages were lit until 8PM, then switched off only gradually. "Otherwise [the birds] would fly about in the dark until they were totally exhausted, and then would have great difficulty in recovering" (G. Steinbacher 1936). Potted flowers were always maintained in the exhibits, and two diets were provided daily. The morning diet was largely Alfred Ezra's with cane sugar added, while in the afternoon they received a diluted solution of only honey and cane sugar. Drosophila, raised at the zoo, were provided continuously. The longest lived of the 1932 hummingbirds died after four years and four days (G. Steinbacher 1937). In 1937, a Swallow-tailed Hummingbird received in April, 1934, died on the 24th of an unfortunately unrecorded month (G. Steinbacher 1937) - immediately after laying an egg. This was the closest any collection had come to propagating hummingbirds outside of Brazil.

Among the first foreign visitors to admire the Berlin hummingbirds was Jean Delacour, who spent several days there in November, 1932, when he noted "many Humming-birds, several in perfect condition" (Delacour 1933a). Shortly thereafter, he had a discussion with his friend David Seth-Smith: "When, some… months ago, M. Delacour told me that a certain German collector was expected to arrive in Europe towards the end of May with a collection of Humming Birds, and that of a similar collection, which arrived a year ago the majority were still thriving in Berlin, I determined that some should be procured for the Zoological Gardens and that, in the meantime, a suitable place should be prepared for their reception" (Seth-Smith 1933).

The last time London Zoo had received hummingbirds had been in 1931, when Cecil Webb made his first expedition to South America: "Mr. Webb's Collection from British Guiana, though a small one, contains some very choice birds. On another page mention has been made of the Hoatzin, of which two specimens arrived though one soon died; but in addition to these there are three Humming Birds, namely the Wood Nymph *Thalurania fissilis*, Lesson's Emerald *Agyrtria fimbriata*, and the Sabre-wing *Campylpterus largipennis*. The Wood Nymph is the smallest and at the same time the most beautiful… Mr. Webb has fed these Hummers entirely on a mixture of equal parts of Horlick's malted milk and Mellin's food, with about five parts of water, and they, or at any rate, the two smaller ones, are in most excellent plumage and condition" (Seth-Smith 1931a)

Also in 1931, the "old Tortoise House at the Zoo, which has for some time been known as the Tropical House, though it has contained nothing more exciting than a young rhinoceros, is now being transformed into a real Tropical House. One side of it will be for birds, and should represent a real bit of a tropical forest. In it, disporting themselves among the foliage and on the ground, will be Sugar-birds, Tanagers, Pittas, and such other birds as may be found to thrive in the hot, moist atmosphere" (Seth-Smith 1931).

It was in this new exhibit that accommodations were prepared for hummingbirds in 1933: "As a site for a Humming Bird aviary in the London Zoo, I could think of no more suitable place than the Tropical House. Here such birds as Sugar-birds, the more difficult Honey-eaters, Sunbirds, and so forth, have done extremely well in a damp and warm atmosphere. So a corner of this was selected for an aviary of about 18 by 12 feet. The front was screened off with plate glass, the open side covered with "Windolite", and the top below the glass roof covered with fine white mosquito netting. A ventilator near the ground admits fresh air while the exhausted air has an exit at the top. As a precaution against the possible failure of the hot-water heating supply an additional electric heating apparatus has been installed. This is thermostatically controlled, so that should the temperature fall below about 70 degrees F, the electric heater comes into action. The place has been planted with flowers and a fountain installed."

The arrival of the inhabitants of this high-tech exhibit has been charmingly documented by Captain H.S. Stokes, a specialist in softbills, the first to breed a *Ptilinopus* Fruit Dove (Stokes 1925). (Delacour (1966) commented "[His] taste in and knowledge of plants are perfect..."). His account in the Avicultural Magazine was entitled "The Cargo of Living Jewels" (Stokes 1933a). Captain Stokes volunteered to assist London's Overseer of Birds and Tropical House Keeper, James Bailey (who was to serve 44 years at the Zoo) in transporting this collection: "The jewel boat docked at Havre on 23rd May [1933], and M. Delacour, Bailey from the London Zoo, and I were the first people up the gangway to see what we had specially gone to France for, and I must readily admit that when I went into the treasure room I felt like the Queen of Sheba among the riches of Solomon - there was no more spirit in me. There they were, rows of emeralds and rubies and sapphires and topaz, flashing and iridescent, and every jewel of them a living Humming Bird, 200 of them! And not Humming Birds with frayed wings and sticky plumage flopping piteously about in dirty cages, but everyone in show condition and perfect. They were the property of a German dealer who had caught them in Brazil for distribution among the fortunate ones of Europe. They traveled in a nice warm room on the boat, in charming little wicker cages about 15 inches square and about four birds to a cage. The cage floors were covered with clean paper, and the birds were fed from glass bottles corked at the top and with a tiny up-turned glass mouth painted red at the bottom.

The food in the bottles was a mixture of honey, sugar, Mellin's food, Leibig's [meat] extract, and water, with a little powdered charcoal to keep it sweet.... The bottles were washed out three times daily with silver sand, and filled each time with fresh food. Every bird in each cage was able to take ample flying exercise, and did so without any stupid banging about or injury to itself."

Jean Delacour provided the services of his aviaries for the next stage of this endeavor: "The next thing was to sort out what we had settled to buy. Twenty each for M. Delacour and the London Zoo and a dozen more for two private aviarists were soon caught and re-caged by the nimble hands of Bailey [then 47], who took upon himself and carried out with utmost care and cleverness their safe transport to their new homes. A three hour drive by motor to Cleres, going very gingerly with all the windows closed: the arrival there just before dark when the side of the car was removed and oil lamps placed by for the treasures to feed by the gradual dimming of the lamps while fifty pairs of little wings were still buzzing; the extinguishing of the lights while four humans listened to hear if every bird was still and safe on its perch for the night; all this was something new and thrilling in bird-keeping and a great experience for me. They were housed in M. Delacour's greenhouse aviary. The next morning was fixed for our departure for London, and 4:30am saw Bailey and Fooks up and working at the birds. M. Delacour and I joined them in our pyjamas, and every bird (and a great many others too besides Humming Birds) were fed and watered. Two hours after sunrise, we were working in a temperature of 115 degrees sun and dire heat, and I fear we were much thirstier than the birds!" (Stokes 1933a). London Zoo's twenty hummingbirds crossed the channel from Dieppe, arriving at Victoria Station that evening, and then liberated into their new display the next morning. Six species, including the Swallowtail and the Ruby-Topaz were included (Seth-Smith 1933).

Jean Dealcour (1933c) had just created a new tropical house at Cleres, connected to the "primitive house" (built in 1928) by a "glass covered passage". Here, in 1933, at least seven species were contained in a special very warm compartment designed especially for hummingbirds, while at least five species lived in a cooler community flight shared with many other birds, among them two species of minivets, three species of Old World flycatchers, and a pair of Hooded Pitas (which were to breed in the near future). Along with the hummingbirds just received from Brazil, were others which had been previously been kept in other indoor accommodations (Delacour 1933b).

While Captain Stokes (1933a) did not, at the time, name the "two private aviarists" who received some of the hummingbirds he helped transport to London, he shortly reported that among "the new arrivals" at Alfred Ezra's collection at Foxwarren was "a Humming Bird from the same consignment" as those displayed at the London Zoo...".

The other recipient was the Hon. (eventually 3rd Viscount) Anthony Freskyn Charles Hamby Chaplin.

I was startled to find he was only born 14 December 1906 (Barclay-Smith, 1969). Aside from his avicultural achievements he was an excellent artist, creating several plates for the Avicultural Magazine (which he helped edit in 1935), as well as a concert pianist and composer, and served as Secretary to the Zoological Society of London from 1952-1955. Perhaps it was a combination of youth and aristocracy that caused him to make a radical break from accepted hummingbird aviculture. In light of what he did, it is amusing to read Captain Stokes' (1933a) admonishment, which concluded his account of the Hummingbirds from Le Havre: "I have since heard from M. Delacour that his are quite happy and doing well in his greenhouse, which, is far hotter than the London Zoo cage. So, let no moist and perspiring humans, eschewing Turkish bath atmospheres, try fresh-air cures or Spartan treatment for the little Hummers, or our treasures will die and our experience and our labour of love be wasted." One of the Hon. Anthony Chaplin's experiments was accidental, the other entirely deliberate.

The accident is described in a letter to the Avicultural Magazine by the distinguished avicultural historian Emilius Hopkinson (1933): "Yesterday (18th July) I was rung up by a gentleman in Cuckfield, about 4 miles away, to say that he had got a Humming Bird which had been about a day or two, and asking if it was mine and what to feed it on. I went over to see it and found it was greedily sucking up Nestle's milk from blossoms stuck into the wire of a canary cage into which it had been put. I, of course, was able to tell the finder that the owner must be Mr. Chaplin, and he at once took it over to him, and I hear this morning that it is none the worse for its long outing and the recent weather. It is a Swallow-tail *Eupetomena mocroura*, one of the two of this species which Mr. Chaplin obtained when the Zoo got theirs at the beginning of June. It escaped the day it was brought here, which was about 8th June, so that it has been on its own for more than a month in spite of the heavy rain and cold and in that time has not only exited but kept fit, and must also have covered quite a lot of ground in the time it has been seen… at Bolney, 7 miles away and other places….Mr. Chaplin promises full details later, but I think it worth while putting on record at once this almost unbelievable (but absolutely proved) adventure of a Humming Bird - a second Balcombe miracle I call it, the first being the same owner's Ruby and Topaz, which has been out-of-doors in a small aviary for about the same time, and that without any worse effects than occasional temporary torpidity."

The Hon. Anthony Chaplin (1933a) provided a table of night-time temperatures from 23 June through 12 July, 1933, during which he kept his Ruby-Topaz in an "all open outdoor aviary". These were mostly in the 50s F, but ranged from 44F to only 58F. He had been prompted to place this bird in this cage after the following experiment: "At 10:45pm on 21st June the two Humming Birds, *Eupetomena macroura* and *Chrysolampis moschitus*, were torpid at a temperature of 66F. Now I have always suspected that this state is not induced primarily or necessarily by a low atmospheric temperature. I therefore deliberately exposed the birds to more cool air (by opening an additional window), but left the electric light burning, after moving the birds gently with a stick so that they hung upside-down like bats, on their respective perches. Within twenty minutes the eyes of each were open, and in a very few minutes more one after the other flew to have a feed. I think this proves that the torpid state is not necessarily brought on by cold, and that it is probably a natural state of rest… When I say that the birds were torpid, I mean that they showed to all appearance no spark of life; they could be moved about and laid out on a table like as many dried skins. But what is, I think, important is that they should be so smooth in feather that they appear to have been stuck dead suddenly while asleep without having had time to alter their positions: all evidence to show that it is a natural condition. I have no doubt that it is due to this habit that the Trochilidae are supposed to need a very high temperature in order to be maintained in health. For my part I have but little hesitation in submitting them to the ordinary atmospheric conditions provided for most foreign birds."

Unfortunately after two months in its aviary, the Ruby-Topaz "escaped through the bars of a cage through which it had been placed temporarily" (Chaplin 1933b). Having given away other birds, Chaplin (1933b) had by late 1933, only a single bird from his share of the London Zoo shipment, a Blue-chinned Sapphire *Chlorestes notatus*, "so robust, owing, I consider, to its treatment on arrival, that it will stand 45F without ruffling a feather and is a veritable atom of ceaseless energy." The treatment in question was Chaplin's "spending a week's toil catching small hedge spiders on wich to feed it, for once its natural standard of health is re-established it can be subjected to a less complicated treatment." His belief was that too sudden a transition from a wild diet was a major source of hummingbird mortality.

The identity of the "German dealer" who landed the 200 hummingbirds in Le Havre in 1933 can now only be guessed at. It might have been Lothar Behrend, based out of Buenos Aires. A detailed and colorful description is given by his protégé, the 1950s celebrity animal catcher Peter Ryhiner (Ryhiner & Mannix, 195: 80-81, 92-93). From 1928 to 1939 Behrend delivered enormous shipments of birds and other animals to Fockelmann and other European dealers, and held a reputation for maintaining sunbirds and hummingbirds. Made bankrupt during the Second World War, he re-established his business and remained active at least until 1951, when he supplied Vermillion Flycatchers to Joachim Steinbacher (1952), the only mention I have found of Behrend in the Avicultural Magazine.

From comments made later by Delacour (1943), it is more likely this "German Dealer" was a Swiss, based in Pernambuco (Recife), Brazil, who, at any rate, quickly became well known to aviculturists in the 1930s. Charles Cordier (1897-1994) was introduced to readers of the Avicultural Magazine in a detailed account by Mrs. N. Wharton-Tigar (1934, 1998), a Fellow of the Zoological Society and an aviculturist who achieved the first breedings of the Vinaceous Firefinch and Black-crowned Waxbill. Mrs. Wharton-Tigar, accompanied her husband to Brazil in September, 1933, and returned in November with a collection of twelve species of hummingbirds, three new to British aviculture. All of these were for the London Zoo, James Bailey meeting the ship at Southampton, although she retained quite a collection of softbills. Cordier had at least a hundred hummingbirds on his premises at the time of her visit, "all booked to go to Germany via Havre in a few days." Cordier was able to supply Mrs. Wharton-Tigar with freshly-caught birds on short notice. She was also able to acquire hummingbirds from a Brazilian whom Cordier had trained.

A few months later, Captain Richard de Quincey (1935, 1960) arrived in Pernambuco, purchased 36 hummingbirds, and landed 33 of them in England, in April, 1934. His only losses were three of the tiny Reddish Hermit *Phaethornis rubber*. A fourth specimen lived five weeks in "a separate case in the Tropical House" at the London Zoo (Seth-Smith 1934c). Captain de Quincey names six other species which he donated to the Zoo, in a 1960 article, where he unfortunately repeatedly gives their date of arrival as 1932. However, it is plain from his 1935 article, as well as from David Seth Smith's notes on the Reddish Hermit, that his collection of Hummingbirds did arrive in 1934.

Two birds from this shipment did not go to the Zoo. The one Long-tailed Wood Nymph *Thalurania watertonii* was given to Alfred Ezra. Commenting on the 1935 Crystal Palace Show, Anthony Chaplin (1935b) noted: "...the star turn was provided by... Mr. Ezra's lovely Waterton's Wood Nymph, a species still rare in museum collections and in Gould's day known only from the type specimens." This bird escaped after Ezra had had it for seven years (de Quincey 1960). Ezra (1943a) himself wrote: "... I was very fond... of a Waterton's Wood Nymph..., which I kept in perfect health for a long time, and only lost him by stupidly letting him out of a window. This bird was the most fearless one I had ever kept, and used to fly and skim over my bath when I was in it, and fly out again like a swallow. When let out of his cage he flew about at a terrific pace, and rushed back into the cage without ever settling on anything in the room. If I held his little bath in my hand he would at once fly out of his cage, and wash himself in this without being the least bit frightened. I could have cried when I let him out, and I was very grateful that I did the stupid thing and no one else." (Apparently all of Alfred Ezra's hummingbirds were caged individually as pets. This is somewhat odd, when one considers that he achieved a truly phenomenal series of first breedings at Foxwarren - Prestwich (1946) lists 23 species bred for the first time in Britain, from 1925 through 1939, and there are some omissions, such as the Bali Mynah.)

The one hummingbird that Captain de Quincey retained from the 1934 Pernambuco collection also lived seven years, then froze to death, having spent all its English winters in an outdoor aviary. Anthony Chaplin (1935a), who of course had kept hummingbirds out of doors during an English summer, went on to successfully keep African Sunbirds out all winter. He suggested "certain hummingbirds, too, would very probably winter in the open in a sheltered aviary", obviously unaware that this had just been accomplished. de Quincey (1935) discussed its first winter in detail. While this Swallow-tail had a heated shelter where its one feeding bottle was maintained, he eventually discovered that it rarely if ever used it, roosting instead high on exposed twigs. "... It could have been said to have lived without artificial heat all that time. I have never had one do this since. Eventually on a very frosty night I think it was scared by something and flew against the wire-netting, and was found frozen there the next morning" (de Quincey 1960).

At the time of his 1934 importation Captain de Quincey did not have a tropical house, but had built one and stocked it by mid 1935. The temperatures in this facility could fall to 45F, and there were several day periods when they stayed in the 50s. He did not note any discomfort from the hummingbirds or other species. In 1935 he listed seven species of Brazilian hummingbirds, most having arrived in a shipment to Europe escorted by Charles Cordier himself (de Quincey 1935). At the same time he was receiving birds from a Mr. Read, an employee on the ship he had traveled on in 1934, who was to make several shipments in the '30s (de Quincey 1960). A species de Quincey donated to London Zoo in late 1934 was the Racket-tailed Hummingbird *Discosura longicauda*, which was also obtained by Delacour for the first time that year (Seth-Smith 1934b), and apparently new to aviculture. Yealland (1956c) notes that London Zoo received five in 1934: "Two were purchased and three presented by Capt. de Quincey. Four of these did not live long, but the fifth lived for two-and-a-half years when it died of senility."

The arrival of hundreds of healthy hummingbirds at once in Europe during the early 1930s of course meant other zoos experimented as well. Scheithauer (1967) describes a specially built "pavilion" at Leipzig, at around this time. When the Rome Zoo was reopened in 1934, with Jean Delacour acting as consultant, Cordier was one of several persons Delacour commissioned to stock it (Delacour 1966, 1962.)

The most well documented European zoo hummingbird aviculture in the 1930s was conducted at Copenhagen. Axel Reventlow (1948), Copenhagen's Director, listed each specimen kept there, from the first consignment of 22 received 17 May, 1935, to a shipment of ten received 8 July, 1938. There were a total of 74 specimens in six consignments, all purchased from Fockelmann. At least three of these shipments had been collected by Cordier. Twelve species, one represented by two subspecies, as well as one unidentified species were represented. In a table, Reventlow (1948) presented the longevities of every one of these, as well as recording the causes of death as best as he was able. After 1940, a number of mysterious deaths were postulated as due to tapeworms. In 1937, while Mellin's Food was unavailable, the substitution of Horlick's Milk appeared to cause mortalities. Behavioral notes were included as well. The best record among these birds was a Glittering-bellied Emerald *Chlorostilbon aureoventris*, which lived at Copenhagen from 18 June, 1937 through 6 May, 1945, 13 days short of eight years. In all, 28 of the 74 hummingbirds lived more than a year (38%). Eight of those lived more than two years, and five lived more than three. Most were kept in large, well-panted community aviaries.

A number of birds were appearing in private aviculture at the same time. David Seth-Smith (1934a), reporting on the Crystal Palace show of 1934, noted: "I have seen one Humming Bird before at a bird show, but at this show, there were no less than seven... The seven Humming Birds were the star of the show, although I hope that Hummers will not be shown too freely, because they are only suitable where special provision can be made for them. Here they were in a small and specially-warmed room, but even so, some of those shown did not look too happy. The two exhibited by Mr. Hopkins had become more or less acclimatized, and they certainly looked very fit. The first prize went to his Ruby and Topaz, the second to a Black-throated Mango." At least one English bird dealer, Percy Hastings, of Portsmouth, imported hummingbirds during this period, having commenced in 1924 (Hastings 1967).

*Meanwhile, in America...*

Across the Atlantic, there was little if any impact from Fockelmann's and Cordier's efforts in the U.S. The National Zoological Park's first species of hummingbird was the Cuban Emerald, obtained in 1930, while the record-breaking Bronx specimen of the same species was still alive. No other species were added to that collection until the 1940s (Tongren 1989). The notes of the late zoo historian Marvin Jones document that the venerable Philadelphia Zoo obtained the wide-ranging Plain-bellied Emerald *Amazilia leucogaster*) and something designated "*Argyais viridissimus*" in 1936.

Lee Crandall (1939) summed up the Bronx Zoo's hummingbird aviculture through the 1930s: "... We have usually exhibited Hummingbirds of one species or another. We kept them singly in glass cases about two feet in each dimension, well ventilated and heated by electric devices, though without accurate thermostatic control. Always we thought of a cage built after Jean Delacour's greenhouse plan but always we hesitated. Hummingbirds are notoriously quarrelsome, and at close quarters will fight like fiends. We could not risk a failure and this seemed a likelihood until the problem of maintaining a friendly and contented group was solved...". As will be later described, a community exhibit of hummingbirds was not established at the Bronx until 1939.

In the meantime, however, other American Zoos were creating hummingbird exhibits. The St. Louis Zoo's Bird House, opened in 1931, was highly innovative in its use of glass-fronted, planted aviaries. By 1936 one of these contained a collection of hummingbirds (Delacour 1937a) as well as honeycreepers (Masure 1937). While I have been unable to find what species were exhibited then, there is ample documentation of the collection in the bird house at the Brookfield Zoo (Chicago Zoological Park), which was opened to the public in 1934. Its elaborate Perching Bird House was opened in 1935.

Brookfield's Curator of Birds, Karl Plath (1937) published a detailed description of its exhibits for the *Avicultural Magazine*: "Before leaving the Perching-bird House mention must be made of the plate-glass cage built for the Humming-birds. Eight species are represented, Allen's, Anna's, Gilt-crest, Ruby, and Topaz, [sic] two species of the Sapphire, and two species of the Emerald. This attractive exhibit is very popular with the visitors. The cage is 13 feet across by 8 feet high and deep and is planted with bamboo and an assortment of flowering vines which impart a tropical jungle effect. The several food bottles are hung in various locations so as to give the birds a choice of feeding places. A mist-like spray is turned on once a day during the winter months; we found it best to watch during this time as the birds sometimes became water-logged, and could not raise again. The temperature is maintained at 80F. While they require a stong light to show up their brilliant refulgence, their feeding bottles are place at eye level so tif the spectator has any patience at all, sooner or later one of these gems will flash in front of him and display his glittereing gorget or crest. These tiny creatures display great skill in capturing the infinitesimal fruit-flies which we release in the cage. While their needs are great comparatively, an aviary of these feathered gems is well worth while judging from the enthusiasm shown by the public, which after all is the criterion".

Aside from the afore-mentioned questionable 1874 record for Philadelphia, the Allen's *Selasphorus sasin* and Anna's Hummingbirds *Calypte anna* at Brookfield appear to be the first California hummingbirds displayed in a public zoo. Both species came from the collection of Eric Kinsey, an aviculturist specializing in Western North American birds in the days when the government was far more liberal in granting permits to private citizens for noncommercial avicultural purposes. Mr. Kinsey and his wife lived in Marin County, near San Francisco, and were active from the 1920s into the '60s (Delacour 1942, 1950, 1961; Prestwich 1957). He claimed the Anna's he sent to Chicago were "the first species of bird to be shipped by airplane". (Kinsey 1939). (Fockelmann flew a shipment of Cordier's Pernambuco hummingbirds from Paris to Hamburg in May, 1935 (Reventlow 1948) ). Kinsey also sent Allen's and Anna's hummingbirds to Japan, the first living hummingbirds in Asia. Two shipments in 1937, and a third in 1938 were all received by the ornithologist, the Marquis Yamashina, who designed a new sort of feeding bottle with a curving "elephant spout" (Kinsey 1938). Kinsey (1939) maintained, at one time or another, six species of California hummingbirds, summing up his experiences with each in an article in the journal of the Avicultural Society of America. They were maintained indoors, in specially designed, rather small glass fronted cages, in units of five. He was usually able to keep at least two to a cage.

Jean Delacour (1937b), visiting Long Beach, California, found "Mrs. S. Tomlinson, who has given up most of her birds after her husband's death, still has, in perfect condition after more than four years, her wonderful male Anna's Humming-bird, which lives in a large sheltered aviary, built along the wall of the house, full of creepers and bushes." Susie Tomlinson (1934) had already maintained Anna's hummingbirds when she received this bird in 1933. Eric Kinsey (1939) referred to "a constant source of insect food in the outdoor vine-shrouded aviary which he occupied."

A curious footnote to the aviculture of California Hummingbirds in the 1930s were the experiments of the wealthy American, Eastham Guild, who released 9,000 birds of at least 54 species (Guild 1944) in Tahiti. Most were estrildids (including 900 Gouldian Finches), but among other species were birds he had collected himself in California (Guild 1940), including Mountain and Western Bluebirds, Lapis Lazuli Buntings, Pygmy Nuthatches and Anna's Hummingbirds (Guild 1944). There is no evidence that the hummingbirds reproduced, though Western Bluebirds did for several years. It appears the only species now established in Tahiti which can be clearly traced to Guild's importations is the Crimson-backed Tanager *Ramphocelus dimidiatus* (Pratt et al 1987).

One might imagine, with all the deliberate, (and government sanctioned) introductions of such birds as North American, Red-crested, and Yellow-billed Cardinals, and the Western Meadow Lark to Hawaii in the 1920s and 1930s, with the result that residential areas of Hawaii resemble a walk-through aviary, that hummingbirds might have been among the attempted species. I have heard rumors to that effect. However, due to fears of cross-pollenation of commercially valuable strains, the powerful Pineapple Research Institute saw to it that the introduction of hummingbirds and other nectar-feeders was discouraged (P.

Breese pers. comm.).

The "Gilt-crests" at the Brookfield Zoo were Antillean Crested Hummingbirds *Orthorhynchus cristatus*, which arrived from Dominica (Chicago Zoological Society 1939). This species had become rather widespread in aviculture, again, thanks to Charles Cordier. Captain de Quincey (1960) reminisced: "Then on one occasion Monsieur Cordier brought a great number [of hummingbirds] over, mostly for the Continent, but his ship docked at Plymouth for a couple of hours and I managed to be there waiting for him, and on that occasion to get some beautiful Garnet-throated [Purple-throated] Caribs *Eulampis jugularis*, and other lovely West Indian and Venezuelan species, Gilt Crests *Orthorynchus exilis* among them". Perhaps this was the shipment from which Axel Reventlow flew nine Antillean Crested Hummingbirds from Fockelman's Hamburg establishment to the Copenhagen Zoo, 22 June, 1936. Reventlow (1948) mistakenly records that Cordier collected these birds in "French Guiana and the Brazils", *Orthorhynchus* being an entirely Caribbean Island species. It is likely by then, that Cordier's days of escorting hundreds of hummingbirds from Pernambuco to Europe were behind him.

*Post-Pernambuco*

Mrs. Wharton-Tigar (1937) describing the hummingbirds then resident at London Zoo, found "the collection... rather low at the present time, comprising only six species", going on to not that "owing to restrictions being placed on catching birds by the Brazilian Government, the supply is likely to be almost cut off from this source, which gave the Zoo many very lovely examples in the past...". Two of these species were Antillean-cresteds and Purple-throated Caribes, which Mrs. Wharton-Tigar specified were collected in Martinique. The others were Brazilian. Among these was a White-throated Sapphire *Hylocharis cyanus* ("a cage and aviary bird... well known to many of us"), at London since November, 1933. Although she did not say so, one would imagine this was the last survivor of Mrs. Wharton-Tigar's Pernambuco shipment. On the other hand, Alfred Ezra donated to London Zoo some of the specimens that Cordier brought to Le Havre around the same time (Wharton-Tigar 1934). Aside from a Swallow-tail in a community flight in the Bird House, these hummingbirds were maintained in the Tropical House (where, by then, "tiny banana flies [were] specially bred and liberated there...").

It is not surprising, considering the volume of hummingbirds alone being exported to Europe in the early 1930s, that Brazil should have enacted a total ban on bird trapping in the middle of that decade. This included not only exports but domestic commerce as well. When the globe-trotting aviculturist, Sydney Porter, traveled to Brazil in 1938, he "visited several bird shops in Rio and Santos but found that their stock-in-trade consisted of mainly of Canaries, Java Sparrows, various species of African Waxbills, and Weavers, and the commoner varieties of Pheasants. One man did show me two Black-headed Siskins and a dying Grackle which he kept under his writing desk for fear the police should see them" (Porter 1938).

Towards the end of 1937 Cecil Webb went to Ecuador, from whence living hummingbirds do not appear to have been previously exported. He wrote two lengthy accounts of this expedition, with detailed discussions of some of the hummingbirds collected (Webb 1939, 1954). Birds were collected both by nets and an Ecuadorian blow-pipe hunter who used clay pellets. Webb and his niece, Delys, assembled a comprehensive collection, only to nearly loose it at the end of the expedition, in early 1938. On the final train journey from Quito to the coastal city of Guayaquil, the special compartment containing the birds was connected to a different train in the middle of the night. Webb discovered this at 5am, but was unable to catch up with his collection until 11am, by which time all the hummingbirds had gone into torpor. In his 1939 account, he recorded: "Some of these I managed to revive, but quite a lot succumbed". On the other hand, in his 1954 account, he wrote: "After Delys and I had worked furiously for an hour, we had revived practically the whole lot."

A more interesting contradiction exists between both of Webb's accounts and the reminiscences of Captain de Quincey (1960). Webb (1939, 1954) made concerted attempts to maintain the Sword-billed Hummingbird *Ensifera ensifera*. He found the standard Melline's Food mixture "acted just like a poison to these specialized feeders and after having a good feed they would roll over dead." (Webb 1939). By "completely changing the diet" (in an unspecified way), he was able to keep one alive for three weeks. One gets the impression from this, and the 1954 account, that he did not bring any back to England. However, Captain de Quincey (1960) wrote the following: "Then Mr. C.S. Webb went collecting in Colombia, on which occasion there was that tragic mishap when two railway vans, containing the greater and rarer portion of his wonderful collection, were shunted into a siding, with a consequent eight-hour delay, during which time many rare things were lost, but he did return with a few Humming Birds, and I had two most beautiful ones from him - a charming Hill Angel *Heliangelus clarrisse* and a Swordbill *Docimastes longirostris* [which on the same page is also listed as "*Docimastes ensifer*"]. The latter, with its beautiful slow wing-beats, was a most exciting thing to watch in an outdoor aviary that summer, but it faded after four or five months when wild-caught insects became scarce. Owing to the upward curve of its enormously long bill, we hung its feeding bottles at an angle, which courtesy it seemed to appreciate. It was very fearless and gentle".

Jean Delacour commissioned Cecil Webb to collect for him in Vietnam in 1927, and Madagascar in 1929 and 1935 (Webb 1954). After Charles Cordier was constrained to give up his base of operations in Pernambuco, following the Brazilian bird trade ban, Delacour sent him on expedition as well, first to Guatemala in 1938, then to Vietnam in 1939 (Delacour 1939a). The main objective of the Guatemalan expedition was Ocellated Turkeys, of which Cordier brought twelve to Cleres. However, seven species of hummingbirds arrived there as well (Delacour 1939a). Among these were the first Magnificent (Rivoli's) Hummingbirds *Eugenes fulgens* in aviculture, and *Lamprolaima rhami*, the Garnet-throated Hummingbird. During these years, the Purple-throated Carib *Eulampis jugularis* was also known as the "Garnet-throated Hummingbird", so that confusion may result. However, one may presume the "pair of Garnet-throated Humming Birds" which "built a lovely nest in an hibiscus tree, but did not go any further", in Delacour's tropical house at Cleres (Delacour 1939b), really were the *Lamprolaima* which Cordier brought from Guatemala.

*Destruction and Creation*

Aside from the afore-mentioned egg produced by a Swallow-tail which promptly died at Zoologischer Garten Berlin in 1937 (G. Steinbacher 1937), this 1939 nest-building at Cleres was as close as anyone had come to breeding hummingbirds in captivity outside of South America. Circumstance prevented anything further. The Second World War commenced in September, 1939. By November of that year, all of Clere's gardeners and four of the bird-keepers had joined the Army. Captain Delacour himself, though on military duty, "by the doubtful privilege of age" (he was then 49) was allowed to remain in the district (Delacour 1939). The tropical house (where an Elliott's Pittas and seven Indo-Chinese Shamas had been hatched and raised that year) was closed down. Delacour (1939b) wrote: "Now, all the birds have been removed from greenhouse to bird-rooms, as it would have been very difficult to obtain coal to heat them during the winter. The rare plants have been deposited at the Rouen Botanical Gardens, and I hope to be able to replant the house when the war is over. It was very heart-breaking to me to close these houses, which were so attractive in every way: this is the only noticeable difference that the war has made to Cleres so far".

On 7 June, 1940, Captain Delacour was ordered away with his Army unit, so saw Cleres for what was to be the last time in more than six years. German troops arrived shortly thereafter and for several days the collection was unattended. By the time the Germans allowed three of the keepers, "who had remained near by, to return and care for what was left of the collection... naturally in the interim all the more delicate species, including the Birds of Paradise, the Humming Birds, Sunbirds, and those of insectivorous habit, had died of starvation" (Dealcour 1941).

In England, rationing was imposed 8 January, 1940. Captain de Quincey, who had maintained at least 22 species of hummingbirds from 1934 through 1939, soon found "honey, Nestle's milk, and Mellin's food became almost unobtainable" (de Quincey 1960). None the less, the Swallow-tail he brought back in 1934, and had kept out of doors ever since, lived until 1941 when it was found frozen against the netting. He still possessed the "*Heliangelus clarisse*" (Amethyst-throated Sunangel *H. amethysticollis*) which Cecil Webb had brought from Ecuador in 1938. Of this bird he wrote: "The Hill Angel was my favourite of all. He escaped from his greenhouse one day when snow was falling, and I imagined him lost forever, but three days later I received a telephone call and learnt that a small bird was hovering round some clumps of "Wanda" primroses in a garden in the village about a mile away. Snow was lying on the ground, but these vivid little flowers were showing just clear of the snow. Hardly daring to hope that it was my most precious *Heliangelus* I hurried there with an all-wire cage and a feeding bottle. 'Helio' came straight to me and started drinking hungrily form the bottle, which I gently advanced to the door of the cage, and the bird, feeding without a break, was gently but surely manoeuvred, flying backwards through the door of the cage, and was safe once more. In winter he would come and pull out tiny pieces of fluffy wool from a scarf I used to wear, and was always friendly and without fear. This little bird outlived all my other Hummers, and was the last nail in the coffin of my bird-keeping activities for a good many years..." (de Quincey 1960).

At the end of 1939 Alfred Ezra (1940a) wrote: "It will interest members to know that I still have two Humming Birds that are still in perfect condition after four and a half years.

One is a hen of the Racket-tailed Humming Bird *Discura longicauda* and the other the "Garnet-throated" *Eulampis jugularis*. Although the "Racket tail" is beginning to show signs of old age, the "Garnet Throated" has just moulted out, and looks more beautiful than ever". The Racket-tail died at the end of July, 1940, having been obtained by Ezra (1940b) in October, 1934. He had never fed it live food. The *Eulampis*, which had also arrived at Foxwarren October, 1934, died 11 November, 1942 (Ezra 1943). This bird had been shown repeatedly at the National Cage Bird Show, where Mrs.Wharton-Tigar, (1938) described it as " a gem of the first water". As of early 1943, Ezra still kept one hummingbird, a Sparkling Violet-ear he had had for nearly five years. This bird died in 1944, having been at Fowxwarren over six years. Ezra (1944a) thought it the last Hummingbird in England.

During a 1940 air raid, the London Zoo's last three hummingbirds escaped through a hole in the roof of the central aviary of the Bird House caused by an" incendiary bomb" which otherwise "spluttered out harmlessly on the floor" (London Times, 1940, Seth-Smith 1943). These were considered the zoo's only air-raid casualties. On the other hand, Zoologischer Garten Berlin's bird house was one of several zoo buildings totally destroyed by allied bombings in 1943. In Japan, where three years of war with China was creating a fuel shortage, the Marquis Yamashina, in 1940, entrusted his California hummingbirds to the Marquis Hachisuka, whose conservatory in Atami was heated by a hot spring (Hachisuka 1940). Lothar Behrend spent the war years wandering through Patagonia eating stolen sheep (Ryhiner & Mannix, 1958). Cecil Webb was stranded in Madagascar, where the local government sided with the collaborationist Vichy French. Until he was picked up by an allied ship in 1942, he largely lived off the land, gathering intelligence and museum specimens (Webb 1954).

Charles Cordier similarly found himself stranded in Venezuela by the outbreak of war in 1939. It eventually came to his attention that Jean Delacour was in New York, employed as Technical Consultant to the New York Zoological Society at the end of 1940 (Delacour 1966; Lindholm 1988), having escaped France through a circuitous route involving Morocco and Lisbon, and being met at the dock by Lee Crandall (Delacour 1966). Cordier contacted him in 1941, and he was soon commissioned to collect for the Zoological Society. Thus began a particularly vibrant chapter in the history of hummingbird aviculture.

As mentioned earlier, the Bronx zoo, after years of keeping individual hummingbirds in smallish cages, opened a planted community aviary for hummingbirds in May, 1939, a donation of Mr. John D. Rockefeller. Its initial inhabitants were five hummingbirds of two species from Venezuela (Crandall 1939). By the following year, keeping this exhibit stocked became a source of anxiety, as foreign species were scarce due to wartime conditions. Through 1941, the zoo largely depended on local donations of Ruby-throated Hummingbirds found exhausted during migration (Bridges & Hollisher 1941, pp.47-49).

On 9 December, 1941, two days after the attack on Pearl Harbor, Charles Cordier arrived in New York with 24 species of birds collected in the highlands of Western Colombia (Barclay-Smith 1942; Bridges 1942; Lindholm 1988). The stars of this collection were a dozen Scarlet Cocks-of-the-Rock, the first Andean Cocks-of-the-Rock to certainly reach aviculture. However, the collection of 22 hummingbirds of eight species stood out in that five were species new to aviculture. These included five Violet-tailed Sylphs *Aglalocercus coelestis*, five Booted Racquet-tails *Ocreatus underwoodii* and two Brown Incas *Coeligena inca*.

On the occasion of Cordier's second Bronx Zoo expedition Jean Delacour (1943) wrote: "For more than ten years it has always been a thrilling experience to see Mr. C. Cordier arrive with a collection of birds. It happens once or twice a year. Until 1939, he used to bring his collections to Cleres, and the surplus material found its way mostly to the London Zoo, to Mr. A. Ezra and Mr. Spedan Lewis in England, and to Dr. E. Beraut and M. Francois Edmond-Blanc in France. These happy days are over, alas! But Mr. Cordier still brings his collections, now to the United States. He brought us, to the New York Zoological Park, a marvelous Colombian collection in December, 1941, and early in October, 1942, he was back again, this time from Costa Rica, with perhaps the finest lot of birds he has ever secured. There are ninety-six birds in the collection, including three Umbrella Birds which have never before been exhibited alive. Fifty-four Humming birds of which the majority have never been imported anywhere, and eighteen Quetzals..."

Regarding the Costa Rican hummingbirds in particular Delacour (1943) noted: "The fifty-four Humming Birds in the collection comprise eleven species, at least seven of which have never been exhibited in any bird collection before. Although larger shipments of Humming Birds, in point of numbers, have been made by Mr. Cordier to Cleres, in its variety and beauty, the present lot is the best that has ever been made from the highlands of the American tropics".

General Curator Lee Crandall considered "perhaps rarest of all" in this shipment to be the single Costa Rican Snow-cap *Microchera albocoronata parvirostris* and the three Black-crested Coquettes *Lophornis helenae* (Ditmars & Crandall 1945). Another avicultural novelty was the Fiery-throated Hummingbird *Panterpe insgnis* of which twelve specimens were imported. Cordier was forced by contingency to contrive a novel method of hummingbird transportation: "Late in September he brought his birds out of the jungle to San Jose. Space was made on a plane for Miami- but at the last minute it transpired that there was no room for all the light but bulky Humming Bird cages. Mr. Cordier thereupon raided shoe stores in San Jose, got together a score of pasteboard shoeboxes, and divided them into compartments. He filled each compartment with loose hay, packed the Humming Birds in the hay, and they traveled with perfect comfort. The hay kept them from fluttering too much and gave them something to cling to. Not a single bird was lost on the trip except by a serious accident at Tegucigalpa, Honduras, where the plane was forced to remain in the full sun for several hours. Some birds suffocated despite everything Mr. Cordier could do. Fortunately, none of the rarer species was lost. All the birds... a list of which follows, arrived in extraordinarily good condition, and more than a fortnight after their arrival none had died." (Delacour 1943).

The arrival of these monumental collections of birds entailed major renovations at the New York Zoological Park. Exhibit design was included among Jean Delacour's responsibilities in his position as Technical Consultant to the New York Zoological Society ("...the first time in my fifty years that I had been paid for my work, except when I was on military service. It felt rather strange, but it was satisfactory to realize that I could live on what I earned if the necessity arose" (Delacour 1966)).

"When the astonishing circumstances under which we have all been living brought me to New York, I found a good chance for experiment. I talked things over with my old bird friends Fairfield Osborn, President of the Zoological Society, and Lee Crandall, Curator of the Zoo... (Delacour 1945). "The Bird House in New York consists of three halls, the largest of which has not been much altered these last years. We have only redecorated the big central flight and some of the compartments. The second room has been completely changed in 1942. The numerous cages and small compartments for parrots and doves have been removed, and five roomy flights have replaced them. They are decorated and planted so that they now form the 'New England Garden', for native species; 'Arid Plain' for desert birds; 'Indo-Malayan Jungle' for Asiatic species; 'Tropical American Jungle Stream' and 'Tropical American Rain Forest' " (Delacour 1946). These pioneering habitat groups were eighteen to twenty feet in length, six feet wide, and eight feet high, standing two and a half feet above the floor (Delacour 1945).

Delacour described some of them in greater detail: "One has a fast-running stream and is called the 'Tropical American mountain stream'. It contains some Blue-headed and Ruddy Buntings, a small South American Barbet, a dozen Manakins and Sugar Birds, a few small tanagers, and a pair of Fire-throated Hummingbirds (*Panterpe*)" (Delacour 1943). "The Arid Plain or desert cage is inhabited by birds adapted to life in dry countries. It contains some Asiatic Pratincoles, African Sandgrouse, Egyptian Plovers, Diamond and Plumed Ground Doves, many species of Australian Grass Finches and African Waxbills, different Larks, some Galapagos Finches, and two Costa's Hummingbirds from the Californian Desert. Succulent Plants... have to be replaced about three times a year, as the glass roof of the house is too high above them and does not afford them sufficient light. But, at that price, the show remains excellent. Needless to say, the birds do exceedingly well and a number have nested successfully. The 'New England Garden' is a little more difficult to keep in good condition, as hardy plants do not last very long indoors. They have to be replaced four times a year. We show in this cage a selection of our local small birds: Ruby-throated Humming Birds, Bluebirds, Hermit, Russet-backed, Veery and Wood Thrushes, Catbirds, Baltimore and Orchard Orioles, Cedar Waxwings, Purple Finches, different Warblers, Sparrows, Nuthatches, small rails, etc. To make and maintain such planted aviaries in suitable condition, it is essential never to place in them birds which would destroy the plants, and also to keep them perfectly tidy. An especially trained gardener waters and tends them every morning with great care and a bird-keeper cleans them thoroughly" (Delacour 1945).

As is often the case, zoogeographical integrity

shortly broke down to an extent. By the spring of 1945, a female Southern Green Violet-ear *Colibri thalassinus cyanotus*, one of the ten Cordier had brought from Costa Rica 9 October, 1942, was living in the "New England Garden". That spring it began stealing material from a Zebra Finch nest in the adjoining "Arid Plain". After the Violet-ear's nest was largely completed, a male was introduced and began to sing regularly. Then the female died from egg-binding. (Anon 1945a,b).

"In the new home of the Humming Birds that is being rushed in the Bronx Zoo's Bird House, they will be exhibited behind glass in small brightly lighted cages, while the public will view them from a black passageway" (Delacour 1943). The first installation of ten small landscaped exhibits, in a small building constructed inside the "Glass Court" of the 1905 Bird House, opened 6 November, 1942, after a preview attended by the consuls of nine Latin American countries 5 November (Anon 1942). In 1945, this was replaced a far more elaborate exhibit, the Jewel Room, which was to remain one of the most popular displays at the Bronx Zoo until 1972, when the Bird House was replaced by the enormous World of Birds.

"The problem of exhibiting cage birds in public zoos is a difficult one. Until recently it has been tackled rather crudely. Too often just rows of wire cages are lined up on shelves. In the better case fixed compartments have been neatly built. But practically never before has it been attempted to show the birds under the best possible conditions of light, which enables one to detect all the usually elusive metallic colours and delicate hues; nor, at the same time, to set them in an artistic frame. For many years I had planned to build a special hall, the walls of which would have glass openings, giving view to birds and fishes. Cages and aquariums would have been decorated and planted. The effect would have been that of so many animated, living pictures... The centuries-old rooms of Cleres, with all their historic interest, did not allow for such a scheme. But I had hoped to build a special house some day in a secluded corner of the park. Fate has decided otherwise... However, I had a chance to achieve at the New York Zoo for the public what I had once dreamed to do at home for my own satisfaction. The result has been what we call the 'Jewel Room' " (Delacour 1946).

"... The third hall was particularly unattractive in its former state: a large room 60 ft by 30 ft, with an ugly glass roof and plainly built compartments all around. They were badly lit, and none of the beautiful colors or metallic reflections of the inmates could be seen at real advantage. It was the more unfortunate that it always housed a wonderful collection..." (Delacour 1946). In restrospect, a description of this hall by the New York Zoological Park's first Director (1896-1926), William Temple Hornaday, is amusing: "In the angle of the main building stands a structure almost wholly composed of metal and glass, which is known as the Glass Court. It was designed especially for North American song-birds... In the Glass Court and around it, the Curator of Birds, Mr. C. William Beebe, has cored a gratifying success in the installation of the Order Passeres. They are arranged by Families, and all of the twenty-one families of eastern North American perching birds are represented" (Hornaday 1906).

"... This hall has been entirely renovated during the winter of 1945, and it was reopened as the 'Jewel Room' a little more than a year ago. The transformation has been comparatively simple and easy - a smaller room has been built inside the hall, entirely dark but for the light which comes through the glass front of the cages that open in the walls; it all looks like a gallery of live bird pictures, as in my original plan. Also it reminds one of modern techniques use in aquariums and terrariums. The hall has two large double doors, one at end of the western wall, the other one occupying most of the smaller southern panel. As a result, the cages form two groups, one occupying the greatest part of the western wall, the other L-shaped, all along the northern and eastern sides. Those of the first group, ten in number and of three different sizes, are dedicated to Humming birds. The others consist of one large (10ft by 11ft) unplanted but nicely decorated aviary mostly for hardbills, of two fair-sized planted compartments (5½ft by 5ft) and seven smaller ones (3½ ft by 3ft). They are at present occupied by a Fairy Bluebird, a Rothschild's Starling, a Cock-of-the-Rock, and a number of Manakins, Sugar-birds, and small tanagers, which are doing exceptionally well in such quarters...

The large hardbill's [finches] aviary and the two adjoining smaller cages are painted a gay bright yellow inside. All the others are pale grayish blue. I am against backgrounds with painted landscapes or other scenes, as they are too difficult to keep clean. The cages are strongly and appropriately lit by natural light form the glass roof, and also from heatless electric tubes disposed at a favorable angle inside the cages, at the top of the corner in front, quite invisible to the public. The metallic reflections of the plumage are thus seen at their best. Troughs are inserted through the bottom, along the back of the floor, and filled with tropical plants and flowers. It looks like Gould's plates, but it is alive" (Delacour 1946).

"The public room, on the other hand, is absolutely plain, painted a dark, neutral grey, so that the visitor's attention is not distracted from the exhibits. We found that in order to give the illusion of adequate space, cages must be very deep, (deeper than wide) and that the back must always be rounded, angles looking very ugly. Proper ventilation is essential; and to ensure this holes have to be pierced at the lower part of the cage, while the top is partly covered by wire netting, as used against flies and mosquitoes. Rather complicated devices had to be conceived for the proper working of the cages. Mr. Lee Crandall, the General Curator and one of the most experienced aviculturists I have known, has invented most of them, with the help of the Head Keeper, G. Scott. For the cleaning of the Humming Bird's compartments, we place on the top a bottomless movable cage. A large trap in the top of the compartment slides outwards so that the Hummer can easily fly up into the movable cage and be confined there while the compartment is cleaned through the whole back side of the compartment, which can be opened. After cleaning, the hummer is easily induced to fly down and re-enter its proper quarters. The larger cages are cleaned through back doors, without shifting the birds, but the keeper operates by placing himself inside movable panels, which constitute a portable cage. If a bird escapes it cannot go far, and it returns almost immediately to its compartment" (Delacour 1946). A photograph of George Scott servicing the hummingbird cages is published in the September-October number of the Zoological Society's magazine (Anon 1946). "Birds behind glass cannot be heard, which is a pity, at least for the largest number of them. We have remedied that by conducting the sound from the cages to the public hall by special devices. There is ample room at the back for spare cages and all sorts of feeding and cleaning facilities. The 'Jewel Room' has proved a great success. It is, we think, an excellent show, and the birds are doing exceedingly well in it...." (Delacour 1946).

As a facility for keeping hummingbirds alive, the Jewel Room could be said to have proved itself through the longevity of a "Lesson's Emerald" (which appears to be some sort of Amazillia), obtained, from some source other than Cordier, on 18 May, 1944, which died 18 June, 1949 (Crandall 1949). It was also the home of a Green-throated Carib *Eulampis holosericeus* and a Purple-throated Carib *E. jugularis* which attained celebrity status when their eighth year at the Bronx Zoo was celebrated in 1961, respectively on 21 July and 21 September (Conway 1961).

The Green-throated Carib died "as the result of an accident" 24 February, 1964 (Prestwich 1965). In two months its captive hummingbird longevity record of ten years, seven months, and seven days was exceeded by its Purple-throated relative, which died after ten years, eight months, and six days in New York (Mobbs 1982). These records were attained despite the necessity of a regularly scheduled procedure: "Adventure among our Zoo hummingbirds is held to the absolute minimum, but once every six months or so a worried old Emerald-throat, and Alarmed old Garnet Hummer, a nervous Head Keeper and a perspiring Curator assemble over a small table behind the Jewel Room for a mutually frightening adventure. It is toenail-clipping time. Hand-catching and holding sixteen year's worth of hummingbirds might best be compared with trying to bunt a Christmas tree ornament with a baseball bat. Yet, if the toenails are left unclipped, the birds will eventually become crippled. So far, the operations have all been uneventful" (Conway 1961), (The trimming of hummingbird toenails appears to be nonexistent practice, at least among present day zoo personal, who do not usually maintain hummingbirds in small cages. While the very experienced hummingbird specialist A.J. Mobbs (1973) emphasized the importance of claw trimming in at least one Avicultural Magazine article, the subject is mentioned only in passing, in some of the species accounts in his authoritative book (Mobbs 1982), and is not listed in the index.)

The Green-throated Carib was part of a collection of birds from Martinique and Surinam, collected by Charles Cordier (New York Zoological Society 1953) along with a Calfbird, four species each of ant-thrushes and manakins, and other species, this consignment included the Bronx Zoo's first specimens of Antillean Crested Hummingbirds, Crimson Topaz, and the Blue-headed Hummingbird *Cyanophaia bicolor*.

Following his 1941 Colombian collection, and his 1942 Costa Rican expedition, Cordier had been kept busy by the New York Zoological Society. From 1943 until sometime after the end of the war, he had been employed in some sort of "defense work" in California, but that did not prevent him from assembling a collection of Western hummingbirds. These Anna's, Costa's *Calypte costae*, Black-chinned *Archilochus alexandri*, Calliope *Stellula calliope* and Broad-tailed *Selasphorus platycercus* Hummingbirds were sent to New York in 1944 (Anon 1944a). On 15 March, 1947, in his official capacity as Staff Collector, he returned from Guatemala with 57 species of birds (Barclay-Smith 1947; Cordier

1947a; Lindholm 1988). These included three species of Trogons, and 33 hummingbirds of eleven taxa, every one of them new to the Bronx Zoo collection. Among this esoterica were eight Magnificent (Rivoli's) Hummingbirds, eleven White-eared Hummingbirds *Hylocharis leucotis*, and one specimen of the Guatemalan subspecies of the Broad-tailed Hummingbird *S. platycercus guatamalae*.

Although Cordier was due to depart for the Congo in October, 1947, he insisted on another expedition in the interim. He wrote: "If there is any one thing a collector of birds dislikes more than another, it is the inactivity of city life when the field work is finished and a collection has been delivered to its destination, a zoological garden. I do not suffer boredom and inactivity in silence, and so the New York Zoological Society was anxious to get rid of me last spring just as soon as I had delivered a collection of birds from Guatemala..." (Cordier 1947). [Jean Delacour told me : "Cordier hated humanity in general, but he liked me!"]. Five-and-a-half months after landing his Guatemalan collection, Cordier arrived from Costa Rica on 31 August, 1947, with 41 species of birds, as well as mammals, reptiles and amphibians (Cordier, 1947). General Curator Crandall (1947) wrote: "The most important collection of birds and mammals that the New York Zoological Society ever received from Central America was first exhibited on 6th September. The stars of the collection were a female Yapock, an almost legendary Central American water opossum that has never before been exhibited alive, and which few white men have ever seen; the Three-wattled Bellbird, a sickle-bill Hummingbird, male Bare-necked Umbrellabirds, Costa Rican Quetzals, and 37 Hummingbirds of 17 species - six of them never before exhibited anywhere, and four more new to the Bronx Zoo".

In his 1942 Costa Rican shipment, Cordier had brought back one Costa Rican Snowcap. This time there were eight. "The most spectacular of all the birds, even though on a small scale, are the new Hummingbirds and particularly eight Costa Rican Snow-caps. These are tiny, wine-rust-colored hummers with snowy white crowns which, in certain light, are afire with opalescent tints. Five male birds are now exhibited in a thorn-planted compartment in the in the Bronx Zoo's 'Jewel Room', together with samples of the other 'first time' hummers. The rarest of all is Salvin's [White-tipped] Sickle-bill [*Eutoxeres aquila salvini*] , a medium-sized Hummingbird with a sharply-curved beak an inch and a quarter long. It has never been exhibited before" (Crandall 1947). The eight Magnificent Hummingbirds in the 1947 Guatemalan shipment were *Eugenes fulgens viridecips*. Cordier included two *E. fulgens spectabilis* in the 1947 Costa Rican collection (Cordier 1947).

When Cordier and his and his wife Emy (who had assisted him in Coast Rica) departed for the Congo, 10 October, 1947, they were expected to be back in "six to eight months" (Anon 1947a). As it happened, Charles Cordier broke his leg in the Ituri Forest, and spent months convalescing in Stanleyville, so that the resulting collection did not arrive in New York until 15 June, 1949 (Barclay-Smith 1949; Bridges 1954; Cordier 1949). Including as it did, the first Congo Peafowl to leave Africa alive, as well as an enormous assortment of birds, mammals, reptiles, and fishes, it remains Cordier's most famous shipment.

Typically, Charles and Emy Cordier left New York for Ecuador on 2 October, 1949, returning 20 January, 1950, with 79 taxa of birds. Lee Crandall (1950b) comments: "The return to New York was made by a special airplane flight, which again demonstrated the efficiency of this mode of transportation when the value of the collection justifies the cost. There were no losses en route and everything was landed in excellent condition". Along with the first, Purple-throated Fruitcrows, Red-crested Cotingas, Long-wattled Umbrellabirds, Eastern Umbrellabirds, and Equatorial Cocks of the Rock in aviculture, were 74 hummingbirds of 29 taxa, 24 of them new to the Bronx Zoo (Crandall 1950a,b). This collection was particularly well documented with photographs (Crandall 1950a). The single Swordbill was considered the prize among the hummingbirds. Crandall (1950b) wrote: "Best of all is the incredible Swordbill, its beak quite 5 inches long and noticeably thick. Fortunately, Cordier's specimen is a female, the beak in this sex being longer than that of the male. In a single day of intensive work Cordier taught the bird to feed from an ordinary bottle, with it continues to do with facility". Among other hummingbirds were Black-tailed Train Bearers *Lesbia victoriae*, both species of *Aglaiocercus* Sylphs, three species of Pufflegs *Eriocnemis nigrivestis*, *E. vestitus*, and *E. luciani*, two subspecies of Chimborazo Hillstar *Oreotrochilus c. chimoborazo* and *O. c. jamesonii*, and five specimens of the Purple-backed Thornbill *Ramphomicron microrhynchum*.

In light of the fanfare with which this collection was received, it is disheartening to note that of these 74 hummingbirds, only seventeen specimens of fourteen taxa were present at the New York Zoological Park as of 31 December, 1950 (New York Zoological Society 1950). The year end report for 1951 (New

York Zoological Society 1951) lists ten hummingbirds of four taxa. If one does the math, obviously, not all of these were birds present at the beginning of that year.

The Cordiers brought another shipment from Ecuador to New York in 1952, including seventy hummingbirds (Cordeir 1952a). In their previous expeditions, Charles Cordier had been employed as Staff Collector. This time, he was not entirely obligated to the New York Zoological Society, and the majority of the two hundred birds in this consignment were purchased by the St. Louis Zoo (Hellman 1952), with others going to Philadelphia (Marvin Jones in litt.). The Bronx Zoo still took the lion's share in regards to jaw-dropping rarities (E. Cordier 1952; Crandall 1952; New York Zoological Society 1952). Along with a Mountain Tapir, a White-backed Water Ouzel, and the first Tapaculo in aviculture, were five hummingbird taxa new to the collection, bringing the total number of taxa ever exhibited at the Bronx Zoo to eighty-five (Crandall 1952). [Through 1943, this had been thirty-one taxa (Ditmars & Crandall 1945)]. The "celebrity" among them was the first Giant Hummingbird *Patagona gigas peruviana* in captivity, which among other things, was profiled in "The Talk of the Town" in The New Yorker (Hellman 1952), where it was mentioned it was maintained in an air-conditioned cage, and that Charles Cordier considered it especially "clever". The Tooth-billed Hummingbird *Androdon aequatorialis* was also landed successfully in New York (Cordier 1952a), but I have found no evidence that any were received by zoos.

Charles Cordier (1952a,b) had been as distressed as anyone over the mortality of his 1950 Ecuadorian collection, and formulated entirely different feeding techniques before embarking in 1952. As will be further discussed, his ideas had rather long-reaching effects.

Jean Delacour, who had forged the partnership of Cordier and the New York Zoological Society, had by this time, long before resigned his position as Technical Consultant. While the majority of animals at Cleres had been destroyed or dispersed by the end of the War, and much damage had occurred from bombs and the stationing of soldiers and civilians, Delacour decided, following a visit there by Frank Fooks in 1945, to restore it (Delacour 1949). Delacour's resignation coincided with the opening of Cleres as a public zoological park, willed to the Natural History Museum of Paris, on 25 May, 1947. At that time a large collection of waterfowl and gamebirds had been assembled (including Blue, Bar-headed and Barnacle Geese from Major Albert Pam at Wormleybury, who also provided wallabies (Delacour 1947)), as well as many pigeons and doves, some psittacines and Senegalese finches and starlings. On the other hand, Delacour found "the tropical houses and indoor bird galleries, however, are a total loss, and it is hardly worth while reconstructing them as long as fuel and special foods are not available; it would also be impossible to procure today the rare small birds for which they had been planned" (Delacour 1949).

At the London Zoo, the Tropical House, which had sustained war damage that "necessitated almost total reconstruction" (Prestwich 1947c), was reopened in 1947. Two hummingbirds were displayed there, in separate compartments (one shared with a Golden-headed Manakin (Prestwich 1947c)). Both of these were Plain-bellied Emeralds *Amazilia leucogaster*, which, with a White-chested Emerald *A. chionopectus whitleyi* were part of a diverse collection of animals from British Guyana, brought back in 1947 by Cecil Webb, then employed by the Zoological Society of London (Prestwich 1947a,b). In October of that year, the Bronx Zoo flew a Ruby-throated Hummingbird and a Costa Rican Quetzal to London Zoo, where they arrived in less than 24 hours, as part of an exchange (Anon. 1947b). For some reason, London reported this hummingbird as "Sp. Inc." (Prestwich 1948).

Inspired by Delacour's achievements at the Bronx, Gerald Iles, Director of the Belleview Zoo in Manchester, opened a "Hall of Living Jewels" in 1949 (Iles 1950). No hummingbirds could then be obtained, so the eleven glass-fronted, planted cages contained sunbirds, whydahs, and various Estrildid Finches. He finally obtained hummingbirds in 1950: "On 11th August last I went to Liverpool to meet Mr. Kenneth Smith, who had arrived from British Guiana with a miscellaneous collection of livestock. On the quayside I noticed a packing case containing five hummingbirds. As nobody appeared to want them I bought the lot and hiring a taxi, conveyed my prizes to Manchester" (Iles 1950). With the assistance of Cecil Webb, these birds were keyed out as White-chested Emeralds. Iles kept them in his office for six weeks before placing any on exhibit, loosing only one in the interrem. At the time of writing, he continued to care for them personally, and presented detailed notes on their diet and care in the Avicultural Magazine.

In June, 1948, Alfred Ezra's collection of bedroom birds consisted only of "a couple of Golden-fronted Fruitsuckers, a glorious shama, and two sunbirds" (Witting 1948). However, in 1950, he exhibited a Swallow-tail at the National Cage Bird Show (Vane 1951). Several days before his death in July, 1955, he received a visit from David Seth-Smith, who found, in the bedroom, "a large stack of cages near the window, containing living jewels, such as Humming Birds, Sunbirds, Shamas, and Tanagers". Seth-Smith (1955) wrote: "I remarked upon the excellent condition of the hummingbirds, and how successful he had been with them. 'Yes,' he said, 'they are some of the easiest birds to keep. I have had the larger of those for seven years'." I imagine this bird was the Swallow-tail which appeared at the National Show.

Although the War had ended in 1945, importing birds commercially, or for private aviculture, into Britain, remained extremely difficult for several years. John Yealland (1949) who would serve as London Zoo's Curator of Birds from 1952 through 1968, reported the following: "A few days ago I went to the Import Licensing Department of the Board of Trade in order to inquire about the regulations governing the importation of birds and was informed that, apart from a possible exception in the case of those required for scientific research, there is a ban on all birds and snakes. The official view seems to be that any small bird will eat seed... It is apparently only the small birds that peck away the foundations of the country's economic structure, for I was told later that permits might be granted for large birds, such as cranes or flamingoes, destined for zoos." Arthur Prestwich (1949), future President of the Avicultural Society, replied: "Mr. Yealland's experience is similar to that of several other members who have approached the board of trade. While steadfastly refusing to grant import licences to any except zoos, they seem entirely unable to give any reason for their attitude. The sympathy of a well-known M.P. has been enlisted and he has taken up the question with the appropriate department, but without getting a satisfactory reply. He has now formulated a question to be put on the Order Paper and answered on the floor of the House".

*Pernambuco Revisited*

By 1951, Jean Delacour was able to report: "It has become much easier to acquire birds in Europe during the past year. Permits for imports have been more liberal in France and in England... while Belgian and Dutch dealers continue to offer large assortments..." (Delacour 1951a). Results of more liberal polices were certainly on display at Britain's National Cage Bird Show: "Visitors were also able to see many welcome reappearances such as the Naked-throated Bell-bird, Touracous, cissas, toucans and toucanettes, pittas, Wilson's Birds-of-Paradise, hummingbirds, sugar-birds, and sunbirds. Definitely, the exotics are coming back, and obviously are once more the center of attention to the casual visitor" (Vane 1952).

In addition, Brazil was once again allowing commercial exports of birds. In 1950, the large, sturdy Swallow-tail which had been one of the mainstays of the 1930s trade, made a multiple re-appearance at the London Zoo. One was received from the Antwerp Zoo (Webb 1950a), and shortly thereafter, more specimens arrived from the Copenhagen Zoo, along with some Plain-bellied Emeralds (Webb 1950b). In 1952, London Zoo was presented with a collection of hummingbirds out of Recife (Yealland 1952). These were two Pygmy Hermits, three Blue-chinned Sapphires, and one tentatively identified as "*Hylocharis lactea*", which does not correspond with any name in current usage. They were from of a shipment of around sixty hummingbirds, with other birds from Brazil, on their way to Dusseldorf, in the care of Dr. Wilhelm Windecker, who had established a reputation as a dealer in South American animals.

In May, 1952, Dr. Windecker was appointed Director of the Cologne Zoo, and in less than two years expanded the collection by more than 200 species, making it the finest collection of birds in West Germany (Jones 1954). His arrival coincided with the remodeling of the interior of the 1890 bird house, which included the creation of "jewel box" exhibits for hummingbirds. Zoologischer Garten Wuppertal had already become the first German zoo to acquire hummingbirds after the war, receiving 12 from Brazil in 1950 (Schuerer, 1983). By the early 1950s, a renaissance for European zoos was in full swing. In 1948, Antwerp opened a startlingly innovative bird house, entirely replacing one destroyed in the War (Barclay-Smith 1948). At opening, the exhibit intended for hummingbirds was inhabited by kookaburras (Barclay-Smith 1948), but, as noted above, by 1950, Antwerp had hummingbirds to spare. By 1951, "Ruby, Topaz, Allen's, Emerald [sic]" could be seen in a planted plate-glass cage 10ft high by 10ft deep, at the bird house at the Wassenar Zoo, and when the famed Louise Hall was opened there in 1953, hummingbirds were exhibited there (de Goedern 1953).

As previously noted, Copenhagen's last hummingbird of the pre-war years had died there 6 May, 1945, after almost eight years (Reventlow 1948). As of 1 August, 1952, "40 Humming Birds of eight or nine different kinds" could be seen there (Reventlow 1953).

Axel Reventlow, who had been an aviculturist since 1902, and a whole-sale birdseed importer before his appointment at Copenhagen, (Prestwich no date), had interesting observations on the state of postwar hummingbird shipments: "According to my experience, the last consignments to Europe of these birds have not, by far, been of the same good quality or consisting of as many different species as before the war. Then the birds were brought to Europe by clever and experienced men such as Charles Cordier, and the German, Daenisch, who generally caught the birds themselves and had time and patience enough to care for and feed them, and to clean the cages properly during the journey from Brazil, a voyage of about 10-12 days by the great passenger liners. The humming birds are now coming to Europe by air, and though this means a quicker journey, it involves the great drawback that the crew of the aeroplanes only in very few cases shows the least interest in the welfare of the birds. During the last few years we have several times got collections of humming birds consisting of big and very small ones all mixed together, and consequently the small ones were constantly chased away from the feeding glasses so that many of the starved to death during the journey. The cages were badly constructed and almost impossible to keep clean. On arrival at our airport late in the evening many of the birds were lying on the bottom of the cage, in a very deplorable state. They were so soiled that they stuck together in ther own feces, and in the food spilled from the feeding-glasses. My wife and I had to wash every single bird several times with cotton-wool in lukewarm water, dry them by means of a heating lamp, and feed them by hand almost the whole night" (Reventlow 1953).

At the end of 1952, John Yealland, London Zoo's new Curator of Birds, noted that twenty-four forms of African sunbirds were then present in the collection (Yealland 1953a). This was at least partially due to Cecil Webb having served as Curator of Birds and Mammals from 1949 to 1951 (Webb 1954). Webb really was an African bird specialist. Three further African sunbird taxa, never before exhibited at London, were added a few weeks later (Yealland 1953b), and two more shortly after that (Yealland 1953c). Karl Plath (1957) at Brookfield Zoo, had, by the 1950s come to the conclusion that: "Generally speaking, it is our opinion that sunbirds do better; they are equally beautiful though, of course, do not have the rapid wing-beat of the hummer". While Brookfield probably maintained the most extensive series of hummingbirds in the U.S. in the 1930s, in the '50s few efforts were made to procure hummingbirds. For instance, when Plath (1953) published a list of all birds acquired in 1952, hummingbirds were entirely absent.

None the less, London entered into a new phase of hummingbird aviculture when it remodeled the Tropical House in 1953. John Yealland (1953c) wrote: "The whole of one side of the Tropical House has been furnished with tropical plants, including orchids, and is now devoted to hummingbirds of which twenty-two specimens of six forms arrived some five weeks ago". These were all species typical of the Recife/Pernambuco commercial trade: "The Golden-throated *Polytmus guainumbi thaumantias* and the Red-throated Sapphire *Hylocharis sapphirina* are new to the collection; the remainder consist of Waterton's Wood Nymph *Thalurania watertonii*); Pucheran's Emerald *Chlorostilbon aureoventris pucherani*; Blue-breasted Sapphire *Chlorestes notatus*, and another not yet identified which might be immature *Amazilia leucogaster*. These birds are being fed on the liquid diet recommended by M. Cordier and plenty of fruit flies. The fateful forty days of which M. Cordier writes have almost passed and there have been no losses so far. This happy state could be attributed as much to the perfect condition of the birds on arrival - due to some excellent packing and to air travel - and to the amount of exercise they are able to take in this large flight as to the food they have received here" (Yealland 1953c).

*Forty Days*

Yealland's (1953c) reference to Charles Cordier's "fateful forty days" requires adding further convolutions to an already convoluted narrative. As mentioned earlier, Cordier was deeply disturbed by the mortality of the Ecuadorian collection he had landed in 1950, as well as that of his previous Bronx Zoo shipments, especially, as has been noted, the birds appeared in excellent condition upon their arrival in New York. In 1952 he wrote in the Avicultural Magazine: "Most of the readers of this magazine are familiar with the longevity records attained by humming birds in captivity ranging from two, four, five, up to eight years. In the light of these records it would seem that the diet for humming birds in captivity has been solved satisfactorily. However, this is far from being the case. The birds that live for years are the exceptions and the others that live from forty days to four or six months are the rule. My own experience over the years has convinced me that practically any humming bird will live on honey or sugar-water alone for abut forty days. Even if fed from the very day of... capture on the standard formula of Mellin's food, condensed milk, meat extract, honey, and a dash of vitamins, some will not live on this antiquated formula for more than forty days..." (Cordier 1952a). It is interesting to note how Cordier's "forty days" corresponds to the "between four and five weeks" (Pam & Pam 1907) that the last bird in the Pam brother's 1907 shipment survived in London.

Cordier (1952a) continues: "Thus the matter has stood for the last seventeen years and the poor collector going in for humming birds simply never knew where he stood after the crucial forty-day test, the period during which these birds live anyway, no matter how strongly they are being fed. Fortunately I have been corresponding over the years with Dr. Beraut of Rio de Janeiro, an ardent hummingbird fan who keeps some forty hummers in thirty species in one single small room with the greatest success. He is keeping in splendid health all the species found in the state of Rio de Janeiro, where I have collected twice, and which I never managed to keep for long. Dr. Beraut simply never like the idea of these birds getting any cereal in their food, as in nature they have no chance to do so. Little by little he evolved his own formula and in view of his enthusiasm over it, it did not take me long to make up my mind to adopt it. The improvement is so phenomenal I do not hesitate in passing on the new formula to our readers. In essence it is simply the known formula from which the Mellin's Food has been left out and essential vitamins added. At this writing I have a collection of seventy specimens consisting mostly of mountain species which are, by far, more delicate than lowland hummers or, rather, I should write of their frailty in the past tense, as losses have been practically nil".

Cordier's (1952a) formula to feed forty hummingbirds was as follows:

- 3g Salted meat extract [such as Armours in glass jars]
- 40g sweetened condensed milk
- 100 to 180g honey
- Water [to complete 1,000cu cm or roughly one quart]
- Vitamins A and D

The Cordiers (1952a) found that the honey they could buy in the U.S. was "so much weaker" then that in Ecuador, that they almost doubled the concentration, hence the above variation. The vitamins were administered in the following way: "The druggist should make up for you the following formula: One ampoule vitamin A of 600,000 units. One ampoule vitamin D of 100,000 units. In 30cm$^3$ of almond oil". Four drops of this formula was beaten into the condensed milk, after some of the water, hot, had been added to it. In the afternoon or early evening, the birds were fed only with a honey and water mixture to which vitamins B and C (ten times as much of the later) were added.

Not everyone greeted Cordier's innovations with the same enthusiasm that London Zoo did. Axel Reventlow (1953) at Copenhagen conceded to "gradually... reduce the quantity of Mellin's Food and to increase the quantity of honey", since he had "what I should call fairly good results... using for many years what Cordier in his article calls 'an antiquated formula' ". On the other hand he did use Cordier's afternoon and evening diet of honey, water, and vitamins A & B. Reventlow (1953) emphasized the importance of fruit flies, having maintained a culture of *Drosophila repleta* originating from 25 specimens he had brought back from Zoologischer Garten Berlin.

At the time the two Caribs at the Bronx Zoo passed their eighth years there in 1961, their diet consisted of " a special liver protein hydrolysate, a cereal-based baby food, beef extract, sweetened condensed milk, pure honey, liquid vitamins and water, all mixed to form a thin, beige-colored, sweet-tasting syrup" (Conway 1961), supplemented with fruit flies. The reversion to a diet containing cereal may have been partially due to continuing poor results despite the new diet. At the end of 1952 the zoo held ten hummingbirds of five taxa (New York Zoological Society 1952), but at the end of 1953, during which, as noted above, at least five taxa (including the two record setting Caribs) were added to the collection, the count was only thirteen specimens of five species (New York Zoological Society 1953).

Dr. Etienne Beraut, who inspired Charles Cordier to adopt a cereal-free diet, had been the recipient, through Jean Delacour, of the some of the hummingbirds Cordier had brought to Europe before the War (Delacour 1943). He left France to live in Rio de Janeiro, from where he wrote in 1951: "I am still very much interested in humming birds, and I begin to have a nice local collection of fourteen species, missing only three or four I have seen in the neighborhood. Contrary to previous experience, they do not fight. I keep birds of various size together, from the tiny *Calliphilox* to large *Rhamphodon*. The terrible *Eupetomena* and *Thalurania* go a little after the others, but without any real viciousness. This happy state of things is probably due to the crowding: I keep twenty-five Hummers in an aviary 11ft by 6ft. They live perfectly and are tremendously active; I practically loose none, but they are looked after very carefully. I secure them with a blowpipe, and it is a most exciting sport. It is difficult and fairly exhausting, as it is necessary to be on the spot at daybreak" (Barclay-Smith 1951).

Several months later, in a letter to Jean Delacour, Dr. Beraut wrote: On my return from Europe in August the weather was not good and I could only make a few trips in search of new humming birds. Following Professor J. Berlioz's directions, I went to the plateau of Minas Geraes to look for *Augastes superbus*, but I did not find it, although it was the exact time when he had seen it there. But on the mountains I found by luck large numbers of *Stephanoxis delalandei* on the Eucalyptus trees in bloom; I now have five in my aviary, it is a beautiful species. I also caught some *Phaetornis pretrei* and *Glaucis tomineo* (*hirsuita*). I now possess twenty-three species in perfect health. I have kept several over a year and some, such as Violet-ears (*Colibri*) and *Heliothrix*, more than six months. I have forty in a space 10 x 7 feet. It is an extraordinary thing that although they fly after one another at terrific speed they do not fight viciously, probably because they are so numerous. Of the Brazilian species, the Coquette *Lopohornis magnificus* is still lacking; I have not yet been able to locate it. I plan to go soon to the hinterland of Matto Grosso and to procure there some interesting specimens. In December, on Monsieur Cordier's advice, I plan to visit northern Brazil, when I should find *Topaza pella* and *Heliactin cornutus*. In this way I hope to add to my collection and to keep it up by means of periodical trips. I have a few true pairs, but they have not yet made any attempt to nest. I believe the perfect condition of my humming birds is due to three main reasons; complete suppression of Mellin's food; daily addition of vitamins A, B, C, D; daily supply of a large number of fruit flies which are indispensable to certain species, in particular *Heliothrix*, and in a still higher degree to all those which wag their tails - *Rhamphodon, Phaetornis, Pygmornis,* and *Glaucis*" (Delacour 1951b).

Several months after that, Dr. Beraut wrote: "My collection of humming birds now includes all the twenty-five species living in the region of Rio de Janeiro; I have added Ruby-Topaz and *Polythmus* [sic] from Pernambuco, and the beautiful *Thalurania baeri* from Matto-Grosso. They all do perfectly well. I have had for 15 months a pair of *Petasophora*... I am soon going to Matto-Grosso and hope to bring back some nice species" (Barclay-Smith 1952).

This collection was fed in the following way: "Early morning and 1p.m.: 190cc of water; 18cc of honey; 7cc of condensed milk; 2cc of meat extract (Wilson's or Armour's), and one drop of a solution of 600.00 units vitamin D2 and 100.000 units vitamin A in 30cc of peanut oil. At 5pm: 190cc water; 18cc honey; 5mg of vitamin C, to which I add twice a week 5mg of vitamin B1. This diet is supplemented by live fruit flies, but these are not eaten by all humming birds. Those who need it most are the species of the *Rhamphodon-Glaucis-Phaetornis-Pygmornis* group which live well with me and are most attractive. They lack bright metallic colours, but their flight is extremely graceful and they become extraordinarily tame" (Barclay-Smith 1952).

By 1954, Dr. Beraut wrote: "My collection of hummingbirds is still doing well. I possess one *Heliactin cornutus* in wonderful plumage, a young bird which has moulted in my aviary, and my best bird. I believe I now know the best way to keep the more difficult species: *Heliactin, Heliothrix*, and particularly *Pygmornis*, which live and moult perfectly, mostly thanks to a massive production of fruit flies (Drossophila). I also use a new way of transportation; instead of putting the humming birds in a cage, which creates difficulties in airplanes, I place each of them in a small bag of cloth with a small hole for the head to stick out. They are thus obliged to keep quiet and do not become exhausted. Every hour, I take them out of the box where I keep them during the trip and make them drink. They can travel for twenty-four hours under such conditions without being tired." (Prestwich 1954b).

Sometime in the following two years, disaster befell this collection. Dr. Crawford Greenwalt, who was assembling the photographs for his classic book *Hummingbirds*, tells the story (Greenwalt 1963): "The day I went to New York to suggest to the American Museum of Natural History that I attempt an illustrated book on hummingbirds..., it just happened that Jean Delacour... was in New York; it just happened he was spending that afternoon at the museum; and it just happened I was taken to meet him. 'If you want to photograph hummingbirds,' Delacour said, 'you should get in touch with my friend Beraut in Rio.'... I turned out that Dr. Etienne Beraut, a charming Frenchman... had an aviary with twenty or more species in his 12th-floor Rio de Janeiro apartment. After much cordial correspondence, a date was made, and I prepared to set out for Rio. Unhappily, about a week before my departure he wrote that his birds had become infected with a fungus disease and that all had died. Dr. Beraut's letter, in French, said his bird had champignons..." .

As a result of this catastrophe Dr. Beraut further

refined his diet. By the time Jean Delacour visited him in 1956, he had again assembled, in a "planted verandah", "a beautiful collection of humming birds, which he had his collectors obtain throughout Brazil" (Delacour 1957). According to Captain Delacour, these birds were fed entirely "on sugar or honey-water and on quantities of fruit flies".

On this same 1956 Los Angeles County Museum expedition to Brazil, Delacour was taken by Beraut to visit another collection that, until then, was little known outside that country."… The visit I made with him to his friend, Mr. F. [sic] Ruschi at Santa Teresa, Espirito Santo (the state just north of Rio) will remain in my memory as one of the greatest thrills that I ever had. Mr. Ruschi is a dedicated Brazilian nature-lover. He has single-handedly organized remarkable wild life preserves in his native state… His property contains museums gardens and aviaries of the greatest interest. In particular he has built an enormous flight, 300ft x 100ft, and 30ft high, where hundreds of humming birds live and breed freely, including the lovely little Coquette. I saw fifteen on a bush, all reared by one original pair. There are about twenty local species in this aviary, and one can also see several hundreds of wild Hummers in the garden. Dozens of them are always buzzing as they drink sugar-water from bottles, hung around the verandah of the house… Furthermore, a beautiful large (100ft. long) house has been built, with a passage for visitors along its front, to accommodate the equatorial species from Amazonia which will not stand the cool nights of Santa Teresa (altitude 3,000 feet)" (Delacour 1957). Like his friend Beraut, Ruschi also fed only sugar-water (de Quincey 1960) and fruit flies.

Augusto Ruschi had commenced breeding hummingbirds in aviaries in 1936, when he was successful with four species (Marden 1963). By 1963, he had propagated 61 species at his Museu de Biolgia Prof Mello Leitao (which he officially founded at his family property in 1949) (Marden 1963).

*Jamaicans*

Preceded by the previously mentioned near-misses at Berlin and Cleres in the 1930s and at New York in 1945, the first captive hatching of a hummingbird outside of South America took place in 1959, at the aviaries of Captain Richard de Quincey.

When the Amethyst-throated Sunangel *H. amethysticollis* which Cecil Webb had brought from Ecuador in 1938 died sometime after 1941, during the Second World War. Captain de Quincey considered it

"the last nail in the coffin of my bird-keeping activities…" (de Quincey 1960). Although he had been a member of the Avicultural Society since 1913, at War's end he was "determined not to start keeping birds in captivity again, but the temptation to do so returned. I don't just remember what started the germ reasserting itself, but it was at one of the post-war National Cage bird show that I fell. I bought a pair of Virginian Cardinals, which I found I could not compare with those pre-war excitements and the next year I bought a pair of Lesser Collared Sunbirds from Mrs. Scamell, which only whetted the appetite to have Hummingbirds again. In that spring, two Hummingbirds, Golden-throated *Polytmus guainumbi*, arrived with Percy Hastings, one for the zoo and one for me, but I changed mine with Raymond Sawyer for a Pucheran, as he wanted the Golden-throat which was new to him, and I was delight with the Puchaeran *Chlorostilbon aureoventris pucherani*, the most charming little fellow I have ever had of that species, very tame and intelligent, and a great vocalist in his own estimation. He was a great favorite and lived about four years. The old greenhouse home of prewar Hummingbirds was restored, and our one and only Percy Hastings had a sort of request to let me know when he had any nice birds. Since then quite a lot have come over an found their way here, some in good form, others very frayed about their wing-feathers, or with that horrible fungoidal tongue trouble… from which there seems to be no recovery. To receive a little parcel of eight Frilled Coquettes - and to find them all going wrong in the tongue from the moment they arrive - is very daunting, as one can do nothing apparently…".

I find it amusing that Percy Hastings , who had ceased operations during the war, re-established his business in 1953 (Scamell 1968) - The same year Richard de Quincey sold the Hereford bull "Vern Diamond" to the Wyoming Hereford Ranch for the then record price of 16,000 Pounds Sterling. The year before, he reduced the world-famous Vern Herd "realizing a record average of 670 pounds for 66 lots offered, breaking a record which had existed since 1918" (Hereford Herd Book Society 1995). The Vern was Captain de Quincey's Herefordshire property, listed in 1086AD in the Domesday Book. His Herefords were considered the finest in the world (Delacour 1966b). So he was not wanting for funds. In less than seven years he had acquired at least 26 taxa of hummingbirds (de Quincey 1960). In 1961, around four years before Richard de Quincey's death at the age of 69, Jean Delacour reported: "My old friend Captain R. de Q. Quincey has in Herefordshire the best collection of Hummingbirds, Sunbirds, Sugar-birds, Tanagers, and various small softbills to be seen in Europe, excellently housed in beautiful greenhouses and planted outdoor aviaries... Among the Hummers, are such rarities as Heavenly Sylphs and a Giant *Patagona gigas*) [neither of which were included in de Quincey's (1960) post-war list] ..." (Delacour 1961).

Raymond Sawyer, presently still keeping birds with Miss Ruth Ezra at Chestnut Lodge, in Surrey, not far from where Miss Ezra's father's aviaries stood at Foxwarren Park, joined the Avicultural Society in 1949. He rather suddenly came to the attention of British aviculture in 1952, at the National Show: "The opening class among softbills was indeed exceptional, there being no less than 15 entries fo Sunbirds and Hummingbirds. Mr. Sawyer's Ruby and Topaz taking first prize, also the best foreign exhibit and supreme champion of the Show. Not content with that, Mr. Sawyer took second and third prize in this class, with a pair of Amethyst Sunbirds, and a pair of Pucheran's Emerald Hummingbirds. He repeated the performance in the next class, for small Tanagers and Sugar Birds, with a team of Black-headed, Yellow-winged, and Blue Sugar Birds. All these exhibits were faultlessly staged in most tastefully decorated surroundings" (Vane 1953). Mr. Sawyer was one of the first private aviculturists to obtain hummingbirds following the relaxation of the afore-mentioned British post-war bird importation restrictions: "I remember around 1949/50 going to collect a crate of Hummingbirds from London Airport. In those days the offices were only in army huts. I bought a crate of... 16 Hummingbirds from a dealer called Randau in Brazil, for 48 pounds, delivered. Twelve birds were supposed to be sent but he put in 16 in case there was any losses on the way. The captain had to be consulted before any birds were taken on the plane, and quite often he would have the Hummingbirds in the cockpit with him. We used to have the same problem with customs that we still get today" (Sawyer 1995).

The dealer G. Randau was based in Recife, Pernambuco, and throughout the 1950s supplied a wide array of Brazilian Hummingbirds. Raymond Sawyer (2002) notes that Percy Hastings "imported a lot of hummingbirds from Randau". Randau donated four Horned Sun Gems *Heliactin cornutus* and a Frilled Coquette *Lophornis magnifica* to the London Zoo in 1955, the first in that collection (Yealland 1955a). The last of these Sun Gems, a female, died in 1957, after one and a half years in London. John Yealland (1957a) considered this an exceptional record for so small a wild-caught bird, noting that it depended on fruit flies more than the other hummingbirds. Randau also donated to London Zoo White-throated Sapphires *Hylocharis cyanus* in 1957 (Yealland 1957a) Rufous-throated Sapphires *H. sapphirina* in 1957 and 1958 (Yealland 1957a & 1958c).

Brazilian Hummingbirds made up the majority of species acquired by the London Zoo during the 1950s. Aside from those previously mentioned, these include Ruby-Topaz and Stripe-breasted Starthroats purchased in 1953 (Yealland 1953d): twelve specimens of four of the typical Pernambuco species purchased in 1954 (Yealland 1954); a Swallowtail received in exchange in 1955 (Yealland 1955a); a collection of seven species, among them the Zoo's first Black Jacobin *Melanotrichilus fuscus*, purchased in 1956 (Yealland 1956a); a Black-throated Mango received in exchange in 1956 (Yealland 1956b); Racquet-tailed Hummingbirds obtained in 1956 (the first to arrive at the zoo since 1934) (Yealland 1956c); and nine species from the aviaries of Augusto Ruschi, "presented by H.E. the Brazilian Ambassador" (Yealland 1958c). These birds were brought to London by Dr. Ruschi himself, and included three species new to the collection.

Not all of London Zoo's 1950s hummingbirds came from Brazil. The second of David Attenborough's joint London Zoo and BBC expeditions was to British Guiana, resulting in an animal collection, including a manatee, arriving in August, 1955. Five hummingbirds of three species were among them. The Roraima Emerald *Chlorostilbon prasinus subfurcatus* was new to the collection. The other two were *Amazillia* (Yealland 1955b).

Although I have net yet investigated, I suspect three acquisitions listed as "exchanges" were from the Bronx Zoo. Three Antillean Crested Hummingbirds and three Purple-throated Caribs were thus obtained in 1953 (The same year the Bronx zoo obtained its first Antilean Cresteds, and its record-setting Carib) (Yealland 1953e). A pair of Ruby-throated Hummingbirds were received in 1956 (Yealland 1955c). And in 1955 London Zoo received its first Red-billed Streamertails *Trochilus polytmus* (Yealland 1955c).

While attempts had been made by the Victorian Naturalist Philip Henry Gosse (the Father of the Saltwater Aquarium) to keep Streamertails in Jamaica in the 1840s (Gosse 1847; Porter 1936), 1955 appears to be the year this species was first exported alive. That year the Bronx Zoo received its first Streamertails, as well as its first Jamaican Mangos *Anthracothorax mango*, Bahaman Woodstars *Pholodice e. evelynae*, and the second smallest species of hummingbird, the Vervain *Mellisuga m. minima*, all collected by Oliver Griswold, Coordinator of the Radio, Television, and Film Department of the University of Miami (New York Zoological Society 1955).

Oliver Griswold (who appears to have been no relation to John "Gus" Griswold, the long-time Curator of Birds at the Philadelphia Zoo) had served on Herbert Hoover's Committee for Employment, and under the administration of Franklin Roosevelt he was successively Assistant Director of Public Relations for the Works Progress Administration, Director of Programming and Distribution of Motion Pictures of the U.S. Department of Agriculture, and Associate Director of Public Relations for the Office of War Information. He joined the University of Miami in 1949, retiring in 1962 (W. Brown 1991). He also served as President of the Tropical Audubon Society in Miami.

Dr. Griswold commenced his career as a hummingbird collector in 1955 in connection with the "science program for the Univeristy of Miami televison show" (O. Griswold, 1960a). For several years he collected hummingbirds in the Caribbean and Costa Rica (Greenwalt 1960; Griswold 1960a). His collecting method was limed twigs (O. Griswold 1960a&b). He also went to much trouble to acclimate Todies, though his experiments with three species did not meet with much success (O. Griswold 1959; Schuerer 1983). Aside from the Bronx Zoo, the birds he collected went to several institutions and individuals, among them Raymond Sawyer and the Cleveland Zoo (O. Griswold 1960a). Streamertails sent to these two collections were destined for fame.

In 1958, the avicultural historian Arthur Prestwich (1958) announced: "One of the most interesting events in the annals of British aviculture is in process in one of Captain R.S. De Quincey's conservatory-aviaries. A pair of Jamaican Streamer-tailed Hummingbirds nested, duly hatched their two eggs and the young ones, according to the latest report, are now two weeks old" (Prestwich 1958).

Captain de Quincey related: "... In the spring of 1957, Raymond Sawyer asked me if I would like some Doctor Humming Birds *Trochilus poytmus* (Linn.), or Streamer-tails. I said I should welcome them gladly, and in due course he received four of these from a friend., three cocks and one hen. He kept a very fine cock bird for himself, and I had the others. I lost one cock bird in the spring of 1958. It had never been as fit as the other two, and was rather an aimless sort of soul. The hen and the other cock moulted out perfectly, and were put out again into my old pre-war greenhouse and open-flight in April. The hen, a very confiding and tame bird of very considerable character, was soon seen to be looking about for cobwebs, and hopes ran high that she might be thinking about breeding, and then a few days later the cock bird, apparently in perfect condition, was picked up dead. The hen seemed determined in her efforts to find nesting material, so cobwebs and lichens and some teased-out sheep's wool and hair were supplied for her to sort out. She built a very flimsy nest, attached to a branch and a leaf of the tender *Rhododendron rhabdotum*. She laid two eggs on two successive days, but the nest toppled over and one egg was broken. She then started building on top of this original nest, but this rather more robust structure was still far from secure, so I wrapped a hairnet round the nest and secured this to the branch somehow. The hen watched the proceedings with interest and no sense of alarm, and seemed grateful for this assistance, and she completed the nest and laid two more eggs, again on two successive days, but somehow she did not seem satisfied with matters. Feeling extremely doubtful by this time that her eggs would be fertile, I managed to persuade Raymond Sawyer to let me have his cock bird, as this was, by this time, known to be the only remaining cock Streamer-tail in the country- but we agreed that if Mr. Sawyer should want this bird at any time to show he could have it back, and it was indeed a very kind gesture on his part to let the bird out of his care at all, seeing the tragedies I had had with mine" (de Quincey 1960)..

"Almost as soon as Raymond Sawyer's cock bird was introduced, there was a lot of excitement and displaying on his part, attentions to which the hen bird was not insensitive, and it is certain that they mated, although this was not actually observed. The hen built another portion of superstructure on top of her second nest and its two now unwanted eggs, and on 23rd May, 1958, laid her first (really her fifth) egg, and on 24th May there was a second (sixth and last) egg in this nest. The cock bird made no attempt to help with the building of the last part of the nest, but, when the hen was on the nest, he perched within about a yard of it, and her, for the first two or three days of incubation, and drove away any other birds that happened to come near. After these first two or three days he took up a position in the outside flight, and took no further notice of the hen or the nest" (de Quincey 1960).

"The first chick hatched out on 10th June, the second on 11th June, not less than twenty-four hours and not more than thirty-six hours after the first one. The hen alone fed the two chicks, which, as they grew older, she did without ceasing. A vast number of fruit flies were bred and released within the greenhouse, and she fed herself on these and the nectar mixture, and then regurgitated it, seemingly pumping it right down into the crops of the two young birds. They grew very fast and were apparently very strong, pushing themselves up to the edge of the nest backwards to excrete. The edge of the nest begin to be pretty unpleasant and sticky but the young birds remained clean, and by 30th June appeared to be fully feathered. On that day I noticed the hen was flying more weakly than usual and unfortunately I had to be away from home just at that moment, although I do not think I could have done any more than was done, had I been at home. I was back on Friday, 4th July. The hen was weak and obviously failing, but still feeding the two young ones. On Saturday, 5th July, I found the hen and one young one dead. I feel that laying six eggs in fairly quick succession may have weakened her, and this, followed by her incessant maternal duties, and probably not exactly the necessary amounts, or proportions, of an attempted compensation for what she would find to feed on in the wild sate was just too much for her constitution. I think she may have pierced the oesophagus of this young one in her weakened state when trying to feed it, for it died with a curious, darkened swelling under the skin on one side of its throat. The other young one climbed on to the side of the nest and started using its wings. We [de Quincey employed two aviculturists] fed it by hand from the nectar bottle several times, and shortly after this it left the nest, flying a short distance. The next time it started to fly it landed on the ground, so I picked it up and put it into one of the usual Hummingbird traveling cages, where it seemed quite intelligent, and fed itself and stretched its wings, flying from its perch and landing on it again. Whether, when it landed on the earth floor of the greenhouse, it picked up some dirt to cause an infection in one eye, or, whether this damage was done in some other way, I don't know, but on 8th July I noticed one of its eyes was gummily closed. I treated this with penicillin cream, but it only scratched this with its foot, and the next morning it was obviously unwell, in fact it was torpid. My record reads: 'On 9th July, the remaining young hummer died.' This young bird left the nest on it own, and fed from a nectar bottle on its own, for a matter of four days, and it died in full, or virtually full, youthful plumage, but I feel myself that it was a 'near miss' rather than a complete success story..." (de Quincey 1960).

The cock bird, which had remained in the aviary throughout this entire period, was returned to Raymond Sawyer, to enter in the 1959 National Cage Bird Show. Donald Risdon, long-time Zoo and Bird Park aviculturist reported: "This year we were honored by the attendance of H.R.H. the Duchess of Gloucester, who spent over half an hour looking round the exhibits and finally presented the Supreme Trophy for the best bird in show to Mr. Raymond Sawyer for his Streamer-tailed Humming Bird... This bird also has the distinction of being the first of its kind ever to attempt to breed in captivity" (Risdon 1960). (It had already won best in the Hummingbird Class at the 1958 National Show (Yealland 1958a), before it sired the chicks at The Vern). Sawyer then retuned it to Captain de Quincey, who had built a new tropical house. De Quincey (1960) wrote: "The Doctor, as Jamaicans always call this bird, since his return, has been flirting rather shamelessly with a hen Wood Nymph *Thalurania glaucopis* to the enraged disgust of her proper mate. Although there will be no hope of breeding' The Doctor' unless some kind person provides me with a mate for him, I do start off this year with two or three different true pairs of Hummingbirds, a thing I have never managed to do before, and I hope that perhaps one of these pairs will do this new place that I have had built for them the honour of trying to breed here". To my knowledge, no further hummingbird breedings took place at The Vern, but other Streamer-tails collected by Oliver Griswold were quick to follow in making avicultural history.

It is an interesting footnote that Charles Cordier, like his counterpart Cecil Webb, made a brief career of zoo administration, serving, for a short time in the 1950s, as Curator of Birds at the Cleveland Zoo before leaving abruptly and establishing a base in the Belgian Congo, from where he supplied animals, primarily to Europe, until the disorders following Congolese Independence in 1960. Cleveland Zoo had built an elaborate Bird House in 1951, promoted as the "most modern in the world", which featured hummingbirds from its inception (Ranney 1951). An Antillean Crested Hummingbird lived at Cleveland for six years and 21 days, dying in 1960 (Prestwich 1960).

Oliver Griswold supplied a male Streamertail to Cleveland on 21 June, 1956, and a female 9 July, 1957, both arriving in adult plumage (O. Griswold, 1960; Reuther & Lamm 1959). They were each placed, shortly after arrival, in 17ft long "habitat cage", a glass-fronted aviary which exhibited a variety of hummingbirds, sunbirds, and other small softbills. It was not until February of 1959, when courtship was noted. Copulation was observed repeatedly during the first week of March, always on a "plant branch" (Reuther

& Lamm 1959). During the second week of March, the female commenced nest-building, but the material was interfered with by yuhinas, resulting in removal of the pair of streamertails to a glass-fronted exhibit "4ft, 8in by 4ft, 3in at the bottom and 4ft, 4in by 3ft, 4 in at the top" and 8ft, 4in high. Copulation was not observed after this relocation, but nest building commenced 15 March, 1959. By 22 March, the female had attempted building at three different locations, and like Captain de Quincey's bird, "the structures she assembled in each case appeared flimsy and unsatisfactory", prompting assistance (Reuther & Lamm 1959). An artificial nest of "clay and strands of fine hemp-string was placed in the third location, a begonia. The female immediately began adding material and laid an egg in this nest 24 March. It continued adding material, while subjected to minor harassment by the male. At 6:30AM the female was found on the ground, apparently egg-bound, but after hand-feeding, returned to its nest and laid a malformed egg by 8AM. The male continued harassing the female and was removed 29 March. The malformed egg was discovered broken 31 March. On 8 April, the begonia branch to which the nest was attached broke, resulting in the breaking of the remaining egg, which contained a well-formed embryo. Up to this point, the female would be off the nest for as long as an hour (Reuther & Lamm 1959).

On 10 April, 1959, the male was returned to the exhibit and the begonia was replaced with a privet. Copulation was observed over the next two weeks, and during the last week of April, the female again placed nest material at various sites, selecting the privet as the third location. An artificial nest was fastened with wire to this shrub. Eggs were discovered 30 April and 2 May. Regular incubation commenced after the second egg was laid. The male was removed 3 May, and died of unknown causes 23 June, 1959. It was noted that "in both nestings the female always incubated in the same position, facing the glass, and thus, the public. This was the position that afforded her head region the most light" (Reuther & Lamm, 1959).

Some time between 6:30AM and 8:20AM, on 18 May, 1959, one of the eggs hatched, establishing an incubation period of sixteen days. The second chick was partially out of the egg at 10:30AM, and fully emerged by noon. The female's feeding routine was to drink nectar first, then catch fruit flies, and then feed the chicks. One chick was found dead in the nest 3 June, with possible trauma due to overcrowding and the rigid structure of the artificial nest. The remaining chick was seen on the edge of the nest for the first time, exercising its wings, 6 June. On 12 June this chick was seen hovering about two inches above the nest, but was not observed to do so again until 14 June, when it flew to a twig. The chick commenced feeding from a bottle 15 June, but the female continued feeding it until it was removed 1 July, after it was observed keeping its offspring away from the bottle. The chick remained in the aviary. The female laid an egg in a holding cage 3 July, and was transferred the same day to the large community aviary where another male was resident (Reuther & Lamm 1959). Oliver Griswold (1960a) noted that the chick had been named "Celia, Jr." (in honor of Lee Crandall's wife) after its sex was confirmed as a female by November, 1959.

In comparison to other diets previously discussed, the remarkably complicated Cleveland Zoo diets of the 1950s are startling. In addition to fruit flies, the morning and mid-day diet was a mixture, prepared in a blender, of meal worms, peeled grapes, raw horse liver, raw carrot and water, to which was added honey and sweet condensed milk. The diet provided at 4PM consisted of honey, Vitamin B, and a mixture of orange, tomato, grape and apricot juices (all canned) (Reuther & Lamm 1959). (In contrast, (Richard de Quincy (1960),after losing birds to "fits' due to diets that were "too fat", provided only fruit flies and a mixture of five parts honey to one part of Percy Hasting's proprietary "Stimulite" paste [of which more will be said], with water).

*The '60s*

Whether or not one defines Augusto Ruschi's Museu de Biolgia Prof Mello Leitao as a "Zoo", Cleveland Zoo's 1959 breeding of Streamertails is certainly the first Zoo breeding of hummingbirds outside of Brazil. And, as can be seen from Table I, it was the last until

1970 - At least no hummingbird hatching between 1959 and 1970 is listed by the International Zoo Yearbook (Zoological Society of London, 1960-1998). In retrospect, this seems perverse, as the 1960s were the decade when the availability of hummingbirds was at its height. Brazil again prohibited export of birds in 1967, but, by then, Ecuador was commercially exporting mass quantities by air, representing a great diversity of species. Several other countries were exporting hummingbirds as well.

It is telling that the cover design of the *Avicultural Magazine* switched from a Greater Bird of Paradise in 1958 to two hummingbirds, one a rather stylized Sparkling Violet-ear, the other apparently intended to be some member of the genus *Coeligena*. Though uncredited, I believe this drawing was by the prolific zoological caricaturist L.R. Brightwell. It remained until 1968, when it was replaced with Diamond Sparrows by Robert Gilmore. By the mid '60s Arthur Prestwich (1965) was able to observe: "The present-day aviculturist can, if he so wishes, obtain humming birds with little difficulty; but before the last war, being great rarities, they were available to only a favored few...".

A series of advertisements from the Cologne-based firm Zoologischer Versand Koeln (run by Victor Franck, who was still active almost forty years later) appeared in the endpapers of the Avicultural Magazine. One running in the November-December 1963 number (Vol.69, No.6) offered more than fifteen species of Hummingbirds, apparently all Ecuadorian, including Giants, the Velvet-Purple Coronet *Boissonneaua jardini*, the Black-tailed Train-bearer *Lesbia victoriae*, and three species of *Heliodoxa*. In the following issue, this firm proclaims: "We are Specialists in Humming Birds and Sunbirds... We have a large number of Humming Birds in stock, both common species and some of the rarer such as *Eugenes fulgens*, etc.... Dealers: We give a discount on ten or more Humming Birds". In Number 4 for 1964, 20 species are listed including both species of *Lesbia*, three *Coeligena*, and two species of *Boissonneaua*. Charles Cordier, having left the disorder of the former Belgian Congo, re-established himself in Cochabamba, Bolivia, and by the mid-1960s was running ads in *International Zoo News*. In the October 1966 number (Vol. 13, No.4), he advertised the "Estellas Hill-Star [*Oreotrochilus estella*] from 13,000 feet altitude and higher" as well as the Red-tailed Comet *Sappho sparganura* which I do not believe was in aviculture before the '60s. On the same page as this ad is one for "P.H. Hastings Ltd". This is startling in that Percy Hastings (repeatedly) advertised his Portsmouth firm to zoo professionals as "Specialists in Mountain and Lowland Gorillas". Hastings (1967) claimed: "...I have specialized in nectar feeders since 1919. I am, as far as I know, the first purveyor to import Humming Birds, around 1924, and since that time, with the exception of the the war period, and some little time after been the sole purveyor of these birds until I dispensed with the services of my collector in Ecuador in early 1963. In all these forty-seven years, I have handled many thousands..."

At this time of large-scale commercial importations of hummingbirds to Europe a now classic work on captive hummingbirds appeared. Originally published in German in 1966 as *Kolibris*, Walter Scheithauer's (1967) *Hummingbirds* became available to English-speaking readers, in a masterful translation by Gwynne Vevers, the London Zoo's Aquarium Curator. Beginning in the 1950s, Walter Scheithauer assembled a wonderful collection of hummingbirds at his house in Munich, including the first Velvet-Purple Coronet in European aviculture, as well as a Swordbill and a White-tipped Sickle-bill. His book has become quite a collector's item due to the pioneering photos he took after much effort in devising techniques and equipment. A wide range of the species available in the 1960s is thus beautifully documented. He was also very thorough in accumulating data. Although at the time his book was published he had not hatched any species, he provided detailed accounts, with pictures, of the nesting of several. Five species built nests in his "communal aviary" and a Brown Inca *Coeligena wilsoni* and a Violet-bellied Hummingbird *Damophila julie* "laid four eggs each in one year, which they incubated for the full period" (Scheithauer 1967).

Scheithauer (1967) also conducted a thorough investigation of diets. He provides a survey of what had been done before, presenting a rather appalling collection of highly complicated diets, including Cleveland Zoo's, and others even more complex. Frankfurt Zoo for instance, included in a single formula seaweed meal, germinating wheat, ant's eggs, locusts (or mealworms), horse heart (or veal), cattle blood serum, raw egg, lettuce leaves, apple, banana, orange and carrot, as well as honey, condensed milk, baby food and vitamins. Scheithauer (1967) commented "My compliments to these little birds which have to swallow it all". After conducting his own experiments on a Brown Inca and a White-eared Hummingbird *Hylocharis leucotis*, he devised his own formula, consisting of a "ready-mixed baby food", Nektar-Mil II (produced by Milupa-Pauly), honey and water, with two sorts of vitamin drops. He was able to restore a sick Sword-bill to health, so that "after an initial illness [it] developed from a feeble flier into a really first-class performer, vibrating very actively and relatively fast" (Scheithauer 1967 & 1961). He also found this diet did not sour overnight, making the provision of a different evening solution unnecessary.

Jean Delacour also came to endorse a highly simplified diet. From 1952 to 1960 he was Director of the Department of History, Science and Art for the County of Los Angeles. He immediately established aviaries at his Los Angeles house. For most of his years there he did not keep hummingbirds. "As the garden is usually occupied by wild hummingbirds, it is not necessary to keep any in confinement where space is very limited" (Delacour 1953a). However, by 1958, he was keeping Black-chinned and Costa's hummingbirds, neither of which are especially common around Los Angeles. He fed these, and a collection of East African sunbirds "a mixture of honey and beef extract, which does not spoil before twenty-four hours, adding to this diet small insects, which are rather scarce in the winter but abundant during the warmer months. It is amusing to witness the incessant chase they give them" (Delacour 1959). He had also recently become an enthusiastic collector of the "beautiful Tanagers and Sugar-birds" that have "lately... arrived by air, in excellent condition, from Colombia and Costa Rica" (1958a), describing them in some detail in an Avicultural Magazine article (Delacour 1958b). He also obtained Long-tailed Manakins *Chiroxiphia linearis* from Costa Rica (Delacour 1959). He made no mention of any hummingbirds arriving with these species, so I gather that the large-scale commercial import of Central American and Andean Hummingbirds had not yet occurred in the U.S. In their comprehensive Finches and Softbilled Birds, the Southern Californian importers Henry Bates and Robert Busenbark (1963) speak only in generalities while discussing hummingbirds, obviously having had little, if any experience with foreign species at the time they wrote it. However, in 1967, they wrote Arthur Prestwich (1967) informing him that, at their Palos Verdes Bird Farm, they were "very busy with quite an number of humming birds. Most are from Ecuador, but we do have two specimens of our native Anna's Humming Bird". The hen Anna's had been reared with an eye-dropper. Apparently, the Government was still fairly liberal in granting permits to private persons to keep native species.

In 1960, Jean Delacour retired at the age of seventy from his directorship of the Los Angeles County Museums. He thus had to sell his California house, but, for the first time since 1939, he was able to keep hummingbirds at Cleres. "As I am now planning to spend a couple of winter months in California and two or three in New York, the collections at Cleres have been considerably increased in the course of the present year, particularly those of small birds. The principle outdoor aviaries have been done up, the greenhouse transformed and replenished. There are a good many indoor cages as well. Among the new birds, are a dozen Hummingbirds, brought from Brazil by Dr. E. Beraut; Sunbirds, Sugar-birds, and various insectivorous and fruit-eating birds, so that it looks much as in previous years" (Delacour 1961).

"Dr. E. Beraut kindly brings me a number of humming birds from Brazil each year. About fifteen in a large outdoor aviary did very well throughout the warm spring and summer of 1964, but, as is inevitable, after a time when they felt fit, they started killing one another, until only one of each species was left in October when they were taken in for the winter. They were Wood-nymph, *Thalurania glaucopis*, Red-tail *Clytolaema rubricauda*, Crested *Stephanoxis verrauxi*, White and Green *Leuchloris albicollis*; but there were still four Black Jacobins *Melanotrichilus fuscus*, a more accommodating species. In 1965, we separated them more carefully. A pair of sunbirds usually share the hummingbird's aviaries but such savage species as Taccazze will attack and kill them, and they have to be watched" (Delacour 1965).

I believe some time later Delacour (1972) devised a system for reducing aggression in mixed collections of hummingbirds: "The writer has successfully kept about 40 hummingbirds for a long time in a n enclosure 16ft long, 3ft high, and only about 20in wide, built in a greenhouse and connected to an outside flight 16ft x 7ft x 3ft, to which the birds were given access during the summer months". He added: "There is no room to collide, especially if the aviary contains plenty of branches, twigs and growing plants" (Delacour 1972). His friend, the master aviculturist Dr. Henri Quinque (1988) described these accommodations thusly: "Hummingbird tunnels were an invention of his own to enable these wonderful, but unfortunately very aggressive birds to live together. Inside his aviaries he set up enclosures 10m long but no more than one metre high and one metre high, filled with exotic plants. In these enclosures the birds loose their territorial instincts and 50 of them can live together without harm". Dr. Quinque (1988) wrote that these "tunnels" were in use at Cleres in the 1930s, but I have found no evidence of this.

In the 1960s, Dr. Beraut moved from his 12th floor Rio de Janeiro apartment to a an extensive property outside the city, where Jean Delacour (1970) found one of the most remarkable softbill collections ever assembled anywhere. Hummingbirds were everywhere, including such foreign species as Red-billed Streamer-tails, and Green-throated and Purple-throated Caribs. Particularly surprising were the species kept in large aviaries with all sorts of other birds. In a single large flight were "Central American and Golden-headed Quetzals, Scarlet Cocks-of-the-Rock, a number of Cotingas; a love and tame Swallow-tailed *Phibalura flaviventra*, a Bare-necked Fruitcrow *Gymnoderus foetidus* and a Black-necked *Tityara T. cayana*; a Swallow-winged Puffbird *Chelidoptera tenebrosa*, various Tanagers, Scarlet-chested Sunbirds; several small Hummingbirds *Sericotes* [*Eulampis*] *holosericeus, Augastes lumachellus, Stephnoxis lalandei*; White-capped Redstarts *Chaimarehornis*, Royal, Splendid, Amethyst Starlings, American Jacanas (nesting), different plovers, etc.".

The hummingbird *Augastes lumachellus*, generally known as the Hooded Visor-bearer, is one of the most famous species Augusto Ruschi worked with. Suspected to be extinct for decades, its habitat was rediscovered by Dr. Ruschi in 1959, after launching a special expedition, during which he collected twelve living specimens (Marden 1963). Confined to a small range in the Brazilian stae of Bahia, it is today considered a Near-Threatened species by the IUCN. Several years later, Dr. Ruschi again achieved celebrity when he discovered the previosuly unknown range of the Marvellous Spatule-tail *Loddigesia mirabilis*, high in the Peruvian Maranon Valley. He brought six living birds back to his Museum, where they were photographed by Crawford Greenwalt (1966), who featured this bird in an article in National Geographic. Augusto Ruschi was also responsible for the inaugural collection for one of the most famous and popular hummingbird exhibits anywhere.

While the San Diego Zoo, founded in 1916, had maintained an important and impressive bird collection since the the 1920s, hummingbirds were not part of it for almost all of the zoo's first fifty years. A few "wounded or sick" Anna's Hummingbird's brought into the zoo's office had been kept for brief periods, then released (Benchley 1935 & 1936), and there had been some discussion of "a hummingbird exhibt built over trees and shrubs" (Benchley 1935), but nothing had come of that.

During his tenure in Los Angeles, Jean Delacour had taken a great interest in the San Diego Zoo. It was his suggestion that the enormous 1922 Scripps be converted into a walkthrough aviary in 1958 (Lint 1958), followed by the transformation of the 1936 bird of prey aviary into the walkthrough Rainforest in 1960 (K.C. Lint pers. comm.). Encouraged by the enthusiasm with which these innovations were greeted by the public, it is only natural that he would have encouraged San Diego to create an outdoor, walkthrough display of Hummingbirds (Heublein 1965). This aviary was opened to much fanfare (Lint, 1965, 1966 & 1970). K.C. Lint, San Diego's long-time Curator of Birds announced: "A spectacular new exhibit was opened October 2, 1964, in the San Diego Zoological Gardens. The round-topped aviary (64 feet long, 24 feet wide, and 12 feet high) houses one of the finest collection sof tropical birds. One hundred and five Brazilian hummingbirds, representing 23 species, were collected by the author and Dr. Augusto Ruschi... A visit with Dr. Ruschi, the world's leading authority on hummingbirds is a memorable and rewarding experience. Without his interest and continual assistance in guiding this expedition into the high country of eastern Brazil, these birds never would have reached San Diego to be seen by zoo visitors (Lint 1965)." Some of the species were Frilled Coquettes, Black Jacobins, and Scale-throated Hemits *Phaethornis eurynome*.

Accompanied by his wife Marie, and his son Roland, K.C. Lint assisted in the capture of these birds. Many were mist-netted, but others were captured as they mobbed the tame Ferruginous Pygmy Owl which Ruschi customarily used as a decoy (Lint 1965; Marden 1960). Ruschi's method of catching hummingbirds thus engaged, using bird lime at the end of a telescoping pole, is famous (Marden 1960). Less well known is the use of a single hair, around nine inches log, made into a noose, and "attached to the end of a piece of No. 26 steel wire, 1-1 ½ inches long, which in turn is mounted on the end of a ten-foot bamboo pole. When the bird hovers for some time to fix its attention on the enemy, the loop is dropped over the bird's head. A slight movement is made so that the loop tightens around the neck. By releasing the loop the bird can be freed quidkly. For the Coquettes *Lophornis* and the Woodstars *Calliphlox*, the smallest species, a strand of human hair is recommended. Hair from the mane or tail of a horse or mule can be used for larger species" (Lint 1965).

The birds were transported over the 16½ hour flight from Rio de Janeiro to San Diego in specially designed travel cases, each bird restrained in its own cloth jacket (Lasiewski 1962; Lint 1965; Marden, 1960). They were easily bottle-fed during the journey. The San Diego Zoo adapted a diet from Dr. Ruschi, consisting entirely of cane sugar, Super Hydramin Powder ("a human protein-vitamin-mineral concentrate") and water, with a continuous supply of fruit flies. Lint (1965 & 1966) provided formulas for both the hummingbird solution and a medium for culturing fruit flies, based on agar, corn meal, yeast, and both dextrose and sucrose.

There was much zoo activity during the 1960s. At London, the decade ended as it began, with a renovation of the Tropical House. In 1960, "the interior of the Tropical House is being improved and flying space much enlarged by the removal of the corridor. Visitors will be able to walk though among the birds, entering by a darkened pouch at one end and going out through a similar darkened porch at the other. Hummingbirds, sugar birds, small tanagers, and perhaps sunbirds will be kept there" (Yealland 1960a). John Yaelland provided further details of this renovation in a separate article (Yealland 1960a). The 1969 remodeling involved the addition of "three small waterfalls" and "the opportunity was taken when the previous occupants were reintroduced to include four new species of Hummingbird, including two Black-throated Mangos *Anthracothorax nigricollis* which had not been in the collection for many years" (Olney 1969). This resulted in sixteen species of birds present in the building.

Among the birds exhibited in London's Tropical House in the 1960s were a Puccheran's Emerald, still present in the building in 1961, having been acquired in 1957 (Yealland 1961). The Brown Vioet-ear *Colibri delphinae* (Yealland 1960c) and the Heine's Hummingbird *Amazilia tzacatl jucunda* (Yealland 1960a), both species that were to be considered "beginner's" standards in the Ecuadorian hummingbird shipments, were new to London's collection in 1960, while another, the Brown Inca *Coeligena wilsoni* was received for the first time in 1965 (Yealland 1965). Another Ecuadorian novelty to the collection was the Napo Sabre-wing *Campylopterus villavicencio*, now classified as a near-theatened species, a gift of Viscount Chaplin in 1965 (Yealland 1965). From the late Captain de Quincey's estate came a small collection of rare softbills including an unspecified Wood Nymph and a "Pied Jacobin", which I take to be *Florisuga mellivora*, in 1967 (Yealland 1967)... Two Red-billed streamer-tails were received in 1968 (Yealland 1968a,b), one from Len Hill's Birdland.

Birdland was one of a number of English bird parks that flourished during the 1960s only to come into difficulties toward's the end of the Twentieth Century. The Avicultural Magazine's long-time editor Phlyllis Barclay-Smith (1968) observed: "A great landmark in the history of Birdland was the construction in 1965 of a new tropical house… The main hall is well planted with trees, shrubs and flowering plants, a stream and waterfall ensure humidity, and the whole forms a perfect setting for the tropical birds, all of which are free-flying. In the outer hall are the "Jewel enclosures" housing several species of beautiful Humming Birds; these enclosures are based on the design of the late Captain de Quincey, that great expert on the keeping of Humming Birds." While it does not appear hummingbirds actually hatched at Birdland, clear eggs were laid (Barclay-Smith 1968). Green-tailed Sylphs *Aglaiocercus emmae* and Pale-bellied Hermits *Phaethornis anthophilus* were among the inhabitants of the new Tropical House, while the Green-crowned Brilliant *Heliodoxa jacula* could be seen in the old tropical house near the front entrance (Barclay-Smith 1968). Len Hill, Birdland's Founder collected Red-billed Streamertails in Jamaica during the making of "The Flying Doctors of Jamaica", a film made for B.O.A.C. and the Jamaican Government to aid tourism. "It included habitat, feeding young, catching and transportation of Doctor Humming Birds" (Prestwich 1969a). Len Hill (1976) discussed these birds in his autobiography. The five Streamer-tails were brought back to England in Ruschi-styled "straight-jackets", in a case. Len Hill invented the plastic hummingbird feeder, which one could buy at the Birdland shops (Pryor 1966), and, in one form or another entered standard usage (Hill & Wood 1976). (They replaced glass feeders, which appear to have been the invention of James Bailey, 1886-1973, 44 years at London Zoo, retiring as Overseer of Birds in 1948 (Silver 1948), having been the first keeper of the Tropical House (Horswell 1973)).

Elsewhere in the 1960s, Zoologischer Garten Berlin opened a huge modernistic bird house in 1963 (Kloes 1963), a replacing the one destroyed in 1944, and Hummingbirds were seen there again (Kloes 1964a, 1964b,1965a, 1965b, 1968). That same year, Tierpark Berlin, which had been created in the Eastern Sector in 1954, opened the enormous Alfred Brehm House, a multi-pupose building, which, among its many species, included hummingbirds (Dathe, 1965). Chester Zoo, in the West of England, included hummingbirds among the birds in its new walk-through Tropical House (Fletcher 1962). Although White-eyes, and eventually a verity of waxbills bred there, hummingbirds did not, and it was proved conclusively that Cocks-of-the-Rock will eat hummingbirds, at least on occasion. (Fletcher 1964)

While no zoo hummingbird breedings appear to have taken place during the 1960s, two master aviculturists did achieve success, and thoroughly documented their achievements. The Californian Alex Isenberg, was, among other things the first person to hatch touracos and birds-of-paradise in the Western Hemisphere, and maintained a marvelous collection of softbills near the San Francisco Bay, from the 1920s until his death in 1971. In 1959, he hand-raised two Allen's Hummingbirds that had fallen from the nest. Both proved to be males, and one killed the other the following year. In the fall of 1960 he caught a female, and after maintaining it for several weeks in a cage in the male's aviary, released it to join the male (who had been displaying during the introduction period). In April 1961, the female commenced building a nest, using "bunches of cotton string" which Isenberg

(1962) had tied to tree branches, also adding cobwebs and lichens. The "flimsy" nest was upset during a windstorm, destroying the two eggs a week after they had been laid.

The female built another nest within four days, this time "much stronger". Both eggs hatched 19 June, 1961, after a nineteen day incubation period (during which time the male was removed because of attacks from the female). After ten days one chick disappeared, the other taking its first flight on 17 July. The female continued to feed the chick for two weeks, then began attacking it, prompting its removal (Isernberg 1962). Thus occurred the first breeding of any North American hummingbird, and only the second hummingbird species captive-bred outside of South America. (Several years before, Isenberg (1961) had a pair of Purple-throated Caribs go to nest and lay two eggs, but the female was "injured before incubation term was up".) The aviary where the Allen's were kept was 40ft log, 8ft wide, and 9ft high, and shared with a breeding pair of Clarinos (Slate-colored Solitaires), a pair of Pygmy Nutchatches and a Japanese White-eye.

Isenberg's (1962) ideas on nectar were quite different from those of his friend Delacour: "Finally a word about our formulas for hummers, sunbirds, etc., may be of interest. The morning formula consists of one part honey to eight to ten parts water... To this we add a generous, overflowing teaspoonful of Thompson's Multi-Vitamin Syrup and then we add four heaping teaspoons of Soyagen Milk Powder (which is a milk substitute) and blend for not more than 30 seconds in an electric blender. At 3PM, all feeders are washed and filled with one part honey to eight to ten parts water, twelve to fourteen drops of ABEDC vitamins for two quarts. We use a little hot water with the honey at the start. Never use beef extract, white sugar, or synthetic vitamins".

The other non-Brazilian to breed Hummingbirds in the 1960s, would become the first person in the Eastern Hemisphere to hatch (but not raise) a Cock-of-the-Rock (Prestwich 1970). Lena Scamell, who was to receive fifteen medals from the Avicultural Society for first British breedings, had already distinguished herself for the first captive breedings of the Daurian Redstart, two species of Rubythroat *Luscinia calliope* and *L. pectoralis*, and the Malachite Sunbird (Sawyer 2002). Raymond Sawyer (2002) recalled her Surrey aviaries thusly: "... The Scamell's garden is best described as having been in a wood with fairly small aviaries designed with breeding in mind. Lena was quite a large person, and when she walked through the long and narrow aviaries, the birds had to fly over her shoulders to get past her; but what success she achieved in them...".

In 1966, Arthur Prestwich announced: "The most exiting avicultural event this year so far is the hatching of two young by Mrs. K.M. Scamell's Violet-eared Humming Birds *Colibri courscans*, on 8th and 9th April. Very unfortunately they did not survive; one died after thirteen days and the other fourteen" (Prestwich 1966).

Lena Scamell (1966), who had been keeping hummingbirds since 1950, wrote a lengthy and detailed account of this breeding. This pair of Sparkling Violet-ears were from Ecuador, the male purchased in late 1964, the female in May, 1965, both from "Mr. Riley of Petcenta, Southall" (who appeared to be able to sex Violet-ears by beak curvature). They had initially been kept in the Scamell's conservatory, but the arrival of hummingbirds from Ettiene Beraut prompted their transfer to a bird room in "the small wood" in August, 1965. Their accommodation was an outdoor compartment 8x2x7' high connected to indoor 3x2x6' high. Until November, they had 24 hour access to the outdoor flight, and after that were given daytime access except during frost. Nighttime temperatures indoors could fall to 38F. Following a successful wintering under these conditions, nest construction was first noted 19 March, 1966, and proceeded rapidly. The site was a "wooden batten 4 ft. 3 in. from the floor, which supports the wire netting" inside the shelter. Courtship flights and copulation were observed the same day. By 22 March, the nest, constructed from fine dried grass, Kapok, cotton, and moss, was an inch deep. An egg was discovered the next morning. Copulation was also observed that day, which was snowy, with a strong north-east wind. When the second egg was laid on the 25th, the male was removed. It was noted on subsequent days that the female consumed far less nectar than the male. On 8 April, the day the first egg hatched (with a fifteen day incubation), the temperature was 59F. The hen, which had been confined to the indoor flight since incubation had commenced was now allowed access outdoors. For several days following the hatching of the second chick the weather was extremely cold, with snow storms, and the shelter temperature was sometimes at 50F, despite artificial heat. Commencing 21 April, the chicks appeared to be weakening, and both died on the 22nd. Mrs. Scamell (1966) suspected that the female, confined to the shelter, had not been able to properly clean its beak and thus impacted the chicks with nest material while feeding them. Attempts to reintroduce the male the next day were disastrous, and the pair were not really compatible until June. In August, another nest was constructed and eggs laid, but the nest was destroyed by the male.

The further doings of this pair were again recorded in great detail in another article (Scamell 1967a). Nest building recommenced in February, 1967, eggs were laid in March, one disappearing, but the other hatched 4 April. (The male had been removed 24 March). For several days after this hatching, the bird room temperature did not rise above 45F. The young bird seemed more disturbed as the temperature rose to about 80F on sunny days. The chick, which underwent torpor during some evenings, did not leave the nest until 4 May. It was transferred to its own aviary 14 May. Further nesting attempts by the parents that year were not successful. A chick hatched 9 February 1968 died after two days (Prestwich 1968c). The bird which had full fledged in 1967 died in 1968 from a ruptured aorta, being in otherwise excellent condition, a male (Scamell 1968).

A peculiar result of Mr's Scamell's highly detailed accounts of her Violet-ear hatchings was some of the most vitriolic correspondence to appear in the *Avicultural Magazine*. It began with Percy Hasting's (1967) response to the Mrs. Scamell's (1966) first article, detailing the unsuccessful first hatching. Her hummingbird diet of choice was that of her friend Ettiene Beraut - The American product Gevral-Protein with water and white sugar, supplemented with copious fruit flies. Hastings, who manufactured and sold his own line of "nectar paste" vigorously attacked this diet, squarely blaming it for the death of the chicks. He had little use for the Gevral-Protein but utterly condemned the use of white sugar, declaring it to be entirely without value as a food source, and a positive danger: "In all these forty-seven years, I have handled many thousands and in that time I have had over 800 post mortems on Nectar feeders, which show about 95 percent die from excess sugar. This results in the almost complete dehydration of the flesh and muscles, and invariably diseases of the liver, extended, perforated, and cancerous conditions, and in most cases also, severe inflammation of the mouth gullet and tongue, the latter in very many instances rotting away altogether." He had long before abandoned the use of Alfred Ezra's Mellin's Food formula: "Many years ago I cut out honey, condensed milk, and prepared invalid foods as they all turn sour overnight; I use only Desxtrine-Maltose ingredients which even in hot weather will not turn sour in 24 hours" (Hastings 1954).

An attempt by Mrs. Scamell (1967b) to reply to this letter led to a more extreme response from Hastings (1968). Other people with a knowledge of chemistry and physiology weighed in (Harrison 1968; Peaker & Peaker 1968), calling into doubt Hasting's arguments of the comparative merits of sucrose versus dexatrine or maltose, and Mrs. Scammell (1968) attempted to defend herself again. Finally, another member called for a pox on all houses (Walton 1968). Hastings then ran ads for his "Stimulite" nectar paste (later granules) that ran in at least eight numbers of the 1968 and 1969 Avicultural Magazine. His arguments continued there. The whole business reminds me of the edginess between the American representatives of the two main proprietary nectars available in the 1980s.

*Sustained Propagation*

From Table I, it can be seen that at least 338 hummingbirds of 16 species were hatched among seventeen zoos, from 1970, through 1996, the last year the International Zoo Yearbook (Zoological Society of London 1998) collected breeding records. Eighty of these were Sparkling Violetears *Colibri courscans*, and forty are indicated to have reached independence. Not included in these statistics are at least seven *C. coruscans* hatched, with at least six reared, at the Wildfowl Trust at Slimbridge from 1983 through 1992 (Hodges 1991, Woolham, 1993). Slimbridge also hatched the Green Violet ear *C. thalassinus*, which does not appear in the IZY list, in 1978 (Harvey 1978), and its first hummingbird hatching, sometime before 1977, was a "Sparkling x Brown Violet-ear which lived for one year" (Elgar 1977). Parc Zoologique de Cleres came close to breeding *C. coruscans* in 1972 when "... a female Violet-eared Hummingbird built several nests without laying" (Delacour 1973).

From Mrs. Scammel's experiences, and as will be seen from the circumstances of San Diego Zoo's hatchings, (and not to forget Albert Pam's 1905 London bird), Sparkling Violet-ears appear comparatively indestructible. They are definitely notorious as one of the more territorial and belligerent hummingbirds, and can be a menace to other species in aviaries. I remember, in 1988, an American zoo, where hummingbirds had not been kept before, acquiring four specimens each of four species, one of which was the Sparkling Violet-ear. They were placed in a fairly small indoor space, and less than a year later only two were alive, one of them a Violet-ear. Be that as it may, this species in particular, and the genus *Colibri* in general, seems more adapted to reproduce in challenging environments. Mary Havey (1978), long-time Editor of the Avicultural Magazine commented: "It is curious that nearly all the captive breedings of hummingbirds have involved the genus *Colibri*".

At times, it has seemed that in exhibits designed to showcase hummingbirds, practically everything else hatched there but hummingbirds. At London Zoo, the first British breeding of the Purple Honeycreeper *Cyanerpes caeruleus* occurred at the Tropical House in 1967 (Yealland 1968a). The first captive breeding of any species of manakin was achieved when Blue-backed Manakins *Chiroxiphia pareola* hatched there in 1972 and 1974 (Olney 1973 & 1976). However, no hummingbirds ever hatched at London Zoo.

Within months after the October, 1964 opening of San Diego Zoo's Hummingbird Aviary, Golden, Golden-masked, and Silver-thoated Tanagers had hatched in this outdoor enclosure, as well as Indian White-eyes and Black-necked Stilts (Lint 1966). The first public zoo hatching of a cotinga took place there in 1965, when an Orange-breasted Cotinga *Pipreola jucunda* was parent-reared (Lint & Dolan 1966). A White-naped Honeyeater *Melithreptus lunatus* hatched in 1968 (Lint 1968) was the first of its family captive-hatched in the U.S. or Europe. Jean Delacour (1972b&c) was especially intrigued by a hybrid produced by a Red-legged Honeycreeper and Masked Tanager *Cyanerpes cyaneus* x *Tangara nigrocincta franciscae* in 1971.

Hummingbirds did hatch at San Diego in 1970. None of the four Brown Violetears *Colibri delphinae* survived to independence (Table I.) However, both of the Sparkling Violetears *C. coruscans* hatched on the 16th and 17th of March, after a seventeen day incubation period, fledged and became self-feeding (though the International Zoo Yearbook indicates one was a juvenile mortality (Table I)). A beautiful series of photos accompanied K.C. Lint's (1970) account of this hatching. 1970 was the year that San Diego Zoo would hold the most species and subspecies of birds in its history (Lindholm 1993 a&b). For 31 December, 1969, San Diego's inventory lists 1,126 taxa of birds. Forty-one of these were hummingbirds, and all lived in the same aviary, a total of 82 specimens (Lint & Dolan 1969). Sparking Violet-ears hatched in this exhibit again in 1971, when, according to the IZY, two hatched and one was reared. In 1973, both of the two chicks hatched survived (Table I). Non-native Hummingbirds would not be hatched at San Diego again until 1983. K.C. Lint (1981), in a privately distributed manuscript, states the Rufous-tailed Hummingbird *Amazilia t. tzacatl*, not listed in the IZY, hatched at San Diego in 1971. On 1 January, 1973 there were 35 hummingbirds of seventeen taxa at San Diego, of which eight were Sparking Violet-ears (Lint & Dolan, 1973). With the exception of two taxa from California and three from Central America, all appear to have been Ecuadorian or Colombian. Among the 41 taxa present at the end of 1969, there were still three Brazilian species, despite the 1967 Brazilian export ban, but already, the Andean birds from Ecuador were dominant.

In 1971, an outbreak of Exotic Newcastle's Disease in California had caused great alarm, resulting in a ban on bird imports to the United States in 1972 (and incidentally resulting in the formation of the American Federation of Aviculture). Subsequently, the importation of birds to the U.S. was allowed only through quarantine stations, which exponentially increased the prices of birds. Prior to 1972 ban, practically all the available species of Ecuadorian and Colombian Hummingbirds could be purchased for less than $25 each, even valued collector's items such as the Velvet Purple Coronet or the sylphs. As previously noted, dealers were quite ready to offer discounts on quantities of hummingbirds, and there are stories, from the pre-Newcastle's days, of $5 hummingbirds.

The immediate effect in Europe of the U.S. Newcastle's import ban was announced by the Belgian aviculturist Johan Ingels (1974) wrote: "The ban on importation of birds into the United States imposed at the end of 1972, has resulted in an increased import of Central and South American birds into west European countries, and large consignments of neotropical birds including tanagers have reached Belgium and the Netherlands. Most of the tanagers were from Ecuador and Colombia, but small consignments were also obtained from Mexico, Brazil, Bolivia, Paraguay and Peru". However, not long after, Ingels (1975) noted: "Due to growing opposition among protectionists, it has become increasingly difficult to export live birds from South and Central American countries. Those which have come in have been mostly from Ecuador and Brazil, with small consignments from Colombia and Surinam also. I find it interesting that birds from Brazil were arriving at all, as a Brazilian export ban had been put into effect in 1967, but it appears such species as the threatened Brazilian endemic, the Superb Tanager *Tangara fastuosa*, continued to be available for years afterwards.

At any rate, a combination of factors began to limit European access to Hummingbirds. By the mid-1970s Newcastle's disease restrictions began to appear in Britain and Europe, and the recently created Convention on International Trade in Endangered

Species was beginning to have some effect (Kear 1976). The '70s were the decade that Costa Rica and Colombia ceased commercial export, and Ecuador greatly curtailed its bird trade. By 1980 only one South American country was supplying hummingbirds commercially: "At present only Peru will allow small numbers of hummingbirds to be exported and under supervision from the authorities whose control is strictly enforced. The birds can only be collected from the Western side of the Andes" Elgar (1982c).

Two German Zoos achieved sustained propagation towards the end of the 1970s. Not surprisingly, one collection did very well with Sparkling Violet-ears. As can be seen from Table I, this species was bred at Tiergarten Heidelberg every year, from 1978 through 1984. A total of 21 were hatched, of which only four are indicated to have died before independence. The most hatched in one year were five in 1981, when it was indicated that a second generation breeding had occurred, the first of which I am aware for any hummingbird.

Dieter Poley (1975), Director of Tiergarten Heidelberg, wrote, regarding sunbirds and hummingbirds: "The best method of keeping these birds is to ensure that they have ample supplies and a variety of live food, and since the breeding of large quantities of fruit flies, etc, is a lot of work, we have our nectar-feeding birds in an aviary that combines an inner shelter with an outdoor flight. The birds can get from the planted and heated shelter measuring 14 x 4 x 3.5m. via an open window to the planted outdoor flight which measures 22 x 2.5 x 3m, where many insects are attracted to the food vessels, while in other parts of the aviary we put receptacles of decaying fish, meat and fruit to attract more insects. In the evening the artificial lighting of the shelter attracts still more insects and the birds are thereby encouraged to come for the night into the warmed shelter. During the day almost all the aviary inmates spend their time in the outside flight..." In 1975, six species of sunbirds were kept in this facility, along with a flock of Scarlet-backed flowerpeckers, Masked Flower-piercers *Diglossa cyanea*, bananaquits, honeycreepers, and un-named hummingbirds. One of the two pairs of Splendid Sunbirds hatched a chick which did not survive (Poley 1975).

It was a different set of aviaries where Sparkling Violet-ears bred at Heidelberg. Dieter Poley (1981) states the indoor planted aviary was "4 x 8m in area and about 3m high", while the outdoor flight was 150 square meters. A male and two females were purchased in 1976, "acclimatized in cages for six months", placed in the indoor facility in March 1977, and given access to the outside flight in May, 1977. "In 1978 five nests were built by the two females, and one nest was used for two successful incubations. Nests were built in both the inside and outside aviaries, always at the base of a fork on a slender twig or a large leaf, at a height of 1.25-2m. They were composed mainly of animal hair (llama, bison, camel, Irish setter), small feather, cotton wool, and a considerable amount of spier's web, which helps to bind the structure together. When the nest is nearing completion the female, which is now ready for fertilization, looks for where the male is singing. As soon as she comes into view the male follows her and copulation is presumed to follow, though it was never observed" (Poley 1981).

In 1978, one female hatched and reared a chick from an egg laid 25 March, while the other reared one chick from and egg laid 15 July and two from eggs laid 9 and 11 September (Poley 1981). It is interesting to note that the clutches laid by one female 15 and 17 May, and the other 24 and 25 May were all infertile The incubation period of was fifteen to sixteen days. The females fed their chicks every twenty minutes. Poley (1981) noted that the female collected the food it ate separately from that which it fed the chicks. Aside from insects, the birds were provided with a rather old fashioned diet: "1 liter water; 250g honey; 50g Aptamil bay food; 25g Boviserin; 25g fish meal" (Poley 1981). Poley (1981) provides a summary of the development of the four chicks hatched in 1978.

In the 1970s European and British private aviculturists were also achieving success with members of the genus Colibri. Mary Harvey (1975) reported that "Dr. Pierre Lamoure has bred the Violet-eared (or Sparkling Violet-eared) Hummingbird *Colibri coruscans* in his aviaries at Saint-Peray, France, two young ones being reared". The British hummingbird specialist Rod Elgar (1977) noted one Brown Violet-ear and two Brown/Sparkling crosses had been bred in Germany in 1977. Elgar himself made extensive documentation of his own propagation on Violetears, commencing with the production of hybrids in 1977 (Elgar 1977). The male of that breeding pair was a Brown Violet-ear obtained in 1974. The female was a Green Violet-ear (initially misidentified as a Sparkling Violet-ear (Yealland 1977)). They were maintained in a 14 ft x 4 ft 6 in indoor flight with nine other hummingbirds. Elgar (1977) presents a lengthy account of the hatching of two chicks and the fledging of one. Neither the male, or any of the other hummingbirds were removed during this time and "apart from a female White-necked Jacobin, which occasionally stole material from the nest, the others showed no interest until the chick left the nest. From this one might conclude that neither *Colibri delphinae* nor *C. thalassinus* display the aggression *C. coruscans* is notorious for.

However, generalities concerning *C. coruscans* are not supported by subsequent happenings in Rod Elgar's Manchester aviaries. In 1978, the male Brown Violet-ear died while the female Green was collecting spider webs. Elgar (1979) relates: "At this time I had been unable to acquire a male Green Violet-ear and the female seemed unwilling to accept display and mating from male hummingbirds to a different genus to herself. After a lot of thought, I introduced a male Sparkling Violet-ear *C. coruscans*. This species is notoriously aggressive and will vigorously defend its singing perch and feeding tube. I have three males of this species, so I introduced the smaller and less aggressive of them and it settled down without causing trouble to the rest of the hummingbirds. This male had been imported from Ecuador in 1976. The female Green mated with it immediately. One chick was reared out of two hatched and fledged in the aviary. Again, this breeding is extensively documented (Elgar 1979), and the account accompanied by excellent photographs.

The avicultural historian David Coles (1980) remarked: "The rearing of hybrid Violet-ear Hummingbirds seems to be an almost annual occurrence", mentioning the successful rearing of a Sparkling X Green hybrid (not listed by the *International Zoo Yearbook*) at the West Country Wildlife Park at Cricket St. Thomas, Sommerset in an indoor walk-through aviary inhabited by a variety of birds including glossy starlings. In 1982, Rod Elgar summed up twenty years of experience with Brown, Green, and Sparkling Violet-ears in another detailed article (Elgar 1982). Of the female green which produced the above mentioned hybrids, he had this to say: "... a female that I acquired in 1973 has been most interesting. She has nested on numerous occasions, in 1977 rearing a hybrid to a male *Colibri delphinae*, in 1979 rearing a further youngster to a male *Colibri coruscans*, and in 1980 having two nests to a male of her own species but failing to rear the young to maturity. Over a period of five years she has hatched twelve chicks, rearing two to maturity" (Elgar 1982). In 1980, a pair of Sparkling Violet-ears in Elgar's collection reared two chicks, and another in 1981. He also mentioned another British aviculturist, K. Dewall, who reared several in 1980.

Rod Elgar was one of two British Aviculturists who wrote extensively on hummingbirds for the *Avicultural Magazine* in the 1970s. The other was A. J. "Tony" Mobbs, who published an extensive series of articles on hummingbirds from 1971 through 1978. A sampling of them may be found in the references of this paper. He deliberately avoided breeding hummingbirds, stating: "My main interest in hummingbirds has always been to study and record their mating displays, songs and (where possible) plumage differences in immature and adult specimens. Because of this I have never attempted to breed with any of the species I have owned, and birds which have shown signs of wishing to go to nest have been deterred where possible, as in most cases such birds would become extremely aggressive and quickly upset the equilibrium of a communal flight. I am fully aware that young have been reared under such conditions, but feel that if a serious attempt at breeding is to be made, it is preferable to give a pair of hummingbirds an aviary or indoor flight to themselves" (Mobbs 1982). In light of this, I find it amusing that when Tony Mobbs gave up hummingbird keeping in the 1980s, due to the lack of diversity of species then imported to Britain (T. Mobbs pers. comm.) he turned to Australian finches, perfecting methods of cage-breeding them, publishing prolifically, both articles and books, on various species, and strongly advocating the establishment of self-sustaining populations in the United Kingdom (Mobbs 1992).

The results of Tony Mobbs years of observations on captive hummingbirds appear in his book *Hummingbirds* (Mobbs 1982), a remarkable compendium of information on the great range of species available to post-war aviculture until the restrictions of the 1970s. Recently republished, it remains a vital reference for anyone interested in hummingbirds in captivity.

The other German zoo which attained remarkable sustained propagation of hummingbirds commencing in the 1970s, was Zoologischer Garten Wuppertal, where Red-billed Streamertails were hatched every year from 1977 through 1984, as well as 1986, at total of 32 chicks (Table I). The five hatched in 1977 were the first hummingbirds successfully reared in any zoo outside the Americas. (A single Sparkling Violet-ear had hatched, but died, at Tierpark Berlin in 1970 (Table I)). This program is described in detail by Dr. Ulrich Schuerer (1983), director at Wuppertal. Two males and six females were collected in Jamaica by Head-Keeper of Birds Juergen Bock in July, 1975. His variation on the Ruschi/Lasiewski "straight jackets" was "paper handkerchiefs into which we make a hole in the center, we then stick the hummingbird's heads and necks through, fasten the tissue with adhesive tape, and stow them away in small boxes originally used for super-8 films... As soon as the handkerchiefs have become soaked from the excrements we change them...." (Schuerer 1983).

What I found especially remarkable is that breeding occurred with multiple adults in a not terribly large "closed, glass-fronted indoor aviary of our bird-house" (Schuerer 1983). This planted exhibited measured 2m x 4m and was 2.8m high. "It is possible to keep two to three adult male and four to five adult female Streamer-tailed Hummingbirds in an aviary of our size. In other species it usually would be less" (Schuerer 1983). The temperature was maintained between 72F and 77F. The diet was a "very old-fashioned honey-and-sugar solution" consisting of one teaspoon each of honey and cane sugar, and "a very small quantity of Liebig's meat extract", dissolved in 100ml of boiled water. This was fed twice daily, supplemented with a "wide range of different vitamin mixtures" which were changed from day to day. The key to using such a simple nectar formula was to always maintain more fruit flies in the exhibit than could be eaten. These Drosophila were bred on a "hidden shelf inside the aviary", where "50L of fruit-fly cultures in different stages" were maintained. This would support twenty hummingbirds. Nesting material consisted of "a synthetic fibre used for aquarium filters". There were no reports of entanglement. The International Zoo Yearbook statistics indicate at least partial second generation breeding took place in 1982 and 1983 (Table I.). Attempts at hand-rearing chicks were not successful past three weeks (Schuerer 1983). Dr. Schuerer informs me that in 1979 two females were sent on loan to Dr. Karl Schuchmann, and one was returned in 1981.

For more than twenty years, Dr. Karl-Ludwig Schuchmann has been a Research Officer in the Department of Ornithology at the Alexander Koeing Zoological Institute in Bonn, and is considered the authority on hummingbirds. Commencing in 1972, at the Zoologisches Institut Frankfurt, he conducted extensive work on captive hummingbirds, beginning with birds he brought back from a field study in California (Schuchmann 1973). Tony Mobbs (1982) was able to conduct observations on an Anna's Hummingbird given to him by Dr. Schuchmann. Mobbs (1982) noted that Schuchmann "housed most of the Anna's in a small aviary along with south American species and, although there was a certain amount of bickering, the birds appeared to get along together remarkably well." Dr. Schuchmann had at least eggs from Jamaican Streamertails before 1977 (Elgar 1977). By the mid-1980s he had maintained at least one hummingbird for thirteen-and-a-half years, and bred ten species. I have in my possession a photocopy of an article by Frank Woolham, published in Cage and Aviary Birds, but unfortunately without citation, given to me in 1985 by the American representative of Schuchmann's company Biotropic. It is stated therein that one of the ten species had been bred to three generations, and second generation hatchings had been achieved for two others. The Rufous-tailed Hummingbird *Amazila tzacatl*, Red-billed Streamertail, and Sparkling Violetears were "regular breeding species". Dr Schuchmann had developed a series of prepared foods in the course of his work, which he then patented, incorporating Biotropic to market them. There was much competition from the Nekton brand of bird diets, created by the German aviculturist Guenter Enderle in the early '70s, and as, mentioned earlier, the presence of agents of both companies at American Federation of Aviculture conventions could lead to an amusing sort of tension. (I also possess a photocopy of an (also undated) Cage and Aviary Birds article by Rosemary Low - endorsing Nekton's products - given to me by Nekton's American agent!). Eventually Biotropic was discontinued and Nekton has become established as the standard "instant" hummingbird formula for zoos. Dr. Schuchmann in the meantime has since concentrated on the systematics of hummingbirds, describing a number of new taxa, otherwise publishing an enormous output. His section on hummingbirds in the *Handbook of Birds of the World* (Schuchmann 1999) remains the definitive monograph of the Trochilidae. In light of his successes in the 1970s and '80s, and his foray into commercial aviculture, it is interesting to note an element of guarded pessimism in his handbook discussions of hummingbird aviculture: "...finding the right balance [of proteins] is still a delicate business; too much remains unknown about the Trochilid digestive system" (Schuchmann 1999).

Aside from Violet-ears and Red-billed Streamertails, a handful of other species were hatched by German and English aviculturists in the 1970s. In Munich, Walter Scheithauer, who at the time his book was written in 1966 had not actually hatched any species, had, by the mid-1970s been successful with the Black-chinned Hummingbird *Archilochus alexandri* (obtained from Mexico or Dr. Schuchmann?) and also produced the first captive inter-generic hybrid I know of, between the Dumerill's Emerald *Amazilia amazilia dumerillii* and the Fork-tailed Woodnymph *Thalurania furcata* (Elgar 1977). Tony Mobbs (1975) described a visit to a Frau Schulze, in Bamberg, Germany, whose Booted Racquet-tails *Ocreatus u. underwoodi* produced two chicks which died after a week, apparently of cotton wool impaction. C.G. Morey (1976), in Basingstoke, kept a pair of Long-tailed Sylphs *Aglaiocercus kingi* in a heavily planted outdoor aviary (heated in winter) 8ft x 6ft x 6ft high, shared by "three sugar birds, a sunbird and five other hummingbirds". The female sylph nested in the sort of box used for Zebra Finches and laid two eggs in December, 1975. These proved to be infertile, as were clutch laid around 24 January, 1976. Eggs laid on the 4th and 6th of July did produce chicks, one of which lived to 21 days. (The other birds were removed when the chicks hatched). "The nest was composed of sheep's wool and fibre glass of the kind used for roof insulation" (Morey 1976). The diet followed Mrs. Scammel, being composed of Superhydramin, sugar, pollen and Bovril, in hot water, with fruit flies in continuous supply.

During the 1970s, the captive propagation of hummingbirds in the U.S. was scant. I am aware of no private breedings. The famed author and entomologist, Marston Bates, who died in 1971, wrote an engaging book about his home greenhouse aviary in Ann Arbor, Michigan, where he kept, among other species, a Napo Sabre-wing, Magnificent Hummingbirds, and Blue-throated Hummingbirds *Lampornis clemenciae* (I imagine his stature both as an author and zoologist on the faculty of the University of Michigan enabled him to obtain permits for native birds), but none of his birds bred (Bates 1970).

From Table I., it can be seen little occurred in U.S. zoos in the 1970s aside from the previously discussed Violet-ears at San Diego. In 1971, two Sparkling Violet-ears hatched, but did not survive at the Philadelphia Zoo. They were to prove the only hummingbirds to hatch in the Eleanor S. Gray Memorial Hummingbird Exhibit, a spacious indoor walkthrough aviary which opened to much fanfare in 1970 (Griswold 1972), with an inaugural collection of 28 Ecuadorian hummingbirds including swordbills and Black-tailed Train-bearers *Lesbia victoriae*, as well as 22 other species of birds (Griswold 1970). Of course the imposition of the 1972 Newcastle's importation ban made it difficult to maintain many hummingbirds there. San Diego Zoo did supply a number of California species in the 1980s, but eventually this aviary was renamed the "Jungle Bird Walk' and hummingbirds were no longer kept there. In 1979, San Diego Zoo hatched an Anna's Hummingbird, a local species which had become a major component of the Hummingbird Aviary, and would remain so for years. This hatching was documented in a profusely illustrated article by Denise Gillen (1979), the Hummingbird Aviary keeper, who with fellow keeper Ruth Wooten (1979) was responsible for "rehabbing" large number of injured and "orphaned" hummingbirds, primarily Anna's, found by local persons. Many were placed on exhibit, and others were sent to various collections, including Wuppertal, where Anna's were hatched every year from 1989 through 1992 (Table I.).

From 1978 until 1980, there were only North American species of hummingbirds at San Diego. It was in 1979 that the first commercial shipments of hummingbirds to the U.S. since the 1972 Newcastle's ban were made. Gail Worth, of the California firm Aves International imported birds from both Jamaica and Peru. A number of U.S. zoos, thus obtained Red-billed Streamertails, among them the Bronx Zoo, Philadelphia, Brookfield, Cincinnati, and Houston. For some reason none of these bred, but one male that arrived at Brookfield 31 August 1979 died there 25 August, 1989, living the whole time in an open-fronted aviary shared with a variety of softbills, some rather robust, such as Scarlet-bellied Mountain Tanagers and a breeding pair of Gray-headed Kingfishers.

This record of just short of ten years is the second shortest longevity from a list of fourteen birds sent to me by Richard Weigl, a primate keeper at the Frankfurt Zoo, who is assembling the captive longevity records of birds for future publication. I requested records of birds that lived ten years or more. Aside from the famous Bronx Zoo Caribs, there were six birds at the Arizona-Sonora Desert Museum, of which more will be said. Only one of these decade-breakers was from Europe, a Versicolored Emerald *Amazilia versicolor*, at the Frankfurt Zoo from 5 August 1977 to 26 September, 1988.

The remaining four were a Rufous-tailed Hummingbird *A. tzacatl* at Cincinnati (where it lived in a butterfly exhibit) from 30 November 1983 through 15 June, 1994; an Anna's Hummingbird, from Arizona, which lived at Philadelphia from 7 September, 1988 until it was noted missing 31 December, 1998; a Black-chinned Hummingbird at the North Carolina Zoological Park, obtained 26 August, 1995 and still alive 21 October, 2005; and a White-bellied Hummingbird *Amazilia chionogaster* obtained for the National Zoological Park's new walk-through Rainforest building in 27 October, 1992, and still alive 21 October, 2005 (as of 2007, this bird is not on ISIS). Neither Richard nor I have found the species or institution of the seventeen year old bird mentioned by Dr. Schuchmann (1999). The longest record I am aware of is of fourteen years for a Planalto Hermit *Phaethornis pretrei* living in the near-natural conditions of Augusto Ruschi's Museum (Skutch 1973).

*Amazilia chionogaster* is one of the hummingbirds from the Western Slope of the Peruvian Andes which were uncommon or unknown in aviculture before the Peruvian government restricted hummingbird exports to this geographical area. Hummingbirds from the coastal desert of Peru were to become quite familiar to aviculturists in the 1980s and '90s, but at the time Tony Mobbs (1982) wrote his book such species as the Oasis Hummingbird *Rhodopis vesper*, the Purple-collared Woodstar *Myrtis fanny*, and the Peruvian Sheartail *Thaumastura cora* were a compete unknown, which he never expected to see. Regarding both the Oasis Hummingbird and the Sheartail, he stated "This species is found in an area where birds are rarely if ever trapped for aviculture". Thanks to the restrictions placed upon Peruvian trappers by their government, this shortly changed drastically. I clearly remember my first glimpse of *Rhodopis* and *Myrtis*, when I was escorted into the holding area of San Diego's Hummingbird Aviary, to see the newly arrived birds in the Summer of 1980. On the other hand, the first Sheartail I saw was in a pet store in Berkeley, California, in 1983. In both places, the birds were from Gail Worth. The energetic Indiana aviculturist and columnist Val Clear (1986) tried his hand, with some success, at importing Peruvian hummingbirds, outlining his experiences of coordinating dealers, airlines, quarantine stations and brokers. Into the 1990s the hummingbirds of Coastal Peru remained fairly available.

Sparkling Violet-ears were included among these shipments, and the Brookfield Zoo hatched two and reared one in 1981 in the large open-fronted central aviary. That was to be the only U.S. zoo breeding of Peruvian Hummingbirds in the 1980s, however. After years of planning, the San Diego Zoo imported a collection of Bolivian birds from Charles Cordier, then 85 years old. As previously mentioned, he'd been advertising hummingbirds from Bolivia since the mid-1960s, and had made at least two shipments for the impending opening of the Bronx Zoo's monumental World of Birds in 1982. In 1971, he sent the first Red-tailed Comets *Sappho sparganura* in the Bronx Zoo's collection (Bell 1971), and in 1972 shipped a novel subspecies of Fork-tailed Woodnymph *Thalurania furcata jelski* (Bell 1972). In 1972, San Diego received Giant Hummingbirds and Red-tailed Comets. The first U.S. hatchings of the Red-tailed Comet took place in 1983 at San Diego Zoo. Unfortunately, the 1984 breeding records for the San Diego Zoo failed to make it into the International Zoo Yearbook, so that the total of hatchings for this species at San Diego is recorded as two, both of which died (Table I). As it happens, the correct statistic is that eight hatched at San Diego. In 1987, Marvin Jones, San Diego's Registrar, provided me with the records for Red-tailed Comets, revealing that six chicks hatched in 1983, of which two lived four months, one three months, one two months, one one month, and one (hatched 29 March) still alive as of 1 July, 1987. Of the two hatched in 1984, one lived almost six months and the other (hatched June 16) was also still living 1 July, 1987, as was the breeding male, received 29 January, 1982.

The only other *Sappho* breedings listed in the IZY are the eighteen hatched at Tiergarten Heidelberg, from 1981 through 1984, of which four survived. I have found no published account of these breedings, but presume they took place in exhibits similar to the ones where Sparkling Violet-ears were hatched (Poley 1981). Brown Violet-ears also hatched at Heidelberg, producing two chicks each in 1983 and 1984. They were the only hummingbird to breed at Cologne (Koeln), two hatching there in 1984. I have found nothing in English about the rather remarkable spurt of breedings at Augsburg, where five species hatched over the years 1986 through 1990, three of them species hatched by no other IZY-listed collection.

In the 1980s, several North American hummingbirds were hatched among three North American zoos. The Costa's at the Bronx Zoo and and the Living Desert would prove to be the only hummingbird hatchings listed for those collections (Table I). On the other hand, the Broad-billed Hummingbirds *Cynanthus latirostris* hatched at the Arizona Sonora Desert Museum in 1989 would prove to be the first of eleven hatched there through 1993. Costa's Hummingbirds would prove even more prolific, 54 hatching (and 45 surviving) from 1990 through 1994 (when second generation hatchings were indicated). As previously mentioned, the Arizona Sonora Desert Museum also was distinguished for longevities, with five birds surpassing ten years and another almost reaching ten years (Weigl pers. comm.). In order of their age, these were an Anna's Hummingbird obtained 2 September, 1988, which died 4 August, 1998; a Magnificent Hummingbird received 1 August, 1984, which died 25 October, 1994; a Black-chinned Hummingbird acquired 22 August 1988, noted disappeared 30 November, 1999; an Calliope Hummingbird *Stellula calliope* (the smallest North American hummingbird) acquired 18 August, 1989, dying 5 January, 2001; a Broad-billed Hummingbird hatched in the collection 29 March, 1990, dying 24 May, 2001; and a Rufous Hummingbird *Selasphorus rufus* received 10 September, 1988, which died 10

November, 2000.

Most of these birds and all of the hatchings took place in the ASDM's "Hummingbirds of the Sonoran Desert Region Exhibit", which opened October, 1988. As many as ten species, all found in Arizona, have been maintained there. Unlike most exhibits of this sort, hummingbirds are the only birds displayed. This outdoor walk-through aviary, with three individual exhibits for large or aggressive males is described in great detail by Karen Krebbs (1992), of the ASDM's Department of Mammalogy and Ornithology. In this same paper, she also discusses in depth the propagation of Broad-billed and Costa's Hummingbirds from 1989 through 1991. Following the period covered by the IZY, propagation of hummingbirds has continued to take place, though with some set-backs (G. Stoppelmoor pers. comm.).

Two other institutions have also recently put together collections of hummingbirds from the American Southwest. Since 2001, the Dallas World Aquarium has worked with hummingbirds, beginning with birds collected in Oklahoma and Texas, then with an extensive series collected in Arizona. In the subtropical indoor environment of the 15,000 sq ft Mundo Maya exhibit (with a 60ft ceiling), opened in 2004, Broad-billed Hummingbirds have hatched chicks repeatedly and a Rufus Hummingbird incubated eggs that were apparently infertile. Altogether around forty hummingbirds are maintained here, representing ten species, including a unique collection of four from the Commonwealth of Puerto Rico. A number have been at The Aquarium for well over three years, including a Magnificent, and several Violet-crowned Hummingbirds *Amazilia violiceps*.

The Henry Doorly Zoo in Omaha opened its enormous indoor Desert Dome in 2003. A series of individual aviaries for North American hummingbirds situated in an artificial canyon are a feature, and hummingbirds are also in this building's freeflight. A Broad-billed Hummingbird was hatched here in 2005.

Since 1993, the acquisition of hummingbirds not native to the U.S. and its territories has been complicated for U.S. zoos and presently impossible for private individuals. That year the 1992 Wild Bird Conservation Act was enacted, prohibiting the importation of almost all birds listed on any appendix of the Convention of International Trade (CITES), except by special permit. In 1987, almost all hummingbirds (except for a few on the more restrictive Appendix I) were placed on CITES Appendix II. Appendix II status in itself means that all international trade must be permitted and documented, but has not in itself necessarily prevented importation. Under the Wild Bird Conservation Act, zoos and related institutions may apply for permits to import restricted species. Several places have done so. Both the San Diego Zoo and Wild Animal Park received Oasis Hummingbirds and Sparkling Violet-ears in 2000 (Ellis 2001a). The Dallas World Aquarium was able to collect and import several Red-billed Streamertails in 2003.

Butterfly World, in Coconut Grove Florida, near Fort Lauderdale, only began to keep hummingbirds after the imposition of the Wild Bird Conservation Act. After some initial work with North American hummingbird, permission was granted to import Sparkling Violet-ears in 1997, which commenced breeding 1999 (Stoppelmoor 2000), as well as other South American species. In 1999, permission was granted for Curator Greg Stopelmoor to collect Red-billed Streamertails. A detailed account of the importation process was published (Stoppelmoor 2002). Ten birds were imported and breeding commenced in 2000 (Stioppelmoor 2002). As of 2006, Streamertails were still present in the collection, along with Jamaican Mangos, Amazilia Emeralds, and Broad-tailed *Selesphorus platycercus* and Rufus Hummingbirds (T. Carney pers. comm.).

The Peruvian government continues to allow an annual quota of coastal and Western Andes Hummingbirds to be exported. To my knowledge, it has been several years since Peruvian hummingbirds have entered the U.S., but the potential for sustained propagation of certain species appears high. The Purple-collared Woodstar was hatched at San Diego Zoo in 1993, and the hatching of two broods is documented in detail by Senior Keeper Amy Kendall Flanagan (1998). This species was also bred by the English aviculturist Kevin Casey, who raised two broods in 1996 (Ellis 1997). The Oasis Hummingbird was hatched at San Diego Zoo in 1994 (Table I). It was bred as early as 1985, by the German aviculturists Maria and Willi Hesse (Ellis 2001b). The first British breeding was achieved by the Merseyside aviculturist Bob McWha in 2001. What makes his breeding outstanding is that the male was acquired in 1992, and the female purchased in 1996. Many infertile eggs were produce before one chicks was reared (Ellis 2001b). It will be noted from Table I that the Peruvian Sheartail was hatched in both 1989 and 1990 at Augsburg Zoo, a chick being fully reared the second year.

While Heidelberg's achievement of hatching 21 Sparkling Violet-ears, and rearing 17 from 1978 through 1984 has been discussed, one must also be impressed by Tierpark Berlin, where 47 were hatched and 18 reared from 1988 through 1996. These hatchings took place in a new crocodile hall. At least partial second generation breeding is indicated for 1992.

The Peruvian Hummingbird with the most apparent potential, however, is the Amazillia Emerald, the most well-represented among the hatchings recorded by the IZY (Table I), with 95 hatched and 46 reared, from 1988 through 1996. It is interesting to note how prolific it is inclined to be in a given collection. Ten were hatched and eight reared between the years 1989 and 1991 at Noorder Dierenpark in Emmen, the Netherlands. In suburban Tokyo's Tama Zoo, 39 were hatched (though only twelve were reared) from 1991 through 1996. At the San Diego Wild Animal Park, 33 were hatched and 20 reared, from 1993 through 1995, with second generation breeding indicated for 1995. It is interesting that in these three collections, the hummingbirds were kept in exhibits emphasizing insects. Tama built an enormous structure for a variety of free-ranging insects. The San Diego Wild Animal Park opened the Hidden Jungle in 1993, and the hummingbirds at Emmen were deliberately introduced to function as spider control in the indoor butterfly garden (Landman 1991)

The first breeding appears to have taken place in 1986, at Rod Elgar's collection in Manchester. His female was housed in an indoor aviary 14ft by 4ft by 6ft high, inhabited by single specimens, mostly female, of eight other species of hummingbirds. The male was introduced for an hour, twice daily until the female completed a nest and laid eggs. Both chicks were reared, their development extensively documented (Elgar 1986).

The introduction of the male only for breeding purpose was carried further by the Dutch aviculturist Jac Roovers. Malcolm Ellis, the Editor of the *Avicultural Magazine*, describes his set-up thusly: "His hummingbird accommodation (in Teteringen, in the Netherlands) measures approx. 56ft x 10ft), with the main room divided into 20 small aviaries, each measuring 115cm$^2$ x 225cm high (approx. 3ft 9in and ½ in$^2$ x 7ft 5in high), and serviced from a central corridor. In these aviaries Jac Roovers has succeeded in breeding the Violet-bellied Hummingbird *Damophilia julie*, Amazilia Hummingbird *Amazilia amazilia*, Fort-tailed Woodnymph *Thalurania furcata* and Violet-tailed Sylph *Aglaiocercus coelestis*. He believes he is the first to breed the last two in captivity, and is convinced that in time he can achieve more first breedings with hummingbirds. One of the secrets of his success may be that he lets the male in with the female just long enough for them to mate a few times, then opens the hatch and allows the male to fly back to his own aviary" (Ellis 1997b).

Jac Roovers still maintains his aviaries in this fashion, and has created a Hummingbird Foundation to attempt to establish various species in captivity. He is presently working with zoos through the European Zoo Association EAZA, providing birds on the condition that they be managed following his guidelines. There is a particular urgency as of this year, as the European Community has banned the commercial import of birds into its member countries. This will restrict private aviculturists to captive-bred birds, though, due to the efforts of Ulrich Schuerer and others, zoos will be able to arrange special importations. So far, a small circle of Dutch, Belgian and German zoos are working with *Amazilia amazilia*, but it is intended to expand as circumstances permit. One of the candidates is London Zoo, which will exhibit hummingbirds for the first time in twenty years, the Tropical House having been demolished in 1985, 54 years after its conversion to a bird facility. It was "in a bad state of repair and [its] destruction was inevitable" (Olney 1986). At the end of 1985, an *A. amazilia* was the sole hummingbird in the collection. Today, plans are underway for a remodeled Bird House, which will include hummingbirds among its inhabitants. It is fitting that the place, which slightly over a hundred years ago, caught the imagination of the public with living hummingbirds should do so once again.

## Acknowledgements

The bequest of archives and materials from my friend, the Late Marvin Jones has been invaluable, and I am grateful to Dr. Kurt Benirschke, Linda Coates, and Tom Schultz for organizing this material, and Ms. Constance Spates, of the E.T. Duncan Foundation for providing for its transport. Linda Coates, Librarian for the Zoological Society of San Diego, Steve Johnson, Manager of the Bronx Zoo Library, and Robert Olley, Information Services Librarian at the Bronx Zoo were generous, as always, in providing me with information and materials. I am also most grateful to Dr. Ulrich Schuerer, Director of Zoolgischer Garten Wuppertal, Dr. Rudolf Reinhard, Curator of Birds at Zoologischer Garten Berlin, John Ellis, Curator of Birds at London Zoo, and Beth Bahner, Animal Collections Manager and Registrar at the Philadelphia Zoo, for quick answers to questions, and to Paul Breese, Richard Weigl, Arthur Douglas, Eric Arndtsen, and Tim Carney for providing me with information and materials. Arden Richardson, Director of Education at the Dallas World Aquarium was most helpful in accessing computerized References. I have deep appreciation for the patience of Myles Lamont, editor of these proceedings, and for my wife, Natalie Mashburn Lindholm, for dealing with a paper that grew far beyond its imagined limits.

## References

ANON. 1942. The "Jewel Room". Animal Kingdom 45:152-153.
ANON. 1945. Behind the scenes. Animal Kingdom 48:52.
ANON. 1945. Behind the scenes. Animal Kingdom 48:88.
ANON. 1946. Bird housekeeping. Animal Kingdom 49:177-180.
ANON. 1947. Off to the Congo. Animal Kingdom 50:176.
ANON. 1947. Zoo housekeeping. Animal Kingdom 50:198.
ANON. 1975. Aviaries past and present. 222-223. IN: Reader's Digest Association. 1975. Our Magnificent Wildlife - How to enjoy and preserve it. Readers Digest Association.
"A FRENCH MEMBER OF THE SOCIETY" [Marquis de Segur]. 1915. My Humming Birds, and how I obtained them. Avicultural Magazine Ser.3, 6:105-109.
A FRENCH MEMBER OF THE SOCIETY. 1915. My Humming Birds, and how I obtained them.. Avicultural Magazine Ser. 3, 6:150-156.
ASTLEY HD. 1906. Humming Birds in Europe. Avicultural Magazine Ser. 2, 4:243.
ASTLEY HD. 1913. Notes. Avicultural Magazine Ser.3, 4:88.
ASTLEY HD. 1914 Powers of resuscitation in Humming Birds. Avicultural Magazine Ser.3, 4: 339-340.
ASTLEY HD. 1915. Humming Birds in Captivity. Avicultural Magazine Ser.3, 7:68-69.
BALDELLI GT. 1906. Humming Birds in Italy. Avicultural Magazine Ser.2, 4:271-272.
BARCLAY-SMITH P. 1941. A collection of Scarlet Cocks-of-the-Rock and other rare Colombian birds. Avicultural Magazine Ser.5, 7:29-32.
BARCLAY-SMITH P. 1947. A collection of Guatemalan birds for the New York Zoo. Avicultural Magazine 53:110-112.
BARCLAY-SMITH P. 1948. The new bird pavilion at the Antwerp Zoo. Avicultural Magazine 54: 114-118.
BARCLAY-SMITH P. 1949. New York Zoological Society's Belgian Congo expedition. Avicultural Magazine 55:195..
BARCLAY-SMITH P. 1951. News from America. Avicultural Magazine 57:108-110.
BARCLAY-SMITH P. 1952. News from America. Avicultural Magazine 58:78-79.
BARCLAY-SMITH P. 1968. Birdland. Avicultural Magazine 74:24-27.
BARCLAY-SMITH P. 1969 Editors. Avicultural Magazine 75:216-221.
BATES H & R Busenbark. 1963. Finches and soft-billed birds. T.F.H. Publications.
BATES M. 1970. A jungle in the house. Walker Publishing Company.
BELL J. 1971. News from the Department of Ornithology at the New York Zoological Park. Avicultural Magazine 77:37-38.
BELL J. 1972. News from the Department of Ornithology at the New York Zoological Park. Avicultural Magazine 78:67-68.
BENCHLEY BJ. 1935. Humming Birds. Zoonooz 8 (No.3): 1-2.
BENCHLEY BJ. 1936. Humming Birds. Zoonooz 9 (No.11): 8.
BERLIOZ J. 1957. Hummingbirds. Avicultural Magazine 63:84-87.
BRIDGES W. 1942. Flame-in-the-Forest. Animal Kingdom 45:24-29.
BRIDGES W. 1954. Zoo expeditions. William Morrow & Company.
BRIDGES W & D Holisher. 1941. Big Zoo. Viking Press.
BROWN W.E. 1991. Oliver Griswold Papers. University of Miami, Otto G. Richter. Library Archives and Special Collections Department. www.library.miami.edu/archives/papers/griswold.html
CARTER C. 1982. Nesting of the Speckled Hummingbird (Adelomyia melanogenys). Avicultural Magazine 88:151-152.

CHAPLIN A. 1933. Torpedity in the Trochilidae. Avicultural Magazine Ser.4, 11:231-232.
CHAPLIN A. 1933. Humming Birds. Avicultural Magazine Ser.4, 11:431-434.
CHAPLIN A. 1935. Sunbirds wintering in the open. Avicultural Magazine Ser.4, 13:58.
CHAPLIN A. 1935. The Crystal Place Show. Avicultural Magazine Ser.4, 12:88.
CHICAGO Zoological Society. 1939. Official illustrated guide - Chicago Zoological Park. Chicago Zoological Society.
CLEAR V. 1986. Thinking of importing? A.F.A. Watchbird 13(3):8-11.
COLES D. 1980. News and Views. Avicultural Magazine 86:114-120.
CONWAY WG. 1961. They show their age - Hummingbirds with Wrinkles. Animal Kingdom 64:151-154.
CORDIER C. 1947. Bird-collecting adventures in Guatemala. Animal Kingdom 50:89-97.
CORDIER C. 1947. A cargo of rarities from Costa Rica. Animal Kingdom 50:171-178.
CORDIER C. 1949. Our Belgian Congo expedition comes home. Animal Kingdom 52:99-114.
CORDIER C. 1952. A better way to feed Humming Birds. Avicultural Magazine 58:143-145.
CORDIER C. 1952. The feeding of Euphonias, Vassor's Tanagers, Red-eared Tanagers, Hooded Mountain Tanagers, and Red-crested Chatterers. Avicultural Magazine 58: 171.
CORDIER E. 1952 Coming from Ecuador: Giant Hummers and another Mountain Tapir. Animal Kingdom 55:50-55.
CRANDALL LS. 1931. A Hummingbird more than three years in captivity. Avicultural Magazine Ser.4, 9:292.
CRANDALL LS. 1931. The New York Hummingbird. Avicultural Magazine 4(9): 350-351.
CRANDALL, L.S. 1939. The Living Jewels called "Hummingbirds". Aviculture Ser. 3, 9:141-147.
CRANDALL LS. 1947. An important collection of birds from Central America for the New York Zoo. Avicultural Magazine 53:233-234.
CRANDALL LS. 1949. Old timers in the Zoological Park. Animal Kingdom 52:175-177.
CRANDALL LS. 1950. Feathered treasurers from Ecuador - Our greatest South American collection. Animal Kingdom 53:144-163.
CRANDALL LS. 1950. Charles Cordier's latest collection. Avicultural Magazine 56:205-208.
CRANDALL LS. 1952. Welcome to the zoo. Animal Kingdom 55:108-113.
DATHE H. 1965. The Alfred Brehm House at East Berlin Zoo. International Zoo Yearbook: 5:230-232
de GOEDEREN G. 1951. A visit to the Zoological Gardens of Wassenaar. Avicultural Magazine 57:139-142
de GOEDEREN G. 1953. The Louise Bird Hall at Wassenaar Zoo. Avicultural Magazine 59:135-136.

DELACOUR J. 1922. Notes on field ornithology and aviculture in Tropical America I. Martinique. Avicultural Magazine Ser.3, 8:101-105.
DELACOUR J. 1922. Notes of a bird-lover in Tropical America. Avicultural Magazine Ser.3, 8:148-157.
DELACOUR J. 1922. Notes of a bird-lover in Tropical America. Avicultural Magazine Ser.3, 8:161-168.
DELACOUR J. 1925. Obituary - Hubert Delaval Astley. Avicultural Magazine Ser.4, 3:179-182.
DELACOUR J. 1925. The late Mr. Hubert Astley. Avicultural Magazine Ser.4, 3:241-243.
DELACOUR J. 1926. Humming Birds. Avicultural Magazine Ser.4, 4:25-29.
DELACOUR J. 1926. The birds at Foxwarren Park. Avicultural Magazine Ser.4, 4:37-41.
DELACOUR J. 1928. A greenhouse aviary. Avicultural Magazine Ser.4, 6:293-295.
DELACOUR J. 1929. Bird notes from Cleres. Avicultural Magazine Ser.4, 7:25-26.
DELACOUR J. 1933. On a recent visit to the Berlin Zoo. Avicultural Magazine Ser.4, 11:21-22.
DELACOUR J. 1933. Bird notes from Cleres for 1932. Avicultural Magazine Ser.4, 11:34-39.
DELACOUR J. 1933. Notes on the small birds in the tropical houses at Cleres. Avicultural Magazine Ser.4, 11:179-181.
DELACOUR J. 1937. American Aviculture (I&II). Avicultural Magazine Ser.5, 2:109-118.
DELACOUR J. 1937. American Aviculture (III). Avicultural Magazine Ser.5, 2:125-139.
DELACOUR J. 1939. M. C. Cordier's collection. Avicultural Magazine Ser.5, 4:267-271.
DELACOUR J. 1939. The birds at Cleres in 1939. Avicultural Magazine Ser.5, 4:347-350.
DELACOUR J. 1941. The end of Cleres. Avicultural Magazine Ser.5, 6:81-84.
DELACOUR J. 1943. A collection of birds from Costa Rica. Avicultural Magazine Ser.5, 8:29-32.
DELACOUR J. (1944. Avicultural Entente Cordiale. Avicultural Magazine Jubilee Supplement: 5-10.
DELACOUR J. 1945. Decorative aviaries in the New York Zoo. Avicultural Magazine Ser.5, 10:57-58.
DELACOUR J. 1946. The Jewel Room in the New York Zoo. Avicultural Magazine 52:123-125.
DELACOUR, J. 1947. Waterfowl at Cleres in 1947. Avicultural Magazine 53:198-199.
DELACOUR J. 1949. The re-birth of Cleres. Avicultural Magazine 55:62-66.
DELACOUR J. 1950. American aviculture 1949- II. Californian aviaries. Avicultural Magazine 56:116-121.
DELACOU, J. 1951. Notes on European aviculture, 1950. Avicultural Magazine 57:5-8.
DELACOUR J. 1951. News from America. Avicultural Magazine 57:234-235.
DELACOUR J. 1953. My California aviaries. Avicultural Magazine 59:115-116.

DELACOUR J. 1953. The Marquess Hachisuka. Avicultural Magazine 59:139-140.
DELACOUR J. 1955. In Memoriam - Alfred Ezra, OBE. Avicultural Magazine 61:218-219.
DELACOUR J. 1957. Bird notes for 1956. Avicultural Magazine 63:19-21.
DELACOUR J. 1958. Bird news, 1957, from California and Canada. Avicultural Magazine 64:31-33.
DELACOUR J. 1958. Tanagers and Sugar Birds from Costa Rica and Colombia. Avicultural Magazine 64:142-46.
DELACOUR J. 1959. Bird news from Cleres and Los Angeles, 1958. Avicultural Magazine 65:18-19.
DELACOUR J. 1961 Bird notes from everywhere. Avicultural Magazine 67:187-192.
DELACOUR J. 1965 The birds at Cleres in 1964-65. Avicultural Magazine 71:170-172.
DELACOUR J. 1966. The living air - The memoirs of an ornithologist. Country Life Limited.
DELACOUR J. (1966. In Memoriam - Captain Richard S. de Quincey (1896-1965). Avicultural Magazine 72:60.
DELACOUR J. 1970. The Birds at Cleres in 1969. Avicultural Magazine 76:24-25.
DELACOUR J. 1970. Brazilian bird collections. Avicultural Magazine 77:71-75.
DELACOUR J. 1972. Suborder: Trochili, Hummingbirds. 266-271. IN: Rutgers, & J. K.A. Norris, eds. (1972) Encyclopedia of aviculture. Vol.2. Blandford Press.
DELACOUR J. 1972. Sugar-bird Tanager hybrids. Avicultural Magazine 78:48.
DELACOUR J. 1972. hybrids Sugar-bird X Tanager (Cyanerpes cyaneus x Tangara nigrocincta franciscae). Avicultural Magazine 78:187-188.
DELACOUR J. 1973. Bird breeding at Cleres in 1972. Avicultural Magazine 79:16-18.
de QUINCEY RS. 1935. The hardiness of a Humming-Bird and other notes on birds at the Vern. Avicultural Magazine Ser.4, 13:139-148.
de QUINCEY RS. 1960. Notes on some humming birds - And the nesting of the Doctor, or Streamer-tailed Humming Bird (Trochilus polytmus). Avicultural Magazine 66:58-66.
DITMARS, R.L., and L.S. Crandall 1945. Guide to the New York Zoological Park. Fourth edition. New York Zoological Society.
ELGAR RJ. 1975 The Ecuadorian Pied-tailed Hummingbird (Phlogophilus hemileucurus). Avicultural Magazine 81:93-95.
ELGAR RJ. 1976. Notes on the Stripe-breasted Starthroat Hummingbird (Heliomaster squamosus). Avicultural Magazine 82:83-84.
ELGAR RJ. 1977. The breeding of a hybrid Hummingbird - Brown Violetear X Sparkling Violetear. Avicultural Magazine 83:130-134.
ELGAR RJ. 1979. A Hybrid Hummingbird - Colibri courscans X Colibri thalassinus Avicultural Magazine 85:71-74.
ELGAR RJ. 1982. Notes on Violet-eared Hummingbirds - The genus Colibri. Avicultural Magazine 88:26-33.
ELGAR RJ. 1982. Nest-building of a female Speckled Humming Bird (Adelomyia melanogenys). Avicultural Magazine 88:153-155.
ELGAR RJ. 1982. Hummingbirds: Their care and management. Avicultural Magazine 88:213-225.
ELGAR RJ. 1986. Notes on the successful breeding of the Peruvian Brown-breasted Amazilia. (Amazilia a. amazilia) (Lesson). Avicultural Magazine 92:177-180..
ELLIS M. 1997. News and views. Avicultural Magazine 103:44-47.
ELLIS M. 1997. News and views. Avicultural Magazine 103:93-96.
ELLIS M. 2001. News and views. Avicultural Magazine 107:42-45.
ELLIS M. 2001. News and views. Avicultural Magazine 107:184-187
EZRA A. 1940. Breeding results at Foxwarren Park. Avicultural Magazine Ser.5, 5:7-9.
EZRA A. 1940. Long-lived Hummingbirds. Avicultural Magazine Ser.5, 5:270.
EZRA A. 1943. Long-lived Hummingbirds. Avicultural Magazine Ser.5, 8:6-7.
EZRA A. 1944. News from Foxwarren Park. Avicultural Magazine Ser.5, 9:117.
EZRA A. 1944. Fifty years of aviculture. Avicultural Magazine Jubilee Supplement: 1-3.
FINN F. 1912. Practical bird-keeping - XXI - Notes on out-of-the-way birds. Avicultural Magazine Ser.3, 4:66-70.
FINN F. 1928. Notes on the Humming-Birds recently exhibited at the Zoological Gardens. Avicultural Magazine Ser.4, 6:92-94.
FLANAGAN AK. 1998 Delicate nests: Breeding the Purple-collared Woodstar. A.F.A. Watchbird 25(4):40-42.
FLETCHER AWE. 1962. News from Chester Zoo. Avicultural Magazine 68:173-174.
FLETCHER AWE. 1964. News from the Chester Zoo bird collection. Avicultural Magazine 70:38.
GILLEN D. 1958. A hatching in the jewel room. Zoonooz. 52(9):8-11.
GOSSE PH. 1847. The birds of Jamaica. John Van Voorst.
GREENWALT CH. 1960. Hummingbirds. Doubleday & Company, Inc.
GREENWALT CH. .1963. Photographing hummingbirds in Brazil. National Geographic 123(No1):100-115.
GREENWALT CH. 1966. The Marvellous Hummingbird rediscovered. National Geographic 126(No7):98-101.
GRISWOLD JA.. 1970. The fauna of the hummingbird exhibit. America's First Zoo 22:18-21.
GRISWOLD JA. 1972. Hummingbird building at Philadelphia Zoo. International Zoo Yearbook 12:89-91.
GRISWOLD O. 1959. Blithe, Bautiful, tiny and - The first Tody to be exhibited alive! Animal Kindgom 62:98-105.

GRISWOLD O. 1960. Streamer-tail expedition - How to catch a shooting star. Animal Kingdom 62:7-12.

GRISWOLD O. 1960 Vervain Hummingbird: Second smallest bird in the world. Animal Kingdom 63:151-153.

GUILD E. 1940. Western Bluebirds in Tahiti. Avicultural Magazine Ser.5, 5:284-285.

GUILD E. 1944. More about birds in Tahiti. Avicultural Magazine Ser.5, 9:104-105.

HACHISUKA M.. 1940. Aviculture in Japan. Avicultural Magazine Ser.5, 5:173-176.

HARRISON CJO. 1968. Nutrition of nectar-feeding birds. Avicultural Magazine 74:36-37.

HARVEY MH. 1975. News and views. Avicultural Magazine 81:231-233.

HARVEY MH. 1978. News and views. Avicultural Magazine 84:235-236.

HASTINGS PH. 1954. The use of nectar in feeding birds. Avicultural Magazine 59:244.

HASTINGS PH. 1967. The "near miss" in rearing Humming Birds. Avicultural Magazine 73:104-105.

HASTINGS PH. 1968. Nutrition of Humming Birds. Avicultural Magazine 74:37-38.

HELLMAN GT. 1952. Clever Giant. The New Yorker July 28, 1952:11.

HEREFORD Herd Book Society. 1995. Early Chronology of the Hereford Breed 1723-1955 www.herefordwebpages.co.uk/herdhist.shtml.

HEUBLEIN E. 1965. Editor's note. Zoonooz. 38(2):3.

HILL L & E Wood. 1976. Birdland - The story of a world famous bird sanctuary. Taplinger Publishing Company.

HODGES JR. 1968. News and views. Avicultural Magazine. 78:179-180.

HODGES JR. 1992. News and views. Avicultural Magazine 98:136-141.

HOPKINSON E. 1933. A Hummingbird's outing. Avicultural Magazine Ser.4, 11:232-233.

HOPKINSON E. 1940. Latham as a bird fancier (Part I). Avicultural Magazine Ser.5, 5:128-134..

HOPKINSON E. 1940. Latham as a bird fancier (Part III). Avicultural Magazine Ser.5, 5:227-230.

HORNADAY WT. 1906. Popular Official Guide - The New York Zoological Park (Eighth edition). New York Zoological Society

HORSWELL HJ. 1973. James Bailey. Avicultural Magazine 79:144.

HOUSDEN JB. 1928. The call of the wild. Avicultural Magazine Ser.4, 6:209-211.

ILES GT. 1950. The bird collection at the Zoological Gardens, Belle Vue, Manchester. Avicultural Magazine 56:233-237.

INGELS J. 1974. Rare tanagers imported into Belgium and the Netherlands in 1973. Avicultural Magazine 80:20-25.

INGELS J. 1975 Rare tanagers imported into Belgium and the Netherlands during 1974. Avicultural Magazine 81:98-104.

ISENBERG AH. 1962. The breeding of Allen's Hummingbird (Selaphorus alleni). Avicultural Magazine 68:86-88.

JONES ML. 1954. Birds of the Cologne Zoological Gardens. Avicultural Magazine 60:26-30.

KEAR J. 1976. New British regulations on import and quarantine of birds. Avicultural Magazine 82:210-213

KINSEY EC. 1938. Hummingbirds in Japan. Aviculture Ser.3, 8:145.

KINSEY EC. 1939. California native bird keeping. Aviculture Ser.3, 9:1-6.

KLOES HG. 1963. New bird house at West Berlin Zoo. International Zoo Yearbook 4:153-155.

KLOES HG. 1964. News from the Berlin Zoological Gardens. Avicultural Magazine 70:147-148.

KLOES HG. 1964. News from the Berlin Zoological Gardens. Avicultural Magazine 70:219-221.

KLOES HG. 1965. News from the Berlin Zoological Gardens. Avicultural Magazine 71:59-60.

KLOES HG. 1968. News from the Berlin Zoological Gardens. Avicultural Magazine 74:68.

KREBBS K. 1992. Hummingbird breeding and nesting success at the Arizona-Sonora Desert Museum. AAZPA Regional Conference Proceedings 1992:410-417.

LANDMAN, W. 1991. Breeding the Amazilia hummingbird (Amazillia amazillia) at the Noorder Zoo, Emmen. International Zoo Yearbook 30:177-180.

LASIEWSKI RC. 1962. The capture and maintenance of hummingbirds for experimental purposes. Avicultural Magazine 68:59-64.

LINDHOLM JH. 1988. Captain Delacour at the Bronx (1940-1947). Avicultural Magazine 94:31-56.

LINDHOLM JH. 1993. Kenton C. Lint, March 19 1912 - December 3, 1992 (Part I). A.F.A. Watrchbird 20(2):55-56.

LINDHOLM JH. 1993. Kenton C. Lint, March 19 1912 - December 3, 1992 (Part II). A.F.A. Watchbird 20(3):46-49.

LINT KC. 1958. Wilderness birds - Within arms reach of visitors. Zoonooz. 31(11):10-14.

LINT KC. 1965. Hummingbirds in zoological gardens… Living jewels. Zoonooz. 38(2):3-9.

LINT KC. 1966. Hummingbird diets and new walk-through hummingbird aviary. International Zoo Yearbook. 6:103-104.

LINT KC. 1968. Breeding of the White-naped Honeyeater (Melithreptus lunatus) in the San Diego Zoological Garden. Avicultural Magazine. 74:172-174.

LINT KC. 1970. Sparkling Violet-ears. Zoonooz. 43(6):6-9.

LINT, K.C. 1981. Birds hatched in San Diego Zoo. Self-published.

LINT KC. & J.M. Dolan 1966. Successful breeding of the Orange-breasted Cotinga (Pipreola jucunda) in the San Diego Zoological Gardens. Avicultural Magazine 72:18-2.

LINT KC. 1969. Bird inventory, 31 December, 1969. Zoological Society of San Diego.

LONDON Times. 1940. Carrying on at the zoo. Zoonooz 12(No.12):5-6.
MARDEN L. 1963. The man who talks to hummingbirds. National Geographic 123(No.1):80-99.
MASURE RH. 1937. The birds of a Caribbean cruise. Avicultural Magazine Ser.5, 2:156-162.
MOBBS AJ. 1971. Notes on the Reddish Hermit Hummingbird. Avicultural Magazine 77:160-163.
MOBBS AJ. 1971. Stretching attitudes in hummingbirds. Avicultural Magazine 77:231.
MOBBS AJ. 1972. Observations on the pre-mating behaviour in the Andean Emerald and the Golden-tailed Sapphire Hummingbirds. Avicultural Magazine 78:173-176.
MOBBS AJ. 1973. The Horned Sungem Hummingbird. Avicultural Magazine 79:49-52.
MOBBS AJ. 1973. Scratching and preening postures in Hummingbirds. Avicultural Magazine 79:200-202.
MOBBS AJ. 1975. Notes on the display of the Booted Racquet-tail Hummingbird. Avicultural Magazine 81:156-159.
MOBBS AJ. 1976. Observations on a female Long-tailed Hermit Hummingbird. Avicultural Magazine 82:32-39.
MOBBS AJ. 1976 Observations on the Purple-throated Carib Hummingbird (Eulampis jugularis). Avicultural Magazine 82:196-201.
MOBBS AJ. 1977. Notes on the Ruby-Topaz Hummingbird. Avicultural Magazine 83:53-56.
MOBBS AJ. 1982. Hummingbirds. Saiga Publishing Co. Ltd.
MOBBS AJ 1992. The status of Australian Finches in the United Kingdom. Avicultural Magazine 98:130-135.
MOREY CG. 1976. Nesting of the Long-tailed Sylph Hummingbird. Avicultural Magazine 82:179.
NEW YORK ZOOLOGICAL SOCIETY. 1950. Birds. Annual report for the year 1950 55:12-15.
NEW YORK ZOOLOGICAL SOCIETY. 1951. Mammals and birds. Annual report for the year 1951. 56:14-16.
NEW YORK ZOOLOGICAL SOCIETY. 1952. Mammals and birds. Annual report for the year 1952. 57:15-18.
NEW YORK ZOOLOGICAL SOCIETY. 1953. Mammals and birds. Annual report for the year 1953. 58:11-15.
NEW YORK ZOOLOGICAL SOCIETY. 1955. Department of birds. Annual report for the year 1955. 58:12-15.
OLNEY PJ. 1969. London Zoo notes. Avicultural Magazine 75:191-192.
OLNEY PJ. 1973. Breeding the Blue-baked Manakin (Chiroxiphia pareola) at London Zoo. Avicultural Magazine 79:1-3.
OLNEY PJ. 1976. 1975 at London Zoo, with brief notes on 1974. Avicultural Magazine 82:54-55.
OLNEY PJ. 1986. Birds. The Zoological Society of London Annual Report 1985-1986: 82:54-55.
PAM A. 1923. The construction of an aviary. Avicultural Magazine Ser.4, 1:2-6.
PAM A & H Pam. 1907. Humming-Birds in captivity. Avicultural Magazine Ser.2, 5:285-289.
Peaker M & SJ Peaker. 1968. Feeding Humming Birds. Avicultural Magazine 74:55-59.
PLATH K. 1937. The birds at the new Chicago Zoological Park (I.). Avicultural Magazine. Ser.5, 2:173-179.
PLATH K. 1937. The birds at the new Chicago Zoological. 1953. A diary for 1952 of new arrivals in the bird department of the Brookfield Zoo at Brookfield, ILL.. Avicultural Magazine 59:104-107.
PLATH K. 1957. News from the Brookfield Zoo. Avicultural Magazine 63:192-195.
POLEY D. 1975. Notes on some nectar-feeding birds. Avicultural Magazine 81:104-106.
POLEY D.. 1981. Breeding the Sparkling violet-ear hummingbird (Colibri coruscans) at Heidelberg Zoo. International Zoo Yearbook 21:121-122
PORTER S. 1936. A West Indian Diary. Avicultural Magazine Ser.5, 1:96-112.
PORTER S. 1938. Notes from South America. Avicultural Magazine Ser.5, 3:207-213.
PRATT HD, PL Bruner & DG Berrett. 1987. A field guide to the birds of Hawaii and the Tropical Pacific. Princeton University Press.
PRESTWICH AA. (No date). Who's who in aviculture - First series. The Avicultural Book Co.
PRESTWICH AA. 1928. The new bird-house at the zoo. Avicultural Magazine Ser.4, 6:102-104.
PRESTWICH AA. 1945. The first importation of hummingbirds. Avicultural Magazine Ser.5, 10:T144.
PRESTWICH AA. 1946. The Society's medal. Avicultural Magazine 52:159-164.
PRESTWICH AA. 1947. British Aviculturist's Club. Avicultural Magazine 53:188-189.
PRESTWICH AA. 1947. Additions to the London Zoo. Avicultural Magazine 53:191.
PRESTWICH AA. 1947. The Tropical House at the London Zoo. Avicultural Magazine 53:193.
PRESTWICH AA. 1948. Additions to the London Zoo. Avicultural Magazine 54:40.
PRESTWICH AA. 1949. (Note regarding Yealland 1949.) Avicultural Magazine 55:40.
PRESTWICH, A.A. 1954. News and views. Avicultural Magazine 60:35-36.
PRESTWICH A.A. 1957. News and views. Avicultural Magazine 63:140-142.
PRESTWICH AA. 1958. News and views. Avicultural Magazine 64:121-122.
PRESTWICH AA. 1960. News and views. Avicultural Magazine 66:121-123.
PRESTWICH AA. 1965. News and views. Avicultural Magazine 71:195-198.
PRESTWICH AA. 1966. News and views. Avicultural Magazine 72:88-91.
PRESTWICH AA. 1968. Council meeting. Avicultural Magazine 72:28.

PRESTWICH AA. 1968. News and views. Avicultural Magazine 74:150-154.
PRESTWICH AA. 1969. British Aviculturist's Club. Avicultural Magazine 75:32-33.
PRESTWICH AA. 1969. News and views. Avicultural Magazine 75:192-195.
PRESTWICH AA. 1970. News and views. Avicultural Magazine 76:248-251.
RANNEY O. 1951. Official illustrated guide to Cleveland Zoological Park. Max M. Axelrod..
REUTHER RT & HG Lamm. 1959. Successful breeding of the Streamer-tailed Hummingbird (Trochilus polytmus). Avicultural Magazine 65:103-106.
REVENTLOW A. 1948. Details on the keeping of Humming Birds in Copenhagen. Avicultural Magazine 54:60-79.
REVENTLOW A. 1953. Experience during fifteen years with the feeding and management of Humming Birds in captivity. Avicultural Magazine 9:1-7.
RISDON DHS. 1960. Some impressions of the National Cage Bird Show, Olympia, 1959. Avicultural Magazine 66:60-79.
RYHINER P & DP Mannix. 1958. The wildest game. J.B. Lippincott Company.
SAWYER RCJ. 1995. Famous aviculturists around 1950. Avicultural Magazine 101:149-154.
SAWYER, RCJ. 2002. Some famous aviculturists I have known. Avicultural Magazine 108:100-109.
SCAMELL KM. 1966. Near misses with Violet-eared Humming Birds (Colibri coruscans). Avicultural Magazine 72:160-167.
SCAMELL KM. 1967. Breeding the Violet-eared Humming Birds (Colibri coruscans). Avicultural Magazine 73:109-115.
SCAMELL KM. 1967. The "Near miss" - which later became a success in rearing Humming Birds. Avicultural Magazine 73:137-138.
SCAMELL KM. 1968. Nutrition of Humming Birds. Avicultural Magazine 74:74-75.
SCHEITHAUER W. 1967. Hummingbirds. Thomas Y. Crowell Company.
SCHUCHMANN KL. 1973. Observations on an albino Black-chinned Hummingbird. Avicultural Magazine 79:203-204.
SCHUCHMANN KL. 1999. Family Trochilidae. 468-535. IN: del Hoyo, J., Elliott, A. & J. Sargatal, eds. 1999. Handbook of birds of the world. Vol.5 (Barn Owls to Hummingbirds). Lynx Edicions.
SCHUERER U. 1983. Maintenance and breeding of softbills - Todies, Streamer-tailed Hummingbirds and Swallows. Proceedings - Jean Delacour/IFCB Symposium on breeding birds in captivity. 567-574.
SETH-SMITH D. 1906. Humming Bird at the Zoological Gardens. Avicultural Magazine Ser.2, 4:109-111.
SETH-SMITH D. 1906. Bird notes from the Zoological Gardens. Avicultural Magazine Ser.2, 4:315-316.
SETH-SMITH D. 1907. Humming Birds at the Zoological Gardens. Avicultural Magazine Ser.2, 5:244-245.
SETH-SMITH D. 1907. Editorial note. Avicultural Magazine Ser.2, 5:289.
SETH-SMITH D. 1927. Avicultural Notes. Avicultural Magazine Ser.4, 5:272-273.
SETH-SMITH D. 1931. Avicultural Notes. Avicultural Magazine Ser.4, 9:203-204.
SETH-SMITH D. 1931. Editor's Note. Avicultural Magazine Ser.4, 9:292.
SETH-SMITH D. 1933. The arrival of Humming Birds. Avicultural Magazine Ser.4, 11:174-176.
SETH-SMITH D. 1934. Foreign Birds at the Crystal Palace. Avicultural Magazine Ser.4, 12:92-94.
SETH-SMITH D. 1934. Rare Birds in France. Avicultural Magazine Ser.4, 12:120.
SETH-SMITH D. 1934. Notes from the Zoological Gardens. Avicultural Magazine Ser.4, 12:190-191.
SETH-SMITH D. 1943. The Bird House in the London Zoo in war time. Avicultural Magazine Ser.5, 8:129-130.
SETH-SMITH D. 1955. In Memoriam - Alfred Ezra, OBE. Avicultural Magazine 61:215-218.
SILVER A. 1948, Retirement of Mr. James Bailey from the London Zoo. Avicultural Magazine 54:54.
SKUTCH AF. 1973. The life of the hummingbird. Crown Publishers, Inc.
STEINBACHER G. 1936 Humming Birds in the Zoological Garden of Berlin. Avicultural Magazine Ser.5, 1:48-49.
STEINBACHER G. 1937, Correspondence, notes, etc. Avicultural Magazine Ser.5, 2:363.
STEINBACHER J. 1952. The Scarlet Flycatcher in freedom and captivity. Avicultural Magazine 58:83-86.
STOKES HS. 1923. Breeding the Lilac-crowned Fruit Pigeon. Avicultural Magazine Ser.4, 1:199-200.
STOKES HS. 1933. The cargo of living jewels. Avicultural Magazine Ser.4, 11:176-178.
STOKES HS. 1933. The President's party, 1933. Avicultural Magazine Ser.4, 11:192-193.
STOPPELMOOR G. 2000. Captive breeding of the Sparkling Violet-ear Hummingbird. A.F.A. Watchbird 27(1):52-55.
STOPPELMOOR G. 2002. Collecting and breeding the Jamaican Red-billed Streamertail (Trochilus polytmus). A.F.A. Watchbird 29(2):4-8.
TOMLINSON S. 1934. Buzz and Whizz. Aviculture Ser.3, 5:19.
TONGREN S. 1989. The birds of the National Zoo - First accessions and hatchings, 1889-1989. National Zoological Park.
TREE I. 1991. The ruling passion of John Gould - A biography of the British Audubon. Grove Weidenfeld. .
VANE ENT. 1951. National Cage Bird Show at Olympia, 1950. Avicultural Magazine 57:36-37.
VANE ENT. 1952. National Cage Bird Show at Olympia. Avicultural Magazine 58:43-44.
VANE ENT. 1953. The National Show. Avicultural

Magazine 59:41-42.
WALTON JT. 1968. The nutrition of humming birds saga. Avicultural Magazine 74:156.
WEBB CS. 1939. A collector in the Andes of Ecuador. Avicultural Magazine Ser.5, 4:237-250.
WEBB CS. 1950. Notes from the London Zoo. Avicultural Magazine 56:186-187.
WEBB CS. 1950. Notes from the London Zoo. Avicultural Magazine 56:267-269.
WEBB CS. 1954. The odyssey of an animal collector. Longmans, Green & Company.
WHARTON-TIGAR N. 1934. An Amateur's experience in importing Humming Birds. Avicultural Magazine Ser.4, 12:122-128.
WHARTON-TIGAR N. 1937. Notes from the London Zoo. Avicultural Magazine Ser.5, 2:62-63.
WHARTON-TIGAR N. 1938. The Crystal Palace Show. Avicultural Magazine Ser.5, 3:53-58.
WHARTON-TIGAR N. 1998. (Republication of 1934 article, with notes by Frank Woolham). Avicultural Magazine 104:65-70.
WITTING R.C. 1948. The President's garden party, Foxwarren Park, 12th June, 1948. Avicultural Magazine 54:123-125.
WOOLHAM F. 1993. News and Views. Avicultural Magazine 99:48-53.
WOOTEN R. 1979. Not all hummingbirds fly. Zoonooz 52(9):12-13.
YEALLAND JJ. 1949. Avian imports. Avicultural Magazine 55:40.
YEALLAND JJ. 1952. London Zoo Notes. Avicultural Magazine 58:114.
YEALLAND JJ. 1953. London Zoo Notes. Avicultural Magazine 59:36-37.
YEALLAND JJ. 1953. London Zoo Notes. Avicultural Magazine 59:67-68.
YEALLAND JJ. 1953. London Zoo Notes. Avicultural Magazine 59:108-109.
YEALLAND JJ. 1953. London Zoo Notes. Avicultural Magazine 59:141-142.
YEALLAND JJ. 1953. London Zoo Notes. Avicultural Magazine 59:216-217.
YEALLAND JJ. 1954. London Zoo Notes. Avicultural Magazine 60:179-180.
YEALLAND JJ. 1955. London Zoo Notes. Avicultural Magazine 61:211-212.
YEALLAND JJ. 1955. London Zoo Notes. Avicultural Magazine 61:254-261.
YEALLAND JJ. 1955. London Zoo Notes. Avicultural Magazine 61:323-324.
YEALLAND JJ. 1956. London Zoo Notes. Avicultural Magazine 62:121-122.
YEALLAND JJ. 1956. London Zoo Notes. Avicultural Magazine 62:159-160.
YEALLAND JJ. 1956. London Zoo Notes. Avicultural Magazine 61:193-194.
YEALLAND J.J. 1956. London Zoo Notes. Avicultural Magazine 62:226-227.
YEALLAND JJ. 1957. London Zoo Notes. Avicultural Magazine 63:71-72.
YEALLAND JJ. 1957. London Zoo Notes. Avicultural Magazine 63:108-109.
YEALLAND JJ. 1958. The National Show. Avicultural Magazine 64:75-76.
YEALLAND JJ. 1958. London Zoo Notes. Avicultural Magazine 64:98-99.
YEALLAND JJ. 1958. London Zoo Notes. Avicultural Magazine 64:146.
YEALLAND JJ. 1959. The National Exhibition of Cage Birds. Avicultural Magazine 65:49-51.
YEALLAND JJ. 1960. London Zoo notes. Avicultural Magazine 66:123-124.
YEALLAND JJ. 1960. London Zoo notes. Avicultural Magazine 66:149-150.
YEALLAND JJ. 1960. London Zoo notes. Avicultural Magazine 66:200-201.
YEALLAND JJ. 1961. London Zoo notes. Avicultural Magazine 67:109-110.
YEALLAND JJ. 1962. London Zoo notes. Avicultural Magazine 68:111.
YEALLAND JJ. 1965. London Zoo notes. Avicultural Magazine 71:156-157.
YEALLAND JJ. 1967. London Zoo notes. Avicultural Magazine 73:17-18.
YEALLAND JJ. 1968. Breeding of the Purple Honeycreeper (Cyanerpes caeruleus) in the London Zoo. Avicultural Magazine 74:21.
YEALLAND JJ. 1968. London Zoo notes. Avicultural Magazine 74:27.
YEALLAND JJ. 1968. London Zoo notes. Avicultural Magazine 74:149.
YEALLAND JJ. 1977. Correction. Avicultural Magazine 75:237.
ZOOLOGICAL SOCIETY OF LONDON. 1960-1998. Birds bred in captivity and multiple generation births. International Zoo Yearbook 1-36.

## Author Biography

From the time he was shown a Guiana-Cock-of-the-Rock at the San Diego Zoo when he was four, Josef Lindholm has been fascinated by soft-billed birds in captivity. Since December, 2004, he has been Senior Aviculturist at The Dallas World Aquarium. The Aquarium's bird collection includes over 100 species and subspecies, with a concentration on Neotropical softbills, among them 26 taxa of toucans and related birds, seven species of cotingas, and twelve species of hummingbirds as well as Puerto Rican Woodpeckers and Todies. Josef's daily responsibilities have focused primarily on Ramphastids and both species of cocks of the rock, but he has recently began working with a unique collection of manakins as well. Birds propagated during his tenure include the first Ivory-billed Aracaris hatched outside of South America, both hand-reared and parent-raised, and now into the second generation. Josef Lindholm was born in Berkeley, California, where he earned a Bachelor's in Zoology from the University of California

He began his full-time avicultural career with Jerry Jennings in 1989, commencing at Walnut Acres Aviary in Woodland Hills, California, and assisting with the transfer of hundreds of toucans and other birds to their current home at Emerald Forest Bird Gardens in Fallbrook in 1990. From 1991 through 1999, he was a keeper at the Fort Worth Zoological Gardens where he developed a breeding program for African Finches. Among other things, more than fifty Golden-breasted Waxbills were hatched in one season, Red-billed Fire Finches were bred in multiple generations for five consecutive years, and Peter's Twinspots were raised for the first time in an American Zoo. Fort Worth is where Josef met his wife Natalie. Immediately after their marriage in 1999, they transferred to the newly-opened Disney Animal Kingdom, then to the Cameron Park Zoo in Waco, in 2000, where they remained until their arrival at The Dallas World Aquarium (where Natalie is the Mammal Supervisor, but also responsible for a variety of birds). He is a founding member of the American Zoo and Aquarium Association's Taxon Advisory Group for Passerines, Hummingbirds, Mousebirds, Frogmouths and relatives, and Trogons (PACCT TAG) and has recently been appointed to its steering committee.

For more than thirty years, he has gathered data on the history of birds and other animals in captivity, authoring more than fifty articles in, among other publications, the Avicultural Magazine, AFA Watchbird, and several regional proceedings of the AZA. He has also written articles for the revised Grzimek's Animal Life Encyclopedia and the Hancock House Encyclopedia of Aviculture. He has been entrusted with the archives of the late Marvin L. Jones, eminent Zoo Historian, which together with his pre-exisitng library and collection of zoo and avicultural ephemera, promise to keep him busy for a very long time.

## Table 1

**THE INTERNATIONAL ZOO YEARBOOK BREEDING RECORDS FOR HUMMINGBIRDS FROM 1959-1996**

*Figures in parentheses indicate pre-fledgling mortalities*
*\*At least one captive-hatched parent*

### 1959
**Red-billed Streamertail** *Trochilus polytmus*
Cleveland Zoo............................................................................................. *Number unspecified*

### 1970
**Sparkling (Gould's) Violetear** *Colibri courscans*
Tierpark Berlin, Germany............................................................ *1(1)*
San Diego Zoo, USA................................................................... *2(1)*
**Brown Violetear** *Colibri delphinae*
San Diego Zoo, USA................................................................... *4(4)*
**White-vented Violetear** *Colibri serrirostriss*
Parque Zoologico Sao Paulo, Brazil........................................ *1*

### 1971
**Sparkling (Gould's) Violetear** *Colibri delphinae*
Philadelphia Zoo, USA............................................................... *2(2)*
San Diego Zoo, USA................................................................... *2(1)*

### 1972
**White-vented Violetear** *Colibri serrirostriss*
Parque Zoologico Sao Paulo, Brazil........................................ *2*

### 1973
**Sparkling (Gould's) Violetear** *Colibri courscans*
San Diego Zoo, USA................................................................... *2*

### 1977
**Red-billed Streamertail** *Trochilus polytmus*
Zoologischer Garten Wuppertal, Germany............................. *5*

### 1978
**Sparkling (Gould's) Violetear** *Colibri courscans*
Tiergarten Heidelberg, Germany.............................................. *4*
**Red-billed Streamertail** *Trochilus polytmus*
Zoologischer Garten Wuppertal, Germany............................. *4*

### 1979
**Sparkling (Gould's) Violetear** *Colibri courscans*
Tiergarten Heidelberg, Germany.............................................. *4(3)*
**Red-billed Streamertail** *Trochilus polytmus*
Zoologischer Garten Wuppertal, Germany............................. *5*

**Anna's Hummingbird** *Calypte anna*
San Diego Zoo, USA................................................................. *1*

## 1980
**Sparkling (Gould's) Violetear** *Colibri courscans*
Tiergarten Heidelberg, Germany............................................*1*
**Red-billed Streamertail** *Trochilus polytmus*
Zoologischer Garten Wuppertal, Germany.............................. *4*

## 1981
**Sparkling (Gould's) Violetear** *Colibri courscans*
Chacago Zoological Park (Brookfield Zoo), USA.................... *2(1)*
Tiergarten Heidelberg, Germany............................................*5\**
**Brown Violetear** *Colibri delphinae*
Tiergarten Heildelberg, Germany...........................................*2(2)*
**Red-billed Streamertail** *Trochilus polytmus*
Zoologischer Garten Wuppertal, Germany.............................. *1*
**Red-tailed Comet** *Sappho sparganura*
Tiergarten Heildelberg, Germany...........................................*6(6)*
**Costa's Hummingbird** *Calypte costae*
The Living Desert, Palm Springs, USA.................................. *2(1)*

## 1982
**Sparkling (Gould's) Violetear** *Colibri courscans*
Tiergarten Heidelberg, Germany............................................*2*
**Red-billed Streamertail** *Trochilus polytmus*
Zoologischer Garten Wuppertal, Germany.............................. *4\**
**Red-tailed Comet** *Sappho sparganura*
Tiergarten Heildelberg, Germany...........................................*6(6)*
**Costa's Hummingbird** *Calypte costae*
The Living Desert, Palm Springs, USA...................................*2*

## 1983
**Sparkling (Gould's) Violetear** *Colibri courscans*
Tiergarten Heidelberg, Germany............................................*3*
**Brown Violetear** *Colibri delphinae*
Tiergarten Heidelberg, Germany............................................*2*
**Red-billed Streamertail** *Trochilus polytmus*
Zoologischer Garten Wuppertal, Germany.............................. *3\**
**Red-tailed Comet** *Sappho sparganura*
Tiergarten Heildelberg, Germany...........................................*4*
San Diego Zoo, USA................................................................*2(2)*

## 1984
**Sparkling (Gould's) Violetear** *Colibri courscans*
Tiergarten Heidelberg, Germany............................................*2(1)*
**Brown Violetear** *Colibri delphinae*
Zooogischer Garten Koeln, Germany...................................... *2*
Tiergarten Heidelberg, Germany............................................*2*
**Red-billed Streamertail** *Trochilus polytmus*

Zoologischer Garten Wuppertal, Germany..............................*1*
**Red-tailed Comet** *Sappho sparganura*
Tiergarten Heildelberg, Germany...............................................*2(2)*

*1986*
**Violet-bellied Hummingbird** *Damophila julie*
Zoologischer Garten Augsburg, Germany................................ *2(2)*
**Red-billed Streamertail** *Trochilus polytmus*
Zoologischer Garten Wuppertal, Germany.............................. *1*

*1988*
**Sparkling (Gould's) Violetear** *Colibri courscans*
Tierpark Berlin, Germany.........................................................*5(1)*
**Amazilia Hummingbird** *Amazilia amazilia*
Zoologischer Garten Augsburg, Germany................................*3(2)*
**Costa's Hummingbird** *Calypte costae*
Bronx Zoo, USA....................................................................... *2(2)*

*1989*
**Sparkling (Gould's) Violetear** *Colibri courscans*
Tierpark Berlin, Germany.........................................................*4(1)*
**Black-thoated Mango** *Anthracothorax nigricollis*
Zoologischer Garten Augsburg, Germany................................ *1(1)*
**Broad-billed Hummingbird** *Cynanthus latirostris*
Arizona-Sonora Desert Museum, Tucson, USA........................*1*
**Red-billed Streamertail** *Trochilus polytmus*
Zoologischer Garten Augsburg, Germany................................*1*
**Amazilia Hummingbird** *Amazilia amazilia*
Noorder Dierenpark/Zoo, Emmen, Netherlands.......................*4(2)*
**Peruvian Sheartail** *Thaumastura cora*
Zoologischer Garten Augsburg, Germany................................*1(1)*
**Anna's Hummingbird** *Calypte anna*
Zoologischer Garten Wuppertal, Germany...............................*1*

*1990*
**Sparkling (Gould's) Violetear** *Colibri courscans*
Tierpark Berlin, Germany.........................................................*12(7)*
**Broad-billed Hummingbird** *Cynanthus latirostris*
Arizona-Sonora Desert Museum, Tucson, USA........................*4*
**Peruvian Sheartail** *Thaumastura cora*
Zoologischer Garten Augsburg, Germany................................*1*
**Anna's Hummingbird** *Calypte anna*
Zoologischer Garten Wuppertal, Germany................................ *3(2)*
**Costa's Hummingbird** *Calypte costae*
Arizona-Sonora Desert Museum, Tucson, USA........................*4*

*1991*
**Sparkling (Gould's) Violetear** *Colibri courscans*
Tierpark Berlin, Germany.........................................................*6(4)*\*

**Broad-billed Hummingbird** *Cynanthus latirostris*
Arizona-Sonora Desert Museum, Tucson, USA.......................*3*
**Amazilia Hummingbird** *Amazilia amazilia*
Noorder Dierenpark/Zoo, Emmen, Netherlands......................*6*
Tama Zoological Park, Tokyo, Japan.........................................*11(11)*
**Anna's Hummingbird** *Calypte anna*
Zoologischer Garten Wuppertal, Germany................................*4*
**Costa's Hummingbird** *Calypte costae*
Arizona-Sonora Desert Museum, Tucson, USA.......................*12(1)*

*1992*
**Sparkling (Gould's) Violetear** *Colibri courscans*
Tierpark Berlin, Germany...........................................................*8(7)\**
**Broad-billed Hummingbird** *Cynanthus latirostris*
Arizona-Sonora Desert Museum, Tucson, USA.......................*2(2)*
**Saphire-spangled Emerald** *Amazilia lactea*
San Diego Zoo, USA....................................................................*1(1)*
**Amazilia Hummingbird** *Amazilia amazilia*
Tama Zoological Park, Tokyo, Japan.........................................*14(11)*
**Anna's Hummingbird** *Calypte anna*
Zoologischer Garten Wuppertal, Germany................................*1*
**Costa's Hummingbird** *Calypte costae*
Arizona-Sonora Desert Museum, Tucson, USA.......................*10(1)\**

*1993*
**Sparkling (Gould's) Violetear** *Colibri courscans*
Tierpark Berlin, Germany...........................................................*1(1)*
**Broad-billed Hummingbird** *Cynanthus latirostris*
Arizona-Sonora Desert Museum, Tucson, USA.......................*1*
**Saphire-spangled Emerald** *Amazilia lactea*
San Diego Zoo, USA....................................................................*1*
**Amazilia Hummingbird** *Amazilia amazilia*
San Diego Zoo, USA....................................................................*2(2)*
San Diego Wild Animal Park, USA...........................................*8(3)*
Costa's Hummingbird *Calypte costae*
Arizona-Sonora Desert Museum, Tucson, USA.......................*5(2)*
**Purple-collared Woodstar** *Myrtis fanny*
San Diego Zoo, USA....................................................................*3(1)*

*1994*
**Sparkling (Gould's) Violetear** *Colibri courscans*
Tierpark Berlin, Germany...........................................................*1*
**Amazilia Hummingbird** *Amazilia amazilia*
San Diego Wild Animal Park, USA...........................................*9(6)*
Tama Zoological Park, Tokyo, Japan.........................................*2*
**Oasis Hummingbird** *Rhodopsis vesper*
San Diego Zoo, USA.................................................................... *1*

**Costa's Hummingbird** *Calypte costae*
Arizona-Sonora Desert Museum, Tucson, USA........................*23(5)\**

*1995*
**Sparkling (Gould's) Violetear** *Colibri courscans*
Tierpark Berlin, Germany............................................................*7(5)\**
**Amazilia Hummingbird** *Amazilia amazilia*
Burger's Zoo and Safari, Arnhem, Netherlands.........................*5(1)*
San Diego Wild Animal Park, USA...........................................*16(4)\**
Tama Zoological Park, Tokyo, Japan.........................................*10(4)*

*1996*
**Sparkling (Gould's) Violetear** *Colibri courscans*
Tierpark Berlin, Germany............................................................*3(3)*
Tiergarten Schoinbrunn, Vienna, Austria...................................*1(1)*
**Red-billed Streamertail** *Trochilus polytmus*
Zoologischer Garten Wuppertal, Germany.................................*5*
**Black-billed Streamertail** *Trochilus scitulus*
Zoologischer Garten Wuppertal, Germany.................................*6*
**Amazilia Hummingbird** *Amazilia amazilia*
Burger's Zoo and Safari, Arnhem, Netherlands.........................*3(1)*
Tama Zoological Park, Tokyo, Japan.........................................*2(2)*

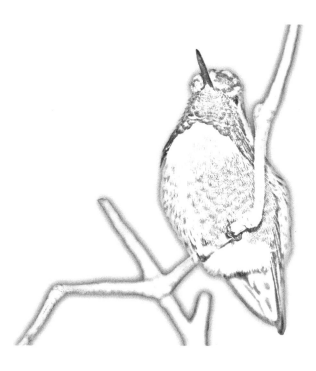

# THE STATE OF CAPTIVE WATERFOWL

MICHAEL R. LUBBOCK

Sylvan Heights Waterfowl Center
4963 Hwy 258, Scotland Neck, NC, USA

## SUMMARY

A review of two surveys conducted by the Conservation Committee of the International Wild Waterfowl Association (IWWA) was undertaken in order to determine which species of ducks and geese were at the greatest risk of disappearing from waterfowl collections in the United States. One survey, published in 2004, focused on ducks known to be in the collections of American zoos and private aviculturists. The other, also released in 2004, reviewed rare captive geese species in American zoos and with private aviculturists. The results of the two surveys clearly indicated that a number of waterfowl species in U.S. collections were quantitatively so low in that their sustainability is in question. The author offers an assessment of the situation using the results from the IWWA surveys and his own insight on the global availability of captive waterfowl. Statistical data of selected species from the 2004 IWWA surveys combined with updated figures from the Sylvan Heights collection is presented as the Rare Captive Waterfowl Species List. The list was established in order to focus attention of North American aviculturists on the species that are on the threshold of disappearing from collections. A current update on each species of the list is presented, citing which species are in critical need of immediate breeding efforts and which require close monitoring of their populations. A review of the decline of large American waterfowl collections and avicultural science is discussed in detail. The role of both private and zoological aviculturalists in the restoration of rare waterfowl species is examined, expressing the importance of the two groups combining efforts to be successful. The discussion concludes with a hopeful plan of utilizing education to revive interest in waterfowl conservation and aviculture.

## INTRODUCTION

Waterfowl collections in many zoos and most private collections are composed largely of species that are

attractive either for their pleasing coloration or for their economic value. Paying the price for this bias are the duller-plumaged and less profitable species of waterfowl, which ironically are often the most significant in terms of their conservation and biological importance. Couple this reality with a continuing decline in the popularity of avicultural science and one could conclude that the outlook for the state of captive waterfowl is currently very bleak. Nevertheless, it may be that a coming resurgence in avicultural interest and an emerging new alliance between the zoological and private collector will ultimately provide hope for the future of endangered waterfowl in both managed populations and the wild.

There is no question that waterfowl collectors have in general overlooked many species that have considerable biological importance. To demonstrate this avicultural prejudice, recent surveys of waterfowl populations in zoological and private collections were conducted by the International Wild Waterfowl Association. The result of these surveys will be presented along with the issuance of the Rare Captive Waterfowl Species List, which is designed to further highlight the species in critical need of intensified managed breeding programs.

Before any improvement in the state of captive waterfowl can be addressed, it is necessary to review the manner in which the artful science of aviculture made its decline. This topic will be presented from the perspective of both the private aviculturist as well as that of zoological organizations.

The real purpose of noting the urgency of any situation is to make suggestions as to how the crisis can be reversed. It is understood that my prescription for preserving rare and endangered species of waterfowl may be considered unlikely, if not impossible. The solution will require renewed cooperation between groups that have distanced themselves from one another over the past thirty years. It will also include a heavy dose of optimism that the general public will become willing supporters in reversing the plight of wild waterfowl. Despite the improbable course to the resolution of this issue, the preservation of waterfowl deserves the best coordinated effort that can possibly be assembled.

## Method and Results of Waterfowl Survey

One concern among aviculturists is the disappearance of certain waterfowl species that once were very common in collections around the country, but are becoming increasingly scarce. In 2003, after having difficulty finding certain birds for my own collection at Sylvan Heights Waterfowl Center (SHWC), I led the effort from within the International Wild Waterfowl Association (IWWA) to conduct a waterfowl survey that would accurately identify the number and location of ducks in captive populations of both private breeders and the Association of Zoos and Aquariums (AZA) members in the United States. The survey was conducted on 72 species of ducks selected by the IWWA Conservation Committee. Questionnaires were completed and returned from 100 members of the AZA and 51 private aviculturists. An additional survey was conducted in 2004 by the IWWA Conservation Committee to determine if similar trends were occurring with captive species of geese. It should be noted that many survey recipients did not respond and some private sector aviculturists were not aware that the survey was being conducted. Therefore the survey results do not represent a complete census of captive birds in the U.S.

The IWWA Waterfowl Survey, which was published in the Spring 2004 issue of the IWWA Newsletter, revealed a list of duck species that were either absent in American collections or had severe weaknesses in their gene pool. The additional report compiled by the IWWA Conservation Committee exposed a similar trend occurring with some species of captive geese. Next, the Conservation Committee determined the most vulnerable species and adopted immediate action to stabilize those populations. The committee initially selected the Hawaiian Duck, Chinese Spot-billed Duck and Baer's Pochard. Funds were budgeted to support a breeding program for these birds and a goal was set to add three species each year to the program.

I took the additional step of identifying and listing all of the vulnerable captive species using not only the IWWA survey results as a guide, but also my extensive experience in obtaining captive bird species from around the world and distributing them within North America. In March 2007, I developed the Rare Captive Waterfowl Species List, which focuses on birds that were once numerous in collections but have recently declined to levels where sustainability is uncertain. While some of the birds on the list are still numerous in the wild, a subset of them is becoming increasingly threatened. The list is designed to focus the attention of American aviculturists on waterfowl species that require immediate breeding programs.

Table 1 shows the goose species of the Rare Captive Waterfowl Species List, the data from the IWWA Conservation Committee Report and the May 2007 census for SHWC. Table 2 exhibits the duck species of the Rare Captive Waterfowl Species List along with the data from the IWWA duck survey. Recent updates of species at Sylvan Heights Waterfowl Center (SHWC) are noted.

Utilizing funds available from the IWWA for obtaining the most vulnerable birds, Sylvan Heights Waterfowl Center imported pairs of Baer's Pochard, Pacific Black Duck and Chinese Spot-billed Ducks.

### Table 1

**GOOSE SPECIES WITH EXTREMELY LOW POPULATIONS IN U.S. COLLECTIONS AS OF 2004***

| Goose Species | Zoo M | Zoo F | Zoo ? | Private M | Private F | Private ? | SHWC M | SHWC F | SHWC ? |
|---|---|---|---|---|---|---|---|---|---|
| Ruddy-headed | 6 | 5 | 0 | 8 | 6 | 0 | 6 | 6 | 0 |
| Ashy-headed | 0 | 2 | 0 | 6 | 6 | 0 | 2 | 2 | 0 |
| L. White-fronted | 9 | 4 | 2 | 10 | 10 | 0 | 4 | 4 | 0 |
| Orinoco | 5 | 6 | 0 | 5 | 5 | 0 | 5 | 5 | 0 |

? Unknown sex
* All data for Private Breeders and SHWC is from May 2007; supplied by the author

Following is an update of the entire Rare Captive Waterfowl Species List.

## Rare Captive Waterfowl Species Assessment as of May, 2007

### Wandering Whistling-Duck *Dendrocygna arcuata*

This bird was never numerous in captivity, but it has declined recently. We have three birds at SHWC that came from San Diego Zoo and are all related; there is one at Palm Beach Zoo and one elsewhere, and that is all that exist in the U.S. It may be possible to obtain some from Europe.

### Spotted Whistling-Duck *Dendrocygna guttata*

Close monitoring is necessary for this species. The captive population tends to go up and down very quickly and we realized two years ago that we were down in the number of breeding pairs after once having as many as 40 birds. When the data is analyzed one might think there is more hope than there actually may be. There are some in zoos, but most of them are really old and could die out in the next few years. Several are already post-reproductive. We have only two breeding pairs at SHWC and there is only a handful around the country. There are not many in Europe, so we need to concentrate on breeding them.

### Cuban Whistling-Duck *Dendrocygna arborea*

Fortunately, several zoos are breeding them, so the captive population is up from where it was a few years ago. Still, a close watch is needed.

### New World Comb Duck *Sarkidiornis m. sylvicola*

The captive population of these birds in North America is hanging by a thread. We have only one male and five females at SHWC, so we may need to import some from Europe where they are more numerous.

### Patagonian Crested Duck *Lophonetta s. specularioides*

This bird tends to be aggressive; therefore, most zoos do not like to keep them. Also, the Patagonian is one of those little brown ducks that zoos generally do not get selected for exhibit. At SHWC, we recognized at the beginning of a recent breeding season that we were down to just two pairs. We had plenty of females, but only two males. There may be less than 20 birds total in the U.S.

### Hawaiian Duck *Anas wyvilliana*

Sylvan Heights Waterfowl Center received a pair of Hawaiian Ducks from a private collector in New York who recently imported some of these birds. In the wild, the Hawaiian Duck has hybridized with mallards to some extent, so there is a question as to how pure the bloodlines are, but what we have is as close to pure one can get today.

### Philippine Duck *Anas luzonica*

This is an example of a bird that became rare in captivity four years ago, but we are beginning to increase the population again. The key is to have a reliable system in place to monitor captive populations so that we do not get to the point of realizing too late that a species is in deep trouble.

### Chinese Spot-billed Duck *Anas poecilorhyncha*

Although there were several in the U.S. a few years ago, they had disappeared until the IWWA imported three pairs in May 2007. We are initiating a breeding program at SHWC. Hopefully, this bird will once again be a part of many collections in America.

### Indian Spot-billed Duck *Anas poecilorhyncha*

This species is slowly coming back after being extremely restricted in its availability. Concentrated breeding at SHWC has contributed to its numbers in the U.S.

### Pacific Black Duck *Anas superciliosa*

These birds nearly disappeared from U.S. collections. The IWWA imported a few two years ago or there probably would not be any around today.

### African Yellow-billed Duck *Anas undulata*

We are concentrating on the African Yellow-bill this year because their numbers are declining rapidly in within the country. Many of the ones in zoos are old and post-reproductive.

### Table 2
**DUCK SPECIES WITH EXTREMELY LOW POPULATIONS IN U.S. COLLECTIONS AS OF 2004\***

| Duck Species | Zoo M | Zoo F | Zoo ? | Private M | Private F | Private ? | SHWC M | SHWC F | SHWC ? |
|---|---|---|---|---|---|---|---|---|---|
| Spotted Whistling | 8 | 9 | 0 | 1 | 1 | 0 | 2 | 2 | 0 |
| Cuban Whistling | 12 | 11 | 0 | 7 | 6 | 2 | 2 | 1 | 0 |
| Wandering Whistling | 5 | 5 | 0 | 0 | 0 | 0 | 1 | 2 | 0 |
| New World Comb | 1 | 2 | 0 | 0 | 0 | 0 | 2 | 6 | 0 |
| Hawaiian (Koloa) | 0 | 0 | 0 | 0 | 0 | 0 | 1 | 1 | 0 |
| Philippine | 4 | 3 | 5 | 3 | 4 | 0 | 4 | 4 | 12 |
| African Yellow-billed | 15 | 11 | 0 | 6 | 6 | 0 | 3 | 4 | 0 |
| Cape Shoveler | 0 | 0 | 0 | 0 | 0 | 0 | 1 | 1 | 0 |
| Australian Grey Teal | 0 | 0 | 0 | 0 | 1 | 0 | 0 | 1 | 0 |
| Chinese Spot-billed | 0 | 0 | 0 | 0 | 0 | 0 | 3 | 3 | 0 |
| Indian Spot-billed | 2 | 2 | 0 | 8 | 4 | 1 | 4 | 4 | 0 |
| Pacific Black | 0 | 0 | 0 | 1 | 1 | 0 | 6 | 5 | 0 |
| Patagonian Crested | 3 | 3 | 0 | 1 | 1 | 0 | 2 | 13 | 0 |
| Red-billed Pintail | 14 | 18 | 0 | 3 | 8 | 0 | 4 | 6 | 0 |
| Garganey | 5 | 2 | 0 | 13 | 9 | 4 | 5 | 12 | 0 |
| Silver Teal | 9 | 10 | 0 | 11 | 11 | 0 | 8 | 6 | 0 |
| Puna Teal | 12 | 10 | 3 | 13 | 10 | 0 | 5 | 5 | 7 |
| Australian White-eyed | 3 | 1 | 0 | 3 | 3 | 0 | 4 | 4 | 0 |
| Baer's Pochard | 0 | 0 | 0 | 0 | 0 | 0 | 2 | 2 | 0 |
| New Zealand Scaup | 1 | 7 | 0 | 4 | 5 | 0 | 7 | 4 | 0 |

? unknown sex          *SHWC data was updated May 2007

### Cape Shoveler *Anas smithii*

These have almost completely disappeared in North America and Europe. Collectors tend to ignore the Cape Shoveler since they are not as striking in color as the Northern or Australian Shovelers. There is one male in the African Aviary at the Sylvan Heights Waterfowl Park & Eco-Center and that is possibly the only one in the country. There is potential for us to import some from Africa.

### Australian Grey Teal *Anas gracilis*

We formally had these birds, but now we are down to a single female. There are few in Europe, so it will be a struggle to get the Grey Teal back. It is not as attractive as the East Indian Grey Teal, which puts this bird at a disadvantage. If we can increase the numbers again, perhaps zoos would be interested in utilizing them in Australia-themed exhibits.

### Red-billed Pintail *Anas erythrorhyncha*

This bird must be monitored closely. There are not many in collections, although they are popular in Africa-themed exhibits at zoos. We may be able to import them from Africa and increase the population again.

### Garganey *Anas erythrorhyncha*

These are starting to disappear in captivity, so they must be carefully watched. There are some in private collections. SHWC is attempting to increase its population this year.

### Puna Teal *Anas puna*

This is another species to monitor closely. Their numbers are decreasing in captivity and there are some inbreeding issues. SHWC focused on breeding Puna Teal in 2007 and already have some young ones.

### Silver (Versicolor) Teal *Anas versicolor*

These birds are not as popular as the similar-looking Hottentot Teal and this may be affecting the captive population of not only the Silver Teal, but also the Puna Teal, which also looks similar. More Silver Teal need to be bred annually.

### Australian White-eyed duck *Aythya australis*

Because this bird is often misidentified in zoo collections, it is difficult to determine if the ISIS data is accurate. Either way, the numbers are low. SHWC is down to two pairs of this duck. Fertility is very poor with our pairs and we do not rear many.

### Baer's Pochard *Aythya baeri*

There are more of these birds in Europe than in North American collections. The IWWA imported birds two years ago and we now have two pairs at SHWC and hope to breed more this year. These are the only Baer's Pochards in the country and they are becoming very rare in the wild as well.

### New Zealand Scaup *Aythya novaeseelandiae*

This bird definitely needs monitoring. No new bloodlines of this species have been brought into Europe or anywhere else in the last forty years. We need to preserve what we have currently. Fortunately they breed very well and there does not seem to be any inbreeding problems. After bringing in some birds from Europe two years ago, we now have four pairs and already have eggs this year.

### Ashy-headed Goose *Chloephaga poliocephala*

This bird needs to be watched with extreme care. As far as zoo collections are concerned, this bird is essentially gone. Last year some were imported from Europe, where the species is more abundant, but the Ashy-headed is in critical need of an expanded breeding effort.

### Ruddy-headed Goose *Chloephaga rubidiceps*

There are more Ruddy-headed Geese in U.S. collections than of Ashy-headed Geese. Although we probably do not need to import anymore, it will be necessary to watch their numbers.

### Lesser White-fronted Goose *Anser erythropus*

This small-sized goose with a beautiful gold eye-ring is fairly popular among collectors, even small ones. They are doing well in the U.S., but monitoring is needed. There are some in Europe. Geese in general are more expensive to maintain due to the large amount of grazing area required, so there are very few geese bred every year compared to the ducks.

Orinoco Goose *Neochen jubata*

This species' population is definitely one that needs to be carefully monitored because they are very close to being lost. There are only a few in U.S. zoos. SHWC and several private breeders are attempting to increase their numbers, but we are all working with the same bloodlines. There are some in zoos in South America from which we might be able to import a couple of males to improve the genetic situation. Wild Orinocos are decreasing rapidly in number due to loss of tropical forest habitat in Venezuela and other areas of their territory. The Orinoco needs a sustainable captive population for protection.

**Waterfowl Species Success Stories**

Cape Teal *Anas capensis*

This is an example of a bird that became rare in captivity three years ago, but we were able to bring the population back up. Private aviculturists and a few zoos noticed the problem in time and began to concentrate on breeding them. Now there is a substantial population of Cape Teal.

Meller's Duck *Anas melleri*

Two years ago, SHWC imported these birds from Europe through the IWWA. Now the U.S. population is coming back.

East Indian Grey Teal *Anas g. gibberifrons*

Two years ago SHWC brought in two pairs and now through our breeding program we have a sufficient number of this Grey Teal sub-species.

Southern Pochard *Netta erythrophthalma*

Ten years ago there were none of these birds in the U.S. Six years ago SHWC imported three pairs and now the numbers are good.

**DISCUSSION**

The results of the IWWA Waterfowl Surveys clearly indicated that certain species of both ducks and geese are rapidly disappearing from managed populations. The most seriously affected are those species that share the characteristics of being unpopular with the viewing public and having low market value. A myriad of social and administrative conditions have combined to cause the decline in birds on the Rare Captive Waterfowl Species List, obscuring the future for these species and the future of avicultural science as well. In order to find our bearings on these issues, it is necessary to review how the current state of affairs came to be in the United States.

**Declining Number of Large Private Waterfowl Collections**

Nearly all of the large private waterfowl collections in North America that were in existence 15 years ago are gone. Many of the owners have died or retired their collections, the end result being the demise of nearly all of the large private waterfowl collections and the birds they once contained. When I arrived in Long Island, New York in 1969 to supervise the sizable collection for Winston Guest, there were seven major collections just on this one island of New York. Now none of them remain. The same thing has happened with the major private collections in New England. Except for the Ripley collection, they have all disappeared. During this time, no new large waterfowl collections have come into existence anywhere in the country. The only one that has increased in size in the last 10 years is the 3,000 birds in the waterfowl collection I founded at Sylvan Heights Waterfowl Center in Scotland Neck, North Carolina.

Membership levels in most avicultural societies have remained roughly the same over the last 20 years, but the size of most of the members' collections is very small, perhaps ten pairs of birds or less, whereas they used to have as many as 60 species. There are no signs that activity in raising waterfowl privately is likely to come back to the level it once was in America during the 1970s.

The same decline in private aviculturists is evident in Europe, where once there was a thriving interest in keeping waterfowl. Most of the small European collections are gone, although some of the large collections remain. However, owners of the large collections tend to breed mostly expensive birds that sell at a high price. Most do not concentrate on raising the birds on the Rare Captive Waterfowl Species List.

## Possible Reasons for Decline in Private Waterfowl Collections

The decline in keeping waterfowl seems to be related to two primary factors: impinging regulatory issues from government agencies and a general lack of interest in aviculture among young people, especially with respect to waterfowl species.

Economics is not a major factor in the decline of large waterfowl collections. The people who once owned these collections, for the most part, did not do so to make money. When they sold birds it was to cover their expenses— not to make a profit from raising them. The activity was mainly a hobby for wealthy individuals that had a property with enough acreage to build ponds, money to invest in the necessary infrastructure and an interest in avian husbandry. I have never advised anyone to start collecting and breeding waterfowl as a business or to make a profit. They may be able to cover their expenses by selling birds, but it is first and foremost a hobby.

There are areas where costs have greatly increased due to additional regulatory requirements. This has particularly affected the remaining small private breeders. For example, the USDA requires a 30-day mandatory quarantine on all imported waterfowl at a current cost of $6.50 per bird per day, which can add up to a large sum if one is importing a great many birds. This regulation is a result over issues the USDA has with the chicken industry. In addition, state agencies have imposed an increasing number of expensive and time-consuming tests for avian-related disease on birds being shipped interstate. Air transportation companies have established onerous restrictions on shipping birds, while simultaneously increasing fares for birds meeting the new criteria. Various environmental regulations in some states affect the testing of water run-off in ponds.

The other major factor in the decline of aviculture is the reduction in the number of people entering the field. Very few young people seem to have an interest in aviculture compared to thirty years ago. Especially in Europe, many young people once learned aviculture techniques by getting jobs at the large waterfowl collections owned by wealthy individuals. Some maintained their interest at older ages by becoming the supervisor for one of these large collections or by establishing their own collections. Not only have most of the large private collections vanished, but jobs at zoological institutions that require avian husbandry skills are also few in number.

## The Declining Number of Large Waterfowl Collections in American Zoos

The Association of Zoos and Aquariums' current database of waterfowl held by zoological entities reflects a marked decline from previous decades. Most of the major waterfowl collections that once existed at many zoos have slowly gone down in numbers. The San Diego Zoo, Bronx Zoo, Philadelphia Zoo, Saint Louis Zoo, and San Antonio Zoo all once had very large and extensive collections of waterfowl. During the past 25 years all of these zoos have all drastically reduced their holdings of waterfowl. SeaWorld San Diego, which once boasted the world's largest waterfowl collection, currently has only a modest number of birds compared to their zenith. Few zoos conduct any substantial breeding program for the few species they maintain. The only American zoo that has added significantly to its collection in recent years is Palm Beach Zoo in West Palm Beach, Florida.

The demise of the major waterfowl collections at zoological institutions has severely reduced the capacity for maintaining the birds on the Rare Captive Waterfowl Species List. The lack of space for these birds is a contributing factor as to why they are so rare today. It also greatly diminished the number of skilled aviculturists and waterfowl biologists, thereby eliminating a major portion of waterfowl expertise in the country. As a result, a valuable piece of avian legacy was lost in the field of natural science.

## Possible Reasons for the Decline of Waterfowl in Zoo Collections

Anecdotal evidence suggests that the reduction of waterfowl on display at zoos corresponds with the general public's relative lack of interest in waterfowl compared to more exotic animals, such as pachyderms, big cats, bears and primates. Space for animals on exhibit is competitive and assigned on the recommendation of Taxonomic Advisory Group (TAG) leaders and other internal decision-making criteria. In the face of this competitive environment for exhibit space, waterfowl often become an after-thought in many zoo collections. Some waterfowl may be utilized as "background interest" within a specific geographical-based display, such as an African exhibit or a tropical rain forest environment, but rarely are they presented at zoos on the basis of educating the public on the importance of preserving or understanding waterfowl species. There are some zoos in the U.S. that display impressive bird collections but have absolutely no waterfowl at all. Many zoos have large natural or man-made ponds within view of visitors. These ponds often become self-populated with Mallards and Canada Geese, providing a ready-made waterfowl exhibit. There is little incentive to absorb the expense to purchase or breed birds and place them among the local inhabitants. In addition, laws have been enacted that make it difficult to remove native birds, even if a zoo or nature center desired to do so in order to display a collection on their ponds.

In addition, there are very few zoological collections in the country devoted exclusively to birds. The only major ones in existence are Tracy Aviary in Salt Lake City and the National Aviary in Pittsburgh. The Sylvan Heights Waterfowl Park & Eco-Center in Scotland Neck, NC, is the only major avian collection which specializes in exhibiting waterfowl. America compares poorly to Europe, where bird gardens are prolific and considered a status symbol for private breeders and royalty. In Asia as well, bird parks attract a large number of visitors as witnessed by the existence of the perhaps the world's largest avian collection at Jurong BirdPark in Singapore, which boasts over 9,000 birds from 600 species.

Compounding the problem is the lack of avian husbandry knowledge among zoo keepers, especially with regard to breeding waterfowl. This is certainly understandable, since as previously noted, zoos keep minimal numbers of waterfowl if any at all. Some of the zoos that display waterfowl do not make any effort to breed them. As a result, the competency level for providing proper nesting requirements, egg collection, incubation techniques and rearing methods for waterfowl are dismally low. For many years zoos have been discouraged from breeding their birds. In 2006, the AZA Anseriformes Taxon Advisory Group implemented the Duck DERP. The acronym stands for Display (to the public), Education and Research Population (for science-oriented and conservation research programs). This program has the potential to revitalize zoological breeding programs for rare waterfowl species, while providing needed display space and public education.

Although money for conservation projects has significantly increased over the years, the commitment among zoological organizations in exhibiting, breeding and educating the public about waterfowl is at a thirty-year low.

**Thinking Beyond the Endangered Species List**

The AZA has a Waterfowl TAG that has taken on the task of identifying and listing species that are vulnerable and need support in order to sustain the population. But the TAG has primarily listed birds that are rare in the wild. The Madagascar Teal, for instance, is on the Waterfowl TAG list and actions have been taken to expand the breeding program for this bird.

While this is undoubtedly beneficial for birds struggling for survival in their native habitat, there are also species of waterfowl that are rare in captivity but are not on the TAG list. Most of these birds are presently not rare in the wild; however, they certainly could become so. If that happens and we do not have a sufficient captive population, then these species are at risk of becoming extinct both in captivity and the wild. The Madagascar Teal had the fortunate advantage of being available in a captive population. No birds or fertile eggs were recently taken from the wild in order to initiate this survival breeding program. In fact, taking wild birds and eggs of endangered species is generally banned internationally, and even when allowed, may occur too late to be successful or sustain appropriate genetic distribution.

With the Waterfowl TAG focused on endangered species, there is little incentive at zoological institutions to make room for the birds on the Rare Captive Waterfowl Species List. In order to prevent the loss of these biologically important species, zoos and private aviculturists have a responsibility to work together in providing space for these birds and their offspring, despite the current trend for these two groups to be separated.

**The Role of Private Aviculturists in the Restoration of Rare Waterfowl Species**

Although Sylvan Heights Waterfowl Center currently has the largest number of waterfowl in North America, there are limitations to how many vulnerable species can be raised there before birds need to be dispersed to other suitable locations. We absolutely need zoos, nature centers and private aviculturists to offer their facilities as destination sites for rare captive birds. Having such a large portion of America's vulnerable species concentrated in one location puts the birds at additional risk due to exposure to weather-related events, such as a tornado or hurricane, as well as exposure to an outbreak of disease.

Developing a new breed of private waterfowl collector is a major goal of the IWWA. The primary motive for breeding any bird species should be conservation-oriented. The profit motive needs to be removed from the equation. All participants should be guided by sound ethical principle and a willingness to cooperate with other aviculturists to help preserve wild birds. The goal of all avian propagators should be to help birds survive the continued loss of habitat in the wild and reduced numbers in managed populations. When private aviculturists get to the point that they gain more satisfaction from raising 50 White-headed Ducks of which 25 are released in the wild (or at least could be) than the profit that could be made from raising those 50 ducks, that is when the private community will become a major asset in waterfowl conservation.

Despite the fact that many aviculturists, both private and zoological, share these altruistic qualities, there are many barriers to conservation-oriented waterfowl breeding. One barrier for the small private aviculturists is the cost associated with raising rare captive birds. The IWWA has established a fund that pays qualified aviculturists for the importation and quarantine expenses of the most vulnerable ducks. A portion of membership dues and other financial activities goes into the conservation fund, so that all the IWWA members are supporting the effort. The conservation fund is utilized to import species that the Waterfowl Survey identifies as vulnerable. These birds are owned by the IWWA and judiciously dispersed to members that want them and have been approved through the Conservation Committee. We must be sure that the aviculturist has the facilities as well as the experience to properly handle and raise these birds.

All breeding projects for birds owned by the IWWA birds allow the member to recoup expenses by selling offspring of the birds provided by IWWA to other aviculturists. The member must maintain records of the birds and inform the IWWA of any dispersal. However, the original breeding pairs remain the property of the IWWA. If deemed necessary, the IWWA may reassign the original birds to other breeding projects. I consider this to be a proactive approach to making sure that a sufficient number of rare captive birds are monitored and sustained. AZA breeding programs do not currently work in this manner. The zoos do not own the birds and all of the birds are dispersed by the TAG. Hopefully, if the birds are not too expensive, some zoos may recognize the logic of this proactive approach and work in cooperation with the IWWA in providing space for these birds.

By adopting a conservation purpose for sustaining the birds on the Rare Captive Waterfowl Species List, many more people can be involved in raising waterfowl. The private aviculturist must first realize that there is little or no money to be made in rearing birds. If the profit motive dominates, then the proper dispersion of the birds will never occur. In the past the profit motive was one reason many people began breeding birds, especially in Europe. It is the responsibility of groups, such as the IWWA, to encourage breeders to act in the best interest of the birds' long-term survival and to instill in its members a commitment to saving species. We also need to assure best breeding practices by offering advanced avicultural training as well as funding for aviculturists willing to engage in conservation management of waterfowl.

If the IWWA model for engaging reliable private breeders can be expanded, then additional space will become available for these vulnerable species to reside and eventually expand in number. I believe the trend among many aviculturists today is to have a rare bird in the collection with the satisfaction of knowing that he or she is helping preserve a species. This constitutes a complete rethinking of the role of the private aviculturist. By replacing the profit motive with a species-conservation motive, the private aviculturist can become a major contributor to the sustainability of rare birds.

**Saving Wild Populations by Utilizing Managed Populations**

By adopting a proactive approach to sustaining the vulnerable captive birds in Rare Captive Waterfowl Species List, we also take a step toward preserving species in the wild. Although only a few of the birds on this list are highly threatened outside of captive populations, many are declining in numbers in the wild. If any of these birds were to reach crisis levels in the wild, we stand a much better chance of preserving the bird, or possibly returning it to the wild, if we first succeed at re-establishing them in managed populations.

Managed populations may also serve as a valuable tool in re-introducing endangered waterfowl populations in the wild. A few breed-and-release programs are in the discussion phase and others may be initiated. To bring these projects to realization, the government agencies and other stakeholders in the host country must first be involved in the decision-making process of the release plan. This requires receiving input and ultimately agreement among all interested parties, which can sometimes drag on for years. However, getting local support for the project is vital. Students or interns from the host country can be trained at an appropriate avicultural location in the U.S. or Europe. The students return to their native country to conduct the necessary field work and education of the local people concerning the project - another crucial step that must be implemented prior to the release of any birds.

Once all preliminary work has been completed, the actual bird release can commence. Captive-bred parent birds from established collections would be sent to the selected release sites in the field. None of these parent birds will be released; instead they become part of a breeding program at the release site, conducted by the interns and supervised by avicultural consultants from the U.S., Europe or elsewhere. The parent birds are clipped, so they cannot fly out of the breeding compound when the netting is removed at the appropriate time. The adolescent birds fly out and eventually choose mates and establish territories in the release area. This type of breed-and-release method was successfully utilized in Spain during the recovery of the White-headed Duck population, for which I was a consultant. By allowing input from all the diverse groups involved, consensus was reached on the essential steps needed to restore the wild population.

Without sufficient numbers of birds in managed collections, wild species will be unnecessarily at risk for extinction. It is imperative that we not overlook the birds of the Rare Captive Waterfowl Species List as potential breed-and-release projects. Some of these birds have declining wild populations that someday may benefit from recovery projects, but only if we have sufficient numbers of these birds in our collections.

## The Role of Education in the Restoration of Rare Waterfowl Species

In 1984 I went to Guatemala on a mission to preserve the Atitlan Giant Grebe (known locally as the Poc) and Black-bellied Whistling-Duck. The ten or so remaining Poc lived on Lake Atitlan, on which were three large Indian villages. It became very difficult for me to go to these villages and ask people not to hunt the birds or take the eggs for food just because the bird was very rare. How do you convince them to follow this advice when they are struggling to feed their families and this bird lives on their doorsteps? Sadly, we did not succeed as the Atitlan Giant Grebe is now extinct.

We tried a different tactic to recover the Black-bellied Whistling Duck, which had dropped significantly in numbers in Guatemala. After putting up nest boxes at specified locations, we told the locals that the birds would nest in these boxes and that they could take the first clutch of eggs to eat. The second clutch of eggs should be left untouched for the birds to rear their young, but if they did this, the following year they might have three times as many eggs to eat. The villagers followed these instructions and the results were astounding. The population of Black-bellied Whistling Ducks grew quickly and the villagers even began making their own nest boxes to attract more birds. A more recent project is being implemented in Venezuela to support the declining Orinoco Goose population by again erecting nest boxes. This project will only be successful if enough local people recognize a benefit from the conservation activities there. In this case, several owners of large ranches located in the llanos, have agreed to the boxes to being erected on their property in order to increase eco-tourism.

The broader point to these examples is that conservation goals are rarely successful when people have no incentive to act in ways that we might deem to be environmentally responsible. Whether it's the native villages, private breeders, zoos or the general public, just telling people to act in a certain way may not produce the result we are seeking. Duck hunters help to protect wetlands because that's how ducks are preserved in order to perpetuate managed hunting. Were it not for the preservation activities associated with duck hunting, waterfowl in this country would be in much worse condition in the wild and have much less habitat. In this case, the hunter has become educated and motivated to preserve wetlands as well as the ducks that live in them. We need to continue to search for more "win-win" situations for a much larger segment of the American population, inducing more reasons for people to take conservation action specifically for waterfowl.

## Advancements in Avicultural Science

One irony in the decline of waterfowl husbandry in North America is that as the number of collections and people breeding them have decreased, advancements in aviculture actually make rearing birds more efficient than ever. Since my entry into aviculture over 40 years ago, there have been remarkable improvements in the feeds, methods, and equipment utilized in avian husbandry. These advancements have resulted in the increased survival rate of hatchlings and the number of birds that can be successfully managed on a site.

While feed is still a sizable expense in breeding waterfowl, nutritional improvements and production advancements not only make feeding more economically efficient, but also contribute to the health and survivability of the birds. The availability of the floating pellet allows for efficient and healthy feeding. The development of the seabird pellet has eliminated the need to purchase fresh fish as a food source. Forced air incubators with more precise temperature control also contributed to the improved hatchling survival rate.

A somewhat serendipitous improvement in the breeding of wild waterfowl is the creation of more captive generations. While we are losing genetic diversity as we produce more captive generations, the later generations are becoming more acclimated to the methods employed by aviculturists, making them easier to breed in captivity. For example, certain species that once nested exclusively in log boxes, now have adapted to porch boxes. Years ago, Black-necked Swans always nested in December or January, but as we get more and more captive generations they are now breeding closer to our spring. While increasing the number of people engaged in aviculture is certainly desirable, it is comforting to note that advancements in avicultural science can efficiently produce more birds from fewer facilities. This alone will not turn the tide for the rare captive waterfowl species, but it does provide an important safety net.

**The Role of Education in the Revitalization of Aviculture**

One enormous challenge to the field of aviculture is the lack of a new generation entering the profession. It is disturbing to note how few young people have any interest in raising birds. The interest of most youth is focused on leisure pursuits, such as computers or electronic games. The prospect of the hard work required to learn aviculture seems to be unattractive to an increasing segment of the population. Of the 360 volunteers or zoo interns that have been trained at Sylvan Heights Waterfowl Center over the past 18 years, only five percent to ten percent are still involved in aviculture at any level. As a result, the next generation of aviculturists will need to be carefully nourished and encouraged.

In order to reverse this trend, hands-on experience with waterfowl and husbandry training need to be offered to those youth expressing interest in this endeavor. Sylvan Heights Waterfowl Center provides a volunteer program that hopefully will attract promising young aviculturists. Our volunteers come from all over North America, as well as countries in Europe, Asia, Africa and South America. Most of them stay at Sylvan Heights for one to three months, giving them plenty of time to learn husbandry techniques and see the end result of their labor. Some of our volunteers are financially supported through scholarships offered by the IWWA and the American Pheasant and Waterfowl Society, which enables us to attract highly qualified young aviculturists who may otherwise need to work at a summer job rather than learn avian husbandry.

Also an increasing number of zoo keepers are no longer receiving significant experience in avian husbandry because zoo protocols have drastically limited the number of birds that can be bred throughout American zoos. In order to address this problem, the Sylvan Heights Avian Husbandry and Management Program was developed. Zoo and wildlife professionals spend two weeks with the Sylvan Heights staff at the Avian Breeding Center, gaining knowledge and practical application of advanced avian husbandry techniques. To my knowledge, this is the only professional level training program in the country that focuses on avian husbandry. It is vital that both young people and the zoological community are provided access to waterfowl husbandry instruction and experience in order to reinvigorate aviculture in America. Sylvan Heights is intensely committed to training the next generation of aviculturists. However, other organizations concerned with biological sciences need to come forward in establishing scholarships and grants that allow youths as well as wildlife professionals to pursue their avicultural interests.

**Cooperative Waterfowl Conservation Efforts**

When one considers the low participation level of zoological organizations in waterfowl breeding and education, it is difficult to postulate how these institutions would be able to lead the way to a brighter future for the birds of the Rare Captive Waterfowl Species List or to a resurgence of interest in waterfowl aviculture. Sylvan Heights has strong relationships with many bird curators at zoos, the result of which is certainly encouraging. Unfortunately, it remains a difficult task to convince most zoos to acquire, display and breed waterfowl species that are not resident on a TAG or SSP. There is also an increasing reluctance at some zoos to develop cooperative relationships, even on a selective basis, with non-AZA accredited organizations, despite the obvious competence and ethical standards of many private waterfowl aviculturists. Zoo education programs rarely feature waterfowl species, even at zoos that have a fair-sized collection. Wetland education programs at zoos and nature centers frequently focus on the plants, mammals and amphibians found in these habitats while practically ignoring the wetlands' importance to the survivability of waterfowl.

As a result, the primary effort to bolster threatened

waterfowl species and educate the American public and wildlife professionals on waterfowl issues (including aviculture) must emanate from a small, but visionary group of zoos and private aviculturists that understand the benefit of blending resources to support waterfowl conservation. Ultimately, this small cooperative group must be widely expanded. In addition, the general public, corporate America and private foundations must become aware of the need to financially support these efforts on behalf of waterfowl populations in the wild as well as in managed collections. In order to achieve this level of institutional, corporate and civic collaboration, these groups must first become aware of the need for waterfowl conservation.

The Sylvan Heights Waterfowl Park and Eco-Center in Scotland Neck, North Carolina, is designed to educate people about waterfowl and the importance of preserving them. Our goal is to tell visitors the story of every species--where it comes from, what habitat it prefers and why the species is important to our world. The Center also immerses visitors into a wetland setting so the feel and scope of a primary waterfowl habitat can be fully experienced. We hope the public will react to what they see and hear by caring more, becoming more aware and wanting take action to help all birds, but especially those on the brink of extinction. The hand-feeding exhibit that is being constructed at the Waterfowl Park is designed to produce close encounters with birds for children and adults. Amazing results are accomplished through this type of one-on-one experience that cannot be duplicated by merely viewing a bird in an aviary. We certainly want the visitor's experience to be enjoyable, but it's not our primary goal to entertain the public. Hopefully, enjoyment will be gained from knowing their admission fee or membership is an investment in the future of waterfowl.

Establishing the Waterfowl Park and Eco-Center in a region of the country rich in waterfowl habitat provides an avenue to reach the general public, civic leaders, conservation groups, and corporate America on matters concerning waterfowl preservation. Equally important, the Eco-Center presents opportunities to form high-level cooperative programs among experts in the fields of aviculture, ecology, veterinary science, behavioral biology, conservation, zoology and education. The resulting actions should be focused on joint projects for breeding endangered species as well as the birds on the Rare Captive Waterfowl Species List. In addition, efforts should be expanded to support species critically endangered in the wild.

## CONCLUSION

Clearly the state of captive waterfowl is at a critical point. Dozens of species are on the verge of being lost in collections and perhaps someday lost in the wild as a result. Despite a decrease in private and zoological collections of waterfowl and the expertise to manage the ones which endure, there still remains reason to maintain hope that the future is not as bleak as it may appear.

A fortunate development is recent years is the increased interest in birds and bird-related activities among the general public, such as bird watching, avian photography and backyard bird feeding. These hobbyist pursuits offer a great opportunity for avian education programs to connect a broad segment of the public with issues concerning waterfowl. Involving and motivating the public also increases the prospect of engaging civic and corporate entities. More of the conservation and educational funding provided by wildlife organizations, private foundations, corporate sponsors and civic groups, needs to be directed toward the cooperative efforts of organizations like the Sylvan Heights Waterfowl Park and Eco-Center and other visionary zoological conservation programs for waterfowl preservation.

I also continue to hold out hope that zoological organizations join with private aviculturists on a broader level than what has occurred in recent years. I look forward to a time when zoos and private waterfowl aviculturists work together on preserving the most susceptible waterfowl species. Time is growing short for many of these birds. Whatever barriers exist that would prevent full cooperation need to be addressed very soon or we will lose these avian treasures.

Waterfowl species in peril deserve the greatest cooperative effort that zoos, conservation societies, breeding facilities, nature centers and the educated public can muster on their behalf. Waterfowl deserve these efforts as much as any other threatened species. Those "little brown ducks", which are often overlooked and under-appreciated, are as precious to mankind as the most colorful macaw, the largest whale or the most endearing baby panda. Internal differences between organizations should never be a reason for losing even one species of waterfowl. I would rather see the last of a species alive in captivity than dead in a museum.

## Acknowledgements

I would like to thank Nicolas Hill, Curator of Aviculture at Sylvan Heights Waterfowl Center, and Brad Hazelton, General Curator at the Sylvan Heights Waterfowl Park and Eco-Center for their contributions to this paper and for their role in preserving wild waterfowl. I would also like to express my appreciation to Arnold Schouten, IWWA Conservation Committee Chair, who compiled the waterfowl survey data and to Dale True, Avian Programs Coordinator at Sylvan Heights Waterfowl Center for his work in editing this paper.

## Author Biography

Mike Lubbock is Executive Director of Sylvan Heights Waterfowl Park & Eco-Center and Founder of Sylvan Heights Waterfowl Center, organizations dedicated to the survival of the world's waterfowl species, in both the wild and in zoological collections. Mike Lubbock serves on the Board of Directors of the International Wild Waterfowl Association (IWWA) and is also Vice President of the Carolinas/Virginia Pheasant & Waterfowl Society as well as the American Pheasant and Waterfowl Society. In April 2007, Lubbock and his wife Ali were named Entrepreneurs of the Year by the NorthEastern Entrepreneur Roundtable (NEER) for the establishment of Sylvan Heights Waterfowl Park & Eco-Center. Mike Lubbock's professional career with birds began at the Wildfowl & Wetlands Trust (WWT) in Slimbridge, England, where he held the positions of Curator and Director of Aviculture. While at the WWT, Lubbock continually refined and improved waterfowl propagation techniques and did a considerable amount of avian field research in some of the world's most remote areas. The ultimate goal of establishing his own avian collection became a reality in 1989, upon arriving in North Carolina and founding Sylvan Heights Waterfowl Center. In addition to receiving 17 World and 13 North American First Breeding Awards, Lubbock was inducted into the International Wild Waterfowl Association's Hall of Fame and is also a recipient of the prestigious Jean Delacour Avicultural Award. This special honor was given to Mike Lubbock at the 2000 IWWA convention along with this introduction:

"Mike Lubbock's avicultural accomplishments on both sides of the Atlantic are legendary. He has brought many new species and new bloodlines in from the wild. He has accomplished many first breedings and he has been a source of bird and breeding advice to many." Mike's passion is the global preservation of threatened waterfowl. In addition to his involvement with the previously mentioned groups and societies, he is active in many conservation efforts worldwide. He assisted in organizing the Venezuelan Waterfowl Foundation, which provides fieldwork and educational support for the preservation of the Torrent Duck, Orinoco Goose and Merida Teal. He is a member of the Brazilian Merganser Recovery Team, developed under the WWT's Threatened Waterfowl Specialist Group to preserve the critically endangered Brazilian Merganser. Also, in 2003, in conjunction with the North Carolina Zoological Society, Mike founded the Sylvan Heights Rare & Endangered Species Breeding Fund, designed to provide financial support to obtain and breed the world's most critically threatened avian species.

# EXPERIENCES WITH THE RESPLENDENT QUETZAL *PHARMACHRUS MOCCINO* IN NATURE AND CAPTIVITY: A BRIEF HISTORICAL REVIEW

JESÚS ESTUDILLO LÓPEZ

Vida Silvestre, Mexico City, Mexico

[*Editors note: The unfortunate passing of Dr. Lopez in 2010 left a significant gap in North American aviculture. Dr. Lopez was one of the foremost experts on cracids in aviculture and established one of the largest bird collections in the Western Hemisphere. His attendance at the ISBBC was possibly the last avicultural event he attended. His presence in the avicultural world will be greatly missed.*]

## INTRODUCTION

I had the fortune of being born into a family in which my father, from the time of my childhood, had passed on to me his love and admiration for nature. He was very much interested in all orchids and bromelias. Wild birds have always been my love: parrots, macaws, toucans, and many other birds like the curassows and guans impacted my interest in a special manner. Searching for one of them, the Horned Guan that in a very fortunate coincidence, lives in the very same habitat of the Resplendent Quetzal in the clouded forest of southern Mexico and the volcanic northern area of western Guatemala, hence when you find one, you often find the other. I was very young when I walked the rain and clouded forests of Mexico in the states of Veracruz, Tabasco, Oaxaca, Chiapas, Campeche, Yucatan, and I had the great luck of observing in the mountains of Los Chimalapas (Oaxaca-Chiapas), the Resplendent Quetzal and the Horned Guan, and other birds like the Harpy Eagle. This was more than fifty years ago, and I still relive those fantastic events in my head. Searching for birds, I have had the fortune of traveling most of the jungle areas of Latin America, Mexico, Central America, the Amazonian rain forest of Venezuela, Colombia, Ecuador, Peru, Brazil, and Bolivia, in the way that my father taught me. By contracting local people from different Indian tribes of the areas I visited (such as Lacandones, Zapotecas, Zoques, Quichés, Motilones, Chocoes, Tucanos, Aucas, Jibaros, Esejas, etc), these people were familiar with their ancestral jungles and knew the birds I was

looking for, some of them part of their diet, so they had intricate knowledge and vital expertise. Needless to say what I have learned from these native people about wildlife in the jungles, and the information they passed on to me, wouldn't be possible to obtain from the best university in the world, beside their moral values and sincere friendship, it is something that people who consider themselves civilized must learn from these primitive people.

### The Most Beautiful Bird in the Americas

The Resplendent Quetzal (R.Q.) is often thought to be the most beautiful bird in the New World and for many the most beautiful in the world. Quetzals are the most ornate members of the Trogon family, there are five species. The Resplendent Quetzal that occurs in Southern Mexico and Central America, is unique in being endowed with a train of long tail plumes. The other four species live in South America, mainly in the clouded forests of the Andes. There are two distinct populations of the Resplendent Quetzal, one from Southern Mexico, Guatemala and Honduras, the other in Costa Rica and Western Panama. The northern birds are the more ornate with longer, wider plumes and a longer crest. Modern bird lovers are not unique in treasuring the quetzal. To the pre-Columbian people of Mexico-middle America, the bird's glorious plumes were more highly treasured than gold, and only royalty where allowed to possess and wear them. Ornaments made of the incredibly long golden-green tail coverts of the male quetzal, adorned the heads of these personages, these feathers are of the color of maize leaves and all the other vegetation that give life and beauty to the earth.

### Historical Importance of the Quetzal

Within the ancient history texts of Mexico resounds the great admiration and respect that the pre-Columbian civilizations had for nature and its manifestations, which created gods that were very important throughout their history. In accordance with these traditions, the origin of Mexico is related with "El Aguila Real", Golden Eagle, a bird that not only flew at great altitude, but possessed feathers with a golden sheen; its descending flight signified sunset, flying with both wings extended represented the center of the universe.

The foundation of the empire had to be built in a place they found the eagle perching on a "tenochtly" (cactus), the tree of life, with its red fruits representeing the hearts of the Aztec warriors who died in battle, the eagle would then eat them to take them to heaven.

In the Mendoza Codex, the eagle appears perching in the sacred tree which is the center of the four cardinal points, pointing to the four sectors for the construction of Tenochtitlan (Mexico City) in 1325a.d. on an island in Texcoco Lake. The beauty of the quetzal, his beautiful colored feathers, elegant shape, rapid flight, and other characteristics, that for 3000 years, assumed a god like status in the form of Quetzalcoatl (the plumed serpent). It obtained its god like status in the form of a number of male and female deities across the pre-Hispanic cultures all over Mexico and Central America, including the Olmecs, Toltecs, Aztecs, Mayas, Mixtecs, and others. Among the Mayan, the deities are variously known as Guacomate or Kukulcan depending on the language and culture groups in question. In Central and Southern Mexico, the Mayan gods are Quetzalcoatl the male deity and Xochiquetzal the female deity. The R.Q. had a profound influence on the old cultures and was so complex that even today, is still not fully understood. An extraordinary number of relevant temples and palaces have been studied by archeologists, such as the temple of Kukulcan in Yucatan, the Palace of Quetzalcoatl, and the temple of the Feathered Serpent in Teotihuacan, (near Mexico City), Xochiquetzal in Cacaxtla, Copan, Tical, Palenque, and several more places in which they worship the quetzal deity Quetzalcoatl-Kukulcan. The mixture of the quetzal and the serpent in Quetzalcoatl represents the union of the majesty and celestial characteristics of the quetzal with the power of the underworld- the serpent. The feathers of the quetzal were used to symbolize royalty and were used as ornaments on the Aztec and Mayan king headdresses. For the Aztec and Mayan warriors the quetzal protected them during battle and if they died the quetzal would also die. The legend of how the quetzal stained its chest with red coloration was originated in 1524, when the Spaniard conqueror Pedro de Alvarado, taking advantage of their superior weapons and horses, massacred 20,000 Mayas, who where commanded by Tekum-Uman, and when he was mortally wounded the Quetzal covered him, thus staining his pectoral feathers with red. Today the name of the place of this battle is Quetzaltenango: place of quetzals and is now one of the main cities in Guatemala. Today the quetzal is featured in many modern pieces of literature. The feathers of the quetzal where more valuable than gold and only individuals that were experts in capturing the quetzal without damaging it, had the permission to collect feathers. Killing a quetzal was punishable by death.

After a victory in battle, a feather was added to the crown; Moctezuma's crown contained 469 quetzals feathers that corresponded to the number of Aztec populations he had conquered. After the conquest of Mexico, the Spaniards along with gold, silver and precious stones took the Moctezuma quetzal headdress, which later appeared in the Viennan museum in Austria. The Mexican government, after a long period of negotiation, claimed it back as its national jewel.

The Quetzal has always been a symbol of freedom for the pre-Hispanic world as demonstrated in the war of independence of 1810 against the Spaniards, and even today the R.Q. is the national bird of Guatemala, featured on both the Guatemalan coat of arms and on its national currency.

**Etymology of the Word Quetzal**

The Aztecs used the quetzal without the "t", which in their language means beautiful green feather, and later added the word 'totol', which means beautiful bird of green feathers. The Mayan term 'gug' means beautiful green feather and the word Que-zalit meaning brilliant, beautiful feathers. It is my believe that any of these terms are much more suitable than the latin name of the quetzal which is derived from pharomacrus: pharos meaning throat and macrus meaning large.

**Biology of the Quetzal**

Taxonomy

Order Trogoniformes
Family Trogonidae
Genus *Pharomacrus*
Species *mocinno*
Subspecies *m. mocinno, m. costarricense*

The Quetzal was first described by Francisco Fernandez in 1651, who was sent by King Felipe the second. In 1825, Temminck painted the beautiful trogon, in 1835 Gould named it as Trogon Resplendent, Juan Pablo de la Llave named it as *Pharmacrus mocinno* after a Mexican ornithologist.

The order Trogoniformes (trogons and quetzals) is constituted by a small group of pantropical birds. Thirty-nine species are included in the order, nine species (24%) are found in Mexico. Trogons are important seed dispersers. They are not found in abundance, so they are important indicators of habitat quality.

Coloration is very different in males and females. Males are more colorful and there is color variation according to the species. There are eight genera of trogons that have pantropical distribution, excluding Australia. The distribution of trogons is wildly extended and includes the Americas, Asia and Africa. None of them are migratory and their flying capacity is limited. The five species of the genus *Pharomachrus* are the most beautiful and the two subspecies of R.Q. belong to that genus. The order Trogoniforms do not have close relatives suggesting that they are a very old group. Fossils of trogons are known date back to the tertiary age, sixty-five millions years ago. All trogons are birds of arboreal habits, and they inhabit rain and clouded forests: from lowlands of 300m to mountains of 3000m. The New World trogons feed mainly on fruits and berries that they collect mid-flight, stopping under the fruit, hovering below and feeding. They will eat rather large fruits and regurgitate the seeds.

## Description of the Resplendent Quetzal

The species is strongly sexually dimorphic, males are more beautiful, they have long tail feathers and a vertical crest on the head, the four covert tail feathers are long and hang down. The head, neck and back are brilliant iridescent green or golden green changing to bluish green or even greenish blue, cobalt green in certain light, and is dependant on the angle that they receive sunlight, the hour of the day and in the area they are perching or flying. The central tail feathers are of dark coloration, and covered by the long tail feathers. The green color of the quetzal feathers contrast with the intense darkening to crimson of the breast and abdomen, the color is produced by the carotenoid feather pigment (zooerythrina) one of the cantaxanthins.

When the quetzal is perched in the jungle without sunlight, it blends perfectly with the green color of vegetation. The beak of the male is yellow, legs are black colored; the measurement of the male from beak to tail is about 1100mm, the body size is about 369mm. The average size of the tail feathers of the northern species is from 965-1075mm and the average of the Costa Rican quetzal feathers are 660mm. These ornamental feathers are much less developed in females and in mature males; it takes about three years for the coverts to reach their maximum length on the males.

The anatomy of the quetzal presents adaptations to ingest big fruits of Lauraceas (Aguacatillo), a wild relative of the avocado that represents an important proportion of the diet of the quetzal; they do not have a crop, the mandibles and clavicle bones are flexible and allows the quetzal to swallow large fruits like Aguacatillos of different species.

## Distribution of the Resplendent Quetzal and Feeding Habits

The original distribution covered almost a continual area from Southern Mexico to Panama which is supported by museums from Europe and USA, institutions that collected several hundred of these birds during the last two centuries. From the origin of the birds it is known 20.4% were originally from Mexico, 17.8% Guatemala, 18.2% Costa Rica, 17.5% Panama and the rest from Honduras, Nicaragua and El Salvador. Of the exported birds 58% corresponded to the northern race and 41% to the southern.

The northern extreme of the distribution of the R.Q. is the clouded forest of the Chimalapas in the state of Oaxaca, Chiapas, Montebello, El Triunfo, Tacana, from 1000-3300m of elevation. In Guatemala the R.Q. occurs from 1000-2000m also in clouded forest, but it has been found at different elevations, 1500-2500m and in other locations at 2800-3300 meters.

Personally, it is not possible to speak of quetzals and Horned Guan in Guatemala without mentioning professor Mario Dary Rivera, a highly appreciated friend of mine. We walked the volcanic axis together, Alta and Baja Verapaz, San Marcos, Sierra de Minas and several other places searching for these birds, we had a friendship that lasted for years. He was one of the pioneers of conservation in Guatemala, he was a scientist, not a politician, and very sadly he was murdered outside of the University San Carlos in Guatemala City without any apparent reason. Today there is a national park named after him- Biotopo El Quetzal Profesor Mario Dary Rivera.

The R.Q. extends from Honduras: Cosusco, Yahoa, La Tigra, Nicaragua: Soslaya, Costa Rica: La Amistad, Chirripo, Cordillera de Tailarán and Panama: en Barú. The actual distribution of the quetzal has greatly diminished and disappeared from great areas and in countries like El Salvador the bird is extirpated. The quetzal inhabits jungles, mountains, and clouded forests in an area of more than 1500km from southern Mexico to Panama. The distribution however is not continuous, in fact, it never has been, it is interrupted in Nicaragua, a country in which existed a great plain along the San Juan River and Nicaragua lake that covers the surface of 8624km$^2$, a distance insurmountable to a bird like a quetzal with a limited flying capacity. Telemetry studies have indicated that the quetzal cannot fly further than 30km without making a stop. This phenomenon can explain the two populations of two different subspecies of R.Q: the northern race from southern Mexico, Oaxaca, and Chiapas to north of Nicaragua, and the southern race from Costa Rica and Western Panama. The R.Q. is an altitudinal migratory bird. During their lifetime they inhabit different environments. During the non-breeding season, they live in a lower altitudinal environment than the one they inhabit during the breeding period. The breeding habitat of the R.Q. from Mexico to Central America is known as cloud forests or mesofil mountain forests that are located at 1500-3000m of altitude and they receive from 3000-5000mm of rain, they are always covered with fog and produce many species of hygroscopic plants like arboreal ferns, hundreds of bromeliads, orchids, lichens, moss, etc.

The vegetation composition of that forest is extremely variable, changing notoriously not only from one place to another, but also depending on the altitude. In the reproductive cycle, three periods are identifiable: the courtship that takes place at the end of December, January, and February, the egg laying in February and March and the raising of the chicks from March to June. During the breeding cycle, both male and female participate, so quetzals are a monogamous species. The R.Q. nests mainly in dead trees at a height of 3-18m above the ground. The nest cavity is about 35cm in diameter, sometimes its very shallow, only 15cm below the entrance, but in many, the bottom is about 30-50cm below the entrance. It is possible that at times the quetzals take over old woodpecker holes enlarging them when necessary. As with other trogons, both sexes help to excavate. Skutch described in Costa Rica ten nests, which ranged from 4-30m above the ground with entrances of 10-15cm in diameter. Many of the trees in which the birds nested where in the last stages of decay, and two collapsed during inspection. The bird sometimes moved their sites downward as the snag rotted from above. One pair of R.Q. began excavating five sites within a month before they finally selected one. Nests are mostly excavated in decaying Ocotea or other Lauraceous trees, 74% of 43 nests sites where in the forests, 12% at the forest's edges and 14% where in the open.

## Eggs and Incubation

The normal clutch size is two eggs, Skutch found that the male incubates in the morning two to four hours, and returns to the nest in the afternoon; the female returns about sunset to take over the night time incubation. The incubation period is very short, from 17 to 18 days. Nestling R.Q., like other trogons studied, hatched with tightly closed eyes and completely naked. Pinfeathers began to appear within two days and began to break open on the 7th day, although retrices and remiges did not expand until the 10th day For the first few days the chicks are fed mainly live animal foods, but from the eleventh day, the adults begin to feed them with fruits. Toward the end of the brooding period mainly the male feeds the chicks, which fledge at 21-31 days. Resplendent Quetzal chicks at 13-14 days are partly feathered, having eyes that are still closed, most developing feathers are blackish. According to Skutch, there was a high rate of nest failure, estimated 67% to 87% of 11 nesting efforts. Winter reported 80% of young quetzals dying before fledging, and few of the remaining survived to adulthood. Some native friends of mine, Zoques from Oaxaca have informed me that after one month, the first brood fledged, some pairs of quetzals laid again in the same hole, probably it is to compensate for the high failure mentioned above.

## Altitudinal Migration

The altitudinal migration of the R.Q. is verified from July to December, descending to warmer jungles. During the 1990s, telemetry observation was practiced with R.Q. in Mexico, Guatemala and Costa Rica and it was found that the pairs migrate separately at different dates and to different places. There is no explanation for such behavior; once the migration period finished, the R.Q. returns to the breeding grounds to meet his partner.

Many authors have tried to explain the altitudinal migration of the quetzal due to the diminution of food and/or the diminution of Aguacatillo (wild avocado), as some authors believe 75-80 % of the quetzal's diet is made up of this fruit. Recent research done by Dr. Solorzano and her group en El Triunfo, Chiapas, do not agree with the anterior hypothesis; the Aguacatillo is an important fruit, but not as important to determine migrational habits of the quetzals. They documented 26 species of fruit in the quetzal diet, furthermore they pointed out that the R.Q. is not a vegetarian species, its diet also includes several small vertebrates like frogs, lizards, mollusks such as snails, and many insect species. In Costa Rica, the R.Q. in June-July, descended into the Pacific where they remained until January-February, where they stayed at an altitude of 800-1500m.

**Resplendent Quetzals in Captivity**

During the early 1990s, I was visiting the clouded forest in Chimalapa, Oaxaca, along Zoques with indigenous friends of mine, we observed some pairs of R.Q. in display; on the last day of my visit, one of my friends found a fallen tree, and he heard some noises so circled the tree and found that it had an opening, it was a quetzal nest with two chicks, that fortunately, were in good condition. I realized I had to save these birds from a certain death and I knew it was going to be a big challenge trying to save them and to maintain them in captivity (and later to breed them in my aviary). From the beginning it wasn't an easy venture, and I was going to face many challenges along the way. At my aviary I kept them in a warm area, feeding them with avocado, fruit, and insects. I informed the wildlife offices of the finding of the birds, and asked for a license to keep them; quetzals are absolutely protected and it is forbidden to take them out of the wild. I knew that quetzals required a special environment, they do not adapt easily to captive conditions like other species of birds, they are very active birds that are almost in continuous movement; so I built a large aviary (50x40x8m high). I planted it heavily with the same plants found in their environment: arboreal ferns, bamboos, palms, avocado trees, bromeliads, orchids, etc. Mexico City valley has the proper altitude for the quetzal, but not the humidity that they require (70-80%), if all the conditions are not met, they develop respiratory diseases, so I adapted my aviary.

Another problem that I faced many years ago, with birds originally from clouded forests, is that they where in good condition for a time, but would suddenly die. I searched for infectious agents, (which is my specialty), but with no results. The histopathological examinations indicated hemosiderosis-hemocromatosis or iron poisoning; at the beginning I couldn't understand the problem, I fed them with the same diet they had in the wild. Many years later, I found that the problem was related with the water they were drinking, all the birds that I had problems with where from clouded forests and they never drink water from rivers or lakes, but water that had accumulated in bromeliads, orchids, etc. Such water contains tannic acid, that diminished the iron absorption and hence they evolved with a low iron requirement. I had no information on the quetzal, but since they have the same habits, we provide low iron diet and drink water in the way they do in nature.

**Breeding of the Resplendent Quetzal in Captivity**

Needles to say that the goal of any effort made with wild animals in captivity is reproduction, if they are provided with the proper conditions, nutrition, habitat, etc. It has been possible not only to breed the species, but to increase it's natural capacity of its production, and in this way it has been possible to avoid the extinction of several endangered species.

When I decided to work with the R.Q, I knew it was not going to be easy, I had the experience of many years of work with almost 500 different species, but I knew that the quetzal was something completely different, just keeping them alive in captivity was a big challenge. Without any success I searched for any possible information on the reproduction in captivity of the R.Q., I found nothing. The quetzals that in the past where kept in captivity didn't stay alive for a long time. At this point I considered that it was necessary to start the program with hand-reared birds, as I guessed that the chances of adaptation in captivity were better than doing it with adult birds that were used to living in the wild. The baby quetzals were kept in a warm room and hand fed for five weeks, time in which they were completely feathered and began eating by themselves a mixture of avocado, tropical fruits, tenebrio worms and day old mice. As they grew, I moved them to larger aviaries that were well planted and had branches so they could practice flying. About 4-5 months, I considered it was time to move them to the large aviary. Fortunately, the two birds turned out to be a male and a female, and incredibly beautiful birds. I considered the possibility of at least another pair of quetzals to avoid endogamy; fortunately, with the support of the National University (UNAM), the government allowed me the opportunity and I received two birds from Chiapas that also turned out to be a pair.

Based on observations in the wild, I put several trunks from one-and-a-half to two meters long, some with two entrances, others with one, and with a cavity of 30-40cm deep. For years I observed some type of display and courtship, males chasing females producing different vocalizations, undulatory flight of the males and entering the nest; these actions occurred at different times, but especially in spring. In 1998, when the oldest pair was now six to seven years old, finally laid two eggs of light blue coloration. I couldn't observe if the eggs had been laid in two, three or four day intervals. Female and male shared incubation duties. The female incubated during the night, and early in the morning came out to eat, while the male incubated for two to four hours and again by evening, sometimes the male incubated during the middle of the day, a period of time that the female usually occupied the nest. Seldom was it possible to observe the change of incubation duties. The substitute perched near the nest or on the nest, while the bird inside of the nest somehow detected its partner, apparently without observing it. The hatching of the baby quetzals occurred after 17-19 days, I tried not to disturb the birds, the first time I could observe the babies was when they were probably three four days old, I did not have a chance before, as one of the adults was always sitting on them. At this time, the babies looked like embryos, completely nude, pink colored skin, no feathers at all, eyes shut. About a week later, the chicks started developing blackish feathers on the back but nothing on the heads, at ten to twelve days, wing feathers began developing.

Authors like Skutch, describe that young quetzals open their eyes after one week of age, this may happen with the southern Costa Rica subspecies, however the ones at my aviary opened their eyes only after sixteen to eighteen days. After two weeks, the chicks were covered with feathers on the back, but not on the head, they had a bulky abdomen and the wing feathers began growing. In the rearing process, like incubation, both sexes participate. During the first two weeks, the female and the male stayed in the nest covering and feeding the chicks, seldom leaving them alone. The feeding routine was typically: seven o'clock the female, at eight the male, nine to ten the female, ten to eleven the male, noon to one the female, three to five the male, several times the adults collected and swallowed the food and remained for half an hour to one hour perching near the nest, then entered the nest and regurgitated the food. Approximately, at eighteen to twenty days, when the chicks had open eyes, they approached the entrance of the nest to receive the food. At four to five weeks, they were completely feathered with gray coloration and greenish shades on the back. They began showing up and perching in the entrance of the nest and if any of the adult birds approached they emitted sounds and if they carried food, immediately took it from the beak of the adult. It is important to note that even at this time, when the young quetzals where capable of eating by themselves, the parents regularly regurgitated fruit for them. In the wild, at four to five weeks, the chicks are ready to leave the nest; they remain near the adults for a few days, who continue to feed the young birds. It is probably the most critical period of the life of the quetzal; it is at this time when they are susceptible to predators, accidents or starvation if they fall from the nest. In the aviary, I was very fortunate in observing one of the chicks that continuously perched at the opening of the nest, and flapping its wings continuously, I suspected the possibility of it falling down, fortunately I had put a net under the nest and after just the second day, the young quetzal fell out, but the net saved him from falling on the floor.

**Juvenile Quetzals**

Young birds are brown-grayish in coloration, somewhat like the color of the female, the coverts of the wings are whitish like the abdomen, tail feathers are barred. At five months, the dorsal feathers and wing coverts are green and the black tail feathers start changing to green. At eight months, males already develop crests, the back is iridescent green, abdomen reddish. At one year they get the adult coloration, the development of tail feathers is not complete, they need a second or third feather molt to obtain full adult plumage.

## CONCLUSION

In the approximately 15 years I have worked with quetzals in captivity, it has been possible to reach some modest goals. In the breeding of the R.Q. five years after the first chicks, the same pair produced two more chicks, and two or three nests with infertile eggs.

I have many doubts for which I cannot find a logical answer; I will mention a few, to begin with: certainly the R.Q. is one of the most beautiful birds of the world, but also one of the most difficult, at least for me. I was able to provide the proper environment, humidity, low iron diet, tannic acid in water, and the R.Q. survived adequately in my aviaries. The success in captive breeding is very modest. I do not understand the low reproductive rate in captivity, as far as I'm concerned the environment and nutrition has been satisfactory, nevertheless in the wild they start breeding after two or three years of age. Obviously, tools that have been successful in many other species of birds, such as artificial insemination and artificial incubation don't work with the quetzal, this bird will not tolerate being handled. Artificial incubation is not possible either as it cannot mimic the needed temperature and humidity requirements for this species, furthermore they have one of the shortest incubation times of any bird, 17 to 18 days and the chicks have to be fed regurgitated food, probably partly digested and with the bacterial flora of the species to survive. One phenomenon that is not easy to understand is the vertical migration of the R.Q. During the warmest months of the year, in summer, it migrates down to the warmest places, some authors have related this with the availability of food. The three year research in El Triunfo, Chiapas by Solorzano and her group, found no relationship between fruit abundance and migration. The quetzal does the opposite of most migratory birds that in winter move to warmer places, and return to breed when cold weather is gone and daylight hours are increasing; a phenomenon that stimulates the endocrine system and reproduction. Such a migration of the R.Q. is very confusing, but must have a reason. Does the lower altitude, high temperature changes in atmospheric pressure produce some physiological stimulation in the endocrine system? Sometimes I wonder if the impossibility of the quetzals in my aviary to practice such migration is part of the problem.

**Conservation of the Resplendent Quetzal**

The fundamental problem of conservation of wildlife is the loss of habitat by human actions, this is especially true in species having a limited distribution, or with specialized territory. In the case of the R.Q, a bird whose territory has been reduced to small island fragments (ie: Guatemala originally had 30,000-35,000km$^2$ of habitat, today they only have 2,500km$^2$ left and the situation is just about the same in all other countries it inhabits). The indigenous farmers that used to have their subsistence on maize fields in the lowlands now have disappeared because the large companies with multinational support took this land for producing big coffee, banana plantations and cattle ranches. As a result, the indigenous farmers have had to migrate to the high lands of clouded forests and they are burning it to grow corn; because of this, the territory of the quetzal has been significantly reduced. These types of corn fields in Spanish are named "Milpas Caminantes" or walking cornfields, because the clouded forests have steep slopes, hence during heavy rains, nutrients are rapidly washed out of the fertile organic soil, so they need to burn more forests to open new cornfields.

The R.Q. has always had natural enemies: the Grey Squirrel, the Emerald Toucanet, different species of jays, etc, that unfortunately discover the quetzals nests and eat the eggs, babies, and even the adults. Most unfortunately, when the male quetzal is incubating or feeding the babies, his long tail feathers hang out of the nest, disclosing himself and the nest to his enemies. Unfortunately, the destruction of the lowlands by the rich agriculture companies has had a multi-faceted effect on the natural enemies of the quetzal. Many animals that are potential predators of the quetzal like the Tayra, Ocelots, Margays and another wild cats, Kinkajous, monkeys, raptors, and reptiles have been forced into the clouded forests and certainly represent a new threat for the quetzal. Due to all of these pressures, many authors consider that not more than 12-15% of young quetzal's survive. Fortunately, national parks have been created for the protection of the quetzal: El Triunfo in Chiapas, Biotopo el Quetzal, Mario Dary Rivera, Sierra de Minas in Guatemala, Volcan Poas, Chyrripo in Costa Rica, and others, but it is very important to remember that the flying abilities of the quetzal are very limited, and they cannot move from one reserve to another. Some authors claim that there are probably not more than one hundred pairs on each reserve, so in the near feature, inbreeding may be another problem with which to deal. Finally it has been a modest step in keeping and breeding a few R.Q. in captivity, but because of the natural complexity of the bird, we are still far from reaching the same success obtained with other species. The most important action to preserve the R.Q. is to stop the destruction of its habitat and let's keep in mind that we can not blame the poor indigenous farmers if we don't provide other ways of living for them. Hopefully in the future, because their knowledge of the forest, we can use them as guardians of the quetzal. We must keep in mind that in the preservation of other species, the purchase or protection of land, forests and jungles, has been the most successful way to protect endangered species. I believe that it would be extremely important to contact large national and international consortiums, and invite them to buy lands to protect the Resplendant Quetzal, jewel of the birds, before it's too late.

## References

AGUILERA C. 1981. Simbolismo mexica del Quetzal I.N.A.H. S.E.P. México. Biblioteca Nacional de Antropología e Historia.

ÁLVAREZ DEL TORO. 1980. Las aves de Chiapas. 2a. ed. Gobierno del Estado de Chiapas. Tuxtla Gutiérrez, Chiapas, México.

BATES H. 1977. Order Trogoniformes. Bradford Press.

GALVEZ G. 1966. El Quetzal Simbolismos Nacionales - Guatemala.

HANSON D. 1982. Distribution of the Quetzal in Honduras. The Auk 93: 385.

JOHNGARD P. 2005. Trogons and Quetzals of the World Smithsonian Institution Press.

LABASTILLE A. 1964. Pharmacos Mocino Monografia.

LABASTILLE A & DG Allen. 1969. Biology and Conservation of the Quetzal. Biological Conservation 1(4): 297-306.

RODRÍGUEZ L. 1979. El Quetzal Historia Natural y Pronatura Guatemala.

RAMÍREZ O. Quetzales en el Sector Volcán Barba Universidad Nacional Heredia, Costa Rica.

ROJAS F. 1964. Presencia del Quetzal en la Cultura Guatemalteca Antropología e Historia de Guatemala.

SKUTCH A. 1944. Life history of the quetzal. Condor 46(5): 213-235.

SKUTCH A. 1982. Resplendent myth. Audubon 84(5): 74-85.

WHEELWRIGHT N. 1983. Fruits and ecology of Resplendent Quetzal. Auk 100: 286-301.

WETMORE A. 1965-1973. The birds of the Replublic of Panama, 1-3. Smithsonian Misc. Coll. 150.

## Author Biography

Jesús Estudillo López is a medical veterinarian from the National Autonomous University of Mexico (UNAM). He obtained his Masters degree on Avian Pathology from Ohio State University. He made numerous important contributions of the development of the poultry industry in Mexico and was director for 25 years at the Avian Pathology Department of the UNAM, where he was also a professor. As a scientist he developed several vaccines for avian infection diseases, particularly Newcastle disease, Laringotracheitis, Gumboro disease and other viral diseases. He has been breeding several endangered species of wild birds in captivity for over 50 years. He also has been recognized as the first aviculturist to breed various species of cracids, parrots, toucans, cranes, and several other birds. He has achieved several national and international awards for preserving wildlife from various organizations including: The World Wild Life Fund, Smithsonian Institute, American Game Bird Breeders Association, Global 500 United Nations, Asociación Nacional de Ciencias Avícolas, Secretaria Nacional del Avicultura (National Reward) and several universities of Mexico and Latin America. He has traveled in the tropical jungles of Mexico, Central America, Amazonian rain forests, and Andean clouded forests of Venezuela, Colombia, Ecuador, Peru, Bolivia, Brazil, Borneo, Malaysia, Vietnam, and Laos, expeditions that last from few weeks to months. Many scientists, conservationists, education specialists and ecologists, from many countries have visited his aviaries to observe and study the 500 species and 4000 individual birds at his facility. He collaborates with local universities and ecological institutions from Mexico, USA and Latin America, on research programs for various wildlife. He is one of the founders of: Pronatura A.C., Naturalia A.C. and Vida Silvestre (Wild life) Jesús Estudillo A.C. All of them Mexican conservation institutions. Recently he opened a part of his aviary to give lectures in ecology, preservation of nature and wildlife to children and visitors who have interest in nature. One of his most resent activities with wildlife has been his work with the Resplendent Quetzal.

# INFECTIOUS AVIAN DNA TESTING: SOME STATISTICAL ANALYSES

Yuri Melekhovets, Tatiana Volossiouk & Alexander Babakhanov

HealthGene Laboratory, 2175 Keele Street,
Toronto, Ontario M6M 3Z4 Canada

## INTRODUCTION

Many bird owners, breeders, and veterinarians have already discovered the large list of avian diagnostic tests offered by HealthGene Laboratory. These DNA-based tests enable practitioners to screen for an assortment of common pathogens known to cause illness and infection in virtually every species of bird. As illustrated in Table 1, HealthGene conducts testing for common avian pathogens such as *Chlamydophila psittaci*, *Avian Polyomavirus*, Psittacine Beak and Feather Disease (PBFD) virus, and Pacheco's Disease virus. However, this list is not exhaustive. Other avian DNA tests include *Aspergillus* spp., *Toxoplasma gondii*, Avian Paramyxovirus, Avian Tuberculosis, *Trichomonas gallinae*, *Bordetella avium*, etc.

In addition to these nine specialty tests, HealthGene offers an exclusive avian general profile that screens for four of the previously mentioned avian pathogens, specifically, for Avian Polyomavirus, PBFD virus, Pacheco's Disease virus, and *Chlamydophila psittaci*. This profile offers veterinarians and bird owners an economical solution to conducting multiple diagnostic tests at once. Bird owners and aviculturists have also realized the benefits of DNA testing since screening all newly acquired birds provides owners with the opportunity to safely and immediately introduce new birds to their collections, while avoiding lengthy quarantine times.

As Table 1 illustrates, a three year-analysis of avian DNA diagnostic testing. This data represents the incidence of positive infection for some of the most notorious avian pathogens known to bird owners and breeders. These numbers denote samples submitted from an array of avian veterinary practitioners and a variety of species, signifying that most pathogens do not discriminate between species and virtually all birds are susceptible. These numbers represent samples from birds displaying clinical symptoms of disease, as well as samples submitted for general screening.

The statistics presented in Table 1 are only valuable to practitioners if they are clinically significant. In order to determine their clinical significance, one must examine the validity of the test underlying the result. To illustrate this idea, let's look at the test used to identify *Chlamydophila psittaci*, as well as the organism itself. Psittacosis *Chlamydophila psittaci* is one of the most common bacterial infections among companion birds. In exotic birds, infection rates are said to vary from 10-90% overall, with reported infection rates approaching 100% in some closely held collections. Diagnosis can often be a problem due to the fact that *Chlamydophila* resides inside the cells of its host and therefore does not stimulate a high production of antibodies for serological testing. As Table 1 illustrates, HealthGene has reported nearly a 10% positive infection rate for *C. psittaci* using a PCR-based test. These positive results are of clinical significance because this test actually detects the genetic material (DNA) of the pathogen itself, thereby accurately

## Table 1
### INFECTIOUS AVIAN DNA TESTING BY SPECIES

| Tests \ Species | Chlamydophila psittaci | | Avian Polyomavirus | | PBFD Virus | | Pacheco's Disease Virus | |
|---|---|---|---|---|---|---|---|---|
| | Positive | Total | Positive | Total | Positive | Total | Positive | Total |
| African Grey | 54 | 545 | 1 | 343 | 47 | 533 | 4 | 305 |
| Amazon | 31 | 301 | 3 | 177 | 4 | 138 | 14 | 193 |
| Budgie | 50 | 246 | 5 | 52 | 11 | 54 | 0 | 31 |
| Caique | 4 | 29 | 3 | 23 | 0 | 17 | 0 | 15 |
| Cockatiel | 85 | 564 | 13 | 154 | 0 | 118 | 5 | 75 |
| Cockatoo | 41 | 332 | 6 | 215 | 12 | 278 | 6 | 187 |
| Conure | 26 | 331 | 8 | 219 | 8 | 218 | 5 | 203 |
| Finch | 8 | 38 | 0 | 4 | 0 | 1 | 0 | 1 |
| Lory/Lorikeet | 12 | 203 | 0 | 118 | 4 | 147 | 0 | 111 |
| Lovebird | 25 | 186 | 16 | 154 | 27 | 157 | 0 | 58 |
| Macaw | 42 | 463 | 12 | 339 | 6 | 300 | 22 | 313 |
| Parakeet, Ringneck | 5 | 82 | 3 | 60 | 2 | 58 | 3 | 40 |
| Parrot, Brown-headed | 0 | 10 | 1 | 12 | 0 | 1 | 0 | 9 |
| Parrot, Eclectus | 4 | 91 | 4 | 68 | 6 | 76 | 1 | 46 |
| Parrot, Jardine's | 1 | 22 | 1 | 16 | 0 | 21 | 0 | 16 |
| Parrot, Meyer's | 3 | 54 | 3 | 32 | 3 | 33 | 0 | 25 |
| Parrot, Quaker | 6 | 98 | 1 | 45 | 1 | 48 | 1 | 36 |
| Parrot, Senegal | 12 | 181 | 27 | 181 | 14 | 182 | 5 | 128 |
| Parrot/Parakeet, Other | 54 | 330 | 19 | 175 | 13 | 194 | 11 | 198 |
| Parrotlet | 5 | 51 | 1 | 34 | 1 | 33 | 0 | 28 |
| Pionus | 10 | 119 | 0 | 65 | 0 | 103 | 6 | 81 |
| TOTAL | 478 | 4276 | 127 | 2486 | 159 | 2710 | 83 | 2099 |
| % of Positive Results | 11% | | 5% | | 6% | | 4% | |

## Table 2

### OTHER AVIAN DNA TESTS

| Test | | Negative | Positive | Total | % Pos |
|---|---|---|---|---|---|
| D321 | Herpesvirus spp. | 366 | 147 | 513 | 28.65% |
| D405 | Giardia spp. | 81 | 5 | 86 | 5.81% |
| D409 | Aspergillus spp. | 167 | 55 | 222 | 24.77% |
| A1181 | Mycobacterium avium | 343 | 7 | 350 | 2.00% |
| A1182 | Mycobacterium genavense | 302 | 22 | 324 | 6.79% |

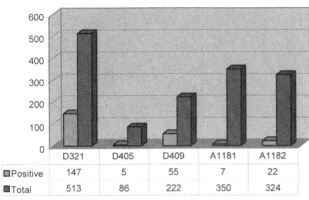

confirming the organism's presence in the biological sample (ex. blood, feces). DNA by itself (i.e. outside of the pathogens protective cell wall) cannot survive for prolonged periods of time since it is quickly degraded by the bird's naturally occurring enzymes. Thus, the detection of a pathogens DNA is a fail-safe marker that indicates the presence of the organism itself.

DNA testing has many other advantages over other diagnostic methods. For example, a pathogen can be detected during any stage of infection even if the host is not showing any clinical symptoms of disease. This is a major advantage over most forms of serological testing, which are based on the detection of antibodies and usually require an immune response. In certain cases, a bird's immune system may be able to overcome or suppress an infection. In this scenario, the bird is considered a carrier of the infection and may actively shed the organism into its environment if exposed to different types of stress. Birds infected with Pacheco's disease virus can display these symptoms. Pacheco's disease is caused by a strain of herpesvirus that is dreaded by aviculturists because of its very devastating and often fatal effects. It was first recognized in Brazil when birds started dying several days after becoming ill.

Any bird that survives an outbreak should be considered a "latent" carrier of the virus and may actively shed the virus under different types of stress, thereby perpetuating the infection. As Table 1 indicates, our laboratory results show a 3.4% positive infection rate from the samples submitted. It should be emphasized that screening all newly acquired birds for the pathogens mentioned in Table 1 should be part of every bird owner standard practice before exposing new birds to their collections. Veterinarians interested in conducting infectious testing should submit 0.2ml blood samples in EDTA / lavender top tubes. Cloacal swabs, environmental swabs, feces and other samples can be submitted in any sterile container such as a red top tube or urine container.

It is recommended to submit multiple samples, which significantly increases the effectiveness of the DNA diagnostic test. Results for infectious testing are typically provided within 2-3 business days following their arrival. The authors recommend that positive results for certain infectious agents (ex. *Chlamydophila psittaci*) be resubmitted two weeks following treatment to ensure that treatment was successful.

## Table 3

### INFECTIOUS AVIAN DNA TESTING BY SUBMITTED SAMPLE

| Sample \ Tests | Chlamydophila psittaci | | Avian Polyomavirus | | PBFD Virus | | Pacheco's Disease Virus | |
|---|---|---|---|---|---|---|---|---|
| | Positive | Total | Positive | Total | Positive | Total | Positive | Total |
| Blood | 436 | 4012 | 122 | 2388 | 140 | 2607 | 72 | 2044 |
| Feces/Cloacal Swab | 162 | 2219 | 12 | 1047 | 19 | 1074 | 23 | 904 |
| Choana Swab | 41 | 345 | 0 | 102 | 0 | 126 | 2 | 83 |
| Feather | 1 | 80 | 1 | 67 | 25 | 148 | 0 | 49 |
| Post-Mortem | 6 | 38 | 5 | 40 | 3 | 25 | 9 | 26 |
| **TOTAL** | 646 | 6694 | 140 | 3644 | 187 | 3980 | 106 | 3106 |
| % of Positive Tests | 10% | | 4% | | 5% | | 3% | |

# IMPROVING THE KEEPING OF PARROTS BY NEW CONCEPTS OF ENVIRONMENTAL ENRICHMENT METHODS

RAFAEL ZAMORA PADRÓN

Loro Parque und der Loro Parque Fundación
Avenida Loro Parque s/n, 38400 Puerto de la Cruz
Spain

## SUMMARY

In the world of aviculture, environmental enrichment has formed the basis for the maintenance and breeding success of species kept in captivity during the last three decades. The definition of "captivity" already implies a series of limits. The best conditions and distractions for captive inhabitants had to be discovered over a period of time. An evident problem arises if the maintained animals need space to fly, and it is made more difficult if these animals have complex social behaviour which requires different stimuli throughout the day. There is an inherent question which is not easily answered when it comes to enrichment. Does it only mean the attempt to replicate the natural environment if we talk about environmental enrichment for parrots? The answer is simply no, because what is meant goes well beyond simply an attempt at replication. Only once this is understood can an optimization of breeding results be reached, which as a consequence, will lead to much progress in the conservation of a species by breeding them in captivity. It is the aim of the Loro Parque Fundación that, by spreading its experience and developments in the field of advanced parrot breeding, it will be possible for all aviculturists caring for birds to experience the joy that their observations and experience can contribute to the conservation of the world's parrots.

## INTRODUCTION

### Understanding the "Environmental Enrichment" Concept

This word implies the necessity for attempting to create a natural environment for the proper care and maintenance of animals in captivity. However, this concept is so broad that it goes far beyond simply to

fill the space built for the development of the biological cycle of a species with aviary furnishings, plants and toys. In the case of parrots, this concept has special importance if it is applied appropriately and if all its meaning has been understood.

In order to have a clear idea of what is applicable to our case, we must keep in mind that environmental enrichment has had a special importance for the maintenance of medium-sized and large mammals in recent decades. We can still recall the bare concrete cells - without any additions or interactive elements - in which the inhabitants continuously repeated abrupt movements. In order to solve this intensively studied problem, the solution arose to add elements which would permit some distractions and avoid these unwanted behaviours.

Arising from the knowledge that animals in the wild use the majority of their time and energy to search for food and water, by contrast in captivity this is presented to them regularly; thus an important amount of time is free, and considerable thought went into how this time could be filled with distractions and entertainment. With cats and primates, this problem was reduced in the beginning by improving and replicating their natural habitats. Later came the realization of how important it is to make efforts in this sense, according to the necessities demanded by the animals. As far as parrots are concerned - according to their life strategy - these efforts need to be more than continued, and go far beyond filling the accommodation with some branches and toys. The needs of parrots demands that enrichment in this sense is continuous, well developed and diversified.

### What does "Environmental Enrichment" for parrots include?

*Preparation of adequate physical space for each group of a species*

If the specific needs of each species are considered according to its behaviour, this helps us prepare the housing and to maximize the potential of an aviary. As contrasting examples, we could mention cockatoos on one hand, which need a lot of space and can often be found on the ground, in contrast to the amazons which avoid longer flights and also avoid being on the ground..

Enrichment with living, non-toxic plants enables us to create a vivid environment which is in constant development. Parrots interact with these surroundings and benefit on all levels. In places with a cooler climate, the plants in outdoor installations function as a weather barrier, and in warmer places they provide shade and depending on the species, potential food sources at all times of the day. A well-developed system of greenery even makes it possible to eliminate the visible physical limits of the wire netting of a cage. Branches which grow from the outside to the inside, diffuse the straight lines which a rectangular housing has by definition. Thus, monotonous patterns are avoided which could potentially harm the parrots psyche and appease the visitor.

Plants also supply the surrounding environment with oxygen, something especially necessary in closed rooms. The greenery contributes to a beneficial balance of the micro-climate, which is desirable in a small environment.

For the species such as cockatoos which might be housed in groups of aviaries, and which are very territorial, especially in the breeding season, the effect of these natural barriers is indispensable. They permit the birds to sense the presence of their conspecifics, but avoid the direct visual confrontation which could result in aggression and which cannot be shown towards the neighbor, being re-directed against the breeding partner. A test, which is very easy to conduct, consists of putting a parakeet into a cage which has no cover provided by plants. Its nervousness is visible from the very outset, as the bird feels unprotected and sees no possibility to hide. However, if the same parrot is put into the same cage with the same dimensions, but with plant cover, aside from the normal nervousness at the beginning it calms down much faster and will sit in the most covered area. From here, it will tend to investigate its territory, but will always return to the point where it feels safest.

*Practical optimization of space*

In most zoological centres, the decision making process regarding physical dimensions of an aviary is determined by the available ground space. This does not mean that we should adapt the available space in order to maintain the maximum number of animals, rather the perspective to be followed is always to place priority on the welfare of the birds.

It is logical to expect that the more space available, the better it is for the parrots, however, if we wish to standardize a zoological facility such that its maintenance is the easiest as possible, the installations have to be planned appropriately. Additional installations must be available which can allow for temporary rotation of some species; this type of maintenance is necessary on some occasions.

*Internal physical barriers*

In the case of aggressive species, the use of mobile and fixed panels has provided good results without the need to capture the birds. In this way, a pair can be kept either together or separated. Furthermore, visual separators make use of elevated surface to increase the usable space, as in the case of very aggressive species which live in groups and defend their territories in the wild.

In some species, during the breeding season, the male will often attack the female, even in very harmonious pairs. When visual separators are installed in the middle of the aviary, allowing for no direct eye contact between the partners, they can still be near to each other if they so choose, which typically reduces the likelihood of attacks. A very excited male usually stops its attack if it cannot see its mate anymore; however, if the female is receptive, they can stay together. These visual separators can be taken out in the course of the year, so that their placement varies according to requirements. The distribution can be changed according to the necessity and the space, such that different combinations can be chosen as needed.

Similarly, in the case of aggressive parrots, these separators that can be combined with a separation of the food zones so that it is not necessary that both parrots must eat together. When we are using the concept "separators", we are thinking about limits, but the consequences of their use permits the available space for the birds to be increased. It creates micro spaces within one aviary.

*Perches*

Without any doubt, perches are essential in an installation and parrots spend more than 80% of their time on them. Thus, the perches as well as their distribution contribute to a passive environmental enrichment of an installation. Renovation of the perches is always stimulating for parrots. Modifications in form and tex-

**Fig. 1.** *An example of an aviary with visual barriers, essential for aggressive species of birds.*

ture invite them to move. In the wild, parrots do not always stay in the same place, although most often they have a preferred site to which they return to sleep at night.

Based on this knowledge, we must always offer them natural perches, above all resistant branches to chewing which have to be variable in size and form to the parrot species for which they are to be used. The diameter of these branches plays an important role for the extremities of the parrots. Broader branches permit that the larger species of parrots can relax on them, and the slimmer and narrower ones promote exercising and use of the wings and toes, even amongst birds which are overweight. The distribution of the perches has an enormous influence on their use. As such, severe mistakes should be avoided, as an example, putting just one short perch in a higher position inside a communal aviary, as the dominant animals will fight for it, especially in the later afternoon. The perches should be varied and well-designed before the animals are put into the aviary. Tree trunks with a broad diameter, and which have smaller branches on them, have proven themselves to be the best perches. This combination results in an increase of functionality and available space for the parrot to use. The distance between the perches is very important. If the distance between them is very small, exercise is not promoted; however, if the distances vary, the birds can fulfil shorter or longer flights at will, depending on the circumstances. With a system of gradual changes of distance, we can make a passive parrot exercise regularly by flying. By installing trunks in different heights, different types of flight are possible. In vertical flight, muscles which are not as commonly used as in horizontal movements, are trained.

We should remember that the more space there is for the birds to perch inside the cage, the more efficiently they will use the offered space and likewise the space available to them is also increased.

*Swings or unstable perches*

These should be offered in addition to the static perches and they form another category for maintaining muscle use of many species. In the case of lories, they constitute part of the toys and are frequently used during the day. The production of different types of swings with various functions has to be carried out extremely carefully to avoid potential accidents. The use of chains or ropes of the right size and material are fundamental in order to make sure that the use of this technique provides only advantages. Perches which improve balance stimulate bigger parrots to fly, and are especially useful for young birds which are still developing their flight muscles.

*Types of soil*

The substrate enriches the environment of all species which can regularly be found on the ground. This is especially applicable for the Australian and some neotropical species which run and play on the ground. It is best to always use the same type of substrate, which contains an appropriate percentage of minerals and permits a fast removal of waste and constant hygiene in order to avoid parasites and diseases. A soil with a porous composition allows us to play with parameters such as temperature and humidity in dry and warmer areas. The use of pebbles or isolated stones gives many species, especially larger ones, enthusiasm to spend hours moving them or simply lifting them up. However, their volume and size should be controlled from the beginning, to avoid accidental trauma and more severe accidents.

*Acoustic stimuli*

The effects of these stimuli are much higher than breeders can imagine. In parrots, being very social birds, this sense is extremely well-developed. Their daily lives in the wild have led them to develop evolutionary strategies in which sounds play a major role. It is absolutely no coincidence that parrots imitate sounds and words. The communication amongst them expresses a whole series of emotions which can categorically determine the emotional state of a parrot in various circumstances.

The accommodation of species and groups within the breeding station defines the security of the individual birds in every aspect. If a cockatoo which is brought to a new installation is accommodated close to animals of the same species, it will adapt itself a lot quicker than an animal which is kept in an isolated room. First it will establish communication with its neighbours and will always know whether a dangerous situation is arising or if the situation is under control. A group of Asiatic parakeets, normally kept as pairs in isolation, can synchronize in the breeding season if they hear the courtship and mating sounds of nearby pairs. It goes without saying that not all species live in groups and some need some isolation when breeding. The stimulation of breeding via tape recordings or a strategically clever distribution of the species within the breeding facility offers an endless number of advantages which should not be underestimated. An isolated or apathetic bird can change drastically if it hears another bird of its own or a similar species.

*Interactions between species*

Enrichment via the introduction of different species is one of the biggest progresses, in quality and in quantity, developed by the Loro Parque Fundación. Detailed knowledge about species which are kept together in generous and modern installations makes it possible to have different species in the same installation. This results in behavioural and breeding advantages, due to factors such as interspecific competition, meaning that the birds are stimulated in new fashion compared to prior to the introduction.
Birds with problems such as feather-plucking change their attitude in this new situation, behave more socially towards their own species and let the necessary feathers grow to be able to perform the indispensable flights so that they can assert their independence within a system searching for balance. Due to interactions between species, inactive pairs have spontaneously started to breed again, which proves that such interaction stimulates the animals which have been inactive for many years, although they had reached the necessary age to breed. Furthermore, animals which were incompatible in the past have proven to be good breeders if they find themselves in a new installation in which there are other species of the same or a bigger size defending their territories. We are, of course, talking about aviaries of a considerable size in which the birds are provided with places to hide and generous flight areas.

*Strategies with nest-boxes*

Nest-boxes form an active part of environmental enrichment for parrots at the reproductive age. Their installation visibly influences mating behavior. Increasing choice in respect of forms and orientation can be decisive for the breeding success of a determined species. Different forms enable the breeding animals to select the appropriate one. With some species which are rare in aviculture, have been in captivity few generations and are very shy in respect of artificial materials, the selection of natural trunks may be the only way to achieve positive results. New natural materials invite the parrots to discover new possibilities. With bamboo, acacia trunks or ceramic jars, incredible results have been achieved.

*Influence of the type and orientation of the aviary*

Big aviaries, in which the young birds are housed during their socialization period, or groups of future breeding birds are formed, are an indispensable tool when maintaining a big collection. In a smaller breeding station, there should be at least one spacious aviary available in order to achieve this aim. Enrichment can be obtained far easier in these spacious aviaries and the furnishings can occupy a bigger volume there. The orientation of this infrastructure is the key to using it one-hundred percent. Orientation towards the east permits the birds to enjoy the first rays of the sun. What can also be achieved is that the individual sections have two environments, one with shade and one with sun. This last point allows the parrots to use the space actively and differently, according to the season.

*The importance of rain*

The rainy season indicates, in tropical regions, the beginning of the breeding period for the majority of species. This stimulus can be very intense, however, this pattern cannot be applied in subtropical regions characterized by mild climate and no real seasons. It can be applied either in cold regions or those with extreme winters. In all cases, rain is an element which is very welcomed by parrots. Therefore, rain forms a further component of good environmental enrichment assuming that the birds are provided with dry sections where they can hide. The installation of sprinklers and showers in every cage contributes to the good health of the parrots and also in maintaining good feather condition.

*Food as enrichment*

Anyone who has ever owned a parrot knows about the tendency of the birds to prefer a special type of seed or fruit. This behaviour shows us how important the composition and the form of the food is which the parrots eat. Apart from the nutrients contained in every meal, the parrots have the urge to explore their environment. Using their beaks and their fleshy tongues, they try flavours and forms, until they know what they like best. By changing the fruit mixtures and varying the days when they are offered, the health of the bird is positively influenced, as it gets used to eating everything and in different combinations.

In the case of parrots, "variety" which also implies nutritional diversity, has a broader meaning. If different food stations are available within an aviary, this brings the development of a new strategy for the parrot, as it has to change its location to feed itself. This enforces an important distraction, which can encourage static animals to move in order to reach their desired aim. The addition of fruits or flowers onto swings makes lories invent incredible artistic tricks in order to get one single drop of nectar or a few seeds.

In the wild, parrots eat different types of foods according to the time of year, which also influences their behaviour accordingly. In captivity, this should also be considered, and the food should be varied according to the season. Soaked and cooked vegetables before and during the breeding season; food which is rich in fruits and vegetables during the hot summer months and the moult and food with a lot of calories and nutrition during the winter. These changes have a stimulating effect and indicate in a natural way the seasons in which a bird is living within an artificial environment. Adding flowers of different colours is also one food addition which can enrich an aviary. The different hibiscus types which are available make it possible to provide our parrots a varied offer of colours. Although they are much more important for the groups of lories, a cockatoo will not miss the opportunity to play for a while and tear a few blossoms apart, which also have a sweet taste. The visual stimulus in the aviary also includes a hidden input of nutrition, which is always welcome. Blossoms such as from *Callistemon* sp. *Grevillea* sp., *Psidium* sp. can complete the offered variety.

The Strawberry and Pineapple Guavas *Psidium cattleyanum* and *Feijoa sellowiana* from Brazil supply a large number of blossoms in a subtropical climate. Blossoms and fruits can both be used for enrichment with various species of lories by introducing an entire branch into the aviary. The supply of local, seasonally dependent herbs can also help to define seasonal differences and to stimulate the parrots.

If we carry out a nutritional analysis of the food for our parrots, we realise that if seed mixtures are used as the main food, they are always offered with the same degree of ripeness, in other words, dry. In the wild, parrots eat the seeds as they are offered according to the season, from green to ripe until, finally, dry. Unripe seeds can only be found very rarely or not at all in most markets. The addition of seasonal herbs is one of a few possibilities to integrate this factor in the food for our parrots. They are useful mainly after the first rains, as they transmit important messages for the stimulation of the breeding birds. The unripe seeds of different herbs are also very rich in various important substances for the birds. Loro Parque Fundación has built up its own ecological cultivation of herbs like Sowthistle *Sonchus* sp., Common Chickweed *Stellaria media*, which has an interesting content of saponine, or Dandelion *Taraxadum officinalis*. The last two plants have a very broad distribution worldwide, which makes their collection in the different countries very easy.

*Toys*

Colourful toys, pieces of leather or plastic on strings, mirrors, bells, etc form part of the repertoire offered in specialised shops. The use of these elements, which are becoming more and more modern and complex, can be very useful, especially in the case of parrots which are kept as pets and have to live in a restricted space.

However, its use can only bring results if a toy is used in an intelligent way and with previous thoughtful planning. It can often be observed that parrots which are kept in cages are surrounded by a huge amount of toys, to which they do not pay any attention, and which hinder their movement. Ocassionally one will see in some aviaries an oxidised gemstone hanging in a corner, which has been especially designed by an animal psychologist, but evidently has not met expectations.

The instinct of the person responsible for the parrot is essential in the question of when these elements are to be used. In the case of parrots which are kept as pets and are used to interactions with their owners, the toy can be offered at a firmly defined hour and over a determined period of time. Thus, it serves its purpose and the parrot uses the time to entertain itself. If different elements can be selected which are used with the same aim according to the occasion, the parrot learns to distinguish them and it will even have favorite toys. In the case of parrots which live in breeding environments, these additions have to be adapted in accordance to the development of psychological receptivity and removed if the enjoyment is over. Some species like the Keas *Nestor notablis* much appreciate it when colourful balls or wooden pieces of different forms are hidden and which can easily be transported. With the Palm Cockatoo *Prosbosciger aterrium*, the availability of little pieces of wood stimulates nest construction before the breeding season and supports the courtship, in which the male knocks on a log with a piece of wood or rock, as if it were playing a drum. Some young birds also try this behaviour if they have the opportunity. Automatic food machines, which deliver sweets or toys if the parrot presses on a dispenser, are the latest inventions to entertain active species. However, their use should be supervised so that they serve their purpose. The list is endless and there are even transparent plastic balls, which include seeds in an internal labyrinth and entice the parrot to turn the ball until the treat falls out off one of the numerous holes. Food blocks with natural substances, which contain seeds or dry fruits, are also very attractive for parrots, which search actively for objects with feet and beaks.

## CONCLUSION

Some individuals who have read this paper might have said to themselves at different points, "Well, I already knew this", and others might even think that this topic has really been exhausted by others previously. These are mainly concepts which will be implemented one step at a time, but which cannot be applied all together. Those who think that everything has already been discovered would be mistaken. We experience almost daily, a member of the Loro Parque Fundación approaching us and telling us about his own successful experience with a specific species, which he could only discover due to his ingenuity and his devotion.

Sometimes it requires more than just trying to think like a parrot and modify things to our liking, it goes much further than this. We are constantly receiving ideas from other fields of aviculture which can be perfectly applied to parrots; thus, we test them, hoping that we can find new and successful protocols for the difficult species.

This concept which we are presenting to you is the application of all these enrichment techniques attempted in constant form over time. We are talking about a "*modus vivendi*" which we must assume with responsibility and obligation. If we can add passion and joy to this idea, not only with our current understanding of behavioral enrichment, but also by keeping an open mind as to future possibilities, we can make great strides in this area of aviculture. Loro Parque Fundación acts as a large data base which accordingly can be consulted by its members at any time. And at the same time members can share their experiences and knowledge by including it into this database. Thus, every time a Spix's Macaw or any other endangered species reproduces, it is not only a success for conservation biologists who are responsible for the wild birds, but it is a victory for each and every individual member sharing this reality together today and into the future.

## REFERENCES

DE WAAL FBM .1993. La Política de los Chimpancés. Alianza, Madrid.

EYRE S. 1997. The Effectiveness of Environmental Enrichment in Preventing And Curing Stereotypic Behaviour in a Sand Cat (*Felis margarita harrisoni*). In Ratel 24 (5): 156-164.

GIL BURMANN CY & F Peláez. 1992. La observación: Selección de datos. En: Psicología Social: Métodos y Técnicas de Investigación. M. Clemente Díaz (ed.). Pp. 224-245. Eudema, Madrid (PS/13/679).

GRZIMEK B. 1977. Grzimeck's Encyclopedia of Ethology, ed. K. Immelmann. Van Nostrand Reinhold Company.

LAW G. 1993. Part I: Behavioural Enrichment. 1993 ABWAK Conference Proceedings.

MICHAULT M. 2001. Enrichment for Brazilian Tapirs and Capybaras. The Shape of Enrichment, Vol.10, No.2, May 2001.

MAZUR JE. Learning and Behavior. Upper Saddle River, NJ: Prentice Hall; 2002.

MEYER-HOLZAPFEL M. 1968. Abnormal behavior in zoo animals. In Abnormal Behavior in Animals, ed. M.W. Fox, 476-504. Philadelphia: W.B Saunders.

RIEDE T. 1991. Examples of Environmental Enrichment for Zoo-Animals, NOD, Amsterdam, NL.

SANDOS A., J. Peeler. 1997. Enrichment is a many splendid thing. Proceedings of the third environmental enrichment International Conference, 1997. Available from: www.aza.org , accessed on September 6, 2005.

# PREVENTION AND TREATMENT OF COMMON MEDICAL CHALLENGES ASSOCIATED WITH REARING AND BREEDING WATTLED CRANES *BUGERANUS CARUNCULATUS* FOR A RECOVERY PROGRAM

J.M. PITTMAN[1], M. BARROWS[2] & S.D. VAN DER SPUY[2]

[1] Johannesburg Zoo, Private Bag x 13, Parkview 2122, Johannesburg, South Africa

## SUMMARY

The Wattled Crane Recovery Program is a captive breeding and release program aimed at preventing local extinction of the Wattled Crane in South Africa. The two main objectives of the program are; the maintenance of a captive breeding flock to serve as a genetic reservoir in the case of catastrophic extinction of the species in the wild and supplementation of the wild population through release of captive-reared fledglings into existing floater flocks. This paper discusses prevention and treatment of the common medical challenges encountered in the rearing and breeding of Wattled Cranes for the recovery program including; infectious diseases, musculoskeletal developmental abnormalities, gastrointestinal tract problems and trauma.

## INTRODUCTION

The Wattled Crane *Bugeranus carunculatus* is one of three crane species found in South Africa and the most critically endangered of the six crane species on the African continent (Meine & Archibald 1996). There has been a 35% decline in its population over the last two decades. A 2004 census found just a few isolated populations remaining, which are now under severe threat from habitat loss, power line collisions and accidental poisonings. These threats are further compounded by a naturally low reproductive rate and a wetland dependent lifestyle. The Wattled Crane has one of the lowest reproductive potentials of all crane species. Recent genetic studies suggest that the Wattled Cranes occurring in South Africa are genetically unique to those occurring in other African regions and relevant conservancies have recommended that South African Wattled Cranes be managed as a distinct population (Jones et al 2006). A Population and Habitat Viability Assessment (PHVA) conducted in July 2000 (McCann et al 2000) resulted in the recommendation of a captive breeding and release program for the Wattled Crane in South Africa and thus the Wattled Crane Recovery Program (WCRP) was created. The main program objectives include: the maintenance of a captive breeding flock to serve as a genetic reservoir in the case of catastrophic extinction of birds in the wild, and supplementations of the wild population through the release of captive reared fledglings into existing wild floater flocks.

Wattled Cranes typically lay a clutch of one or two eggs. However, even when two eggs are laid, the parents only ever rear one chick. Nests containing two eggs are closely monitored and when the first egg pips, the second egg is collected and artificially incubated. Ezemvelo KwaZulu Wildlife allows for the removal of the second egg, as this has no detrimental effect on the wild population. The resultant chicks are puppet-reared to avoid human imprinting. Throughout the history of the program, a variety of medical challenges have arisen in the captive breeding flock, including infectious diseases (parasitic infestation, mycobacteriosis, botulism) musculoskeletal developmental abnormalities (digit deformity, angular limb deformity, tibiotarsal and tarsometatarsal deviation), GIT problems (foreign body ingestion, heavy metal toxicosis, impaction), and trauma (self-induced and intraspecific trauma, exertional myopathy). While these challenges have not created a high level of morbidity or mortality, measures were taken to address future occurrence. A series of neonatal and flock management protocols have been developed to address husbandry related medical issues, along with a comprehensive Preventative medicine program to minimize potential disease risk.

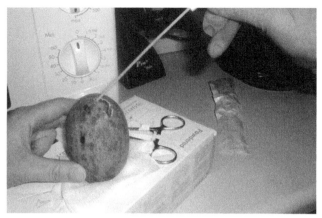

Fig. 1. *Culturing the air cell of a non-viable Wattled Crane egg. (Michelle Barrows)*

## Discussion

### Incubation

Due to their porous nature, eggs are very susceptible to bacterial contamination, which can easily result in the death of the developing embryo. Some of the more common pathogens isolated from avian eggs include, *Salmonella*, *Staphylococcus* and *Escherichia coli* (Olsen and Clubb 1997). Contamination generally occurs either in the female's oviduct before the egg is laid, in the environment after laying, or in the incubator (Ellis et al 1996). Culturing the air cell and/or egg contents can confirm the diagnosis (Fig 1). Additional culture swabs can be made from incubators, equipment and the laying female's cloaca to determine the source of the contamination. Prevention is the best course of action as treatment is largely unsuccessful once a contaminant enters the egg. Preventative measures include; maintaining healthy laying females, keeping nesting areas clean and dry, removing non-viable eggs from nests and incubators, wearing gloves when handling eggs and routine disinfection of incubators and equipment. Post-mortems should be carried out on all fertile non-viable eggs.

### Malposition

Malposition of the chick in the egg can lead to rapid death of the chick if unassisted. When the chick is ready to pip, it should be positioned with its head at the large end and curled downward directly under the abdomen, the beak tucked underneath the right wing and the legs and tail filling the small end of the egg. The most common method of determining malposition is through candling of the egg; however in Wattled Cranes, the dark colouration of the shell makes candling difficult and other methods such as radiography or opening the shell should be considered. Radiography is an excellent non-invasive method of determining malposition (Fig 2). Radiographic images should be taken of all four quadrants for ease of evaluation. Care should be taken to avoid damaging the developing embryo by over rotating the egg. Alternatively, the large end of the egg can be opened under sterile conditions and the beak or other body parts visualized by moistening the eggshell membrane with sterile water. Repeated occurrence of malposition may be related to improper egg position or rotation during incubation (Ellis et al 1996). As in chickens, continuous elevation of the large or small end of the egg during incubation can increase the incidence of certain types of malposition (Hutt & Pilkey 1934). Elevation of the large end may result in failure of the chick to tuck its head properly; tucking of the head under the left wing, and tucking of the head with the beak over the wing. Elevation of the small end of the egg may result in the chick being positioned backwards with its head in the opposite direction to the air cell (Talmadge 1977). Preventative measures for reducing the incidence of malposition include increasing the frequency of turning and/or changing the position of the egg (Ellis et al 1996). Eggs should either be incubated lying horizontally with the large and small ends at the same distance from the tray or alternatively, the egg can be placed at a 20-30% angle so that the large end is elevated when the tray tilts forward and the small is elevated when the tray tilts back (Ellis et al 1996).

### Assisted hatching

Failure of a chick to hatch on its own accord may result from weakness, malposition, dehydration of the

Fig. 2. *Egg radiography to determine embryonic positioning. (Michelle Barrows)*

shell membranes, an overly thick egg shell, or developmental deformity of the chick due to improper incubation (Ellis et al 1996). When a chick has failed to hatch 48 hours after pipping, it should be assisted as failure to do so will likely result in death of the chick (Ellis et al 1996). Assisted hatching is a delicate procedure and extreme caution as well as appropriate hygiene should be implemented to avoid injuring the chick or exposing it to potential pathogenic organisms. The process begins by gently peeling away the shell around the air cell and drawing the bill and nares through the shell membrane to allow the chick to breathe. Keeping the shell membranes moist, the chick is then slowly encouraged to hatch by calling to it, and incrementally working the shell and membrane away. Care should be taken to avoid damaging the yolk sac and any blood vessels that may still be intact (Ellis et al 1996).

**Neonatal Challenges**

*Exteriorized yolk sac*

The yolk sac represents a major source of nutrition for newly hatched chicks during their first few days of life. Spasmodic contraction of the abdominal muscles withdraws the yolk through the umbilicus prior to hatching. Failure of the yolk sac to be withdrawn into the abdominal cavity and resultant exteriorization can result from, incorrect incubation temperatures, excessive humidity during incubation or assisting the chick to hatch too early (Ellis et al 1996). The exteriorized yolk sac is very delicate and excessive handling can cause bleeding, infection or rupture. It is best to seek advice from a veterinary professional prior to attempting this procedure as failure to apply appropriate disinfection methods could result in the introduction of bacteria into the abdominal cavity and subsequent death of the chick due to peritonitis. Gentle attempts can be made to manipulate the yolk sac into the abdominal cavity and the umbilical opening closed with suture, using a purse-string pattern. Alternately, if manipulation fails, the yolk sac can be surgically removed by placing a ligature of absorbable suture around the stalk (Hartman et al 1987).

*Musculoskeletal developmental problems*

Digit and leg deformities are some of the most common medical challenges encountered in captive neonatal Wattled Cranes. Early intervention is necessary to prevent the development of irreversible abnormalities.

**Fig. 3.** *Hobbles placed on a four-day old Wattled Crane chick to correct splay-leg. (Jeanne Marie Pittman)*

Common deformities seen include, splay leg, curled and/or rotated digits, and tibiotarsal or tarsometatarsal rotation (also known as angular limb deformity). Prevention of developmental abnormalities can be achieved by thorough dietary analysis, close monitoring of weight gain, use of appropriate substrates, limited careful handling and a rigorous exercise program tailored to the developmental stage of the chick (Ellis et al 1996).

*Splay-leg*

Splay-leg is a condition where one, or more commonly both legs deviate laterally from the hip. This condition is most often seen in newly hatched chicks as a result of improper incubation or slippery substrates, but may also occur spontaneously (Ellis et al

**Fig. 4.** *A sling is used to support a three-day old Wattled Crane with severe splay-leg. (Jeanne Marie Pittman)*

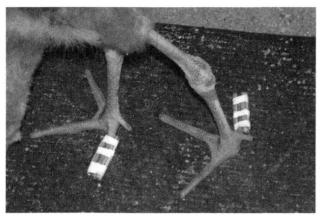

**Fig. 5.** *Samsplint® applied to the digits of a Wattled Crane chick to correct digit rotation.*
*(Jeanne Marie Pittman)*

1996). Treatment involves the application of hobbles using a light bandage material such as Micropore© above and/or below the hocks (Fig 3). Care must be taken to ensure that hobbled chicks have adequate substrate for good footing to prevent them from compounding the problem by tripping and injuring themselves. In severe cases where chicks are unable to stand even when hobbled, slinging may be necessary in addition to taping (Fig 4).

*Digit deformities*

Curled toes are most often seen immediately post-hatching, while rotated or crooked toes tend to occur throughout the rearing process (Ellis et al 1996). When treated early these conditions generally resolve quickly but if left for extended periods, may result in severe irreversible deformity. Samsplint© is an ideal product for splinting deformed/deviated digits. This product consists of a thin sheet of aluminum bonded on either side to thin sheets of resilient foam padding. It can be cut to any shape and due to its lightweight, flexibility and water resistance; it is an ideal material for splinting young cranes. Older birds may require the application of a more rigid material to the Samsplint® for added strength. Splint material is applied to the medial and lateral aspects of the toes and attached with waterproof bandage tape (Fig 5). Curled toes at hatching are generally caused by improper incubation methods or genetic predisposition (Ellis et al 1996). Prevention includes fine-tuning incubation protocols and selective breeding if a genetic component is suspected.

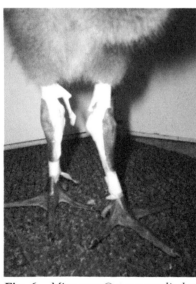

**Fig. 6.** *Micropore® tape applied to the medial aspect of tarsometatarsus to correct lateral deviation.*
*(Jeanne Marie Pittman)*

Rotated or crooked digits in older chicks may be due to a number of factors including poor substrate, lack of exercise and poor nutrition (Ellis et al 1996). In ratites, digit rotation appears to be a variation of angular limb deformity (Donley 2006). Corrective splinting is similar to that used for curled toes whereby the Samsplint® is applied to the medial and lateral aspect of the toes, except that the waterproof bandage tape is applied to the digit prior to applying the splint. One continuous strip of tape is applied to the medial or lateral aspect of the digit on the opposite to the direction of the rotation. Light tension is then applied to the tape as it is wrapped ventrally the full circumference around the digit, rotating the digit back into its normal position. The splint is then placed on the medial and lateral aspects of the digit and the tape wrapped around the digit a second time over the splint material. Preventative measures to reduce the incidence of digit rotation include good nutrition, the use of appropriate substrates, and an age appropriate exercise program (Ellis et al 1996).

*Tibiotarsal/tarsometatarsal rotation (angular limb deformity)*

Angular limb deformity occurs when the growth plate of the tibiotarsus or the tarsometatarsus develops too rapidly on one side resulting in a rotation along its long axis and a lateral (or occasionally medial) rotation of the limb (Ellis et al 1996). The exact cause of the condition has not been determined but it appears to be multifactoral. Diets high in protein or fat, rapid growth rates, calcium/phosphorous imbalances and genetic predisposition have all been implicated (Ellis et al 1996; Donley 2006). Studies conducted on Sandhill Cranes *Grus canadensis* indicate that feeding crane chicks animal based protein may increase the incidence of leg and wing problems (Serafin 1980, 1982). Serafin (1980) suggests that a diet containing no more than 24% protein and 0.73% sulfur amino acids can reduce the incidence of limb deviation. Chicks should be evaluated daily for signs of limb rotation. When the chick is standing still, the middle toe should point forward. If the toe is pointing medially or laterally, this is a good indication that the limb may be rotated. Chicks with angular limb deformity will also exhibit a "windmill" gait when running.

Tension taping can be effective in resolving mild cases of angular limb deformity by slowing growth on one side of the growth plate (Haffner 1988). This process involves applying a strip of bandage tape to the limb on the rapidly growing side (the side opposite the direction the limb is deviated) and securing it with circumferential bandage strips at the top and bottom (Fig 6). Surgical techniques such as transverse, oblique, wedge and dome osteotomies have been successful in correcting angular limb deformities in Psittaciformes, Falconiformes and Strigiformes (Martin & Ritchie 1994), but have been shown to have limited success in cranes (Ellis et al 1996). Hydrotherapy may be helpful if the condition has developed due to lack of exercise (Ellis et al 1996). For long-term successful treatment, all physical corrective measures (taping or surgery) must be accompanied by management changes that address the contributing causes. Preventative measures include, thorough dietary analysis, close monitoring of weight gain, a rigorous exercise program tailored to the developmental stage of the chick and selective breeding where genetic factors are suspected.

*Beak abnormalities*

Developmental abnormalities of the beak can occur as a result of trauma during the growth phase. Even fairly minor injuries can result in severe deviation. In 2006, as the result of pen mate aggression, two juvenile Wattled Cranes sustained fairly minor injuries to the maxilla. Both injuries appeared to heal without incident. Over time however, lateral deviations became evident and in the more severe case, surgical repair became necessary to correct the deviation. Surgical correction was achieved by passing a heavy gauge surgical wire through mandible and adhering polymethyl methacrylate (PMMA) around it to raise the beak surface and push the maxilla back into normal alignment (Fig. 7 & 8). The PPMA brace was left in place for several months and resulted in a partial correction of the deviation.

In the second case, the chick sustained a 2.5cm fracture to the tip of the maxilla, which was eventually removed. Post-injury swelling created a lateral deviation, which did not resolve when the swelling subsided. Preventative measures to avoid the development of beak abnormalities include; housing juvenile cranes in individual units to reduce pen mate aggression, close

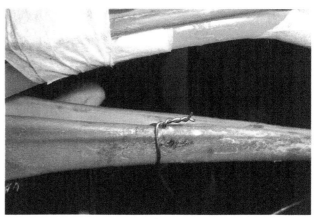

**Fig. 7.** *Heavy gauge surgical wire passed through the mandible of a Wattled Crane to act as a base for attaching PMMA. (Michelle Barrows)*

supervision of juvenile cranes during socialization and frequent physical examination post-injury.

*Impaction*

Wattled Crane chicks routinely ingest a variety of substances, which in small volumes generally pass through the intestinal tract uneventfully but in large volumes can cause impaction. These substances most commonly include substrates such as sand, wood chips or loose strands of carpeting, and dietary grit that has been offered ad libitum rather than being mixed into the feed (Fig. 9). Impaction may also result from ingesting large quantities of uncooked vegetables such as beets or carrots. Signs of impaction may include, anorexia, lethargy, dyspnea and esophageal distention. Contrast radiography using barium sulfate can confirm the diagnosis. Fluids should be administered when using barium sulfate due to its hydroscopic nature (Flammer & Clubb 1994). For mild cases, treatment may be successful through the administration of laxa-

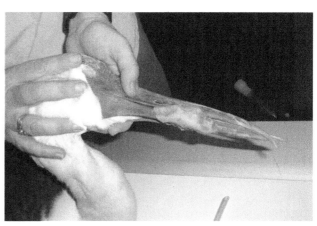

**Fig. 8.** *PMMA brace used to correct lateral deviation of the maxilla. (Michelle Barrows)*

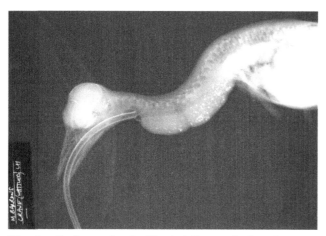

**Fig. 9.** *Radiograph of a Wattled Crane chick with severe distention of the esophagus due to impaction caused by the ingestion of a large volume of raw vegetables and grit. (Michelle Barrows)*

tives such as dioctyl sodium sulfosuccinate, psyllium, digestive enzymes or mineral oil (Flammer & Clubb 1994). Care must be taken when administering medication by gavage to birds with impaction, as there is an increased potential for regurgitation and subsequent aspiration. More severe cases may require gastric lavage or proventriculotomy but these procedures carry a much more guarded prognosis (Flammer & Clubb 1994). Preventative measures include, selective use of substrates (especially in the first two weeks of age), limiting quantities of raw vegetables, finely chopping raw vegetables to increase digestibility and eliminating grit from the diet or mixing small amounts in the feed vs. offering ad lib.

**Fig. 10.** *Head trauma in a six-week old Wattled Crane caused by a younger sibling. (Michelle Barrows)*

## Trauma & Related Conditions

### Self inflicted injury

Wattled Cranes are very territorial and will injure themselves if housed too closely to other pairs, especially during breeding season. Males tend to become so aggressive toward keepers or other pairs, that they entangle their beaks, wings and legs on the enclosure wire and even sustain internal injury or exertional myopathy by repeatedly flailing against the fence in an attempt to ward off intruders. Beak injuries are most common, and can range in severity from minor fractures to the tip of the beak to a full fracture of the mandible or maxilla. Depending on the severity, treatments range from the application of hemostatic agent to surgical repair or removal. Pinioned males have a great tendency to damage the carpus during breeding season and application of a light, bandage material such as Microfoam© may help to alleviate continual abrasion. Preventative measures include the use of shade cloth to provide visual barriers, creating turnstile devises that allow food and water bowls to be accessed without entering the pen and leaving a considerable distance between enclosures of breeding pairs.

### Pen mate trauma

In the wild, Wattled Cranes will lay a clutch of one or two eggs but will only rear one chick. When the first egg hatches, the parents often abandon the second egg. In cases where both eggs hatch simultaneously, the stronger chick generally exhibits cainism, resulting in the death of the weaker sibling (Fig 10). This trait proves challenging when attempting to rear several Wattled Cranes simultaneously. The propensity toward cainism varies between individuals. Some chicks are content to accept multiple pen mates, while others will persist in actively seeking to injure or kill their siblings. Prevention methods include housing neonates in individual units and utilizing a central socialization area whereby social interactions can be closely monitored until such time that aggression is reduced to an acceptable level. When housed as a cohort (group of sub-adults of mixed sexes living together), Wattled Cranes must be monitored closely and separated at the first sign of sexual maturity to prevent fatal aggression (Fig 11). Additionally, it is not uncommon for the male of an adult pair (even birds that have been paired for years with no sign of aggression) to suddenly attack and kill their mate. Prevention of cage mate trauma amongst pairs requires close observation to detect subtle changes in behaviour.

*Capture injury*

Leg injuries, including fractures, tendon rupture and tibial cnemial crest avulsion are a common result of incorrect restraint methods. A variety of restraint methods have been developed for chicks and adult cranes. Young chicks are generally restrained in the palm of one hand with the legs dangling between the fingers and the second hand placed gently over the back (Fig 12). Older chicks and adults can be restrained by facing the crane, wrapping the left arm around the body and simultaneously grasping the legs around the hocks with the right hand. The crane is then lifted off the ground and held close to the handler's body just above the hip (Fig 13). Never force the legs to stay in a flexed position as this may result in tendon rupture or tearing of the cnemial crest when the bird attempts to kick. While proper restraint will reduce the incidence of injury, all attempts should be made to take a hands-off approach to management. Many procedures such as SC and IM injections, body score indexes and screening radiographs for metal ingestion can be done with the bird standing. Puppet-reared chicks can be trained to step onto a scale for weight monitoring greatly reducing the need for handling.

*Myopathy*

Exertional myopathy (EM) also referred to as capture myopathy, is a non-infectious disease characterized by skeletal and cardiac muscle necrosis and severe metabolic disturbance following extreme exertion, struggling and/or stress. While exertional myopathy has been most extensively documented in ungulates, the occurrence of EM has been reported in a wide variety of avian taxa (Williams & Thorne 1996) including several crane species (Brannian et al 1981; Hartman 1983; Windingstad et al 1983; Carpenter et al 1991). While EM in cranes has most frequently been associated with capture and restraint, cases in captive cranes have been reported subsequent to serious traumatic injury or after prolonged restraint in a sling (Ellis et al 1996). Factors which may contribute to the occurrence of exertional myopathy include fear, anxiety, over exertion, excessive handling, prolonged capture or transport, constant muscle tension, and general anesthesia (Ellis et al 1996). Clinical signs may include; ataxia, paresis, paralysis (Hayes et al 2003), stiff painful movements, hard swollen muscles that are warm to

**Fig. 11.** *Severe head trauma in a two-year old Wattled Crane caused by a pen mate. The bird subsequently died from severe trauma and exertional myopathy. (Jeanne Marie Pittman)*

**Fig. 12.** *Correct restraint method for young Wattled Crane chicks. (Jeanne Marie Pittman)*

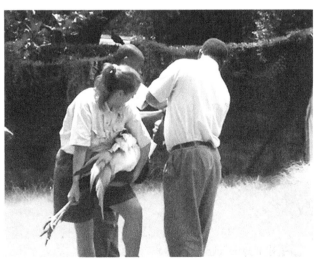

**Fig. 13.** *Correct restraint method for adult cranes. (Jeanne Marie Pittman)*

Fig. 14. *The use of a sling to support a severely injured adult Wattled Crane suffering from exertional myopathy. (Jeanne Marie Pittman)*

the touch, cardiac failure and sudden death (Ellis et al 1996). Serum analysis may show high elevations in creatinine kinase, lactic dehydrogenase, and aspartate aminotransferase (Brannian et al 1981; Hayes et al 2003). Treatment of exertional myopathy includes the administration of intravenous fluids, vitamin E, selenium, corticosteroids and antibiotics (Ellis et al 1996). Cranes that are unable to stand should be maintained in a sling and given intensive physiotherapy (Fig 14). The prognosis for recovery from exertional myopathy is poor.

**Non-infections Disease**

*Foreign body ingestion*

In the wild, Wattled Cranes feed by continual probing. This adaptation proves problematic in captivity, as Wattled Cranes tend to readily find and eat foreign material in their enclosures. Care should be taken to

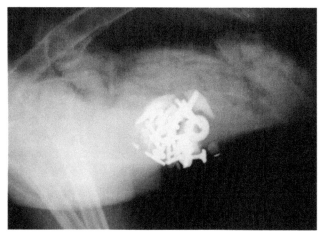

Fig. 15. *Foreign body ingestion in a juvenile Wattled Crane. The hardware was left behind by building contractors when the pen was constructed. (Michelle Barrows)*

remove all bits of scrap metal when building new enclosures, especially metal that contains zinc, lead or other toxic substances. The vast majority of ingested foreign bodies go undetected. Periodic radiographic screening may be useful in detecting radiopaque materials such as bolts, nuts and bits of wire (Fig 15). Puppet-reared chicks can be trained to stand for radiographic procedures (Fig 16), while more fractious chicks and adults can be placed in a crate constructed of radiolucent material eliminating the risk of handling and anaesthesia. While many objects such as small pieces of glass, plastic or other non-toxic materials can remain in the proventriculus or ventriculus for long periods without complication, materials that have the potential to cause toxicity or obstruction, as well as any sharp objects that might puncture gastrointestinal structures, should be removed. Several procedures for the removal of foreign bodies in avian patients have been developed including, endoscopic retrieval (Fig 17 & 18), gastric lavage, and proventriculotomy (Flammer & Clubb 1994). Preventative measures include, careful selection of enclosure furnishings and enrichment items and scanning new exhibits with a metal detector (limited use on small metal fragments) or high-powered magnets to detect metal left behind by building contractors.

*Heavy metal toxicosis*

Heavy metal toxicosis as the result of ingesting zinc or lead based foreign bodies can be fatal if untreated. Removal from the gastrointestinal tract is essential in preventing serious toxic effects (Bauck & LaBonde 1977). A number of substances have been used to aid in the passage of smaller pieces of heavy metal including, mineral oil, corn oil, peanut butter, barium sulfate and high fiber substances such as psyllium and whole grain products (Dumonceaux & Harrison 1994). Larger pieces of metal that are unable to pass through the lower intestinal tract may need to be removed by endoscopic retrieval, gastric lavage or proventriculotomy. Additional measures to reduce the toxic effects of heavy metals include supportive treatment and the removal of heavy metals from body tissues through the use of chelating agents (Bauck & LaBonde 1977).

*Zinc*

Zinc may be found in hardware such as wire, screws, nuts and bolts as well as coins (USA pennies) and galvanized metal used in enclosure wire, feed trays and water buckets. Clinical signs of zinc toxicity include general weakness, anemia, excessive urination, excessive thirst, loose faces, weight loss, cyanosis, seizures and death (Richardson 2006). Various chelating agents used to treat zinc toxicosis include, calcium disodium ethylene diamine tetracetate (Ca EDTA), D-penicillamine and Succimer (2,3 dimercaptosuccinic) (Richardson 2006; Dumonceaux & Harrison 1994).

*Lead*

Sources of lead ingestion may include galvanized metals, paint, linoleum, enrichment items and ceramic glazing on feed bowls. Clinical signs of lead toxicity include general weakness, anorexia, lethargy, polyuria, ataxia, circling and convulsions (Richardson 2006). Chelating agents used to treat lead toxicity include Ca EDTA, penicillamine, Succimer, and diethylene triamine pentaacetic acid (DTPA) (Richardson 2006). Sodium sulfate and activated charcoal are also used to bind lead in the gastrointestinal tract and prevent absorption (Dumonceaux & Harrison 1994).

*Note: In the authors' experience, oral administration of penicillamine to juvenile Wattled cranes resulted in anorexia in several birds. One birds also suffered from depression and regurgitation. Symptoms disappeared shortly after the medication was discontinued.*

**Infectious Disease**

*Avian mycobacteriosis*

Avian mycobacteriosis commonly referred to as avian tuberculosis, is an infectious chronic wasting disease transmitted by the ingestion or inhalation of *Mycobacterium avium* from fecal contaminated soil or water (Pollock 2006). Due to their aquatic nature and propensity for probing in wet soil, Wattled Cranes are highly susceptible to avian mycobacteriosis (Pollock 2006). Three forms of avian mycobacterium have been described; classical, paratuberculous and diffuse (Gerlach 1994). Some of the more common clinical signs include, chronic wasting, depression, diarrhea, polyuria, abdominal distention, lameness and the development of subcutaneous masses (Pollock 2006). Euthanasia is generally recommended for birds infected with *Mycobacterium avium*, due to the zoonotic potential of the disease, the low efficacy and high cost of treatment and the potential for infected birds to shed large numbers of organisms into the environment. Diagnosing avian mycobacteriosis in live animals is difficult due to the wide variety of clinical signs and

**Fig. 16.**
*A juvenile Wattled Crane is taught to stand still for a radiographic procedure.*
*(Judith Lazek)*

physical findings and the lack of reliable diagnostic tests (Pollock 2006).

Over the course of four years, three Wattled Cranes managed by the Wattled Crane Recovery Program succumbed to avian mycobacteriosis in three separate facilities (Fig. 19 & 20). Screening the remaining population for the presence of *Mycobacterium avium* proved problematic due to the lack of reliable testing methods. A full medical work-up including radiographs, hematology, biochemistry (including bile acids and fibrinogen), endoscopic examination and liver biopsy was conducted on individuals that were potentially exposed to the disease. A strict long-term quarantine was instituted. Preventive medicine protocols were developed which include; annual physical examinations, hematology (total white blood cell count, white blood cell differential), biochemistry, radiographic and laproscopic examinations and/or liver biopsies on birds with suspicious findings, strict screening and quarantine of new individuals, and post-

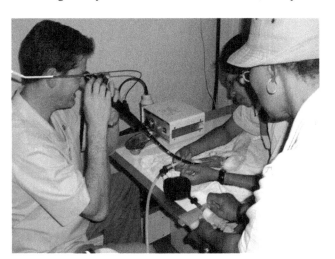

**Fig. 17.** *A flexible endoscope is used to retrieve foreign bodies in a juvenile crane. (Mike Harman)*

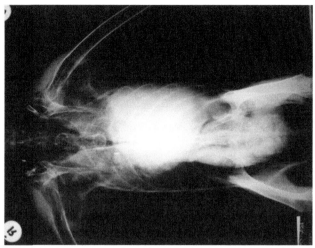

Fig. 18. *A zinc-coated washer and screw are removed from a juvenile Wattled Crane using the grasping tool from a flexible endoscope. (Mike Harman)*

mortem examinations on all birds that die. Husbandry protocols were also developed to lessen the potential for exposure to *Mycobacterium avium* and include the following recommendations; the implementation of high standards of hygiene, keeping Wattled Cranes separate from other avian species, especially waterfowl, using flight netting to prevent wild and feral birds from entering the enclosure and refraining from serial filling of ponds between enclosures.

### Botulism

Botulism is a paralytic condition caused by the ingestion of *Clostridium botulinum* toxin. Outbreaks of botulism caused by *Clostridium botulinum* are associated with prolonged warm weather, stagnant shallow water, alkaline conditions and an accumulation of rotting vegetation coupled with decaying animal carcasses (Dorrestein 1977). Botulism mortality has been seen in many avian species, including cranes. Clinical signs include, flaccid paralysis of legs, wings, neck and head (Bennet 1994), green diarrhea, chemosis, swelling of the eyelids and nictitans, ocular discharge and hypersalivation (Cambre 1988; Foreyt & Abinanti 1980; Harrison 1974). Treatment includes supportive care and the administration of laxatives, cathartics and drenches to flush the toxin from the gastrointestinal tract (Bennet 1994). Antitoxin has been used to treat botulism but its efficacy is uncertain (Cambre 1988; Harrison 1974; LaBonde 1992). In 2003, an outbreak of botulism resulting in the death of waterfowl and various other species including a Blue Crane *Anthropoides paradiseus* and several psittacines, occurred at one of the facilities housing cranes for the WCRP. An annual botulism vaccination program was implemented for the collection, including the Wattled Cranes using *C. botulinum* vaccine (Bio Onderstepoort). To date, no additional cases of botulism have been reported in the collection. Husbandry practices aimed at controlling outbreaks of botulism include vaccination, good sanitation, preventing the accumulation of stagnant water and rotting vegetation in enclosures, and refraining from serial filling of ponds between enclosures.

### Endoparasites

Twenty-five species of endoparasites, including acanthocephalans, cestodes, trematodes, and nematodes, have been reported to infect cranes (Carpenter 1993). The most commonly found parasites amongst the birds in the WCRP have included, *Capillaria* sp. *Eimeria* sp. and *Ascaridia* sp. Clinical signs of parasitic infection may include, weight loss, diarrhea, lethargy and dyspnea. While mild parasitic infections do not generally present a serious health risk, heavy parasitic burdens can lead to malnutrition, decreased reproduction and increased susceptibility to other diseases. A formulary indicating therapeutic doses for common antiparasitic medications used in cranes can be found in Ellis et al (1996) Cranes: Their Biology, Husbandry and Conservation chapter 8, Medicine and Surgery. A comprehensive parasite monitoring and control program should be implemented including, routine fecal analysis (wet prep, float and sedimentation), treatment of positive individuals, pen rotation and good sanitation.

### Disseminated visceral coccidiosis

The protozoal parasite *Eimeria*, which normally infects the intestinal tract, can migrate to other parts of the body causing a condition known as Disseminated Visceral Coccidiosis. While adult cranes appear to tolerate extra-intestinal coccidia, infection in young captive cranes can represent a significant health problem resulting in bronchopneumonia, hepatitis, myocarditis, splenitis and enteritis (Olsen & Carpenter 1997). Preventative measures include; pen rotation, good sanitation, routine fecal analysis, treatment of positive individuals, and prophylactic treatment of young cranes where the presence of *Eimeria* has been demonstrated.

## CONCLUSION

One of the best methods of preventing common medical problems is the establishment of a neonatal and annual health-screening program. All birds in the Wattled Crane Recovery Program undergo annual physical examinations and routine laboratory testing to detect and treat potential medical problems before they become life threatening. The health and physical development of chicks is monitored through a series of health checks and husbandry protocols designed to minimize disease risks and reduce the incidence of developmental abnormalities. Many common medical challenges can be avoided through proper enclosure construction, good hygiene practices and fine tuning incubation and rearing protocols. The incidence of developmental abnormalities can be controlled through the implementation of strict dietary monitoring together with an age appropriate exercise program. The spread of parasites and other infectious diseases can be reduced through the use of pen rotation and good hygiene practices. Self-inflicted injuries and pen mate trauma can be mitigated with intensive behavioural monitoring and the construction of enclosures designed to reduce intra and interspecific interactions. Finding creative alternatives to husbandry and medical management that reduce the need for disturbance and handling can help prevent capture related mortality and morbidity. And finally, closely monitoring neonatal health and development and implementing standardized annual health checks will ensure early detection of and treatment of medical challenges, reduce the potential severity and spread of disease and ensure a greater potential for successful recovery.

## ACKNOWLEDGEMENTS

We would like to thank Jenny Gray, CEO of the Johannesburg Zoo for her support of the Wattled Crane Recovery Program; the staff of the Johannesburg Zoo (especially Mike Harman, Thoko Mesina and Thapelo Maserumule for their tireless hours of chick-rearing); our partners the Endangered Wildlife Trust's South African Crane Working Group, Ezemvelo KwaZulu Wildlife and the African Association of Zoos and Aquaria; the participants of the Wattled Crane Recovery Program; and our sponsors Airlink and the Bateleurs.

## REFERENCES

BAUCK L & J LaBonde. 1977. Toxic Diseases In Avian Medicine and Surgery, Altman, Clubb, Dorrestein, & Queenberry ED. W.B. Saunders Co. Philadelphia, PA. p. 606.

BENNET RA. 1994. Neurology In Avian Medicine: Principles and Application, Ritchie BW, Harrison GJ, Harrison LR. ED Wingers Publishing Inc, Lake Worth, Fla. pp. 728-747.

BRANNIAN RE, DL Graham and J Creswell. 1981. Restraint associated myopathy in East African Crowned Cranes. In M. E. Fowler, editor. Proceedings of the American Association of Zoo Veterinarians, Seattle, Wash. pp. 21-23.

CAMBRE RC. Consecutive avian botulism outbreaks in a zoological garden. 1988. Proc Am Assoc Zoo Vet, 1988, pp. 43-47.

CARPENTER JW. 1993. Infectious and parasitic diseases of cranes. In Fowler M.E., (ed): Zoo and wild Animal Medicine, Current Therapy III, Philadelphia, PA, WB Saunders Co. pp. 229-237.

CARPENTER JW, NJ Thomas and S Reeves. 1991. Capture myopathy in an endangered Sandhill Crane (*Grus canadensis pulla*). Journal of Zoo and Wildlife Medicine 22: 488-493.

DONLEY B. 2006. Management of Captive Ratites In Clinical Avian Medicine Volume II, Harrison & Lightfoot ED. Spix Publishing Inc. Palm Beach Fla. pp. 957-989.

DORRESTEIN GM. 1977. Bacteriology In Avian Medicine and Surgery, Altman, Clubb, Dorrestein, & Queenberry ED. W.B. Saunders Co. Philadelphia, PA. W.B. Saunders Co. Philadelphia, PA. p. 275-278.

DUMONCEAUX G & G Harrison. 1994 Toxins In Avian Medicine: Principles and Application, Ritchie BW, Harrison GJ, Harrison LR. ED Wingers Publishing Inc, Lake Worth, Fla. p. 1038.

ELLIS DH, GF Gee and CM Mirande. 1996. Cranes; Their biology, husbandry, and conservation. US Department of the Interior, National Biological Service, Washington D.C. and International Crane Foundation, Baraboo, Wisconsin.

FLAMMER K, SL Clubb. 1994. Neonatology In Avian Medicine: Principles and Application, Ritchie BW, Harrison GJ, Harrison LR. ED Wingers Publishing Inc, Lake Worth, Fla. p. 828.

FOREYT WJ & FR Abinanti. 1980. Maggot-associated type C botulism in game farm pheasants. J Am Vet Med Assoc 177(9): 827-828.

HAEFFNER, S. 1988. Correcting leg and joint abnormalities in long-legged precocial birds, with notes on prevention. Pages 550-563 In Proceedings of the Regional Conference of the American Association of Zoological Parks and Aquariums, Pittsburgh, PA.

HARRISON GJ. 1974. *Clostridium botulinum* type C infection on a game fowl farm. Proceedings of the American Assoiation of Zoo Veterinarians. Pp. 221-224

HARTMAN LM. 1983. Summary of mortality of 14 species of cranes at the International Crane Foundation, 1972-1982. Pages 555-570 In G.W. Archibald and R.F. Pasquier, editors. Proceedings 1983 International Crane Workshop, Bharatpur, India. International Crane Foundation, Baraboo, Wisconsin.

HARTMAN LM, S Duncan and G Archibald. 1987. The hatching process in cranes with recommendations for assisting abnormal chicks. Pages 387-396 In Proceedings 1985, Crane Workshop. Platte River Whooping Crane Habitat Maintenance Trust and U.S. Fish and Wildlife Se rvice, Grand Island, Nebr., J. C. Lewis, ED.

HAYES MA, BK Hartup, JM Pittman and JA Barzen. 2003. Capture of sandhill cranes using Alpha-chloralose. Journal of Wildlife Diseases 39(4): 859-868.

HUTT, F. B. & A. M. Pilkey. 1934. Studies in embryonic mortality in the fowl. Relationships between positions of the egg and frequencies of malpositions. Poultry Science 13: 3-13.

JONES K, L Rodwell L, McCann K, Verdoorn H, Ashley M. 2006. Genetic conservation of South African Wattled Cranes. Biological Conservation 127: 98-106.

LABONDE J. 1992. The medical and surgical mangement of domestic waterfowl collections. Proceedings of the Association of Avian Veterinarians pp. 223-233.

MARTIN H & BW Ritchie. 1994. Orthopedic Surgical Techniques In Avian Medicine: Principles and Application. Wingers Publishing Inc, Lake Worth, Fla. p. 1161.

MCCANN K, A Burke, L Rodwell, M Steinacker and US Seal. 2000. Population and Habitat Viability Assessment for the Wattled Crane (Bugeranus carunculatus) in South Africa. IUCN/SSC Conservation Breeding Specialist Group, Apple Valley, Minnesota.

OLSEN GH & JW. Carpenter. 1977. Cranes In Avian Medicine and Surgery, Altman, Clubb, Dorrestein, & Queenberry ED. W.B. Saunders Co. Philadelphia, PA. W.B. Saunders Co. Philadelphia, PA. p. 973-991.

OLSEN GH & SL Clubb. 1977. Embryology, Incubation, and Hatching In Avian Medicine and Surgery, Altman, Clubb, Dorrestein, & Queenberry ED. W.B. Saunders Co. Philadelphia, PA. W.B. Saunders Co. Philadelphia, PA. p. 65.

POLLOCK CG. 2006. Implications of mycobacteria in clinical disorders p. 681-689 In Clinical Avian Medicine Volume II, Harrison & Lightfoot ED. Spix Publishing Inc. Palm Beach Fla.

RICHARDSON JA. 2006 Toxic substances in clinical disorders p. 711-719. In Clinical Avian Medicine Volume II, Harrison & Lightfoot ED. Spix Publishing Inc. Palm Beach Fla.

SERAFIN JA. 1980. Influence of dietary energy and sulfur amino acid levels upon growth and development of young Sandhill Cranes. In Proceedings of the Annual meeting of the American Association of Zoo Veterinarians, Washington, D.C. p. 30.

SERAFIN JA. 1982. The influence of diet composition upon growth and development of Sandhill Cranes. Condor 8(4) :427-434.

TALMADGE DW. 1977. The effect of incubating eggs narrow end up on malposition II and hatchability. Poultry Science 5(6): 1046-1048.

WILLIAMS ES & ET Thorne. 1996. Exertional Myopathy (capture myopathy). In Non-infectious diseases of wildlife, 2nd edition. A. Fairbrother, L.N. Locke and G. L. Hoff (Eds). Iowa State University Press, Ames, Iowa, pp. 181-193.

WINDINGSTAD RM, SS Hurley and L Sileo. 1983. Capture myopathy in a free-flying greater Sandhill Crane (*Grus canadensis tabida*) from Wisconsin. Journal of Wildlife Diseases 19(3): 289-290.

MEINE C & G Archibald. 1996. The Cranes: Status Survey and Conservation Action Plan. IUCN, Gland, Switzerland.

GERLACH H. 1994. Bacteriology In Avian Medicine: Principles and Application, Ritchie B.W., Harrison G.J., Harrison L.R. ED Wingers Publishing Inc, Lake Worth, Fla. p. 1038.

## Author Biography

Jeanne Marie Pittman CVT, is an American veterinary nurse currently residing in South Africa and employed by the Johannesburg Zoo as the hospital supervisor. She is also the coordinator for the Wattled Crane Recovery Programme, a conservation initiative aimed at preventing local extinction of the Wattled Crane in South Africa. Jeanne Marie's long-standing interest in conservation and wildlife has lead to her involvement with a number of conservation projects including the Vulture Study Group, the Okavango Delta Crocodile Research Project and the Ground Hornbill Research & Conservation Project. Prior to moving to Africa, Jeanne Marie was employed by the International Crane Foundation in Baraboo Wisconsin as part of the support team for the Whooping Crane Eastern Partnership's Ultralight Migration Project; a conservation project aimed at the re-introduction of the Whooping Crane in North America. She has spent the last 30 years gaining experience in avian and exotic medicine both in private practice and in zoological institutions. Jeanne Marie passed the United States National Veterinary Technicians Examination in 1984 and again in 2002 and is currently registered by the South African Veterinary Council.

# SOUTHERN CASSOWARY *CASUARIUS CASUARIUS JOHNSONII* IN THE WILD AND CAPTIVITY IN AUSTRALIA

LIZ ROMER

The Australian Museum, Sydney, Australia.
*(formerly Currumbin Wildlife Sanctuary)*

## SUMMARY

One of the most threatened members of the Ratite family and the largest extent vertebrate of Australia, this iconic genus of birds has long fascinated both ornithologists and aviculturists alike. Despite being much sought after by zoological facilities, they have historically remained difficult to reproduce reliably in captivity. More recent avicultural techniques and management practices have resulted in improved breeding results however careful attention has to be paid to these birds in order to maintain viable captive populations. The captive history, behavioral aspects, dietary requirements, artificial rearing and general husbandry are covered in this paper.

## INTRODUCTION

Cassowaries, a familiar rainforest species, are widely held in captivity for educational displays as they represent both a flagship species for rainforests and the unique group of birds known as Ratites. The thick glossy black plumage combined with the striking reds, blues and purples of the head and neck, topped with a bony casque make it an unmistakable and outstanding looking animal. The word cassowary is of Papuan origin, from "kasu" - horned and "weri' - head (Boles 1987). Three species of cassowary are currently recognised. All three species occur in Papua New Guinea with only the Southern Cassowary occurring in Northern Australia. The Australian sub-species of the Southern Cassowary is also known as the Double-wattled Cassowary *Casuarius c. johnsonii* referring to the two wattles at the base of the neck. This sub-species will be the focus of this paper.

### Conservation in the Wild

The Southern Cassowary (cassowary henceforth) is the largest native vertebrate in the Australian rainforests. There are two populations in Australia. The southern (wet tropics) population includes the area between Cooktown and Townsville west to the extent of the rainforest and the northern (Cape York) population is found in the eastern Cape York Peninsula (Garnett 2000). The Southern Cassowary is important to the survival of many rainforest plants as they are a major long distance dispersal agent for about 150 species. The cassowary is seen as the flagship species for the conservation of wet tropics rainforest conservation and attracts regional, national and international attention as a unique and impressive bird. Unfortunately, this species is under extreme pressure. At this stage about 85 per cent of its lowland habitat and 75 percent of its upland habitat has been cleared (Garnett pers. comm. 2007). They have been classified by the Commonwealth Government as Endangered, the Queensland Government has classified the wet tropics population as Endangered and the Cape York population as vulnerable. It's thought that as few as 1200 individuals survive in Australia. Apart from the loss of habitat a number of other factors have contributed to the decline of the cassowary in Australia. Roads and traffic kill a significant number especially in the Mission Beach area. Given that cassowaries are a long lived, slow-reproducing animal each road death of an adult bird may potentially influence population dynamics and the population's reproductive fitness (QPWS 2002). Hand-feeding has also been implicated in their decline as it makes them more vulnerable to dog attack and road mortality. This can also endanger humans as the birds can become aggressive when accustomed hand-outs are not forth coming. Dogs have been known to attack birds and dog attacks are classed as the second most important source of cassowary mortality (QWPS 2002). Pigs have been documented destroying nests and causing decline in the Daintree and Atherton tablelands area (Crome and Moore 1988). Diseases such as Aspergillosis and Avian Tuberculosis and parasites have additionally been implicated in the decline of the cassowary (Latch 2007).

**Fig. 1.** *A Southern Cassowary.*
*(L. Romer)*

Catastrophic events such as the recent cyclone in March 2006 also play a part in their plight. After the last cyclone, a number of cassowaries moved in search of food resources and subsequently entered urban areas where they were at greater risk of vehicle strike and dog attack. Following Cyclone Larry, 32 deaths of cassowaries were reported, 70 percent from vehicle strike and 22 percent from dog attack. Another 15 birds were taken into care including eight orphaned juveniles, six sub adults showing signs of malnutrition and one injured adult. Most were housed at the Queensland Parks & Wildlife Garners Beach Cassowary Rehabilitation Facility and two at Hartley's Crocodile Adventures. The six sub adults have since been released to the wild and the adult euthanized after assessment by a veterinarian. Three juveniles have been moved into captivity permanently due to unsuitability for rehabilitation. The remaining birds are due to be released shortly following assessment of potential release areas. The Queensland Parks and Wildlife Service (QPWS) also has undertaken a supplementary feeding program using up to 61 feeding stations. Approximately 1000kg of fruit is purchased weekly supplemented by approximately 100kg donated by community members. These feeding stations are used to draw the birds away from urban areas where they are more at risk and will be reduced over time (Gayler 2007).

Habitat protection is being seen as a high priority for conserving the cassowary but still 16 percent of remaining habitat in the wet tropics remains unprotected (QPWS 2002). Corridors linking patches of habitat are also important in assisting dispersal of birds. Many community groups have become actively involved in the restoration of habitat for cassowaries.

Recovery plans have been prepared for this species detailing a number of actions. These include protection of habitat as well as public education programs relating to road mortalities, dog control and hand feeding. Feral pig control programs are also part of the actions listed. Captive breeding for release is not seen at this stage as a conservation action as there are too many concerns with habitat protection that need to be solved first. Cassowaries in captivity do however give us a great tool to use for education about cassowary conservation and thus help support the wild conservation in that way.

## Life History

Although occurring primarily in rainforest and associated vegetation mosaics, the cassowary also uses woodland, swamp and disturbed habitats as intermittent food sources and as connecting habitat between more suitable sites (Crome & Moore 1993; Bentrupperbäumer 1998). It requires a high diversity of fruiting trees to provide a year-round supply of fleshy fruits (Latch 2007).

Their diet includes fleshy fruits of up to 238 plant species, including seven exotics (Westcott et al 2005). While fallen fruit is the primary food source, cassowaries also eat small vertebrates, invertebrates, fungi, plants and carrion (Marchant & Higgins 1990). They forage for about 35 percent of the day, mainly early morning and late afternoon (Latch 2007; Westcott et al 2005). The cassowary is both territorial and solitary, with contact between mature individuals generally only tolerated during mating. Sexes will maintain independent but overlapping home ranges with female home ranges encompassing those of one to several males (Bentruperbäumer 1998). Home ranges fluctuate depending on season and availability of fruit, with estimates of between $0.52km^2$ to $2.35km^2$ recorded (Bentrupperbäumer 1998; Moore & Moore 2001). Cassowaries may also tolerate each other in areas of very abundant fallen fruit and have been known to congregate in areas when artificially fed on a regular basis (QPWS unpublished; Latch 2007). Females lay three to five olive-green eggs, generally between June and October. Males incubate the clutch for about 50 days before raising the young alone for about a year (Bentrupperbäumer 1998). Young birds must then seek their own home range, but with limited opportunities, particularly due to high fragmentation and loss of habitat, the sub-adult mortality rate is probably high (Latch 2007). Chicks when hatched are striped but begin to lose the stripes at about three months of age. By six months their sub-adult brown plumage has developed and the neck and head are beginning to colour. It takes from about two to three and a half years of age for the glossy black plumage to develop fully. They may mature before this as brown birds will sometimes attempt to mate with adult females. In captivity the youngest recorded breeding was a male of three and a half that successfully fertilized eggs with an older female at Currumbin Sanctuary. This contrasts with Denver Zoo whose breeding male was 31 years old when he first successfully mated (Romer 1997).

**History in Captivity**

The Southern Cassowary has a long history in captivity. In 1596, the first bird seen by Europeans was when a *Casuarius casuarius* from Banda Island was given as a gift to a captain of a Dutch merchant vessel. The captain was subsequently murdered by the bird (which only goes to show how dangerous cassowaries can be!) but the live bird of the now dead captain was taken back to Amsterdam where it arrived in 1597 and was presented to the Holy Emperor Rudolf II as a gift and was put on public display for some years. Thus, both discovery and captivity arrived at the same time for the cassowary (Moore 1994). Records of breeding show London Zoo produced single chicks in 1862 and 1863 although neither survived. The first successful rearing (one chick) was probably San Diego Zoo in 1957. The male had lived there for 31 years. Taronga Zoo hatched a chick in 1962 but it is thought to have died soon after. From 1973 to 1982 the Australian Reptile Park in Gosford NSW was relatively successful in breeding them. Edinburgh Zoo (Scotland) documented breeding from a four year old pair. Since then results have been sporadic with Denver Zoo being perhaps the most productive with 98 chicks bred from 1977 - 1992 (ISIS Studbook data). Airlie Beach Wildlife Park has been the most prolific breeders in Australian Wildlife Parks with over 28 birds being produced to this present time. Some pairs have had one or two good breeding seasons then produced no more while other pairs have still to breed.

**Current Status in Captivity**

Currently cassowaries are reasonably numerous in captivity in Australia with many zoos and wildlife parks maintaining them. While the general husbandry of this species appears to be generally well established breeding is still sporadic. In the current cassowary studbook run by the Australasian Species Management Program (ASMP) fifteen institutions hold cassowary - mostly pairs. This number however represents probably only half of the current number of institutions known to be holding cassowaries. Many of Australian wildlife parks are not members of the ASMP. In the past breeding season only 2-3 institutions have been known to be successful in breeding cassowary, of these, only one was a member of the ASMP. As general rule cassowaries in captivity are for display purposes only. As mentioned earlier they represent a great opportunity to tell visitors about conservation of rainforests and act as a flagship species for the Wet Tropics.

**General Husbandry**

As a rainforest species, cassowaries require enclosures which provide shade at all times. Anecdotal evidence shows that cassowaries maintained in enclosures without suitable shade are prone to the development of cataracts. Shelter from extremes of cold is required in areas that fall below the minimum temperatures in their natural environment. In coastal areas this would rarely be below 10°C. Philadelphia Zoo did however keep a bird outside with only a roof for shelter for a number of winters. The provision of a hut or shelter may also encourage the pair to nest within or against the side of the structure.

A formal shelter may be required for cassowaries depending on the climate and amount of shade trees in the enclosure. In very cold climates there may be the need to provide heat in these holding areas. It is recommended that the enclosures be heavily planted with large shady rainforest trees. Those providing natural food sources are recommended over those that are strictly ornamental. Fresh water should be available at all times for drinking. Cassowaries also have a particular fondness for bathing and should be supplied an area that allows them to do this. A mud bath is highly enjoyed and sprinklers appreciated. Cassowaries swim well and will wade chest deep in water given the opportunity. A large stream running through the exhibit is ideal; sprinklers can also be installed.

Enclosure furniture (logs, rocks, trees and shrubs) should be available in pens used for pairing to allow birds to have visual separation if required. These obstacles should be avoided around the fence boundary. Rocks and logs can provide homes for prey items, they may also encourage the birds to nest as they provide a suitable backing. The aggressive nature of cassowaries can necessitate the separation of pairs. The enclosure should be designed to allow isolation of one animal from the other if the need arises but should still allow visual or auditory contact to be maintained.

Additionally, enclosures need to be designed to enable servicing of the enclosure with the birds being locked into holding areas to eliminate the risk of injury to staff servicing them. Ideally routine feeding and watering should be possible without the need to enter the enclosure. It is also useful to provide suitable access for a vehicle to the enclosure to facilitate transportation of birds. Cassowaries have been bred in enclosures as small as 200m$^2$ though this was divided in half to separate the pair and was additionally heavily planted (Hopton 1992). It has been suggested that the small size of cassowary enclosures has contributed to the poor breeding record experienced in captivity. The size of an enclosure for a breeding pair is not as important as obtaining a compatible pair. However, for introduction purposes a larger pen is required. If the pen is large then a pair can possibly remain together at all times including while incubating and raising young. Rundel recommends a good size pen for breeding would be 10m x 30m combined with good cover facilities. Queensland Wildlife Parks Association (QWPA) minimum standards require 200m$^2$ for an individual bird and 300m$^2$ for a compatible pair.

A suggested minimum size enclosure for a pair of cassowaries is two adjacent enclosures each measuring 18m x 12m with a safety passage 1.5m wide designed so that during the breeding season the birds have access to both enclosures. There should also be a separate feeding area that can be shut off from the birds for servicing while also acting as a holding facility while the main enclosure is serviced. A minimum suggested size is 5m x 4m. Chain link fences are commonly used as a boundary. Foliage, particularly climbing plants, planted along the perimeter fences provide cover and protection for the birds. Solid concrete rendered walls and tin fences have both been successfully used. Fence heights have varied from 1.4m to 1.8m. Several jump outs have occurred at the lower height while one recorded jumpout has occurred at 1.8m. These jump outs have generally been associated with cassowaries attacking other cassowaries or people. QWPA standards require a minimum height of 1.8m. A double fence may be necessary if the birds are aggressive. This can contain a planted barrier to improve the appearance. Chicken wire is not recommended as toes can easily get caught in the small mesh and is not strong enough to contain an adult bird if aggressive or frightened. Vertical panels must also be avoided due to similar injury potential (whole legs and heads may get caught). Mesh size needs to be considered carefully as large mesh can lead to heads being pushed through causing irreparable damage to the casque with a potential for the bird to break its neck. Additionally kicks through the mesh can lead to torn and peeled leg scales and infections. QWPA recommended mesh size is 50mm x 50mm.

The fence needs to be of solid construction, not easily lifted and should follow the contour of the land as well as being secured into the ground to a depth of 40cm or of similar stability as birds have been known to go through a fence if they can't go over it.

Fences should be free of obstacles or loose wires. All fence posts and straining wires should be positioned on the outside of the enclosure. Where the park perimeter forms part of the boundary provisions need to be made to protect the birds from outside disturbances such as the public, traffic and dogs. Moats were tried unsuccessfully at Los Angeles Zoo. Melbourne Zoo was successful in the use of a wet moat. Careful thought into the design needs to be made. The moats should allow the birds to use it for swimming. This would be beneficial from a display point of view. Heat is only required in very cold climates where temperatures regularly fall below approximately 0°C. In these cases, a heated shed may need to be provided. This is a necessity in any area where snow is known to fall or regular heavy winter frosts occur. Cassowary chicks in particular are not cold hardy (see rearing section).

Cassowaries regularly eat their faeces in captivity as well as in the wild. It is however still important to remove the faeces especially in small enclosures. Lack of removal of faeces may contribute to a high worm burden. It is important that clean water and feed containers be provided. Small scatter feeds may add behavioural enrichment thus making it important to keep the enclosure clean. Appropriate measures need to be taken to ensure keepers can enter the enclosure

**Fig. 2.** *Male cassowary shows its chick how to feed.*

safely - thus routine cleaning and watering should be possible without the need to enter the enclosure with the birds present. Care should also be taken not to disturb an incubating male. Regular checks of the enclosure with particular attention given to problems that cause leg trauma eg. broken fencing, need to be carried out.

## Captive Behavioural Notes

Cassowaries are solitary animals in the wild, normally only coming together to breed. Some institutions have been able to keep a pair together year round even when the male is incubating. Other institutions need to keep the pair apart in the non-breeding season and when the male is incubating. Males with chicks can be particularly aggressive towards the female (and keepers). Groups of young birds have been kept together but is not recommended as breeding is often interfered with by the other birds. Cassowaries are a particularly aggressive species. In some institutions a pair can easily be kept together year round with no apparent aggression problems (except to the keeper). In other institutions the birds may only be introduced to each other at the beginning of the breeding season and need to be separated once the clutch has been laid. Often aggression problems can be solved by ensuring the enclosure is large enough and well landscaped with areas where the birds can avoid visual contact.

Some institutions have experienced difficulty even getting a pair together without severe fighting. Aggression towards the keeper can be a particularly dangerous situation. This can be solved by good exhibit design - allowing the animals to be penned up separately to where the keeper needs to work. Additionally being able to feed & water the birds without entering the enclosure can assist and protect the keepers from injury! It is of utmost importance that the keeper must not incite or encourage aggressive behaviour in the cassowary. In the early stages of breeding the birds are often noticed in close contact - feeding and foraging together. Before copulation the male displays to the female by "dancing" around her in a circle, his throat trembling and swelling and emitting a series of low "boos". The male then leads the female a short distance where she squats and allows him to mount (Crome 1993). The female has also been seen to display. The female stands with her neck upright and her head tucked into the neck. The neck is then filled with air all the way down to the breast - the neck diameter stretched to twice the size. She then emits a low frequency rumble - the whole neck is seen to vibrate. This

**Fig. 3.** *Cassowary can be extremely dangerous and need to be managed accordingly.*

is done for a few minutes (Z. Gubler pers. comm).

Cassowary mating is somewhat prolonged. Often the male ruffles his plumage and pecks the ground whilst slowly approaching a resting female. If she remains still, he pecks her neck, adopts a squatting stance and eventually mounts from behind. Copulation continues between the laying of each egg (Whitehead & Masson 1984). Crome & Moore (1988) state copulation is brief. Richard Rundel's observations of mating are described as follows: "Breeding commonly lasts 30 minutes from the time the male starts circling behind the female, preening her to induce her to lay down and present herself, to the final moment of copulation. In order for their respective cloaca's to touch the male must lock his legs under the female and lean way back. Since they copulate frequently he is used to the ritual and knows that once contact is made she will jump to her feet, causing him to fall over backwards. At this point he is desperately trying to get to his feet and run because she by now is performing a Mexican Hat dance on top of him." In some institutions where the pair must be kept apart the birds are kept in visual

**Fig. 4.** *A silkie bantam used to teach a cassowary chick how to eat independently.*
*(L. Romer)*

contact. When the birds are seen interacting together at the start of the breeding season they are introduced to each other. As a general rule it is found that the best way to establish breeding pairs is to start with young birds that will bond as they mature. Some adult birds put together may take up to a year or two to establish a viable pair.

**Diets and Supplements**

Cassowaries are mainly frugivorous though they also eat some animal matter and have a very short digestive system and hence a fast passage of ingesta. This results in a high dietary requirement which can reach 10% their body weight in daily intake. A large variety of fruits are fed including tomato, banana, apple, pear, paw paw, watermelon, grapes, mango, plums, nectarines, cherries, kiwi fruit, figs, rock melon, eggplants and cooked sweet potato and carrot. Mice, rats, day old chickens and some fish are readily taken. One of the cassowaries at Adelaide Zoo also considered House Sparrows a delicacy.

As a general rule most institutions feed between four and five kilograms of fruit per day per bird, though amounts vary seasonally. Some institutions also have a number of fruiting trees in their enclosures which supplement the captive diet. Supplements of calcium carbonate and Petvite multivitamin supplement have been successfully used in some institutions that have bred cassowaries. Wombaroo insectivore mix sprinkled on their food has also been used. Wheat germ is sprinkled daily on the fruit at Fleay's Wildlife Park. Other dietary items used include brown bread, boiled rice, turkey pellets, trout chow, hydroponics (seven day oat grass), cotoneaster berries, canned raspberries, *Eugenia* berries, *Pyracantha* berries, soaked raisins, horse cubes, kangaroo pellets, water-

**Fig. 5.** *Chicks begin to lose their stripes from three months of age.*
(L. Romer)

fowl pellets and horse meat. Rocks are also eaten to aid digestion. It is noted that cooked vegetables are utilized (metabolically) better than raw, though if overcooked and mushy they will clog around the bill and be flicked away. Between March and June soft fruits are decreased at Fleay's Wildlife Park to coincide with the normal dry season in North Queensland, with increased protein foods offered daily eg: rats, mice, day old chicks and fish. In the USA, a pelleted diet is used on birds from one month of age. The primary nutritional problem is providing sufficient level of vitamins A and E. Sonoma bird farm use a diet formulated by Purina Mills and sold under their Mazuri label that is specifically blended for cassowaries with additional A, E, and D vitamins.

At this stage it is recommended that more analysis is done on native fruits for assistance in formulating diets & trigger diets. It is known that Laurels are high in protein and when they don't fruit in the wild breeding drops off. Toxins in unripe fruit also needs to be investigated. The National Capital Botanical Gardens Papua New Guinea hold both Double-wattled Cassowary *Casuarius c. scalterii* and Dwarf Cassowary *C. bennetti*. The adult birds are fed the following daily: two apples, one whole pawpaw, two hands of sweet banana, one loaf wholemeal bread, two mangoes, Pandanus fruit when available, cooked Kaukau and rice when available (J. Tkatechenko pers. comm).

**Breeding**

Breeding ages appear to cover a wide span. One breeding male first mated when 31 years old at San Diego Zoo (Whitehead & Mason 1984). The youngest mature male known to breed was a three-and-a-half years old at Currumbin Sanctuary fertilized a clutch. Most breeding activity appears to start around four years. A female at Rockhampton is still breeding at 37 years of age. The best combination appears to be an older hen with a younger cock bird. Breeding season in the wild appears to be from May to October/November with the main period being June to October (Crome & Moore 1988) coinciding with the maximum availability of fruit. In the northern hemisphere breeding appears to begin around March (Fisher 1968). Females in the wild may take more than one mate with up three being seen (Crome & Moore 1988).

Many institutions have experienced difficulty getting a pair together. Some years there are no problems while other years the amount of aggression causes the pair to remain segregated. Other pairs appear very compatible and even if separated there is not a problem getting them together later. The best approach appears to be pairing birds as juveniles allowing the bond to develop as they mature. If this is not possible and the birds are already adults it can take up to a year to get the birds together. Cassowaries generally don't like change and it takes time for them to adjust. It is best to try putting them together after an interest is shown in each other but no aggression. If the birds are not compatible and aggression is being displayed after 15 minutes separate them and attempt introduction again later.

If worried about aggression the introduction pen can be lined with plastic to avoid injuries. As the breeding season nears both birds are likely to exhibit more aggression towards the keepers. Some institutions find the female shows more aggression whilst others find the male more aggressive. At this time the male will sometimes be seen sitting in a favoured nesting spot to show his parental suitability. In the wild nests are made up of leaves, grass and debris, and are about a metre in diameter and up to 2-5cm thick, supposedly placed in a sheltered position such as between buttresses or beside a log. Nests in captivity generally are in strange and problematic places, often near paths, Brush Turkey *Alectura lathami* mounds and on the sides of hills. The use of dummy eggs can be of use to change nest site position. The provision of a hut or shelter will often encourage the pair to nest within it. This can be of use for management purposes.

High humidity is thought to be essential for incubation. Irrigation may be needed in drier climates during the day provided it is not too cold. Increasing the protein and vitamin B may be necessary prior to the breeding season. A chick which hatched at Taronga Zoo and died shortly after showed a vitamin B deficiency. Supplementation with Petvite solved this in the subsequent clutch. Fleay's Wildlife Park increase the protein levels between March and June with the female feeding keenly on the protein. After July the interest in high protein food drops. Adelaide Zoo supplement with calcium prior to breeding.

### Table 1
**INCUBATION PARAMETERS USED AT TWO INSTITUTIONS**

|  | Dry Bulb (°C) | Wet Bulb (°C) |
|---|---|---|
| Currumbin Sanctuary | 35.5 - 36.7 (Av 36.4) | 27.2 - 30.6 (Av 28.3) |
| Melbourne Zoo | 36.5 | 31.1 - 31.7 |

**Fig. 6.** *Pair of cassowaries mating.*

A possible vitamin E deficiency was seen in a nearly formed chick at Currumbin Sanctuary. The muscles at the back of the neck used for pipping the egg were not developed. Cassowaries will breed on a fruit only diet however. More research into wild animal food preferences would be beneficial to assist in formulating breeding diets in captivity. Generally four but up to nine eggs have been recorded in captivity. Double clutching is possible by pulling the chicks at one - two weeks. Pulling eggs as laid will also increase the amount of eggs laid- up to 28 being recorded. However not all were fertile. Generally the more laid the lower the fertility. Incubation periods of 49-57 days have been recorded. The smaller eggs tending to hatch sooner than larger eggs and the male will brood the eggs until the full clutch is laid. A male incubating eggs for more than 65 days after the last egg is laid calls for consideration of a decision to take the eggs for examination.

Chicks being parent raised will require food to be provided in smaller portions though the males are known to break up the food and drop it in front of the chicks; live insects such as mealworms can be provided for the chicks. The cock bird will pick up food items and drop it in front of the chick followed by a short burst of beak clapping to gain the chicks attention and direct it to the food. Once the chick understands what happens at feed time it will feed independently taking food only occasionally from the father though the cock bird would continually offer food indicating that it was simply a progressive behaviour; this will happen for many months (Wexler 1995).

Chicks can remain with the male up to 18 months of age, but are likely to be independent before then. If any aggression is displayed by the male or female towards the young, they should removed immediately. Ideally, the young should be removed prior to the next breeding season to allow the pair to breed again. In the wild, the chicks can remain with the male resulting in his breeding every two years (D. Westcott pers. comm. 2007).

Emu's *Dromaius novaehollandiae* have been used to hatch cassowaries at Wildworld Cairns (Bullen pers. comm. 2007). Three eggs apparently late in incubation were placed under an Emu that was incubating eggs at that time. The eggs successfully hatched and the chicks were hand-raised.

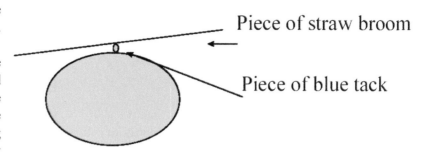

**Fig. 7.** *Simple method for detecting a heartbeat in an egg.*

## Artificial Rearing

There is a wealth of information on artificial incubation techniques. This section will be dealing with specifics relating to cassowaries. One critical factor influencing hatching results appears to be incubator humidity. Cassowaries naturally occur in areas of high humidity and their eggs may be particularly susceptible to embryo desiccation following excessive water loss. Temperature does appear to be critical also. A temperature that is too high results in premature hatching and the chicks will not survive. It is recommended that the temperature not go above 36.1°C dry bulb. Late deaths are often related to high temperature. Denver Zoo had success with a dry bulb reading of 36.1°C and a wet bulb reading of 30.6°C, or relative humidity of 68% (Birchard et al 1982). Ten days before hatching the humidity increased to a wet bulb reading of 31.7-31.8°C and the temperature decreased from a dry bulb reading of 36.5°C to 36.2°C. Weight loss is generally desired to be in the 12-15% range for most eggs. Rundel (pers. comm. 2007) finds that 12-15% is most successful but successful hatches have occurred from 12% to 20% weight loss; Currumbin Sanctuary successfully hatched a chick with an 11% weight loss.

Assisted hatch is sometimes necessary. Male birds have been observed assisting chicks out of the shell. Rundel has had a problem with mal-positioned chicks with an estimated 20% exhibiting this problem. This can be seen with an infrared candler - the chick expanding through the air cell. It is not always a disaster. For an assisted hatch create a hole in shell at the top where the air cell is- do not break the membrane. The movement of the chick can then be checked by holding a torch against the membrane. Prick the membrane with a dull probe if required. If membrane starts to bleed stop immediately. Exposing the nostrils of the chick will increase the activity of the chick. The head can be located by x-ray or by flicking the shell with a fingernail while holding the egg to the ear to listen for the location of the call.

Eggs can be left for a number of days on a table in a well ventilated area before being set in the incubator. It is better to set a clutch together for synchronised hatching and easier hand-rearing. To detect movement in eggs you can place a piece of straw from a straw broom on the top of the egg attached with a piece of blue tack. By watching for movement you can detect the heartbeat. When hatching the eggs can be placed on the floor of the incubator on a towel or similar solid surface to give them adequate secure footage when hatched. Brooder boxes measuring approximately 2m x 1m x 0.8m high have been used by Currumbin Sanctuary, Taronga Zoo, Fleay's Wildlife Park and Rundel. All four had solid sides with an overhead light - Rundel uses a 100 watt bulb and a heating pad at one end. Currumbin used a heat lamp at one end to create a slight temperature gradient. Cassowary chicks are not cold tolerant and must be kept reasonably warm. It is best to maintain the chicks initially at 30°C. A slight temperature gradient is advisable.

## Feeding

Chicks generally do not eat for 2-6 days as the yolk is absorbed. To start the chick feeding a day old chicken tutor may be used. This also provides any single chicks with company which appears to be important (Rundel & R. Dunn pers. comm. 2007). Otherwise tapping the feeding dish with a pen or similar object will stimulate the chick to show interest. A low bowl of water should be provided always and will usually be used before food is eaten. It is important to weigh birds daily at first to ensure weight gain is not too great. As with other ratites too much weight and too much protein coupled with a lack of exercise can cause leg problems. The diet is fairly similar to the adults. Soaked sultanas and diced tomato and banana are good starting foods. This can be sprinkled with Wombaroo Insectivore. Nekton S has also been added to supply trace vitamins and minerals. Calcivet has also been added to their drinking water. Insectivorous food is necessary for the first two weeks of the cassowary chicks life (Lint 1981). Birds should be fed 2-4 times daily. Pinkies can be offered from two weeks of age and grit is offered in a separate dish by some breeders (Lint 1981).

As with all ratites, it is of vital importance that the chicks are exercised daily preferably two to three times. The chicks will follow staff as they jog around a suitable area. Thirty minutes two to three times daily is a suggested starting point. Company is important for the chicks, either other cassowary chicks or a young chicken can substitute. The provision of a chick also assists in stimulating exercise. A dirt bath is appreciated from about two weeks of age. Dig out a depression in the soil and run water into the depression and the chicks will roll on their backs in the mud. Brooder box substrate should be a non slip surface such as astro turf. Do not use soft sand as a substrate. Indoor/ outdoor carpet is a good substrate or rubber matting which can be easily cleaned. The use of a mirror in the brooder and a cassowary head hand puppet when teaching the chick to eat may reduce imprinting, though imprinting does not appear to present a great problem. Aggressiveness to other chicks can start at 12-14 months though has been seen in chicks under six months of age. Aggressiveness towards staff may start under two years.

## ACKNOWLEDGEMENTS

Much of the information contained here is the result of a workshop held in 1996 on cassowary management and compiled into a Cassowary Husbandry Manual. The manual can be obtained by contacting me at email slromer@bigpond.com. Thanks to all who participated in the workshop especially Richard Rundell who made the long trip to Australia to share his experience. Thanks also to Maurice Gayler for the up to date information on the latest after Cyclone Larry.

## REFERENCES

BENTRUPPERBÄUMER JM. 1998. Reciprocal ecosystem impact and behavioural interactions between cassowaries, *Casuarius casuarius* and humans, *Homo sapiens* exploring the natural human environment interface and its implications for endangered species recovery in north Queensland, Australia. Unpubl. PhD thesis. James Cook University of North Queensland. Townsville.

BOLES W. 1987. Alias Emu. Australian Natural History 22: 215-216.

CROME FHJ. 1993. A catalogue of important Cassowary populations in the Wet Tropics. Unpublished report to the Wet Tropics Management Authority.

CROME FHJ & J Bentrupperbaumer. 1991. Management of Cassowaries in the fragmented rainforests of north Queensland. Unpublished report to the Endangered Species Program of the Australian National Parks and Wildife Service.

CROME FHJ & LA Moore. 1988. The Southern Cassowary in North Queensland - a pilot study. Vols I-IV. Unpublished report prepared for the Queensland National Parks and Wildlife Service and the Australian National Parks and Wildlife Service.

CROME FHJ & LA Moore. 1990. Cassowaries in north-eastern Queensland: report of a survey and a review and assessment of their status and conservation and management needs. Australian Wildlife Research 17: 369-86.

CROME FHJ & LA Moore. 1993. Cassowary populations and their conservation between the Daintree River and Cape Tribulation. II. Background, Survey Results and Analysis. Unpublished report to the Douglas Shire Council.

GARNETT S & G Crowley. 2000. The Action Plan for Australian Birds 2000. Environment Australia, Canberra.

GAYLER M. 2000. Exit Strategy for the post-Cyclone Larry cassowary supplementary feeding program. Queensland Parks and Wildlife Service.

HOPTON D. 1992. Breeding Southern Cassowaries *Casuarius casuarius* at Adelaide Zoo. Proceedings of the Saving Wildlife Conference - ARAZPA & ASZK.

LATCH P. 2007. National recovery plan for the Southern Cassowary *Casuarius casuarius johnsonii*. Draft. http://www.environment.gov.au/biodiversity/threatened/publications/pubs/c-c-johnsonii.pdf

LINT KC & AM Lint. 1981. Diets for birds in captivity. Blanforde Poole.

MARCHANT S & PJ Higgins (eds). 1990. The Handbook of Australian, New Zealand and Antarctic Birds, Part A and B, Ratities to Ducks, Oxford University Press, Melbourne.

QPWS. 2002. Queensland Parks and Wildlife Service). Recovery Plan for the Southern Cassowary *Casuarius casuarius johnsonii* 2001-05. Environmental Protection Agency, Brisbane.

ROMER L. 1997. Cassowary Husbandry Manual, Proceedings of February, 1996 Workshop. Currumbin Sanctuary, Currumbin.

WESTCOTT DA. 1999. Counting cassowaries: what does cassowary signs reveal about their abundance? Wildlife Research 26: 61-8.

WESTCOTT DA & K Reid. 2002. Use of medetomidine for capture and restraint of cassowaries (*Casuarius casuarius*). Australian Veterinary Journal 80: 150-3.

WESTCOTT DA, J Bentrupperbäumer, MG Bradford and A McKeown. 2005. Incorporating patterns of disperser behaviour into models of seed dispersal and its effects on estimated dispersal curves. Oecologia 146: 57-67.

WEXLER P. Breeding the Double Wattled Cassowary at Birdworld. In press.

WHITEHEAD M & G Masson. 1984. Notes on the Double-wattled cassowary at Twycross Zoo, U.K and elsewhere. The Management of Cranes, Storks and Ratites in Captivity. Proceedings of Symposium 9 of the Association of British Wild Animal keepers.

**Products Mentioned in text**

Petvite (vitamin & mineral powder)
Marrickville Holdings Ltd
I.G.Y. Veterinary Products,
74 Edinburgh Road,
Marrickville NSW

Wombaroo Insectivore Rearing Mix
Wombaroo Food Products
P.O. Box 151
Glen Osmond SA 5064

Nekton S
Nekton Produkte
75177 Pforzheim
Germany

Calcivet
Vetafarm
3 Bye Street
Wagga Wagga, NSW 2650

**Author Biography**

Liz Romer has had along association with cassowary working with them in a number of institutions including Taronga Zoo and Currumbin Sanctuary in Queensland over 20 years. While at Currumbin Sanctuary she was studbook keeper and coordinated a workshop on cassowary which resulted in a comprehensive husbandry manual being compiled. She is currently working at the Australian Museum as an exhibition development researcher and runs a consultancy business.

# FIRST CAPTIVE BREEDING OF THE PHILIPPINE EAGLE-OWL *BUBO PHILIPPENSIS PHILIPPENSIS*

Leo Jonathan, A. Suarez & Cristina Georgii

Negros Forest Ecological Foundation, Inc. South Capitol Rd, Bacolod City 6100, Phillipines

## Summary

The Philippines harbors a great diversity of owls. Habitat loss and intense hunting pressure has, however, significantly reduced there numbers in the wild. With the aim of protecting Philippine owl species, the Philippine Owls Conservation Program (POCP) was created and agreed among the Department of Environment and Natural Resources-Protected Areas and Wildlife Bureau, World Owl Trust and UK Owl Taxon Advisory Group. One of the current projects of the POCP is the conservation breeding and research program for the Philippine Eagle-owl *Bubo p. philippensis* which has resulted to the first captive breeding of this species. The pair of owls that successfully bred in 2005 was maintained in a fairly large enclosure that was landscaped with natural vegetation. Their diet consisted of rats, mice and beef. Routine health monitoring was done to ensure the owls' health. Several interesting behaviors were observed during the courting ritual, nest preference, nesting and the rearing of the owlet. Some of the most interesting behavior observed during rearing was that skills that are important for its survival in the wild such as hunting and techniques in eating prey of different sizes were being taught to the young owl. Natural breeding is therefore very important consideration in the captive breeding of this species, especially if the owls bred will be released in the future.

## Introduction

### Philippine Eagle-Owl

The Philippines harbors a great diversity of native owls. There are currently 24 subspecies and 16 species from eight genera and three different families recognized in the country. The largest owl occurring in the country is the endemic Philippine Eagle-owl *Bubo philippensis* which has two subspecies; the *B. p. philippensis* occurring in the islands of Luzon and Catanduanes and the *B. p. mindanensis* occurring in the islands of Samar, Leyte, Bohol and Mindanao

The Philippine Eagle-owl occurs in the lowland forests edges, often near rivers and lakes, and in coconut plantations with patches of secondary forests (Kennedy et al 2000). However, recent surveys made at the Mt. Pulag National Park in Luzon have recorded this species at altitudes between 2100-2300m, suggesting that it also occurs at higher altitudes (Ogbinar 2005). There had been very few recorded sightings of this species in recent years suggesting that it is relatively rare (Collar et al 1999). Because of its scarcity, virtually nothing is known about this species in the wild (T. Warburton pers. comm. 2005). With the continued destruction of its habitat, it is inevitable that so much loss will have a major effect on its populations (Collar et al 1999). The Philippine Eagle-owl is one of the threatened endemic owl species of the Philippines and is categorized as vulnerable in the 2004 IUCN Red List.

### Philippine Owl Conservation Project

To protect the Philippine Eagle-owl along with the other equally important endemic owl species of the country, the Philippine Owls Conservation Program (POCP) was agreed among the Department of Environment and Natural Resources & Protected Areas and Wildlife Bureau (DENR-PAWB), World Owl Trust (WOT) and UK Owl Taxon Advisory Group (O-TAG). In 1998, a Memorandum of Agreement (MOA) was signed among the three parties to facilitate Philippine owl field research, education and captive breeding projects.

One of the projects of the POCP is the conservation breeding and research program for the Philippine Eagle-owl *B. p. philippensis* which is located at the Negros Forests and Ecological Foundation, Inc.- Biodiversity Conservation Center (NFEFI-BCC) in Bacolod City. The NFEFI-BCC acquired its stock through a breeding loan of six unsexed individuals from the Avilon Montalban Zoological Park on the 29th of November, 2002 which constitutes the first ever such breeding loan between DENR-accredited institutions in the country. These birds were all from Luzon and were kept on display in Avilon for viewing purposes.

Fig. 1. *Inside of the owl aviary.*

## Management

Since the six eagle-owls had been kept together at Avilon in a large aviary together with other individuals for several months prior to the transport, they were also placed together in a large enclosure (9m long x 4m wide x 4m high) upon arrival at the NFEFI-BCC to monitor their adjustment to their new surroundings and to encourage natural bond formation. From the flock, bonds were observed a few days after arrival. As aggression was observed among the formed pairs, they were transferred to individual enclosures. One of the pairs, however, had to be separated because the female began displaying aggression towards the male.

### Enclosure

The breeding pair was housed in an enclosure in the breeding center that was 6m x 4m x 2.5m high and a 0.7m wide inner enclosure at the back, which also served as service corridor. The three walls of the enclosure are made of welded wire mesh and the wall at the back is made of concrete. The enclosure floor is composed of soil and has a shallow pool of 10cm deep where the owls bathe and drink. Several wooden perches of different diameters were secured along the walls and corners while trunks from dead trees were set in the middle of the enclosure. The enclosure was landscaped with a variety of non-poisonous and thornless plants and shrubs to serve as visual barriers, giving the enclosure a more natural look.

### Diet

The recommended diet for owls in captivity is generally a variety of whole prey such as rodents, poultry and fresh freshwater fish, depending on the species. Nothing is clear about the diet of the Philippine Eagle-owl in the wild although the fact that it is often found near lakes and rivers suggests that it obtains its food in these areas (Collar et al 1999). The NFEFI-BCC offers the eagle-owls beef, rats and mice at 75g/day per bird, in the late afternoon, for six days a week. The bulk of the diet of the breeding pair was beef, due to the limited number of rodents being harvested, though rats and mice were offered once a week. When breeding behaviors were observed a pure whole rodent diet was offered to the pair to stimulate breeding (P. Hospodarsky pers. comm. 2005) since the amount of food consumed by breeding owls is normally more than when they are in non-breeding state. Rats and mice were offered four to five times a day with a higher amount offered in the late afternoons. The total amount of food offered would vary from approximately 100-200g for the two birds, depending on the amount of food they consumed the previous day. One day a week was for fasting, which is normally practiced, was no longer done at this stage.

### Health

The health of the eagle-owls was assessed through daily observations of the individuals (general appearance and consumption of food). The eagle-owls were

Fig. 2. *Inside covered partition.*

checked for gastrointestinal parasites twice a year by direct sedimentation and flotation and were routinely dewormed once a year with pyrantel embonate (Combantrin®) at the recommended dosage of 4.5mg/kg orally (D.C. Plumb pers. comm. 1995).

**Breeding**

*Courting & mating*

The first courting behavior was observed in November, 2004. Courting was characterized by the male making gentle squeaking sounds while gently touching and biting the feathers of the female's face as if encouraging her to accept his offer of a piece of meat from his talons. The female would then grab and take the piece of meat from the male while gently touching and biting the feathers of his face and produce the same gentle squeaking sounds. The "sharing" of the meat between the two was repeated several times before they stopped when they were disturbed by the presence of people. Courtship normally proceeds to mating among owls (P. Hospodarsky pers. comm. 2005) but copulation was not observed after this ritual.

In June 2005, the pair was observed perching at the edge of the nest platform producing similar sounds as observed in November 2004, while gently biting the feathers of the face of the other. The two were apparently disturbed by the human presence and flew out of the nest. The following day, however, the pair was observed mating at approximately 9am, which proceeded to the first attempt of the female to breed. In September of the same year, similar sounds were also heard from within the nest platform for approximately 15 minutes. Four days later, the male attempted to mate with the female at midday. In both instances of courting, however, neither of the birds was grabbing a piece of meat. This observation suggests that the offering of food is a part but not an essential element in the courtship ritual of this species.

*Nesting & incubation*

Since nothing is known about the breeding biology of the Philippine Eagle-owl (Collar et al 1999), several nesting options were made available to the pair so they could choose in which area they want to breed: one on the ground and two platforms, 1m and 2m above the ground, which were located inside the "inner enclosure." The nests were filled with nest substrates such as dried leaves, sprigs and wood shavings. Dried banana leaves were hung from the roof in front of the nest platforms to provide a visual barrier from the zookeeper during feeding.

The first nesting attempt of the female was on July 21, 2005, three weeks after mating was observed. The female began frequenting the topmost of the three nest platforms setup in the enclosure. The area around the enclosure was cordoned and closed for public viewing to minimize disturbances to the breeding pair. The female eventually spent most of the time inside the nest platform and only went outside for a few minutes either to feed or just to perch. The male was also observed occasionally bringing mice and rats inside the nest to the female. While these activities were being monitored, nest inspection was not carried out, to minimize disturbance to the female. Three weeks later, the female abandoned the nest. During inspection, an egg was observed lying on a poorly arranged nest of dry leaves. The egg was taken out for artificial incubation, but candling revealed that it was infertile. Renovation of the nest was not conducted because of the pair's very aggressive behavior towards the keeper.

On August 29, 19 days after the female abandoned the nest, she once again began frequenting the nest platform. Six days later, she remained inside and only left occasionally as before. The male was also observed bringing in mice and rats as he did before. Two weeks later the nest was inspected and an egg was discovered. The female continued to incubate the egg but abandoned the nest four days later. The egg was taken out of the nest but it was again found to be infertile. This time, however, while the egg was being candled, the nest platform was renovated. Fitted plywood was placed over the welded wire mesh bottom of the platform and was filled with layers of dry sand, soil, wood shavings and sprigs. A small log was also placed at the edge of the nest to prevent the substrates from falling out of the nest, as well as to serve as a perch. The egg was returned to the nest to observe for any interesting behavior. The female was then observed visiting the nest occasionally but only for a few minutes. A week later, the nest was inspected and the egg was found to have been crushed.

The third attempt of the pair to breed was on September 28, four days after the discovery of the crushed egg, when the female once again started frequenting the nest. On one occasion the female and the male were heard producing similar squeaking sounds in the platform. Two days later, the male attempted to mate the female. On October 6, the female remained in the nest. The nest was inspected a week later and an egg was discovered, which the female was incubating. On November 21, while the female was bathing in the

**Fig. 3.** *Owlet at 19 days old.*

**Fig. 4.** *Owlet at 20 days of age.*

**Fig. 5.** *Owlet at 21 days of age.*

pool, an owlet was discovered in the nest. It was uncertain when the egg hatched since neither the female nor the chick made chirping noises but the owlet appeared to be not more than three days old. Tony Warburton, chairman of the World Owl Trust, predicted that the incubation period of this species is 35 days, making it four days old when discovered. This theory ties quite nicely with the keeper's observation when it was observed that the male brought a rat to the nest four and two days before the owlet was discovered.

*Owlet Rearing*

Despite the inexperience of the pair, both were very good parents and reared the owlet very well. Feeding usually involved the male bringing a mouse or a rat to the nest and offering it to the female. The female 'tenderizes' the mouse by biting on it several times before offering it to the owlet, which swallowed it whole. When the male offered a rat to the female, she tore the rat into small pieces before offering it to the owlet. On one occasion, the male brought a rat and tore it to smaller pieces while the female was watching. It was not observed, however, if the male gave the torn pieces of rat to the female and she to the owlet, or directly to the owlet, since the presence of the observer disturbed the pair. Both parents are very aggressive and protective of the nest and even attacked the keeper once during feeding. While rearing the chick, several interesting behaviors were observed. Once, when the owlet was three weeks old, the female was observed gently pulling the tail of a mouse from the mouth of the owlet then releasing it. The female did this three times, seemingly checking to see if the mouse was stuck in its throat, before the owlet swallowed it completely. On one occasion, when the owlet was five weeks old, the female appeared to be teaching the owlet how to tear small pieces of meat from a rat. The female was observed grasping a piece of twig and biting on it while the owlet looked on. The owlet then proceeded to awkwardly tear a piece of meat from the rat it was grasping, apparently imitating the female parent.

Our presumption that the male owl's role in rearing the chick was only to bring food to the nest was proven incorrect when he was once observed bringing a mouse to the nest and, instead of giving it to the female, he tenderized the rodent and offered it directly to the owlet. The amount of food consumed by the owls was increased when the egg hatched. The quantity of food offered was still dependent on how much the owls consumed during the previous days and the frequency depended on the size of rat or mouse consumed. Figure 13 summarizes the amount of food offered. The red dots indicate leftovers from the total amount of food offered on the day. The figure does not include the food amounts offered before and nine days after hatching of the chick. Between one and four weeks old, the owls consumed approximately 222g of rats and mice a day, which is 32% more than the average amount they consumed before the egg hatched. Between four and eight weeks, during the period when the owlet fledged out of the nest, they consumed

**Fig. 6.** *Owlet at 25 days old.*

**Fig. 7.** *Owlet at 32 days old.*

**Fig. 8.** *Owlet at 46 days old.*

approximately 241g of rats and mice or 38% more before hatching. And between eight and ten weeks, they consumed 211g of rats and mice or 29% more before hatching and 13% less before fledging.

*Owlet growth & development*

The growth and development of feathers of the owlet was carefully monitored and recorded. When the owlet was first observed, its size was similar to a tennis ball. Its eyes were still closed and its body covered with fine white hairs. Two days later its size appeared to have doubled. Needle feathers began growing on the wings of the owlet at 11 days old but its eyes were still closed. The following day, the owlet was observed to react to the beam of the flashlight but it was not clear whether the eyes were already open. At 13 days, needle feathers started to grow in the lumbar area and at 16 days in the head and neck and its eyes already opened. At 21 days, the feathers of the belly were beginning to grow while the wings were already covered with feathers. At 24 days, the primary wing feathers were already coming out of the pins. At 30 days, the owlet was already fully feathered and was about three quarters the height of the female, and at 45 days the owlet already had a full but lighter adult plumage.

*Fledging*

On December 31, at 45 days old, the owlet was discovered on the ground at the bottom nesting area previously setup just below the nest. It is not clear

**Fig. 9** *Nest site used by the breeding pair.*

**Fig. 10.** *Both parents taking part in feeding the chick.*

Fig. 11. *Female owl teaching the chick how to eat.*

whether the owlet attempted to fly out of the nest but fell or whether the owlet accidentally fell down from the nest. Two days before it was discovered on the ground, the owlet had been observed exercising its wings and perching at the edge on the nest platform. On January 13, 2006, at 58 days old, the owlet was discovered perching in the enclosure. Other species of eagle-owl generally fledge at 60 days (P. Hospadarsky pers. comm. 2005) which matches with the time the Philippine Eagle-owlet fledged.

The passing of knowledge and skills from the parents to the fledgling appear to extend even after fledging. Three weeks after fledging, the female was observed swooping to the ground after a piece of mouse was thrown into the enclosure. Instead of grabbing the prey, it grabbed a piece of twig, and looked up at the fledgling which was perching on a branch. The female seemed anxiously, waiting for the fledgling to swoop down and grab the prey. After a few seconds, the female took the mouse, tenderized it and offered it to the owlet. This behavior of apparently teaching the owlet how to hunt was observed a few more times. At 12 weeks old (and at the time this paper is written) both the male and the female were still feeding the owlet.

## Conclusions

Captive breeding programs can provide important information for conservation initiatives that are often very difficult to obtain from the wild. The breeding of the pair of Philippine Eagle-owl at the NFEFI-BCC proves that this species can be bred in captivity and has provided us with the first information on its breeding biology.

The breeding of this species in captivity is observed between June and November. This, however, does not mean that they only breed during these months. Further breeding of this species in captivity may eventually identify its breeding period. Courtship ritual is characterized by the "offering/sharing" of meat/prey between the male and the female while producing a gentle squeaking sounds and gentle biting and touching of the feathers of the face of the pair, which lasts for approximately 15 minutes. Food offering/sharing is a part but not an essential element of the ritual.

This species prefers nests high above the ground in captivity. It lays only one egg per clutch and incubation period is approximately 35 days. Development of the owlet is fast which becomes fully feathered at 45 days and fledges at around 58 days after hatching.

The rearing of the owlet mainly involves the male offering the female with prey. The female tenderizes small pieces of whole prey or tear small pieces of meat from larger prey and offers it to the owlet. Although the male's main role is to bring food to the nest, it is also involved, at a lesser extent than the female, in directly feeding the owlet.

Apart from feeding, the parents, especially the female, has a very important role of passing knowledge and skills to the owlet such as tearing of meat from large prey and hunting. The achievement of finally breeding this species in captivity has raised the

Fig. 12. *A comparison of both the juvenile and adult Phillipine Eagle-owl.*

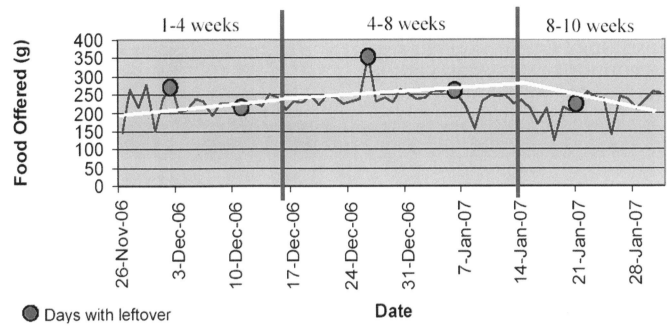

Fig. 13. *Amount of food consumed by the breeding pair of owls.*

idea of doing artificial incubation and rearing to increase its hatching and survival rate. However, the skills and knowledge that the parents pass onto the young is a very important point to consider in the captive breeding of this species. Natural breeding is the most practical method at this stage since more information is still needed to further understand the biology of these birds and especially if these are to be released back into the wild in the future.

### ACKNOWLEDGEMENTS

The authors would like to thank the following institutions that have supported and initiated the Philippine Owls Conservation Project, without this paper would not have been written: World Owl Trust, UK-Owl Taxon Advisory Group, Department of Environment and Natural Resources - Protected Areas and Wildlife Bureau, Fauna and Flora International-Philippines Biodiversity Conservation Program, Avilon Montalban Zoological Park, Negros Forests and Ecological Foundation, Inc., including the German Avicultural Society and the German Development Service (DED). Special thanks is likewise extended to William Oliver and Tony Warburton for making this whole project possible, Pavel Hospadarsky for sharing his experience and expertise that lead to this success and Dr. Maripi Diaz and Edmund Marcella for their hard work and dedication to the project.

### REFERENCES

COLLAR NJ, NAD Mallari and BR Tabaranza Jr. 1999. Threatened Birds of the Philippines. Bookmark, Inc. pp. 291-295

KENNEDY RS, PC Gonzales, EC Dickinson, HC Miranda and TH Fisher, Jr. 2000. A Guide to the Birds of the Philippines. Oxford University Press. pp. 173-182.

OBGINAR LL. 2005. Species Diversity of Sensitive and Indicator Avian Species Used as Biological Monitors in Mt. Pulag National Park. Unpubl. Baguio City, Philippines

PARRY-JONES J. 2001. Understanding Owls. STIGE SpA.

PLUMB DC. 1995. Veterinary Drug Handbook, 2nd Ediiton. Iowa State University Press. pp 597-599

WARBURTON T. 1997. The World Owl Trust Story: 25 Years of Owl Conservation. Print express (Cumbria) Limited, Whitehaven.

# FACING THE CHALLENGE OF SPIX'S MACAW *CYANOPPSITTA SPIXII* MANAGEMENT: OPTIMIZING THE HUSBANDRY, VETERINARY CARE AND REPRODUCTIVE MANAGEMENT OF SPIX'S MACAW AT THE AL WABRA WILDLIFE PRESERVATION

Ryan Watson[1], Richard Switzer[1] §, Simon Bruslund-Jensen[2] & Sven Hammer[1]

§ Speaker
[1] Al Wabra Wildlife Preservation, P.O. Box 44069, Doha, Qatar
[2] Vogelpark Walsrode Fonds e.V. Am Rieselbach, D-29664 Walsrode, Germany; *formerly Al Wabra Wildlife Preservation*

## SUMMARY

The Al Wabra Wildlife Preservation (AWWP) is currently home to 51 individuals of one of the world's most threatened parrots, the Spix's Macaw *Cyanopsitta spixii*. These birds represent 65% of the birds listed in the international studbook for the official captive breeding program, so the AWWP population is very significant and valuable as a conservation resource. Accordingly, AWWP considers many factors in optimizing the management of the birds. Many of the birds initially acquired by AWWP have suffered from medical conditions which compromised their health and reproductive status; numerous birds were potential carriers for psittacine diseases. Consequently, it has been essential to implement a strict health-screening, disease management and quarantine regimen, in an attempt to combat the potential risk of disease. In parallel with the veterinary issues, every attempt has been made to encourage the birds to breed successfully, by optimizing housing conditions, husbandry and genetic management. Reproductive potential is also maximized through artificial incubation and hand-rearing techniques, when necessary. Furthermore, attempting to undertake this entire program in a remote desert location, within a small, but rapidly-developing country, presents an additional set of challenges.

## INTRODUCTION

### Introduction to Spix's Macaw

Spix's Macaw was first discovered and collected in 1819 by two European explorers, Dr. Johan Baptist Ritter von Spix and Dr. Carl Friedrich Philip von Martius; first described six years later by Johan Wagler, who was Spix's assistant and named the bird in his honor. The bird's documented range has been limited to the dry, riverine forest of the Caatinga in northern Bahia, Brazil. Ornithological records suggest that Spix's Macaw was never considered to be abundant in the wild. Avicultural records suggest that the species has always been rare in captivity and highly-prized by collectors.

The demise of the Spix's Macaw has been well documented (Juniper 2002). The primary threat to the species' survival in wild has been poaching of nests for the illegal pet trade. This threat was compounded by a limited species distribution, habitat destruction and the alteration of floral diversity by grazing livestock, as well as invasive bees occupying nesting cavities. These constrictions finally resulted in one lone male surviving in the wild by 1987. A female (probably the original mate of the lone wild male) was released back into the wild in March 1995, but she died approximately two months later after colliding with power lines. The last sighting of the lone male, and consequently the species in the wild, was on 5th October 2000; Spix's Macaw is now considered critically endangered and likely to be extinct in the wild (BirdLife International 2007).

In captivity there are currently 79 birds (34.42.3)

### Table 1

**INSTITUTIONS HOLDING SPIX'S MACAWS REGISTERED IN THE OFFICIAL INTERNATIONAL STUDBOOK, JULY 2000**

| Location | Number of birds held |
|---|---|
| Al Wabra Wildlife Preservation, Qatar. | 21.29.1 |
| Association for the Conservation of Threatened Parrots, Germany. | 7.5.2 |
| Loro Parque, Tenerife. | 1.5.0 |
| Lymington Foundation, Brazil. | 3.2.0 |
| Sao Paulo Zoo, Brazil. | 2.1.0 |
| **Total** | **34.42.3** |

registered in the official studbook. The official international studbook is overseen by the Brazilian Institute for the Environment and Natural Renewable Resources (IBAMA). Five institutions are involved in cooperatively managing the registered birds - Lymington Foundation (Brazil), Sao Paulo Zoo (Brazil), Loro Parque Foundacion (LPF; Tenerife), Association for the Conservation of Threatened Parrots (ACTP; Germany) & AWWP. Although a significant number of undisclosed birds are also thought to exist in private, covert hands throughout the world, the total world population is considered to be less than 130 birds.

**Historical Perspective of Spix's Macaw at AWWP**

The Al Wabra Wildlife Preservation (AWWP) is the private collection of His Excellency Sheikh Saoud Bin Mohammed Bin Ali Al-Thani, located in the state of Qatar. The AWWP collection comprises 52 species of birds, 46 species of mammal and three species of herpefauna. There are two main focuses of the bird collection - Birds-of-Paradise and parrots, with Spix's Macaws being top priority (See Appendix 4).

The first Spix's Macaws arrived at AWWP in February 2000, when two pairs arrived from Birds International Incorporated (BII; Antonio de Dios), Philippines. Subsequent shipments arrived from Roland Messer's Swiss collection and a further 25 birds from BII. Since these Philippine birds represented a significant proportion of the world's population, it was necessary to transfer them to Qatar in four shipments, to ameliorate the potential impact of logistical difficulties, or even a plane crash.

In January 2004, AWWP took over the management of Roland Messer's collection of 13 birds. In March that year, AWWP took ownership of 11 of these birds; the ownership of one pair from these birds, with the only significant breeding potential, was retained by Messer. This pair produced two offspring in June 2004 - the first Spix's Macaws bred under AWWP management. Subsequently in August 2005, this pair plus one of their male offspring was transferred to the ACTP's collection in Germany. In September 2005, an attempt was made to break into the Swiss Spix's Macaw facility. Under the pressure of inadequate security, AWWP explored options for these 12 birds. In the absence of suitable alternatives, these birds were finally transferred to the AWWP's main facilities in Qatar in October 2005.

**Veterinary Management: Disease Monitoring & Quarantine**

When AWWP received its Philippine consignment of 25 Spix's Macaws in late-2003/early-2004, it was fully recognized that these birds could be potential carriers of disease. At BII, the Spix's Macaw flock had suffered from diseases, which had also caused an impact on the breeding success of their birds. Similarly, soon after AWWP acquired its Swiss birds, it became apparent that all 11 of the birds acquired by AWWP suffered from some kind of behavioral problem or physiological ailment (including Proventricular Dilatation Disease in one bird).

In the interest of bio-security and population management, AWWP continued to manage the Swiss population separately from the Qatar population. However, following the attempted theft of the birds in Sept 2005, and a failure to find suitable arrangements elsewhere, the only viable option was to bring the birds to Qatar. Upon their arrival in Qatar, all 12 of the Swiss birds were installed directly into a temporary quarantine facility between 6 to 12 months.

When AWWP took over the management of the Philippine and Swiss birds, it was already acknowledged that disease control and treatment were going to be present immense challenges.

Upon arrival at AWWP, each bird was initiated on a rigorous program of health screening. Proventricular Dilatation Disease (PDD) had already been identified as a problem in the newly-acquired population, and crop biopsies were essential for more thorough diagnosis. Other diseases of particular concern included Avian Herpes Virus (AHV), Avian Paramyxovirus (APMV), Avian Polyoma Virus (APV), Avian Chlamydia/Psittacosis, Psittacine Beak and Feather Disease and Avian Tuberculosis. Additionally, all birds were endoscoped to assess breeding condition/gonad maturation and abnormalities in internal organs. The health checks revealed that birds from both the Swiss and Philippine populations had been exposed to PDD, APV, APMV, AHV and Avian Chlamydia; there also appeared to be a high incidence of *Pseudomonas aeruginosa* in choanal or cloacal swabs, even in birds displaying no clinical symptoms.

On the basis of an assessment of the risk of virus transmission that each bird presented for others in the captive population, each individual was assigned the status of one of three colors - green, yellow or red (see Tables 3a & 3b.). Since the purpose of AWWP's Spix's Macaw program is captive propagation as a conservation tool, it was inherently necessary to start establishing birds as breeding pairs, with major consideration given to their risk of virus transmission. As a general rule, 'green' birds should be paired with other 'green' birds, and never 'red' birds; 'red' birds should be paired with other 'red' birds; 'yellow' birds' should be paired with other 'yellow' birds but for the sake of maximizing reproduction within the flock, could potentially be paired with either 'green' or 'red' birds.

Whilst the newly acquired birds were being kept in the quarantine facility and health checks undertaken, two purpose-built, 10-aviary breeding complexes were being constructed at AWWP to house the birds, whose population now numbered more than 40. One of the primary considerations in the design of these complexes was the strategy to maintain each individual aviary as a form of quarantine unit. (See 'Housing and husbandry' below.) Even when set up as breeding pairs, each pair can be kept in isolation from other birds. This even allows for 'red' birds to be kept in the same complex as 'green' ones, although naturally in separate aviaries and pairs.

The health-screening program has continued, with all birds receiving an examination every 12 months. Diagnostic analysis of blood (including serum chemistry, complete blood counts, serology and PCR for detection of viruses), cloacal and choanal swabs for bacteriology and x-ray examination being the minimum. Notably, it is possible for a bird's disease transmission risk assessment to change from year to year, with subsequent checks revealing or eliminating symptoms that were not recorded or suspected in previous examinations.

The health-screening program has also enabled us to assess the overall condition of an individual bird's health, with a particular focus on the health of its reproductive system. Each bird is assigned a score which reflects the bird's clinical or apparent health condition and its viability to be reproductively successful - score 1 for very good, score 5 for poor. (Tables 3a & 3b). This individual score provides another guide for the selection of pairs, enabling us to select pairs with the greatest breeding potential.

The health-screening and physical examinations have also provided some interesting discoveries. Endoscopy revealed that the AWWP Spix's Macaws appear to have unusual air-sac physiology, with certain air-sac walls appearing in random locations and occasionally missing all together in some individuals. This may be a natural phenomenon, or may even be the result of frequent endoscopies or previous infections. The air-sacs of the Philippine birds were very cloudy and many of them exhibited an anthracosis of the lungs, the incidence of which was higher in older birds - a possible consequence of the heavy air pollution around Manila. Anthracosis was also found in several organs and skeletal tissue of ribs and skull on post mortem examination. The Philippine population all had extremely dark irises, whilst all the Swiss birds have white irises. One can speculate that the dark iris represents a heavy metal accumulation, as certain heavy metals are known to deposit in the eyes; these heavy metals may also be found in high concentrations in the polluted air around Manila. There is now evidence that suggests that the degree of anthracosis in the birds' lungs may reduce with time as pollutants are purged from the bird's body in cleaner air; it is also a possibility that the color of the iris may do the same.

**Veterinary Management: the Impact of Proventricular Dilatation Disease**

Proventricular Dilatation Disease (PDD) has been implicated in five of the eight deaths of Spix's Macaws at AWWP. Three of these PDD-related deaths occurred in 2004, in birds we had accepted knowing that they were not healthy, and possibly suffering from PDD.

It has been noted at AWWP that Spix's Macaws affected with PDD do not develop the typical massively dilated proventriculus or pass undigested seeds in the droppings, as is typical for the disease (Hammer 2005). Instead, they suffer progressive damage to the central cervous system (CNS) and, despite the presumptive diagnosis from the results of the crop biopsy, PDD may often only be confirmed when necropsy samples from the CNS, adrenals, heart and digestive organs are sent for histopathological examination.

Clearly the diagnosis of PDD is a key consideration in the monitoring of the Spix's Macaw population. A crop biopsy is considered to have a sensitivity

## Table 2

### HISTORICAL OVERVIEW OF AWWP SPIX'S MACAW POPULATION

| Year | AWWP Recruitment | AWWP Immigration | AWWP Mortality | AWWP Emigration | Comments | AWWP End-of-year Total | Registered World Total |
|---|---|---|---|---|---|---|---|
| 2000 | 0.0 | 2.2 | 0.0 | 0.0 | Feb 2000: 2.2 arrive from BII, Philippines. | 2.2 | 53 |
| 2001 | 0.0 | 0.0 | 0.0 | 0.0 | | 2.2 | 57 |
| 2002 | 0.0 | 1.3 | 0.0 | 0.0 | Nov 2002: 1.3 arrive from Roland Messer, Switzerland. | 3.5 | 55 |
| 2003 | 0.0 | 3.3 | 2.0 | 0.0 | Aug 2003: Permits issued for BII's 25 remaining birds to be legally transferred from Philippines to Qatar. Nov 2003: First shipment of 3.3 birds arrives from BII, Philippines. | 4.8 | 53 |
| 2004 | 2.0 | 18.14 | 2.1 | 0.0 | Jan 2004: Second shipment of 4.3 birds arrives from BII, Philippines. Jan 2004: AWWP takes over the management of Roland Messer's remaining 6.7 birds, still in Switzerland. March 2004: Third shipment of 4.5 birds and fourth shipment of 3.0 birds arrive from BII, Philippines. March 2004: AWWP takes ownership of 11 of the 13 Swiss-based birds. | 22.21 | 54 |
| 2005 | 0.3 | 0.0 | 2.1 | 2.1 (managed only by AWWP) | Aug 2005: 2.1 birds (breeding pair & 1 offspring still owned by Messer) sent to Martin Guth (ACTP) in Germany. Sept 2005: Serious attempt to break into Swiss breeding facility. Oct 2005: 6.6 birds transferred from AWWP Swiss facility to Qatar. | 18.22 | 54 |
| 2006 | 2.5 | 0.0 | 0.0 | 0.0 | | 20.27 | 74 |
| 2007 | 1.2.1 | 0.0 | 0.0 | 0.0 | | 21.29.1 | 79 |
| **Total** | **5.10.1** | **24.22** | **6.2** | **2.1** | | | |

## Table 3a
### AWWP HEALTH STATUS SCORING SYSTEM

| Virus transmission risk assessment: based on the results of viral tests and biopsies. | | Individual health assessment: based on health check (x-ray, blood screens, clinical signs, etc) | |
|---|---|---|---|
| Green | No risk: all pairing possible | score 1 | Very good: If no abnormalities are detected in all tests & examinations. |
| Yellow | Less risk: virus transmission not ruled out; restricted pairing | score 2 | Good: If it is positive for one test but no clinical signs. |
| Red | High risk: virus carrier; restricted pairing, never with green color. | score 3 | Fair/average: If it is positive for two tests but no clinical signs. |
| | | score 4 | Below average: If it is positive for two tests and clinical signs. |
| | | score 5 | Poor/bad: Most of the tests positive and clinical signs. |

## Table 3b
### OVERVIEW OF THE HEALTH STATUS OF AWWP SPIX'S MACAW FLOCK, JULY 2007

| Number of birds in each category | | Virus transmission risk assessment | | | | | | | | |
|---|---|---|---|---|---|---|---|---|---|---|
| | | Green | | | Yellow | | | Red | | |
| | | Adult breeding | Adult non-breeding | Juvenile | Adult breeding | Adult non-breeding | Juvenile | Adult breeding | Adult non-breeding | Juvenile |
| Individual health assessment | (score 1) Very good | 1 | - | 7 | 3 | - | 1 | - | - | - |
| | (score 2) Good | 5 | 1 | 2 | 13 | - | - | 2 | - | - |
| | (score 3) Fair | 2 | - | - | 6 | 1 | - | 2 | - | - |
| | (score 4) Below average | 1 | - | - | 1 | - | - | - | 1 | - |
| | (score 5) Poor | - | - | - | - | - | - | - | 2 | - |

N = 36 adult breeding
N = 5 adult non-breeding
N = 10 juveniles

of between 60-70% for detecting the changes in nerve cells caused by PDD. A positive test for PDD is said to be definitive, but a negative (no obvious lesions) test is no guarantee that a bird is free of the disease even if they present otherwise healthy indications. Since 2004, 73 crop biopsies have been taken from 46 individual Spix's Macaws at AWWP, with several individuals having had as many as three biopsies during this period.

Sometimes a bird may present suspicious crop biopsy result for PDD, indicated by signs of lymphoplasmacytic inflammation surrounding the ganglia, but not within the ganglia. This bird is managed as though it is PDD positive in order to mitigate the risk to the rest of the breeding population, and is initially assigned the risk assessment color red. Notably, seven of the nine living birds to test suspicious for PDD, have done so in a single crop biopsy and have not exhibited obvious lesions in subsequent crop biopsies, thereby raising their status to the colour yellow.

One bird (marked with * in Table 4) tested positive for PDD by crop biopsy in 2004, but subsequently has had no obvious lesions (indicating a negative result) from biopsies in 2006 and 2007. Radiographs show that although the bird does have an enlarged proventriculus in comparison to other Spix's Macaws, the size does not appear to have changed over the subsequent three years. This bird has never shown any of the classical signs of the disease, nor any of the CNS symptoms now associated with the disease in the Spix's Macaw. This may suggest that this bird has overcome a PDD infection.

### Table 4
SURVIVABILITY OF SPIX MACAWS FOLLOWING THE RESULTS OF PDD CROP BIOPSIES, JULY 2007

| Crop biopsy results for PDD | Number of birds | | |
|---|---|---|---|
| | Total tested | Dead | Alive |
| Positive at least once | 3 | 2 | 1* |
| Suspicious at least once, but never positive | 12 | 3 | 9 |
| Negative always | 31 | 0 | 31 |

N = 46 birds
* Positive for PDD

## Veterinary Management: Disease Treatment & Disease Prevention.

Avian Polyoma Virus (APV) does not appear to be a significant cause of mortality in the Spix's Macaw population at AWWP. We have detected APV antibodies in 26 out of 50 Spix's Macaws tested, but no mortalities have been directly attributed to the disease (Deb 2007). However, it is known from other species that adult parrots with sub-clinical infections may transmit the virus to chicks and fledglings, especially hand-reared ones which are immunologically naïve when introduced to the breeding population, and therefore much more susceptible to acute disease and mortality as a result. Consequently, all hand-reared psittacines at AWWP are vaccinated against APV at five weeks of age followed by a booster three weeks later using a commercially available killed APV vaccine (PSITTIMUNE® APV, www.baymunecompany.com, USA). Parent-reared psittacine chicks are not vaccinated, since it is assumed that they would have already had sufficient exposure to the virus in small quantities, if carried by their parents. In this way parent-reared chicks will receive a natural immunization against APV. To date, however, no Spix's Macaw chicks have been parent-reared.

None of AWWP's psittacines, including the Spix's Macaws, are vaccinated against any Avian Paramyxovirus (APMV), since it has been suggested that there is a link between the APMV vaccine and the onset of PDD. It has even been suggested that APMV Strain I can be used as a preliminary indicator for the presence of PDD in a psittacine collection (Grund 1999). Until the link between the APMV-1 virus, the vaccine, and PDD are fully explored, we will not vaccinate against APMV. This is supported by the evidence that AWWP has experienced no mortality attributable to APMV, while PDD is a major threat (Table 5). Additionally, Qatar has so far not experienced any incidents of Highly Pathogenic Avian Influenza H5N1. Although Avian Influenza remains a serious threat to our birds, it is currently illegal to vaccinate birds against the disease in Qatar. Fortunately, we have the capability to permanently shut a large proportion of our bird collection, including all the Spix Macaws, into indoor aviaries (see 'Housing'). This allows us to isolate each of the bird buildings from the outside world, by the use of quarantine measures, such as footbaths, clothing changes, special trained staff and other sanitary measures, allowing sufficient protection from any external Avian Influenza threat.

Whilst we continue to take every measure to limit

### Table 5

**SUMMARY OF DISEASE EXPOSURE AND MORTALITY IN THE AWWP SPIX MACAW POPULATION**

| Disease | Number of birds | | |
|---|---|---|---|
| | Total tested | Positive result for antibodies | Mortality due to disease |
| Avian Polyoma Virus | 50 | 26 | 0 |
| Avian Herpes Virus | 57 | 10 | 0 |
| Avian Paramyxovirus Type 1 | 57 | 17 | 0 |
| Psittacosis / Avian Chlamydia | 48 | 20 | 0 |
| Avian Tuberculosis | 45 | 0 | 0 |

transmission of disease to and through our Spix's Macaw population, we still have to manage a flock of birds which contains many compromised birds. With the goal of improving the health of these birds, as well as ensuring the healthy birds remain in good condition, we undertake routine monthly fecal checks to evaluate any parasite infestation and to monitor the balance of bacteria in the gut. Weekly weighing allows us to monitor the weight of each bird. (Each Spix's Macaw is conditioned to perch on an electronic weighing scale.) The annual health checks enable us to diagnose and monitor any chronic conditions. Lastly, AWWP maintains a staff of four permanent veterinarians and one laboratory staff to manage ongoing veterinary cases and provide immediate treatment, if necessary.

## Housing

AWWP's two primary Spix's Macaw breeding facilities have been constructed, combining strategies that attempt to:

1. Optimize husbandry & well-being
2. Optimize breeding success
3. Minimize disease transmission between different pairs
4. Optimize aviary maintenance and civil engineering services

Each of the 20 uniform aviaries is intended for keeping no more than one pair. Due to the brutal heat and sunshine of the Qatari summer, each aviary has both an indoor and outdoor compartment - the indoor compartment is air-conditioned, and the roof of the outdoor compartment is covered in shade netting. The birds have free access between the indoor and outdoor compartments (except when bird management prohibits it) through two windows big enough for the birds to fly directly.

The indoor compartment of each aviary measures 4m x 4m x 2.8m, and is surrounded on two sides by an L-shaped service corridor. The indoor aviary is serviced through the wire front, where two externally-mounted feeding stations facilitate food and water provision, whilst minimizing the time spent inside the aviary itself for cleaning. Along the side of the aviary, a brick wall limits visual disturbance to the birds, whilst the service corridor accesses the outdoor section and nest boxes.

The nest boxes are all located within the indoor compartment, with external hatches enabling access for nest inspections without the need to enter the aviary itself. A transparent plastic box is designed to house surveillance cameras. The air-conditioning unit is strategically placed to avoid draughts blowing directly onto the birds. Lighting is controlled by automatic timers and adjusted according to the length of the day. Glass windows are also installed above mesh roof level of the indoor compartment so as to allow more daylight. The substrate inside the indoor compartment is a 5cm layer of fine dune sand, with solid concrete floor beneath.

The outdoor compartment measures 5.6m x 5.6m x 2.8m, and is surrounded on three sides by solid walls, with a wire mesh front. A 25cm distance separates two layers of mesh roofing, in between which lies an artificial rain sprinkler system. As well as sun-protection, the shade-netting also doubles as consider-

## Table 6

### ANNUAL MORTALITY ACCORDING TO AGE CLASS SINCE 2002

| Year | Annual mortality | | | Comments/Causes of death |
|------|------------------|---|---|--------------------------|
|      | Nestlings & fledglings | Juveniles & Adults | Total* | |
| 2000 | -<br>(n = 0 from 0) | 0%<br>(n = 0 from 4) | 0%<br>(n = 0 from 4) | |
| 2001 | -<br>(n = 0 from 0) | 0%<br>(n = 0 from 4) | 0%<br>(n = 0 from 4) | |
| 2002 | -<br>(n = 0 from 0) | 0%<br>(n = 0 from 8) | 0%<br>(n = 0 from 8) | |
| 2003 | -<br>(n = 0 from 0) | 14.3%<br>(n = 2 from 14) | 14.3%<br>(n = 2 from 14) | 1.0 - Kidney failure<br>1.0 - Stroke, kidney failure |
| 2004 | 0%<br>(n = 0 from 2) | 6.8%<br>(n = 3 from 44) | 6.5%<br>(n = 3 from 46) | 0.1 - PDD confirmed<br>1.0 - Lung bleeding, kidney failure<br>1.0 - Euthanized; (PDD confirmed after death) |
| 2005 | 0%<br>(n = 0 from 3) | 6.9%<br>(n = 3 from 43) | 6.5%<br>(n = 3 from 46) | 1.0 - PDD confirmed<br>0.1 - PDD confirmed<br>1.0 - Auto-mutiltion, lesions of peripheral nerves, kidney failure. |
| 2006 | 0%<br>(n = 0 from 7) | 0%<br>(n = 0 from 40) | 0%<br>(n = 0 from 47) | |
| 2007 | 0%<br>(n = 0 from 4) | 0%<br>(n = 0 from 47) | 0%<br>(n = 0 from 51) | |

\* Sample size includes all birds maintained at AWWP throughout the year, including hatches and mortality, immigration and emigration. Consequently, the total may be larger than the actual total on any particular moment of the year.

able protection against droppings from wild birds. The substrate of the outdoor aviaries is either solid concrete or 15-20cm of washed sand; both provide a clean, well-drained surface. The aviaries are furnished with a variety of perches, ropes and living plants, such as grasses, shrubs and small trees. An outdoor service corridor allows for observation of the birds, as well as occasional access to the outdoor aviary.

Not only do the solid concrete walls provide visual barriers between pairs, but crucially they also act as a physical barrier to the transmission of pathogens between adjacent aviaries. This enables each aviary to be considered an individual unit, in isolation from all others. Equally important, each aviary unit has its own individual access door, with no connecting access to neighbouring aviaries. Similarly, each aviary is equipped with its own footwear and tools for cleaning, to avoid the spread of pathogens between aviaries.

Pest control considerations include the use of 2" x ½" mesh which should prevent the passage of rats and adult mice. Similarly the use of solid concrete throughout the construction should eliminate cavities which might harbor mice. Finally, a narrow moat around the entire complex helps to prevent the passage of ants into the feeding stations.

**Husbandry & Diet**

Strict dietary provisioning is a very important consideration in maintaining healthy Spix's Macaws (Watson 2006). A high fat content in their diet, due to nuts and sunflower seeds, can easily lead to obesity. One bird arrived in Qatar from Switzerland, weighing 546g and was so obese he could not fly. In comparison, the average weights for healthy Spix's Macaws are 318g for males and 288g for females (AWWP unpublished data). This male has now decreased in weight to 346g, thanks to a strict maintenance diet, is now healthy and can fly perfectly.

Like many species, it appears that Spix's Macaws are not capable of nutritional wisdom. Consequently they are provided with a varied and balanced diet, but in a very limited quantity. This ensures that each bird consumes the majority of the contents of its food dish, ensuring a nutritionally balanced food intake.

All the birds (in the entire AWWP bird collection) are fed twice per day. Every day the Spix's Macaws receive a morning food dish at 6am containing mostly fruit salad, vegetables, pulses and sprouted seeds. Between 3-4pm, the birds receive an afternoon food dish, which contains a mixture of pellets, seeds and nuts. This food dish is left in the feeding station overnight, and is often presented in ways which encourage more active foraging behaviours and prolong the time spent feeding (such as hidden amongst wood shavings). Tables 7a & 7b present details of the current Spix's Macaw diet at AWWP.

The process of providing adequate supplies of fresh, clean water in the middle of the desert is not without its challenges. At AWWP, drinking water is brought in from a local town by tanker-truck; before being given as drinking water to the birds, the water is triple-filtered to eliminate bacterial contamination.

The desert heat can lead to rapid decomposition and bacterial contamination of fresh diet items, so specific precautions are taken to limit this threat. During the summer months, the morning food dish containing fruit and vegetables is removed between 8:30-9am; during the winter the food is removed in early afternoon. At this time, the spilled food is also picked up off the aviary floor, as necessary. Similarly, the process of sprouting seeds involves soaking them in a dilute (1:500) concentration of F10® (www.healthandhygiene.net, South Africa) disinfectant and triple filtered water, and leaving them in cool room at 16°C for 36 hours whilst they germinate. Followed by repeated washing before it is offered to the birds.

The sandy substrate of indoor aviaries is sieved clean every other day. Minimal food remains are discarded in the outdoor aviary, but these are also

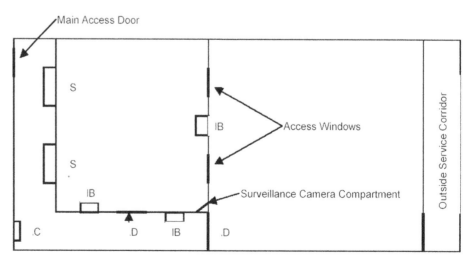

**Fig. 1.** *Plan of an isolated aviary unit in AWWP's Spix's Macaw breeding complex*

### Table 7a
### SPIX'S MACAW DIET AT AWWP

| Food Items | Spix's Macaw Maintenance Diet | | Spix's Macaw Breeding Diet | |
|---|---|---|---|---|
| | Morning: per pair | Afternoon: per pair | Morning: per pair | Afternoon: per pair |
| Parrot soft food mix | 2 table spoon (30g) | - | 3 table spoon (45g) | - |
| Fruit salad | 2 table spoon (30g) | - | 2 table spoon (30g) | - |
| Zeigler pellets maintenance®: | 4 pieces (5g) | 8 pieces (10g) | - | - |
| Zeigler pellets breeder®: | - | - | 4 pieces (5g) | 12 pieces (10g) |
| Milk thistle seed | 8 seeds (1g) | 8 seeds (1g) | 8 seeds (1g) | 8 seeds (1g) |
| Spix's Macaw dry food mix | - | 1 table spoon (15g) | - | 1 table spoon (15g) |
| Almonds: Mon/Wed/Fri/Sun | - | 2-pieces | 2-pieces | 2 pieces |
| Walnuts: Tues/Thur/Sat | - | 2-halves | - | 2 halves |
| Hard boiled egg | - | - | 1 table spoon (10g) | - |
| Crushed mineral block | - | (1g) | - | (1g) |

removed when necessary. The use of the feeding stations ensures that approximately 90% of the food discarded by the Spix's Macaws falls on the base of the feeding station or the floor of the service corridor beneath it. This greatly reduces the amount of time that birds are disturbed inside their aviaries by the cleaning process. In order to ensure high levels of food hygiene, the feeding station itself is scrubbed every other day

Since the majority of the Spix's Macaws arrived in Qatar in 2004, there have been significant improvements in the general health of the flock. Anecdotally, we can state that the feather condition of the birds has improved and there has been a significant increase in their activity level. At the time of writing we had suffered no Spix's Macaw mortality since 2005 and the incidence of illness has been negligible during this time. We attribute the improved health of the flock to both the rigorous program of veterinary monitoring and management, as well as high standards of housing and diet.

However, it is acknowledged that since the Spix Macaws started breeding in earnest in 2004, we have not experienced a significant improvement in either fertility or hatchability, nor in the fecundity or productivity of the flock. Although the reproductive potential of the flock is undoubtedly better than it was prior to their assemblage at AWWP, we are still presented with some huge challenges to maximize productivity.

**Optimizing Breeding Success**

When the majority of the birds were acquired into the AWWP flock in 2004, it was already recognized that the physical, behavioral or reproductive condition of many of the birds was so poor that the potential for reproduction was going to be limited. It was even anticipated that a significant number of birds might never breed.

Table 8 displays the summary of the reproductive output of the flock since the AWWP program began in 2003. Of the 124 eggs laid, only 37 have had confirmed fertility (30% fertility), of these fertile eggs, only 16 eggs have hatched (43% hatchability). The one piece of good news is that all 16 chicks that hatched have been raised successfully.

### Table 7b
**COMPOSITION OF SPIX'S MACAW DIETARY MIX**

| Parrot soft food mix | | Spix's Macaw dry food mix | | Fruit salad | |
|---|---|---|---|---|---|
| Mixed frozen vegetables | 5 parts | Harrison maintenance pellets (fine)® | 5 parts | Apple<br>Banana<br>Broccoli<br>Endives<br>Mango | Papaya<br>Red chili<br>Pear<br>Carrot<br>Orange |
| Boiled seed | 3 parts | NutriBird - P15 pellets 50:50 original tropical® | 2 parts | Celery heads<br>Red bell-peppers<br><br>*Mixture varies on a weekly cycle* | |
| Sprouted seed | 2 parts | Prestige - Parrot premium mix®: | 2 parts | | |
| 100% cranberry concentrate: | 2ml per cup (250ml) of mix | Prestige - Tropical finch mix®: | 1 part | | |

### Table 8
**REPRODUCTIVE STATISTICS OF AWWP FLOCK FROM 2003-2007**

| | 2003 | 2004 Qatar | 2004 Swiss | 2005 Qatar | 2005 Swiss |
|---|---|---|---|---|---|
| Number of egg-laying females | 1 | 2 | 3 | 3 | 2 |
| Number of eggs laid | 6 | 12 | 28 | 15 | 19 |
| Number off eggs broken pre-fertility | 0 | 1 | 2 | 0 | 1 |
| Percentage confirmed fertility | 0% (n = 0 from 6) | 25% (n = 3 from 12) | 46% (n = 13 from 28) | 20% (n = 3 from 15) | 5% (n = 1 from 19) |
| Fertile but no development | - | 0% (n = 0 from 3) | 8% (n = 1 from 13) | 0% (n = 0 from 3) | 0% (n = 0 from 1) |
| Percentage early embryo deaths | - | 0% (n = 0 from 3) | 23% (n = 3 from 13) | 0% (n = 0 from 3) | 0% (n = 0 from 1) |
| Percentage mid embryo deaths | - | 67% (n = 2 from 3) | 31% (n = 4 from 13) | 0% (n = 0 from 3) | 0% (n = 0 from 1) |
| Percentage late embryo deaths | - | 33% (n = 1 from 3) | 23% (n = 3 from 13) | 0% (n = 0 from 3) | 100% (n = 1 from 1) |
| % Fertile egg broken | - | 0% (n = 0 from 3) | 0% (n = 0 from 13) | 0% (n = 0 from 3) | 0% (n = 0 from 1) |
| Percentage hatchability | - | 0% (n = 0 from 3) | 15% (n = 2 from 13) | 100% (n = 3 from 3) | 0% (n = 0 from 1) |
| Number of chicks surviving | - | - | 2 | 3 | - |
| Percentage chick survivability | - | - | 100% (n = from 2) | 100% (n = 3 from 3) | - |

The problems of fertility and hatchability are further compounded by a lack of egg-laying potential of the flock. Of the 30 females that have been kept at Al Wabra, only ten females have ever laid eggs. Of these ten females who have laid eggs, only six have produced fertile eggs, and only four females have produced offspring. Similarly, of the 28 males that have been kept at Al Wabra, only six males have fertilized eggs, and only four males have produced offspring (see Tables 9a & 9b).

The current situation is still rather frustrating. During the 2007 breeding season, whilst there were a total of 27 females in the flock, only five females laid eggs and only two females produced fertile eggs. Similarly, during the 2007 breeding season, whilst there were a total of 20 males in the flock, only two males fertilized eggs. At this time, there were eight females and three males who were too young to breed, but there were a total of 14 other females and 10 males of breeding age, who made no contribution to the reproductive effort (see Table 9a & 9b).

Appendix Table A1 shows the egg production and fertility of the individual females; Appendix Table A2 shows the hatchability of eggs from the individual females. Clearly there are some major challenges faced in increasing the reproductive effort of the flock. Listed below are some explanations for the problems so far encountered:

* The potential fecundity, fertility and productivity of the AWWP flock were reduced by the 2005 transfer of the Swiss female #4329 to ACTP in Germany. Unfortunately, she has not yet produced offspring in Germany, but she has recently been paired with a male transferred from Loro Parque Foundation and there are optimistic signs that she will breed again in the future.

### Table 8 (cont'd)
**REPRODUCTIVE STATISTICS OF AWWP FLOCK FROM 2003-2007**

|  | 2006 | 2007 | Total |
|---|---|---|---|
| Number of egg-laying females | 2 | 5 | 10 |
| Number of eggs laid | 12 | 32 | 124 |
| Number of eggs broken pre-fertility | 0 | 0 | 4 |
| Percentage confirmed fertility | 75% (n = 10 from 12) | 22% (n = 7 from 32) | 30% (n = 37 from 124) |
| Fertile, but no development | 0% (n = 0 from 10) | 0% (n = 0 from 7) | 3% (n = 1 from 37) |
| Percentage early embryo deaths | 0% (n = 0 from 10) | 14% (n = 1 from 7) | 11% (n = 4 from 37) |
| Percentage mid embryo deaths | 0% (n = 0 from 10) | 0% (n = 0 from 7) | 16% (n = 6 from 37) |
| Percentage late embryo deaths | 20% (n = 2 from 10) | 29% (n = 2 from 7) | 24% (n = 9 from 37) |
| % Fertile egg broken | 10% (n = 1 from 10) | 0% (n = 0 from 7) | 3% (n = 1 from 37) |
| Percentage hatchability | 70% (n = 7 from 10) | 57% (n = 4 from 7) | 43% (n = 16 from 37) |
| Number of chicks surviving | 7 | 4 | 16 |
| Percentage chick survivability | 100% (n = 7 from 7) | 100% (n = 4 from 4) | 100% (n = 16 from 16) |

### Table 9a

**CONTRIBUTION BY FEMALES TO REPRODUCTIVE EFFORT 2003-2007**

|  | Breeding Season | | | | | Total |
|---|---|---|---|---|---|---|
|  | 2003 | 2004 | 2005 | 2006 | 2007 | 2003-2007 |
| Number of females in AWWP flock | 5 | 21 | 20 | 22 | 27 | 30 |
| Number of females set up in a breeding pair | 2 | 8 | 15 | 14 | 18 | 22 |
| Number of females laying eggs | 1 | 5 | 5 | 2 | 5 | 10 |
| Number of females laying fertile eggs | 0 | 3 | 3 | 2 | 2 | 6 |
| Number of females successfully producing offspring | 0 | 1 | 2 | 2 | 1 | 4 |

- Female #3254 who was an egg-laying female in 2003, was later diagnosed as being suspicious for PDD, and has consequently been kept in isolation. Two subsequent crop biopsies in 2006 and 2007 revealed no obvious lesions. Other health parameters indicate that she is in very good health, so in March 2007 she was cleared to return to breeding status and moved to the breeding center to be paired with a new mate.

- Female #4332 laid 18 eggs in 2004 & 2005, all of which were infertile. Although this was a behaviorally compatible pairing, the male has since been evaluated by endoscopy and testis biopsy as having very low chances of fertility. The female has since been re-paired with another male, but is yet to produce eggs.

- Female #4365 and #4366 both laid their first eggs in 2007, but neither female produced fertile eggs. #4365 was endoscoped prior to the 2007 breeding season, and she was assessed at having poor reproductive potential - the outcome of eggs being laid is progress in itself.

- Female #4369 produced 7 fertile eggs and 5 offspring in 2006. In 2007 all 7 of her eggs were infertile, despite being observed mating frequently and over prolonged copulations. The reasons for the lack of fertility in 2007 are currently unknown.

- Female #3253 has produced 10 eggs, of which 6 were fertile. However, the hatchability of these eggs has sadly been 0%. In 2004, her 3 fertile eggs died as embryos; similarly her 3 fertile eggs died as embryos in 2007, and 2 of these were found to be deformed. Recent micro-satellite DNA analysis to evaluate the gene pool of the IBAMA-managed breeding program has now shown that she shares a very high mean kinship with her mate (Presti et al, in prep.) This exemplifies the difficulty in managing a population with undocumented sources for founders. However, this result is no great surprise, since it is known that the final remnant population of Spix's Macaws in the wild was very small, and it is probable that many of the founders of the breeding program were originally poached as chicks from the very last wild nest cavity along the Melancia Creek. This bird and her former mate have been re-paired in time for the 2008 season.

- Furthermore, we are discovering a trend of high incidences of immature gonads in both males and female at ages when they should be considered sexually mature - older than 6 years. For example #3252 was 9 years-old during the 2006 season, and this was her third season as a breeding arrangement. In this time she failed to lay any eggs, despite

### Table 9b
**CONTRIBUTION BY MALES TO REPRODUCTIVE EFFORT 2003-2007**

| | Breeding Season | | | | | Total 2003-2007 |
| --- | --- | --- | --- | --- | --- | --- |
| | 2003 | 2004 | 2005 | 2006 | 2007 | |
| Number of males in AWWP flock | 3 | 20 | 18 | 18 | 20 | 28 |
| Number of males set up in a breeding pair | 2 | 8 | 15 | 14 | 18 | 22 |
| Number of males fertilizing eggs | 1 | 5 | 5 | 2 | 5 | 6 |
| Number of males successfully producing offspring | 0 | 1 | 2 | 2 | 1 | 4 |

showing all other typical breeding behaviors such as allo-feeding, allo-preening, frequent copulation and regular activity in the nest cavity. When she was endoscoped in January 2007, her primary ovary was still found to be in a stage of early maturation. Conversely, a Qatar-bred female #5158 was found to have an ovary at an advanced stage of maturation despite being only 20-months old. We hypothesize that birds in large communal aviaries as youngsters rapidly reach a level of maturity, due to increased stimulation of the pituitary gland. Meanwhile, those youngsters maintained in small aviaries for long periods with insufficient exercise and mental stimulation, lack physical fitness, have reduced endocrine function and are slow in sexual maturation. Alternatively, the late maturation might be attributable to an inbred population, a limited gene pool or prior health issues.

Naturally every attempt is being made to increase the reproductive potential of the AWWP flock. Four criteria are taken into account when selecting potentially compatible new pairs:

1) Reproductive ranking, based on a scoring system for gonads when viewed via endoscopy during a health check, combined with the individual's breeding history, if it has one.
2) Genetic compatibility - the new study of microsatellite analysis will provide a valuable, accurate perspective on genetically-compatible pairings. Previously, genetic compatibility had been based on historical evidence and assumptions.
3) Health status, according to the red-yellow-green system, to avoid combinations that present a high risk of disease transfer.
4) Age of birds - pairing birds of a similar age can maximize the breeding life span of the pair.

During the selection process, females are graded according to their reproductive ranking, and then mates are selected according to criteria 2-4. Of course, the final evaluation of a pair's compatibility is based upon the temperament and behavior of the pair - this decision is obviously made by the birds, but we must be flexible to re-evaluate behaviorally-incompatible pairs. Conversely, disease management can impede pair selection on the basis of all the other criteria, particularly behavioral. Consequently, many of the current pairings are forced pairings because we have not been able to flock birds for self selection due to our strategy to control PDD. Unfortunately, Spix's Macaws also appear to be very amenable to their cage-mate, whether they are a bonded pair or not. This makes the assessment of breeding compatibility more difficult. With young, presumably PDD-negative birds soon to be recruited into the breeding population, it is hoped that in the near-future we can use the strategy of off-season flocking of small groups, to enable the pairs to self-select. We have already been keeping juvenile Spix's Macaws in large flocking aviaries (15-20m long) with natural substrates, perching and vegetation, with an artificial rain system. We hope that these conditions will also provide sufficient stimuli to encourage maturation of the gonads, and will ensure that birds will be recruited into the breeding population at a younger age in the near future.

Optimizing reproductive output is a balance between short-term breeding success and long-term fecundity. We encourage females to lay a maximum of two clutches of eggs during the spring breeding season; the typical clutch size is four eggs. At AWWP we often experience a secondary breeding season in the autumn - if pairs choose to lay another clutch again then we do not prevent them. Whenever possible, if the pair have proven themselves as reliable parental incubators, we leave the eggs under the pair until 1-2 days before each is due to externally pip. This ensures that the incubation process is as natural as possible, and hopefully maximizes hatchability. At this point the eggs are placed in an incubator/hatcher for the hatching process. After hatching, the chicks are hand-reared. Crucially, hand-rearing significantly helps to minimize the risks of disease transmission from parent to offspring, thereby increasing the chances of PDD-negative youngsters into the captive flock, and thereby providing more options for mate-selection in the future. In order to monitor incubation and other breeding behaviors, all 20 of the indoor aviaries in the Spix's Macaw breeding complex are designed with the option of installing surveillance cameras. Similarly, in the future we aim to install nest box cameras which will provide a more detailed insight into the parents' behavior at the nest, and hopefully enable us to respond even more rapidly in emergency interventions.

To date, the hand-rearing effort has been completely successful, with all 16 chicks hatched being raised to independence. Although the principles and process of hand-rearing are very similar to that of most other psittacines, it is noticeable that Spix's chicks have a particularly mild-mannered begging response. Additionally, in comparison to other larger macaws, the chicks are susceptible to gaining too much weight, and a lower fat formula is more suitable. If Spix's Macaws are overfed, they counter-balance this by regurgitating food. We utilize observations of post-feeding regurgitation as an indicator for when to reduce the number of daily feedings.

For the first time, in 2007 one fertile egg was left to hatch under the parents. This embryo was the sibling of three others already hatched that season, with three others from previous seasons (dam is female #4269 listed in Appendix Table A1 & A2). Although the parents were not considered to be totally disease-free - both had a disease transmission risk status as 'yellow' - it was considered a calculated risk to leave this one fertile egg from their second clutch to be hatched and raised parentally. This strategy was adopted due the female's reproductive history. In three consecutive seasons, she had produced a total of 18 eggs, and we felt that such intelligent birds would benefit from the stimulus of raising an offspring, and perhaps ensure that they retained an enthusiasm for breeding in subsequent seasons. Additionally, having already produced three offspring in three years, any additional offspring would not be considered particularly genetically valuable. The nest box was checked daily shortly before, during and after hatch to monitor the progress of the egg/chick, since we were fully prepared to rescue the chick if its progress was less than ideal. Unfortunately, despite the initial encouraging signs of full crops and good weight gains, on day 9 the chick was found to have injuries on its toes, including the complete removal of one toe-nail, caused by parental mutilation. At this point, the chick was removed to the nursery for hand-rearing, where it has since recovered from its injuries and is making good progress.

In the incubation and hand-rearing rooms we take a number of extra precautions to ensure the safety and welfare of the Spix's Macaw eggs and chicks:

- As a component of the health-screening program, swabs are taken for bacteriological/fungal cultures from the cloaca and buccal cavity of each chick. These swabs are taken at age three days, seven days, 14 days, and every seven days until two weeks after the chicks have weaned.
- As another disease-prevention measure, no staff is allowed in the Spix's Macaw hand-rearing room if they have been in contact (or in a building) with any other psittacines, unless the member of staff has showered and completely changed their clothes.
- Located in the middle of the Qatari desert, it would be fair to say that Al Wabra does not experience the most reliable electrical supply. Consequently, AWWP has back-up generators to provide power in the occurrence of a power cut. Even so, as soon as there is a power cut to the incubation and hand-rearing rooms, senior members of bird staff receive an automated cell phone message announcing the concern and respond immediately.

Similarly, power surges are not uncommon occurrences, so each incubator and brooder is connected to a power surge protector to prevent damage to machines, both acute and chronic.

## CONCLUSION

Despite the challenges faced by AWWP in breeding Spix's Macaws, significant progress is being made in the propagation and management of the flock. A rigorous program of disease control, monitoring and treatment is limiting the incidence and transmission of clinical disease through the population, with no mortalities since 2005. Although there are still significant hurdles in achieving high levels of fertility and hatchability in the population, strategies to ensure high standards of husbandry, housing and artificial propagation have led to the production of 16 healthy chicks over 4 breeding seasons. Figure 2 displays the progress being made in increasing the population of Spix's Macaws, both at AWWP and throughout official collaborative institutions.

With recently-acquired genetic knowledge, we can look forward to improving our genetic management of the population, both at AWWP and with our collaborators throughout the studbook. Furthermore, as the 16 healthy youngsters are recruited into the breeding population, we hope to enhance our reproductive success. Although the current progress is slow, AWWP is optimizing the husbandry, veterinary care and reproductive management of its flock. We hope to provide a solid foundation for the future of the Spix's Macaw.

## ACKNOWLEDGEMENTS

Special thanks to His Excellency Sheikh Saoud Bin Mohammad Bin Ali Al-Thani, the founder and owner of Al Wabra Wildlife Preservation, for making our program of conservation and science for the Spix's Macaw possible. Thanks also to our collaborators for their advice, expertise and cooperation:

IBAMA, particularly Yara Barros & Onildo Marini Filho, consultant veterinarians Dr. Marcellus Burkle, Dr. Lorenzo Crosta, Prof. Dr. Helga Gerlach and Dr. Susan Clubb. Prof. Dr. Christina Miyake (University of Sao Paulo) for her genetic analysis of the captive Spix's Macaw population. Matthias Reinschmidt & David Waugh of the Loro Parque Foundacion; Martin Guth of ACTP; the staff of the Sao Paulo Zoo.

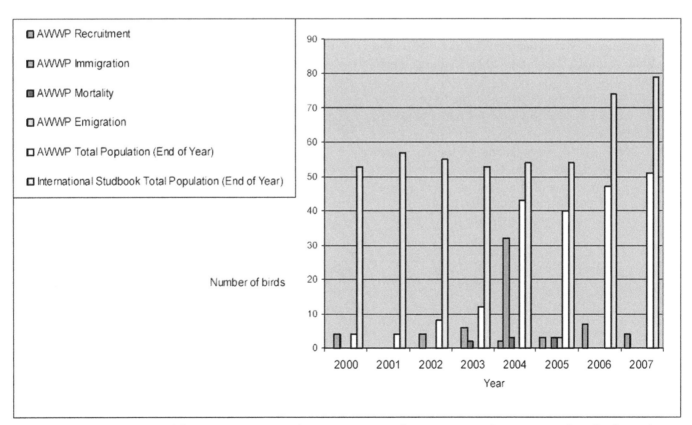

**Fig. 2.** *Historical Overview of the Spix's Macaw Population at AWWP, with comparison to the international studbook population.*

Monalyssa Camandaroba, Patrick Hodgens and Heidi Groffen for their contributions to hand-rearing and scientific research. Thanks also to the staff of AWWP who have contributed in numerous ways to the Spix's Macaw program, notably the AWWP Veterinary Department, particularly Dr. Amrita Deb, Dr. Julia Schulz, Dr. Abdi Arif, Dr. Raffy Borjal and Abid Sharif. The AWWP Bird Department, particularly Julius Celis, Romeo Castillote & Dean Tugade.

## References

BIRDLIFE INTERNATIONAL. Spix's Macaw species fact sheet accessed 22/06/07; www.birdlife.org.

DEB A, U Foldenauer, R Borjal, R Johne, S Hammer. 2007. Detection of persistent antibody titres to Avian Polyomavirus in a captive population of Spix's Macaws (Cyanopsitta spixii), Proceedings of the 43rd International Symposium on Diseases of Zoo and Wild Animals, 16-20th May, Edinburgh.

GRUND C, F Grimm and J Klosters. 1999. Serological studies on persistent PMV-1 infection associated with PDD. Proceedings of the Association of Avian Veterinarians. Pp 19-23.

HAMMER S, H Gerlach, M Buerkle and J Schulz. 2005. Proventricular Dilatation Disease (PDD) in Spix's macaws (*Cyanopsitta spixii*). Forty-second International Symposium on Diseases of Zoo and Wild Animals, May 08, 2005, Prague, Czech Republic.

JUNIPER T. 2002. Spix's Macaw- The Race to Save the World's Rarest Bird. Fourth Estate Publishers, London, UK.

WATSON R. 2006. Introduction to the Al Wabra Wildlife Preservation and its Spix's Macaw breeding program, presented at "Parrots 2006" hosted by the Parrot Society of Australia, Brisbane.

Explanation of proprietary nutritional products mentioned in text:

HARRISON ADULT LIFETIME FORMULAS®
Harrison's Bird Foods
Unit 7 Windmill Road
Loughborough
Leics
LE11 1RA
Telephone 01509 265557
info@hbf-uk.co.uk

NUTRIBIRD®
Versele-Laga nv
Kapellestraat 70
B - 9800 Deinze
Belgium
Tel: +32 (0)9 381 32 00
Fax: +32 (0)9 386 85 13
http://www.versele-laga.com

ZEIGLER®
Parrot Maintenance and Parrot Breeder - For Macaws and Large Parrots
Zeigler Bros., Inc.
P.O. Box 95, Gardners, PA 17324 USA
Tel :800-841-6800
Fax: 717-677-6826
www.zeiglerfeed.com sales@zeiglerfeed.com

## Appendix Table 1

**TOTAL NUMBER OF EGGS LAID AND PERCENTAGE CONFIRMED FERTILITY OF INDIVIDUAL BREEDING FEMALES**

| Breeding female | 2003 Eggs laid | 2003 % Fertility | 2004 Eggs laid | 2004 % Fertility | 2005 Eggs laid | 2005 % Fertility | 2006 Eggs laid | 2006 % Fertility | 2007 Eggs laid | 2007 % Fertility | Total Eggs Laid | Mean % Fertility |
|---|---|---|---|---|---|---|---|---|---|---|---|---|
| 3253 | - | - | 5 | 60% (n = 3) | 1 | 0% (n = 0) | - | - | 4 | 75% (n = 3) | 10 | 60% (n = 6) |
| 3254 | 6 | 0% | - | - | - | - | - | - | - | - | 6 | 0% (n = 0) |
| 4180 | | | 7 | 0% (n = 0) | 7 | 29% (n = 2) | - | - | - | - | 14 | 14% (n = 2) |
| 4269 | | | - | - | 7 | 14% (n = 1) | 4 | 75% (n = 3) | 7 | 57% (n = 4) | 18 | 44% (n = 8) |
| 4329* | | | *15* | *80% (n = 12)* | *10* | *10% (n = 1)* | - | - | - | - | 25 | 52% (n = 13) |
| 4332 | | | *9* | *0% (n = 0)* | *9* | *0% (n = 0)* | - | - | - | - | 18 | 0% (n = 0) |
| 4336 | | | *4* | *25% (n = 1)* | - | - | - | - | - | - | 4 | 25% (n = 1) |
| 4365 | | | - | - | - | - | - | - | 6 | 0% (n = 0) | 6 | 0% (n = 0) |
| 4366 | | | - | - | - | - | - | - | 8 | 0% (n = 0) | 8 | 0% (n = 0) |
| 4369 | | | - | - | - | - | 8 | 87.5% (n = 7) | 7 | 0% (n = 0) | 15 | 47% (n = 7) |
| Total | 6 | | 40 | | 34 | | 12 | | 32 | | 124 | |
| Mean | | 0% (n = 0) | | 40% (n = 16) | | 12% (n = 4) | | 83% (n = 10) | | 22% (n = 7) | | 30% (n = 37) |

*Bolded text indicates birds maintained in Switzerland; all other birds maintained at Qatar*

### Appendix Table 2

**PERCENTAGE HATCHABILITY, NUMBER OF CHICKS PRODUCED AND MEAN HATCH WEIGHT OF CHICKS ACCORDING TO EACH INDIVIDUAL BREEDING FEMALE FROM 2004-2007**

| Breeding female | 2004 | | 2005 | | 2006 | | 2007 | | Total Chicks | Mean % Hatch. | Mean chick hatch weight |
|---|---|---|---|---|---|---|---|---|---|---|---|
| | % Hatch. | Mean chick hatch weight | % Hatch. | Mean chick hatch weight | % Hatch. | Mean chick-hatch weight | % Hatch. | Mean chick hatch weight | | | |
| 3253 | 0% (n = 0) (N = 3) | - | | | | | 0% (n = 0) (N = 3) | | | | |
| 4180 | - | - | 100% (n = 2) (N = 2) | 12.87g | | - | - | -- | 2 | 100% | 12.87g |
| 4269 | - | - | 100% (n = 1) (N = 1) | 11.90g | 67% (n = 2) (N = 3) | 11.45g | 100% (n = 4) (N = 4) | 12.98g | 7 | 88% | 12.39g |
| 4329 | *17% (n = 2) (N = 12)* | *16.27g* | *0% (n = 0) (N = 1)* | - | | - | | - | 2 | 15% | 16.27g |
| 4332 | - | - | - | - | - | - | - | - | - | - | - |
| 4336 | - | - | - | - | - | - | - | - | - | - | - |
| 4365 | - | - | - | - | - | - | - | - | - | - | - |
| 4366 | 0% (n = 0) (N = 1) | - | - | - | - | - | - | - | - | - | - |
| 4369 | - | - | - | - | 71% (n = 5) (N = 7) | 11.67g | - | - | 5 | 71% | 11.67g |
| Mean | 12.5% (n = 2) (N = 16) | 16.27g | 75% (n = 3) (N = 4) | 12.54g | 70% (n = 7) (N = 10) | 11.61g | 57% (n = 4) (N = 7) | 12.98g | 16 | 43% (N = 37) | 12.71g |

% Hatchability: n = number off eggs that hatch; N = number of fertile eggs.
Note: Bolded text identifies birds maintained in Switzerland; all other birds were maintained in Qatar.

### Appendix Table 3
### EGG BIOMETRICS OF INDIVIDUAL BREEDING FEMALES FROM 2004, 2006 & 2007

| Breeding female | 2004 | | 2006 | | 2007 | |
|---|---|---|---|---|---|---|
| | Mean length (mm) | Mean diameter (mm) | Mean length (mm) | Mean diameter (mm) | Mean length (mm) | Mean diameter (mm) |
| 3253 | n.d.a. | n.d.a. | - | - | 41.42 (N = 4) | 30.3 (N = 4) |
| 3254 | - | - | - | - | - | - |
| 4180 | n.d.a. | n.d.a. | - | - | - | - |
| 4269 | - | - | 37.8 (N = 2) | 29.2 (N = 2) | 37.8 (N = 7) | 30.2 (N = 7) |
| 4329 | *40.2 (N = 9)* | *31.3 (N = 9)* | - | - | - | - |
| 4332 | *41.4 (N = 9)* | *31.0 (N = 9)* | - | - | - | - |
| 4336 | *39.8 (N = 4)* | *31.4 (N = 4)* | - | - | - | - |
| 4365 | - | - | - | - | 36.6 (N = 6) | 30.3 (N = 6) |
| 4366 | - | - | - | - | 39.8 (N = 8) | 30.0 (N = 8) |
| 4369 | - | - | 39.8 (N = 4) | 28.9 (N = 4) | 39.1 (N = 7) | 29.6 (N = 7) |
| Mean | 40.7 (N = 22) | 31.2 (N = 22) | 39.1 (N = 6) | 29.03 (N = 6) | 38.8 (N = 32) | 30.1 (N = 32) |

*Note: Sample size of eggs (N) reflects number of eggs measured, not necessarily the number of eggs laid by each female.*
*n.a. = no data available, although eggs were laid.*
*Note: Bolded text identifies birds maintained in Switzerland; all other birds were maintained in Qatar.*

## Appendix Table 4
### A SUMMARY OF THE AWWP AVIAN COLLECTION AS OF JULY, 2007

| Family | AWWP Order / Group | Number of Species | Number of individuals | Notable species |
|---|---|---|---|---|
| Psittaciformes | Blue Macaws | 3 | 70 | Spix's Macaw, Lear's Macaw |
| | Other Macaws | 3 | 22 | Blue-throated Macaw, Blue-headed Macaw |
| | Other Parrots | 3 | 15 | Pesquet's Parrot; Golden Conure |
| | Cockatoos | 6 | 16 | 3 spp of Black Cockatoo |
| Passeriformes | Birds of Paradise | 6 (7 taxa) | 80 | 2 ssp of Greater BOP, 12-Wire BOP |
| | Bowerbirds | 1 | 6 | Flame Bowerbird |
| | Cotingas | 1 | 2 | Long-Wattled Umbrella Bird |
| Struthionidae | Ratites | 1 (2 taxa) | 7 | Sudanese Red-neck Ostrich |
| Pelecaniformes | Pelicans | 1 | 10 | |
| Ciconiformes | Storks, Ibis & Spoonbills | 7 | 97 | Breeding group of Marabou |
| Phoenicopteriformes | Flamingos | 2 | 36 | |
| Anseriformes | Waterfowl | 2 | 5 | |
| Falconiformes | Raptors | 1 | 1 | |
| Galliformes | Pheasants & Guineafowl | 4 | 200+ | Bulwer's Pheasants, Somali Tufted Guineafowl |
| Gruiformes | Cranes & Bustards | 3 | 3 | |
| Pteroclidae | Sandgrouse | 3 | 100+ | Chestnut-bellied Sandgrouse |
| Columbidae | Pigeons | 2 | 9 | Pheasant Pigeon |
| Cuculiformes | Turacos | 2 | 9 | Violet Turaco |
| Coraciformes | Kingfishers | 1 | 1 | |
| **Total** | | **52 (54 taxa)** | **707** | |

# Conservation & Re-introduction

Re-introduction, Avian Conservation & Small Population Management

# ROLE OF CAPTIVE BREEDING CENTRES IN REDUCING IMPACTS OF BIRD TRADE IN SUB-SAHARAN AFRICA

ABDOU KAREKOONA[1], JASON BECK[2]
& RAYMOND KATEBAKA[3]

[1] Nature Uganda, The East African Natural History Society
83 Tufnell Drive, Kampala, Uganda
[2] Blue Lagoon Sanctuary, Central City
Nebraska 68826, USA
[3] Institute of Environment and Natural Resources Makerere University, Box 7062, Kampala, Uganda

## SUMMARY

There is growing concern in most Sub-Saharan African countries that the traditional "protectionist" approaches to conservation is rather expensive and fails to deliver the desired outcomes. "Supply side" policies and captive breeding facilities as measures to stop the illegal bird trade and thus conserve endangered species continue to draw support. Supply side policies, however, are based on a weak legal system and on a naïve representation of the institutional framework within which the wildlife trade takes place, and neglect the potential strategic responses of economic agents. In this paper, we review the status of bird trade in four Sub-Saharan Africa countries and the role of captive breeding centres in counterbalancing the effects of the illegal trade with particular interest to cranes.

## INTRODUCTION

For centuries African cultures co-existed with biodiversity but these ages are now long gone. The fast growing human population requires increasingly more resources, and reconciling the fulfillment of basic human needs with the conservation of biodiversity is now among the biggest challenges for Africa. In Sub-Saharan Africa, bird trade refers to the movement of birds between captive facilities, or removal from the wild for any reason, usually with some form of financial or barter transaction and includes live or dead birds or their parts. Cranes and several water birds mainly Shoebills *Balaniceps rex* Saddle-billed Storks *Ephippiorhynchus senegalensis* and Greater Flamingos *Phoenicopterus roseus* continue to be traded illegally in the four countries that our study has sampled.

Protecting endangered species is increasingly becoming an expensive preoccupation in Sub-Saharan Africa largely due to poverty levels. While the opportunity cost of conserving habitat may be very high and could worsen as the human population in the vicinity of protected areas increases, protecting animals from illegal trade involves substantial additional costs. Recent estimates of the amounts required to prevent poaching range from US$200 to $500 per hectare in Africa (Parker & Graham 1989), and far exceed actual expenditures on enforcement (Dublin et al 1995). Since many of the world's threatened high-profile species are found in developing countries with limited resources and tight budget constraints (Baker & Baker 2002), it is not surprising that effective enforcement expenditures lag far behind recommended rates. As a result, many species especially water birds in Sub-Saharan Africa, have been victims of illegal trade.

Given the foregoing circumstances, conservationists in Sub-Saharan Africa like in other parts of the world continue to look for alternative ways to conserve wildlife. Economists and liberal conservationists in the region have proposed that trying to keep poachers/illegal traders out of protected areas at enormous cost, which in turn drives up prices on black markets and only increases a poacher's incentive to hunt. They therefore support the "supply-side" policies aiming at flooding the market for wildlife commodities with captive-bred varieties and other alternatives. This will depress prices and make illegal trade and poaching unprofitable. Traders will in turn search for alternative employment allowing wild populations of endangered species to recover. The supply side solution appears to have gained considerable popular support in South Africa relative to other Sub-Saharan Africa countries.

Understanding the whole concept of illegal bird trade Sub-Saharan Africa is, in our opinion, essential for the effective implementation of conservation actions to save endangered species like Wattled Cranes.

Captive breeding has been recommended by several conservationists as a way to ensure a steady supply of eg: bear bile, tiger bones and rhino horns, thus protecting their wild counterparts (Brown & Layton 2001; Mills et al 1995).

## RESULTS

In our preliminary surveys and visits to seven random breeding centre's in four Sub-Saharan African countries, namely Tanzania (three centres), Uganda (two centres) Botswana (one centre) and South Africa (two centres) respectively. We agree with farmers and conservationists that captive breeding is a viable solution to illegal bird trade in Sub-Saharan Africa and more specifically to Tanzania where the trade is still rampant and where prices have been depressed by developing close substitutes for birds obtained from the wild. This seems especially relevant for species that are primarily hunted for their food in some countries (e.g. ducks and geese in southern Tanzania).

At the moment it seems the major source of the wild-bird trade in Africa is Tanzania (Karekoona unpublished). Not only does its own birds get transferred to South Africa but, because of its slack regulation, it has also become the preferred export hub for birds from other countries.

One of the most over-traded bird in the region is the Grey Parrot *Psittacus erithacus* which is favoured for its beauty and its exceptional ability to mimic the human voice. Though not yet considered endangered, our preliminary surveys agree with Birdlife International (2005), that its formerly large numbers are declining rapidly as a result mainly of trapping for the wild-bird trade. Interviews with concerned Birdlife partners in the respective countries suggest that the trade is controlled by a relatively small number of cartelized traders, while poaching and trapping are carried out by subsistence village dwellers (most especially impoverished youth) under open access conditions.

We characterize this situation to be developing a stylized three-staged market. In the first stage, the traders of wild birds set the remuneration to be paid to poachers. In the second stage, the poachers determine the harvest of birds, taking as given the prices set by the traders. In the final stage the traders sell the birds through dealers who are largely based in South Africa.

Like any other trade in illegal commodities, the case of imperfect competition on the market is well-documented and appears to be among the most important challenges for all those involved in the trade. More surprising, perhaps, is evidence that captive-breeding of some species is posing a big challenge to the poachers is that is has increased returns in South Africa where the two breeding centers centre's we visited are doing well.

In general, the impact of the legal trade in various species of birds in the four sampled countries is vastly compounded by the ravages of the equally lucrative and even crueller illegal trade.

**Overview of Illegal Trade in Cranes**

Approximately 2000 Blue Cranes *Anthropoides paradiseus* are believed to be illegally kept in South Africa, 43 of which were confiscated by government officials in 2003/2004 (Karekoona unpublished), Black Crowned Cranes *Balearica pavonina* have declined from >100,000 to <15,000 in the wild over the past decade due to illegal trade and domestication. Forty-one Grey Crowned Cranes *Balearica regulorum* were confiscated in South Africa in 2006 (Endangered Wild life Trust, in press). Five-thousand Wattled Cranes *Bugeranus carunculatus* are thought to have been exported illegally out of Tanzania in the last ten years, with only 47 Wattled Cranes exported on CITES from Tanzania in four years (African Wattled Crane Program, unpublished).

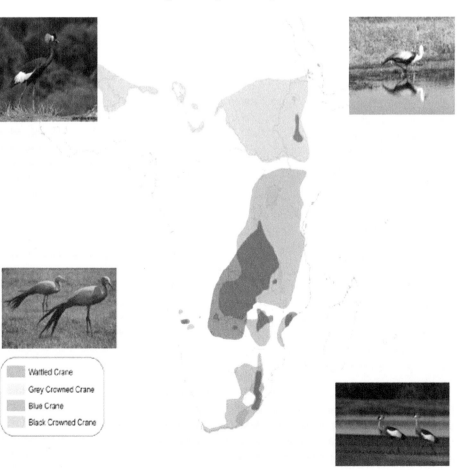

**Fig. 1.** *Geographical location of crane species involved in the illegal trade.*

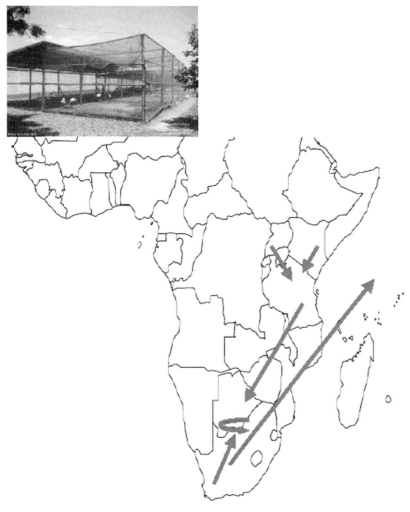

**Fig. 2.** *Trade syndicates within and between four African countries.*

## DISCUSSION

We agree that when captive bred birds are introduced to the market, the demand for poached wild birds is modified. The introduction of captive bred birds into the market place will lower the level of poaching relative to that which occurs in the absence of farming for any given wildlife stock. If the demand for wild birds is either unchanged or declines, competition in the retail market will occur through quantity setting, of which these act as perfect substitutes.

In light of the uncertainties that surround the messy situation of the illegal trade in Sub-Saharan Africa, we recommend and promote captive breeding programs and believe they will affect the "supply-side" and that they will eventually counterbalance the illegal trade. We observe that since trade in most endangered species is either regulated or banned by CITES, a key decision (International) in trade in commodities should be legalized so that the legal trade can overwhelm that of the illegal one. If captive breeding is to contribute to conservation in Africa, it may call for some form of government intervention to ensure optimal supplies of captive-bred animals/birds.

From our preliminary data, we observe that for any given population of wild birds that have a lucrative market, the illegal trade and poaching levels in equilibrium with captive breeding will still be lower than those without captive breeding. If the supply of captive bred birds has no impact on the parameters of the original inverse demand function for wild birds, the introduction of farmed birds will automatically lower the level of poaching relative to that which occurs in the absence of farming.

If a captive breeding program is to contribute effectively to conservation in Sub-Saharan Africa, it may call for some form of government intervention to ensure optimal care and supply of captive-bred animals. Imperfect competition continues to be at the heart of commercial endangered species poaching, and failure to acknowledge this fact could have detrimental consequences for wildlife. This realization is particularly important now that growing dissatisfaction with the conventional protectionist approach to conservation is prompting calls for the adoption of supply side policies. Captive breeding programs may however be detrimental in the region if they will induce aggressive competition, since this drives output levels towards the perfectly competitive outcome. The picture becomes more complex when we allow for the possibility that consumer preferences are likely unstable and that transaction costs of illegal trade are affected when a parallel legal trade develops. As a result if the traders earn excess profits and use prices as the strategic variable, then the trader's best response to a lower price set by a rival is also to lower its own price. This in turn necessitates an increase in wild supplies.

A number of issues have not been addressed in this paper. There remains the possibility that one of the players may have a strategic first-mover advantage in the retail market. It is well established in the literature that with quantity setting competition, a Stackelberg leader can limit the output of the follower while, conversely, the leader is at a disadvantage relative to the follower under price competition (Dowrick 1986).

Thus, the effects of captive breeding on poaching levels in Sub-Saharan Africa will depend on the instruments of competition.

In general, it is still unclear to us whether a strategic advantage would accrue to traders or breeders. On the one hand, the benefits of incumbency might confer a strategic benefit upon traders. However, it is possible that the legal status of farmed birds might give breeders a strategic advantage. Thus the introduction of sequential moves in retailing does little to mitigate the ambiguity identified in this paper. There remains the possibility that captive breeding may either promote or undermine conservation efforts. Finally, if at any given price all consumers prefer wild birds to captive birds, then it may be more appropriate to model the goods as being vertically differentiated in product quality space (Tirole 1988). The literature on vertical product differentiation suggests that when the quality of goods is given, the higher quality product always captures a larger share of the market (Shaked & Sutton 1983).

In our ongoing research we intend to analyze the role of captive breeding centres in reducing illegal bird trade in four Sub-Saharan African Countries under quantity competition.

## REFERENCES

BAKER NE & EM Baker. 2002,. Important Bird Areas in Tanzania: A first inventory. Wildlife Conservation Society of Tanzania, Dar es Salaam, Tanzania.

BROWN G & DF Layton. 2001. A Market Solution for Preserving Biodiversity: The Black Rhino. In "Protecting Endangered Species in the United States: Biological Needs, Political Realities, Economic Choices" (J. Shogren and T. Tschirhart, ed), Cambridge: Cambridge University Press.

DOWRICK S. 1986. von Stackelberg versus Cournot, Rand J. Econom., 17(2): 251-60.

DUBLIN HT, T Milliken and RFW Barnes. 1995. Four Years After The CITES Ban: Illegal Killing Of Elephants, Ivory Trade And Stockpiles. A report of the IUCN/SSC African Elephant Specialist Group. Gland, Switzerland

MILLS J, S Chan and A Ishihara. 1995. The Bear Facts: The East Asian Market for Bear Gall Bladder, Cambridge: Traffic East Asia.

PARKER I & AD Graham. 1989. Elephant Decline: Downward Trends in African Elephant Distribution and Numbers, Part I, Int. J. Environ. Stud. 34: 287-305.

SHAKED A & J Sutton, 1983. Natural Oligopolies, Econometrica, 51(5): 1469-83.

TIROLE J. 1988. The Theory of Industrial Organization, MIT Press, Cambridge, MA.

# CONSERVATION OF THE CRITICALLY ENDANGERED NEGROS BLEEDING-HEART PIGEON *GALLICOLUMBA KEAYI* ON THE ISLAND OF NEGROS, PHILLIPINES

APOLINARIO CARIÑO[1], ANGELITA CADELIÑA[2], RENE VENDIOLA[1], JOSE BALDADO[3], CHARLIE FABRE[4], MERCY TEVES[3], PAVEL HOSPODARSKY[5], EMILIA LASTICA[5] & LOUJEAN CERIAL[5]

[1] Pederasyon sa Nagkahiusang mga Mag-uuma nga Nanalipud ug Nagpasig-uli sa Kinaiyahan Inc. (PENAGMANNAKI), and Mt. Talinis Peoples Organization Federation Inc. (MTPFI) Sibulan, Oriental Negros, Philippines
[2] Biology Department, Silliman University, Dumaguete City, Philippines
[3] Province of Oriental Negros, Philippines
[4] Community Environment and Natural Resources - Dumaguete
[5] Fauna and Flora International-Philippines Biodiversity Conservation Program, Silliman University Center for Tropical Conservation Studies

## SUMMARY

On Negros Island and elsewhere in the Philippines, where over 90% of its original forest has been removed and implementation of existing wildlife laws remain weak, populations of many wildlife species including birds, are rapidly declining and may eventually become extinct. Among these birds is the Critically Endangered Negros Bleeding-heart Pigeon. Field surveys revealed for the first time the current distribution and population of the species on the island. The species distribution showed it to be very restricted on lowland areas with forests. It is also shown to survive in areas with agricultural development (small-scale farming). Observations on nesting habits were recorded and documented at the vicinity of Malangwa River, Mt. Talinis (480m) and at Landay, Siaton (780m). Feeding areas and food plants were documented in Calinawan, Mantikil, Twin Lakes and Mt. Talinis areas. Breeding observations on the behavior in captivity at the A.Y. Reyes Zoological and Botanical Garden were documented. A province-wide environmental public awareness campaign on the protection of this and other wildlife species was initiated by the provincial and local governments of Oriental Negros. This has given way to a sustainable conservation program of the species all throughout the Island of Negros. It is hoped that these efforts may lead to activities that could be done to mitigate the threats to the sustainability of its population and the preservation of its habitat where conservation bred species can be reintroduced.

## INTRODUCTION

The Philippines is fortunate to have at least five of the seven species of the genus *Gallicolumba* pigeons known worldwide. Except on Palawan, the Bleeding-heart Pigeons are unique to the Philippines. These are: the Luzon Bleeding-heart Pigeon *Gallicolumba luzonica*, Mindoro Bleeding-heart Pigeon *G. platenae*, Mindanao Bleeding-heart Pigeon *G. criniger*, Sulu Bleeding-heart Pigeon *G. menagei* and Negros Bleeding-heart Pigeon *G. keayi*. Bleeding-heart pigeons or doves are characterized by a golden, red patch or vertical streak on the breast. All of the seven species are terrestrial, living singly or in pairs, and feeding on seeds, berries and invertebrates (Gibbs et al 2001). On Negros, *G. keayi* are often observed feeding on invertebrates (Gibbs et al 2001) and fruits of *Pinanga, Arisaema*, and *Ficus* species on lowland dipterocarp forest grounds, near agricultural and in degraded areas (Cariño pers. obs.). They are very difficult to observe in the wild but oftentimes observed walking searching for food in small bodies of water like streams, brooks and/or along river banks (Gibbs et al 2001). They are classified as Critically Endangered by Birdlife International (2004) and the IUCN (2004 CD database) and that the bird has been regarded as "an extremely rare species" (Collar et al 1999). Its population and ecology are poorly known except for some notes on the nesting observations done by Slade et al. (2005) in Sibaliw, Buruanga, Aklan, Panay on the north-west Panay peninsula mountain range. Also, a detailed observation of an individual bird caught in the wild during banding operations by the Philippine Endemic Species Conservation Project (PESCP) which was held in a rehabilitation cage in the Panay forest, a part of a one and a half year project (Curio 2001). According to Kennedy et al. (2000) and Mallari et al. (2001), this is a very cryptic, elusive and very difficult bird to observe. The continuing destruction of our lowland forests as well as the upland primary forests of Negros, the imminent large habitat loss of the species, poses a serious threat to the population (Brooks et al 1992). Local trapping, hunting, and trade are chronic problems that exacerbate the effects of deforestation (Cariño et al 2006 in prep.). This study aims to gather information on its distribution and ecology (breeding and feeding behavior) in the wild. The ultimate goal is to protect the species by

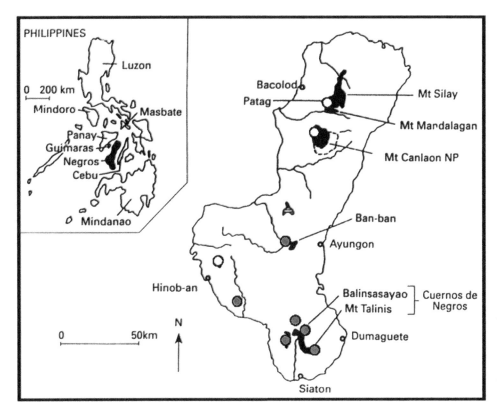

Fig. 1. *Map showing confirmed present distribution of the Negros bleeding-heart pigeon (in red) and unconfirmed reported sites (in yellow) on the Island of Negros. Map adopted from Brooks et al. 1992.*

formulating conservation measures involving the local government units (LGUs) and academia by initiating a community-based education program at the grassroots level.

## METHODS

### Study Areas

Negros Island has five major forest patches mainly located along its north-south mountain formations (Figure 1). Most of the remaining populations of the study species, which include other endangered endemic species, are in the lowland forests of these areas. Vegetation of the surrounding areas include; Cogon *Imperata cylindrica* grasslands, wild thickets and various crops on farms (corn, root crops, fruit trees, etc.) which are mostly for the subsistence of their owners. The farms and cleared areas are actually encroachments into public forest lands that are supposed to be inalienable and protected. The remaining forests are also pock-marked with small clearings due to small-scale logging and slash-and-burn farming. The survey covered the remaining lowland forests where hunting activities are quiet rampant. The hunters included in the survey were: those who live in or near the hunting areas, and are mostly farmers; and those from urban areas who are not farmers and some few are in fact affluent.

### Mt. Talinis-Twin Lakes Balinsasayao and Danao Area

Quarterly field monitoring was conducted by the Mt. Talinis-Twin Lakes Biodiversity Monitoring and Evaluation Team between April 2004 and December 2006, the objective of which is to make a survey on whether or not the species is one of the hunter's bird targets among others. The team surveyed six established transects (2km) and explored all existing trails in all forest types of the area. This is located at the southeastern mountain range of Negros Island and is also called Cuernos de Negros range. It rises 1,800m above sea level (Dolino et al 2004). To the north of Mt. Talinis are substantial areas of primary and secondary lowland dipterocarp forest. Around the Twin Lakes at Balinsasayao and Danao Natural Park, are patches of secondary growth in recently cleared areas. The Twin Lakes Balinsasayao Natural Park was officially declared as a Natural Park by virtue of Proc. No. 414 series of 2000 pursuant to RA 7586. The lakes are about 850m elevation and the submontane forest surrounding it rises to 1,050m. Species of the genus *Ixonanthes* are the dominant tree species found at the center of the Lake Kabalin-an, some 18km away from the Twin Lakes going north. Although Mt. Talinis rises some 6,000m, the lake area can be reached by a motor vehicle such as a motorcycle or a non-four wheel drive vehicle due to improved road system. One can also proceed to the area by taking a hired vehicle, going 15km northbound and to the west's all-weather roads 17km off the highway. The Mt. Talinis-Twin Lakes (MTTL) Forest Reserve is one of the most important but critically endangered ecosystems in the Philippines (Cadeliña et al 2004). This is also known popularly as the Southern Negros Forest Reserve, and is within the jurisdiction of the 133,000-hectare Philippine National Oil Company (PNOC) geothermal reserve.

### Calinawan Community-based Wildlife Sanctuary

Field surveys were conducted between May 2004 to November 2006 in an old-growth lowland dipterocarp forest (500-1000m). The species was reported to be hunted in the area and most of the reports from poachers and people who have bought the species as pets were all trapped from this place. The forest is dominated by White and Red Lauan *Shorea polysperma* and *S. negrosensis* and oftentimes with Almaciga *Agathis philippinensis* and *A. dammara* comprising about 300ha. This area is characterized by low-lying hills with gentle slopes within the jurisdiction of the Municipalities of Sibulan and Sta. Catalina, Oriental Negros. The area is also popularly known as "Moratorium"- where two of the logging companies; Pan Oriental Lumber Company (POLCO) and Phil-American Timber Company (PATIC) made a 1,000ha reservation in the past. Now it is reduced to 300ha due to local timber poaching and "kaingin" or slash and burn farming (Cariño 2002 & 2004; Tiempo et al 2002). This forest extends towards Barangay Dobdob, Valencia and rolling towards the road project of the province of the Tamlang Valley Zone for Peace and Development. Field visits were done along the remaining lowland and secondary forests surrounding the Dobdob - Tamlang Valley areas. Most of the valley areas are filled with Cogon *Imperata cylindrica* grass interspersed with some agricultural crops such as corn, carrots, onion leaves, Cassava *Manihot esculenta*, Camote *Ipomoea batatas* and other root crops.

### Canaway, Mantikil, Siaton

The study was also conducted along the Canaway river headwaters of Mantikil, ridge tops, and agricultural portions of the area from March 2004 to May 2006. The lowland forest (750-850m) is a dipterocarp forest dominated by *Shorea* sp. and some species of *Lithocarpus* and *Ficus* sp. clinging on boulders and rocks. Between 750m to more than 1,300m, tree species, *Agathis* and *Podocarpus* interspersed with *Schizostachium* sp., often times in high densities along steep ravines and between ridges, dominated the area. Epiphytes observed were composed of several species of orchids *Vanda lamellata*, ferns *Lycopodium* sp., climbing vines, hoyas, lianas, bromeliads, *Medinilla* sp. and lipstick vines *Aeschynanthus* sp. In most of the forest clearings, agricultural crops planted include coconuts, coffee, carrots, corn, green onions, cassava, and sweet potato.

Field surveys were also conducted in nearby forests (planted with mahogany trees *Swietennia macrophylla* and *Cassia mangium*) privately owned by Mr. Carding Teves and near Sitio Atimon of Barangay Mantikil and Barangay Balastro of Siaton, Negros Oriental on December 2006, where the species was observed mostly in riparian areas. Most of the planted mahogany trees are now fully grown with other indigenous species in between vacant areas and saplings of *Shorea* sp. were left to grow through Assisted Natural Regeneration (ANR). Sugarcane plantations and cattle ranching encroached and threaten these habitats.

### Landay, Mantikil, Siaton

Direct observations were carried out along a mature secondary forest and a mossy forest dominated by dipterocarp species (*Hopea, Pentacme,* and *Shorea* spp.), along river banks, and from 756m up to 950m on ridge tops in Landay, Mantikil, Siaton, Negros Oriental dominated by *Agathis* sp., *Podocarpus* sp., and many varieties of *Ficus* sp. during occasional field visits between April and May 2004 to 2006. Canopy is oftentimes closed and sometimes with as little as 30 percent light penetration. Fallen logs were observed as fairly common and shrubs, tree ferns *Cyathea* sp., and bird's nest ferns *Asplenium nidus* clinging on trunks of trees dominate most of the ridges and slopes. Species of *Pandanus, Araceae*, ground ferns, and *Calamus* sp. constitute the ground cover. Climbing lianas, hoyas, drynarias, and lycopodiums were observed on tree trunks of most emergent trees in the area (Cariño 2004). Moss cover is common even on dead logs and along rocks and riverbanks, and tree buttresses. Common epiphytes observed were orchids, bromeliads, and lipstick vines and *Medinilla* sp. (also observed on understorey and riverbanks). The species is oftentimes encountered along Landay river banks.

### Mabato-Candanaay, Banban, Ayungon

This is considered one of the Important Bird Areas (Mallari et al 2001) and is located at the north of the Province of Negros Oriental close to the border separating the latter with Negros Occidental. The survey was conducted between August 2005 and March 2006 on the lowland forests and by communities living around Mounts Tihol-tiholan, Katungaw-tungawan and Manlawaan in Barangay Banban, Mabato, Candanaay, Maaslum, and Jandalamanon of Ayungon, Negros Oriental from elevation 750-896m. These five barangays harbor a mature secondary lowland dipterocap forest dominated by red and white laua-an and tangile trees *Pentacme contorta*, *Shorea negrosensis* and *S. polysperma* (Cariño 2004). People who commute to the other municipalities such as Bindoy and Mabinay regularly use this road. Farm-lots, plantations (rice-fields and sugarcane), and abandoned agricultural fields surround this watershed area (Paguntalan et al 2002). The species was reportedly caught using indigenous traps (lit-ag) while feeding on palms (*Pinanga* sp.) and placed in small baskets and sold to nearby flea markets in Mabilog, Ayungon, Negros Oriental.

### Sipalay - Cauayan - Hinobaan Area

The forest in these areas are fragmented and of lowland limestone forest type (Alcala et al 2004) just like the Sipalay-Cauayan Area. The Calatong Forest, the more popularly known forest, is located in Manlucahoc where most of the hunters reportedly hunt the species, although the species was not encountered during the survey in July 2005. This forest is a logged-over dipterocarp forest with many patches of agricultural clearings located in the area within an elevation of 20- 270m above sea level. The Hinobaan-Candoni forest area is relatively small and is considered the last remaining forest patch in southern Negros Occidental (Mallari et al 2001). Most of the respondents interviewed hunt around the communities surrounding the Sipalay (Manlucahoc and Calatong), Cauayan and Hinobaan areas, although some hunters extend their hunting activities towards the forests of Candoni; Hinobaan; Damutan; Cabatuanan, Basay; Bayawan and Kabankalan. The field survey was conducted along Manlukahoc area in Sipalay and in Hinobaan-Damutan areas and extended towards the forests of Sitio Cabatuanan of Barangay Tangtang, Basay, Oriental Negros.

### Mt. Canlaon - Guintubdan and Mambukal Area

Field surveys were conducted in lowland forests of Guintubdan, La Castellana and Mambukal, Minoyan, Murcia and interviews were conducted in the communities surrounding barangays and municipalities of Mt. Canlaon (including Barangay Tagbino of the Municipality of Vallehermoso), and communities of Guintubdan in La Castellana and Mambukal of Murcia. The species was not encountered during the survey in September 2005. Mt. Kanlaon National Park was first proclaimed as a natural park by virtue of Proclamation 721 on 8 August 1934, and revised by Proclamation No. 1005 on 8 May 1997 and was a GEF-CPPAP (full first) site (Mallari et al 2001). Mt. Canlaon has the highest peaks on Negros Island, and lies c. 35km southeast of Bacolod City. Several volcanic craters and peaks are found in the Canlaon range, the highest so far reaches 2,435m. These mountains are subsumed by the Mt. Canlaon National Park, and the forest within the park is estimated to cover about 11,475ha or 46.75% of its total area. Other habitats include open grassland and cultivated lands occupied by settlers (Mallari et al 2001). Most of the forests are montane, including mossy forest above 1,700m to the bare peaks of Mt. Canlaon where active volcanoes and low shrubby vegetation and grassland of the inactive peaks are found. Some of the lowland forests are located at Guintubdan and Mambukal ranging in height from 400-1,050m.

### Mt. Silay - Patag - Mandalagan Area (North Negros Natural Park)

Field surveys and interviews of hunters were conducted in September, 2005 along the vicinity of Mts. Silay and Patag in Mandalagan area. The species was never encountered during the survey. The area is one of the Important Biodiversity Areas (IBAs) located within the North Negros Natural Park (NNNP), which was declared formerly as (North Negros Forest Reserve), a protected area for birds by Administrative Act No. 789 on 28 April 1935 (Mallari et al 2001). It was declared as a Natural Park by the National Integrated Protected Areas System last August, 2005 (Bibar pers. comm. 2005). This area lies to the north of Mount Canlaon. It is an old forest reserve, mostly logged except for the two mountains, Mts. Silay and Mandalagan (Mallari et al 2001). The area of NNNP covers 80,454ha but only 16,687ha is forested of which c. 75% is old growth and c. 25% is secondary growth.

## Field research

Field observations were done through forest exploration and purposive searches for the species, transect walks and visits to reported hunting and feeding areas. Market day places were also visited for possible trading of the species. Interviews were also conducted with known hunters and traders of the study species in the area.

## RESULTS

The results and discussion of the activities accomplished and on-going to date are enumerated below starting from recording of its distribution on the Island of Negros (past and present distribution); nesting and breeding observations in the wild; taking notes on its feeding ecology in the wild and observations of breeding activities of the species in captivity; and public awareness campaign.

## Historical Background of the Species Distribution on Negros Island

Recorded data of previous studies conducted by Sharpe (1877) were obtained from the northern tip of Negros. Other data was taken in San Carlos City when a male was obtained alive in or around 1926 by McGregor (1927). Other data was taken from Panyabunan, Bais, Negros Oriental on May 1949, where two bird specimens were found at 300m. Three female specimens are kept at the Delaware Museum of Natural History (DMNH) and Field Museum of Natural History, Chicago (FMNH). Another individual of the species was recorded in Basay, Bayawan, Negros Oriental, in December 1959 (female individual in YPM). In April and May 1950 individuals were recorded in Candomao, Naliong and Balangbang of Tolong, Negros Oriental and the first two were recorded at 600m (FMNH). Alcala and Carumbana (1980) also reported the species in the forests surrounding Lake Balinsasayao and Danao and were recorded from between January 1977 to July 1978. The species was also recorded in Mambukal, Murcia, Negros Occidental based on either observation or interviews with reliable witnesses between July-August 1991 by Evans et al. (1993) and from Diesmos and Pedregosa (1995) and by Robson and Davidson (1995) at 1,005m in March 1994 (in Collar et al 1999).

Unconfirmed reports about the species were obtained from Mt. Mandalagan, Mt. Patag, North Negros Forest Reserve, Guintubdan, Mt. Canlaon, Mt. Talinis (Cuernos de Negros), and Mabato, Ayungon (in Collar et al 1999). Anecdotal reports claimed that they are seen commonly in the 80s by the locals and are locally called "pugngan" meaning "restrain".

## The Current Survey (2004-2006)

During this survey, reports on the most recent confirmed distribution of the species were recorded either through direct field observations, from anecdotal reports of hunters, trading activities (previously trapped birds for pets), and documented reports on rescued birds from bird poachers. At least three individuals were recorded (only one was rescued alive the rest died of starvation) in Calinawan, Enrique Villanueva, Sibulan, Negros Oriental at 536m. One individual was caught dead in a "batayan" (an indigenous trap) set on a *Pinanga* sp. fruiting palm at the vicinity of the Twin Lakes Balinsasayao and Danao.

Another three individuals were trapped from the same plant variety in Mabato-Candanaay, Ayungon, Negros Oriental and were traded and brought to Dumaguete City as pets although, the rest of the three individuals died in captivity a few days after. Solitary individuals were observed calling and feeding along the Malangwa Creek, Timbao, Bacong, Negros Oriental (with mountain ridges going towards Mt. Talinis) at 480m (see Figure 1).

Solitary individuals were observed in three different occasions at Mangudkod Creek, Sitio Nawakat, Barangay Kasalaan, Siaton, Negros Oriental which is approximately five kilometers from Barangay Mantikil (21km away from Siaton town proper). Recorded observations were also made along Naubo River (27 km from Siaton town proper). At least a single individual was observed walking along Naubo River. This was also observed in Atimon, Mantikil at around five o'clock in the afternoon at 677m. The species was also recorded in Canaway, Mantikil, Siaton, Negros Oriental (750-850m) and Landay, Mantikil, Siaton, Negros Oriental from elevations between 756m up to 950m on ridge tops). Another individual was observed at around 5:30pm, 5 February 2007, in Bongalonan, Basay, Negros Oriental perching on a branch of an unidentified tree. The bird stayed on the branch until night fall.

The species was not encountered in Mambukal, Murcia, Sipalay (Manlucahoc and Calatong), Cauayan and Hinobaan, Mt. Kanlaon area and Guintubdan, La Castellana, Negros Occidental. However, reports from the fighting cock caretakers revealed that the species feeds on some of the left over food of the cocks in their poultry yard and observed to be walking along the river of the Guintubdan falls. Reports from the locals in this area revealed that the reported bleeding heart is actually the Yellow-breasted Fruit-dove *Ptilinopus occipitalis* with the yellow colored patch on its breast. This was verified based on descriptions the locals made concerning the species.

**Feeding Observations**

Several species of plants (Araceae, Palmae, and Moraceae) were personally observed as food plants which correlated with anecdotal reports of hunters and farmers who encountered them accidentally or intentionally trapped for trade. Fruit of these plants are used by hunters as baits for traps: *Arisaema* sp. and *Amorphophallus campanulatus* both from the Araceae family; *Ficus* sp. (*F. cumingii* var. *terminalifolia*, *F. obscura* var. *scaberrima* and *F. virgata*) and *Pinanga cf. philippinensis*, a common forest palm. Fruits from these plants were utilized by the Negros Beeding-heart Pigeon and other species of birds (*Ducula*, *Phapitreron*, *Penelopides* and *Aceros*). Most of the traps were also laid down by hunters on ground level at the base of each of the plant varieties mentioned. Although, the bird was not encountered feeding on ground invertebrates, claw prints indicating removal of fallen dried leaves from surrounding tree trunks were evident of this type of feeding activity. Therefore, further studies on this aspect are still needed to fill the gaps on our knowledge concerning feeding preferences of this species in the wild.

**Nesting and Breeding Observations in the Wild**

Notes and observations on its nesting activities were documented for the first time on Negros Island in Timbao, Bacong, Negros Oriental in March, 2005. The nest was discovered while conducting a river trekking activity to fix the pipes for the water supply of Barangay Liptong, Bacong, Oriental Negros. The nest was built and well camouflaged by vines and branches of some shrubby plants found in the area. The nest sat on the fronds of the *Dicksonia* sp. (a giant fern) at about 0.75m from the ground along the Malangwa River at 480m. Its nest is made from twigs and sticks of plants (loosely arranged) found around the area. It measures about 8-10 inches in diameter. The inside structure of the nest measures at least eight inches where the birds lay their eggs and sat on them. Enclosing this structure is an extension which measures at least two inches. At least two glossy white eggs were observed during the first day of observation. However, the eggs might be two to three days old already. One week after, the eggs hatched and during the first three hours the mother guarded her two chicks and then left to find food for the hatchlings. Most of the observations were made using a Bushnell spotting scope at least a 20m distance away to avoid disturbance to the female tending the hatchlings. Occasional photographs of the hatchlings in the nest were taken while the female is out to find food for the hatchlings.

Another nest was recorded in Landay, Mantikil, Siaton from an elevation of 800m along a ~70° angle slope on a Bird's Nest Fern *Asplenium nidus* at about seven meters away from the Landay River. The nest is located at least three to four feet from the sloping ground. This paralleled the observations made by Slade et al. (2005) in Sibalom, Panay. The fledgling birds tried their wings and hopped after 14 days in Landay.

**In Conservation Breeding Facilities**

In captivity, the red streak on its breast becomes brighter and deeper in color during breeding season. This is well demonstrated by the male individuals in captivity at the A.Y. Reyes Zoological and Botanical Garden. This occurs during the months of February (starting on the second week) towards the last week of April. Then it loses its radiant color and becomes pale red and in thinner streak on non-reproductive months. However, a more thorough observation can be made on its natural habitat. Information on mating and other activity patterns in the wild are still wanting.

Notes on the activity patterns, courtship, and breeding activities of the Negros Bleeding-heart Pigeon in captivity at the A.Y. Reyes Zoological and Botanical Garden

The three individuals of the Negros Beeding-heart Pigeon were observed for the first time in captivity at the A.Y. Reyes Zoological and Botanical Garden. The conservation facility is managed by the Center for Tropical Conservation Studies, a research arm of the Biology Department of Silliman University.

**Activity Patterns**

In most times of the day, the birds were observed walking, resting, grooming, and perching on available tree branch perches at different vertical levels from the ground inside the enclosure. The birds can be observed walking around the enclosure as early as 6 o'clock in the morning and drinking at the water pools provided; during sunny mid days and in the afternoon (2-3pm). Feeding on provided food (beans, pigeon meal, concentrates, and fruits) and grooming takes place anytime of the day. However, schedule of feeding for these birds are at 8-9 o'clock in the morning. Resting while sitting on provided dried leaves inside the enclosure was observed anytime of the day. One individual was also observed perching occasionally on branches of Ficus plants and other artificial perches made of wood branches at different levels (one foot to seven feet in height). At least 60 percent of its time during the day was spent walking around the enclosure and the rest include; drinking, feeding, and resting either by sitting on dried leaves or on provided perches. The birds were also observed submersing its head and sometimes dipping its entire body inside the water pools during hot and sunny days. The pool is about two feet in diameter and the water is at least four to five inches deep. Then finally, all these activities ceased at around 6 o'clock in the evening. The next moment, they were found sleeping while perching at the highest level of the branch provided (about six feet in height).

**Courtship, Mating and Breeding Observations in Captivity**

Notes on the courtship, mating, and breeding behavior of a pair of Negros Bleeding-heart Pigeon were also observed and recorded for the first time in captivity at the A.Y. Reyes Zoological and Botanical Garden between 15 February 2007 and still going on (at this writing-April 2007). A suspected female and male Negros Bleeding-heart Pigeon were the subject of observation. During this study, the two individuals were placed in an enclosure but were separated with a fine plastic net (used typically in gardens) as partition to separate the two birds since the sex of each is still uncertain.

Courtship was observed since 15 February when the male displayed its shoulders, gradually shaking them upwards and presented them to the female accompanied by cooing calls. During this time, they were still separated in the enclosure until the female laid the first unfertilized egg on ground on 21 February 2007 at 6:45am. The second egg was laid on 22 February 2007 at 4 o'clock in the afternoon. Upon removing the partition and the unfertilized eggs the following morning, the birds were observed copulating for the first time at around 3:18pm for about 3-4 seconds and again about 8-10 seconds at 3:58pm. Copulation occurs when the male is on top of the female parallel to its body and with the latter mounting the female's cloaca sideways. From then on, both mated two to five times in a day. Mating occurred in the morning at anytime between 6am to 10am and then at around 2pm to 4pm in the afternoon.

On 25 February 2007, the male was observed selecting, gathering and/or picking up twigs (provided in the enclosure) while courtship continued. Allopreening was also very evident during most times of the day. Mating would sometimes occur after sharing of food and pecking each other before or after copulation between four to eight seconds. On this day, the birds were observed copulating at 9:45am and at 3:06pm.

Most of the activities on 26 February, 2007 were spent preening, for both individuals. Mating was also noted at around 3:45pm. Frequent drinking of water was also observed throughout the day. The male was observed gathering twigs on the ground which was brought to the newly constructed man-made nest framework while the female arranged them in the nest.

Courtship was observed again at 7:29am on 27 February, 2007. This was initiated by both male and female while doing intermittent calls amid the activity. This activity lasted for three minutes while simultaneous allopreening one another. They then both walked together to search for twigs for the nest to be arranged on the mounted nest framework. Between 7:50am to 9:31am regular drinking of water for both individuals was observed. This also occurred in the afternoon along with preening, feeding, and searching and the piling of twigs on the nest.

The female laid another egg, this time on the nest at around 6:30am on 1 March, 2007. The female sat on the nest the entire day while intermittent calls and drinking of water were made by the female during this day. However, the male still wanted to copulate with the female, but the female was no longer receptive. Between 7:45-10:00am, searching for food, feeding, grooming, walking around the enclosure and gathering of twigs for the nest were the activities observed during this time initiated by the male pigeon. Similar activities were also observed in the afternoon. On the following day 2 March, 2007, the female went down from the nest to feed and went up again to its nest after five minutes. The activities continued until 2pm. The second egg was finally laid on the nest and now with a laying interval of about $31\pm1$ hours.

Between 3 March and 12 March, 2007 both birds took turns in incubating the eggs. The activity occurred during either mid day at 1pm or early morning and sometimes at 3 o'clock in the afternoon and in the evening. While the other is incubating, the other individual is feeding, preening, drinking water and /or walking around the enclosure. But on 13 March 2007, signs of courtship and mating behavior were still observed. Intermittent calls by the male were heard. On 14 March 2007, the female no longer incubated the eggs, although the male tried doing the incubation, but left after five minutes. After both birds totally abandoned the eggs they continued their activities of courtship, mating, grooming, and feeding.

Then on 28 March 2007, at around 6:15am the female laid a new egg and laid the second one the following day (29 March 2007) at around 3:15 in the afternoon. This time, no behavioral observation activities were allowed so as not to disturb the breeding pair but intermittent checking was done in the morning, mid-day, mid and late afternoons, and during night time. Similar nesting and breeding activities happened during this time and the pair took turns in incubating the eggs in the nest. About 15 days later, on 12 April 2007, the first egg hatched at around 7:20am and the second egg at 3:45pm the following day. After 10 and 11 days, the hatchlings left the nests alternately, the first hatchling that was hatched on 12 April 2007 left the nest first then followed by the second hatchling the following day. Feeding was facilitated by both parents. So far, the two hatchlings in captivity were unsexed and are now separated from the parents as they are capable of feeding on their own.

**Public Awareness Campaign**

Between November 2005 and November 2006, a wildlife conservation education caravan was driven through every municipality and city. All members of the municipal council, the barangay captains, local government officials, and line agency heads would meet to show support for the caravan, once every two weeks among the 24 municipalities and cities of Oriental Negros. This activity was spearheaded by the Vice Governor and Governor of the Province of Oriental Negros along with the Oriental Negros Wildlife Conservation Coordinating Committee members (ONWCCC-ENRD, PNP, DENR) of which the primary author is also a member. During the caravan, status and importance of the threatened vertebrate wildlife of Negros Island (of which the Negros Bleeding-heart Pigeon is one of the highlighted species for conservation) and penalties for violations were discussed. Each member of the ONWCCC contributed their expertise during the open forum. The LGU officials were targeted as the main audience during the caravan so that these local officials eventually help disseminate the information before their respective barangay general assemblies. The team (comprising of LGUs and NGOs) provided and distributed information materials on the status and penalties for violation of the target species in all of the municipalities and cities visited. However, the environmental education campaign was only limited to the province of Oriental Negros. So in order to cover and include Occidental Negros in the campaign, the province of Oriental Negros initiated the Summit on Environmntal Awareness held last November, 2006 in Bayawan City which included representatives from Negros Occidental. This was held in conjunction with the celebration of the Province's Annual Wildlife Conservation Month that is celebrated every year in November. The main highlight of the summit was the presentation of the status of the vertebrate wildlife on Negros Island and its priorities for conservation. Each municipality and city head officials (Mayors) were requested to sign a Manifesto assuring their commitment and support for the conservation of wildlife and their habitats, along with the institutionalization of the Technical Working Group or the Wildlife Conservation Coordinating Team of each Province.

Aside from this, the Province of Oriental Negros initiated a province wide Buglasan Festival that showcased environmental themes along with their respective cultural ingenuities of each municipality and city participating in the festival. This move has also endorsed the species to be the official mascot of the festival and all the future festivals to come. On-going community-based mobile environmental education activities were also conducted in collaboration with the Philippine Biodiversity Conservation Project on the Island.

## CONCLUSION

Since the Negros Bleeding-heart Pigeon is a critically endangered endemic species of the Philippines, a country-wide public awareness campaign is in order.

To do this the following are recommended:

1. The print and broadcast media can be utilized to address the conservation problems of the species.
    a. target hunting by affluent hunters should avoid this small-bodied pigeon.

b. subsistence hunters should not hunt this small-bodied pigeon as their meat is too little to be used for food.

   c. during the weather-watch tv segments, this bird species can be included in the discussion since the anchor of this portion of television usually includes animal studies as an interest-grabbing adlib.

2. The combined efforts of managers, conservation specialists and bird technicians will be useful in community-based educational campaign which can be integrated in an extension program initiated by colleges and universities. The latter program is usually required of universities before earning the center of development or excellence status to be granted by the Commission on Higher Education (CHED).

3. Conservation facilities and bird sanctuaries staff should initiate conservation awareness after-school activities which may consist of essay contests about the bird species, poster drawing or walk-through and "assist feed preparation activities". The same activities could also be initiated for the out-of-school youth who are residing in rural areas.

4. Assisted natural regeneration of bird food plants should be encouraged among the locals.

5. Policies on wildlife conservation and during formulation of protective legislation should be initiated at the local level.

6. Committee chairmen on the Environment (Sanggunian Bayan or Municipal Council) and the Police Environmental Desk Officers (PEDO) should be involved as focal persons in the campaign for all environmental conservation efforts.

## ACKNOWLEDGEMENTS

We thank the Rufford Small Grants Program for their support and giving us the opportunity to undertake this project. To Bristol Zooloogical Gardens for supporting the rearing of confiscated Negros Bleeding-heart Pigeons in captivity at the A.Y. Reyes Zoological and Botanical Garden and the island wide ethnobiological survey; the Foundation for the Philippine Environment (FPE) for providing support for the community-based biodiversity conservation activities for the Federation of POs from Mt. Talinis-Twin Lakes areas; Chester Zoo for supporting the Mobile Environmental Education Unit; Fauna and Flora International for logistical support, and the local government units of the provinces of Occidental and Oriental Negros. Our special thanks to Mr. Tiny and Mrs. Marilyn Suasin of Valencia for donating the birds in captivity; Governor George Arnaiz of Oriental Negros for the legislative support; to Mr. Henry Abancio of PENAGMANNAKI and Mr. Servano Enid of MTPOFI and their PO members for all the assistance in the data collection; to Prof. Mirasol N. Magbanua of the S.U. Biology Department for the use of equipment and facilities; Mr. Leonard Co of the Philippine Museum of Natural History for the plant identification; Dr. Thomas Brooks of Conservation International; Dr. Jon Ekstrom of Birdlife International; Dr. Nigel Collar of Cambridge University; Dr. Eberhard Curio of Bochum University, Germany; and Ms. Myrissa L. Tabao of FPE for their technical inputs, advice, and provision of available literature materials.

## REFERENCES

ALCALA AC & E Carumbana. 1980. Ecological observations of game birds in South Negros, Philippines. Silliman Journal 27: 91-119.

ALCALA EL, RB Paalan, LT Averia and AC Alcala. 2004. Rediscovery of the Philippine Bare-backed Fruit Bat (*Dobsonia chapmani* Rabor) in southwestern Negros Island, Philippines. Silliman 45(2): 123-136.

BIRDLIFE INTERNATIONAL. 2004. Threatened birds of the world 2004. CD-ROM. Cambridge, U.K: Birdlife International.

BROOKS TM, TD Evans, GCL Dutson, GQA Anderson, DC Asane, RJ Timmins and AG Toledo. 1992. The conservation status of the birds of Negros, Philippines. Bird Conservation International. 2:273-302.

CADELIÑA AM, AB Cariño and CN Dolino. 2004. Saving a Physically Challenged Ecosystem: Who Takes Charge of the Mt. Talinis - Twin Lakes Forest Reserve. Silliman 45(2): 237-250.

CARIÑO AB. 2004. Studies of Fruit Bats on Negros Island, Philippines. Silliman 45(2): 137-159.

CARIÑO AB, AM Cadeliña and FA Tiempo. 2006. Ehtnobiological Survey of Vertebrate Wildlife Hunters on Negros Island, Philippines. Paper in preparation.

COLLAR NJ, NAD Mallari and BL Tabaranza, Jr. 1999. Threatened Birds of the Philippines. The Haribon Foundation, Birdlife International Red Data Book pp. 558.

CURIO E. 2001. Taxonomic status of the Negros Bleeding-heart *Gallicolumba keayi* from Panay, Philippines, with notes on its behaviour. Forktail 17: 13-19.

DIESMOS AC & MDG Pedregosa. 1995. The conservation status of threatened species of bleeding-hearts (Columbidae) and hornbills (Bucerotidae) in the Philippines. Wildlife Biology Laboratory, IBS - CAS, U.P. Los Baños.

DOLINO CN, AB Cariño and AM Cadeliña. 2004. Threatened Wildlife of the Twin Lakes Balinsasayao and Danao Natural Park, Negros Oriental, Philippines. Silliman 45(2): 160-208.

EVANS TD, GCL Dutson and TM. Brooks. 1993. Cambridge Philippines Rainforest Project 1991: final report. Cambridge, U.K: Birdlife International (Study Report 54).

GIBBS D, E Barnes and J Cox. 2001. Pigeons and doves. Robertsbridge, Susses, U.K: Pica Press.

IUCN. 2000. IUCN Red List of Threatened Species. International Union for the Conservation of Nature, Cambridge, U.K. (compact disc).

KENNEDY RS, PC Gonzales, EC Dickinson, HC Miranda Jr. and TH Fisher. 2000. A Guide to the Birds of the Philippines. Oxford University Press.

MALLARI NAD, BR Tabaranza Jr and MJ Crosby. 2001. Key Conservation Sites on the Philippines. Haribon Foundation and Birdlife International. Bookmark, Inc., Makati City, Philippines.

MCGREGOR RC. 1927. New or noteworthy Philippine birds. V. Phil. J. Sci. 32: 513-527.

PAGUNTALAN LMJ. 2002. Bird Abundance and Diversity in Forest Fragments in Sothern Negros, Philippines. Unpublished thesis. Silliman University, Dumaguete City.

PAGUNTALAN LMJ, JCT Gonzales, MJC Gadiana, ATL Dans, MDG Pedregosa, AB Cariño and CN Dolino. 2002. Birds of Ban-ban, Central Negros, Philippines: Threats and Conservation Status. Silliman 43(1): 110-136.

ROBSON C & P Davidson. 1995. Some recent records of Philippine birds. Forktail 11: 162-167.

SHARPE RB. 1877. On the birds collected by Professor JB Steere in the Philippine Archipelago. Trans. Linn. Soc. Lond. Zool. 1: 307-355.

TIEMPO FA, ST Villegas, SMB Villagante and AB Cariño. 2002. Management of a non-NIPAS protected area by a People's Organization: The case of the Calinawan sanctuary. Building on Lessons from the Field, Conference on Protected Area Management in the Philippines. Haribon Foudnation, DENR pp. 98-102.

# A COMPARISON OF BEHAVIOR AND POST-RELEASE SURVIVAL OF PARENT-REARED VERSUS HAND-REARED SAN CLEMENTE LOGGERHEAD SHRIKES *LANIUS LUDOVICIANUS MEARNSI*

Susan M. Farabaugh, Ania Bukowinski, Susan Hammerly, Christine Slocomb, Angela Sewell, Kathleen De Falco, Lynne Neibaur & Jeremy Hodges

Conservation and Research for Endangered Species, Zoological Society of San Diego, 15600 San Pasqual Valley Road, Escondido, CA, 92027-7000, USA.

## Summary

The San Clemente Loggerhead Shrike is an endangered subspecies endemic to Navy-owned San Clemente Island (SCI), one of the Channel Islands off the coast of southern California. The US Navy funds a recovery project for this bird that involves monitoring the wild shrike population, managing potential predators, restoring habitat, and augmenting the wild population through captive breeding and release.

The captive program began in 1991 when three clutches of wild eggs were brought in and artificially incubated and hand reared. Releases of captive-hatched birds began in 1992, but the early releases (i.e., 1992-1996) were not successful, and none of the released birds were recruited to the breeding population. In 1999, the cooperative groups participating in the shrike's recovery developed a new plan for releases of various types, including releases of captive-hatched juveniles, bonded adult pairs, family groups, and solo adult releases to unpaired wild birds. Since 1999, a proportion of each year's captive releases have survived to become breeders in the wild population, and this has resulted in a dramatic increase in the size of the wild population. Survival of captive hatched birds varies with age of the bird and rearing (hand versus parent). Juveniles survive better than adults, and parent-reared juveniles survive better than hand-reared. Different levels of vertical space are considered to represent different levels of predation risk for young birds. We found that parent-reared juvenile shrikes spend significantly more time in high safe locations than do hand-reared juveniles. Conversely, hand-reared juveniles spend significantly more time than parent-reared juveniles perched low or on the ground. These hand-reared juveniles may have more poorly developed anti-predator skills than parent-reared juveniles and may be more at risk for predation when released into the wild.

*No manuscript submitted*

# THE SCIENCE AND ART OF MANAGING CAPTIVE BREEDING FOR RELEASE: SAN CLEMENTE LOGGERHEAD SHRIKES *LANIUS LUDOVICIANUS MEARNSI*

SUSAN HAMMERLY, SUSAN M. FARABAUGH, TANDORA GRANT, CHRISTINE SLOCOMB, ANGELA SEWELL, KATHLEEN DE FALCO, LYNNE NEIBAUR & JEREMY HODGES

Conservation and Research for Endangered Species, Zoological Society of San Diego,
15600 San Pasqual Valley Road, Escondido, CA, 92027-7000, USA.

## SUMMARY

The San Clemente Loggerhead Shrike *Lanius ludovicianus mearnsi* is a nonmigratory endangered subspecies endemic to Navy-owned San Clemente Island (SCI), one of the Channel Islands off the coast of southern California. The US Navy funds a complex recovery project that involves monitoring the wild shrike population, managing potential predators, restoring habitat, and augmenting the wild population through captive breeding and release.

The captive breeding program has an annual commitment to provide juveniles, bonded pairs for paired or family release, and solo adults to unpaired wild birds. Annual breeding goals must produce the right number of juveniles (no more than can be accommodated by the available release sites, cages, and staff), of the right age (52-80 days), at the right time (mid-June to mid-July), plus a few more to replace adults lost through release or death. Due to cage availability, the captive flock is limited to 60-65 birds, and thus making a fairly tight target, not too many or too few. The shrike captive husbandry protocols are designed to imitate the basic biology of the species. For example, captive females are placed next to potential mates when the wild females are searching for their mates, and captive birds are fed live vertebrate and insect prey throughout their lives. Another important part of captive management is behavioral monitoring. Standardized daily behavioral observations are conducted to assess mate compatibility before pairing, to monitor the pairs during breeding, and to assess the foraging and flight skills of the captive candidates for release. Finally, the genetics and demography of both the captive and wild populations are managed intensively, by optimizing the genetic diversity in the captive flock through selective breeding, as well as providing sufficient numbers of genetically diverse individuals for release to overcome the demographic challenges facing the wild population.

*No manuscript submitted*

# POPULATION HISTORY AND MITOCHONDRIAL GENE POLYMORPHISM IN BIRDS: IMPLICATIONS FOR CONSERVATION

AUSTIN L. HUGHES & MARY ANN K. HUGHES

Dept. of Biological Sciences, University of South Carolina, Columbia SC 29205 USA

## SUMMARY

The maintenance of genetic diversity has been considered a priority for captive breeding programs, but there is little knowledge of general factors shaping genetic diversity of avian species in nature. We examined nucleotide sequence diversity at mitochondrial protein-coding loci from 72 species of birds from different geographic regions was analyzed in order to test the hypothesis that past population histories have affected genetic diversity. Temperate zone species showed reduced nucleotide diversity in comparison to tropical mainland species, suggesting that the former have reduced long-term effective population sizes due to population bottleneck effects during the most recent glaciation. This hypothesis was further supported by evidence of an unusually high estimated rate of population growth in species breeding in North America and wintering in the New World tropics (Nearctic migrants), consistent with population expansion after a bottleneck. Nearctic migrants also showed evidence of an abundance of rare nonsynonymous (amino acid-altering) polymorphisms, a pattern suggesting that slightly deleterious polymorphisms drifted to high frequencies during a bottleneck and are now being eliminated by selection. Because the extensive glaciation in North America limited the area available for refugia during glaciation, the bottleneck effects are predicted to have been particularly strong in Nearctic migrants, and this prediction was supported. The reduced genetic diversity of Nearctic migrants provides an additional basis for concern for the survival of these species, which are threatened by loss of habitat in the winter range and by introduced disease.

## INTRODUCTION

Conservation biologists have argued that maintaining genetic diversity in natural populations can be an important factor in assuring their continued survival, both because polymorphisms at certain key loci (such as immune system loci) may be essential for survival and because genome-wide polymorphism underlies the additive genetic variance enabling short-term adaptive responses to environmental change (Hughes 1991; Templeton 1994; O'Brien 1994; Maillard & Gonzalez 2006). Thus an understanding of global patterns of genetic polymorphism and their causes can play an important role in devising comprehensive conservation strategies, including those involving captive breeding.

The level of polymorphism present in a species largely reflects the species' demographic history. For organisms inhabiting the earth's temperate zones, a major factor in that history was the most recent glaciation between 60,000 and 10,000 years ago, which caused population bottlenecks and a consequent loss of genetic diversity in numerous temperate zone species (Avise et al 1988; Bucklin & Wiebe 1998; Leonard et al 2000; Lessa et al 2003; McCusker et al 2004). In birds, species which breed in the temperate zones but winter in the tropics are particularly likely to have undergone bottlenecks during glaciation, as a result of the restriction of available breeding habitat (Steadman 2005; Williams & Webb 1996). Because of their inability to survive low winter temperatures, migrants are expected to have experienced more severe range reduction due to glaciation than non-migrants (Williams & Webb 1996).

Most nucleotide sequence polymorphism are believed to result from genetic drift affecting selectively neutral or nearly neutral variants (Kimura 1983; Nei 1987). Since the rate of fixation of neutral or nearly neutral variants is inversely related to effective population size, the extent of polymorphism provides an index of a species' long-term effective population size (Fuerst et al 1977). There is abundant evidence that by far the most prevalent form of natural selection in nature is not positive selection but purifying selection; that is, selection acting to eliminate selectively deleterious alleles (Kimura & Ohta 1974; Ruiz-Pesini et al 2004). The pattern of purifying selection also provides information regarding population history. In populations of small effective population size, selection is expected to be inefficient at removing slightly deleterious mutations, which can then drift to high frequency or even become fixed (Ohta 1976, 2002). On the other hand, if a species experiences population growth after a severe bottleneck, there will be a number of slightly deleterious alleles that drifted to high frequencies when population size was small but are subject to purifying selection when the effective population size becomes larger (Hughes et al 2003, 2005).

We analyzed an extensive database of protein-coding gene sequences from the mitochondrial genomes of a worldwide sample of avian species in order to test for correlations between geographical distribution and the pattern of genetic polymorphism (Hughes and Hughes 2007). In these analyses, we assumed that biogeographic categories are correlated with population history. On average, we assumed that tropical mainland species, should have larger long-term effective population size than temperate-zone mainland species because of the longer lasting climatic stability of the tropics in comparison to the temperate zones over the past 700,000 years, during which seven well-defined glaciations occurred (Barron 1984: Kastner & Goñi 200; Lessa et al 2003), and we predict a particularly strong impact of glaciation on migrants. Within the tropics, mainland species are predicted on average to have greater effective population size than island species because of the larger average range sizes of mainland species (Stattersfield et al 1998) and because founder effects may reduce effective population size of island species (Estoup & Clegg 2003; Miller & Lambert 2004).

## METHODS

### Sequence Data

Our analyses used 103 data sets, each of which consisted of a set of four or more aligned allelic partial or complete sequences for one of five mitochondrial protein-coding genes (*COI, ND2, ND3, cytb,* and *ATP6*). These data sets included a total of 2377 individual sequences and represented 72 species (Table 1); 19 of these species were represented by two or more data sets (Hughes & Hughes 2007). A total of 3237 sites were polymorphic within species. Species were placed in four biogeographic categories: (1) Nearctic migrant, including species breeding in the Nearctic region and wintering in the Neotropics; (2) other temperate zone, including both year-round residents of the Nearctic or other temperate regions (e.g., Australian) and temperate-to-tropic migrants of the Old World; (3) tropical island, including species whose life cycle is confined to one or more small (<40,000 km$^2$) oceanic islands; (4) tropical mainland, including species resident on tropical continental areas or larger tropical islands (e.g., Madagascar). Nearctic migrants were analyzed separately from other temperate zone species because of the availability of data from a substantial number of these species and because of the expectation that the effects of glaciation might have been particularly acute in Nearctic migrants.

### Statistical Analyses

The number of synonymous substitutions per synonymous site and the number of non-synonymous substitutions per non-synonymous site were estimated by Li's (1993) method, using the MEGA2 software (Kumar et al 2001). This method was used because it takes into account transitional bias; and, as is typical in vertebrate mitochondrial genomes, there was a strong transitional bias in the present data, with the transition:transversion ratio (R) in all data sets estimated at 7.2:1. Within each data set, the mean for all pairwise comparisons of the number of synonymous substitutions per synonymous site provided an estimate of nucleotide diversity at synonymous sites ($\pi_S$); and the mean for all pairwise comparisons of the number of non-synonymous substitutions per non-synonymous site provided an estimate of nucleotide diversity at synonymous sites ($\pi_N$) (Nei & Kumar 2000). Weighted averages of $\pi_S$ and $\pi_N$ for the 19 species represented by more than one data set were obtained by weighting with the numbers of synonymous and non-synonymous sites, respectively, estimated by the modified Nei-Gojobori method (Zhang et al 1998), assuming an R of 7.2.

Gene diversity (Nei 1987, p.177) was estimated separately at each polymorphic site, as in Hughes et al. (2003); where $x_i$ is the frequency of the *i*th allele (nucleotide) at a given locus (site), the gene diversity is 1- $\Sigma x_i^2$. Polymorphic sites were classified as synonymous or non-synonymous, based on the coding effect of the nucleotide change. There were 29 sites (0.9%) that could not be so classified; in these cases, because of multiple polymorphic sites within a single codon, the coding effect of a given substitution depended on the pathway taken by evolution. These 29 sites were therefore excluded from analyses of gene diversity at individual polymorphic sites.

In order to examine the relative frequency of rare alleles at synonymous and non-synonymous sites, we compared the average number of nucleotide differences and the number of segregating sites (Tajima 1989) separately for synonymous and non-synonymous sites (Hughes 2005). We computed separately for synonymous and non-synonymous polymorphisms, the difference $k - S/a_1$, where $k$ is the mean number of nucleotide differences for all pairwise comparisons

## Table 1
### SPECIES AND GENES USED IN ANALYSES

| Species | Genes | Biogeographic Category |
|---|---|---|
| *Somateria mollissima* | COI | Other Temperate |
| *Picoides villosus* | COI | Other Temperate |
| *Picoides tridactylus* | ND2, ND3, cytb | Other Temperate |
| *Opisthocomos hoazin* | COI, cytb, ATP6 | Tropical Mainland |
| *Tringa solitaria* | COI | Nearctic Migrant |
| *Empidonax trailli* | cytb | Nearctic Migrant |
| *Empidonax difficilis* | ND2, ND3, cytb | Nearctic Migrant |
| *Empidonax occidentalis* | cytb | Nearctic Migrant |
| *Thamnophilus caerulescens* | cytb | Tropical Mainland |
| *Glyphorhynchus spirurus* | ND2, ND3, cytb | Tropical Mainland |
| *Malurus leucopterus* | ND3 | Other Temperate |
| *Ptilonorhynchus violaceus* | ATP6 | Other Temperate |
| *Vireo cassiniii* | cytb | Nearctic Migrant |
| *Vireo solitarius* | cytb | Nearctic Migrant |
| *Vireo gilvus* | COI | Nearctic Migrant |
| *Cyanocitta cristata* | COI | Other Temperate |
| *Corvus macrorhynchos* | cytb | Tropical Mainland |
| *Corvus corax* | COI, cytb | Other Temperate |
| *Phainoptila melanoxantha* | ATP6 | Tropical Mainland |
| *Turdus olivaceus* | ND3, cytb | Tropical Mainland |
| *Turdus smithi* | ND3, cytb | Other Temperate |
| *Stiphrornis sanghensis* | cytb | Tropical Mainland |
| *Enicurus leschenaulti* | ND2, ND3 | Tropical Mainland |
| *Mimus polyglottos* | ATP6 | Other Temperate |
| *Mimus gilvus* | ATP6 | Tropical Mainland |
| *Cinclocerthia ruficauda* | ATP6 | Tropical Island |
| *Margarops fuscatus* | ATP6 | Tropical Island |
| *Toxostoma curvirostre* | ND2, cytb | Other Temperate |
| *Toxostoma redivivum* | cytb, ATP6 | Other Temperate |
| *Cistothorus palustris* | COI | Nearctic Migrant |
| *Henicorhina leucophrys* | ATP6 | Tropical Mainland |
| *Baeolophus inornatus* | cytb | Other Temperate |
| *Cettia diphone* | cytb | Other Temperate |
| *Nectarinia olivacea* | ND3 | Tropical Mainland |
| *Nectarinia oritis* | cytb | Tropical Mainland |
| *Nectarinia dussumieri* | ND4 | Tropical Island |
| *Nectarinia sovimanga* | ND4, ATP6 | Tropical Mainland |
| *Nectarinia humbloti* | ND4 | Tropical Island |
| *Nectarinia mediocris* | ND4 | Tropical Mainland |
| *Nectarinia moreaui* | ND3 | Tropical Mainland |
| *Nectarinia notata* | ND4 | Tropical Mainland |

## Table 1 (cont'd)
### SPECIES AND GENES USED IN ANALYSES

| Species | Genes | Biogeographic Category |
|---|---|---|
| *Motacilla alba* | ND2, cytb | Other Temperate |
| *Motacilla citreola* | ND4 | Other Temperate |
| *Motacilla flava* | ND4 | Other Temperate |
| *Fringilla coelebs* | cytb, ATP6 | Other Temperate |
| *Loxia curvirostra* | cytb | Other Temperate |
| *Emberiza cioides* | cytb | Other Temperate |
| *Passerella iliaca* | ND3 | Nearctic Migrant |
| *Melospiza melodia* | cytb | Nearctic Migrant |
| *Zonotrichia leucophrys* | cytb | Nearctic Migrant |
| *Zonotrichia atricapilla* | cytb | Nearctic Migrant |
| *Passerculus sanswichensis* | ND2, ND3 | Nearctic Migrant |
| *Vermivora pinus* | ND2 | Nearctic Migrant |
| *Vermivora chrysoptera* | ND2 | Nearctic Migrant |
| *Parula pitiayumi* | ATP6 | Tropical Mainland |
| *Parula gutteralis* | cytb | Tropical Mainland |
| *Dendroica nigrescens* | COI, ATP6 | Nearctic Migrant |
| *Dendroica townsendi* | COI, ATP6 | Nearctic Migrant |
| *Dendroica occidentalis* | COI, ATP6 | Nearctic Migrant |
| *Dendroica adelaidae* | COI, ATP6 | Tropical Island |
| *Dendroica vitellina* | ND2 | Tropical Island |
| *Dendroica plumbea* | COI, ATP6 | Tropical Island |
| *Basileuterus fulvicauda* | COI, ND2, cytb, ATP6 | Tropical Mainland |
| *Basileuterus rivularis* | COI, ND2, cytb, ATP6 | Tropical Mainland |
| *Coereba flaveola* | ATP6 | Tropical Island |
| *Chlorospingus opthalmicus* | ATP6 | Tropical Mainland |
| *Tangara gyrola* | ND2, cytb | Tropical Mainland |
| *Tangara cayana* | ND2, cytb | Tropical Mainland |
| *Tangara cucullata* | ATP6 | Tropical Island |
| *Certhidea olivacea* | cytb | Tropical Island |
| *Certhidea fusca* | cytb | Tropical Island |
| *Passerina cyanea* | cytb | Nearctic Migrant |

### Table 2

**MEANS (± S.E.) OF MEASURES OF SYNONYMOUS AND NON-SYNONYMOUS POLYMORPHISM IN MITOCHONDRIAL PROTEIN-CODING GENES OF BIRDS**

| Measure | No. species | Synonymous | Non-synonymous |
|---|---|---|---|
| Nucleotide diversity | 72 | 0.0385 ± 0.0058 | 0.0027 ± 0.0004[1] |
| Gene diversity at polymorphic sites | 58[3] | 0.2940 ± 0.0160 | 0.2582 ± 0.0185[2] |

[1] Test of the hypothesis that synonymous and non-synonymous nucleotide diversity are equal ($P < 0.001$; paired t-test).
[2] Test of the hypothesis that gene diversity at synonymous sites equals that at non-synonymous sites ($P = 0.001$; paired t-test).
[3] The data for 14 species included no polymorphic non-synonymous sites.

among $n$ allelic sequences, $S$ is the number of segregating sites, and $a_1$ is the sum from 1 to $n-1$ of $1/n$, which provides an adjustment for sample size (Tajima 1989). We then computed the ratio of this difference to the absolute value of the minimum possible value of the difference, which would occur if all polymorphisms were singletons (Schaeffer 2002). We designate this ratio $Q_{syn}$ in the case of synonymous polymorphisms and $Q_{non}$ in the case of nonsynonymous polymorphisms. Comparing $Q_{syn}$ and $Q_{non}$ provides an index of the relative abundance of rare alleles at synonymous and non-synonymous sites, with a strongly negative value indicating an abundance of rare alleles (Hughes 2005). We applied this method to all data sets (N = 82) that included both synonymous and non-synonymous polymorphisms.

### Results

Mean $\pi_S$ and $\pi_N$ at mitochondrial protein-coding loci within each of 72 avian species were compared by paired-sample t-test (Table 2). Overall mean $\pi_S$ was over an order of magnitude greater than overall mean $\pi_N$; and the difference was highly significant (Table 2). Moreover, in the case of 58 species for which there were data on both synonymous and non-synonymous polymorphisms, the mean gene diversity at synonymous polymorphic sites was significantly greater than mean gene diversity at non-synonymous polymorphic sites (Table 2).

There was a significant difference with respect to mean $\pi_S$ values among the four biogeographic categories of species (Table 3). Individual comparisons with a family error rate of 5% revealed a significant difference between mean $\pi_S$ for Nearctic migrants and that for tropical mainland species and a significant difference between mean $\pi_S$ for other temperate zone species and that for tropical mainland species (Table 3). Likewise, there was a significant difference with respect to mean $\pi_N$ values among the four biogeographic categories (Table 3). Mean $\pi_N$ values for all other categories were significantly different with a family error rate of 1% from that for tropical mainland species (Table 3).

We used the Q statistic, computed separately at synonymous and non-synonymous sites, as a measure of the relative abundance of rare alleles in the 82 data sets that included both synonymous and non-synonymous polymorphisms. There was no significant difference among biogeographic categories with respect to $Q_{syn}$ (based on synonymous sites) (Table 4). By contrast, there was a highly significant difference among biogeographic categories with respect to $Q_{non}$ (based on nonsynonymous sites; Table 4). By individual comparisons with a 5% family error rate, both Nearctic migrants and other temperate species showed significant differences from tropical mainland species with respect to mean $Q_{non}$ (Table 4).

### Discussion

Analysis of polymorphism at mitochondrial protein-coding loci from 72 avian species revealed a substantial impact of purifying selection. The fact that the mean nucleotide diversity at synonymous sites greatly exceeded that at non-synonymous sites (Table 2) is evidence that purifying selection has acted to eliminate a substantial fraction of non-synonymous mutations occurring at these loci. Furthermore, reduced genetic diversity at non-synonymous polymorphic sites (Table 2) is evidence that many non-synonymous polymorphisms are subject to ongoing purifying selection, acting to reduce population frequency of slightly deleterious alleles (Hughes et al 2003, 2005).

Biogeographic categories differed with respect to nucleotide diversity (Table 3) and the abundance of rare non-synonymous variants, as measured by $Q_{non}$ (Table 4). Tropical mainland species showed the highest nucleotide diversities, supporting the hypothesis that these species have had relatively large long-term

## Table 3

MEAN (± S.E.) SYNONYMOUS ($\pi_S$) AND NON-SYNONYMOUS ($\pi_N$) NUCLEOTIDE DIVERSITIES OF MITOCHONDRIAL PROTEIN CODING GENES 72 AVIAN SPECIES CATEGORIZED BIOGEOGRAPHICALLY

| Biogeographic category | No. species | $\pi_S$ (± S.E.) | $\pi_N$ (± S.E.) |
|---|---|---|---|
| Nearctic migrants | 19 | 0.0231 ± 0.0082[1] | 0.0014 ± 0.0005[2] |
| Other temperate | 19 | 0.0273 ± 0.0055[1] | 0.0016 ± 0.0004[2] |
| Tropical island | 11 | 0.0285 ± 0.0070 | 0.0011 ± 0.0003[2] |
| Tropical mainland | 23 | 0.0653 ± 0.0146 | 0.0054 ± 0.0010 |

[1] Significantly different from value for tropical mainland species ($P < 0.05$; Dunnett's test). One way ANOVA for differences among categories, $P = 0.014$.
[2] Significantly different from value for tropical mainland species ($P < 0.01$; Dunnett's test). One way ANOVA for differences among categories, $P < 0.001$.

effective population sizes in comparison to the other categories. Several previous studies have showed increased within-species genetic diversity in the tropics in a variety of organisms (Alvarez-Bullya et al 1996; Hewitt 2004; Martin & McKay 2004). Increased genetic diversity in the tropics is frequently associated with population subdivision (Martin & McKay 2004). Such genetic differentiation among sub-populations requires long-term stability and a high effective population size for the species as a whole (Wakeley 2000).

In comparison to tropical mainland species, temperate zone species showed significantly reduced gene diversities at both synonymous and non-synonymous sites, a result consistent with the hypothesis that these species have reduced long-term effective population sizes as a result of population bottlenecks caused by glaciation. The loss of breeding habitat may have been particularly acute for species breeding in the Nearctic region and wintering in the Neotropics (Nearctic migrants). As a consequence of the funnel-like shape of the North American land mass, a much greater proportion of North American than of Eurasia was covered by glaciers at glacial maxima; and the availability of refugia from glaciation was thus much reduced in the Nearctic in comparison to the Palearctic (Hewitt 2004). A particularly pronounced effect of glaciation on genetic diversity is not surprising in Nearctic-to-Neotropical migrant birds is not surprising in light of these biogeographic considerations.

Nearctic migrants also showed strongly negative values of $Q_{non}$, significantly different from those of tropical mainland species, whereas there were no significant differences among biogeographic categories with respect to $Q_{syn}$. Strongly negative $Q_{non}$ indicates that there are abundant rare non-synonymous polymorphisms. Under a population bottleneck, slightly deleterious mutations can drift to high frequencies because purifying selection cannot eliminate them effectively in a small population (Ohta 1976, 2002). When a bottlenecked population subsequently increases in size, purifying selection is predicted to act to remove such slightly deleterious alleles, leading to a

## Table 4

MEAN (± S.E.) $Q_{syn}$ AND $Q_{non}$ IN DATA SETS AVIAN SPECIES CATEGORIZED BIOGEOGRAPHICALLY.

| Biogeographic category | No. data sets | $Q_{syn}$ (± S.E.) | $Q_{non}$ (± S.E.) |
|---|---|---|---|
| Nearctic migrants | 15 | 0.125 ± 0.378 | -0.603 ± 0.331[1] |
| Other temperate | 21 | -0.032 ± 0.126 | -0.431 ± 0.163[1] |
| Tropical island | 10 | -0.002 ± 0.244 | 0.208 ± 0.463 |
| Tropical mainland | 36 | 0.060 ± 0.151 | 0.490 ± 0.215 |

[1] Significantly different from value for tropical mainland species ($P < 0.05$; Dunnett's test). One way ANOVA for differences among categories, $P = 0.007$.

decrease in gene diversity at such sites in comparison to linked neutral sites. Consistent with this prediction, in Nearctic migrants we observed a significant abundance of rare polymorphisms at non-synonymous sites, where deleterious mutations are likely to occur, but not at synonymous sites. Note that, in mitochondrial genomes, the elimination of slightly deleterious mutations is slowed by the lack of recombination (Lowe 2006; Weinreich & Rand 2000), although the latter can be compensated to some extent by back-mutation due to the higher mutation rate and strong transitional bias in mitochondrial genes.

It might be hypothesized that the reduced genetic diversity in the temperate zones and in Nearctic migrants in particular can be attributed to some factor other than glaciation, such as human activities. However, other than a few well-known cases of extinction, temperate zone birds have not experienced unusually severe population reduction. Species with restricted ranges have increased incidence of extinction and endangerment (Hughes 2004), and restricted range species occur much more frequently in the tropics than in the temperate zones (Stattersfield et al 1998). Moreover, human impacts are mostly too recent to have had a substantial impact on genetic diversities of avian species, since, in order to have a significant impact, a bottleneck must last for many generations (Nei et al 1975).

Under the hypothesis that temperate zone species will show genetic effects of past glaciation, species that breed in North America and winter in the tropics might be expected to show strong bottleneck effects because the shape of the North American land mass severely restricted available breeding habitat during the last glaciation. Consistent with this prediction, the overall strongest evidence of genetic bottleneck effects was observed in Nearctic migrants. These species have been subjects of conservation concern because of recent population declines, some but not all of which may be attributed to wintering habitat loss due to clearing of Neotropical forests (Askins et al 1990; Pimm & Askins 1995). That these species have relatively reduced genetic diversity and, potentially, an elevated frequency of slightly deleterious mutations adds further urgency to conservation efforts, since these genetic characteristics may make them particularly vulnerable to recently introduced infectious agents such as West Nile virus (McLean 2006; Van der Meulen et al 2005).

Given the goal of preserving genetic variability in populations, information on genetic diversity in nature is a key component of any effective captive breeding program. For example, if there is evidence that a species has low genetic variability in nature, prudence suggests that special care be taken to preserve that diversity as part of an effective management scheme. Our results showed that relatively limited amounts of genetic information from easily amplified mitochondrial protein-coding genes can provide sufficient information to detect major trends in the genetic diversity within species. Data on polymorphism at these loci thus is a straightforward and cost-effective means of surveying overall genetic diversity that can be applied to any species of conservation concern.

## ACKNOWLEDGMENTS

This research was supported by grant GM43940 from the National Institutes of Health to A.L.H.

## REFERENCES

ALVAREZ-BUYLLA ER, R García-Burrios, C Lara-Moreno and M Martínez-Ramos. 1996. Demographic and genetic models in conservation biology: applications and perspectives for tropical rain forest tree species. Annual Review of Ecology and Systematics 27: 387-421.

ASKINS RA, F Lynch and R Greenberg. 1990. Population declines in migratory birds in eastern North America. Current Ornithology 7: 1-57.

AVISE JC, RM Ball, J Arnold. 1988. Current versus historical population sizes in vertebrate species with high gene flow: a comparison based on mitochondrial DNA lineages and inbreeding theory for neutral mutations. Molecular Biology and Evolution 5: 331-344.

BARRON EJ .1984. Ancient climates: investigation with climate models. Reports of Progress in Physics 47: 1563-1599.

BUCKLIN A, PH Wiebe. 1998. Low mitochondrial diversity and small effective population sizes of the copepods *Calanus finmarchicus* and *Nannocalanus minor*: possible impact of climatic variation during recent glaciation. Journal of Heredity 8: 383-392.

ESTOUP A, M Clegg. 2003. Bayesian inferences on the recent island colonization history by the bird *Zosterops lateralis*. Molecular Ecology 12: 657-674.

FAY JC, CI Wu. 1999. A human population bottleneck can account for the discordance between patterns of mitochondrial versus nuclear DNA variation. Molecular Biology and Evolution 16: 1003-1005.

FUERST PA, R Chakraborty and M Nei. 1977. Statistical studies on protein polymorphism in natural populations. I. Distribution of single-locus heterozygosity. Genetics 86:455-483.

HEWITT GM. 2004. The structure of biodiversity - insights from molecular phylogeography. Frontiers in Zoology 2004 I: 4.

HUGHES AL. 1991. MHC polymorphism and the design of captive breeding programs. Conservation Biology 5: 249-251

HUGHES AL. 2004. A statistical analysis of factors associated with historical extinction and current endangerment of non-passerine birds. Wilson Bulletin 116: 330-336.

HUGHES AL. 2005. Evidence for abundant slightly deleterious polymorphisms in bacterial populations. Genetics 169: 533-538.

HUGHES AL, B Packer, R Welch, AW Bergen, SJ Chanock and M Yeager. 2003. Widespread purifying selection at polymorphic sites in human protein-coding loci. Proceedings of the National Academy of Sciences of the USA 100: 15754-15757.

HUGHES AL, B Packer, R Welch, AW Bergen, SJ Chanock, M Yeager. 2005. Effects of natural selection on inter-population divergence at polymorphic sites in human protein-coding loci. Genetics 170: 1181-1187.

HUGHES AL, MA Hughes. 2007. Coding sequence polymorphism in avian mitochondrial genomes reflects population histories. Molecular Ecology 16:1369-1376.

KASTNER TP, MA Goñi. 2003. Constancy in the vegetation of the Amazon Basin during the late Pleistocene: evidence from the organic matter composition of Amazon deep sea fan sediments. Geology 31: 291-294.

KIMURA M. 1983. The Neutral Theory of Molecular Evolution. Cambridge University Press, Cambridge.

KIMURA M., T. Ohta. 1974. On some principles governing molecular evolution. Proceedings of the National Academy of Sciences of the USA 71: 2848-2852.

KUMAR S, K Tamura, IB Jakobsen and M Nei. 2001. MEGA2: molecular evolutionary genetics analysis software. Bioinformatics 17: 1244-1245.

LEONARD JA, RK Wayne and A Cooper. 2000. Population genetics of Ice Age brown bears. Proceedings of the National Academy of Sciences of the USA 97: 1651-1654.

LESSA EP, JA Cook and JL Patton. 2003. Genetic footprints of demographic expansion in North America, but not Amazonia, during the Late Quaternary. Proceedings of the National Academy of Sciences of the USA 100: 10331-10334.

LI WH. 1993. Unbiased estimates of the rates of synonymous and nonsynonymous substitution. Journal of Molecular Evolution 36: 96-99.

LOWE L. 2006. Quantifying the genomic decay paradox due to Muller's ratchet in human mitochondrial DNA. Genetical Research 87: 133-159.

MAILLARD JC, JP Gonzalez. 2006. Biodiversity and emerging diseases. Annals of the New York Academy of Science 1081: 1-16.

MARTIN PR, JK McKay. 2004. Latitudinal variation in genetic divergence of populations and the potential for future speciation. Evolution 58: 938-945.

MCLEAN RG. 2006. West Nile virus in North American birds. Ornithological Monographs 60: 44-64.

MCCUSKER MR, E Parkinson and EB Taylor. 2000. Mitochondrial DNA variation in rainbow trout (*Onchorhynchus mykiss*) across its native range: testing biogeographical hypotheses and their relevance to conservation. Molecular Ecology 9: 2089-2108.

MILLER HC & DM Lambert. 2004. Genetic drift outweighs balancing selection in shaping post-bottleneck major histocompatibility complex variation in New Zealand robins (Petroicidae). Molecular Ecology 13: 3709-3721.

NEI M. 1987. Molecular Evolutionary Genetics. Columbia University Press, New York.

NEI M & S Kumar. 2000. Molecular Evolution and Phylogenetics. Oxford University Press, New York.

NEI M, T Maruyama and R Chakraborty. 1975. The bottleneck effect and genetic variability in populations. Evolution 29: 1-10.

O'BRIEN S. 1994. A role for molecular genetics in biological conservation. Proceedings of the National Academy of Sciences of the USA 9: 5748-5755.

OHTA T. 1976. Role of very slightly deleterious mutations in molecular evolution and polymorphism. Theoretical Population Biology, 10, 254-275.

OHTA T. 2002. Near-neutrality in evolution of genes and gene regulation. Proceedings of the National Academy of Sciences of the USA 99: 6134-16137.

PIMM SL & RA Askins. 1995. Forest losses predict bird extinctions in eastern North America. Proceedings of the National Academy of Sciences of the USA 92: 9343-9347.

RUIZ-PESINI E, D Mishmar, M Brandon, V Procaccio and DC Wallace. 2004. Effects of purifying and adaptive selection on regional variation in human mtDNA. Science 303: 223-226.

SCHAEFFER SW. 2002. Molecular population genetics of sequence length diversity in the Adh region of *Drosophila pseudoobscura*. Genetical. Research 80: 163-175.

STATTERSFIELD AJ, MJ Crosby, AJ Long and DC Wege. 1998. Endemic Bird Areas of the World: Priorities for Biodiversity Conservation. BirdLife International, Cambridge.

STEADMAN DW. 2005. The paleoecology and fossil history of migratory landbirds. In: Birds of Two Worlds: the Ecology and Evolution of Migration (eds R. Greenberg & P.P Marra), pp 5-17. Johns Hopkins University Press, Baltimore.

TAJIMA F. 1989. Statistical method for testing the neutral mutation hypothesis by DNA polymorphism. Genetics 125: 585-595.

TEMPLETON AR. 1994. Biodiversity at the molecular genetic level: experiences from disparate macrooorganisms. Philosophical Transactions of the Royal Society of London Series B 345: 59-64.

VAN DER MEULEN KM, MB Pensaert and HJ Nauuwynck. 2005. West Nile virus in the vertebrate world. Archives of Virology 150: 637-657.

WAKELEY J. 2000. The effects of subdivision on the genetic divergence of populations and species. Evolution 54: 1092-1101.

WEINREICH DM & DM. Rand. 2000. Contrasting patterns of nonneutral evolution in proteins encoded in nuclear and mitochondrial genomes. Genetics 156: 385-399.

WILLIAMS TC, TH Webb III. 1996. Neotropical bird migration during the Ice Ages: orientation and ecology. Auk 113: 105-118.

ZHANG J, HF Rosenberg and M Nei. 1998. Positive Darwinian selection after gene duplication in primate ribonuclease genes. Proceedings of the National Academy of Sciences of the USA 98: 3708-3713.

# CONSERVATION EFFORTS TO RESTORE THE CAPTIVE-BRED POPULATION OF HOUBARA BUSTARD *CHLAMYDOTIS MACQUEENII* RELEASED IN THE WILD IN THE KINGDOM OF SAUDI ARABIA

M. Zafar-ul Islam, P. Mohammed Basheer, Moayyad Sher Shah, Hajid al-Subai & Mohammad Shobrak

National Wildlife Research Centre, P.O. Box 1086
Taif, Kingdom of Saudi Arabia

## Summary

The breeding program of Houbara Bustard was started in Saudi Arabia in 1986 to undertake the restoration of native species such as Houbara through a program of reintroduction, involving the release of captive bred birds in the wild. Two sites were selected for Houbara reintroduction such as Mahazat as-Sayd and Saja Umm Ar-Rimth, both protected areas in 1988 and 1998 respectively. Both the areas are fenced, fairly level and sandy plain with a few rock outcrops. Captive bred Houbara have been released in Mahazat since 1992 by NWRC and those birds have been successfully breeding since then. The nesting season of the Houbara at Mahazat recorded from February to May and on an average 20-25 nests are located each year but no nesting recorded in Saja. Houbara are monitored using radio transmitters through aerial tracking techniques and also by vehicle for terrestrial tracking. Total populations of Houbara in Mahazat is roughly estimated around 500 birds. In Saja, only 25 Houbara have survived since 2001 because most of the birds are predated immediately after release. Mean annual home range was calculated using Kernel and Convex polygons methods with Range VII software. The minimum density of Houbara was also calculated. In order to know Houbara movements or their migration to other regions, two captive-reared male Houbara were released into the wild and one wild born female were fitted with Platform Transmitter Terminals (PTT). The home range showed that the wild-born female had greater movements than two captive bred males. More areas need to be selected for reintroduction programs to establish the network of sites to provide easy access to move birds and associate with the wild Houbara. Some potential sites have been proposed which require more surveys to check the habitat suitability.

## Introduction

For thousands of years the falconry and the hunting of Houbara Bustard *Chlamydotis undulate* have been deeply embedded in the Arabian culture and traditions as depicted by many poets and story tellers. Historically, Houbara used to arrive in the Arabian region in large numbers from Central Asia, falconers used to hunt on camel back and considered the Almighty's recompense to those who have endured the summer heat. Since the advent of four wheel drive vehicles, it became easier for people to chase and poach Houbara even on sand dunes. These days Houbara are not hunted for food but for sport as the bird provides a challenge for hunters. Houbara is also illegally trapped to supply the demand for falcon training. The populations of all sub-species of Houbara have been declining at an alarming rate and the main threats are habitat loss and degradation as desert areas are developed for agriculture and infrastructure projects; these are compounded by high hunting pressure from falconers, with new areas in Central Asia, close to breeding grounds, increasingly being exploited (BirdLife International 2001). There are no reliable data for rates of decline, but given the substantial threats, population declines are likely to be significant and possibly widespread; moreover, they may accelerate if hunting pressure in Central Asia increases (Alekseev 1980; Collar 1979; Shams 1985).

### Distribution and Population of Houbara

The Houbara Bustard is a globally threatened species, which is listed as Vulnerable by the IUCN. This medium sized, arid and semi-arid area inhabiting bird is found in open or scrub-covered plains and occurs over a huge range from Canary Islands, Spain, across North Africa to the Middle East and Central Asia via South Asia to mainland China (Ali & Ripley 1987; del Hoyo et al 1996; Roberts 1985). The population has been estimated at 49,000-62,000 individuals, but it is likely to exceed 100,000 birds (BirdLife International 2001).

There are three sub-species recognized (1) *Chlamydotis undulate undulata* (9,800 birds) resident in North Africa where it has declined in Libya, Egypt and Tunisia, and probably also in Algeria, Mauritania, Morocco and Sudan; (2) *C. u. fuertaventurae* (700-750 birds) occurs on the Canary Islands, Spain (Goriup 1997); and (3) *C. u. macqueenii* [now given full species status] is thought to occupy six sub-regions: resident and migratory birds occur in the Middle East

(Turkey, Jordan, Israel, Iraq, Kuwait, Bahrain, Oman, Qatar, Saudi Arabia, United Arab Emirates, Syria, Yemen), and in Russia (including in the Asian region), Iran, Pakistan, India, Afghanistan, Uzbekistan, Tajikistan, from western Kazakhstan to Turkmenistan, and on the Mongolian plateau and in the Gobi desert of Mongolia and western China (BirdLife International 2001; Goriup 1997).

The population of *C. macqueenii* is estimated at 39,000-52,000 individuals, mostly breeding in Kazakhstan (30,000-40,000) (BirdLife International 2001), although numbers in mainland China are likely to be much higher than the current estimate of 500 birds (BirdLife International 2001). Declines are reported from Bahrain, Jordan, Iran, Iraq and India (BirdLife International 2001; Goriup 1997). Populations from some sub-regions are thought to mix on the wintering grounds.

The breeding program for Houbara Bustard started in Saudi Arabia to undertake the restoration of native species such as Houbara as the breeding population was virtually absent and the restoration program could realistically only be achieved through a program of reintroduction, involving the release of captive bred birds. From 1986-1988 fertile eggs were collected, under government permit, from resident populations in Baluchistan in Pakistan. By 1990s, through the application of artificial insemination techniques the NWRC was able to produce enough Houbara chicks to replace losses in the breeding unit and also were ready to release.

Two sites were selected for Houbara reintroduction those being the Mahazat as-Sayd and Saja Umm Ar-Rimth protected areas in Saudi Arabia, bearing in mind the IUCN resolutions on reintroduction.

The IUCN Guidelines for Reintroductions (1998) states that: "If captive or artificially propagated stock is to be used (for reintroductions), it must be from a population which has been soundly managed both demographically and genetically, according to the principles of contemporary conservation biology. [...]. Reintroductions should not be carried out merely because captive stock exists, nor solely as means of disposing of surplus stock".

### Wild Houbara Distribution in Saudi Arabia

As per the previous field studies on Houbara in Saudi Arabia the number and distribution of breeding birds have had a steep decline. The Houbara had breeding records that extended from the Jordanian border in the north, down in a band encompassing the north-west and eastern regions as far as the Rub-al-Khali but the Houbara is now an uncommon breeding visitor and restricted to north-western region of Saudi Arabia specifically Harrat al-Harrah, al Hammad and Al Nafud protected areas. Within the last five years there has been a number of recorded incidences from these reserves of people disturbing and poaching Houbara.

### Reintroduction Sites in Saudi Arabia

Mahazat as-Sayd protected area in Makkah province of about 219,000ha of area with fairly level, sandy plain at 900-1,100m elevation with a few rock outcrops. Mahazat is a special nature reserve established

**Fig. 1.** *Houbara bustard fitted with radio transmitter in the Mahazat as-Sayd protected area.* *(NWRC)*

in 1988, especially to re-introduce oryx, gazelle and houbara. Mahazat as-Sayd is about 175km northwest of Taif and south of al-Muwayh and other nearby town are Zalim and al-Khurmah. It is fenced and moderately to well vegetated with *Acacia totilis*, *Indigofera* sp. and *Salsola* sp. as dominant shrub/trees. The substrate at Mahazat may be sand, gravel, or alluvial clays, and is usually loose, but not shifting, forming an even surface.

Saja Umm Ar-Rimth is another protected area that was established as an extension of the Mahazat as-Sayd protected area in March 1994 by HRH Prince Saud Al Faisal as a possible reintroduction site for Houbara Bustards. In 1998, a ~6000km$^2$ area was proposed by the NWRC for the reintroduction of Houbara and of that 10% of the total protected area was fenced to release Houbara, a 400km$^2$ (20x20km) area of Jabal Barah, which is 5km east of the Zalim-Afif road.

In these protected areas the elevation ranges from 900-1100m, temperature varied from 8°C in winter to 46.3°C in summer with mean rainfall 6.2±8.2mm (min-max: 0.3-60.5).

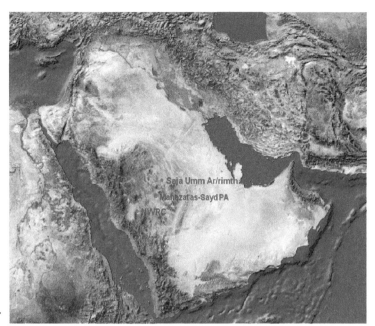

**Fig. 2.** *Houbara reintroduction sites in Saudi Arabia.*

## Habitat in the Study Area

In Mahazat as-Sayd PA, during the breeding season most of the females made nests on the open area and avoided more vegetative cover but males were seen in the vegetative cover consisting of moderate or sparse perennials, primarily grasses, herbs and shrubs but some times including larger bushes and trees such as *Acacia* spp. They were also recorded foraging in the green vegetation in wadis and small silty depressions but found roosting in the elevated boulder fields at night (Seddon & van Heezik 1994; pers. obs. 2006). Similarly in Saja Umm Ar-Rimth, Houbara are regularly seen in more open areas but seen resting under the shade during the hottest hours of the day in the summer months.

## Breeding in the Wild

The nesting season of the Houbara at Mahazat starts from February to May. There were 29 nests located in Mahazat as-Sayd protected area in 2006 but no nest was recorded in Saja Umm Ar-Rimth protected area. These nests were found close to sandy wadis, in areas of small basaltic boulders and in good vegetation cover. We also found large numbers of ant colonies around Houbara nests, which could be a reason for nests locations as there is an easy accessibility of food for females and their broods. The mean date of the first laid clutch was (5 April 2006 ±3 days) and breeding success per female (0.67±0.83 fledging chicks per breeding female, (n=29)), with breeding females from the 1993, 1994, 1995, 1996, 1998, 1999, 2002, 2003 and one from 2004 release cohorts.

## Reintroduction Methods

We have been releasing Houbara in Mahazat as-Sayd and Saja Umm Ar-Rimth protected areas using techniques of captive bred juveniles of 4-6 months, translocated to long, tunnel shaped cages and after three to four weeks release them in the enclosure.
Since 1991, in the Mahazat as-Sayd protected area, a total of 781 Houbara were released, of them 374 were males and 407 females. Out of 781 Houbara released in Mahazat, 133 died within a span of one month after the release and 648 survived. These mortalities are due to mammal predation and some are because of starvation. In the Saja Umm Ar-Rimth protected area, the reintroduction program of Houbara was started in 2003 and a total of 145 Houbara Bustard were released until 2006 of which 67 were females and 78 males. In Saja four are alive from 2003 to 2005 cohorts and around 30 Houbara are alive from 2007 cohort. Most of the birds died due to predation by mammals (foxes and cats), some cases of starvation and poaching were also documented. We recommended that we should have a predator free enclosure of 4km$^2$ to get Houbara acclimatized to the natural environment then the survival rate will also likely increase.

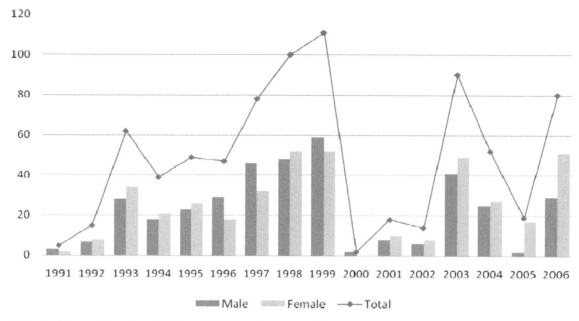

**Fig. 3.** *The total number of released Houbara in protected areas.*

## Total Population of Houbara in Mahazat and Saja Protected Areas

The Houbara population in Mahazat is the only population in the world of known size, sex and age ratio. Total population size (N) is estimated as follows:

Houbara released, radio tagged & regularly checked (monitored) = $n^1$
Radio-tagged missing birds = $n^2$
Wild born chicks not recorded = $n^3$
Wild born chicks recorded, but not radio-tagged = $n^4$
Immigrating Houbara = $n^5$
Birds deceased after release = $n^6$
Total population can be expressed as
$$N = (n^1 + n^2 + n^3 + n^4 + n^5) - n^6$$

This estimation is on annual basis. Roughly estimated, the population is 500 individuals surviving in Mahazat as-Sayd Protected Area. Mean annual home range was $467.7 \pm 352.6 km^2$ (n=59) using Kernel and Convex polygons methods with Range VII software.

## Houbara Density

Densities were assessed on three different ways:
* Observations during car driving transects
* Observation on circular points
* Number of Houbara radio tracked during the month divided by the size of the all area they used (Kernel 95 method)

The minimum density of Houbara in 2006 was 0.367 individuals per $km^2$.

## Challenges Ahead

Regular monitoring of Houbara is necessary to determine population status of Houbara in Saudi Arabia. Some aspects that need to be considered include:

* Need to support national legislation concerning the Houbara and other species of birds for conservation.
* Hunting should be avoided during the breeding season and no hunting should occur in and around reintroduction sites.
* Regular review of conservation planning protocols, and make necessary modification when needed.
* Falconers/hunters should also participate in the conservation programs initiated by the NWRC (National Wildlife Research Centre) in Taif, and provide information about their activities.
* All the Houbara range countries should exchange information regularly.
* There should be international team of researchers to share data and strategize conservation planning in consultation with government bodies.
* All breeding sites should be strictly protected.
* Large-scale habitat conservation programs are among the most promising steps for Houbara conservation, particularly if hunting is prohibited or strictly controlled is such areas.
* Additional areas can be identified for Houbara release and protect these sites to assist in bringing back a stable population.

# Conservation & Re-introduction

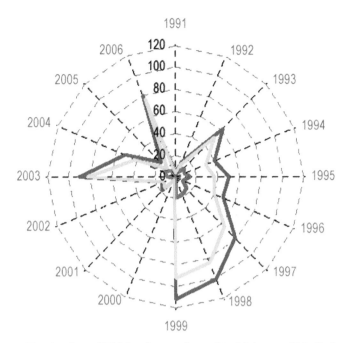

Fig. 4. *Out of 781 houbara released in Mahazat, 133 died within a span of one month after the release and 648 survived. These mortalities are due to mammal predation and some are because of starvation.*

* Conservation awareness programs should be initiated.
* Regular articles in local news papers should be published.

**Awareness Programs**

NWRC has carried out the educational and awareness programs to meet these challenges for long-term survival and conservation of endangered Houbara Bustard in Saudi Arabia.

## CONCLUSION

If hunting trends continue, Houbara populations may eventually collapse over most of its range, reaching such low levels that recovery would be difficult. Such a decline would certainly mean Houbara would no longer be able to be harvested for falconry purposes and this ancient Arab tradition would likely die. If the disappearance of Houbara from many countries is to be avoided, then conservation action must be taken immediately to ensure the protection of habitats within the Houbara's range, and to reduce the loss of animals from hunting.

## ACKNOWLEDGEMENTS

We want to extend our thanks and gratitude to HH Prince Bandar bin Saud Al Saud (Secretary General, NCWCD) for his leadership, generosity and continuous support towards the research and conservation work by the NWRC in the Kingdom. We also want to thank Mr. Abdulrahman Khoja, Mr. Ali Zahrani, Mr. Chukkans, rangers in the protected areas for their continuous support and encouragement for the field work.

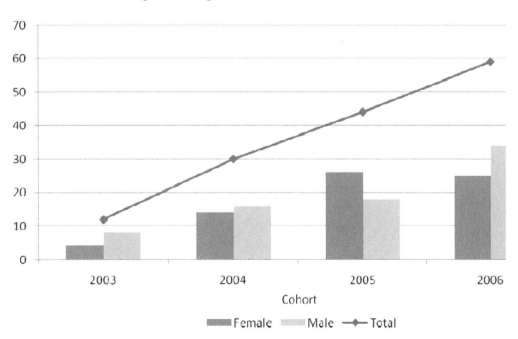

Fig. 5. *Number of surviving cohorts after release in to the Saja region.*

## REFERENCES

ALEEKSEEV AF. 1980. The Houbara bustard macqueenii in the North-West Kyzylkum (USSR). Zoologicheskij Zhurna 59: 1263-1266.

ALI S & SD Ripley. 1987. Handbook of the Birds of India and Pakistan. Compact Edition. Oxford University Press.

BIRDLIFE International. 2001. Threatened Birds of Asia: Red Data Book. BirdLife International, Cambridge, UK.

COLLAR NJ. 1979. The world status of the Houbara: A preliminary review. Proceedings of the Houbara Bustard Symposium, Athens, Greece.

COMBREAU O & T Smith. 1998. Release Techniques and Predation in the Introduction of Houbara Bustards in Saudi Arabia. Biological Conservation 84: 147-155.

JALME SM & VY Heezik (eds). 1994. Propagation of the Houbara Bustard. Kegan Paul International, London & NCWCD Riyadh. 112 pp.

MENDELSSOHN H, U Marder and M Stavy. 1979. Captive breeding of the Houbara (*Chlamydotis undulate macqueenii*) and a description of its display. XIIth Bulletin ICBP: 134-149.

ROBERTS TJ. 1985. The Houbara bustard in Pakistan in relation to conservation. Bustard Studies 3: 35-37.

SEDDON P, J Judas, RF Maloney, F Combreau. 2000. The reintroduction of Houbara Bustards in the Kingdom of Saudi Arabia. Pp. 22- 24. In: Soorae, P. S. & Seddon, P. J. (eds). 2000. Special Bird Issue. Reintroduction News, 19. IUCN/SSC Reintroduction Specialist Group, Abu Dhabi, UAE. 56 pp.

SHAMS KM. 1985. Occurrence and distribution of bustards in Baluchistan. Bustard Studies 3: 51-53.

HEEZIK VY, PJ Seddon and RF. Maloney. 1999. Helping Reintroduced Houbara Bustards Avoid Predation: Effective Anti-predator Training and the Predictive Value of Pre-release Behaviour. Animal Conservation 2: 155-163.

# THE ROLE OF SCIENCE IN AVIAN CONSERVATION: EXAMPLES FROM PACIFIC ISLAND KINGFISHERS

DYLAN C. KESLER

Fisheries and Wildlife Department
103 Anheuser-Busch Natural Resources Building
University of Missouri-Columbia
Columbia, MO 65211

## SUMMARY

Island species and populations appear to be more susceptible to extinction than their continental counterparts. Massive conservation efforts have been directed at rescuing island populations, and the results of those efforts have been mixed. One way to improve success with such conservation efforts is through thorough scientific investigations aimed at gathering information that will be useful to conservation practitioners. These studies have typically taken two forms, including research of declining and almost-extinct populations, and studies of surrogate populations and subspecies. I present examples of investigations that were conducted with the intent of improving conservation success for the critically endangered Guam Micronesian Kingfisher and the critically endangered Niau Kingfisher. To address conservation needs for the Guam Kingfishers, which are extinct in their native range, I worked with others to study a surrogate population on the island of Pohnpei, Federated States of Micronesia. Investigations focused on improving study techniques, nest site resources and selection, microclimate, and movement and territoriality. In summaries of these studies, I present the original needs of conservation practitioners, summarized results and conclusions, and important conservation implications. I also outline an ongoing *in-situ* investigation of the critically endangered Niau Kingfisher. The project's history, preliminary findings, and plans for future research are presented. Finally, I introduce a recently published idea of conducting research so that results can be applied to suites of similar species across the Pacific.

## INTRODUCTION

Studies of insular biota have profoundly influenced our understanding of the biological world. Islands are

simple microcosmic versions of their larger continental counterparts, and as such, investigations of insular fauna and ecosystems have inspired some of the most fundamental theories in ecology and evolutionary biology. Studies of island community structure (Simberloff & Wilson 1968; Komdeur 1994; Komdeur & Pels 2005), natural and human-caused changes and catastrophes (van Riper et al 1986; Savidge 1987; Steadman 1989; Steadman 1995), and evolutionary time (Darwin 1859; Wallace 1881; Grant 2001) have become the backbone of modern ecology. That Charles Darwin and Alfred Russell Wallace first developed the theory of evolution after observing assemblages of unique insular species is no coincidence (Darwin 1859; Wallace 1881). And although it has been aptly applied to continental "habitat islands," MacArthur and Wilson's (1967) Theory of Island Biogeography was the result of much time studying oceanic islands.

Island systems and species are highly susceptible to extinction (Moors 1993). Both stochastic and deterministic events profoundly affect island populations because of their small size and an ecological naiveté that stems from an evolutionary history often lacking competition and predation (King 1993). Although some refute the assertion that insular species are predisposed to extinction (Simberloff 1995), astounding losses of biodiversity have led most to conclude otherwise (MacArthur & Wilson 1967; Myers 1983; Steadman 1989; Johnson & Stattersfield 1990; Wiles et al 2003). For example, only one-fifth of the world's bird species occur on islands, yet more than 90% of the avian extinctions witnessed during historic times were island forms (Johnson & Stattersfield 1990).

On some islands, endemic populations decline so rapidly that captive breeding efforts are initiated with hopes that reintroductions to native ranges can occur some time in the future (e.g. Guam, see Wiles et al 2003). Captive breeding programs are usually necessitated by extreme circumstances, however, there is often little time to conduct thorough investigations of natural history before wild populations go extinct (e.g.

*Gallirallus owstroni* and *Todiramphus cinnamomina cinnamomina*). Thus, managers of captive populations of endangered species are often left with little information from which to design programs for their new guests. Basic information about nest sites, nutrition, breeding behavior, population demography, and climate conditions are often lacking (e.g. Bahner et al 1998). Similarly, those working to plan reintroductions of the endangered birds back into their native ranges are left without basic guidance about resources and habitats that might be best for introductions, and about the amount of area that should be protected in order to provide for minimum viable populations.

One way to provide the information needed to facilitate recovery in captivity, and simultaneously enhance reintroduction planning, is to undertake scientific investigations. The roles of science, conservation biology, and wildlife management have been described in detail previously (Meffe & Carroll 1997). When it comes to on-the-ground conservation situations, however, managers sometimes question the need for thorough scientific investigations, and instead choose to make decisions based on ad-hoc observations or advice. While these data can be quite useful, results cannot be reproduced, reconsidered, or re-evaluated without disclosed methodologies, and factors with the potential to bias conclusions that might not be identified. Scientific methodologies attempt to address these issues by outlining the exact methods used and the reasoning behind conclusions.

Two primary types of scientific studies have frequently been used for insular species that were suffering from population declines, or that were extinct in their native ranges. Surrogate species investigations focus on closely related or ecologically similar species on nearby islands. Investigations have also focused on remnant populations that remain in their native ranges. Both types of studies have the potential to provide profoundly important information, and both also have the potential to be biased and to mislead conservation practitioners.

In this paper, I will begin by discussing the relationship between science, conservation biology, and endangered species management to set a framework for the case examples that follow. The examples focus on scientific investigations of two species of Pacific *Todiramphus* kingfishers from Micronesia and Polynesia, and include a surrogate species study and an ongoing investigation of a declining population. The Guam subspecies of Micronesian Kingfisher *Todiramphus c. cinnamominus* is currently extinct in its native range and only exists as a captive population in U.S. zoos and one breeding facility on Guam. The second species discussed is the Niau Kingfisher *Todiramphus gambieri niauensis*, from the small isolated atoll of Niau in French Polynesia. Both studies were initiated with the intent of gathering information that would be useful to conservation efforts. I then conclude with a description of recent research that was conducted to provide broad results for suites of endangered species in the Pacific, which focuses on the *Todiramphus* group of kingfishers.

## Science and Conservation

Science, conservation, values, and wildlife management are all important concepts that come into play when working to keep a species from going extinct. However, each of these concepts is also somewhat amorphous and not always well defined. Science has been defined in many ways throughout history. For the purposes of this paper, I will embrace some common threads that run through nearly all of the definitions. In general, science attempts to identify simplified truths and realities of the natural world by employing unbiased techniques and synthesizing new results with existing information. Or as the Nobel Prize Laureate, Linus Pauling succinctly defined it, "Science is the search for truth" (Pauling 1958). With aims of being unbiased, science is fundamentally different from other value-laden endeavors, including conservation biology, endangered species conservation, and wildlife management. Perhaps the architects of conservation biology described it most concisely when they stated that conservation biology is a "discipline" that is "value laden" (see Meffe & Carroll 1997).

Science is vitally important to endangered species conservation, despite the differences in approach. In the recovery of endangered species, new information often is needed and science is the way to obtain unbiased and reliable new data. While few facts can ever be conclusively known as "truth," scientific methodologies are good at providing the most reasonable explanations. Statistical methods employed by most scientists also provide indicators and measures of reliability. Thus, conservation biologists using information derived through scientific investigations often have a means by which to evaluate the quality of the information they use.

## Surrogate Species Investigations

Research substitutes, in the form of closely related or ecologically similar species, have been used to gather

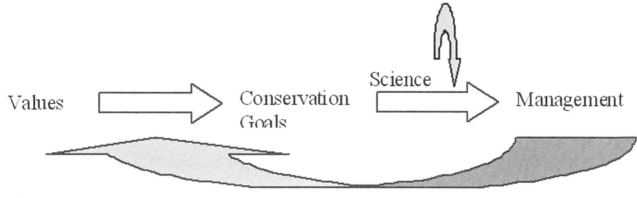

Fig. 1. *Conceptual relationship between values, conservation goals, science, and conservation management.*

important information needed for the conservation management of endangered organisms. Studies of these substitutes, or surrogate species investigations, are often used when the populations of interest have already gone extinct within native ranges, or when declining populations are so affected by their predicament that they might not provide useful biological information. These conservation surrogate species have the potential to be particularly effective in the Pacific, and on islands, where many species are ecological generalists and where speciation among islands has more likely been a result of isolation by watery distances than by specialized adaptations.

Surrogate species data should be used with caution, however, when they are used to guide endangered species management. Consideration should be given to differences in the evolutionary forces that shaped the ecology of the primary species of interest, and then comparisons should be made with the situations underlying the natural history of surrogate species. Additionally, the problem of bias takes on a new dimension when information from one population is generalized to other populations in regions with different resource and environmental conditions. Although natural patterns are not always closely correlated with taxonomic distance (Heppell et al 2000), perhaps one way to mitigate effects of bias is to use a level of caution that is directly related to both taxonomic and ecological distance from the source population.

## Surrogate Species Example - The Guam Micronesian Kingfisher

Like the nameless town in Rachel Carson's (1962) Silent Spring, Guam's forests are silent because of an environmental legacy that has become the case-study example of ecological problems encountered by insular biota (e.g. McCoid 1991; Meffe & Carroll 1997, p. 225-228). Guam is a U.S. territory with a land area of 550km$^2$, and it is situated at the southern end of the Mariana Archipelago in the central Pacific Ocean. The island historically hosted a diversity of habitats and associated endemic species that included eleven birds, two mammals, and countless plants, reptiles, and invertebrates (Baker 1951). In recent years, however, the island's fauna have suffered from reductions in available habitat, invasive species, and over-harvest.

Major military development began on Guam during World War II, when it was central to the U.S.'s Pacific campaign. After the war, the island continued to serve as a military hub and it simultaneously became a vacation destination. Guam's increasing human population and infrastructure severely impacted native habitats, and by the 1980s some of the only remnant blocks of native forest occurred on U.S. military-controlled lands in the north and along coastal benches and cliffs (Engbring & Ramsey 1984; Mueller-Dombois & Fosberg 1998).

Guam's fauna also suffered severe declines following the introduction of the Brown Tree Snake *Biogza irregularis*. The effects of introduced alien species are among the major conservation issues currently facing island systems (Elton 1958; Carlquist 1974; Drake et al 1989; Brown 1989; Richardson 1992; Atkinson 1993; D'Antonio & Dudley 1995), and the snakes were no exception. Over the course of several decades, the snake spread across the island and rapidly devastated local wildlife populations (Savidge 1987; Wiles 1987; Rodda & Fritts 1992). Seven birds and two bats were listed as endangered under the U.S. Endangered Species Act in 1984, and by 1985 three bird species were extirpated from Guam and two birds and one bat were extinct. In the end, the snakes severely affected seventeen of eighteen native bird populations, and twelve were likely extirpated as breeding residents (USFWS 1984; Wiles et al 2003). The Guam Micronesian Kingfisher exemplifies the decline in Guam's avifauna. The Guam Kingfisher population plummeted over several decades, and threats to the birds were finally recognized in 1984 when they were listed as endangered (USFWS 1984). Twenty-nine individuals were subsequently captured and placed in a captive breeding program in U.S. institutions as they went extinct in the wild (reviewed in Haig & Ballou 1995; Bahner et al 1998). Attempts to breed the birds in captivity have met with limited success since that time and fewer than 100 individuals are currently extant (B. Bahner pers. comm. 2006). Reasons for the stymied captive breeding program are not altogether clear, but some have suggested that a lack of knowledge about nutrition, microclimate, nesting resources and breeding behavior were hindrances (Bahner et al 1998; Baltz 1998; Kesler 2002; Kesler & Haig 2004; Kesler & Haig 2005a, Kesler & Haig 2005b). Few publications addressed the birds before their extinction from the wild, so those charged with the kingfisher's recovery had little information upon which to base management. The Micronesian Kingfishers were known to occupy mature forest, agricultural forest, mangrove and open habitat before their demise (Marshall 1949; Jenkins 1983; Pratt et al 1987). They nested in tree cavities or cavities excavated from the soft material of arboreal termite nests, or termitaria (Marshall 1989). In addition to the Guam Micronesian Kingfisher, the islands of Pohnpei and Palau also host endemic subspecies of conspecific kingfishers *T. c. reichenbachii* and *T. c. pelewensis*, respectively. Pohnpei Micronesian Kingfishers inhabit several small islands approximately 1,700km southeast of Guam, and the Palau Kingfishers occur on the Palau islands 1,300km to the southwest. The extant Micronesian Kingfisher subspecies are similar to the Guam birds in that recent information indicates drastic population declines, 63% on Pohnpei (Buden 2000). In an effort to provide information vital to the recovery of the captive population of Guam Kingfishers, as well as data useful for managing the wild populations of Pohnpei and Palau Micronesian Kingfishers, Dr. Susan Haig and I began a surrogate species investigation of Pohnpei Kingfishers in 1998. In the following section I present summaries of the studies that we undertook to address conservation concerns. The studies summarize data collected during six field seasons and more than twenty-four months of fieldwork on the island of Pohnpei.

## Micronesian Kingfisher Investigations

### Sex determination in Micronesian Kingfishers

Literature about the natural history of North American and other continental species is usually very complete and deep. However, investigators working with endangered species in other parts of the world are often confronted with a dearth of information at the onset of their studies. Our surrogate subspecies work in Micronesia was no exception, as challenges were omnipresent from the beginning of the Pohnpei Micronesian Kingfisher investigations. For example, understanding the bird's behavioral ecology in a way that would be useful to conservation efforts required that we at least know the sex of the study subjects. While the Guam Kingfisher subspecies is sexually dimorphic as an adult, the Pohnpei Kingfisher is not. Thus, one of our first goals with the surrogate species investigation was to develop and evaluate methods for sex determination in Pohnpei Micronesian Kingfishers. In 2006, Iara Lopes, Susan Haig, and I published a paper comparing morphological and molecular genetic techniques for sex determination in Micronesian Kingfishers (Kesler et al 2006). Tissue samples (blood) and morphological measures were collected throughout fieldwork with the hope that they would be useful for deriving methods for sex determination in the field. First, we developed methodology for sex determination in the lab by modifying existing molecular genetic techniques that had been developed for a suite of birds (Jensen et al 2003). Our new technique worked well for the Pohnpei Kingfishers and we were able to determine the sex of nearly every study individual. Molecular genetic techniques are not always available to field biologists, however, so we also attempted to develop a second method for determining the sex of study individuals. We derived a discriminant function - a mathematical equation that would yield the sex of study individuals when appropriate morphological measures were included. For each Micronesian Kingfisher that was captured during the five field seasons, we measured tarsus length, exposed culmen, wing cord, and total weight (see Baldwin et al 1931). Along with sexing results from the molecular genetic analyses, we used the morphological measures to derive a discriminant function that yielded the probability that a particular individual was male. The resulting equation follows:

$$\pi_i = \frac{e^{(35.34 - 0.07(\text{wing cord}_i) - 0.72(\text{tarsus length}_i) - 0.06(\text{culmen length}_i) - 0.19(\text{weigh}_i))}}{1 + e^{(35.34 - 0.07(\text{wing cord}_i) - 0.72(\text{tarsus length}_i) - 0.06(\text{culmen length}_i) - 0.19(\text{weigh}_i))}}$$

where morphometric measures for individual $i$ are used, and $\pi i$ is the probability that the bird is male (from Kesler et al 2006). To use the equation, one can enter the morphological measures for a single bird and solve for $\pi$. The resulting percentage will then indicate the predicted probability that an individual is male. For example, when the measures from a bird with a tarsus of 18.4mm, a wing cord of 103mm, exposed culmen of 41.4mm, and a weight of 56.5g are entered into the equation, it predicts that there is an 83.5% probability that a bird is male. Alternatively, the equation only yields a 17.7% probability if the measurements from a female are entered (20.0mm tarsus, a wing cord of 101mm, exposed culmen of 44mm, and weight of 67g). Molecular genetic techniques showed that the former was actually male and that the latter was a female.

The discriminant function equation was not perfect, however, because it correctly predicted the sex of only 73% (30 of 41) of the birds from the set used to derive the equation. The function is especially difficult to trust for individuals with predictions in the 40-60% range. For example, it is difficult to say that an individual with a 45% chance of being male is extremely different from one with a 55% chance of being male. However, the equation can be made more trustworthy with some adjustment in expectations. There is a much better chance that the predictions are correct if the equation yields predicated values that approach the extremes of 0 and 1, which is illustrated by an improvement in accuracy to 86% (18 of 21) for the sex of the individuals with $\pi i < 30\%$ and $> 70\%$. Nonetheless, we concluded that for a high degree of accuracy, field investigators should rely on molecular genetic techniques, which are now generally available from commercial facilities for relatively low costs.

**Nest Site Selection**

Many bird species have been driven by evolutionary forces to select nest sites that exclude predators or provide shelter from extreme environmental conditions (Martin 1995). At the same time, information about reproduction, including nest site use, is one of the most fundamental pieces of knowledge needed to manage captive and wild populations of birds. Little was known about the nesting habits of the Guam Micronesian Kingfishers at the onset of the captive breeding program (see Marshall 1989), so institutions managing the population requested data about the types of materials to supply to birds in aviaries. Similarly, the U.S. Fish and Wildlife Service, which was charged with designing a reintroduction plan for the birds, had little upon which to base evaluations of potential reintroduction sites on Guam. In an attempt to provide information about resources that might help increase breeding success in captive institutions, and to provide those planning the reintroduction of the birds to their native range with tools for site evaluations, we studied nesting materials and nest site selection in Pohnpei Micronesian Kingfishers (Kesler & Haig 2005b). Pohnpei Micronesian Kingfishers excavate nest cavities from the nests of arboreal termites *Nasutitermes* spp. or termitaria. The termite nests resemble bulbous paper mache structures attached to the sides of trees, or in areas where trees have been damaged. Micronesian Kingfishers excavate cavities from the soft material by flying into termitaria and colliding with their bills extended. They use the cavities for nesting. For this analysis, we traversed forests in the three kingfisher study areas on Pohnpei, and we measured the characteristics of all termitaria encountered. We assessed 25 termitaria that were used by Micronesian Kingfishers for nesting, and 242 termitaria that went unused. The physical dimensions and placement of the termitaria were measured, along with proximity to territory boundaries, and the characteristics of surrounding forest vegetation (Table 1).

We then employed a logistic regression analysis and modeled the features of the termitaria that were used by Micronesian Kingfishers for nesting, by comparing them to termitaria that went unused. Results of the analysis illustrate that nest termitaria were higher in the forest canopy, larger in volume, and occurred in areas with more contiguous canopy cover than unused termitaria. However, nest termitaria were selected independent of their proximity to forest edges and territory boundaries, and we found no difference in characteristics of termitaria on territories occupied by groups of birds and those on territories occupied by breeding pairs (Table 1). We further evaluated the relative abundance of termitaria with characteristics like those used for nesting to determine whether they were abundant in Pohnpei forests or rare. Results indicated that termitaria with nest-like characteristics were not limited in abundance, however, and that Micronesian Kingfishers thus had ample nesting materials on Pohnpei. We then used results from the investigation to provide guidelines for those managing the captive populations of Micronesian Kingfishers in the U.S. zoos. Along with basic measures of nest cavities, results suggested that the birds preferred nesting substrates that were larger in volume (>10 liters), and that the birds might prefer substrates that were placed

### Table 1

**MEAN SITE CHARACTERISTIC VALUES (+ SE) OF NEST AND NON-NEST TERMITARIA SELECTED BY MICRONESIAN KINGFISHERS ON POHNPEI, FEDERATED STATES OF MICRONESIA**

| Site Characteristic | Nest (n = 25) | Non-nest (n = 242) | Probability of equal means |
|---|---|---|---|
| Volume (l)* | 28.7 (1.1) | 11.4 (1.1) | $t_{67.17}$ -8.5 $P<0.001$ |
| Height in canopy (m)* | 4.3 (0.41) | 2.5 (0.12) | $t_{28.02}$ -4.21 $P<0.001$ |
| Open canopy (m$^2$)* | 1.5% (0.01) | 3.4% (0.03) | $t_{45.11}$ 3.06 $P=0.004$ |
| Mature forest (m$^2$)* | 97.4% (0.02) | 91.5% (0.14) | $t_{30.76}$ -2.12 $P=0.043$ |
| Distance to foraging area (m)* | 53 (6) | 39 (2) | $t_{28.54}$ -2.19 $P=0.037$ |
| Distance to forest edge (m) | 99 (14) | 77 (4) | $t_{27.16}$ -1.54 $P=0.135$ |
| Distance to territory (m) | 30 (5) | 21 (1) | $t_{27.70}$ -1.77 $P=0.089$ |

*P-values are presented for unequal variance t-test of nest vs. non-nest termitaria*
*\* denotes variables with statistically significant differences (Kesler & Haig 2005b)*

higher in aviaries. We also outlined a set of guidelines for nest site evaluations on potential reintroduction sites on Guam.

### Thermal Characteristics

Microclimates have the potential to affect reproductive success in many species of birds, so we studied the thermal characteristics of nesting sites used by Micronesian Kingfishers in the wild, and those provided to birds in the captive breeding program for Guam Micronesian Kingfishers. On Pohnpei, we assessed whether Micronesian Kingfishers selected sites based on microclimate (Kesler & Haig 2005a). Then in a second study, we investigated whether the birds in the captive breeding institutions were provided with microclimates resembling those used by wild Micronesian Kingfishers for nesting. For each of the microclimate investigations, we compared temperatures at nesting sites with the thermoneutral zone of Micronesian Kingfishers. The thermoneutral zone includes the range of temperatures within which extra metabolic energy is not needed to thermoregulate (Calder & King 1974). We estimated that the zone for Micronesian Kingfishers included the range of temperatures between 23.8°C and 38°C.

We compared the microclimate in areas where termitaria occurred to the microclimates at random forest sites to determine if the areas with termitaria might be more amenable to nesting birds. We also compared termitaria that were actually used by the kingfishers with those that were not used to determine if the birds were selecting from among potential breeding sites based on microclimate. Automated temperature loggers were placed at termitaria and at random sites in the forest, and we used their records to compare four thermal metrics: 1) daily maximum temperature; 2) daily minimum temperature; 3) the proportion of time when temperatures were below the thermoneutral zone for Micronesian Kingfishers; and 4) the overall amount of temperature fluctuation. Results showed that termitaria occurred in areas of the forest where the temperature fluctuated less. However, the magnitude of the differences was not extreme, and there were no statistical differences between the temperatures at termitaria used for nesting and those that went unused.

In a second set of analyses, we used automated temperature loggers to compare the microclimates in aviaries where Micronesian Kingfishers were breeding with those where the birds did not breed (Kesler & Haig 2004). We used the same four metrics that are

### Table 2

**LEAST SQUARES MEANS AND SE FOR ESTIMATED TEMPERATURES IN MICROHABITATS SURROUNDING POHNPEI MICRONESIAN KINGFISHER NEST TERMITARIA, NON-NEST TERMITARIA, AND FOREST LOCATIONS WITHOUT TERMITARIA ON POHNPEI IN SEPTEMBER 2001**

| Thermal Metric | Nest termitaria (n = 18) | Non-nest termitaria (n = 19) | Forest (n = 21) |
|---|---|---|---|
| $T_{max}$ (°C) | 28.05 (0.41) | 27.71 (0.40) | 28.70 (0.39) |
| $T_{min}$ (°C) | 22.86 (0.11) | 22.91 (0.11) | 22.72 (0.11) |
| $T_{flux}$[a] | 7.32% (0.02%) | 6.57% (0.02%) | 8.26% (0.02%) |
| % below $T_{lc}$[b] | 28.7% (2.6%) | 27.3% (2.6%) | 31.7% (2.5%) |

[a] *arcsine square root transformed for analysis and back transformed for presentation as CV*
[b] *proportion of 6-minute temperature observations below the lower critical temperature for adult Micronesian Kingfishers (Kesler & Haig 2005a)*

described above, and results were quite striking. Micronesian Kingfishers were breeding in warmer aviaries, whereas those in cooler aviaries were not (Figure 2). Daily maximum temperatures were 2.1°C higher in captive environments in which birds bred, and temperatures were within the thermoneutral zone 25% more often than in aviaries where the kingfishers did not breed. Then, we compared the temperatures in aviaries with those at nest sites used by Pohnpei Micronesian Kingfishers in their native range, and with temperatures at a set of aviaries in Guam, which were not being used to breed kingfishers at that time. Again, results showed that daily cyclical temperatures at the nest sites used by Pohnpei birds, and aviaries in Guam, were much higher than temperatures at aviaries that were part of the captive breeding program in U.S. zoos. The difference was even greater when cast in the light of physiological requirements, because temperatures in Guam and Pohnpei were within the thermoneutral zone of the birds for a much greater proportion of time. Compared to aviaries, habitats used by wild Pohnpei Kingfishers had daily maximum and minimum temperatures that were 3.2°C higher and the proportion of time when temperatures were in the birds' thermoneutral zone was 45% greater (Figure 2).

The conservation implications of the work were evident. To resemble microclimates like those at nest sites used by the birds within their native range, captive breeding aviaries needed to be warmer. Results suggested that captive breeding facilities should provide aviaries with daily ambient temperatures ranging from 22.06°C to 28.05°C to reduce microclimate-associated metabolic stress and to replicate microclimates used by Micronesian Kingfishers. Additionally, if temperature was a factor determining whether birds bred, Guam aviaries were more likely to be successful because of their temperatures.

Since that time, institutions within the Guam Kingfisher Species Survival Plan have worked to increase aviary temperatures, and birds are now being bred at Guam aviaries. Of course, many other improvements have also been made to the Guam Kingfisher breeding program, and the kingfisher population has increased dramatically in the last several years (B. Bahner pers. comm. 2007).

### Movement and Territoriality

Conservation practitioners planning the reintroduction of the Guam Micronesian Kingfishers back into their native ranges lacked information about how the birds used landscape resources. Managers asked, "Would a reintroduced population require large tracts of continuously forested areas?", "What if those forests were broken by roads?", "Would urban development prevent the birds from successfully breeding in certain areas, or would they avoid developed areas?", "How much area do a pair of Guam Micronesian Kingfishers need to breed?" and "What size is a territory?" Answers to these questions were key to designing a reintroduction plan for the birds, and to identifying areas on Guam that are most likely to support a reintroduced population of the Guam Kingfishers. Thus, we focused on evaluating where Micronesian Kingfishers occurred on the island of Pohnpei, and what the characteristics of those areas were. Then we assessed the movements of Micronesian Kingfishers where they did exist. We used island-wide surveys and radio telemetry to complete the study, and results of the investigation of multiscale occurrence and resource use were published in 2007 (Kesler & Haig 2007b; Kesler & Haig 2007c).

Point transect surveys were conducted on Pohnpei to determine where Micronesian Kingfishers occurred. Previous investigations documented the kingfishers throughout the island, with the highest densities in mangrove areas (Engbring et al 1990; Buden 2000). Additionally, results from the previous works documented kingfisher population declines of 63% between 1983 and 1994 (Buden 2000). Although they remained 15% to 40% lower than 1983 indices, our transect survey results were not so dire, because they showed kingfisher detection frequencies that were higher than those reported in 1994. We also used a combination of global information system analyses to evaluate correlations between landscape characteristics and kingfisher occurrences. We compared the vegetation characteristics of areas where birds were detected with the characteristics of areas where birds were not detected. Results indicated Micronesian Kingfisher detections were positively associated with the amount of wet forest (e.g. mangrove and swamp forests) and grass-urban vegetative cover (e.g. open grassy areas) and the birds were negatively associated with agricultural forest (e.g. banana, coconut, and papaya), secondary vegetation (e.g. areas of re-growth after forest clearings) and upland forest cover types (e.g. native upland "dwarf" forests).

We also used radio telemetry and high-resolution remote-sensing vegetation data to evaluate habitat use by individual kingfishers at the home range scale. At the most basic level, we described how the birds divide space among individuals. Micronesian Kingfishers are territorial. The home ranges of individuals within territories overlapped substantially, but there was very little overlap among territories. We then compared the vegetation in areas used by the birds with the vegetation in randomly located areas of similar size and shape. This allowed us to discern which habitat features the birds selected for use, and which features were randomly available in the immediate surrounding area. Results of the analyses illustrated that birds used more forested areas than were randomly available. The results also suggested that both open areas and lowland forest vegetation were extremely important to the birds (Table 3). Further, members of cooperatively breeding family groups included more forest in their home ranges than birds in pair-breeding territories and forested portions of study areas appeared to be saturated with territories (Table 3).

Together, results of the investigation suggested that both early and late succession forest habitats were limited for Micronesian Kingfishers on Pohnpei. Thus, protecting and managing forests is important for the

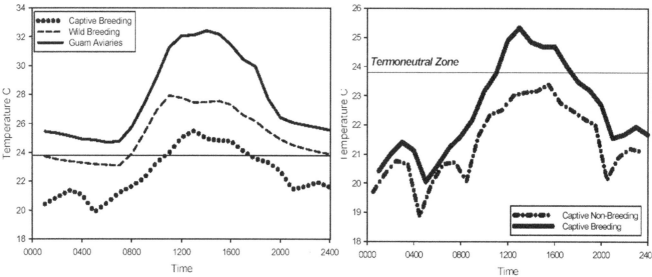

Fig. 2. *Mean temperatures throughout the diurnal cycle during the 2001 breeding season for wild Micronesian Kingfishers on Pohnpei and for Guam Kingfishers in the captive breeding program during the 2002 breeding season (left; from {Kesler & Haig 2005b}). Mean temperatures for Micronesian Kingfishers in nine of eleven Guam Micronesian Kingfisher captive breeding institution aviaries during the 2002 breeding season (right).*

restoration of Micronesian Kingfishers to the island of Guam, where they are currently extirpated, as well as to maintaining kingfisher populations on the islands of Pohnpei and Palau. Results further indicated that limited forest resources may restrict dispersal opportunities and therefore play a role in delayed dispersal behaviors in Micronesian Kingfishers.

**Investigations of Population Remnants**

Remnant and small populations can provide profound insight into the behavior and natural history of a species. Drawing information from remnant populations may also present problems, however, if they continue to be impacted by the forces that initially drove declines, or if they exist only in suboptimal situations where extinction forces have not yet arrived. For example, California Condors *Gymnogyps californianus* once ranged along the west coast of North America from Mexico to Canada (Snyder and Schmitt 2002). The North American explorers, Lewis and Clark, noted that the birds were abundant and numerous in the extremely wet Douglas Fir forests of the Columbia River valley, and along the Pacific coasts of Oregon and Washington (Lewis & Clark 1905). By the time conservation practitioners were ready to study the birds, however, they remained in only a small portion of their original range: the deserts of the American Southwest.

In 2002, the Société d'Ornithologie de Polynésie (SOP) contacted me with questions about a critically endangered kingfisher that they were attempting to recover on the island of Niau, in French Polynesia. The birds had been so affected by their situation that SOP personnel suspected that the entire population might be fewer than 50. We immediately began a cooperative project that is currently ongoing. In the following section, I provide a synopsis of recent work with our *in-situ* investigation of the critically endangered Niau Kingfishers.

**Population Remnant Example - Niau Kingfishers**

The Niau Kingfisher *Todiramphus gambieri niauensis* is one of the most endangered birds from the South Pacific region, and the world. Despite its location on the remote and sparsely populated atoll island of Niau in the UNESCO Tuamotu Man and Biosphere Reserve, there are indications that the birds are at extreme risk of extinction. Some estimates of the total Niau Kingfisher population have been lower than 50 individuals (Gouni & Sanford 2003; Gouni et al 2004). Additionally, the birds are listed as being in critical danger of extinction by the IUCN (2006 IUCN Red List Category, Critically Endangered; BirdLife International 2006) and they have become the focus of local governmental concerns. The reason for the Niau Kingfisher population decline is not altogether clear, however, because they remained largely unstudied throughout the Twentieth Century (Bruner & Dykes 1972; Pratt et al 1987; Fry et al 1992). Nonetheless, some have speculated that declines were caused by

**Table 3**

**VEGETATIVE COVERAGES (Ha) OBSERVED IN THE 95% KERNEL DENSITY HOME RANGES OF MICRONESIAN KINGFISHERS ON THE ISLAND OF POHNPEI, FEDERATED STATES OF MICRONESIA, 1999, 2000 AND 2002**

| Social Status | Birds | Short grass | Tall grass | Early forest | Old Forest | Forest Edge |
|---|---|---|---|---|---|---|
| Dominant Pair territory | 18 | 0.75 | 1.48 | 0.67 | 2.24 | 2.76 |
| Family territory | 16 | 0.96 | 1.96 | 1.08 | 2.94 | 4.25 |
| Helper Family territory | 11 | 0.80 | 3.46 | 1.19 | 3.35 | 6.72 |
| Juvenile Pair territory | 9 | 0.30 | 1.02 | 0.56 | 2.08 | 2.07 |
| Family territory | 7 | 0.54 | 1.27 | 0.51 | 1.55 | 1.95 |

habitat changes (Bruner & Dykes 1972), nest-site limitations (Gouni & Sanford 2003; Gouni et al 2004), and human persecution (Bruner & Dykes 1972).

Lack of information has hindered conservation efforts. BirdLife International, IUCN, and local conservation authorities (Société d'Ornithologie de Polynésie "Manu") have requested additional data about the natural history and population status of the birds (Gouni & Sanford 2003; Gouni et al 2004; BirdLife International 2006). In response, work began on the Niau Kingfisher project in 2002 (Gouni & Sanford 2003; Gouni et al 2004). After detecting only a few birds during site visits in 2002 and 2003, the project was expanded and a plan was developed to study the birds on Niau and use results for a two-pronged conservation strategy. Results would direct *in-situ* conservation efforts on Niau, and they would simultaneously be used to identify suitable locations to establish a second population of Niau Kingfishers on another island within the same biosphere reserve.

Our objectives included the following:

1) Conduct a thorough survey of Niau using point-transect surveys and make an empirical population estimate.
2) Investigate habitat use and breeding behavior of Niau Kingfishers with radio telemetry and direct observations.
3) Study potential translocation sites on nearby islands.
4) Design and conduct a translocation of the Niau Kingfishers to an appropriate site.
5) Study habitat use and demography of a post-translocation Niau Kingfisher population.

**Niau Kingfisher Current Project Status**

In addition to preliminary censuses and nest site augmentation in 2003 and 2004, more than four months of fieldwork was undertaken during two field seasons in 2006 (February-March and October-November). The island of Niau was thoroughly surveyed for Niau Kingfishers, empirical population estimates were made, and the distribution of birds was compared to habitat coverages and rat densities. During fieldwork, 171 variable-distance point-transect surveys were conducted. High-resolution Ikonos images were orthorectified, and used in combination with remote sensing and more than 132km of vegetation transects to develop a hyper-accurate vegetation habitat map of the island. Trap-lines were distributed throughout Niau (more than 432 trap nights) to assess the densities of introduced rat species. Kingfisher survey data were then related to vegetation compositions and rat densities using a global information system (GIS; ArcView, ESRI, Redlands, CA). Results yielded a Niau Kingfisher population estimate, and they illustrate the relationship between bird occurrence, habitat coverage, and rat densities. In short, preliminary results are showing that Niau Kingfishers are strongly associated with both native forests and coconut plantations, and that their occurrence is not associated (positively or negatively) with introduced rats. Nonetheless, populations remain extremely small and the birds are in grave danger of extinction.

Radio telemetry was also used during 2006 to investigate Niau Kingfisher breeding and social behavior, territory-scale habitat use, and dispersal. In February and October, 16 Niau Kingfishers were located, captured with mist nets and fitted with radio telemetry packages (Holohill Systems Ltd.). The birds were then tracked for approximately 45 days each. Preliminary results suggest that the birds maintain long-term territories, approximately eight hectares in size, multiple co-territorial occupants share space and resources, and yearling birds range widely when seeking dispersal destinations. We also located nests for five breeding pairs, tracked breeding phenology, and observed nestling and fledgling feeding and rearing.

Through this work, we have provided the first documentation of nest site characteristics, movement, home range size, territory structure, habitat use, nestling nutrition, and breeding phenology for this critically endangered bird. All of these data have been provided to local conservation practitioners, and much will soon be summarized in peer-reviewed publications. We intend to return to Niau in November 2007 and July 2008 for additional research. With the completion of fieldwork, we will then advise conservation practitioners about methods that might be most useful for increasing Niau Kingfisher populations. Additionally, we will develop evaluation criteria to assist with the identification of potential translocation destinations, so that a second population can be established.

**Suites of Species**

In addition to investigations of declining populations and surrogate species studies, conservation practitioners have begun discussions about conducting scientific investigations to derive information that might be useful in managing suites of species. Susan Haig and I recently published a paper aimed at addressing the needs of the many endangered *Todiramphus* kingfishers of the Pacific (Kesler & Haig 2007a). As research funds continue to dwindle while the numbers of endangered birds from the Pacific region continue to increase, conservation practitioners will need to reach across species boundaries for desperately needed information. Recent discussion has included investigations that are aimed at addressing problems for suites of species. This is a new idea, however, and one that needs additional development because there are numerous potential risks associated with extrapolating information.

## CONCLUSIONS

Scientific investigations have been used to address some of the most fundamental questions asked by those managing the captive population of Guam Micronesian Kingfishers. Investigations of basic breeding biology, nest site selection, habitat use and population biology have clarified a picture of the birds' resource requirements. Moreover, microclimate investigations of nesting sites give indications of the similarities and differences in conditions in captive facilities and those in wild environments. These same investigations provided information important for designing and implementing a reintroduction plan for the Guam Micronesian Kingfishers. However, many questions posed by captive breeding specialists and reintroduction biologists remain unanswered. With continued work, perhaps those too can be addressed through scientific investigations.

Niau Kingfishers are an example of a species just now beginning to be an *in-situ* study subject, and information about them has been gathered relatively quickly. In the course of just over a year, we obtained critical data relating to the social structure of the birds, movement and space use patterns, habitat use, and the potential influences of introduced rat populations. Undoubtedly, lessons learned during the Micronesian Kingfisher surrogate subspecies investigations bolstered our ability to work with the Niau birds. Hopefully this pattern will continue, and future investigations of these and other *Todiramphus* species can yield results with even greater efficiency.

## REFERENCES

ATKINSON IAE. 1993. The spread of commensal species of Rattus to oceanic islands and their effects on island avifaunas. Pages 35-81 in Conservation of island birds (P. J. Moors, Ed.). International Council for Bird Preservation, Cambridge.

BAHNER B, A Baltz, and E Diebold. 1998. Micronesian Kingfisher Species Survival Plan Husbandry Manual. Zoological Society of Philadelphia, Pennsylvania, USA.

BAKER RH. 1951. The avifauna of Micronesia, its origin, evolution, and distribution. University of Kansas Museum of Natural History Publication, Lawrence.

BALDWIN SP, HC Oberholser and LG Worley. 1931. Measurements of Birds. Cleveland Museum of Natural History, Cleveland, Ohio.

BALTZ AP. 1998. The assessment of reproductive potential in Micronesian Kingfisher pairs. Zoo Biology 17:425-432.

BIRDLIFE *International*. 2006. Species factsheet: *Todiramphus gambieri*. Downloaded from http://www.birdlife.org on 18/1/2007.

BROWN JH. 1989. Patterns, modes and extents of invasions by vertebrates. Pages 85-109. in Biological invasions: a global perspective (J. A. Drake, H. A. Mooney, F. di Castri, R. H. Groves, F. J. Kruger, M. Rejmanek, and M. Williamson, Eds.). Wiley, New York, USA.

BRUNER PL & OG Dykes. 1972. The birds of French Polynesia. Pacific Science Information Center, Bernice P. Bishop Museum., Honolulu, HI.

BUDEN DW. 2000. A comparison of 1983 and 1994 bird surveys of Pohnpei, Federated States of Micronesia. Wilson Bulletin 112:403-410.

CALDER WA & JR King. 1974. Thermal and caloric relations of birds. Pages 259-413 in Avian Biology vol. 4. (D. S. Farner, and J. R. King, Eds.). Academic press, New York, New York, USA.

CARLQUIST SJ. 1974. Island biology. Columbia University Press, New York, New York, USA.

D'ANTONIO CM & TL Dudley. 1995. Biological Invasions as agents of change on islands versus mainlands. Pages 103-121 in Islands: Biological Diversity and Ecosystem Function (P. M. Vitousek, L. L. Loope, and H. Adsersen, Eds.). Springer Verlag, New York, New York, USA.

DARWIN C. 1859. On the origin of species by means of natural selection. John Murray, London, England.

DRAKE JA, HA Mooney, F di Castri, RH Groves, FJ Kruger, M Rejmanek and M Williamson. 1989. Biological invasions: a global perspective. John Wiley & Sons, Chichester.

ELTON CS. 1958. The ecology of invasions by animals and plants. Chapman & Hall, New York.

ENGBRING J & FL Ramsey. 1984. U.S. Fish and Wildlife Service, Washington, DC, USA.

ENGBRING J, FL Ramsey and V Wildman. 1990. Micronesian forest bird surveys, the Federated States: Pohnpei, Kosrae, Chuuk, and Yap. Report to the U. S. Fish and Wildlife Service. U.S. Fish and Wildlife Service, Honolulu, Hawaii, USA.

FRY CH, K Fry and A Harris. 1992. Kingfishers, bee-eaters, & rollers. Princeton University Press, Princeton, New Jersey, USA.

GOUNI A & G Sanford. 2003. L'avifaune de Naiu (Polynésie française) en février 2003, cas particulier: Martin-chasseur, Todiramphus gambieri niauensis. Société d'Ornithologie de Polynésie Manu, Papeete, Polynésie Française.

GOUNI A, C Noiret, T Tehei and JB Tahua. 2004. Etude du Martin-chasseur de Niau, Todiramphus gambieri niauensis. Société d'Ornithologie de Polynésie Manu, Papeete, Polynésie Française.

GRANT PR. 2001. Reconstructing the evolution of birds on islands: 100 years of research. Oikos 92:385-403.

HAIG SM & JD Ballou. 1995. Genetic diversity in two avian species formerly endemic to Guam. Auk 112:445-455.

HEPPELL SH, H Caswell and LR Crowder. 2000. Life histories and elasticity patterns: perturbation analysis for species with minimal demographic analysis. Ecology 81:654-665.

JENKINS JM. 1983. The native forest birds of Guam. Ornithological Monographs. American Ornithologists' Union, Washington, D.C. No 31.

JENSEN T, FM Pernasetti and B Durrant. 2003. Conditions for rapid sex determination in 47 avian species by PCR of genomic DNA from blood, shell-membrane, blood vessels, and feathers. Zoo Biology 22:561-571.

JOHNSON TH & AJ Stattersfield. 1990. A global review of island endemic birds. Ibis 132:167-180.

KESLER DC. 2002. Nest site selection in cooperatively breeding Pohnpei Micronesian Kingfishers *Halcyon cinnamomina reichenbachii*: Does nest-site abundance limit reproductive opportunities? Masters Thesis, Oregon State University, Corvallis, OR.

KESLER DC & SM Haig. 2004. Thermal characteristics of wild and captive Micronesian Kingfisher nesting habitats. Zoo Biology 23:301-308.

KESLER, D.C. & S.M. Haig. 2005a. Microclimate and nest site selection in Micronesian Kingfishers. Pacific Science 59:499-508.

KESLER DC & SM Haig. 2005b. Selection of arboreal termitaria for nesting by cooperatively breeding Pohnpei Micronesian Kingfishers. Ibis 147:188-196.

KESLER DC & SM Haig. 2007a. Conservation biology for suites of species: population demography for the Todiramphus kingfishers. Biological Conservation 136:520-530.

KESLER DC & SM Haig. 2007b. Multi-scale resource use and selection in cooperatively breeding Micronesian Kingfishers. Journal of Wildlife Management 71:in press.

KESLER DC & SM Haig. 2007c. Territoriality, prospecting, and dispersal in cooperatively breeding Micronesian Kingfishers *Todiramphus cinnamominus reichenbachii*. Auk 124: in press.

KESLER DC, IF Lopes and SM Haig. 2006. Sex determination in the Pohnpei Micronesian Kingfisher using morphological and molecular genetic techniques. Journal of Field Biology 77:229-232.

KING WB. 1993. Island birds: will the future repeat the past? Pages 3-15 in Conservation of island birds (P. J. Moors, Ed.). International Council for Bird Preservation, Cambridge.

KOMDEUR J. 1994. Experimental evidence for helping and hindering by previous offspring in the cooperative-breeding Seychelles Warbler *Acrocephalus sechellensis*. Behavioral Ecology and Sociobiology 34:175-186.

KOMDEUR J & MD Pels. 2005. Rescue of the Seychelles warbler on Cousin Island, Seychelles: The role of habitat restoration. Biological Conservation 124:15-26.

LEWIS M & W Clark. 1905. Original journals of the Lewis and Clark expedition 1804-1806. in Vol. 3, pt. 2, vol. 4, pts. 1 and 2. (R. G. Thwaites, Ed.). Dodd and Mead and Co., New York, NY, USA.

MACARTHUR RH & EO Wilson. 1967. The theory of island biogeography. Princeton University Press, Princeton, New Jersey, USA.

MARSHALL JT Jr. 1949. The endemic avifauna of Saipan, Tinian, Guam, and Palau. The Condor 51:200-221.

MARSHALL SD. 1989. Nest sites of the Micronesian Kingfisher on Guam. Wilson Bulletin 101:472-477.

MARTIN TE. 1995. Avian life history evolution in relation to nest-sites, nest predation and food. Ecological Monographs 65:101-127.

MCCOID MJ. 1991. Brown Tree Snake *Bioga irregularis* on Guam: a worst case scenario of an introduced predator. Micronesica 3:63-69.

MEFFE GK & RC Carroll. 1997. Principles of conservation biology, 2nd ed. Sinauer, Sunderland, Massachusetts, USA.

MOORS PJ. 1993. Conservation of island birds. International Council for Bird Preservation, Cambridge.

MUELLER-DOMBOIS D & FR Fosberg. 1998. Vegetation of the tropical Pacific islands. Springer, New York.

MYERS N. 1983. The sinking ark : a new look at the problem of disappearing species. Pergamon Press, New York.

PAULING L. 1958. No More War! Dodd Mead, New York.

PRATT HD, PL Bruner and DG Berrett. 1987. A field guide to the birds of Hawaii and the tropical Pacific. Princeton University Press, Princeton, New Jersey, USA.

RICHARDSON AM. 1992. Altitudinal distribution of native and Alien Landhoppers *Amphipoda talitridae* in the Ko'olau Range, O'ahu, Hawiian Islands. Journal of Natural History 26:339-352.

RODDA GH & TH Fritts. 1992. The impact of the introduction of the Colubrid Snake *Bioga irregulairs* on Guam's lizards. Journal of Herpetology 26:166-174.

SAVIDGE JA. 1987. Extinction of an island forest avifauna by an introduced snake. Ecology 68:660-668.

SIMBERLOFF D. 1995. Why do introduced species appear to devastate islands more than mainland areas? Pacific Science 49:87-97.

SIMBERLOFF DS & EO. Wilson. 1968. Experimental zoogeography of islands: the colonization of empty islands. Ecology 50:861-879.

SNYDER NFR & NJ Schmitt. 2002. California Condor *Gymnogyps californianus*. in The Birds of North America, No. 610 (A. Poole and F. Gill, Ed.). The Birds of North America, Philadelphia, PA.

STEADMAN DW. 1989. Extinctions of birds in eastern Polynesia: a review of the record, and comparisons with other Pacific island groups. Journal of Archaeological Science 16:177-205.

STEADMAN DW. 1995. Prehistoric extinctions of Pacific island birds: biodiversity meets zooarchaeology. Science 267:1123-1131.

U.S. FISH AND WILDLIFE SERVICE. 1984. Endangered and threatened wildlife and plants: determination of endangered status for seven birds and two bats on Guam and the Northern Mariana Islands. Federal Register 50 CFR Part 17 49:33881-33885.

VAN RIPER CI, SG van Riper, ML Goff and M Laird. 1986. The epizootiology and ecological significance of malaria in Hawaiian land birds. Ecological Monographs 56:327-344.

WALLACE ARR. 1881. Island life. Harper & Bros., New York, USA.

WILES GJ. 1987. The status of fruit bats on Guam. Pacific Science 41:148-157.

WILES GJ, J Bart, RE Beck and CF Aguon. 2003. Impacts of the Brown Tree Snake: Patterns of Decline and Species Persistence in Guam's Avifauna. Conservation Biology 17:1350-1360.

## Author Biography

My research focuses on the interaction between landscape resources, social behavior, movement, and population biology in endangered birds. Specifically, I am interested in conservation biology, behavioral ecology, spatial ecology and movement and population biology. I am interested in conducting research on the interface of science and conservation. I intend for results from my work to address why species go extinct and I hope they provide insights that can be used to stop extinction. I also participate in recovery groups to ensure that results from my work are available to population managers and conservation practitioners. I have been fortunate to work with many talented scientists and conservation biologists. Most notably, Dr. Susan Haig provided guidance, inspiration, and focus as my doctoral program advisor during Micronesian Kingfisher studies. As a postdoctoral sponsor, Dr. Jeff Walters lent his keen insight and deep experience to recent projects focused on Red-cockaded Woodpeckers and Niau Kingfishers. Anne Gouni is among the most enthusiastic and dedicated conservation practitioners in the Pacific, and I appreciate the research opportunities that she has provided. Numerous other individuals provided valuable insights, including Director Mark Ryan, Department Chair Jack Jones, UMC faculty, graduate students and staff, the Haig and Walters labs, friends around the dinner table in French Polynesia, and at Sakau en Pohnpei.

# TRANSLOCATING SPECIES: THE VALUE OF THE ENVIRONMENT VERSUS THE VALUE OF THE SPECIES

Pete McClelland

Department of Conservation
P O Box 743
Invercargill
New Zealand

## Summary

The translocation of species has been carried out extensively in New Zealand for many years. It has saved a number of species from extinction and provided important back up populations for other species. It should be noted, however, that most of the earlier translocations and transfers were done with little thought to the effect the species involved would have on the receiving environment, be it direct, such as predation or competition, or indirect such as the effect of ongoing monitoring or management. This has now changed with the introduction of the IUCN Reintroduction Guidelines and other operating procedures which require any proposal to take into account all possible impacts so that these can be judged against the likely benefits of establishing a new species at the site and a considered decision made on what is best overall.

## Introduction

Many of New Zealand's unique animal species have been negatively affected by the introduction of a suite of introduced predators to which they have no defence, with many species becoming extinct and many more on the verge. Fortunately, the fact that there are over a hundred smaller islands around the coast of New Zealand, outside the swimming range of introduced predators, including some in the 500-2000ha range, has proved the saviour for a number of species.

The first species transfers in New Zealand occurred when Maori, the initial Polynesian settlers of New Zealand, transferred species with the primary if not sole purpose of providing additional food sources at important locations. The most common and obvious of these was the introduction of Weka *Gallirallus australis*, a large flightless rail valued as a food source to the early Maori, to many islands around New Zealand.

Wildlife managers in New Zealand have in the last five decades rescued many threatened species, particularly birds, from likely extinction by removing what are often the last individuals from the causes of their decline and marooning them on predator free islands (Williams & Mills 1978). They have also established backup populations of some island species to reduce the risk of losing a species to a stochastic event (e.g: predator invasion).

Translocation was first used for conservation reasons by Richard Henry - New Zealand's first wildlife ranger, who, in the 1890s after observing that many of New Zealand's flightless birds were rapidly disappearing in the face of the introduced mammalian predators, transferred over 400 flightless Kakapo *Strigops habroptilus* as well as Kiwi *Apteryx spp.* and Weka to Resolution Island off the Fiordland coast (Hill & Hill 1987). While the transfers were successful, the program was abandoned when Stoats *Mustela eminea*, one of the main culprits for the decline on the mainland, were found to be able to swim the 520m to the island.

More recently, the technique has been used to save the Saddleback *Philesturnus carunculatus* (Hoosen & Jamieson 2003) of which there are two subspecies, one found in each of the North and South Islands. The North Island subspecies died out on the mainland and most offshore islands until a single population of <500 birds remained on 500ha Hen Island. Since 1964, additional populations totalling approximately 6000 birds have been established on 12 other islands using direct transfer techniques.

The South Island subspecies came even closer to extinction when its sole remaining refuge, Big South Cape Island (Taukihepa), was invaded by Ship Rats *Rattus rattus* in the early 1960s. In a last ditch attempt to save the subspecies, 36 birds, all that could be caught, were transferred to two nearby islands. These populations grew to the point where they could be harvested for further transfers until now the birds are established on 14 different islands with an estimated population in excess of 1000 birds. Saddlebacks proved to be sturdy birds, able to be held in captivity and readily taking artificial food, but also able to handle a hard release into new habitat. Of the several hundred South Island Saddlebacks that have now been transferred, some over hundreds of kilometers, and some being held in captivity for over a week due to bad weather, only one has died during transfer. Unfortunately, while the Saddleback was saved, two species of bird, the Bush Wren *Xenicus longpipes* and Stewart Island Snipe *Coenocorypha aucklandica* and one species of mammal, the Greater Short-tailed Bat *Mystacina robusta*, were lost to the rats when insufficient numbers could be caught to establish populations on other islands.

There were a range of other transfers carried out in the intervening years- some to safeguard species, some for site restoration projects, and yet others for aesthetic reasons. Of those that were successful, many have had, often unintentionally, significant conservation benefits. They included the transfer of Little Spotted Kiwi *Apteryx owenii* to Kapiti Island off the southwest coast of the north island (Jolly & Colbourne 1991). Another species to have since died out on the mainland, Little Spotted Kiwi, have thrived on Kapiti and have since been transferred to several other islands, although the size of their territories approximately two hectares on islands with good habitat and likely much larger at less productive sites requires larger islands to hold viable long term populations. There are many other examples where species that have been at risk of becoming extinct, or in some cases on the verge of extinction, on the mainland have been introduced to offshore islands to try and protect them at least in the short-to-medium term. Yet others, which have been relatively safe on a limited number of islands, in some cases only one, have been introduced to additional islands to safeguard them in case of a stochastic event such as a predator invasion or disease. The one thing that, until relatively recently, has been common to most transfers was the limited or often total lack of thought about what impact the translocated species would have on its new home. There is little doubt, although equally little hard data for most species, that at least some translocated species would have had a serious negative impact on their release site. This is especially true in the case of conservation introductions to sites where the species was not previously present, as opposed to reintroductions, translocations of a species back to a site from which they had at some stage in the past been extirpated. Most of New Zealand's island translocations have been conservation introductions to help safeguard the species (i.e. species to new sites where the possible effects were little understood). Although in New Zealand there have really been no true top-level predator translocations, there have been many middle order predator transfers. This includes the Weka mentioned previously, which is known to predate smaller seabirds and ground dwelling birds, as well as lizards and invertebrates (Department of Conservation 1999). The introduction of Weka has in fact caused the local extinction of a range of bird species, including Banded Rail *Rallus philippensis* and Fernbird *Bowdleria punctata* on Kundy Island (R. Trow pers. comm. 2007) and significant declines in others, as observed in the decline of the Cook's Petrel *Pterodroma cookii* on Codfish Island/Whenua Hou from an estimated 20,000+ burrows in 1934 to as few as 100 breeding pairs by the early 1980s (Taylor 2000). Following the removal of Weka in 1984, they have increased to over 3000 pairs (Rayner & Parker in prep). This has lead to a conflict among managers as to whether the Weka, itself a threatened species, should be removed from islands to safeguard the other species. This is usually handled on a case by case basis depending on the known and potential values of the island, and the threat status of the particular subspecies of Weka.

Although it has not been documented, Saddlebacks are also likely to have had a significant ecological effect on their release sites, as they are active predators on a wide range of invertebrates and almost certainly small lizards. Until recently, the transfers of this species were carried out with little or no regard for other species and based on recent work in the area, many of the islands on to which they were released are likely to have had threatened, or at least had species with limited distributions, present. One anecdotal example for the impact of Saddleback is for the Herekopare or Foveaux Weta *Deinacrida carinata* a large threatened othopteran only recorded on three islands off southern New Zealand. Herekopare Weta were recorded on Kundy Island, the most remote of the sites, in the 1970s. Following the introduction of South Island Saddlebacks in 1978 only one specimen has ever been found there. Big Island, which is close to Kundy, and may have had the Weta present in the past, was one of the first two islands to have Saddlebacks released on it in the 1960s before any invertebrate surveys had been carried out on the island. More recent surveys have failed to find any sign of the Weta on the island so we can never be sure if they were there or not. This may have meant that saving the Saddleback from extinction (which would almost certainly have happened without the transfer to Big Island), may have lead to the Weta becoming locally extinct before it was even recorded on the island. Big Island also has a large skink species *Oligosoma chloronoton*, adults of which are commonly seen but no juveniles have been recorded in recent years possibly due to predation by Saddlebacks.

The impact on lizards and particularly invertebrates from birds such as Weka and Saddleback is hard to detect, let alone prove, since on very few islands is there sufficient baseline information to compare any contemporary survey with; however, comparisons between islands with and without Weka and/or

Saddleback indicate that there is an impact although, it has frequently been confused by the presence of rats.

The impact of many translocated species would be even less obvious than that of Saddlebacks, especially insectivores that feed on smaller invertebrates and herbivores that may compete with existing species. There is also the risk of introducing diseases especially in transfers over greater distances or involving birds from captive institutions, although this can usually be more easily managed through appropriate disease screening.

An additional consequence of many translocations is that of the ongoing impact on the release site from humans who visit to either manage or monitor the species. Two New Zealand examples of this are the Kakapo *Strigops habroptilus* and the Chatham Island or Black Robin *Petroica traversi*. The Chatham Island Robin was the subject of an extremely intensive management program in the 1980s and 1990s to try and build up the population from a low of five birds, including only one breeding female (Butler & Merton 1992). The work involved inter-island transfers as well as nest manipulations, and required personnel to be on the islands for extended periods. The islands are heavily burrowed by seabirds, meaning that the track network that had to be put in place initially for the management work destroyed a lot of burrows, albeit the impact is likely to have been small relative to the total number on the island. Later tracking was kept to a minimum, and was put in with consideration for the impact on other species. There would also have been less obvious impacts simply from having people present on the island for extended periods, including disturbance to other species, waste disposal, biosecurity risks and the effect of structures that were erected (e.g. the hut), although the later would have largely been limited to the area around the structure itself.

Kakapo are another species which has required intensive hands on management to try and bring it back from the brink of extinction, (Department of Conservation 1996) although unlike the robins which now number over 200 and require minimal further management, Kakapo, which have a very low productivity, are likely to require assistance for some time to come. The level of management they require involves having personnel present on the islands on which they are held all year, some times, e.g. when they breed which is not every year, in large numbers (20+). This impacts not only via the tracking on the island, which is slowly increasing on at least one island as access is required to new areas, but also the infrastructure that is required to provide for that number of people- waste disposal, electricity, water supply, etc., as well as using aircraft to resupply the staff on the island which has a noise impact and quarantine risks.

There can also be the cultural aspects of a translocation, where access to a site is restricted due to the presence of a species. For Kakapo this is not straightforward as the major site was a protected area with restricted access before Kakapo were introduced, although some people attribute the restrictions to the presence of Kakapo. Other cultural considerations could include the impact of the translocated species on other culturally important species and in some cases the moving of a species between areas can have cultural implications that need to be considered. In the case of Kakapo these impacts have been considered for the islands it is currently present on and balanced against the risk of losing the sole remaining representative of its genus. Kakapo have the added advantage over many other species in that they can, if required, all be removed from an island at any stage, at least during this current intensive management phase.

## Discussion

The past dominance of the larger, higher profile or charismatic species in translocation projects reflects the past emphasis on all aspects of management on those species, often to the detriment of the smaller and less conspicuous species. Colourful, attractive and charismatic birds versus small, scaly and innocuous invertebrates - not much of a competition for the public vote! For New Zealand in the past, this has usually lead to the hierarchy of protecting birds, then reptiles, followed by the larger invertebrates with small plants and invertebrates often forgotten.

While the public profile of the larger species, has been, is, and is likely to continue to be, higher than the other taxa, this will have an effect on management projects due to public expectations and public support often guiding political support and funding. The emphasis on the charismatic species above all else is changing in New Zealand. This is reflected in the revised translocation Standard Operating Procedures, which have been prepared by the Department of Conservation; they require consideration to be given to the impact any translocated species will have on its release site, most often islands in the case of New Zealand, before the transfer is approved. This does not prejudge the values of either the site or the species but simply allows the best overall outcome for conservation. Important considerations in the evaluation of any transfer proposal include:

- What impact the species will have on its new habitat - competition, predation, habitat modification, disease risk, aesthetic, human impacts from monitoring, etc?
- What is the risk to the species if it isn't transferred to the new island?
- What are the long term benefits to the species from being released into the site?
- What is likely to be the public perception, including transfers to open sanctuaries, primarily to give the public the chance to see the species in the wild which would not other wise be possible?

The public factor should not be underestimated as often landowners or local groups want particular high profile species released at sites to act as a focus for their project, as they believe they will get more attention and often funding, by having kiwi at your site rather than some unique but obscure invertebrate.

An example of the system in action is a proposal to trial having Kakapo on Campbell Island, 700km south of New Zealand (Department of Conservation 2002 unpublished). Prior to 2001, the presence of rats and cats meant the island was not suitable for receiving any species, but following the removal of all introduced mammals there have been a number of suggestions put forward. These included species that are likely to have been present before the rats invaded, as well as species that are in need of additional habitat to safeguard their long-term future.

Due to their breeding and ecological behaviour, Kakapo require large predator free islands with some specific habitat criteria such as suitable food plants, in order to establish long term viable populations. Understandably, suitable islands are few and far between and the managers of the species are keen to make the best use of any that are available. However, not only have Kakapo never been on Campbell Island there is unlikely to have been any ecological equivalent, so the impact is unknown. It is also unknown if Kakapo would survive, let alone breed, on the island, therefore, the proposal was to trial "spare" males and see if they came into breeding condition. This set the species managers at odds with those who see the value of the island and its natural fauna as paramount. Due to another option being identified for Kakapo in the short term at least, the issue has yet to be worked through but it will be watched keenly as the debate of how one ranks an iconic species against a unique island still needs to be examined in the New Zealand setting.

For reintroduction programs, it is important to consider that some environmental factors may have changed since the species was previously present and not simply thought, that since it was there in the past, returning it has priority over all other factors; after all something happened to cause the local extinction in the first place, so has the causal factor been clearly identified, understood and solved?

There can often be conflict between species managers over which species should be released at a site first, especially in the case where there is limited habitat available and where the two species may compete or where one predates on the other. This situation can occur for both restoration projects, or where there are two species both requiring new safe locations. Common sense indicates that ecologically it is better to release species lower down the food chain first giving them a chance to establish themselves before introducing their predators. This can, however, take many years with no obvious changes compared to releasing more conspicuous species such as birds that people can see and hear straight away. Managers have to weigh a number of factors when setting priorities, including scientific advice from relevant disciplines as well as community interest, as this may lead to decisions that go against what is believed to be best ecologically speaking. While considering a translocation proposal against the Translocation Standard Operating Procedures will usually not directly answer the main question of which species should be released first, they do force managers to consider all the relevant factors so that an informed decision can be made. In New Zealand, this has lead to a more coordinated approach to translocations, especially in the case of restorations where returning the site to as near to natural as possible is as important, if not the, paramount concern. The SOP also allows an informed decision to be made on whether the overall conservation benefits from an introduction outweigh the possible negative impacts and whether a reintroduction is likely to have an acceptable chance of success.

## CONCLUSION

There is no hard and fast rule for deciding which, if any species should be released at a particular site. What is important is that consideration is given not only to the benefit to the species to be transferred but also to the impact they will or may have at the release site. This includes any possible effect on other species either currently present or which may be introduced in the future. As a general rule this means introducing species lower down the food chain first and giving them a suitable chance to establish before introducing possible predators. The development of restoration plans for sites can go along way to ensuring that all the factors are considered when considering the translocation of species. Restoration plans should take into account the requirements of individual species as laid out in the Species Recovery plans and compare them with other considerations such as community interest and environmental factors. A restoration plan will also allow factors such as: what ecological state do you want the site returned to and what habitat management (e.g. wetland establishment), needs to be done, in addition to what species should be released and in which order. If the program is carried out with suitable consultation from multiple stakeholders, it also means that the managers can focus on getting the tasks done and not have to debate each action as they arise.

# REFERENCES

BUTLER D & D Merton. 1992. Saving the world's most endangered bird. Oxford University Press Auckland ISBN 0-19-558260-8.

DEPARTMENT OF CONSERVATION. 1996. Kakapo recovery plan 1996-2005 Threatened Species Recovery Plan 21. Department of Conservation: Wellington.

DEPARTMENT OF CONSERVATION. 1999. Weka (*Gallirallus australis*) recovery plan 1999- 2009. Threatened Species Recovery Plan 29 Department of Conservation: Wellington.

DEPARTMENT OF CONSERVATION. 2002. National Kakapo Team. Proposal for the transfer of Kakapo *Strigops habroptilus* from Pearl Island to Campbell Island/Motu Ihupuku. Department of Conservation Unpublished.

DEPARTMENT OF CONSERVATION. 2002. Standard Operating Procedure for the Translocation of New Zealand's indigenous terrestrial flora and fauna. Department of Conservation Electronic file OLDDM-718296.

HILL S & J Hill. 1987. Richard Henry of Resolution Island: John McIndoe in association with the New Zealand Wildlife Service, 364pg.

HOOSEN S & I Jamieson. 2003. The distribution and current status of New Zealand Saddleback *Philesturnus carunculatus*. Bird Conservation International 13:79-95.

JOLLY J & R Colbourne. 1991. Translocations of little Spotted Kiwi (*Apteryx owenii*) between offshore islands of New Zealand. Journal of the Royal Society of New Zealand: Vol 2 No 2 pp: 143-149.

RAYNER M & K Parker. 2007. Cook's petrel population survey, Codfish Island:. File NHS 03-07-04 Department of Conservation, Invercargill.

TAYLOR GA. 2000. Action plan for seabird conservation in New Zealand, part a: threatened seabirds. New Zealand Department of Conservation Threatened Species Occasional Publication 16, Wellington.

WILLIAMS G & J Mills. 1978. Marooning - a technique for saving threatened species from extinction. International Zoo yearbook 17:103-106.

# HOW DISEASE CAN AFFECT CAPTIVE BREEDING FOR CONSERVATION: A NEW ZEALAND EXPERIENCE

Kate McInnes[1], Richard Jakob-Hoff[2], Emily Sancha[3] & Jack Van Hal[4]

[1] Department of Conservation, Wellington, New Zealand
[2] New Zealand Centre for Conservation Medicine, Auckland, New Zealand
[3] Department of Conservation, Twizel, New Zealand
[4] Department of Conservation, Christchurch, New Zealand

## Summary

Captive breeding programs are becoming an increasingly important part of reintroduction programs. Among the criteria of the IUCN reintroduction guidelines is the necessity to screen release candidates for any transmissible disease. This paper looks at the history of disease risk in New Zealand avian breeding programs and the importance of these screening protocols in protecting *in-situ* populations of threatened taxa. Examples from the New Zealand Dotterel, Black Stilt and Orange-fronted Parakeet conservation programs are reviewed and discussed.

## Introduction

Due to its long history of geographic isolation, New Zealand has a unique fauna dominated by avian species, with only three species of native terrestrial mammal in the form of bats. As a result, the avifauna is highly susceptible to predation by introduced mammals such as mustelids, rodents and cats which has contributed to the serious decline in numbers and range of many avian species.

Conservation of the wild populations of many species in New Zealand relies heavily on controlling the predator numbers by trapping and poisoning, or by translocating birds to predator-free off-shore islands. In some situations, the threat of extinction is so immediate that captive breeding is essential to the survival of the species, or in other cases it may be used to supplement existing wild populations and/or establish new populations in newly created predator-free habitats. Captive breeding is not without its own suite of potential problems including inbreeding, breeding for "domestic traits", nutritional deficiencies of captive diets and infectious diseases. This paper examines three breeding programs in New Zealand and discusses the disease issues which have affected the programs. Each program had a slightly different goal and different husbandry techniques and these differences are highlighted in their unique disease issues ranging from reduction of breeding success/offspring survival, to deaths in captivity, to progeny being unable to be released into the wild due to infectious disease. The range of issues faced in these programs emphasizes the need for establishing clear goals and objectives from the beginning of the program, continued monitoring to ensure goals are achieved and to detect potential problems quickly with a flexible adaptive approach to problem solving to achieve the best outcomes.

## New Zealand Dotterel (Tuturiwhatu) *Charadrius obscurius*

Captive rearing and release trials of Northern New Zealand Dotterel *Charadirus o. obscurius* were undertaken as an analogue species to test the technique prior to use in the critically endangered Southern New Zealand Dotterel *Charadrius o. aquilonius*. Eggs were collected from the wild birds, hatched and the chicks reared in two captive facilities. Pre-release health monitoring was undertaken to ensure only healthy birds were released and to protect wild populations form the introduction of new diseases. This monitoring detected avian pox lesions on 70% of the birds, and Plasmondium-like organisms (avian malaria) in blood smears from 46% of the birds. Interestingly there had been no overt signs of illness from the malaria infections and without blood testing this infection would have remained undetected.

Under IUCN guidelines, animals chosen for rein-

**Fig. 1.** *New Zealand Dotterel eggs.*
*Photo: Department of Conservation.*

Fig. 2. *New Zealand Dotterel.*
Photo: D. Merton

Fig. 4. *Black Stilt or Kaki.*
Photo: Dick Vietch

troduction should be free of infectious disease that are absent at possible release sites. Disease investigation of the wild populations was undertaken to determine the prevalence of these diseases in the wild. Neither was detected during surveillance and sampling work, suggesting either an absence or very low prevalence of both diseases. It was decided that the birds should be maintained in captivity where a number subsequently succumbed to the effects of avian malaria. Disease investigation suggested a local source of infection would spread to the captive Northern Dotterel via biting insects. Prevention of future infections was achieved through modification of the aviaries to prevent access by insect vectors and subsequent rearing of the birds was unaffected by either disease.

This case highlights the need to undertake routine testing of apparently healthy birds prior to release to ensure diseases are detected and managed, and also illustrates that, once a disease risk has been identified, practical measures can be taken to manage this risk.

## Black Stilt (Kaki)
*Himantopus novaezelandiae*

An important component of the restoration strategy for the critically endangered Kaki or Black Stilt is captive breeding for release. Since 1981, 1,879 eggs were collected from wild and captive pairs, incubated artificially and most chicks reared by hand until released at either 60 days of age, or 9-10 months of age (van Heezik et al 2005). A significant increase in hatching and late-incubation deaths during two breeding seasons (1997 & 1998) prompted an investigation into the potential role of iodine deficiency (Sancha et al 2004). An initial analysis of the data compared the survival rates of wild and captive laid eggs collected and incubated in the same environment, and revealed a poorer rate for captive laid eggs. Necropsies of the dead birds revealed abnormal thyroid structure in 30% (1997) and 81% (1998) of birds examined. Necropsies of wild birds of Black Stilt, Pied Stilt and hybrids were undertaken to determine if the problem was also occurring in the wild and revealed no thyroid abnormalities. A presumptive diagnosis of iodine deficiency in the captive diet was made. Further investigations looked at serum levels of thyroxine (T4) and these were higher in wild stilts than in captive birds. Supplementation of the diet with iodine resulted in an increased T4 levels in the captive birds. As a result of inclusion of iodine supplementation in the captive Kaki diet, the survival of birds raised for release has improved and the wild population has increased dramatically.

## Orange-fronted Parakeet (Kakariki)
*Cyanoramphus malherbi*

There has been much debate over the species status of the Orange-fronted Parakeet. First described by

Fig. 3. *Black Stilt/Kaki breeding aviary.*
Photo: D.P.Murray

Fig. 5. *Orange-fronted Parakeet chicks.*
Photo: J. van Hal

Souance in 1857, it was declared a colour morph of the more common Yellow-crowned Parakeet *Cyanoramphus auriceps* based on cross-breeding experiments with Yellow-crowned Parakeets (Taylor et al 1986 & 1998). However, subsequent genetic work (Boon 2000) confirmed the separate species status.

The Orange-fronted Parakeet has always been rare compared with other mainland parakeet species, however it is now considered by the Department of Conservation to be a critically endangered species. Its wild populations are under threat from mammalian predators and potentially from hybridisation with the more numerous conspecific Yellow-crowned Parakeet. Part of the recovery plan for the species involves captive breeding for release into new predator-free habitats, which is being achieved by breeding captive pairs and through collection of eggs from the wild to be raised by Yellow or Red-crowned Parakeet *Cyanoramphus novaezelandiae* in captivity. The Red-crowned Parakeet used as surrogate parents were sourced from captive holdings rather than from wild birds. The advantage of this process was the use of known breeding pairs and birds which are accustomed to captive holding. The disadvantage was the risk of introduction of disease via the parents. A disease screening/quarantine program was designed to reduce this risk and required the collection of blood, faecal, feather and microbial samples to test for a suite of diseases. During the process, psittacine beak and feather disease (PBFD, psittacine circovirus) was detected in two birds. This disease has the potential to cause high mortality rates in neonatal pstittacines and also immunosuppression with subsequent bacterial and viral infections. The infection is not able to be treated and there is currently no vaccine available. Because this disease had been identified as a risk, it had been targeted with a specific test. Routine blood testing would not have detected this disease. It is fortunate that the captive management team followed the departmental procedures for assessing disease risk and were able to detect and exclude these two birds from the captive program.

## DISCUSSION

Each program was faced with its own unique challenges for employing captive breeding to contribute to the conservation of threatened species.

In the case of the Northern Dotterel, husbandry techniques and captive management were able to produce a high success in hatching and rearing of birds, but infectious disease from outside the facility prevented the birds from entering the wild population. This problem was detected quickly because the facility maintained a high level of disease management through pre-release screening. This early detection allowed the problem to be contained quickly and well managed for the future of the program through redesign of the aviaries to exclude the disease vectors. Without this vigilance and awareness of disease issues, two new diseases might have been spread into the wild population, with severe consequences.

The Black Stilt program was faced with a very different problem. Poor survival of eggs was a significant hindrance to production of birds for release. Investigation of the problem relied on good data keeping (source of each egg, clutch size, egg management records, etc) for the problem to be pinpointed to a difference between captive and wild laid eggs. Routine necropsy of dead birds to determine cause of death then provided further information in the puzzle of why these embryos were failing to hatch successfully and pointed to the iodine deficiency. The investigation could then expand to find more data to support this by looking at wild birds T4 levels, and by supplementing

Fig. 6. *Orange-fronted Parakeet after release.*
Photo J. van Hal

the captive birds and measuring the response in their T4 levels. This was a challenging situation which brought together the efforts of the captive facility and wildlife managers and the veterinary pathologists in the necropsy room and the laboratory, but without the initial good record keeping and monitoring of results, this problem might have taken longer to diagnose. In the case of the Orange-fronted Parakeet, the preparation of a good disease risk assessment and testing/quarantine program enabled the managers to exclude a potentially devastating disease. If PBFD had entered the captive population it would have had severe consequences, both for the production of healthy offspring and in limiting the options for release back into the wild. The diligence of the managers and the preparation of a comprehensive methodology for assessing disease risk was essential in avoiding this unwanted outcome.

## Acknowledgements

Massey University Veterinary School and in particular Maurice Alley for Iodine work in Kaki. Carter Atkinson and Karrie Rose (USGS) for diagnostic work in Dotterels.

Department of Conservation staff and independent researchers involved in captive management programs in New Zealand.

## References

BOON WM. 2000. Molecular systemics and conservation of the *Cyanoramphus* parakeet complex and the evolution of parrots. Unpubl. PhD thesis, Victoria University of Wellington, Wellington.

REED CEM .1997. Avian Malaria in New Zealand Dotterel. Kokako 4(1): 3.

SANCHA E, Y van Heezik, R Maloney and P Seddon. 2004. Iodine deficiency affects hatchability of endangered captive Kaki (Black Stilt, *Himantopus novaezelandiae*). Zoo Biology 23(1): 1-13.

VAN HEEZIK Y, P Lei, R Maloney and E Sancha. 2005. Captive breeding for reintroduction: influence of management practices and biological factors on survival of captive Kaki (black stilt). Zoo Biology 24(5): 459-474.

TAYLOR RH. 1998. A reappraisal of the Orange-fronted Parakeet (Cyanoramphus sp.) - species or colour morph? Notornis 45: 49-63.

TAYLOR RH, EG Heatherbell and EM Heatherbell. 1986. The Orange-fronted Parakeet *Cyanoramphus malherbi* is a colour morph of the yellow-crowned parakeet *Cyanoramphus auriceps*. Notornis 33:17-22.

# POPULATION BIOLOGY: THE SCIENCE OF POPULATION MANAGEMENT FOR CAPTIVITY, REINTRODUCTION & CONSERVATION

COLLEEN LYNCH[1,2]

[1] University of South Dakota, Department of Biology
  414 E. Clark Street, Vermillion, SD 57069
[2] Association of Zoos and Aquariums Population Man. Center
  8403 Colesville Rd., Suite 710
  Silver Spring, MD 20910-3314

## SUMMARY

Conservation biology is concerned with the fate of populations, which in turn are greatly influenced by demographic and genetic characteristics of those populations. Such characteristics determine the potential for adaptation to changing conditions and population persistence. In captivity, populations may be managed to ensure that viable populations of selected species are available into the foreseeable future for the purposes of continued captive breeding and exhibition as well as for the support of *ex-situ* and *in-situ* conservation efforts. Using methods initially developed for zoological captive breeding programs, population biologists and animal managers utilize demographic and genetic data to inform management strategies for populations from their founding, to reintroductions, to the integrated management of captive and wild populations for conservation. Such methods have been essential in recovery efforts for several taxa including Black-footed Ferret, Island Fox, Andean Condor, California Condor, San Clemente Loggerhead Shrike, Guam Rail, Micronesian Kingfisher, and Mississippi Sandhill Crane. Methods applied in these and other programs will be discussed.

## INTRODUCTION

Captive populations, as well as wild populations requiring conservation action, are typically small. As a result, biological and logistical challenges to their successful management exist. Small populations experience high levels of demographic stochasticity (random variation in individual reproduction, mortality and sex ratio) and deleterious genetic effects, which combine to greatly enhance extinction risk. As genetic effects influence reproduction and mortality rates, populations decline further and demographic stochasticity increases. These declines lead to smaller populations and increased inbreeding and loss of gene diversity. Genetic and demographic effects are therefore highly synergistic. Populations of sizes less than 200 are generally considered to be especially susceptible to synergistic dynamics of demographic and genetic risk (Frankham et al 2002).

Captive populations are often initiated with a limited number of individuals, and initially, husbandry knowledge may not be sufficient to encourage rapid population growth. High mortality and low reproduction due to lack of husbandry knowledge exacerbate the random variation in these vital rates and the resulting age structure of the population may not be stable. Skewed sex ratios, empty age classes and weak bases to age structures result in inconsistent population growth and potential population decline. Even as husbandry is perfected and the potential for population growth is enhanced, limited holding capacity for the many species of concern will ultimately constrain population growth.

Genetic effects due to small population size include accelerated rates of inbreeding and higher potential for exhibition of inbreeding depression (Crnokrak & Roff 1999). Loss of adaptive potential occurs with increasing homozygosity (Avise & Hamrick 1996) as genetic diversity decreases and adaptive potential has been linked to both individual and population fitness. While mutations generate gene diversity, mutations do not occur at rates sufficient to influence population structure within management time scales. Genetic changes, such as those associated with population fragmentation, isolation, and reduction in population effective population size, are, in turn, inherently linked with population viability.

Through the application of standardized methods for population management, these potentially detrimental demographic and genetic effects can be mitigated, and optimal management strategies for captive populations can be devised. Examples of such strategies include those employed by the Species Survival Plan® (SSP) of the Association of Zoos and Aquariums

(AZA) and the EEP of European Association of Zoos and Aquariums (EAZA). While zoo populations may be dispersed among several holding institutions across a geographic span, animals are often transferred between institutions and individuals are managed across space in cooperative population management programs.

The goals of these programs include rapid growth of founding populations to achieve demographically stable populations at program carrying capacities, creation of stable age structures and limiting extinction risk. Management strives to maintain the genetic variation present in the founder stock to the greatest extent possible, avoiding loss of heterozygosity due to genetic drift and inbreeding and thus maintaining adaptive potential. These populations are managed both for genetic health in captivity and as genetic reservoirs in the event of future reintroductions to wild populations.

To ensure the suitability of these captive populations as genetic reservoirs, management goals also include avoidance of artificial selection, including both the unintentional selection of animals with characteristics "well-suited" to captivity, and the intentional selection for or against specific traits. The role of selection in captive populations is poorly understood and maintaining maximum genetic variation is therefore prioritized.

**Data for Population Management**

A prerequisite to the development of a population management plan is the compilation and maintenance of basic data required for analysis. Typically this data is maintained in the form of a studbook. This record of a population includes parentage information and life history events (births, deaths, transfers, etc.) on individuals within the population from the time of population founding to the present. If a studbook does not exist, data may be compiled from a variety of sources, including the in-house records of institutions holding specimens currently and historically, the International Species Information System (ISIS) and the International Zoo Yearbook.

Studbook data include detailed information about individuals within a population including the following: specimen identification numbers including local ID, tags, bands and tattoos; specimen parentage; sex; birth date; birth origin (captive or wild); if wild, possible relationships to other wild-caught specimens; specimen location and transfer history; specimen reproductive status and history; death date and circumstances of death; and other miscellaneous information including but not limited to behavioral notes, taxonomic notes, and husbandry notes. Data entered into the studbook should be as complete as possible, but unknown or missing data is to be expected. "Analytical studbooks" incorporating potential or assumed values for missing data can be created and evaluated, but should never replace the "true studbook", composed only of factual information.

Studbooks are typically computerized databases, facilitating data manipulation and analysis. The most commonly utilized software for this purpose is the Single Population Animal Record Keeping System, SPARKS (ISIS 1991). This DOS workhorse has been the standard for many years, though more user-friendly systems are currently under development, including PopLink (Faust et al 2006) and Zoological Information Management System (ZIMS).

**Population Status**

Once compiled, studbooks can be analyzed using a variety of software tools including Population Management 2000 (Pollack et al 1999), Vortex (Lacy 2000), ZooRisk (Earnhardt et al 2004), and MateRx (Ballou et al 2001). Studbook analysis is used to examine population history, evaluate current population status, and predict future population status under varying management conditions. Demographic and genetic analyses are conducted as part of informed population management planning.

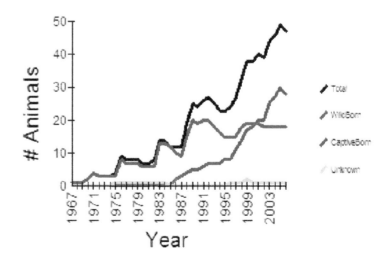

Fig. 1. *Population census of Rhinoceros Hornbills in AZA facilities.*

## Table 1

### LIFE TABLE FOR GUAM RAILS *GALLIRALLUS OWSTONI* IN AZA FACILITIES

| Age | Males | | | Females | | |
| --- | --- | --- | --- | --- | --- | --- |
| | Qx | Lx | Mx | Qx | Lx | Mx |
| 0 | 0.330 | 1.000 | 0.050 | 0.330 | 1.000 | 0.040 |
| 1 | 0.060 | 0.670 | 0.470 | 0.090 | 0.670 | 0.760 |
| 2 | 0.040 | 0.630 | 0.650 | 0.070 | 0.610 | 0.840 |
| 3 | 0.050 | 0.605 | 0.720 | 0.110 | 0.567 | 0.850 |
| 4 | 0.040 | 0.574 | 0.630 | 0.110 | 0.505 | 0.920 |
| 5 | 0.080 | 0.551 | 0.820 | 0.150 | 0.449 | 1.000 |
| 6 | 0.050 | 0.507 | 1.010 | 0.200 | 0.382 | 0.740 |
| 7 | 0.060 | 0.482 | 1.120 | 0.150 | 0.305 | 1.090 |
| 8 | 0.110 | 0.453 | 1.230 | 0.240 | 0.260 | 0.360 |
| 9 | 0.060 | 0.403 | 0.410 | 0.200 | 0.197 | 0.250 |
| 10 | 0.110 | 0.379 | 0.550 | 0.200 | 0.158 | 0.000 |
| 11 | 0.170 | 0.337 | 0.030 | 0.170 | 0.126 | 0.000 |
| 12 | 0.130 | 0.280 | 0.000 | 0.440 | 0.105 | 0.000 |
| 13 | 0.360 | 0.244 | 0.000 | 0.200 | 0.059 | 0.000 |
| 14 | 0.400 | 0.156 | 0.000 | 0.500 | 0.047 | 0.000 |
| 15 | 0.000 | 0.094 | 0.000 | 0.000 | 0.023 | 0.000 |
| 16 | 0.470 | 0.094 | 0.000 | 0.500 | 0.023 | 0.000 |
| 17 | 0.000 | 0.050 | 0.000 | 0.000 | 0.012 | 0.000 |
| 18 | 1.000 | 0.050 | 0.000 | 1.000 | 0.012 | 0.000 |
| 19 | 1.000 | 0.000 | 0.000 | 1.000 | 0.000 | 0.000 |

*This life table indicates first year mortality (qx) for males and females has been observed to be 33%. Both sexes have a high probability of breeding (mx) in their first year. While life span has been observed to be as great as 18 years, reproduction has not been observed in individuals of ages greater than 13 years.*

## Demography

Demographic analyses are used to investigate population size, structure, and distribution. The population census is used to examine historic population growth and decline in captivity (Figure 1). Knowledge of these trends, in combination with biological and husbandry history, can identify planned versus unplanned changes in population size and provide insights into threats to population viability.

Age distributions, illustrating the number of individuals in each age and sex class, are plotted to examine structure in populations. This structure determines the population's overall potential for reproduction and growth. Often referred to as "age pyramids" due to the desirable triangular shapes, age distributions facilitate the visualization of pre-reproductive, reproductive, and post-reproductive composition and sex ratio of a population. Population growth is predicted by distributions with strong bases of many individuals in younger age classes (Figure 2, left). Distributions can also warn managers of potential growth slow downs or population declines, illustrated by shortages of individuals entering breeding age classes (Figure 2., right) or severely skewed sex ratios, limiting the number of individuals available for breeding.

Population vital rates are calculated for males and females using individual life histories; age-specific

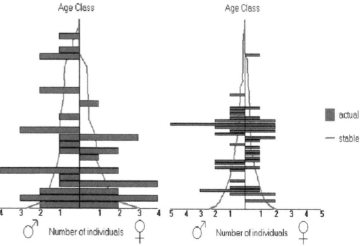

**Fig. 2.** *Age distribution of Blue-faced Honeyeaters and of White-naped cranes in AZA facilities in 2006.*

birth rates (*mx*), mortality rates (*qx*), and survivorship (*lx*) are used to construct life tables (Table 1.). These values, based on historic studbook data, can be used to inform decisions about the management of living specimens as they indicate expectations for the likelihood of reproduction and mortality for individuals in given age and sex classes. Applying historic vital rates to current age distributions allows predictions of potential population attrition and growth. These rates may also be used to parameterize population viability analysis (PVA).

The accuracy of life tables and their value as predictive tools are dependent on both the quantity and quality of studbook data; rates based upon the observation of few individuals will lack the robust nature of larger data sets. For species lacking the data necessary to construct life tables, "surrogate" species can be utilized to provide vital rates for the purposes of population modeling. Data "surrogates" should be closely related species with similar life histories.

Population growth and declines are described using the metric lambda, annual rates of growth. This value is calculated as:

$$\lambda_t = \frac{N_t}{N_{t+1}}$$

Values of lambda greater than 1.0 indicate population growth while values less than 1.0 indicate decline. This value is determined by birth and death rates as well as by immigration (e.g. acquisition of additional founders) and emigration (e.g. reintroduction to wild populations).

## Genetics

Genetic analysis of populations begins with founders. A founder is an individual entering a population in which it is assumed to be unrelated to all other individuals. Wild-caught specimens are usually considered to be founders, as are specimens entering a managed population from another unrelated managed population. If "founders" of known relatedness (e.g. nest mates captured together, or individuals linked through molecular genetic applications), hypothetical founders can be created within an analytical studbook database to represent these known relationships.

To be considered founders, animals must also contribute descendants to the population; they are considered to be potential founders until offspring have been produced. The number of founders contributing to a breeding program is of great concern, as higher numbers of founders will more accurately represent the

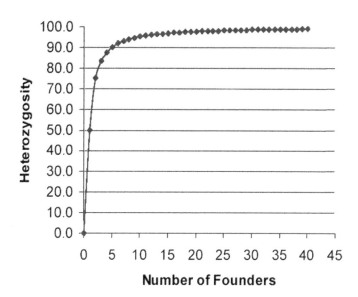

**Fig. 3.** *The proportion of heterozygosity of a source population available to a captive population depends on the number of founders sampled.*

genetic composition of the source population (Figure 3).

Genetic parameters are measured relative to a base population in which founders are assumed to be unrelated unless data indicates founder relationships. Analyses generate values of gene diversity (*GD*), inbreeding (*F*), individual mean kinship (*mk*), population or descendant mean kinship (*MK*), and founder representation of the descendant population. These values are generated through pedigree analyses.

Such analyses include gene drop analysis, which estimates changes in genetic variation relative to that found in the source population (Mace 1986; MacCluer et al 1986), and results in an estimate of heterozygosity (Wright 1969) commonly referred to as gene diversity (*GD*) (Lacy 1989). Gene diversity is equal to the proportion of the source population's heterozygosity found in the living descendant population (Lacy 1995). Calculated from gene drop models, the frequencies of founder alleles (*pi*) present in the extant population (averaged over 10,000 iterations) are used to estimate heterozygosity where:

$$H = 1 - \sum p_i^2.$$

Gene diversity is lost over time and rate of loss depends on several genetic and demographic factors: number of founders, population growth rate and size, effective population size (the proportion of breeders in the population), generation time (average age at reproduction), and immigration/emigration rates. Many of these factors can be manipulated through population management to slow rates of GD loss.

From the pedigree, values of pair-wise kinship

coefficients (Falconer 1981) also are derived, and individual mean kinship is calculated. Individual mean kinship ($mk_i$) is equal to the average kinship between individual $_i$ and all individuals in the population including itself:

$$mk_i = \frac{\sum_{j=1}^{N} f_{ij}}{N}$$

This value indicates the relative rarity of an individual's genome within the population (Table 2). Individuals with low *mki* relative to the population mean kinship ($MK$ = mean $mk_i$) have few relatives within the population and are most likely to carry rare alleles. Individuals with high relative *mki* have many relatives within the population and are less likely to carry rare alleles. Mean kinship is utilized in the creation and prioritization of breeding pairs, with pairs preferentially consisting of individuals with low and similar *mki* values (Ballou & Lacy 1995). This breeding strategy is the most effective way of retaining gene diversity (Lacy 1995). The inbreeding coefficients ($F$) of individuals and their potential offspring are also calculated from the pedigree and used in population management; $F$ is equal to the probability that alleles at a locus sampled in two individuals are identical by descent (Ballou 1983). Inbreeding is expected to result in reduced fitness (Ralls & Ballou 1983) and is avoided or minimized in most managed breeding programs (Ballou & Foose 1996). Founder representation is the proportion of the genes in the living, descendant population that are derived from a given founder (Lacy 1995). Gene diversity is maximized when founder representation is equalized (Figure 4). Equalization of founder representation is achieved through a mean kinship breeding strategy, matching *mki* within pairs to avoid linkage of rare and common alleles and preferentially breeding individuals of low *mki*.

## Population Management Plans

Using demographic and genetic analyses, population biologists create population management plans suited to needs of specific programs. These needs vary from the creation of breeding and transfer recommendations for individual animals to overall strategies for long-term management of healthy populations in captivity and the wild.

### *Ex-situ* Breeding Programs

Many conservation programs include the management of *ex-situ* breeding programs. These programs serve many roles including long-term public exhibition and education, research, redundant populations as insurance against catastrophe, or producing surplus individuals for reintroduction programs both in the long term and in emergent situations. Whatever their role, captive populations must themselves be stable and secure, being managed for demographic and genetic health, to support these functions. Mean kinship management strategies are generally employed, though they may be customized to meet specific program needs. Monogamous, polygamous, and colonial breeders can all be accommodated. Populations can be managed to maintain at carrying capacity or to supply surplus individuals for reintroduction or for export to other managed programs.

### *In-situ* Breeding Programs

Much like *ex-situ* breeding programs, *in-situ* captive populations can be carefully managed for genetic and demographic health. Advantages to *in-situ* breeding programs are numerous in the case of many conservation programs, especially those involving research and reintroductions. Managing captive individuals within their native range mitigates many risks, including limiting exposure to novel diseases, maintenance of natural environmental conditions, minimizing potential adaptation to captive environments with shorter time in captivity and reduced costs of quarantine and transport. *In-situ* programs also have inherent challenges that need to be addressed. The relatively isolated locales of many *in-situ* captive breeding programs

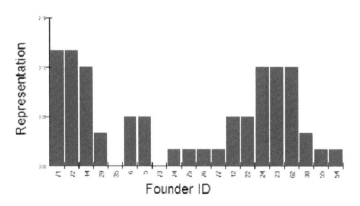

Fig. 4. *Founder representation in the Wrinkled Hornbill population in AZA illustrates unequal representation including unrepresented founders. Gene diversity is maximized when founder representation is equalized. Breeding efforts should focus on specimens from under-represented founder lines; those individuals have the lowest values of mki in the population.*

## Table 2

### INDIVIDUAL MEAN KINSHIPS OF GUAM RAILS *GALLIRALLUS OWSTONI* IN AZA FACILITIES

| Males | | | Females | | |
| --- | --- | --- | --- | --- | --- |
| SB# | *MK* | Age | SB# | *MK* | Age |
| 1292 | 0.104 | 6 | 853 | 0.08 | 7 |
| 1293 | 0.104 | 7 | 727 | 0.104 | 9 |
| 1314 | 0.138 | 5 | 508 | 0.106 | 5 |
| 1591 | 0.14 | 1 | 747 | 0.113 | 8 |
| 1593 | 0.14 | 1 | 1460 | 0.133 | 3 |
| 1594 | 0.14 | 1 | 1595 | 0.14 | 1 |
| 1596 | 0.14 | 1 | 1459 | 0.154 | 3 |
| 988 | 0.142 | 6 | 1564 | 0.154 | 2 |
| 985 | 0.144 | 6 | 1412 | 0.164 | 4 |
| 1313 | 0.148 | 5 | 1584 | 0.167 | 1 |
| 872 | 0.152 | 7 | 1319 | 0.168 | 5 |
| 986 | 0.154 | 6 | 1410 | 0.171 | 4 |
| 981 | 0.161 | 7 | 871 | 0.174 | 7 |
| 1579 | 0.167 | 1 | 1580 | 0.174 | 1 |
| 1582 | 0.167 | 1 | 1556 | 0.175 | 3 |
| 1583 | 0.167 | 1 | 1557 | 0.175 | 2 |
| 994 | 0.168 | 6 | 1559 | 0.175 | 2 |
| 1322 | 0.168 | 5 | 993 | 0.178 | 6 |
| 1569 | 0.174 | 2 | 995 | 0.178 | 6 |
| 1578 | 0.174 | 1 | 1573 | 0.183 | 1 |
| 1581 | 0.174 | 1 | 1574 | 0.183 | 1 |
| 1558 | 0.175 | 2 | | | |
| 1318 | 0.178 | 5 | | | |
| 1565 | 0.183 | 2 | | | |
| 1572 | 0.183 | 1 | | | |

*The population mean kinship is equal to 0.1576.*

may result in limited resources and expertise available to these programs. Limited facilities, in both number and size, may create smaller carrying capacity for breeding programs, resulting in greater loss of gene diversity and increased vulnerability to stochasticity. A small number of participating facilities may increase risk to a population should a disease outbreak or natural catastrophe occur. The importance of genetic management and demographic monitoring in these situations therefore becomes even more critical. In response to environmental risks such as typhoon (Guam Rail), wild fires (San Clemente Loggerhead Shrike) and other risks, several programs have utilized genetic strategies to develop evacuation protocols. In the event that a rapid evacuation becomes necessary, a group of individuals has been identified as priorities for evacuation. These individuals have been selected based on genetic criteria to create a subset of the population most representative of the original founder stock. Should a large portion of the population be lost to natural disaster, this subset would serve as "founders" for the surviving population. In the case of Loggerhead Shrikes, in 2001, an evacuation protocol was created for an *in-situ* breeding population of 54 specimens. It was determined that, in the event of an emergency, only eight individuals could be removed from San Clemente Island and be accommodated in a mainland quarantine station. Four males and four females were prioritized for evacuation. The captive population at the time had *GD* equal to 92.39% of that present at the start of the captive breeding program; the evacuated subset had a *GD* of 90.47 %; thus, while the population size would be reduced by 85%, the population *GD* would be reduced by less than 2%.

### Reintroduction Management

In the selection of individuals for reintroductions, it is critical that the genetic and demographic integrity of the source population be maintained. Indiscriminately selecting animals for reintroductions can have serious genetic and demographic effects, such as skewed founder representation or loss of founder lines, and destabilization of age structure through over-harvest of targeted age classes. Selection of animals for reintroduction must consider the maintenance of the remaining captive population's ability to serve its conservation role into the future, for the perceived extent of the program. Demographic analyses can be utilized to evaluate harvest rates (the number of individuals removed from the population annually) and their effect on the remaining population. Risk assessment software packages such as ZooRisk (Earnhardt et al 2002) and Vortex (Lacy 2000) can be used to examine changes in extinction risk to populations under differing management strategies (Figures 5) or to determine the number of surplus specimens available for reintroduction (Figure 6). The ages of individuals selected to be released should be carefully considered. While young of the year may have had less time to adapt to captivity and may have high post-release survival rates, enough young must be retained in captivity for future breeding (Figure 7). Thought must be given to the fate of post-reproductive individuals; individuals may not be contributing to the program as breeders but may be occupying limited housing resources. Some programs, such as Black-footed Ferrets, choose to release individuals well before reproductive senescence, but do release adult specimens after several years in captivity. Once the number and age structure of individuals available for release is determined,

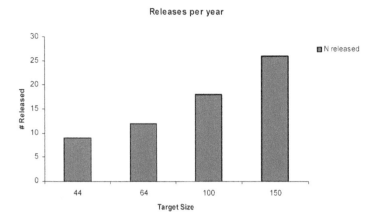

**Fig. 6.** *The captive population of Santa Cruz Island Fox is managed in a facility with a capacity of 44 specimens. At this size less than 10 surplus animals are likely to be produced per year.*

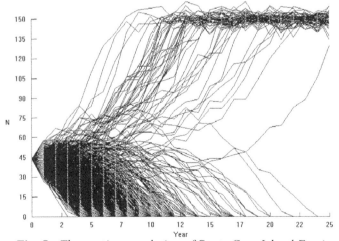

**Fig. 5.** *The captive population of Santa Cruz Island Fox is managed for a reintroduction program. Projections indicate that if the captive population has a capacity of 44 animals and 16 animals are released per annum, there is a 95% likelihood of extinction of the captive population within 25 years from present. Given the current demographic parameters, the population is incapable of producing 16 surplus animals per year.*

specimens may be selected using genetic criteria. Typically those specimens having a positive genetic effect on the captive population are prioritized for retention in captivity and continued breeding. From those specimens not required for captive breeding, release specimens are chosen. Several approaches to these selections exist and the three most common are discussed. The first approach is often referred to as "dumping". This approach involves the release of those specimens which are simply the most over-represented in captivity and will maximize gene diversity in the captive source population. This method is appropriate in the case of rapidly reproducing species with large unmonitored release cohorts, such as the Guam Rail. However, this method may not optimize gene diversity in small, slowly growing recipient populations as it may result in the release of several related individuals having redundant genetic backgrounds. The next approach is that of representative releases. This release strategy is appropriate in cases where wild population monitoring is not sufficient to allow tracking of a large portion of the released population. When the fate of previously released specimens is unknown, release selections may be made to ensure that the total releases over the program lifespan are genetically representative of the captive population. Founder representation, once it is equalized in the captive population, is then duplicated in the release selection. All founder lineages are released in equal frequencies under the assumption that maximum genetic variation will be available for natural selection. Release planning for Andean Condors has utilized this method, and recommended releases are projected to result in gene diversity increases in both captive and recipient populations (Table 3). The last strategy is that of selective release. This strategy is employed in the case of

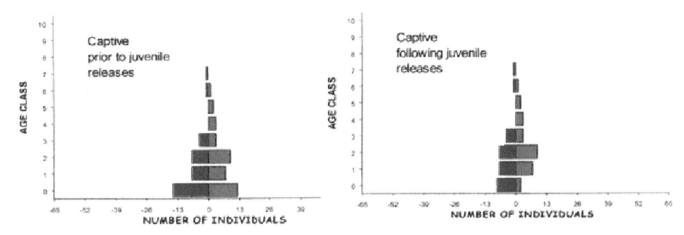

**Fig. 7.** *Potential result of juvenile releases on age structure in Loggerhead Shrikes may reduce reproductive potential in the captive population with too few young animals entering the breeding age classes and several older individuals occupying breeding enclosures.*

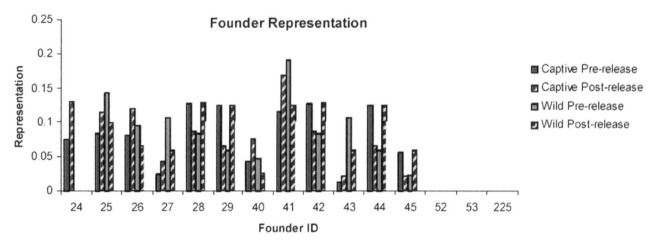

**Fig. 8.** *Recommended breeding and releases of Island Foxes on Santa Rosa Island influence founder representation of several founder lines. Active efforts to breed potential founders continue.*

### Table 3
### SUMMARY OF OUTCOME OF RECOMMENDED REINTRODUCTIONS

|  | Captive Source | Wild Recipient Range 1 | Wild Recipient Range 2 | All |
|---|---|---|---|---|
| **Prior to recommendations** |  |  |  |  |
| N | 81 | 52 | 7 | 143 |
| Founders | 34 | 28 | 12 | 34 |
| Gene Diversity Retained (%) | 97.54 | 96.87 | 91.33 | 97.76 |
| **Following recommendations** |  |  |  |  |
| N | 80 | 63 | 14 | 157 |
| Founders | 38 | 30 | 26 | 38 |
| Gene Diversity Retained (%) | 97.72 | 97.34 | 95.34 | 97.92 |

*Results of recommended releases within two wild Andean Condor populations have positive impacts on founder representation and gene diversity in both source and recipient populations.*

highly monitored recipient populations. When the fates of previously released specimens and wild-born specimens are tracked, selections may be made that equalize founder representation and maximize gene diversity in the recipient population. Release selections for San Clemente Loggerhead Shrikes and Channel Island Foxes are conducted in this manner. As a result of this strategy and after three years of releases of Loggerhead Shrikes, gene diversity in the wild (92.05%) approximated gene diversity in captivity (92.42%), and 16 of 17 founders to the captive population were represented in the wild population. Island Fox populations on Santa Cruz Island and Santa Rosa Island experienced similar results.

## References

AVISE JC & JC Hamrick, Eds. 1996. Conservation Genetics: Case Histories from Nature. Chapman & Hall, New York, USA. Pp 512.

BALLOU JD. 1983. Calculating inbreeding coefficients from pedigrees. In Genetics and Conservation: A reference for managing wild animal and plant populations. Pp. 509-520. CM Schonewald-Cox, SM Chambers, B McBryde, WL Thomas, eds. Benjamin/Cummings. Menlo Park, CA.

BALLOU JD, J Earnhardt J and SD Thompson. 2001. MateRx Software V 1.9. Lincoln Park Zoo.

BALLOU JD & TJ Foose. 1996. Demographic and genetic management of captive populations. In Wild Mammals in Captivity. Pp 263-283. DG Kleiman, M Allen, K. Thompson, S. Lumpkin, H. Harris, eds. University of Chicago Press. Chicago.

BALLOU JD & RC Lacy. 1995. Identifying genetically important individuals for management of genetic diversity in pedigreed populations. Pp 76-111 In Population Management for Survival and Recovery. JD Ballou, M Gilpin, TJ Foose, eds. Columbia University Press, New York.

CRNOKRAK, P. & D.A. Roff. 1999. Inbreeding depression in the wild. Heredity 83: 260-270.

EARNHARDT JM, A Lin, LJ Faust, SD Thompson. 2005. ZooRisk: A Risk Assessment Tool. Version 2.53. Chicago, IL: Lincoln Park Zoo.

FAUST LJ, YM Bergstrom and SD Thompson. 2006 . PopLink Version 1.0. Chicago, IL: Lincoln Park Zoo

FRANKHAM R, JD Ballou, DA Briscoe. 2002. Introduction to Conservation Genetics. Cambridge University Press, Cambridge, UK. Pp 617.

ISIS. 1994. SPARKS. International Species Information System. Apple Valley, MN.

LACY RC. 1995. Clarification of genetic terms and their use in the management of populations. Zoo Biology. 14:565-578.

LACY RC. 2000. Structure of the VORTEX simulation model for population viability analysis. Ecological Bullitin 48:191-203.

FALCONER DS. 1981. Introduction to Quantitative Genetics. New York: Longman.

MACE G. 1986. Genedrop: computer software for gene drop analyses. Zoological Society of London. London.

MACCLUER JW, JL Vandeberg, B Read and OA Ryder. 1986. Pedigree analysis by computer simulation. Zoo Biology 5:147-60.

POLLACK, JP, RC Lacy and JD Ballou. 2002. PM2000: Population Management Software. Cornell University. Ithaca, N.Y.

RALLS K & JD Ballou. 1983. Extinction: lessons from zoos. In Genetics and Conservation: A reference for managing wild animal and plant populations. Pp. 164-184. C.M. Schonewald-Cox, S.M. Chambers, B. McBryde, W.L. Thomas, eds. Benjamin/Cummings. Menlo Park, CA.

RALLS K & JD Ballou. 1992. Managing genetic diversity in captive breeding and reintroduction programs. Wildlife and Natural Resources Conference, 57:263-282.

WRIGHT S. 1969. Evolution and the Genetics of Populations, Vol. II: The Theory of Gene Frequencies. University of Chicago Press, Chicago.

## Author Biography

Colleen Lynch received her B.S. from the Illinois Benedictine College and her M.S. from Northern Illinois University. She is currently pursuing a Ph.D. in conservation genetics at the University of South Dakota. Lynch is a population biologist with the Association of Zoo and Aquariums Population Management Center in Chicago and conducts research in applied population biology. She blends ten years of experience as a zoo keeper and aviculturist with her skills as a population biologist and has produced over 250 breeding and transfer plans for cooperatively managed breeding programs. She participates in several reintroduction programs, providing genetic and demographic advising and developing management strategies tailored to the special needs of captive breeding for endangered species recovery in the wild.

# REARING TO RELEASE: MANAGING RISKS IN THE REINTRODUCTION OF CAPTIVE-BRED BIRDS

PHILIP J. SEDDON & YOLANDA VAN HEEZIK

Department of Zoology
University of Otago
P. O. Box 53 Dunedin
New Zealand

*"It is not enough, any more, simply to keep animals alive—or even alive and breeding... the ultimate aim is conservation, of which it is a condition that at some time the animals in captivity might return to the wild. Animals in zoos must be encouraged to retain enough of their natural behavior to make it possible for them to go back to the wilderness; or enough at least of their native wit to enable them to relearn the necessary skills.*
— *Colin Tudge 1992*

## SUMMARY

Captive management is a key and growing component in the restoration of threatened bird species. Currently there are at least 112 bird species globally that are being captive bred or reared for release to re-establish or reinforce wild populations. However, any period of captivity will pose a set of risks to captive birds, risks that may compromise both their survival in captivity and their fitness after release. The key challenge facing those charged with the captive management of birds in support of population restoration programs is the production of individuals with the best possible chance of post-release survival and successful breeding in the wild. To achieve this requires management of the potential risks posed by trauma, stress, disease, and behavioral and genetic effects. Each risk factor will have an associated risk function, the cumulative probability of a deleterious effect with time spent in captivity, and each will necessitate specific mitigation measures. Drawing on some two decades of published research in reintroduction biology and captive management, this chapter reviews the diverse and innovative approaches developed to overcome many of the deleterious effects of captivity on birds destined to contribute to *in-situ* restoration projects. Wider application and further development of such techniques will allow reintroduction practitioners to capitalize on the inherent advantages of releasing birds from captivity, and will lead to improvements in the success rates of bird reintroduction programs worldwide.

## INTRODUCTION

Recent estimates indicate that one in eight bird species is globally threatened with extinction due to habitat loss, hunting, trade and the impacts of invasive species (Birdlife International 2007). As these threats increase, and are exacerbated by the looming impacts of global climate change, so too has the need for and prevalence of intervening conservation measures increased. One of the most intensive and expensive options in the conservationist's toolkit is that of the restoration of extirpated populations through reintroduction (see Box 1 for definition) from captivity. As conservation options run out for rapidly declining wildlife populations trapped in dwindling habitat patches, recourse to some form of captive management is seen as a necessary step. A review of 314 approved Recovery Plans for threatened and endangered wildlife in the US found that 70% recommended wild-to-wild translocation and >64% called for the establishment of a captive population (Tear et al 1993). Zoos are increasingly seeking and finding a role as conservation partners (Sheppard 1995; Stanley Price & Soorae 2003). In 2006, the 216 accredited institutions of the Association of Zoos and Aquariums (AZA) reported on their participation in 1,807 conservation projects, and over a five-year period AZA member institutions funded 3,693 conservation projects in over 100 countries, an average spending on conservation of nearly US $70 million per annum (AZA 2006). However, zoos will only ever house a small proportion of all threatened species (Magin et al 1994) and increasingly the majority of animals for reintroduction will come from specialized facilities in their native country (Beck et al 1994; Stanley Price & Soorae 2003). The number of species of bird that are the focus of reintroduction projects increased from 62 in 1993 (Wilson & Stanley Price 1994) to 77 in 1998, and to 138 by 2005 (Seddon et al 2007). Combining the lists of current bird reintroductions (derived from the Reintroduction Specialist Group (RSG) database) and the species not on RSG records but for which some form of captive management is underway, planned for, or proposed as a conservation action (Birdlife International online database) we get a total of 204 species restorations that are or will involve wild-to-wild translocation, or translocation from captive rearing or captive breeding facilities (Fig. 1) to meet population restoration goals. Over 60% (112/172) of these programs, where the methods are clearly defined, primarily involve captive breeding or captive rearing. Clearly then captive management is a key and growing component in the restoration and conservation of threatened bird species. Despite its prevalence however, few would naively embark on full-scale captive breeding and reintroduction projects as a sure-fire fix. The costs, logistic requirements and institutional commitment needs aside, and assuming the technical expertise has been gained to hold, sustain and to breed a given species in captivity, there are a number of other challenges facing any captive management program that aims to produce birds that are suitable reintroduction founders. The available evidence suggests that the success of restoration projects releasing captive-reared or captive bred animals is significantly less than for projects using wild-caught founders (Griffith et al 1989; Beck et al 1994; Wolf et al 1996). This paper seeks to examine what these challenges are and how the various risk factors may operate over different time scales across a spectrum of captive management ranging from short-term captivity involved in wild-to-wild translocations, through to multi-generation captive breeding. We explore some approaches that have been used to mitigate the effects of captivity-related risks to birds, and conclude by considering the tangible opportunities that captive management can provide for well planned and resourced reintroduction projects. We seek to make the point that the form of captive management applied must be appropriate to the ultimate goals of a project.

## Types of Captivity

Three broad categories of captive management in support of bird conservation efforts can be described (see Box 1 for definitions).

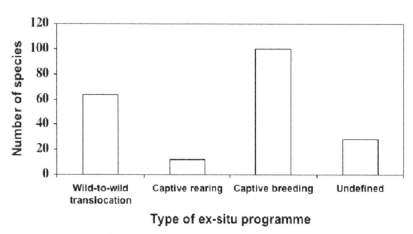

Fig. 1. *Numbers of bird species that are the focus of restorations involving wild-to-wild translocations, captive rearing and captive breeding (see Box 1 for definitions).*

### (1) Wild-to-wild translocation

This entails the live capture of wild free-ranging birds, their transportation to a release site under some form of restraint and their release with or without some period of acclimation. It has been suggested that the release of wild-caught animals has a greater likelihood of success as indicated by post-release survival, primarily due to the higher probability that wild animals will be better able to cope with the necessities of finding food and shelter and avoiding predators than would captive animals that may have lost or never acquired the requisite skills (Griffith et al 1989; Beck et al 1994). While post-release survival may be higher, there are significant risks associated with the process of live capture and transport of wild birds (e.g. Leech et al 2007), and levels of pre-release mortality should be carefully considered in any cost-benefit analysis. Wild-to-wild translocations are the most cost-effective means of re-establishing populations where a suitable source of wild founders exists. In some cases, particular source populations may be harvested repeatedly to provide founders for reintroductions. In New Zealand, localised populations of South Island Saddleback *Philesturnus carunculatus*, North Island Saddleback *P. rufusater*, and Stitchbird *Notiomystis cincta* have been harvested up to 12 times for wild-to-wild translocations (Lovegrove 1996; Taylor et al 2005; Hooson & Jamieson 2003; Armstrong 1999-2006). One critical consideration is whether the removal of founders for releases elsewhere will endanger the viability of the source population. In the past this aspect has received little or inadequate attention, but recent work provides a good model for how such assessments can be undertaken with rigour (Dimond & Armstrong 2006), and increasingly the need to estimate sustainable "harvest"

> **Box 1. Definition of Terms**
>
> The IUCN Position Statement on Translocation of Living Organisms (IUCN 1987) defines translocation as movement of living organisms from one area to another, and recognises three types of translocation according to the status of the area into which animals are released:
>
> **Introduction:** movement of an organism outside its historically known native range.
>
> **Reintroduction:** intentional movement of an organism into a part of its native range from which it has disappeared or become extirpated in historic times.
>
> **Re-stocking:** movement of individuals to build up an existing population.
>
> Note that, strictly speaking, after the first releases for a reintroduction, subsequent releases become a form of population augmentation or supplementation. Effectively however, we can refer to all releases that form part of a single coherent reintroduction program at a given site as reintroductions.
>
> It is also be useful to differentiate between the different sources of animals used in a translocation. Animals to be released into an area may come from free-ranging natural populations or from some form of captive management facility. Three types of captivity may therefore be involved in a reintroduction project:
>
> **Wild-to-wild translocation:** entailing capture and a period of relatively short-term captivity during transport and release; this may be extended in the case of soft-release protocols requiring holding of birds at the release site for a period of acclimation.
>
> **Captive rearing:** most often involving the collection of eggs or young chicks to be raised in captivity in order to avoid periods of high risk of mortality in the wild; captive rearing thus involves only one generation, with no production of offspring by captive animals.
>
> **Captive breeding:** is multi-generational captivity whereby captive birds are induced to breed and consequent generations of offspring provide founder animals for releases.

levels for source populations is being considered (e.g. Greaves 2007). Wild-to-wild translocations may become an important component of a species restoration program in the latter part of a project, enabling the establishment or reinforcement of new sub-populations, or the manipulation of existing populations to maximise productivity. For example, since the reintroduction of Peregrine Falcons *Falco peregrinus anatum* ceased in 1992 the focus of management has turned to the translocation of wild young away from sites with high fledgling mortality (Kauffmann et al 2003).

*(2) Captive-rearing*

In some cases the high risk of mortality of eggs or dependent chicks will justify the collection of eggs or chicks for hatching and raising in captivity, thereby markedly increasing early survival. Birds may then be released back into the population from where they were taken, or used as reintroduction founders to establish a new population within vacant habitat elsewhere in the species' distribution range. An example of captive-rearing for population restoration is provided by the Black Stilt or Kaki *Himantopus novaezelandiae* a critically endangered New Zealand endemic. Black Stilts are braided river specialists that, as a result of habitat loss, predation by introduced mammals and disturbance were reduced by 1983 to only 22 free-ranging adults within a single population in the central South Island of New Zealand (Pierce 1996). High rates of egg and hatchling loss to a suite of exotic mammalian predators have been countered by the collection of eggs from the wild, for artificial incubation and the rearing of hatchlings in a purpose-built facility (Keedwell et al 2002). Captive-reared Kaki are released as juveniles (~3 months old) or sub-adults (~9 months old) to supplement the existing wild population. High rates of hatching success and adequate, though variable post-release survival have increased the free-ranging Kaki population to over 100 adults by 2007 (NZ Department of Conservation, unpublished data). An additional benefit of the collection and captive-rearing of wild-laid eggs is the use of egg-pulling techniques to stimulate relaying and thus maximizing the productivity of wild pairs; up to four clutches of up to four eggs may be collected from a wild pair. The potential costs of repeated re-clutching on adult survival and productivity are not known.

*(3) Captive-breeding*

Captive-breeding entails the management of multiple generations and carries not only a greater cumulative risk of effects that may compromise future fitness, but also two basic challenges: (i) the replication (or simulation) in captivity of the normal processes of mate selection, copulation, laying, hatching and rearing; and (ii) the selection and preparation of offspring derived from captive breeding to be released into the wild for population supplementation or reintroduction. In some cases it may not be possible, for a number of reasons, to induce captive birds to mate. For example, the breeding system of a given species may be incompatible with replication under captive conditions. The Houbara Bustard *Chlamydotis undulata*, now also recognised as the separate species *Chlamydotis macqueenii*, is a drab desert bird that has been driven close to extinction in parts of its range by the dubious virtue or being the premier quarry for Arab falconers (Seddon & van Heezik 1996). As a consequence, over the last two decades a number of large-scale captive-breeding facilities have been created, variously to provide birds for put-and-take hunting operations to relieve pressure on vulnerable wild populations, and/or to restore natural populations through releases of founder birds for reintroduction and supplementation projects (Seddon et al 1995). The Houbara breeding system is described as an exploded lek, whereby males make nuptial displays over a dispersed area that is visited by females who choose a mate, then withdraw to undertake all egg-tending and chick rearing duties alone. The large areas and the special social interactions involved have meant it is neither easy, nor efficient, to try to replicate natural breeding conditions (van Heezik & Seddon 2001). A system of sperm collection from males imprinted on their human handlers, and artificial insemination, incubation, and chick rearing was developed during the 1990s (Saint Jalme & van Heezik 1996) and is now the basis for the annual captive production of thousands of Houbara Bustards from captive breeding facilities in Morocco, Saudi Arabia and the United Arab Emirates. This completely artificial captive breeding process, while extremely labour intensive and heavily reliant on suitably tame sperm donors, enables aviculturists to have full control over all aspects of breeding, from the selection of geneti-

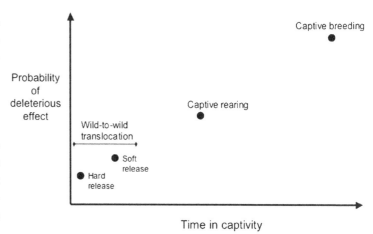

**Fig. 2.** *Relative risk of deleterious effects faced by birds within projects involving releases from wild-to-wild translocations, captive rearing, or captive breeding.*

cally appropriate parents, through to the pre-release rearing environment of offspring.

**Risk Factors**

Any period of captivity will impose a new set of risk factors on captive birds that may compromise both their survival in captivity and their future viability after release. We can group these into five categories: stress, trauma, disease and behavioral and genetic effects. The shapes of risk functions seem intuitively likely to differ in important ways of relevance to reintroduction managers. There are little data available with which to model the shapes of the various risk functions. In the following sections we speculate on the types of curves that may best describe the cumulative risk posed by different factors; we invite aviculturists to gather data necessary to test these predictions.

The longer the period an animal spends in captivity, the greater the likelihood that there will be a deleterious effect. We can envisage that short-term capture and holding will carry some not insignificant risk of stress or trauma-related mortality, but behavioral and certainly genetic effects will be negligible or absent. However, other sub-lethal effects of capture and transport of wild birds, such as weight loss (e.g. Wanless et al 2002) could influence post-release survival and may necessitate some period of acclimation to allow improvement of body condition before release [see also hard vs. soft release protocols]. Captive rearing may have a lower risk of stress-related impacts, retain a risk of trauma, but in addition have potential impacts from disease and behavioral modification. Finally, the multi-generational management of captive breeding will carry an additionally risk of deleterious genetic effects (Fig. 2). The risk that a given bird will experience an event that will reduce its post-release fitness increases with time spent in captivity, so at one level the best thing to reduce the risk of captivity-related reduction in fitness would be to minimise the time spent in captivity. This is a sensible guiding principle for wild-to-wild translocations and may even be a key factor in captive-rearing, where the longer a bird is held before release the greater the risk of trauma, disease, or behavioral adaptation to captive conditions. But reduction of the time in captivity seems an unhelpful suggestion for managers of captive-breeding facilities, where potentially the most serious fitness reductions for captive-born founders due for release may arise due to genetic selection of captive parents and the absence of opportunities to learn appropriate behaviors, such as predator avoidance, or conversely, the presence of opportunities to acquire inappropriate behaviors, such as tameness to humans. However, a number of innovative techniques have been developed by resourceful aviculturists to ensure the captive production of the best possible candidates for release into the wild.

### Trauma and Stress

Causes of death of captive birds can be divided into those that occur almost entirely within a few months of hatching (neonatal, impactions and perforations, and congenital), and those that occur throughout the rest of a bird's time in captivity. Of 457 recorded causes of death of captive Houbara Bustards, the single most significant was trauma (16%) (van Heezik & Ostrowski 2001). Trauma was also the primary cause of mortality among adult, and sub-adult Black Stilts (van Heezik et al 2005). Injuries incurred in captivity are primarily the result of collisions with built structures, particularly during capture operations or when birds respond to the proximity of predators. Stress may result from short-term effects of capture and transport, or the physical characteristics of a captive environment and can be a major factor limiting captive reproductive success. Increased heart rate is part of the stress response to stimuli that are perceived by an animal as being novel, challenging or threatening. Elevated heart rate can occur independently of any overt behavioral reaction to perturbation and is one manifestation of the vertebrate stress response that is activated by the hypothalamic-pituitary-adrenal axis and mediated by an increase in glucocorticosteroids from the adrenocortical tissue- the adrenocortical stress response (Romero 2004). Short-term increases in circulating levels of glucocorticosteroids enable individuals to escape from or cope with adverse conditions, however, long-term elevation of stress hormones can be physiologically damaging to individuals resulting in higher susceptibility to disease, reduced fertility and lower life expectancy (e.g. Walker et al 2005). Some of the most detailed recent work on quantifying the costs of human disturbance on birds has been conducted on wild penguins, and has used experimental blood sampling protocols to measure changes in corticosterone (the glucocorticosteroid in birds) in response to standardised experimental disturbance. Magellanic Penguins *Spheniscus magellanicus* have significantly elevated levels of corticosterone in response to a person visible nearby for only five minutes (Fowler 1999), but can habituate to human disturbance as long as the stimulus is short, intense and consistent (Walker et al 2006).

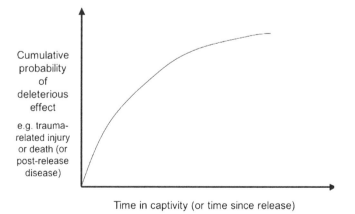

Fig. 3. *Cumulative risk of a deleterious effect due to trauma (or post-release disease) with time spent in captivity (or time since release).*

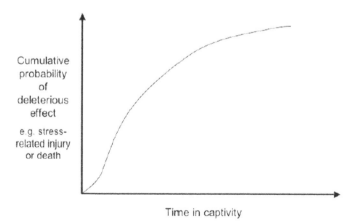

Fig. 4. *Cumulative risk of a deleterious effect due to stress with time spent in captivity.*

*Risk Function*

The instantaneous probability of trauma-related death will be highest around the time of initial capture and decline as a bird adapts to a captive situation. The cumulative risk of trauma-related death will therefore describe an exponential association curve that will approach an asymptote (Fig. 3). The risk of stress-related injury or death may differ from that of the trauma risk function if there is some time delay before the deleterious effects of an elevated stress response are expressed. Therefore, the cumulative risk of stress-related death could take the form of a sigmoidal curve (Fig. 4), with the probability of mortality building exponentially over some initial period, before the rate of increase declines to approach an asymptote as a bird habituates to captivity.

**Mitigation of Trauma and Stress**

Reduction of traumatic injuries is most easily achieved through appropriate modification of the captive-breeding environment to lessen the likelihood of birds accidentally colliding with hard surfaces, for example through the use of soft-walled cages and seclusion from visitors (van Heezik & Ostrowski 2001), and the development of appropriate capture techniques (van Heezik et al 2005). Environmental enrichment is one method to reduce captive stress through prevention of undesirable behaviors such as stereotypies and aggression (McDougall et al 2006). In poultry extra handling and the introduction of novel objects in rearing cages results in calmer birds, with faster growth rates and increased resistance to disease (James & Hughes 1981; Gross & Seigel 1982; Jones & Waddington 1992). Handling of Orange-winged Amazon Parrots *Amazona amazonica* increased tameness (Aengus & Millan 1999), but extra handling and novel objects had no measurable effect on the tameness of captive-bred Houbara Bustards (van Heezik & Seddon 2001). The stress-reduction benefits of extra handling as an environmental enrichment technique need to be balanced against an increased risk of birds imprinting on their human keepers, and is therefore not likely to be appropriate for birds scheduled for release (see Behavior, below).

**Disease**

The captive environment introduces two related disease risks: (i) risk of contagious disease spread through close contact with conspecifics, and (ii) risk of acquiring disease from wild birds coming into contact with the captive facility, or from other species in captivity, keepers and visitors. Infections and infectious disease were the second most common cause of death among captive sub-adult and adult Black Stilts (van Heezik et al 2005) and Houbara Bustards (van Heezik & Ostrowski 2001). While there is an extensive and well-established literature on the management of disease risks within a captive population, only in the last two decades has significant interest been focussed on the disease risks associated with releases of captive-bred animals back into the wild, and in particular those releases of endangered species that aim to supplement or restore wild populations (Cunningham 1996). Again two issues arise: (i) diseases present in wild populations at a release site that may compromise the survival of released animals, and (ii) diseases that may be transmitted by captive animals released into contact with extant wild populations.

*(i) Wild to captive disease transmission*

Captive animals due for release may have failed to acquire immunity to the suite of relatively common pathogens they would encounter in the wild because of lack of exposure to low level infection during captivity (Mathews et al 2006). In addition, captive-bred and wild-caught animals may have reduced immunocompetence and thus increased susceptibility to disease as a result of nutritional, social and other stresses (Viggers et al 1993; Mathews et al 2006). Failure of the Hawaiian Goose *Branta sandvicensis* reintroduction was due to their occupancy only at the higher altitudes of their historic range as a result of their exclusion from lower altitude breeding areas by infestations of introduced mosquitoes carrying avian pox virus (Kear 1977). Exotic doves are believed to be reservoir hosts to trichomoniasis, leucocytozoonosis and avian pox and thus posed a risk to the successful recovery of Pink Pigeons *Columba mayeri* and Echo Parakeets *Psittacula eques* in Mauritius (Swinnerton et al. 2005).

*(ii) Captive to wild disease transmission*

Animals bred in captivity will acquire local infections, possibly including alien parasites, and may become symptomless carriers of pathogens that may pose a risk to wild populations (Woodford & Rossiter 1994), and risk causing even ecosystem level effects (Cunningham 1996). It is insightful to consider a released captive-bred animal as not just a single spe-

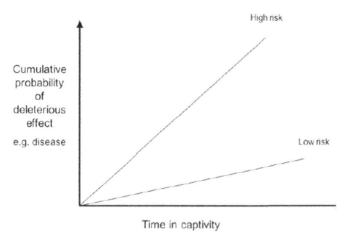

Fig. 5. *Cumulative risk of a deleterious effect due to pre-release disease with time spent in captivity.*

cies, but rather an assemblage of species, a "biological package containing a selection of viruses, bacteria, protozoa, Helminths, and arthropods" (Nettles 1988, cited in Woodford & Rossiter 1994). Captive conditions may bring endangered species into contact with new diseases that could pose a serious risk to populations of wild conspecifics. Sandhill Cranes *Grus canadensis* used to rear chicks of the endangered Whooping Crane *Grus americana* carried infections of visceral coccidiosis and it is believed that transmission of infection by foster parents resulted in this being a significant cause of Whooping Crane mortality. Similarly, Pink Pigeons were infected by Herpes virus via foster Domestic Pigeons *Columba livia*, while avian malaria and avian pox co-introduced by other birds contributed to the reproductive failure and declines in the endangered Hawaiian Crow *Corvus hawaiiensis* (Viggers et al 1993).

*Risk Function*

In any well-managed captive facility the instantaneous risk of disease will be constant, but not zero. Therefore the cumulative risk with time in captivity might be expected to be a straight line, the slope of which will depend on the base level of risk (Fig. 5). Disease risks posed to released birds will initially be high, but will reduce as surviving founders acquire the appropriate immunity; the cumulative risk function may therefore be best described as an exponential curve declining to some asymptote, as for risk of trauma-related death in captivity (Fig. 3).

**Mitigation of Disease**

Three processes can be applied to reduce the risk of disease transmission between captive founders and wild populations:

- Evaluation of the release site will allow identification of the disease agents and their vectors present in wild populations, so that specific risks to founders can be anticipated and addressed.
- Pre-release selection and health screening of founders will ensure that only healthy, pathogen-free birds are scheduled for release (Black 1991). Such screening should consider parasite and pathogen loads, but also include other health indices, such as haematological parameters, body conditions indices, and markers of immunocompetence (Mathews et al 2006). Pathogen loads of captive birds can be minimised by strict rearing protocols, for example, captive-bred Masked Bobwhite *Colinus virginianus ridgwayi* chicks are reared separately from the breeding flock to reduce potential disease exposure (Carpenter et al 1991).
- Pre-release preparation of founders may include ringing or other forms of individual marking to facilitate post-release analyses of survival against known pathogen loads and histories of veterinary treatments; reduction of parasite loads, e.g. deworming; and possibly vaccination against specific pathogens. In 1984, seven of 39 Whooping Cranes died in a captive facility in Maryland, USA, due to infection by eastern equine encephalitis (EEE). Following identification of the causal agent, serological monitoring of both captive and wild birds was started, and an inactivated EEE virus was developed to protect susceptible birds (Woodford & Rossiter 1994). Both captive-reared and wild-caught Campbell Island Teal *Anas nesiotis*, due for release back onto Campbell Island following the eradication of Norway Rats *Rattus norvegicus*, were vaccinated against Erysipelas after an outbreak of the disease amongst Kakapo *Strigops habroptilus* on a holding island (McClelland 2007).

Timing of releases may be an additional factor affecting the likelihood of survival. For example, captive White-tailed Sea-eagles *Haliaeetus albicilla* released in central Europe in mid-summer would come into contact with *Clostridium botulinum*, a common cause of mortality of wild waterbirds; releases delayed until autumn avoid the possibility of sea-eagles feeding on infected carrion (Viggers et al 1993).

Finally, plans should be in place for more than standard captive and pre-release veterinary care and the feasibility of post-release necropsy should be considered in order to understand ongoing disease risks and their role in project success or failure.

**Behavior**

One of the more complex captivity-related effects most applicable to captive-rearing and especially captive-breeding, is the production of birds with behavioral traits that may reduce post-release fitness. Some inadvertent selection for tameness and adaptation to the captive environment is regarded as inevitable in captive populations (Frankham et al 1986), e.g. the birds that are most likely to breeding in captivity are those that are behaviorally most suited to captive conditions, while trauma, stress and failure to breed will select against individuals that are most flighty and least tameable. This form of selection will have a genetic basis and a behavioral phenotypic expression (Håkansson & Jensen 2005), and may reduce post-release survival in offspring. In addition, for many species some key behaviors are not innate and may need to be learned from parents or other conspecifics. Thus deleterious behaviors may result from two processes: (i) acquisition of inappropriate behaviors, and (ii) lack of opportunities to acquire appropriate behaviors.

*(i) Development of inappropriate behaviors*

Close and regular contact with humans during captive management may result in the loss of wariness of people, both within and between generations, that could compromise post-release survival or hinder normal social interactions with conspecifics. Hand-reared Mississippi Sandhill Cranes *Grus canadensis pulla* proved to be unsuitable for reintroduction because of their reluctance to associate with wild conspecifics (Ellis et al 2000). In addition, current understanding of the ontogeny of habitat selection suggests that the natal environment experienced by an individual may increase post-dispersal preference for comparable environmental cues (Stamps & Swaisgood 2007). Thus captive-reared birds may develop specific habitat preferences related to their rearing environment and actively seek similar environments after release. Excessive tameness and curiosity of California Condors *Gymnogyps californianus* towards humans and urbanized areas, drawing them particularly towards rectangular human structures and the sounds of civilization familiar to them from their rearing environment, has contributed towards high post-release mortality rates during early releases (Meretsky et al 2000).

*(ii) Absence of appropriate behaviors*

As a result of the predictable and controlled environment, captive birds may either lose or fail to acquire the appropriate range of behaviors that would enable them to respond to unpredictable events in the wild (McPhee 2003). Post-release mortality of captive-reared or -bred founders is frequently due to behavioral deficiencies, affecting such things as locomotor skills, spatial orientation, the recognition of both natural foods, and appropriate responses to predators (Kleiman 1989; Miller et al 1994; Biggins et al 1999; van Heezik et al 1999). The social environment in captivity will be important for the development by young of species-typical behavior (Håkansson & Jensen 2005). Captive-bred hacked Aplomado Falcons *Falco femoralis septentrionalis* had lower survival and recruitment rates than wild-reared falcons due to the inability of captive-bred birds to out-compete rivals for vacant territories (Brown et al 2006). Reintroduction failures resulting from behavioral deficiencies are most common in species that must learn most of their behavioral repertoire (Snyder et al 1996). Attempts to reintroduce Thick-billed Parrots *Rhynchopsitta pachyrhyncha* failed, with high mortality rates due to poor food-processing ability, inability to avoid predators, and other "deficiencies in basic survival skills" (Snyder et al 1994).

*Risk function*

Behavioral changes will accrue with successive generations in captivity as captive conditions act as a selection pressure for traits such as tameness, and as learned behaviors are lost. The cumulative probability of there being some degree of fitness reduction related to a behavioral change over time is likely to take the form of an exponential curve, similar to that for genetic effects (Fig. 6).

**Mitigation of Behavior Effects**

*Pre-release behavioral selection*

Pre-release behavior has been proposed as a means to select appropriate captive-bred animals for release, whereby the behavior of wild-bred and captive-bred animals in identical novel environments is compared, using the wild-bred animals as a normal baseline from which to identify behavioral deficiencies in captive-bred animals and thereby rank potential candidates for release (Mathews et al 2005). This approach has been further developed with the recognition that individual animals will express consistent behavioral types that respond differently to the same stimuli, and thus population resilience may be enhanced if not all members exhibit the same responses to changing selection pressures (Watters & Meehan 2007). Assessment and categorization of potential reintroduction founders on the basis of behavioral responses measured in captivity is a potentially fruitful new area of investigation. For example, studies of domestic birds have shown consistent tonic immobility responses of individuals suggesting that each bird has a characteristic level of innate fearfulness (Jones et al 1994), implying that perhaps tonic immobility could be used to select behavioral types and rank suitability for release. However, care needs to taken to ensure that behaviors expressed to stimuli in captivity are relevant for post-release survival. Pre-release responses to a model predator and the degree of tonic immobility were not predictors of post-release survival in Houbara Bustards (van Heezik et al 1999).

*Predator recognition/avoidance training*

In general hand-reared birds will show less appropriate responses to predators than do parent-reared or wild birds (Robertson & Dowell 1990; Zilletti et al 1993; Anttila et al 1995). Given the evidence that anti-predator behavior in birds can be learned (Curio et al 1978; Maloney & McLean 1995; McLean et al 1999) there is increasing attention on the use of pre-release training to increase the predator awareness and hence post-release survival of released captive-bred animals (Kleiman 1989; Beck et al 1994; Wallace 1994; Holzer et al 1996). Pre-release anti-predator training can take several forms, including the use of classical conditioning whereby animals learn that model predators are predictors of some aversive event (Griffin et al 2000). Use of a model of a predator, while logistically easiest, may be ineffective in enhancing predator awareness in naïve captive-bred animals, and can even lead to habituation of responses. A more effective stimulus may be provided through the use of a live predator under a controlled, but appropriately realistic situation. Captive-bred Houbara Bustards that were trained using a live Red Fox in the presence of wild conspecifics displaying appropriate alarm responses, had increased post-release survival up to breeding age compared with untrained birds (van Heezik et al 1999). Although providing a potentially richer stimulus, the use of a live predator in pre-release training carries a number of logistic and ethical concerns (Griffin et al 2000). The challenge remains in understanding what the most appropriate anti-predator response will be in the wild, and to stimulate this response effectively in a captive environment.

*Other types of pre-release training*

A wide variety of conditioning and learning experiences may be applied to overcome the limitations imposed by captive conditions. Environmental enrichment is a well-established technique for minimizing the chronic stress and maximizing normal behavioral development in captive animals (Shepherdson 1994). Enrichment can take the form of the introduction of novel objects into enclosures, but of more relevance in the preparation of animals for release would be to provide opportunities for captive animals to acquire future survival skills, such as predator avoidance (see above), and foraging skills to utilise natural food sources. For example, the enclosures for White-winged Guans *Penelope albipennis* include the native trees and bushes that are the natural food source of wild guans (Pratolongo 2003). Milky Storks *Mycteria cincerea* are encouraged to catch live fish in ponds within their captive enclosures (Sebastian 2005), and pre-release training for Tarictic Hornbills *Penelopides panini* included provision of bundles of fruit-laden branches so that they could exercise manoeuvring in dense foliage (Hembra et al 2006). Often more specific pre-release training may be necessary to compensate for the effects of captivity, or to prepare birds for release procedures and for post-release conditions. Captive-bred Bali Starlings *Leucospar rothschildi* were given a six-week pre-release training period to accustom them to transport boxes, to retain a fear of humans and to develop foraging skills (Balen & Gepak 1994). Captive-reared Puerto Rican Parrots *Amazona vittata* were fitted with dummy transmitters to acclimate them to wearing radio-transmitters, and subjected to flight conditioning to develop and maintain physical stamina before release (White et al 2005). Innovative techniques using micro-light aircraft have been used to teach directed migration movements or to establish new migration pathways, for example for captive-bred Whooping Cranes in the USA (Hartup et al 2005), and for hand-reared Northern Bald (Waldrapp) Ibis *Geronticus eremita* moving between Austria and Italy (Fritz 2007).

## Rearing and Release Environment

### Hand-reared versus hen-reared

Rearing techniques can influence the behavioral repertoires of released animals. One method, hand-rearing, was developed to supplement or to overcome the challenges of parent-rearing in captivity, however, hand-rearing methods have been criticized for developing abnormal post-release behaviors (Metetsky et al 2000). Hazel Grouse *Bonasia banasus*, Black Grouse *Hyurus tetrix* and Gray Partridge *Perdix perdix* developed retarded crouching responses to aerial predators and failed to respond to conspecific alarm calls (Curio 1998, cited in Kreger et al 2005). Attempts to minimize human contact and the risks of sexual or filial imprinting during hand-rearing have involved the use of costumes or puppets that resemble adult birds (Wallace 1994). Puppet-reared Common Ravens *Corvus corax* (used as a surrogate for the endangered Hawaiian crow *Corvus hawaiiensis*, showed appropriate social and dispersal behaviors, and association with conspecifics after release and were more vigilant and fearful pre-release (Valutis & Marzluff 1999). There is concern that, although tameness towards humans may be avoided and association with conspecifics facilitated, hand-rearing does not replicate the learning experience young birds would gain from their parents (Snyder et al 1996). To address this need Eastern Loggerhead Shrike *Lanius ludovicianus migrans* fledglings are reared in large enclosures with their parents to enable learning of anti-predator behavior and hunting skills (Woolaver 2005). However, the assumed advantage of parent-rearing has been challenged by recent work indicating that if hand-reared animals are provided with appropriate experience then post-release survival may be equivalent to that of parent-reared birds (van Heezik et al 1999; Ellis et al 2000). For example, the treatments of parent-rearing, hand-rearing, and hand-rearing with exercise had very little long-term effect on the behavior of Whooping Cranes released to the wild (Kreger et al 2005). The ultimate assessment of the performance of hand-reared birds will be in comparison with wild-reared birds. There was no difference in breeding success between hand- versus wild-reared Snowy Plovers *Charadrius alexandricus* (Quinn 1989), and hand-reared Takahe *Porphyrio mantelli* raised with minimal human contact by keepers using glove puppets survived at least as well as wild-reared birds (Maxwell & Jamieson 1997). Growth and behavior of captive-reared and wild-reared Piping Plover *Charadrius melodus* chicks was similar, but fledging rates of captive-reared plovers were actually higher than those of their wild-reared conspecifics (Powell et al 1997).

### Hard versus soft release protocols

The term "soft release" applies to any release strategy designed to ease transition into a new habitat (Scott & Carpenter 1987), including post-release support such as provision of supplementary food. More commonly however, it is used in specific reference to some form of delayed release, whereby animals are held captive at the release site for some period. Considerable management and research attention has been paid to the question of how or whether to ease the transition of founder animals to full independence at a release site. At its simplest a hard release protocol would entail transporting founder animals to the chosen site, releasing them on arrival and providing no subsequent interventive care or support. At the other extreme, a soft release protocol may require that founder animals are held at the release site in some form of captivity or semi-captivity for an extended time (days-months) during which they are provided to some extent with sustenance, shelter and even veterinary care. This period of pre-release confinement is intended to allow acclimation to the new surroundings and thereby increase the probability of post-release survival and decrease the likelihood of post-release long-distance dispersal. Less intensive forms of soft release for birds include the post-release provision of supplementary food, veterinary care, shelter, or localised control of potential predators. Captive-bred Mauritius Kestrel *Falco punctatus* chicks are fledged from nest boxes at the release site (hacked), with food provided until full independence (Nicol et al 2004). Captive-reared Yellow-shouldered Amazon Parrots *Amazona barbadensis* are provided with supplemental food 1-2 times per day for up to one month before release (Sanz and Grajal 1998). Wild-caught flightless Aldabra Rails *Dryolimnas cuvieri aldabranus* were kept in enclosures at the release site and fed daily for between 6 and 14 days to enable compensation for weight loss that occurred during capture and transport; this soft release acclimation period is believed to have been instrumental in the high post-release survival rate (Wanless et al 2002). Probably as a result of published success stories there has been a tendency for reintroduction practitioners to assume that de facto a soft release protocol is preferable (Wanless et al 2002), hence it tends to be the first option considered for improving reintroduction programs. However, there is increasing experimental evidence that indicates, for some species at least, soft release may reduce survival or have no obvious effect (e.g. Fancy 1997; Hardman & Moro 2006). Decisions concerning the best type of release protocol should be made on a project-by-project basis, taking into account the duration, type and possible impacts of pre-release captive management and transport, and with reference to any relevant taxon-specific experience or guidelines that may be available.

## Genetics

Loss of genetic diversity may increase the extinction risk for small populations (Frankham 2005), especially when genetic factors such as inbreeding depression negatively impact population growth rates and make populations more vulnerable to either deterministic (e.g. habitat loss) or stochastic (e.g. environmental variation or catastrophic events) factors (reviewed in Jamieson 2007). Consequently the genetic management of a captive population poses a number of challenges, since for endangered species a captive population must often be derived from only a small number of founders, some of which may be related. Reduction in fitness of individuals resulting from the mating of close genetic relatives has been shown in captive populations of several species (reviewed in Swinnerton et al 2004). The concentrating of deleterious recessive alleles seems to be the main cause of this inbreeding depression; for example, a high frequency (9%) of the recessive allele causing chondrodystrophy (a lethal form of dwarfism) was found in the captive population of Californian Condors due to a founder effect (Ralls et al 2000). Because of a low molecular diversity of founders inbreeding depression reduces egg fertility and squab, juvenile and adult survival in Pink Pigeons being bred for reintroduction, with the strongest effects in the most highly inbred birds (Swinnerton et al 2004).

Maximizing the genetic unrelatedness of the founder population may entail the collection of individuals from different geographic areas, but this will also introduce the possibility that local parental adaptations will disadvantage offspring if they are released into sufficiently different ecological regions. Such outbreeding depression needs to be assessed on a species by species basis. The founder captive population of Peregrine Falcons *Falco peregrinus*, for example, was derived from individuals of widely differing genetic stocks, yet their hybrid offspring have successfully re-established wild populations following releases in the Mid-Western United States (Tordoff & Redig 2001).

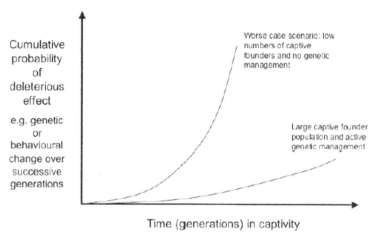

**Fig. 6.** *Cumulative risk of a deleterious effect due to behavioral or genetic changes with time spent in captivity.*

Selection of individuals from a captive population for reintroduction is effectively a harvest of that population that will inevitably alter the remaining genetic composition and structure of the captive population. Releases from a captive population can function like a second population bottleneck in reducing genetic diversity (Thévenon & Couvet 2002). Thus there is an inherent conflict between reintroduction and captive breeding programs that is not always well managed. Earnhardt (1999) sets out five demographic and genetic-based strategies for release founder selection, including the not uncommon strategy of "genetic dumping" entailing the release of animals over-represented in the captive population. While genetic dumping is most beneficial for the captive population, this strategy would provide the least genetic diversity for the reintroduced population. Earnhardt further suggests that there is no one ideal strategy, but that evaluation and selection of appropriate individuals for release will depend on the type (reintroduction or supplementation) and phase (early versus late) of a program, and the availability of wild-caught founders to increase captive genetic diversity.

### Risk function

Genetic diversity is lost with each subsequent captive generation, thus a projected change in the heterogeneity of a captive population will follow a negative trajectory, and the key issue is the predicted rate of loss for a given population size (Earnhardt et al 2004). The cumulative risk of an individual acquiring a genetic load that may reduce fitness will therefore follow an exponential curve (Fig. 6) the steepness of which will depend on the efficacy of management to maintain genetic diversity and avoid inbreeding depression (see below).

*Mitigation of genetic effects*

The advantage of a captive population is that managers are able to monitor each individual and to manage changes in genetic diversity, population size and population structure by controlling the frequency and pairings of matings. Modern breeding programs avoid loss of genetic diversity, genetic drift and inbreeding depression by prioritizing the breeding of individuals with the lowest relatedness, prevention of breeding of individuals with a high mean coefficient of kinship, maximizing founder representation, and maximizing the size of the captive population (Kalinowski & Hedrick 1999; Earnhardt et al 2004). It is often the case that the full genetic variation of an original population is no longer available, and one approach to produce a "locally adapted genotype" has been to maximize genetic heterogeneity by facilitating hybridization among populations or even subspecies; this approach has been used with varying success in the reintroduction of some mammals and lizards (reviewed in May 1991), but no avian examples are apparent in the literature. Genetic drift models are used to project predicted rates of loss of genetic diversity against the current operative standard for genetic objectives within captive breeding programs of retention of 90% of expected heterozygosity for 100 years (Earnhardt et al 2004). When these models indicate a decline in population heterozygosity below the standard, managers of captive breeding programs have three options: increase the effective population size; import additional founder lineages, or phase out the captive program (Earnhardt et al 2004). The costs, difficulty, and impacts on free-ranging populations of taking new animals from the wild, and the space limitations in captive facilities, will place restrictions on the feasible responses of managers to predicted losses of genetic diversity.

The best course of action will take place when captive management is being planned, to ensure that any reintroduction captive program is set up from the least inbred founder population possible, and that provisions are in place to maximise both the initial population size and the carrying capacity of the captive facility to overcome the possible deleterious effects of declining genetic diversity and any new bottleneck (Thévenon & Couvert 2002).

## CONCLUSIONS

The type of captive management used in the restoration of a given bird species will be dictated by the current status of wild populations and their habitats, risks to extant populations and habitats, technical expertise, availability of facilities, and funding. For populations that face a high risk of extinction in the wild, captive breeding may be the only appropriate action in order to sustain the species until the causal factors of their decline have been addressed. High rates of egg and chick losses may be avoided with egg collection and captive rearing, enabling the release of older, less vulnerable age classes. Lack of technical expertise or appropriate facilities for the captive management of a particular species may necessitate use of only wild-to-wild translocations, provided suitable founder populations exist. Adequate funding underpins any restoration effort; funding not just for any captive component, but also to support post-release monitoring and assessment of the success of a release program. Clearly, there are many reasons other than reintroduction why birds may be held in captivity: as pets, as private collections, for public awareness and education within zoological collections, for training for recreation and sport (e.g. falconry), as insurance against the loss of wild populations, and even as surrogates for the development of husbandry techniques for rarer species. None of these objectives requires the selection of individual birds for release into the wild. Even programs that do focus on the release of captive birds may not necessarily be concerned with the restoration of free-ranging populations of native species. Motivations for releasing captive birds include introductions of species outside their natural distribution ranges (today thankfully rare, but a major undertaking early last century in countries such as New Zealand (Green 1997)), put-and-take hunting, and restocking of hunted populations. Conservation projects requiring the release of birds after some, even brief, period of captivity fall into three categories: benign introductions (e.g. to restore a key ecosystem component through the release of the taxonomically closest extant form in the face of the extinction of the original species (Seddon & Soorae 1999)); reinforcements, i.e. the supplementation of an existing population, and reintroduction.

The objectives of any captive management program must be clearly defined a priori as this will determine the type of management that is appropriate and necessary. Projects using birds as pets and for advocacy may seek to promote tameness and even imprinting on human keepers, whereas wildness might be valued in put-and-take hunting operations but long-term survival and breeding in the wild would not be a focus. Management of insurance populations must keep one eye to the future possibility of using captive-bred founders to restore free-ranging wild populations. Reintroduction programs must place the emphasis on the post-release persistence of founders and their wild-hatched offspring. If the goal is to restore wild populations, then all stages of the captive-management process must be aimed at producing suitable founder birds for release. There is a danger that, given the necessary intensity of focus on captive management and captive production, that the needs of the captive component of a given project overwhelm the ultimate aims of the program. The birds allocated for releases to support reintroductions should not be those that are genetically, physically or behaviorally unsuitable for, or surplus to the needs of captive breeding. The emphasis should be on the quality of release birds, not simply the quantity of birds released in the hope that this will overcome un-sustainably high post-release mortality (Meretsky et al 2001; Beres & Starfield 2001). Animal ethics concerns are relevant in this regard also; just because unnecessary starvation, predation or other causes of mortality of naïve and ill-prepared individuals occurs out of sight once released birds are on their own does not mean it can be ignored or accepted in seeking population-level goals. The animal ethics of wildlife reintroductions warrant wider discussion. We propose the application of a simple three-stage process to guide releases of birds from captivity to support species conservation: SPA - selection, preparation, and assessment.

### Selection

Whether pulling passerines out of a mist net in preparation for a wild-to-wild translocation, or choosing captive-bred chicks, a careful selection process should be applied considering: health, through physical checks and disease screening; behavior; genetics (where known), such as level of inbreeding or potential to add new blood lines to a released population; and the composition of the release cohort since captive management will allow managers to select certain age classes or compositions, sex ratios, and social groups, or even groups that represent a variety of inherent behavioral types (sensu Watters & Meehan 2007).

### Preparation

At a minimum some form of marking, either batch or individually identifiable, will be required to facilitate post-release monitoring. For captive-reared and captive-bred birds pre-release preparations are likely to be more extensive, including inoculations against potential diseases in the wild and behavioral training or conditioning, for example development of anti-predator responses (Griffin 2000); or natal habitat preference induction training that seeks to expose animals in captivity to cues they will encounter at the release site and thus equate with their natal habitat, reducing their propensity to undertake long-distance post-release movements (Stamps & Swaisgood 2007).

### Assessment

For the last twenty years the reintroduction literature has called for rigorous post-release monitoring to form an essential component in the assessment and reporting of reintroduction success. Only very few recent projects have not made provision for adequate post-release assessment; by far the majority of reintroduction projects at least attempt the estimation of vital rates (survival, productivity, dispersal) that will enable a robust evaluation and reporting of the efficacy of the selection and preparation phases, and allow adaptive adjustments to be made for future releases. Captive management of birds in support of population restoration programs is a challenging undertaking, with the overriding objective being the production of individuals with the best possible chance of post-release survival and successful breeding in the wild. The captive environment poses a number of risks, from the dangers of trauma and disease, through to detrimental behavioral and genetic changes, all of which will reduce individual fitness and lower the probability that releases will result in viable free-ranging populations. However, the last two decades have seen the development of a number of innovative techniques to overcome many of the potentially deleterious effects of captivity and thereby better prepare birds for release into the wild. Refinement and wider application of such techniques will enable reintroduction practitioners to capitalize on the inherent advantages of releasing birds from captivity, and will do much towards improving the success rates of reintroduction programs worldwide.

### Acknowledgements

We are grateful to Ross Curtis, Glen Greaves, Ian Jamieson, Jeanne Marie Pitman, Emily Sancha and Micky Soorae for providing images and information, sometimes at short notice.

# References

AENGUS WL & JR Millam. 1999. Taming parent-reared orange-winged Amazon parrots by neonatal handling. Zoo Biology 18: 177-187.

ARMSTRONG DP. 1999-2006. Reintroduction projects in New Zealand. Available from http://www.massey.ac.nz/~darsmtro/nz_projects.htm . Accessed April 2006.

ANTILLA I, A Putaala and R Hissa. 1995. Tarhattujen ja villien peltopyn poikasten kättäymisestä. Suomen Riista 41: 53-65.

AZA (Association of Zoos and Aquariums). 2006. Conservation Impact Report. 43 pp. Available from http://members.aza.org/departments/ConScienceMO/ARCS/

BALEN B van & VH Gepak. 1994. The captive breeding and conservation program of the Bali Starling (*Leucospar rothschildi*). In: Olney, P. J. S. Mace, G. M. and Feistner, A. T. C. (Eds). Creative Conservation: Interactive Management of Wild and Captive Animals. Chapman & Hall, London, pp. 420-430.

BAUER GB. 2005. Research training for releasable animals. Conservation Biology 19: 1779-1789.

BECK BB, LG Rapaport, MR Stanley Price and AC Wilson. 1994. Reintroduction of captive-born animals. In: Olney, P. J. S. Mace, G. M. and Feistner, A. T. C. (Eds). Creative Conservation: Interactive Management of Wild and Captive Animals. Chapman & Hall, London, pp. 265-286.

BERES DL & AM Starfield. 2001. "Demography of the California Condor" and management decisions based on modelling: a reply to Meretsky et al Conservation Biology 15:1445-1448.

BIRDLIFE International. 2007. http://www.birdlife.org/datazone/species/index.html. Accessed May 2007.

BLACK JM. 1991. Reintroduction and restocking: guidelines for bird recovery programs. Bird Conservation International 1: 329-334.

BROWN JL, MW Collopy, EJ Gott, PW Juergens, AB Montoya and WG Hunt. 2006. Wild-reared Aplomado Falcons survive and recruit at higher rates than hacked falcons in a common environment. Biological Conservation 131: 453-458.

CARPENTER JW, RR Gabel and JG Goodwin Jnr. 1991. Captive breeding and reintroduction of the endangered Masked Bobwhite. Zoo Biology 10: 439-449.

CUNNINGHAM AA. 1996. Disease risks of wildlife translocations. Conservation Biology 10: 349-353.

DIMOND WJ & DP Armstrong. 2006. Adaptive harvesting of source populations for translocation: a case study with New Zealand Robins. Conservation Biology 21: 114-124.

EARNHARDT JM. Reintroduction programs: genetic trade-offs for populations. Animal Conservation 2: 279-286.

EARNHARDT, J.M., Thompson, S.D. and Schad, K. 2004. Strategic planning for captive populations: projecting changes in genetic diversity. Animal Conservation 7: 9-16.

EBENHARD T. 1995. Conservation breeding as a tool for saving animal species from extinction. Trends in Ecology & Evolution 10: 438-443.

FANCY SG, TJ Snetsinger and JD Jacobi. 1997. Translocation of the Palila, an endangered Hawaiian Honeycreeper. Pacific Conservation Biology 3: 39-46.

FRANKHAM R, H Hemmer, OA Ryder, EG Cothran, ME Soule, ND Murray and M Snyder. 1986. Selection in captive populations. Zoo Biology 5: 127-138.

FRANKHAM, R. 2005. Genetics and extinction. Biological Conservation 126: 131-140.

FRITZ J. 2007. The Scharnstein Waldrapp Ibis Migration Project after four years: bird leave the micro lights behind. In: Boehm C, Bowden CGR, Jordan MJR. and King C. (Eds). Proceedings of the International Advisory Group for the Northern Bald Ibis (IAGNBI) meeting Vejer, Spain, September 2006. Royal Society for the Protection of Birfds, The Lodge, Sandy, Bedforshire, UK. Pp. 67-76.

GREAVES G. 2007. Species reintroduction as a tool for the conservation of takahe, New Zealand. In: Soorae, P. S. (Ed.). Reintroduction News. Newsletter of the IUCN/SSC Reintroduction Specialist Group, Abu Dhabi, UAE. No. 26: 16-17.

GREEN RE. 1997. The influence of numbers released on the outcome of attempts to introduce exotic bird species to New Zealand. Journal of Animal Ecology 66: 25-35.

GRIFFON AS, DT Blumstein and CS Evans. 2000. Training captive-bred or translocated animals to avoid predators. Conservation Biology 14: 1317-1326.

GROSS WB & PB Siegel. 1982. Socialization as a factor in resistance to infection, feed efficiency, and response to antigen in chickens. American Journal of Veterinary Research 43: 2010-2012.

HÅKANSSON J & P Jensen. 2005. Behavioural and morphological variation between captive populations of Red Junglefowl (*Gallus gallus*) - possible implications for conservation. Biological Conservation 122: 431-439.

HARDMAN B & D Moro. 2006. Optimising reintroduction success by delayed dispersal: is the release protocol important for hare-wallabies? Biological Conservation 128, 403-411.

HARTUP BK, GH Olsen and NM Czekala. 2005. Fecal corticoid monitoring in Whooping Cranes *Grus americana* undergoing reintroduction. Zoo Biology 24: 15-28.

HEMBRA SS, B Tacud, E Geronimo, J Villanueva, J Jamangal, E Sanchez, N Bagac and E Curio. 2006. Saving Phillipine hornbills on Pany Island, Philippines. In: Soorae, P. S. (Ed.). Reintroduction News. Newsletter of the IUCN/SSC Reintroduction Specialist Group, Abu Dhabi, UAE. No. 25: 45-46.

HOOSON S & IG Jamieson. 2003. The distribution and current status of New Zealand Saddleback *Philesturnus carunculatus*. Bird Conservation International 13: 79-95.

IUCN 1987. IUCN position statement on the translocation of living organisms: introductions, reintroductions, and re-stocking. IUCN World Conservation Union, Gland, Switzerland.

JAMIESON I. 2007. Has the debate over genetics and extinction of island endemics truly been resolved? Animal Conservation 10: 139-144.

JONES RB & BO Hughes. 1981. Effects of regular handling on growth in male and female chicks of broiler and layer strains. British Poultry Science 22: 461-465.

JONES RB & D Waddington. 1992. Modification of fear in domestic chicks, *Gallus gallus domesticus*, via regular handling and early environmental enrichment. Animal Behaviour 43: 1021-1033.

KALINOWSKI ST & PW Hedrick. 1999. Detecting inbreeding depression is difficult in captive endangered species. Animal Conservation 2: 131-136.

KAUFFMANN MJ, MF Frick and J Linthicum. 2003. Estimation of habitat-specific demography and population growth for Peregrine Falcons in California. Ecological Aspplications 13: 1802-1816.

KEEDWELL RJ, RF Maloney and DP Murray. 2002. Predator control for protecting Kaki *Himantopus novaezelandiae* - lessons from 20 years of management. Biological Conservation 105: 369-374.

KLEIMAN DG. 1989. Reintroduction of captive mammals for conservation: guidelines for reintroducing endangered species to the wild. BioScience 39: 152-161.

LEECH TJ, E Craig, B Beaven, DK Mitchell and PJ Seddon. 2007. Reintroduction of rifleman *Acanthisitta chloris* to Ulva Island, New Zealand: evaluation of techniques and population persistence. Oryx (in press).

LOVEGROVE TG. 1996. Island releases of Saddlebacks *Philesturnus carunculatus* in New Zealand. Biological Conservation 67: 135-142.

MAGIN CD, TH Johnson, B Groombridge, M Jenkins and H Smith. 1994. Species extinctions, endangerment and captive breeding. In: Olney PJS, GM Mace and ATC Feistner (Eds). Creative Conservation: Interactive Management of Wild and Captive Animals. Chapman & Hall, London, pp. 3-31.

MATHEWS F, M Orros, G McLaren, M Gelling and R Foster. 2005. Keeping fit for the ark: assessing the suitability of captive-bred animals for release. Biological Conservation 121: 569-577.

MATHEWS F, D Moro, R Strachan, M Gelling and N Buller. 2006. Health surveillance in wildlife reintroductions. Biological Conservation 131: 338-347.

MAXWELL JM & IG Jamieson. 1997. Survival and recruitment of captive-reared and wild-reared Takahe in Fiordland, New Zealand. Conservation Biology 11: 683-691.

MAY RM. 1991. The role of ecological theory in planning reintroduction of endangered species. Zoological Symposium No. 62: 145-163. The Zoological Society of London.

MCCLELLAND P. 2007. Reintroduction of the Campbell Island Teal, New Zealand. In: Soorae PS. (Ed.). Reintroduction News. Newsletter of the IUCN/SSC Reintroduction Specialist Group, Abu Dhabi, UAE. No. 26: 19-21.

MCPHEE ME. 2003. Generations in captivity increases behavioural variance: considerations for captive breeding and reintroduction programs. Biological Conservation 115: 71-77.

MERETSKY V, NFR Snyder, SR Beissinger, DA Clendenen and JW Wiley. 2001. Demography of the California Condor: Implications for reestablishment. Conservation Biology 14: 957-967.

MERETSKY V, NFR Snyder, SR Beissinger, DA Clendenen and JW Wiley. 2001. Quantity versus quality in California Condor reintroduction: Reply to Beres and Starfield. Conservation Biology 15: 1449-1451.

MILLER B, D Biggins, D Hanebury and A Vargas. 1994. Reintroduction of the Black-footed Ferret (*Mustela nigripes*). In: Olney, P. J. S. Mace, G. M. and Feistner, A. T. C. (Eds). Creative Conservation: Interactive Management of Wild and Captive Animals. Chapman & Hall, London, pp. 455-464.

NICOLL MAC, CG Jones and K Norris. 2004. Comparison of survival rates of captive-reared and wild-bred Mauritius kestrels (*Falco punctatus*) in a re-introduced population. Biological Conservation 118: 539-548.

PIERCE RJ. 1996. Ecology and management of the Black Stilt *Himantopus novaezelandiae*. Bird Conservation International 6: 81-88.

POWELL AN, FJ Cuthbert, LC Wemmer, AW Doolittle, and ST Feirer. 1997. Captive-rearing Piping Plovers: Developing techniques to augment wild populations. Zoo Biology 16: 461-477.

PRATOLONGO FA. 2003. Reintroduction of the white-winged guan in north-west Peru. In: PS Soorae (Ed.). Reintroduction News. Newsletter of the IUCN/SSC Reintroduction Specialist Group, Abu Dhabi, UAE.

No. 23: 19-20.

QUINN PL. 1989. Comparison of the breeding of hand- and wild-reared Snowy Plovers. Conservation Biology 3: 198-201.

RALLS K, JD Ballou, BA Rideout and R Frankham. 2000. Genetic management of chondrodystrophy in California condors. Animal Conservation 3: 145-153.

ROBERTSON PA & SD Dowell. 1990. The effects of hand-rearing on wild gamebird populations. In: The Future of Wild Galliformes in the Netherlands, pp. 158-171. Lumeij, J. T. and Hoogeveen, Y. R. (Eds). Amersfoort, The Netherlands.

ROMERO LM. 2004. Physiological stress in ecology: lessons from biomedical research. Trends in Ecology and Evolution 19: 249-255.

SAINT JALME M and Y van Heezik. (Eds). 1996. Propagation of the Houbara Bustard. Kegan Paul International, London.

SANCHA E, Y van Heezik, R Maloney, M Alley and P Seddon. 2004. Iodine deficiency affects hatchability of endangered captive Kaki (Black Stilt, *Himantopus novaezelandiae*). Zoo Biology 23: 1-13.

SANZ V & A Grajal. 1998. Successful reintroduction of captive-raised Yellow-shouldered Amazon Parrots on Margarita Island, Venezuela. Conservation Biology 12: 430-441.

SCOTT JM & JW Carpenter. 1987. Release of captive-reared or translocated endangered birds: what we need to know. Auk 104: 544-545

SEDDON PJ, M Saint Jalme, Y van Heezik, P Paillat, P Gaucher and O Combreau. 1995. Restoratiuon of Houbara Bustard populations in Saudi Arabia: developments and future directions. Oryx 29: 136-142.

SEDDON PJ & PS Soorae. 1999. Guidelines for subspecific substitutions in wildlife restoration projects. Conservation Biology 13: 177-184.

SEDDON PJ, DA Armstrong & RF Maloney. 2007. Developing the science of reintroduction biology. Conservation Biology 21: 303-312.

SHEPPARD C. 1995. Propagation of endangered birds in US institutions: How much space is there? Zoo Biology 14: 197-210.

SHEPHERDSON D. 1994. The role of environmental enrichment in the captive breeding and reintroduction of endangered species. In: PJS Olney, GM Mace and ATC Feistner. (Eds). Creative Conservation: Interactive Management of Wild and Captive Animals. Chapman & Hall, London, pp. 167-177.

SNYDER NFR, SR Derrickson, SR Beissinger, JW Wiley, TB Smith, WD Toone and B Miller. 1996. Limitations of captive breeding in endangered species recovery. Conservation Biology 10: 338-348.

SNYDER NFR, SE Koenig, J Koschman, HA Snyder and TB Johnson. 1994. Thick-billed Parrot releases in Arizona. Condor 96: 845-862.

STAMPS JA & RR Swaisgood. 2007. Someplace like home: Experience, habitat selection and conservation biology. Applied Animal Behaviour Science 102: 392-409.

STANLEY-PRICE MR and Soorae P.S. 2003. Reintroductions: whence and whither? International Zoo Yearbook 38: 61-75.

SWINNERTON KJ, JJ Groombridge, CG Jones, RW Burn and Y Mungroo. 2004. Inbreeding depression and founder diversity among captive and free-living populations of the endangered Pink Pigeon *Columba mayeri*. Animal Conservation 7: 353-364.

SWINNERTON KJ, AG Greenwood, RE Chapman and CG Jones. 2005. The incidence of the parasitic disease trichomoniasis and its treatment in reintroduced and wild Pink Pigeons *Columba mayeri*. Ibis 147: 772-782.

TEAR TH, JM Scott, PH Hayward and B Griffith. 1993. Status and prospects for success of the Endangered Species Act: a look at Recovery Plans. Science 262: 976-977.

THÉVENON S and D Couvet. 2002. The impact of inbreeding depression on population survival depending on demographic parameters. Animal Conservation 5: 53-60.3

TORDOFF HB & PT Redig. 2001. Role of genetic background in the success of reintroduced Peregrine Falcons. Conservation Biology 15: 528-532.

TUDGE C. 1992. Last Animals at the Zoo. Island Press, Washington DC. 266 pp.

VALUTIS LL & JM Marzluff. 1999. The appropriateness of puppet-rearing birds for reintroduction. Conservation Biology 13: 584-591.

VAN HEEZIK Y, & PJ Seddon. 1998. Ontogeny of behavior of hand-reared and hen-reared captive Houbara Bustards. Zoo Biology 17: 245-255.

VAN HEEZIK Y, PJ Seddon and RF Maloney. 1999. Helping reintroduced houbara bustards avoid predation: effective anti-predator training and the predictive value of pre-release behaviour. Animal Conservation 2: 155-163.

VAN HEEZIK Y & S Ostrowski. 2001. Conservation breeding for reintroductions: assessing survival in a captive flock of Houbara Bustards. Animal Conservation 4: 195-201.

VAN HEEZIK, Y. and Seddon, P.J. 2001. Influence of group size and neonatal handling on growth rates, survival, and tameness of juvenile Houbara Bustards. Zoo Biology 20: 423-433.

VAN HEEZIK Y, P Lei, R Maloney and E Sancha. 2005. Captive breeding for reintroduction: influence of management practices and biological factors on survival of captive Kaki (Black Stilt). Zoo Biology 24: 459-474.

VIGGERS KL, DB Lindenmayer and DM Spratt. 1993. The importance of disease on reintroduction programs. Wildlife Research 20: 687-698.

WANLESS RM, J Cunningham, PAR Hockey, J Wanless, RW White and R Wiseman. 2002. The success of a soft-release reintroduction of the flightless Aldabra rail (*Dryolimnas* [cuvieri] *aldabranus*) on Aldabra Atoll, Seychelles. Biological Conservation 107: 203-210.

WATTERS JV & CL Meehan. 2007. Different strokes: can managing behavioural types increase post-release success? Applied Animal Behaviour Science 102: 364-379.

WHITE TH, JA Collazo and FJ Vilella. 2005. Survival of captive-reared Puerto Rican Parrots released in the Caribbean National Forest. The Condor 107: 424-432.

WILSON AC & MR Stanley Price. 1994. Reintroduction as a reason for captive breeding. In: PJS Olney, GM Mace and ATC Feistner (Eds). Creative Conservation: Interactive Management of Wild and Captive Animals. Chapman & Hall, London, pp. 243-264.

WOODFORD MH & PB Rossiter. 1994. Disease risks associated with wildlife translocation projects. In: PJS Olney, GM Mace and ATC Feistner (Eds). Creative Conservation: Interactive Management of Wild and Captive Animals. Chapman & Hall, London, pp. 179-200.

WOOLAVER L. 2005. Captive propagation and experimental release of the Eastern Loggerhead Shrike in Ontario, Canada. In: PS Soorae (Ed.). Reintroduction News. Newsletter of the IUCN/SSC Reintroduction Specialist Group, Abu Dhabi, UAE. No. 24: 29-32.

ZILLETTI B, E Venturato and L Beani. 1993. Comportamento anti-predatorio nella pernice rossa (*Alectoris rufa*): influenza dell'allevamento. Supplemento alle Richerece di Biologia della Selvaggina 11: 661-667.

## Author Biography

Philip Seddon completed a doctorate in zoology at the University of Otago, in Dunedin, New Zealand. His work on the conservation of the endangered Yellow-eyed Penguin led to a postdoctoral position as Senior Scientific Officer at the FitzPatrick Institute of African Ornithology, at the University of Cape Town. Between 1989 and 1991 he worked within the Benguela Ecosystem Program, studying the behavioural ecology and conservation of seabirds in South Africa and Namibia. In late 1991 he took up a position as field ecologist for the National Commission for Wildlife Conservation in the Kingdom of Saudi Arabia, and during 1993 to 2000 was Research Coordinator at the National Wildlife Research Center in Taif. Work in the Middle East concentrated on the creation and management of a network of wildlife protected areas within the Kingdom, and the restoration of critically endangered species, such as the Arabian Oryx, through intensive captive breeding and reintroduction programs. During this period Dr. Seddon earned a MSc in Protected Landscape Management by distance learning from the University of Wales, basing thesis work around his concurrent efforts to develop sustainable nature-based tourism in Saudi Arabia. In 2001 he returned to New Zealand to take up the position of Director of the Wildlife Management Program at the University of Otago. His current research relates to the reintroduction and population establishment of threatened native species in New Zealand; mitigation of the effects of introduced mammalian predators on native bird species, and assessment of the impacts faced by wildlife due to a burgeoning ecotourism industry. Dr Seddon is involved in a number of World Conservation Union (IUCN) groups; he is a member of the World Commission on Protected Areas' Tourism Taskforce, and the chair of the Bird Section of the Species Survival Commission's Reintroduction Specialist Group.

# CALIFORNIA CONDOR RECOVERY: A WORK IN PROGRESS

Michael Wallace

Applied Animal Ecology Division
San Diego Zoo Institute for Conservation Research
15600 San Pasqual Valley Road
Escondido, California 92027

## SUMMARY

One of the most endangered birds in the world, the California Condor, has become a flagship species symbolizing rare bird recovery efforts. While great strides have been made in captive propagation of condors, returning the species to its former range has been challenging. Without an existing population for released birds to emulate in the wild, program managers have had to develop unique methods for training this highly social species to acquire adaptive behaviors. Researchers have used aversion training designed to teach condors to use appropriate perches in the wild and addressed the continual process of reshaping chick-rearing techniques. Two significant challenges that remain hindering the recovery of the species are lead poisoning and the feeding of micro trash to nestlings by parent birds. Only by correcting these problems can the long-term recovery of this species be achieved.

## INTRODUCTION

### History

Only a few hundred years ago, the California Condor ranged from British Columbia to Baja California, Mexico. As European pioneers settled within its range, the species declined dramatically to near extinction in the mid-1980s. Chemical analysis of historic and recent museum study skins indicate that in the early 1800s the species diet shifted from principally one of a marine animal, shoreline based food supply to one of domestic ungulate carcasses inland. The marine mammal populations were drastically reduced by human over-exploitation and the availability of livestock carcasses steadily increased. The continued decline throughout the 1900s, from a few hundred birds to a few dozen condors through the 1970s, brought conservationists to consider a more intense scientific effort to recover the species.

As population estimates for California Condors fell to 25 to 35 birds, The American Ornithologists Union, National Audubon Society, California Fish and Game Commission (CFGC), and U.S. Fish and Wildlife Service (USFWS) agreed on a more aggressive research and captive breeding program.

The recovery strategy beginning in 1980, when an estimated 19 to 23 birds remained, was to conduct field research, marking individuals and using radio-telemetry, to identify the principal causes of the decline and, hopefully, correct the issues. Concurrently, by harvesting first laid eggs (inducing a replacement egg), from three to five nests per year and selectively removing some young from the wild, a captive flock could gradually be established as back up to the wild population.

Even with this concerted effort the decline of California Condors continued due to anthropogenic factors such as shooting, lead poisoning, collisions with man-made structures and, historically, egg collecting. By 1987, the controversial decision had been made to trap the few remaining wild condors into the custody of the Los Angeles Zoo (LAZ) and the San Diego Zoo's Wild Animal Park (SDWAP) placing the program's hopes entirely on the concept of captive breeding and re-establishment of wild populations through release of captive produced progeny.

### Captive Breeding

The captive breeding program was initiated when wild nestlings were brought into captivity, Xolxol in 1982 to the SDWAP and Cuyama in 1983 to LAZ. An adult condor at the LAZ (Topatopa) had been brought in from the wild as a recently fledged juvenile in 1967. Through the 1980s, 16 eggs were taken from wild nests and brought to the SDWAP where 13 successfully hatched, creating nearly half of the initial captive flock of 27 birds in 1987. The sex of every individual in the program was then, as it is today, determined by the genetics laboratory at the Zoological Society of San Diego's department of Conservation and Research of Endangered Species (CRES) enabling the program to form appropriately matched breeding pairs. Outbreeding is important for the long term health of small populations and a genetic analysis at CRES using mitochondrial DNA early in the program allowed managers to determine the degree of family relatedness between individuals and the formation of minimally related pairs. With the last remaining wild birds captured in 1987, the captive flock of 27 condors was split almost evenly between the LAZ and the SDWAP according to sex, age and genetic background. A genetic Master Plan for the population is conducted through the American Zoo and Aquarium Association (AZA) biologists every two to three years in order to update the genetic information as mortalities and hatchings alter the genetics and demographics over time. Owing to their social, plastic behavior, the wild condors settled into captivity relatively quickly. The first breeding was by a young pair of wild birds that had shown courtship behavior in the wild a year previous to capture. That first egg produced in 1988, at the SDWAP, was fertile and successfully hatched. In 1989, a second pair successfully reproduced at the LAZ.

Captive production of eggs was artificially increased four to six times over that seen in the wild by removing eggs and inducing female condors to lay one to two more than their usual single egg per season. With an aim to achieve a 14% egg weight loss over the 57- day incubation period, incubator temperature was held constant at 98.0F while humidity was adjusted between 12% and 95% as determined by the weight taken daily on the egg. The eggs were rotated several times a day as indicated by observations on both captive and wild incubating parents. Initially, because of the relative inexperience of most captive condor pairs, all eggs were removed and incubated artificially. Keepers using human isolation techniques such as blinds with one-way glass and video camera systems, reared all chicks that hatched from these eggs. The method also includes socializing the chicks with a condor-like hand puppet during feeding time and when management necessitates handling. Hatchability rose from a historic high of 60% to 80% over time mainly due to the increased care of malposition embryos and greater attention to embryonic behavior by the keepers and radiographing aberrant eggs close to hatching. Eventually the more experienced pairs could be trusted to rear their own young.

**Release Program**

In order for condor releases to begin, the California Condor Recovery Plan called for at least 96% of the heterozygosity of a family line be represented in captivity before any offspring of that line could be considered a candidate for release to the higher risk it would face in the wild. This amounted to six (later changed to seven) young held back in the relative safety of the zoos. By 1992, this criterion was met by two of the 14 family lines and a male from SDWAP and female from LAZ were the first condors to be released back into the species former range. Since condors are social, and release experiments on Andean Condors showed that they release better in groups, two California Condors were raised and released with two young, female Andean condors of similar age to round out the group. The Andean's were trapped back into captivity later as more California Condors were released in subsequent years.

During the first two years of condor releases (1992–1994) most aspects of the program went well except for one critical issue. As the young condors explored their new environment they tended to use power-line pole cross-arms as day and nighttime perches. Because the cross-arms support insulators and lines the birds were at higher risk of collision with the lines as they landed and flew from the perches. The program suffered four mortalities of the 13 birds released over the first two-year period from colliding with lines. Expecting other forms of mortality throughout the life of the program the power-line issue was unacceptable and the remainder of the birds were returned to captivity. The program experimented with the use of aversion training at the Los Angeles Zoo where a mock power pole was constructed and the perching surface on the cross-arm was electrified to give a mild shock. All episodes were recorded via time-laps video as to time, date and bird reaction. The birds learned quickly to stay off the poles in captivity and retained the behavior after release. All subsequently released condors subjected to the power-line aversion training ultimately proved it successful at virtually eliminating condor use of power poles for perches and significantly reduced line collision mortalities.

The California Condor Recovery Plan (Revised in 1996) calls for the establishment of condors wherever feasible and practical in their former range. The down-listing criteria, moving the bird from endangered to threatened status, as outlined in the Recovery Plan lists three disjunct populations numbering 150 birds per site. Each population would have at least 15 breeding pairs and a positive rate of increase. One wild population is being established in the Grand Canyon area while the other would be in the bird's more recent range from Baja north through western California. The third population in captivity serves for preservation of the original 14 founding family lines and support for the two wild populations with both genetic and numerical replenishment as needed.

The Peregrine Fund joined the program in the early 1990s bringing their expertise with rearing birds of prey and release programs and, in addition to building a breeding facility in Boise Idaho, in 1996 they began the release program at the Vermilion cliffs north of the Grand Canyon, Arizona. Concurrently, The Ventana Wilderness Society started their first releases in Big Sur, California, where they had been successful in re-establishing Bald Eagles in that part of coastal California. Where elk, deer and cattle form the main food base for released condors in Arizona, cattle in inland areas and marine mammals washed up on the beaches, form the natural food supply at Big Sur. It takes several years of negotiations, public hearings and logistics management to develop and supervise a release site. Program support from the local ranchers and residents is essential. In order to begin condor releases, on an international scale, in northern Baja California, Mexico, CITES permits were needed for export out of the US and import for a bird's return back into the US should it be necessary because of injury or death. The same CITES import/export is required on the Mexican side. Both the United States Department of Agriculture (USDA) and The Secretariat of Agriculture, Livestock, Rural Development, Fisheries and Food of Mexico (SAGARPA) require permits and quarantine as well. After nine years of discussions, the international program in Baja began releases in 2002.

In 2003, releases began at the Pinnacles National Monument in Central California, which is managed by the National Park Service. This site, as well as the other two in California and the Baja release site will eventually integrate to form the California population. Also in 2003, condors were shipped to Oregon Zoo to establish a fourth captive breeding facility for the program. Their off site secluded pens form the most secluded breeding environments in the program. To establish a condor population in an area they once inhabited we, in effect, train them to a site with behavioral management, food and topographic features that will enhance their ability to become competent flyers and learn how to function as condors in their new "wild" environment. Early on in life they are completely dependent on us for food and water, as they would be on their parents for an extended period of months after fledging, hence we provide large animal carcasses in and immediately outside the release pen where they can watch the smaller scavengers and previously released condors feed.

**Foraging Behavior**

Condors rely heavily on each other, particularly more experienced birds to consistently find food. Without the olfactory ability of the smaller Turkey Vulture, condors use observation of the activities of the more numerous smaller scavengers as well as other condors to find a carcass. Historically, they had the benefit of learning the tried and true foraging traditions from the local populations of birds that indicated where food was likely to show up in any given period of the year within their potential foraging distance of several hundred miles. As expected, it is taking decades for our released condors to develop a functional food finding tradition in areas we are attempting to re-establish them. To manage the process we subsidize the inexperienced population with carrion offered at various times and locations in their environment thus encouraging natural foraging. Interesting characteristics of this highly endangered species are its intelligence and social behavior. As a large carcass scavenger it must cover great distances, often at high altitudes, conserving energy by soaring. Unlike the Turkey Vulture that uses olfaction to find food, condors observe the behavior of other scavengers on the landscape attracted to the activity around carcasses large enough to persist for some days or weeks in time. Having a highly developed sense of orientation along with good associative learning abilities allow the development of long term survival skills. With condors, like primates or social carnivores, each bird has a specific ranking within the pecking order of the local population. This social status is based on age, experience, sex and physical attributes like weight and physical ability. To survive, condors need to conserve energy and an ordered social hierarchy reduces the amount of time individuals spend in altercations over resources. While continuous, mild, non-violent displacements occur throughout the day by dominant birds towards subordinates over resources such as food, perches and even air space, higher intensity aggression is usually seen between two birds of nearly the same rank. Because the birds return to the feeding sites every two to four days, we tally the winners and losers of their interactions and assess their hierarchal relationships from camouflaged blinds as they feed. By better understanding the social rules under which this gregarious species functions we gain predictability and can tune our management for better program results. Radio transmitters, along with more sophisticated GPS-satellite transmitters attached to each wing, allow us to keep track of short and long distance movements. Trailing from each transmitter, vinyl tags display numbers large enough for the birds to be identified at some distance with binoculars, whether the bird is sitting or flying. These "studbook" numbers are each bird's personal ID and are never repeated.

**Lead Poisoning**

Within a few years, as the released condors began to venture further from the release sites, the birds increasingly feed on natural carcasses in the environment. Some of these carcasses of deer, coyote, and elk have been shot legally or poached and some still contained intact lead bullets or fragments that the birds inadvertently ingested. Recent ballistics tests show that as much as a third of the lead bullet can be shed as a fine powder throughout the wound, even if the bullet passes through the animal without hitting a bone and fragmenting. The entry and exit wounds provide and easy access point in the thick hide of a large carcass, attracting condors and increasing the odds of lead contamination. When ingested, lead is poisonous to both people and wildlife by compromising the nervous system; it can reduce mental faculties (human studies) in minor exposures and cause debilitation and death in more severe cases.

Lead poisoning was identified as a major condor mortality factor during the decline of the original wild population. Although few (less than a dozen) deaths have occurred since releases began, this is only holds true because of the intensive management of the free flying condors in the form of clean proffered food and chelation therapy of affected birds. Each released condor, per program protocol, is trapped on a yearly basis and tested for lead and other toxicants. Field test kit levels up to 15 micrograms per deciliter (ug/dl) are considered background and not a concern. Levels from 15 (ug/dl) to 65 (ug/dl) (test kit limits) indicate exposure and may need chelation therapy in the field. Higher values require a lab test and chelation treatment in captivity at the LAZ, SDWAP or The Phoenix Zoo. Although therapy, consisting of daily injections of EDTA, a chelating agent that binds the lead in the blood allowing elimination from the blood stream, is very effective and in most cases enabling a successfully treated bird to be re-released within a few weeks or months, the process is resource consuming and stressful on the birds. Nor is it clear if the dozen or so condors chelated, program wide, each year are adversely affected in the long term by the heavy metal exposure or the therapy. Ongoing efforts over the last four years to educate hunters within the condor range to remove or make carcass refuse unavailable to condors, as well as to switch to non-lead ammunition, have not yet resulted in a reduction of blood lead levels in released condors. Efforts are mounting to legally regulate lead, in California, in the current condor range.

**Micro-trash Ingestion**

Another hindrance to the long-term success of the program has been with the feeding of "micro-trash" to wild chicks by the parents. Since the first breeding in the wild by released condors in 2001, thirty eggs had been laid by the end of 2006. Sixteen hatched, but only six chicks fledged. Two chicks had been taken into the LAZ for rehab and eight died in the field. The 96% chick survivorship after hatching in captivity is a drastic difference to the 37% survivorship seen in the wild during the first few years of breeding. While we expected reduced fertility and hatchability in the first time breeders in the wild, we also expected better survivorship of nestlings. While nutrition and diet could have been a factor, (whole still born calf carcasses were almost exclusively used in the beginning of the program) almost all of the parents fed their young some amount of foreign (man-made) material. The micro trash consists of small pieces of glass, bottle caps and plastic that condors bring into the nest, apparently mistaking it for bone fragments. Condors specifically forage for bone chips probably as a calcium source to feed their chicks. As a large carcass scavenger, ingestible size bone fragments can apparently warrant a specific search behavior in contrast to raptors that feed on smaller prey, even small enough to swallow whole, to get their calcium needs. Ingesting micro-trash can cause damage to the GI tract, impactions that stunts growth and poisoning from ingesting zinc coated nuts, bolts and washers. Two wild-hatched chicks have had micro-trash surgically removed at the Los Angeles Zoo; one chick contained 36 bottle-caps.

After studying several nests, and particularly one intensely in 2006 by making four nest visits and making continuous observations from a distance, we have been able to gain some insight into the problem and formulate some ideas for corrective measures. By placing a large amount (15lbs) of bone chips at the nest at every visit we noticed some encouraging behaviors. The parents fed the proffered bone chips to the chick, the prevalence of trash was reduced by the end of the five month nestling phase and the parents began bringing in natural bone they discovered away from the nest cave. Through observations at other nests, it appears that once a chick has left the nest, re-ingestion of regurgitated trash is not likely since cliff or tall tree perches allow the offending items to drop some distance away. This would also explain why we have never seen an impaction issue in flying juveniles or adults. Once the chick makes it out of the nest they seem to do fine in regards to the micro-trash issue. In 2007, we began studying six nests in California closely and hope to, not only duplicate the successful fledging of last year's chick but, gain more insight on how to shift breeding, released condors, to a more appropriate search image for bone fragments.

**Tame Behavior**

Multi-clutching of breeding pairs by removal of first eggs to induce replacement eggs, has been highly successful in increasing productivity of birds, providing many more young for potential reintroduction to the wild than would otherwise be possible. Initially, because of the relative inexperience of most captive condor pairs, all eggs were removed and incubated artificially. All chicks that hatched from these eggs between 1983 and 1990 were hand-reared by keepers using human isolation techniques that included socializing the chicks with a condor-like hand puppet. Beginning in 1991, "puppet-reared" nestlings were placed in pairs or trios and reared as small nestling groups. These groups were gradually introduced to each other during the six-month nestling period until 5-10 member cohorts were formed for release to the wild. This rearing system worked well for Andean Condors in South America where release candidates were set free to integrate into existing natural populations. However, in California and Arizona the first released California Condors did not have the benefit of an existing wild population to impose social rules and teach condor foraging traditions since the entire wild population had been removed for captive breeding. In the wild the natural curiosity of un-chaperoned juvenile condors, especially when dehydrated or food stressed, led some individuals to seek out novel stimuli including, at times, human contact. After several years of releases of only puppet-reared condors to the wild, parent-reared birds became available as captive pairs and gradually became more experienced and could be trusted to rear their own young. Although the effects of rearing method (i.e. puppet vs. parent-reared) did not appear to significantly influence survivorship in the wild. Young released from captive parents, who raised their own chicks for three to six months in captivity, appeared to be better adjusted socially and initially approached people less often than puppet-reared chicks. While these differences were sometimes overstated by critics of the program, parent-reared birds tended to be more shy in novel situations and focused more on the social activities of other condors than did the puppet-reared birds. This observation led us to experimentally modify puppet-rearing techniques to mimic more closely the social experience of parent-reared chicks. Modifications in the captive-rearing environment included nestlings reared singly instead of in cohorts; less traumatic handling during necessary zoo procedures; reduced but more assertive and realistic use of rearing puppets; the use of mentors in fledging pens; food restriction to induce better feeding responses; reduced moves during the nestling phase; greater isolation from humans throughout the captive phase; and more chick-oriented determination of readiness for transfer to release sites.

Releasing condors in the Sierra San Pedro Mártir region of Northern Baja California, Mexico, in the southern portion of the species' former range, afforded an opportunity to evaluate the effects of these modifications to the puppet-rearing technique. Since the growing Baja population would remain isolated from condor populations released in California, USA, for several years, we took the opportunity to release only condors puppet-reared using the modified technique. Preliminary results with 16 free flying condors over a five-year period indicated that the changes in socialization made during the rearing and release process may have a positive effect on the behavior of released condors. Condors in Baja conduct their activities more restricted to remote areas and display a greater wariness of people as they expand their range. As condors released as young juveniles at the US release sites neared sexual maturity, and they spent more time in reproductive behaviors, their tendency to exhibit curious behavior associated with people reduced dramatically. With the older condors in Baja entering their reproductive phase, (one pair hatched an egg in 2007 that unfortunately died within a month likely due to an un-naturally high parasite infestation) they are displaying exceptionally wild condor behavior that will likely benefit the younger birds as they interact at feeding sites and mutual roosts.

## Conclusion

From the beginning, the California Condor program has been a research and development exercise with episodic progress spanning one challenge after another. The species extreme life history pattern of a long lifespan (50 to 60 years), delayed sexual maturity (5 to 8 years) and prolonged parental dependency period (4 to 8 months), have complicated and prolonged efforts for its recovery. Knowledge of avicultural techniques and behaviors has greatly assisted the recovery of the species from extinction in captivity. With a population of only 27 birds in 1987 to 306 birds in 2007, the species has been virtually saved from certain extinction. The salient challenge today is to re-establish the species in as many parts of its former range as possible. With 147 condors flying in the wild (2007) in Arizona, California and Baja the program is slowly but surely approaching that goal. Lead and the presence of micro-trash remain powerful challenges to the condor recovery program and require solutions for ultimate success in recovering this species.

## Author Biography

Dr. Michael Wallace serves the Zoological Society of San Diego as a Wildlife Scientist in the Applied Animal Ecology Division of the Conservation and Research of Endangered Species (CRES) since 1989. He is responsible for coordinating reintroduction projects for endangered species such as the California Condor having over 28 years experience with that species. He directs the effort to re-establish California Condors in Baja California, Mexico for the Zoological Society. He continues to serve as the team leader for the U.S. Fish & Wildlife Service's California Condor Recovery Team, responsible for inter-agency coordination and the overall condor program. He is also Species Coordinator for the American Zoo and Aquarium Association's (AZA) California Condor Species Survival Plan. He earned his master's (1979) and Ph.D. (1985) degrees in Wildlife Ecology at the University of Wisconsin at Madison. He has authored or co-authored more than 30 scientific articles and papers and has lectured frequently throughout his career on various wildlife topics.

# ISBBC Awards

The ISBBC Awards are a continuation of those presented during previous symposia.

The awards are broken down into three categories:

**1) The ISBBC Conservation Award**

*Presented to individuals, institutions or organisations contributing to the preservation of both in-situ and ex-situ populations of threatened taxa.*

**2) The ISBBC Avicultural Award**

*Presented to individuals, institutions or organisations who have made significant contributions to the progress and advancement of the aviculture and conservation breeding.*

**3) The Jean Delacour Lifetime Achievement Award**

*The most prestigious of the three awards, the Jean Delacour Lifetime Achievement Awards are reserved for those individuals who have dedicated a lifetime worth of efforts to the avicultural community and made significant avicultural achievements throughout their career.*

Nominations were accepted from February through August 2007 and were made by various individuals from both North America and Europe. Nominations were put forward in front of a panel of four judges, from both Europe and North America consisting of:

Ms. Nancy Clum
*Wildlife Conservation Society*

Mr. Josef Lindolm, III
*Dallas World Aquarium*

Mr. John Ellis
*Zoological Society of London*

Mr. Grenville Roles
*Disney Animal Kingdom*

Nominations were reviewed and voted upon by the panel.

The ISBBC Awards recipients were as follows:

**1) The ISBBC Conservation Award**

*Presented to Jurong Bird Park for their Oriental Pied Hornbill conservation program.*

**2) The ISBBC Avicultural Award**

*Presented to Mrs. Martine Van Havere of Belgium, founder, with her husband Luuc Van Havere, of IBISRING, for her development and promotion of sustainable avicultural management techniques for the Threskionithidae family.*

**3) The Jean Delacour Lifetime Achievement Award**

Three award recipients were presented:

Mr. Lynn Hall
*California, USA*

Mr. Robert Berry
*Texas, USA*

Mr. Michael Lubbock
*North Carolina, USA*

All award recipients received commissioned, hand blown glass statuettes of the Bali Mynah, representing the significance that aviculture can have to help prevent species extinctions.

# Author Index

| | |
|---|---|
| al Subai, Hajid | 417 |
| Babakhanov, Alexander | 332 |
| Baldado, Jose | 396 |
| Barrows, M. | 341 |
| Beck, Jason | 392 |
| Bell, Diana | 127 |
| Bell, Martin | 16, 27 |
| Bird, David | 169 |
| Böhm, Christiane | 38 |
| Bowden, Christopher | 38, 47 |
| Brader, Kathy | 186 |
| Bruslund-Jensen, Simon | 191, 370 |
| Bukowinski, Ania | 406 |
| Cadeliña, Angelita | 396 |
| Cariño, Apolinario | 396 |
| Cerial, Loujean | 396 |
| Cline, Heidi | 69 |
| Combreau, Olivier | 76 |
| Connors, Neville | 199 |
| Conrad, Laurie | 55 |
| Cornejo, Juan | 209 |
| Coym, Mollie | 61 |
| Cristinacce, Andrew | 127 |
| Cunningham, Andrew | 47 |
| Cuthbert, Richard | 47 |
| De Falco, Kathleen | 406, 407 |
| de Ravel Koenig, Frederique | 127 |
| Estudillo López, Jesús | 324 |
| Fabre, Charlie | 396 |
| Farabaugh, Susan | 406, 407 |
| Gailband, Charles | 55 |
| Georgii, Cristina | 363 |
| Grant, Tandora | 407 |
| Green, Rhys | 47 |
| Hammer, Sven | 370 |
| Hammerly, Susan | 406, 407 |
| Handschuh, Markus | 127 |
| Hodges, Jeremy | 406, 407 |
| Hollmen, Tuula | 69 |
| Hospodarsky, Pavel | 396 |
| Hughes, Austin | 408 |
| Hughes, Mary Ann | 408 |
| Hunt, Samuel | 108 |
| Jakati, Ram | 47 |
| Jakob-Hoff, Richard | 441 |
| Jonathan, Leo | 363 |
| Jones, Carl | 127 |
| Jordan, Lara | 127 |
| Jordan, Mike | 38 |
| Karekoona, Abdou | 392 |
| Karsten, Peter | 219, 227 |
| Katebaka, Raymond | 392 |
| Kesler, Dylan | 423 |
| Khan, Uzma | 102 |
| King, Amy | 108 |
| King, Catherine | 232 |
| Ladkoo, Amanda | 127 |
| Lastica, Emilia | 396 |
| Leon, Olivier | 76 |
| Lieberman, Alan | 88 |
| Lina, Bardo | 169 |
| Lindholm III, Josef | 252 |
| Lindsay, Nick | 47 |
| Lubbock, Michael | 311 |
| Lynch, Colleen | 445 |
| McClelland, Pete | 93, 436 |
| McInnes, Kate | 441 |
| McNair, Tyna | 158 |
| Melekhovets, Yuri | 332 |
| Mohammed Basheer, P. | 417 |
| Murn, Campbell | 102, 108 |
| Neibaur, Lynne | 406, 407 |
| Nichols, Rina | 116 |
| Owen, Andrew | 127 |
| Padrón Zamora, Rafael | 334 |
| Pain, Deborah | 47 |
| Parry-Jones, Jemima | 47 |
| Perschke, Mario | 191 |
| Pittman, Jean-Marie | 341 |
| Prakash, Vibhu | 47 |
| Rahmani, Asad | 47 |
| Rojek, Nora | 69 |
| Romer, Liz | 353 |
| Sancha, Emily | 441 |
| Seddon, Philip | 455 |
| Sewell, Angela | 406, 407 |
| Sher Shah, Moayyad | 417 |
| Shobrak, Mohammad | 417 |
| Slocomb, Christine | 406, 407 |
| Smith, Ashley | 108 |
| Smith, Ken | 38 |
| St. Leger, Judy | 55 |
| Steiner, Jessica | 116 |
| Suarez, A. | 363 |
| Switzer, Richard | 127, 370 |
| Teves, Mercy | 396 |
| Tuininga, Ken | 116 |
| van der Spuy, Stephen | 341 |
| van Hal, Jack | 441 |

| | |
|---|---|
| van Heezik, Yolanda | 455 |
| Vencatasawmy, Vanessa | 127 |
| Vendiola, Rene | 396 |
| Volossiouk, Tatiana | 332 |
| Wallace, Michael | 471 |
| Ward, Gary | 142 |
| Watson, Ryan | 370 |
| Wilkinson, Roger | 127 |
| Williams, Elaine | 116 |
| Williams, Paul | 158 |
| Woolaver, Lance | 116 |
| Zafar-ul Islam, M. | 417 |
| Zembal, Richard | 55 |

CPSIA information can be obtained at www.ICGtesting.com
Printed in the USA
BVOW10*2006091214
377187BV00003B/1/P

9 780888 397317